Sensory Processing in Aquatic Environments

Springer
New York
Berlin
Heidelberg
Hong Kong
London
Milan
Paris
Tokyo

Illustration of three organisms that rely heavily on sensory processing in a range of habitats: from the deep-sea (anglerfish), temperate reef areas (leafy sea dragon), and the intertidal zone (galatheid crab). The First International Conference on Sensory Processing of the Aquatic Environment was held on Heron Island on the Great Barrier Reef in March 1999 and was the inspiration for this book. Design and illustration by Shaun P. Collin.

Shaun P. Collin
N. Justin Marshall
Editors

Sensory Processing in Aquatic Environments

Foreword by Ted Bullock

Introduction by Jelle Atema, Richard R. Fay, Arthur N. Popper, and William N. Tavolga

With 140 Illustrations, 8 in Full Color

Springer

Shaun P. Collin
Department of Anatomy and Developmental Biology
School of Biomedical Sciences
University of Queensland
Brisbane, Queensland 4072
Australia
s.collins@uq.edu.au

N. Justin Marshall
Vision Touch and Hearing Research Centre
School of Biomedical Sciences
University of Queensland
Brisbane, Queensland 4072
Australia
justin.marshall@uq.edu.au

Cover illustration: Background, scanning electron micrograph of the hair cells from the lateral line organs of deep-sea fish *Anoplogaster cornuta* (Photograph by Justin Marshall). Inset photographs, left to right; bathypelagic crustacean *Cystisoma latipes* (Photograph by Edie Widder and Harbour Branch Oceanographic Institute), the eye of coral reef fish *Oxymonocanthus longirostris* (Photograph by Justin Marshall), deep-sea anglerfish *Phrynichthys wedli* with stud-like lateral line organs (Photograph by Justin Marshall and Harbour Branch Oceanographic Institute).

Library of Congress Cataloging-in-Publication Data
Sensory processing in aquatic environments / editors, Shaun P. Collin, N. Justin Marshall.
p. cm.
Includes bibliographical references (p.).
ISBN 0-387-95527-5 (alk. paper)
1. Aquatic ecology—Congresses. 2. Senses and sensation—Congresses.
I. Collin, Shaun P. II. Marshall, N. Justin.
QH541.5.W3 S46 2003
577.6—dc21

2002070736

ISBN 0-387-95527-5 Printed on acid-free paper.

Printed in the United States of America.

9 8 7 6 5 4 3 2 1 SPIN 10883078

www.springer-ny.com

Springer-Verlag New York Berlin Heidelberg
A member of BertelsmannSpringer Science+Business Media GmbH

Foreword

Since this volume has both an introduction and a preface, which explain the relation to a previous volume and the scope and organization of the book, my role, as author of the foreword, is to be as cryptic as an oracle and be detached, not to say Olympian. This is the view of one who was not at the Heron Island Conference but cares passionately for the bridging of research—from descriptive natural history of what animals do to reductionist analysis of how they do it, through all the grades of complexity from nerve nets to human brains.

My angle, like that of the conference organizers and book editors, is the input phase that determines and guides behavior. This phase offers great freedom to manipulate conditions and to choose among modalities, to discriminate stages and levels of information processing, and to compare ontogenetic, experiential, evolutionary, and mood factors. The hardware and software of sensory processing give us particularly good shots at the brass ring of how nervous systems have evolved from exceedingly simple to unbelievably complex.

It continues to be true, as it has been for a long time, that sensory neurobiology enjoys unique advantages. It turns up new organs, like the lateral line analogs in cephalopods, and new functions for old organs, like the infrared specialization of the facial pits of crotalid snakes. It sparks revolutions in understanding of integrative mechanisms, like the submodalities in photoreception, or the organization of brain divisions, like the roles of the cerebellum. It breaks the brain-mind barrier by easing us into genuinely cognitive "higher functions," such as expectations and attention arousing. Hence, a flood of new research and the need for a new book.

Minirevolutions at many levels, with the explosion of new methods, change basic concepts of brain and receptor operation. Thus, the approaches and organization of this volume depart from the past. I welcome especially the frequent introduction of multisensory analysis and the bringing together of diverse senses involved in common ethological domains, like navigation, communication, and finding food. I commend the authors for taking on and dealing with such slippery and demanding questions as parallel evolution, plasticity of tuning, and translation of indoor controlled experiments into outdoor life in the raw. I am pleased to see that the editors have insisted on informative summaries that help the browsing reader to choose what to peruse next.

Now, dear reader, it is your turn. If you are not astonished and amused by something within the first hour, it will not be the fault of the authors, the editors—or the animals.

La Jolla, CA *Ted Bullock*

Preface

Our understanding of the way in which animals see, hear, smell, taste, feel, and electrically and magnetically sense the aquatic environment has advanced a great deal over the last 15 years. In March 1999, after successfully enticing many of the leaders in the field of sensory processing to converge at Heron Island on the Great Barrier Reef in Australia, a wonderful week of intellectual exchange ensued. During that week, the idea was hatched to present an update of the landmark text, *Sensory Biology of Aquatic Animals* by Atema et al. (Springer-Verlag, 1988). This earlier volume raised the idea of considering the sensory systems of an animal as an integrated whole, rather than studying one sense and its capabilities separately. In planning this book, we aimed to follow this idea through, addressing specific problem-based tasks set by the physics of the world and the animals within it, and then arranging chapters that examine their biological solutions. The tasks identified form the five sections of the book.

Navigation and Communication in the aquatic medium presents a set of problems quite different from those in air, and Part 1 examines some of these. Not surprisingly, as water is often turbid, vision may become secondary to other sensory modalities in solving such problems, and this is reflected by the auditory, olfactory, and magnetic senses included in this section. Navigation using polarized light patterns from the sky is a visual solution employed by many terrestrial animals and some from the aquatic realm, an additional problem for water dwellers being the destruction of the information at the air/water interphase. Some aspects of this are also examined in Chapter 13.

Finding Food and Other Localized Sources, such as potential mates, rivals, or predators, is one of the vital day-to-day tasks for any animal. Part 2 takes up this theme and, again because of the physics of life in water, many solutions employ senses foreign to us such as the detection of electrical or magnetic field disturbances or vibration detection. Of course, vision is used by some aquatic organisms, and some adaptations for visual targeting are discussed in the final chapter of this section.

The evolution of any sensory system is limited and guided both by the physical attributes of habitat and medium and the biological building blocks that construct the animal. This is the subject of Part 3, entitled, *The Coevolution of Signal and Sense*, a section that samples recent work in vision, audition, and olfaction. Not included here are some of the wonderful discoveries in recent years within the field of electrosense. These are discussed elsewhere in the book in other contexts (Chapters 4, 5, 20, and 22), and the reader is directed to a recent special issue of the *Journal of Experimental Biology* (volume 204, 2001) on the subject.

Organisms living in the deep-sea environment have received much attention over the past 15 years. This research has been driven by advances in sampling techniques, using ingenious devices deployed from both submersibles and ships, bringing deep-sea creatures to the surface in almost perfect condition. The deep-sea is a natural laboratory, where one can explore sensory thresholds such as visual sensitivity. Therefore, we thought it appropriate to include a part dedicated to the challenges of vision in the deep, Part 4, *Visual Adaptations to Limited Light Environments*. Traveling deeper in the ocean, sunlight is replaced by bioluminescence and the visual systems of the inhabitants there have undergone incredible changes to optimize light capture, sensitivity, and tuning of the visual pigments to the ambient light. The final chapter in this part examines the visual adaptations of crustaceans from both deep and shallow oceans, notably including the mantis shrimps (stomatopods), the beautiful but violent possessors of the world's most complex color vision system. Stomatopods seemingly examine the color world with the same coding principles used by the ear, emphasizing the need for an integrated approach to our understanding of sensory processing.

Part 5, *Central Coordination and Evolution of Sensory Inputs*, focuses on the evolution of the central nervous system, and some of the ways the large input of sensory signals are processed and filtered to allow behaviorally important signals to be sorted from noise. Particularly within the relatively large brains of vertebrates, this is a complex area with which we struggle, and it is certainly one of the major challenges for the future of sensory biology. We hope that those undertaking this rewarding task will remember to keep their angle of attack both comparative and integrated. The final chapter returns to the periphery and the convergent evolution of a bill-shaped electrosensory organ. A very Australian structure, it is now shared with the southern United States. Different animals solving the same tasks in the sensory world often come up with similar solutions, the classic example previously being the convergence in design of cephalopod and vertebrate eyes. As visual animals, we naturally tend toward working on visual systems and that bias is represented in this book. The convergence of lateral line systems between cephalopods and vertebrates is pointed out in the Foreword and Chapter 14 and, along with this final example of electrosensory convergence, it is clear that working on senses other than vision is worth the effort. Although both vision scientists ourselves, we encourage students to think beyond the sense of vision and to consider the other senses. Olfaction or chemosense, the first sensory system to evolve on earth, still remains more important to most animals on earth than vision. Perhaps in 15 years, when the next update on aquatic sensory systems is published, the bias of chapters will reflect this?

We would like to sincerely thank the generous support of the University of Queensland and the University of Western Australia since the inception of this project. We would also like to thank the staff of Springer-Verlag, especially Robin Smith and Janet Slobodien, for their patient cooperation and all the contributing authors for sharing their ideas and presenting their exciting research in such an integrated way.

We hope that this book will challenge students and established scientists alike to learn more about sensory processing and the novel and astonishing ways in which aquatic animals survive in an environment occupying over nine-tenths of this planet.

Brisbane, Queensland
Australia

Shaun P. Collin
N. Justin Marshall

Contents

Contributors

Jelle Atema
Marine Biological Laboratory, Woods Hole, MA 02543–1015, USA.

Andrew H. Bass
Department of Neurobiology and Behavior, Cornell University, Ithaca, NY 148653, USA.

Curtis C. Bell
Neurological Sciences Institute, Portland, OR 97209–1595, USA.

Horst Bleckmann
Institut für Zoologie, Universität Bonn, 53115 Bonn, Germany.

David Bodznick
Department of Biology, Wesleyan University, Middleton Court, CT, 60457, USA.

James K. Bowmaker
Department of Visual Science, Institute of Ophthalmology, University College London, London EC1V 9EL, UK.

Christopher B. Braun
Parmly Hearing Institute, Loyola University, Chicago, IL 60626, USA.

Ann B. Butler
Krasnow Institute for Advanced Study and Department of Psychology, George Mason University, Fairfax, VA 22030, USA.

Shaun P. Collin
Department of Anatomy and Developmental Biology, School of Biomedical Sciences, University of Queensland, Brisbane, Queensland, 4072 Australia.

Sheryl Coombs
Parmly Hearing Institute, Loyola University, Chicago, IL 60626, USA.

Thomas W. Cronin
Department of Biological Sciences, University of Maryland, Baltimore County Campus, Baltimore, MD 21250, USA.

Guido Dehnhardt
Institut für Zoologie, Universität Bonn, 53115 Bonn, Germany.

Leo S. Demski
Division of Natural Sciences and Pritzker Marine Biology Research Center, New College of Florida, Sarasota, FL 34243, USA.

Carol E. Diebel
Auckland War Memorial Museum, Auckland, NZ.

Ronald H. Douglas
Department of Optometry and Visual Science, City University, London, EC1V 7DD, UK.

Kjell B. Døving
Division of General Physiology, Department of Biology, University of Oslo, N-0316 Oslo, Norway.

Gerhard von der Emde
Institut für Zoologie, Universität Bonn, Edenicher Allee 11-13, AVZ 1, 53115 Bonn, Germany.

Richard R. Fay
Parmly Hearing Institute, Loyola University, Chicago, IL 60626, USA.

Tamara M. Frank
Harbor Branch Oceanographic Institution, Bioluminescence Research Group, Fort Pierce, FL 34946, USA.

Roger T. Hanlon
Marine Biology Laboratory, Woods Hole, MA, 02543–1015, USA.

Craig W. Hawryshyn
Department of Biology, University of Victoria, Victoria, British Columbia V8W 3N5, Canada.

David M. Hunt
Department of Molecular Genetics, Institute of Ophthalmology, University College London, London, EC1V 9EL, UK.

Ad. J. Kalmijn
Faraday Laboratory, University of California, San Diego, Scripps Institution of Oceanography, La Jolla, CA 92093–0220, USA.

Joseph L. Kirschvink
Geology Division, California Institute of Technology, Pasadena, CA 91125, USA.

Friedrich Ladich
Institute of Zoology, University of Vienna, 1090 Vienna, Austria.

Michael F. Land
Sussex Centre for Neuroscience, School of Biological Sciences, University of Sussex, Brighton BN19QG, UK

N. Adam Locket
Department of Anatomical Sciences, University of Adelaide, South Australia, 5005 Australia.

Lars G.C. Lüthje
Physiologisches Institut, Universität Göttingen, 37073 Göttingen, Germany.

N. Justin Marshall
Vision Touch and Hearing Research Centre, School of Biomedical Sciences, University of Queensland, Brisbane, Queensland, 4072, Australia.

Joachim Mogdans
Institut für Zoologie, Universität Bonn, 53115 Bonn, Germany.

John Montgomery
Level 1, School of Biological Sciences, University of Auckland, Auckland, NZ.

Christa Neumeyer
Institut für Zoologie III, J. Gutenberg-Universität, Mainz 55099, Germany.

R. Glenn Northcutt
Department of Neuroscience, University of California San Diego, Scripps Institution of Oceanography, La Jolla, CA 92093-0201, USA.

John D. Pettigrew
Vision Touch and Hearing Research Centre, School of Biomedical Sciences, University of Queensland, Brisbane, Queensland, 4072 Australia.

Christopher Platt
Program Director, Sensory Systems, Natural Science Foundation, Arlington, VA 22230, USA.

Arthur N. Popper
Department of Biology, University of Maryland, College Park, MD 20742–4415, USA.

Olav Sand
Department of Biology, Division of General Physiology, Faculty of Mathematics and Natural Sciences, University of Oslo, Blidern, N-0316 Oslo, Norway.

Julia Shand
Department of Zoology, School of Animal Biology, University of Western Australia, Crawley, Western Australia, 6009 Australia.

Nadav Shashar
Interuniversity of Elat, Elat 88103, Israel.

Ole B. Stabell
Department of Natural Sciences, Faculty of Mathematics and Sciences, Agder University College, N-4604 Kristiansand, Norway.

Timothy C. Tricas
Department of Zoology, University of Hawaii at Manoa, Honolulu, HI 96822, USA.

Misha Vorobyev
Vision Touch and Hearing Research Centre, School of Biomedical Sciences, University of Queensland, Brisbane, Queensland, 4072 Australia.

Michael M. Walker
School of Biological Sciences, University of Auckland, Auckland, NZ.

Eric J. Warrant
Department of Zoology, University of Lund, S-22362 Lund, Sweden.

Lon Wilkens
Department of Biology, University of Missouri, St. Louis, MO 63121, USA.

Brian D. Wisenden
Department of Biology, Minnesota State University, Moorhead, MN 56563, USA.

H. Peter Zippel
Physiologisches Institut, Universität Göttingen, 37073 Göttingen, Germany.

Introduction

This volume began with a most delightful and exciting meeting of almost the same name that took place at Heron Island, the Great Barrier Reef, Australia, in March 1999. The topic of the Heron Island meeting, and of this volume, had been previously explored in a meeting that took place in Sarasota, Florida, in June 1985 and in a subsequent edited volume *Sensory Biology of Aquatic Animals* (New York: Springer-Verlag, 1988) that is now out of print. As organizers and editors of the 1985 meeting and 1988 volume, we are very pleased to take part in this new look at the topic.

The 1988 book was organized to foster conceptual and intellectual collaboration, in the broadest sense, among investigators who studied all aspects of sensory biology of aquatic animals. It started with a section on the physical properties of the underwater stimulus world as a background against which animal senses evolved. The subsequent sections were organized by sensory modality, from peripheral receptors to the central nervous system (CNS) processes, including some of the better-known models from different taxa. Broad topics included chemoreception, vision, hydrodynamic reception, hearing, equilibrium, and electroreception. It was designed by the editors to treat specific topics in an integrated and synthetic way, in contrast to conference proceedings, which are often organized around specific authors and the stories they have to tell arising from their own work. The editors of this new volume adopted our strategy wholeheartedly. As a consequence, this volume will stand for a longer time as a synthesis of an entire field, rather than as a series of related papers that might be more appropriate for specific journals. As such, the present volume could well serve as a text for a graduate course on the topic, just as our earlier volume did. It should be also noted that many of the papers from the 1999 conference were published as a special issue of the *Philosophical Transactions of the Royal Society* in 2000 (volume 355).

In *Sensory Biology of Aquatic Animals*, we identified a general impediment to understanding animal sensory behaviors. We argued that in studying one sensory modality, as most of us do, one is led to a rather intense concentration on one modality and to a loss of appreciation that, in nature, the functions of behavior with respect to a given object or event are likely served by multiple modalities. One example of this is in auditory neuroscience, where many investigators work to determine the mechanisms by which animals determine the location of the sound source. Of course, the focus of the investigator is on the source, as an object, and the requirement facing the organism is to behave appropriately with respect to it. While investigators often concentrate solely on the involvement of hearing in localization, it seems unlikely, except in exceptional circumstances, that behaving appropriately with respect to an object that happens to produce sound would engage only auditory processing. Yet, theories and experiments on sound source localization tend to investigate source determination and localization by hearing alone. In doing this, we may be asking too much of one sensory system. Generalizing from human behavior, we know that visual information and estimates of the plausibility of a location, at least, contribute to our perceptions of a source's location, particularly when auditory cues alone are ambiguous. Why would any other organism function essentially differently? Similar examples can be easily found for sensory behavior mediated by any other modality. A different type of multimodal sensory processing is known from "tasting" food, which involves not only the taste sense but simultaneously the tactile sense, allowing animals to sort out edible from nonedible particles inside the mouth. Odor plumes are composed not only of odor but also of the eddies of flow that disperse them. To detect these flavored eddies requires both olfactory and hydrodynamic receptors.

One of the features of this new volume, and something rather different from our earlier work, is that its design and organization may help remedy this sort of sensory myopia. Rather than organizing the volume according to modality, it is divided into parts that encompass the sensory requirements for general types of behaviors, such as Navigation and Communication, Finding Food and Localizing Sources, and Coevolution of Signals and Senses. In 1988, we expressed the hope that in a subsequent treatment of the subject, a more integrated multisensory story could be told. In the present volume, although individual chapters do not necessarily analyze behaviors in a multimodal context, the current book's content and organization fosters the view that behaving competently with respect to environmental objects and events is often a multimodal enterprise. Indeed, we would encourage readers not only to look at the chapters that deal with their own specific sensory interest, but also to look at the other chapters in each section to glean an integrated view of a topic and gain an appreciation for the complex sensory world of marine organisms. The chapters of the present volume tend to be longer and more general or expansive in topic, attempting to synthesize a larger field, often combining the physics and chemistry of stimulus signals with sensory biology and behavior. As in the earlier volume, evolution is an important theme here, either in explicit essays, or implicitly in the consideration of comparative work on sensory behavior and processing in many diverse species.

In summary, as organizers and editors of the 1988 book we are very pleased and honored that Drs. Collin and Marshall would think enough of our earlier work to follow up on it rather than take a totally new tack in exploring the sensory world of aquatic organisms. This volume will stand as a benchmark for future examinations of marine sensory biology.

Jelle Atema
Richard R. Fay
Arthur N. Popper
William N. Tavolga

Part 1
Navigation and Communication

Arthur N. Popper

Marine organisms have evolved a plethora of ways to sense their environment and then use these senses to provide information that allows them to communicate and to find their way. The three chapters in this section illustrate this perfectly. Chapter 1, by Popper, Fay, Platt, and Sand, discusses detection movement of the body through the vestibular senses and the evolution of hearing, which is the ability to detect signals at some distance from the fish, and outside of visual range. To the best of our knowledge, hearing by aquatic species is primarily confined to vertebrates, though there are many very noisy aquatic invertebrates and the lack of data on invertebrates is sufficiently large so as to make any broad generalization of their inability to hear very precarious.

Chapter 2, by Doving and Stabell, continues with the theme of long distance communication in discussions of one of the most fascinating of all abilities of fishes, the homing sense of salmon. Exactly how fishes find their natal streams is still not fully understood. However, Doving and Stabell provide an interesting new hypothesis that is based on use of a variety of stimuli, including infrasound (which is also covered in Chapter 1).

In Chapter 3, Walker, Diebel, and Kirschvink discuss a third sensory capability of fishes (and other aquatic vertebrates), the ability to detect and use magnetic fields for orientation and navigation. They provide insight into the use of this sense, and its physiological basis, and show its presence in diverse species.

Taking into account the other chapters in this book on vision, electroreception, and the lateral line, this volume illustrates once again the amazing abilities of marine organisms to sense and communicate in their environment.

1
Sound Detection Mechanisms and Capabilities of Teleost Fishes

Arthur N. Popper, Richard R. Fay, Christopher Platt, and Olav Sand

Abstract

This chapter, written from the perspective of four authors who have been studying fish bioacoustics for over 120 years (cumulative!), examines the major issues of the field. Each topic is put in some historical perspective, but the chapter emphasizes current thinking about acoustic communication, hearing (including bandwidth, sensitivity, detection of signals in noise, discrimination, and sound source localization), the functions of the ear (both auditory and vestibular, and including the role(s) of the otoliths and sensory hair cells) and their relationships to peripheral structures such as the swim bladder, and the interactions between the ear and the lateral line. Hearing in fishes is not only for acoustic communication and detection of sound-emitting predators and prey but can also play a major role in telling fishes about the acoustic scene at distances well beyond the range of vision. The chapter concludes with the personal views of the authors as to the major challenges and questions for future study. There are still many gaps in our knowledge of fish bioacoustics, including questions on ear function and the significance of interspecific differences in otolith size and shape and hair cell orientation, the role of the lateral line vis-à-vis the ear, the mechanisms of central processing of acoustic (and lateral line) signals, the mechanisms of sound source localization and whether fishes can determine source distance as well as direction, the evolution and functional significance of hearing specializations in taxonomically diverse fish species, and the origins of fish (and vertebrate) hearing and hearing organs.

1. Introduction

Both Aristotle and Pliny appreciated that fishes produce sounds (cited in Moulton, 1963). However, it was not until the mid-1800s that various investigators started to write about fish sounds and hearing. Controlled experiments to actually try to determine what fishes can hear were not performed until the early part of the twentieth century (G.H. Parker, 1902, 1903, 1904, 1909; and see Moulton, 1963 and Tavolga, 1971a, 1976b, 1977 for historical introductions to the field of fish bioacoustics). Following the seminal work of Parker, such notables as Karl von Frisch and one of his students, Sven Dijkgraaf (e.g., von Frisch, 1923, 1936, 1938a,b; Dijkgraaf, 1932, 1947, 1952; von Frisch and Dijkgraaf, 1935), provided evidence not only supporting the idea that fishes can detect sounds but also showing that sound detection is done by the saccule of the ear (one of the inner ear end organs—see below). Subsequent work provided significant insight into acoustic communication of fishes and demonstrated that many species produce, and use, sounds for a variety of behaviors (e.g., Tavolga, 1956, 1958; Demski et al., 1973; Myrberg, 1980, 1981; Zelick et al., 1999). Tavolga and Wodinsky (1963), followed by a number of other investigators (reviewed in Fay, 1988), measured hearing in many fish species and showed that at least some species can discriminate between different frequencies and intensities and detect the presence of a sound within substantial background noise. During these investigations, one of the real enigmas of fish hearing has been whether fishes can determine the location of a sound source (e.g., von Frisch and Dijkgraaf, 1935; van Bergeijk, 1964). More recent studies have shown that localization is indeed possible (e.g., Schuijf et al., 1972; Schuijf, 1975; Hawkins and Sand, 1977) and may also involve peripheral processing that provides directional information directly to the brain (Fay, 1984; Lu et al., 1998; Edds-Walton et al., 1999; Lu and Popper, 2001).

In this chapter, we provide some overview of how and what fishes hear and also comment on a number of recent investigations that have broadened our understanding of the hearing mechanisms and capabilities of fishes since the first conference of this type in 1988 (Atema et al., 1988).

2. Use of Sound by Fishes—Communication and Sense of the Environment

This chapter is directed at questions of sound detection, so we will only briefly discuss sound production and the use of sound by fishes for normal behaviors (see Demski et al., 1973; Myrberg, 1981; Zelick et al., 1999). Teleost fishes produce sound in several ways, none of which involves a larynx or syrinxlike structure as used by terrestrial vertebrates. Instead, fishes use a variety of different methods to produce sounds that range from simply moving two bones together to more complex mechanisms involving exceptionally fast muscles connected to the swim bladder. In this latter instance, the muscles contract at high frequencies to produce the fundamental frequency of the sound (reviewed in Tavolga, 1971b, 1976b; Demski et al., 1973; Myrberg, 1981; Hawkins and Myrberg, 1983; Zelick et al., 1999). The gas-filled swim bladder (or gas bladder) in the abdominal cavity may serve as a sound amplifier (although it has other functions as well—see Steen, 1971). Sounds produced in this way usually have most of their energy below 1,000 Hz, and most often below 500 Hz.

Fishes use sounds in behaviors including aggression, defense, territorial advertisement, courtship, and mating (reviewed in Tavolga, 1971a; Demski et al., 1973; Myrberg, 1981; Zelick et al., 1999). Some marine catfish have been suggested to use a form of "echolocation" to identify objects in their environment by producing low-frequency sounds and listening to the reflections (Tavolga, 1971b, 1976a).

The temporal pattern of fish sounds, rather than their frequency spectrum, has been considered the most important communicative feature of sounds generated by fishes (Winn, 1964). Temporal patterns of the sounds from different species vary, but there are still only limited data suggesting that fishes use this temporal patterning in discrimination. For example, Spanier (1979) showed that four species of damselfish (*Stegastes* spp.) discriminate between the species based on number of pulses and pulse rate in a sound. In addition, Crawford et al. (1997) found that two species of African mormyrids (*Pollimyrus* sp.) are able to discriminate between species using several acoustic parameters of gruntlike sounds, including pulse repetition rates. On the other hand, other sounds ("groans") were discriminated on the basis of fundamental frequency. Similarly, Myrberg and Riggio (1985) demonstrated that bicolor damselfish, *Stegastes partitus*, can discriminate between individuals of the same species, and they speculated that the discrimination was based on frequency components of the sounds.

While there are data on sound production

by several hundred fish species (e.g., Fish and Mowbray, 1970; Tavolga, 1971a, 1977) and data on use of sound for communication by perhaps 20 to 30 species, very little is actually known about acoustic communication in fishes, in part due to difficulties in studying underwater acoustic behavior (see Zelick et al., 1999 for a discussion of this issue). It is likely that many more fish species make and use sounds than currently reported in the literature. Indeed, Tavolga (1971a) pointed out that sound is probably the best channel for underwater communication for fishes since it is rapid, directional, and not impeded by the presence of visual barriers (e.g., coral heads) or darkness.

We have recently argued that fishes are likely to use sound for more than interspecific communication (Sand and Karlsen, 1986, 2000; Popper and Fay, 1997; Fay and Popper, 2000). It is now widely thought that terrestrial vertebrates glean a good deal of information about the general nature of the environment from biological and nonbiological sounds making up the auditory scene (Bregman, 1990). From this concept, we have suggested that vertebrate hearing evolved in aquatic ancestors of fishes as an adaptation to gain information about the environment in ways that were not obtainable by vision or the chemical senses, and especially about the environment beyond the range of these senses (Popper and Fay, 1997; Fay and Popper, 2000). It was only after hearing evolved that fishes are likely to have adapted the general sound-processing capabilities for communication. Thus, while all fishes probably detect sounds and are likely to use sounds to learn about their environment, a smaller set of fishes have actually evolved use of sound for communication.

3. Inner Ear Anatomy

3.1. Structure of Fish Inner Ears

Teleost fishes, like other vertebrates, have a bilateral pair of inner ears that lie inside the cranium on either side of the head at about the level of the hindbrain and are supplied by cranial nerve VIII (Fig. 1.1). The inner ears of fishes share many features with those of other vertebrates ranging from jawless fish to mammals (Retzius, 1881).

The inner ear is a complex structure of enclosed membranous tubes and pouches, often called the labyrinth. It has three dorsally located semicircular canal ducts, which are small looping tubes in nearly orthogonal planes. Each canal has a swelling at the base called the ampulla that contains sensory hair cells on a transverse ridge called the crista ampullaris. Ventrally, at the base of the canals, are three fluid-filled otolith organs (utricule, saccule, and lagena), each containing a dense calcified matrix (the otolith) overlying a sensory epithelium (or macula) containing hair cells. The utricle is located at the base of the canals, with a sensory macula as a bowl in the ventral part. The saccule is beneath the utricle and has a macula on the medial wall. The lagena is a vertically flattened pouch typically developing off the caudal part of the saccule and having a macula on its medial wall. There also may be a small macula neglecta near the utricle in some species (Retzius, 1881).

Unlike the pasty aggregation of otoconial crystals found in the otolith organs in other vertebrates (Carlström, 1963), each calcified mass in teleosts is usually solid and can be formed in highly specific shapes and relative sizes that are different in the three otolith organs and different in different fish species (Lychakov and Rebane, 2000; Popper and Lu, 2000).

Inner ear anatomy in nonteleost fishes is generally similar to that in teleosts, with some exceptions. For example, some chondrichthyans (sharks and rays) have a huge macula neglecta (Corwin, 1981), while nonteleost actinopterygians (ray-finned fishes) and sarcopterygians (lungfishes, coelacanth, and tetrapods) have otoconial masses that are not always solid (Retzius, 1881; Carlström, 1963). Some stem actinopterygians (sturgeons, bichir) have only two otoconial masses, with an apparently fused single sacculolagenar mass (Popper, 1978; Popper and Northcutt, 1983). The inner ear of chimeras (Holocephali) is not well known and may have only two otolith organs (Gauldie et al., 1987).

Fossils of isolated otoliths and of cranial remains showing parts of the bony labyrinth in extinct bony fishes do not suggest major differences in these species from the gross morphology of the ears of extant fishes, although it remains unclear where one, two, or three otolith organs occur (see Maisey, 1988; Schultze, 1990; Clack, 1996; Coates, 1999). The retention of a basic plan of the labyrinth in modern fishes,

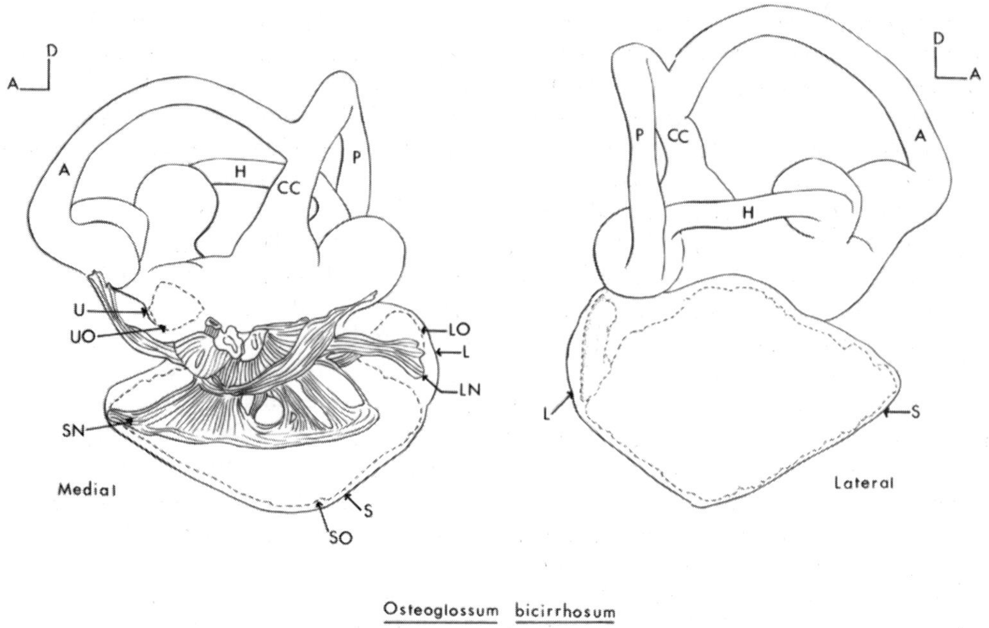

FIGURE 1.1. Drawing of the right ear of *Osteoglossum bicirrhosum*, the arawana (family Osteoglossidae), with medial on the left and lateral on the right. A—anterior semicircular canal; CC—common canal of semicircular system; H—horizontal semicircular canal; L—lagena; LN—lagenar branch of eighth cranial nerve; LO—lagenar otolith; P—posterior semicircular canal; S—saccule; SN—saccular branch of eighth cranial nerve; SO—saccular otolith; U—utricle; UO—utricular otolith. (From Popper, 1981. Reprinted with permission from *Journal of Comparative Neurology*, Copyright 1981 Wiley-Liss, Inc.)

such as teleosts, as well as in terrestrial tetrapods suggests that structural adaptations for postural control, and possibly even for hearing, evolved early to a quite successful design (see also Platt, 1988; Popper and Fay, 1997; Fay and Popper, 2000).

In a classic pair of volumes on the anatomy of the ears of vertebrates, Retzius (1881) noted that variations in the structure of the otolith organs of fishes were most evident in the saccule. Subsequent research has confirmed this idea. The utricle tends to be conservative in shape and sensory epithelium ultrastructure and very similar to the utricles of most other vertebrates (Platt, 1983; Lewis et al., 1985). An ariid catfish, *Arius felis*, has a utricle that is greatly hypertrophied and this structural modification may be associated with exceptionally sensitive low-frequency hearing and a narrow bandwidth (Popper and Tavolga, 1981; Tavolga, 1982). The clupeiform fishes (herrings, shads, anchovies, and relatives) have a unique utricle divided into three separate sensory maculae, one of which sits atop a gas-filled chamber that is an extension from the swim bladder (reviewed in Blaxter et al., 1981; see Section 4.5). Several clupeiforms in the genus *Alosa* are able to detect ultrasonic signals to over 180 kHz (Mann et al., 1997, 1998), and it has been suggested that the specialized utricle portion may be the end organ involved with ultrasonic hearing (Mann et al., 1998, 2001; Popper, 2000).

In most teleosts, the lagenar sensory macula is far smaller in area than that of the saccule and even the utricle. However, exceptions are found in the otophysan fishes (goldfish, catfishes, zebrafish, and relatives—all of which have a series of bones, the Weberian ossicles, connecting the swim bladder to the inner ear, Section 4.5) where the lagenar macular area has become equal to or greater than the sensory area in the saccule in these species (Platt, 1977, 1993). A similar enlarged lagena is found in the mormyrid fishes (elephant-nosed fishes), a group that is taxonomically far from the otophysans, suggesting that the enlarged lagena evolved separately in the two groups (McCormick and Popper, 1984). The mormyrids have a broad hearing bandwidth (McCormick and Popper, 1984) and an air bubble directly associated with the saccule (Stipetić, 1939; Koslowski and Crawford, 2000; Yan and Curtsinger, 2000; Fletcher and Crawford, 2001).

The saccule shows the most interspecific structural variation. Associated with wide variation in the overall size and shape of the saccular chamber are obvious differences in the shape and size of the otolith (reviewed in Popper, 1983; Popper and Lu, 2000) and in the orientation patterns of the sensory hair cells (Fig. 1.2) (Popper and Coombs, 1982; Popper and Platt, 1993). While the functional differences associated with different sizes and patterns of saccules are not known, it has been proposed that hair cell orientation patterns and perhaps overall saccular size (and particularly the size of the rostral end of the saccular epithelium) may correlate with hearing specializations (e.g., Popper and Platt, 1993).

3.2. Vestibular and Auditory Functions

The inner ear has two major sensory functions. One, the "vestibular" sense, is related to posture and balance, and the other "auditory" sense is hearing. For postural control, the fluids and masses of the inner ear allow inertial detection of acceleration vectors from head movements including rotations and linear movements, with the otoliths also detecting the static directional acceleration vector of gravity as a vertical reference cue. This sense can be essential for postural orientation of an aquatic animal in open water where there are no tactile cues

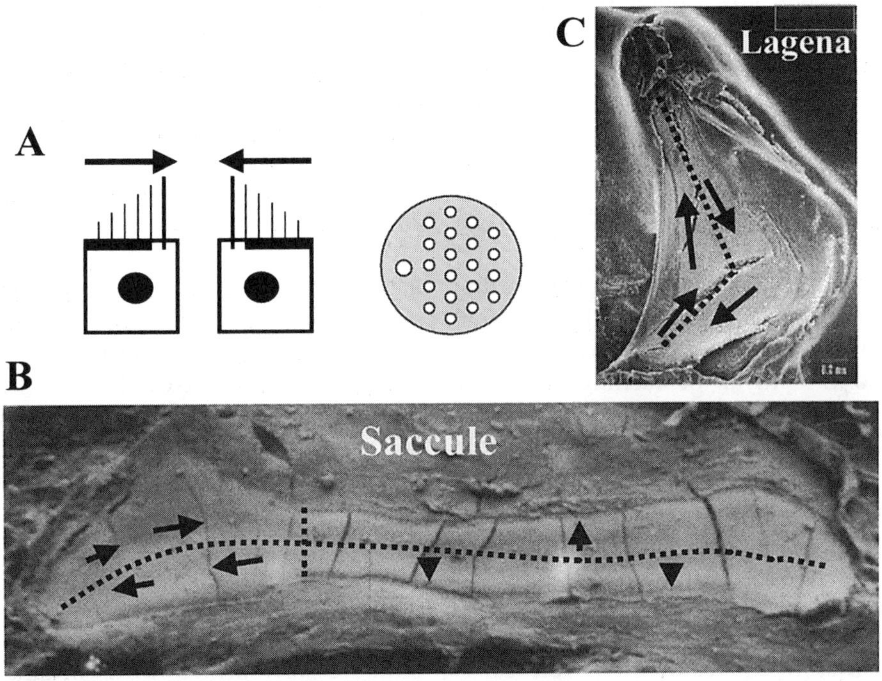

FIGURE 1.2. Saccular and lagenar maculae showing hair cell orientation patterns. (A) Schematic illustration showing two hair cells in side view and the top of one ciliary bundle. The arrows point from the stereocilia to the kinocilium in each case. These arrows illustrate the orientation of the hair cells on the epithelia shown in B and C. Each ciliary bundle has a single kinocilium (the larger circle in the top view) and a series of stereocilia that are graded in size. (B) Saccular macula from a tuna. Rostral to the left, dorsal to the top. This saccule, as in most teleosts, has ciliary bundles oriented in four directions. (C) Lagena macula from a butterfly fish. As in most other lagenae, there are two major hair cell orientation groups.

from a substrate and in turbid or deep water where there may be no light cues for the vertical. Gravity is also essential for normal development of the inner ear (Moorman et al., 1999).

Hearing is based on the detection of oscillatory movements at a range of frequencies. In teleosts, the otolith organs are stimulated directly by the particle motions associated with underwater sound fields and can be stimulated indirectly by particle motions created when sound pressure fluctuations are transformed into particle oscillations by gas-filled accessory organs such as the swim bladder. This sense allows perception of the surrounding spatial world (especially where visibility is poor), enhances detection of predators and prey, and serves as a communication channel.

The otolith organs of teleost fishes thus function both as vestibular and as auditory sensory organs. It remains unclear how these functions are distinguished by peripheral and/or central processing mechanisms. Signal detection for these two different senses requires very different peripheral tuning mechanisms involving high dynamic order, beyond simply frequency sensitivity for appropriate stimuli (Cortopassi and Lewis, 1998). In teleost fishes, the evidence that the utricle is primarily a vestibular organ and the saccule primarily an auditory organ derives primarily from experiments by von Frisch (1938b) and von Holst (1950) using lesions and behavioral responses. Physiological recordings from the isolated labyrinth of the skate (a chondrichthyan or cartilaginous fish) supported these roles (e.g., Lowenstein and Roberts, 1950, 1951). No such primary role has been established in teleosts and chondrichthyans for the lagena, and this end organ may participate in both vestibular and auditory functions (see reviews by Lowenstein, 1971; Platt, 1983).

It may be that every otolith organ has some degree of "multimodal" capability, even if it is "predominantly" a vestibular or an auditory end organ (Platt and Popper, 1981; Platt et al., 1989). Some evidence for this view comes from the apparent evolutionary adaptation of the saccule as a primary end organ for vestibular function in flatfish (Schöne, 1964; Platt, 1973) and the utricle as a primary end organ for hearing in the catfish *Arius* (Popper and Tavolga, 1981) and in clupeid fishes (herrings and relatives) (e.g., Denton and Blaxter, 1976; Blaxter et al., 1981; Popper, 2000).

3.3. Sensory Hair Cells—Overview of Basic Structure and Heterogeneity

The sensory receptors of the inner ear epithelia are hair cells, a class of ciliated mechanosensory cells found in the ear throughout the vertebrate lineage as well as in the lateral line of fishes and aquatic amphibians (reviewed in Lewis et al., 1985). The apical end of each cell faces the endolymph in the lumen of the end organ and has an apical bundle containing a single true cilium, the kinocilium, at one side of an array of several rows of stereovilli (Fig. 1.2A). The bundle is generally tapered, with the tallest stereovilli next to the kinocilium, and each subsequent row of stereovilli is shorter. The stereovilli are relatively stiff and pivot at a tapered base where they insert in the cuticular plate beneath the cell surface. The rows of graded height provide a many-tiered lever structure that is exquisitely sensitive to deflections; bending the bundle in the direction toward the kinocilium causes depolarization of the cell, giving directional sensitivity for the excitatory nerve signal (summarized in Hudspeth and Corey, 1977; Hudspeth, 1989). The bundle of stereovilli is necessary for the cellular response, while the kinocilium might provide some added mechanical coupling to movement of the surrounding fluid or accessory masses. Responses can be to static or dynamic bending. For gravitational detection, static bend would be an adequate stimulus, while for sound reception, a range of frequencies of oscillation would be an adequate stimulus.

Within this basic form, detailed structures of hair cells vary in different end organs or even regions of end organs. Teleosts and other nonamniote vertebrates were originally characterized as having only a single ultrastructural type of sensory hair cell that is cylindrical and innervated by both afferent and efferent neurons of the eighth cranial nerve (Wersäll, 1961). However, substantial diversity has been discovered in the ultrastructure and physiology of hair cells within individual end organs of teleosts (Sento and Furukawa, 1987; Steinacker and Rojas, 1988; Steinacker and Romero, 1991, 1992; Chang et al., 1992; Popper et al., 1993; Saidel et al., 1995; Lanford et al., 2000; Popper, 2000), including considerable variation within single end organs in bundle sizes (both lengths and numbers of stereovilli) (e.g., Platt and Popper, 1981; Popper and Platt, 1983), suggesting functional differences that are not yet understood.

3.4. Hair Cell Orientation Patterns

Because each hair cell has a directional sensitivity vector for maximal depolarization/excitation, maps can be constructed of each sensory surface showing the array of directional sensitivity vectors across whole epithelial sheets (Flock, 1964). All the otolith organ maculae show cells organized into populations of opposing orientation, with a boundary line that can be drawn between groups of different polarization vectors (Fig. 1.2). In most teleost fishes so far studied, the utricular pattern is like that in other vertebrates, with an inner group of cells with vectors facing "outward" and a broad band of hair cells around the rostral and lateral margin that face "inward" (Platt and Popper, 1981; Platt, 1983).

In the saccule, though, there are about five distinct patterns that have been found in a large range of teleosts (Popper and Coombs, 1982) (Fig. 1.3). The most common is called the "standard" pattern, in which the rostral half of the macula has a dorsal group of cells with vectors facing caudally and a ventral group with vectors facing rostrally. In addition, the caudal half of the macula has a dorsal group with vectors facing dorsally and a ventral group facing ventrally. In the otophysan fishes, the saccule has only two hair cell orientation groups: cells in the dorsal half have vectors facing dorsally and in the ventral half facing ventrally. This dual pattern, lacking a different rostral group, is like that in the saccule of tetrapods. The three other patterns have some type of elaboration in the rostral end of the saccular epithelium, and, whenever examined, it is apparent that fishes with such patterns have other specializations in the auditory system that appear to enhance hearing sensitivity (Popper and Coombs, 1982; Popper and Fay, 1999).

There is evidence (Section 5.3) that the array of directionally sensitive hair cells in the ear, oriented in a wide range of directions, enables fishes to directly determine the vectorial component of a sound field.

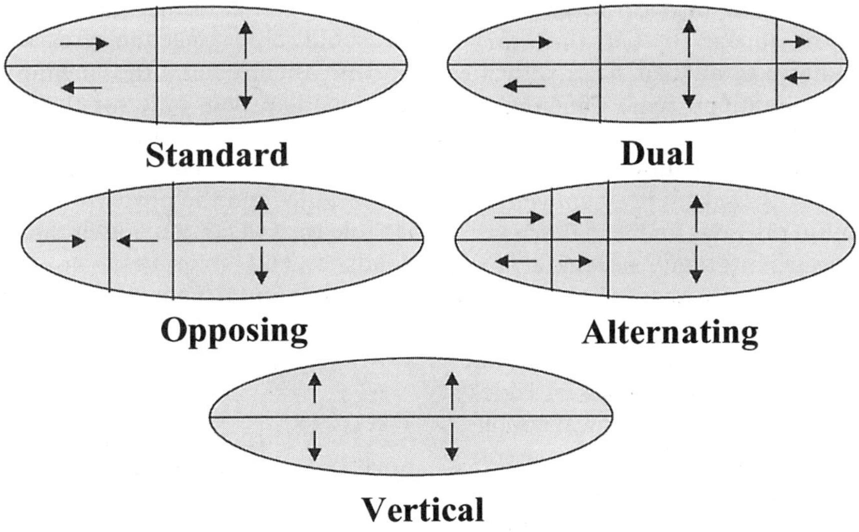

FIGURE 1.3. Saccular hair cell orientation patterns in different species. These can be divided into approximately five different patterns (see Popper and Coombs, 1982). Each pattern is distributed among different taxa, suggesting that each pattern arose several times. The most common pattern is the "standard." All but the vertical pattern (which is found in otophysans and in the unrelated mormyrids) have hair cells oriented both vertically and rostrocaudally. In all cases, the major variations from the standard pattern are found in the hair cells oriented rostrocaudally.

4. Mechanisms of Inner Ear Stimulation

4.1. Mechanics of the Sensory Organs—Accelerometers

The three semicircular canal organs are specialized for detecting angular accelerations produced by head rotations in the three rotational axes of roll, yaw, and pitch. When the head rotates, the inertial lag of fluid in the canal deflects a transverse partition called the cupula, which lies over a narrow transverse ridge, the crista ampullaris. Hair cells in the crista have their cilia, which are among the tallest of any hair cells, extending into the cupula, so deflection of the cupula results in deflection of the ciliary bundle and detection of head rotation. The structure of a cupula on top of a sensory crista in a semicircular canal is very similar to the structure of a cupula on top of a neuromast in the lateral line organs. All bundles within a single crista have their morphological orientation in the same direction so that they all are excited by head rotation in one direction and inhibited by the opposite rotation (Lowenstein et al., 1964). Structural details of each canal's shape and the duct cross section and the structure of each cupula have effects on spatial and temporal aspects of the physical response, and the canal organs appear to send the brain a signal related more to angular head velocities rather than accelerations (e.g., Oman et al., 1987; Rabbit et al., 1994).

The otolith organs detect linear accelerations produced by gravity and by head movements. The dense otolith provides an inertial mass that is pulled downward by gravity and that lags relative to the walls of the organ when the head moves. In a sound field, the entire fish is subject to accelerations from oscillations in the water mass, and the otolith's inertial mass provides a stimulus for the hair cells. Mechanical coupling to the hair cells depends partly on the viscosity of the surrounding fluid and the stiffness of the hair cell ciliary attachments. Morphological polarity of an individual cell acts as a directional filter for maximal response along one preferred axis so that each cell detects the component of the vector of shearing force in that direction within the plane of the local epithelium. A single macula typically has an array of cells giving sensitivity to a wide range of directional vectors, and for curved or cupped maculae, this sensitivity will be to vectors in all three dimensions.

Frequency filtering also occurs in the otolith organs. Because the force of gravity produces a sustained acceleration, detecting the vector of gravity for postural control requires hair cells that detect maintained spatial deflections without adaptation. Detecting transient linear motions, such as accelerations in all three major axes from swimming and other movements, requires units sensitive to both static and dynamic components of a stimulus. Detecting sound at a range of frequencies may require hair cells that can respond to oscillations at frequencies up to several kilohertz. Anatomical studies of tetrapod ears suggest that cells with taller ciliary bundles act as filters more sensitive to lower frequencies, and those with shorter ciliary bundles are more sensitive to higher frequencies (see Lewis et al., 1985). Some teleost fishes show a gradient of bundle heights across otolith organ maculae (see Popper and Platt, 1983), but a direct link of this gradient to auditory sensitivity in fishes has not been established.

The filtering of various frequencies and static sensitivity is currently not well understood but could rely on mechanical factors in the end organs themselves, the high-order tuning of individual hair cells for detecting tonic or phasic classes of stimuli, the specific modes of functional attachments between stereocilia and the otoliths, and the properties of the transmitter release and conduction in the octaval nerve dendrites and fibers (e.g., Boyle et al., 1991; Steinacker and Romeo, 1992; Fay, 1997; Fay and Edds-Walton, 1997b; Cortopassi and Lewis, 1998).

4.2. Peripheral Accessory Structures and Their Contribution to Hearing

The existence of a swim bladder or other gas-filled compartments in teleosts may provide an auditory advantage from the high compressibility of gas compared to water. When a volume of gas is exposed to oscillating pressure changes, it will display larger volume pulsations than a comparable volume of water. Thus the surface of a gas-filled chamber underwater will show larger radial motion amplitudes to pressure changes than the water particles in the absence of the chamber. These amplified motions may then be transmitted to the inner ear, thereby providing an auditory gain to the fish. By transforming sound pressure into particle motion in a sound field, a gas-filled chamber may make the fish sensitive to sound pressure, while the otolith organs remain sensitive to particle motion. The otolith organs may even be able to distinguish between the particle motions of the incident sound and those re-radiating from the gas-filled structures, making the fish sensitive to both the kinetic and pressure components of sound. Such ability may be essential for discrimination of distance (Schuijf and Hawkins, 1983) and direction (Buwalda et al., 1983) to a sound source (Section 5.3).

Teleost fishes may be roughly divided into three nontaxomic groups depending on their utilization of the swim bladder or other gas-filled structures as accessory hearing organs (Fig. 1.4; Fay, 1988; Popper and Fay, 1999). The *hearing specialists* either have a bony connection between the anterior part of the swim bladder and the inner ear or possess gas-filled vesicles in close or direct contact with the inner ear otolith organs. Species lacking gas-filled structures constitute the other extreme, while fishes possessing a swim bladder but lacking specialized connections fall in between. The latter two groups are commonly termed *hearing nonspecialists* (or hearing "generalists"). The hearing specialists have both higher sensitivity in the optimal frequency range and higher

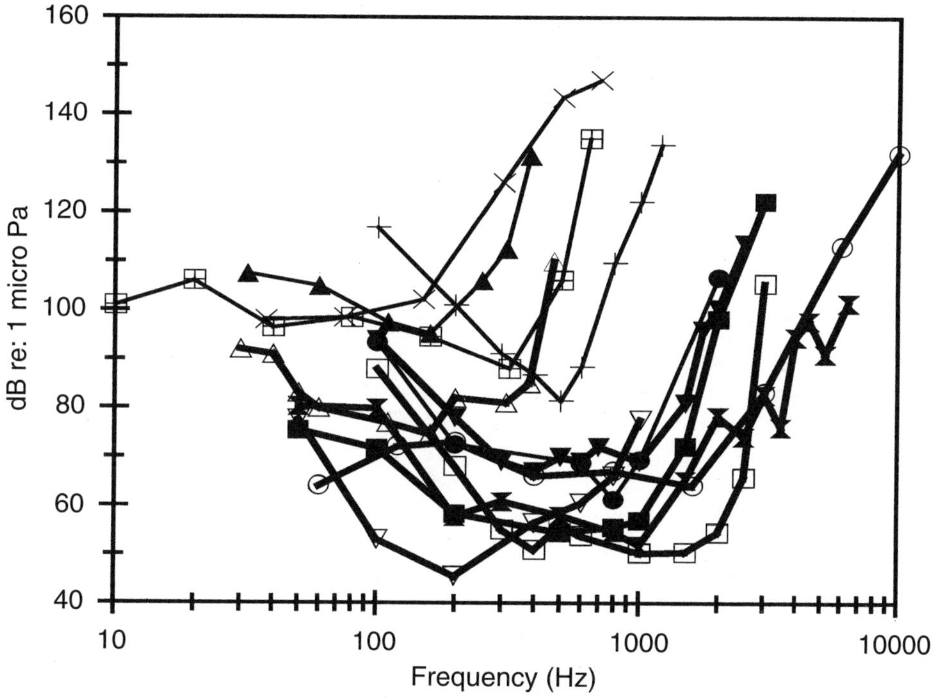

FIGURE 1.4. Audiograms for selected fish species illustrating specialists (thick lines) and nonspecialists (thinner lines). The species are: ■ *Carassius auratus* (goldfish) (Jacobs and Tavolga, 1967); ○ *Amiurus nebulosus* (catfish) (Poggendorf, 1952); ▽ *Arius felis* (marine catfish) (Popper and Tavolga, 1981); ⧖ *Astyanax jordani* (Mexican blind cave fish) (Popper, 1970); □ *Myripristus kuntee* (soldierfish) (Coombs and Popper, 1979); ▼ *Gnathonemus petersii* (elephant nose) (McCormick and Popper, 1984); △ *Gadus morhua* (Atlantic cod) (Chapman and Hawkins, 1973); X *Opsanus tau* (oyster toadfish) (Fish and Offutt, 1972); ▲ *Salmo salar* (Atlantic salmon) (Hawkins and Johnstone, 1978); + *Eupomacentrus dorsopunicans* (a damselfish) (Myrberg and Spires, 1980); ◫ *Negaprion brevirostris* (lemon shark) (Banner, 1967); ● *Equetus acuminatus* (chubbyu) (Tavolga and Wodinsky, 1963).

upper-frequency cutoff than the other groups (Fig. 1.4). However, for frequencies below 30–50 Hz, hearing sensitivity probably converges in all groups. This convergence occurs because the free-field particle motion oscillations will be exceeded by the pulsation amplitudes of a gas-filled bladder only above a certain frequency, which depends on both swim bladder volume and depth (Sand and Hawkins, 1973). Therefore, gas-filled bladders provide no auditory gain in the very low frequency range, where all species are insensitive to sound pressure (Section 4.4).

4.3. Swim Bladder Structure and Acoustic Properties

The swim bladder develops embryonically from the roof of the foregut. The connecting duct between the swim bladder and the gut is retained in adults of species called physostomes, but the duct degenerates in fishes called physoclists. In its simplest form, the swim bladder is an oval-shaped sac located in the abdominal cavity, but in many species the shape is more elaborate. It may, for instance, be divided into an anterior and a posterior chamber, as in many otophysans, or possess horns extending rostrally, as in the gadids and chaetodontids.

The functions of the swim bladder in sound production and as an accessory hearing organ are usually secondary to its function as a hydrostatic organ controlling buoyancy. For neutral buoyancy of the fish, the swim bladder volume should be between 5% and 7% of the body in freshwater and marine species, respectively, and similar values are found in depth-adapted specimens (Jones and Marshall, 1953). Notable exceptions where the bladder is small or even absent include bottom-dwelling species, like flatfish, and some fast pelagic predators that make frequent and rapid excursions between different depths, like mackerel. In most species the swim bladder is located so that its positive buoyancy acts at the center of gravity of the fish, thus avoiding pitch or roll of the body.

Physoclist species charge the swim bladder by gas secretion (see Steen, 1971), while physostomes can fill the bladder by gulping air at the surface and also may be able to refill the bladder below the surface by gas secretion (Sundnes and Sand, 1975). With the exception of the otophysans where the swim bladder gas may have an excess pressure of about 10 mmHg (Alexander, 1959a), the pressure of the swim bladder gas in depth-adapted fishes is generally similar to the surrounding water pressure, and the swim bladder compliance is very high (Alexander, 1959b; Sand and Hawkins, 1974). During rapid excursions to shallower depths, the swim bladder volume in physoclists is therefore inversely related to the hydrostatic pressure. Rapid dives to greater depths cause both reduced swim bladder volume and a pronounced negative pressure of the swim bladder gas relative to the surroundings (Sundnes and Gytre, 1972). During a prolonged stay at the new depth, gas secretion or absorption will readjust the swim bladder size toward its adapted volume. Both lasting (constant gas volume) and dynamic (constant gas mass) depth changes will change the acoustic properties of the swim bladder and hence its function as an accessory hearing organ (Sand and Hawkins, 1973). However, data describing the depth dependence of the auditory sensitivity in fishes are lacking.

A gas bubble in water can be regarded as a simple mass/spring system, where the spring factor is provided by the low elastic modulus of the contained gas and the mass results from the high inertia of the surrounding water (Minnaert, 1933). Compared to a free bubble, the swim bladder is heavily damped at the adaptation depth, with a Q value of about 1 in the Atlantic cod (*Gadus morhua*) (Sand and Hawkins, 1973). Swim bladder resonance will therefore have only moderate influence on the shape of the audiogram when the resonance frequency falls within the audible range. The resonance frequency of the swim bladder is inversely related to its linear dimensions and the hydrostatic pressure.

In the Atlantic cod, the resonance frequency at the adaptation depth is considerably above the value predicted from acoustic theory, possibly due to increased shear modulus of the surrounding tissues (Sand and Hawkins, 1973). The maintenance of a resonance frequency above the audible range ensures that the relative sensitivity to different frequencies is independent of adaptation depth.

4.4. Auditory Function of the Swim Bladder in Species Lacking Specialized Bladder–Ear Connections

The auditory potential of gas-filled compartments can be shown by using species that lack such structures, such as flatfish. Employing sound fields with different ratios between sound pressure and particle motion, Chapman and Sand (1974) measured auditory thresholds in the flatfish *Limanda limanda* and showed that this bladderless species is sensitive to particle motion throughout its hearing range. However, after being fitted with a small gas-filled balloon beneath the head, the fish became sensitive to sound pressure and the hearing range was extended to higher frequencies (Fig. 1.5A). At 200 Hz, the sound pressure threshold dropped about 20 dB after introduction of the balloon. Obviously, specialized structures connecting a gas-filled bladder with the inner ear are not essential in order to obtain at least some auditory gain.

Conversely, the value of an existing bladder can be shown by deflating it to show auditory deficits. In the Atlantic cod, a hearing nonspecialist possessing a swim bladder, hearing sensitivity is dependent on the presence of gas in the swim bladder (Sand and Enger, 1973). Figure 1.5B shows relative audiograms based on measurements of extracellular receptor potentials from the saccule of a cod suspended in the sea at a depth of several meters. Emptying the bladder through a hypodermic needle drastically reduced the hearing sensitivity and shifted the upper-frequency cutoff toward lower frequencies. However, in the low-frequency range, the hearing sensitivity was independent of gas content, as expected.

In the cod, the anterior part of the swim bladder is fairly close to the inner ear, so the amplified swim bladder motions are conducted through ordinary tissue to the inner ear efficiently enough to provide an auditory gain. The gas-filled bladder acts as a secondary near-field source, as a monopole that reradiates particle

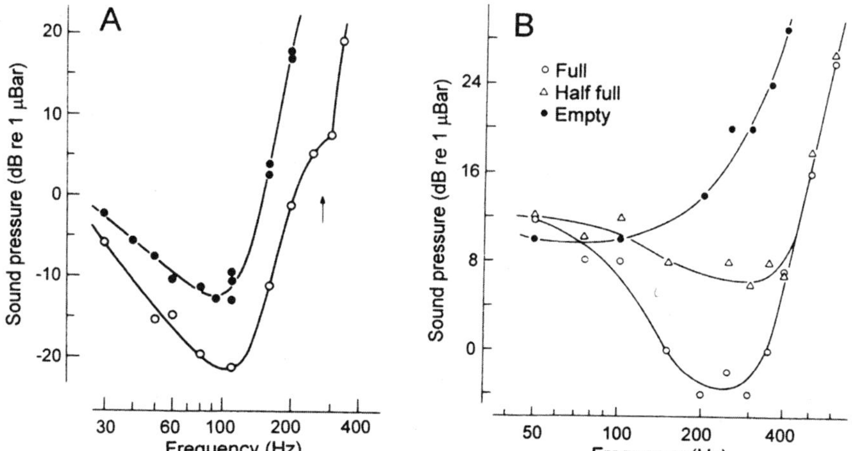

FIGURE 1.5. A gas-filled compartment provides an auditory advantage. (A) The normal audiogram (●) for a dab (*Limanda*), which lacks a swim bladder, compared with the auditory thresholds obtained after supplying the fish with a gas-filled balloon close to the head (○). The arrow indicates the balloon resonance frequency (from Chapman and Sand, 1974). (B) Relative audiograms in the Atlantic cod (*Gadus morhua*) for different swim bladder volumes (from Sand and Enger, 1973). In both A and B the gas-filled compartments enhance the auditory sensitivity in the upper part of the audiogram and shift the upper frequency cutoff to higher frequencies.

motions that will decay with the square of the distance to the bladder, if the transmission channel behaves like water. The auditory advantage of the swim bladder in hearing nonspecialists would therefore be expected to greatly depend on the distance between the ear and the swim bladder. At a moderate distance, the reradiated motions would reach values well below the particle motions in the incident sound, seemingly precluding an enhancement of hearing. For an elongated bladder, the particle motions parallel to the major axis are exaggerated, but the drop-off with distance is steeper than for a spherical bubble (see Schellart and Popper, 1992, for a more detailed discussion of the relationship between bladder shape and re-radiated particle motions).

It is not simple to estimate the auditory gain based on anatomy alone since the physical properties of the transmission channel between the bladder and the ear are essentially unknown and may well deviate from those of water. European eels (*Anguilla anguilla*) with ears about 10 cm from the swim bladder were still sensitive to sound pressure in the upper part of the audiogram (Jerkø et al., 1989), although the estimated reradiated particle motions at this long distance should have been too attenuated to provide an auditory gain. This indicates that the transmission channel between the swim bladder and the ear in hearing nonspecialists can be more efficient than water. It is thus possible that a swim bladder could confer an auditory advantage even in species with a relatively long distance between the bladder and the ear, although lacking other specialized linkages from the bladder to the ear.

4.5. Auditory Function of Gas-Filled Chambers in Hearing Specialists

In the otophysans, the anterior part of the swim bladder is mechanically coupled to the inner ear by an intervening chain of ossicles, the Weberian ossicles. Weber (1820) first described this anatomical structure and suggested an auditory function of the ossicles. In his classic paper, Poggendorf (1952) showed that the otophysan catfish *Ictalurus nebulosus* is sensitive to sound pressure and that surgical disruption of the Weberian ossicles reduced the hearing sensitivity by up to 30–40 dB. This estimate is supported in a recent theoretical analysis of the functions of the Weberian ossicles (Finneran and Hastings, 2000). Poggendorf (1952) noted that the experimental animals were still sensitive to sound pressure after impairment of the Weberian ossicles, and he was the first to also suggest an auditory function of the swim bladder in hearing nonspecialists.

Although experimental data to directly support the hypothesized mechanical function of the Weberian ossicles are scant (reviewed by Alexander, 1966), many studies have clearly demonstrated that otophysans display lower auditory thresholds in the optimal frequency range and higher upper-frequency cutoffs than hearing nonspecialists (see Fay, 1988; Finneran and Hastings, 2000).

The simplest way to mechanically couple a gas-filled chamber to the ear is to arrange these structures in close physical contact. However, shifting the swim bladder far forward can make the body unstable. An alternative is to position small, paired, gas-filled vesicles or bullae in direct contact with the auditory part of the skull or the perilymphatic space of the inner ear. Such arrangements have evolved apparently independently in more than 10 families that are not closely related (e.g., Jones and Marshall, 1953; Alexander, 1966; van Bergeijk, 1967; Tavolga, 1971a). Thin tubes may connect the vesicles to the swim bladder, as in the Clupeidae (Enger, 1967; Blaxter et al., 1981), or the paired gas vesicles may be completely separate from the swim bladder, as in the Anabantoids (Saidel and Popper, 1987; Yan et al., 2000) and mormyrids (e.g., Stipetić, 1939; Koslowski and Crawford 2000; Yan and Curtsinger, 2000; Fletcher and Crawford, 2001). In the clupeids, the coupling to the bladder through the thin tube (inner diameter of less than 10 µm) is of little significance during sound-induced volume oscillations but is apparently necessary for the functional state of the bulla in fishes making vertical excursions, where the more compliant swim bladder is thought to act as a gas reservoir for the bulla (Blaxter et al., 1979, 1981). Within the anabantoids (Saidel and Popper, 1987), all members possess gas-filled vesicles close to the ear, while in the holocentrids (Coombs and Popper, 1979) the presence of such vesicles varies by species. The species with gas vesicles have a wider auditory frequency range and greater sound pressure sensitivity than related species without vesicles, indicating at least circumstantially that such gas-filled vesicles can improve hearing ability (see also Yan et al., 2000).

4.6. Coupling Between Gas-Filled Vesicles and the Lateral Line

In the clupeids, the gas compartment of the auditory bulla is coupled to the lateral line canal system, which in this family is restricted to the head (Denton and Blaxter, 1976). A small window in the wall of the skull is covered by a flexible membrane and separates the perilymphatic space from the fluid in the lateral line canal system. Pressure-induced pulsations of the gas-filled bullae not only stimulate the otolith organs but also cause fluid motions in the lateral line canals (see Blaxter et al., 1981). A functionally similar, but anatomically different, coupling between swim bladder and the lateral line has recently been described in chaetodontid species (Webb, 1998; Webb and Smith, 2000). In these species, rostrolateral swim bladder "horns" extend anteriorly toward the otic capsules and make direct or indirect contact with the lateral line canal of the supracleithrum. This system is likely to impart sound pressure sensitivity to both the inner ear and a subdivision of the lateral line. The clupeids and the chaetodontids are certainly impressively equipped to analyze complex combinations of hydrodynamic and acoustic signals, although the functional implications of this ability are poorly understood. At excessive pressure oscillations, pulsations of an abdominal swim bladder may stimulate even the trunk lateral line, as observed in the cyprinid *Rutilus rutilus*, but the sensitivity of the lateral line to sound-induced swim bladder pulsations was too low to be significant under natural conditions (Sand, 1981).

4.7. Particle Displacement Versus Pressure

It is widely believed that each otolith organ of the ears of all fishes functions primitively as particle motion detectors, potentially in both the near- and far-fields. For any species in which fluctuations of the swim bladder or other gas-filled cavities can stimulate the otolith organs by reradiated particle motion, the question to be answered is whether this second, indirect mechanism actually is used. In addition, the two mechanisms may operate simultaneously in the same or different otolith organs and the relative contribution of each mechanism may be frequency and level dependent.

Sound pressure thresholds and audiograms can be interpreted only for the pressure-specialized species and have little or no meaning for unspecialized species (Fig. 1.4). Nevertheless, it is often said that the sound pressure hearing specialists hear with greater sensitivity and over a wider frequency range than hearing nonspecialists. For most sound sources (vibrating bodies) and under many environmental conditions, specialists will be able to detect the sound at lower source levels of motion or energy, at greater distances, and at higher frequencies than nonspecialists. Specialists detect lower source levels and a given source at greater distances because of the auditory gain provided by the swim bladder, and they have a higher frequency range of hearing than nonspecialists because the underwater acoustic particle motions are smaller at the higher frequencies for a given sound pressure level.

These considerations may help us understand some of the adaptive advantages of sound pressure specializations. The propagation of sound underwater depends on water depth with respect to sound wavelengths (e.g., Rogers and Cox, 1988). In shallow water, high frequencies with shorter wavelengths generally propagate better than lower frequencies. Thus, species in shallow water that evolved specializations for detecting sound pressure would gain a hearing advantage by listening at higher frequencies. However, there are wide differences in hearing specializations among shallow-water species, and it is not yet clear why some species have developed sound pressure specializations while others have not. Whether a species uses sound communication in social and other behaviors does not seem to predict the presence or absence of sound pressure detection specializations (Ladich, 2000).

5. Hearing Capabilities of Teleosts

5.1. Range of Hearing

A fundamental description of the hearing capabilities of any animal begins with the audiogram or a plot of the lowest detectable level for tones in quiet as a function of frequency. Most audiograms have been obtained by manipulating and measuring sound pressure level under the assumption that the ears respond in proportion to sound pressure, as they do in most terrestrial vertebrates. As discussed in Section 4.1, however, it is thought that the primitive mode of otolith organ stimulation is as an accelerometer responding to a kinetic component of underwater sound (acoustic particle displacement, velocity, or acceleration) and not to sound pressure per se. For a progressive sound wave in an ideal acoustic free field, pressure and the kinetic components of sound are simply related and one can be calculated from the other through the value for the specific acoustic impedance of the medium. However, due to the long wavelengths of underwater sound at frequencies detected by fishes, a progressive sound wave is extremely difficult or impossible to produce in the usual laboratory tanks (e.g., Parvulescu, 1964, 1967). Under these conditions, the effective impedance of the test tank medium cannot be reliably predicted, and thus the relationships between sound pressure and the kinetic components cannot be calculated with certainty. In general, the acoustic impedance of water in small laboratory tanks is lower than the ideal, and particle motion amplitude is thus greater than under many free-field conditions. Thus, a measurement of sound pressure at the location of the fish would not readily predict particle motion amplitude. With few exceptions, published studies of fish audiograms have not included independent measurements of acoustic particle motion. For those species that are not highly pressure sensitive (the hearing nonspecialists), a sound pressure threshold may be inadequate and misleading to describe or predict the animal's sensitivity to sound sources in its usual environment.

There have been several attempts to deal with these issues in the experimental literature. These include manipulating acoustic wave impedance using standing waves (e.g., Poggendorf, 1952; Fay and Popper, 1975) and conducting experiments in the field (in free fields or in environments that are usual for the species) (e.g., Chapman and Hawkins, 1973; Popper et al., 1973). Another approach has been to measure sound pressure thresholds as a function of distance between the source and the receiving animal (e.g., Chapman and Hawkins, 1973). In this case, a finding that the sound pressure thresholds do not vary with source distance suggests that the animal responds in proportion to sound pressure.

In the face of these considerable difficulties in specifying the effective component of underwater sound for fishes, most investigators have nevertheless forged ahead with descriptions of sound pressure sensitivity without first determining the validity of this approach. One reason for this is that pressure-sensitive hydrophones are readily available to investigators while transducers and methods for measuring acoustic particle motion are not. We are left, then, with many published audiograms of unknown validity (reviewed in Fay, 1988). The only behavioral thresholds for fishes that can be interpreted with confidence are sound pressure thresholds for "hearing specialists" that are known or reasonably assumed to be highly pressure sensitive (Fig. 1.4, thick lines) and particle motion thresholds for several hearing nonspecialist that have been demonstrated to be sensitive to particle motion (Fig. 1.6).

Figure 1.4 illustrates that hearing specialists detect sound pressure, with the lowest thresholds between 50 and 75 dB re: 1 μPa and in the frequency range between about 100 and 2,000 Hz. Pressure sensitivity generally declines

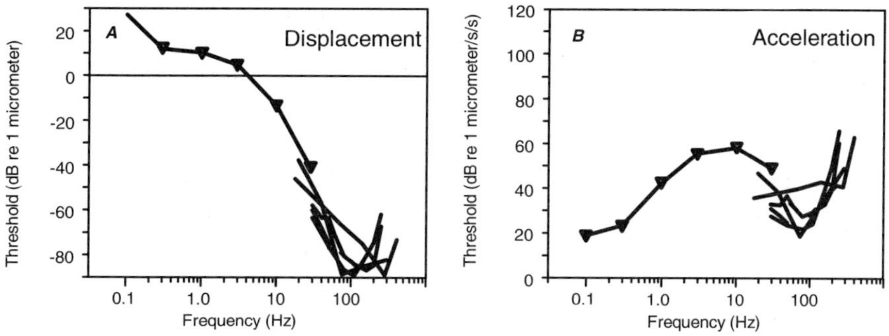

FIGURE 1.6. Behavioral particle motion audiograms for cod (Buerkle, 1969; Chapman and Hawkins, 1973; Offutt, 1973; Sand and Karlsen, 1986 [▼]); plaice (Chapman and Sand, 1974); dab (Chapman and Sand, 1974). Panels A and B plot the same thresholds in terms of displacement and acceleration, respectively.

at frequencies below 200–300 Hz and above 400–1,000 Hz. The most sensitive hearing specia-lists have approximately the same sensitivity as the most sensitive mammals and birds (see examples in Fay, 1988) when signal level at threshold is specified in units of acoustic intensity. The nonspecialists (including sharks) shown in Figure 1.4 (thin lines) generally hear best below 500 Hz and have poorer sensitivity than the specialists. Note, however, that the nonspecialist thresholds plotted here may not be quantitatively valid since these species probably do not respond to sound pressure (except, possibly, *Equetus acuminatus*). However, it is likely that the nonspecialist's exclusively low-frequency range of best sensitivity is reasonably accurate. Furthermore, their relatively poor sensitivity is probably qualitatively correct, in the sense that detecting particle motion at a level of about 0.1 nm would mean that the detectability of a given source in a usual environment would be poorer than for specialists that can detect sound pressure at 50–75 dB re: 1 μPa.

As discussed in detail in Section 5.2.2, it has been recently shown that at least some clupeid fishes are able to detect sounds to over 100 kHz (Mann et al., 1997, 1998, 2001). Other studies have shown that a number of species are able to detect sounds substantially below 50 Hz, in the infrasonic range (e.g., Sand and Karlsen, 1986).

Figure 1.6 shows the audiograms for species that respond in proportion to particle motion amplitude at threshold. These audiograms can look different depending on whether the thresholds are given in terms of particle displacement (panel A) or acceleration (panel B). At 100 Hz, displacements of 0.04 to 1.0 nanometers (rms) can be detected. These displacements are very small and correspond to the acoustic particle motion in an ideal free field at a sound pressure level below about 80 dB re: 1 μPa. This motion sensitivity also corresponds to the amplitude of motion of the mammalian basilar membrane at the threshold of hearing at best frequency (Allen, 1997). This same extreme displacement sensitivity of 0.1 nanometer or less has also been measured electrophysiologically for primary saccular afferents of goldfish (*Carassius auratus*: Fay, 1984) and oyster toadfish (*Opsanus tau*: Fay and Edds-Walton, 1997a). The lowest displacement thresholds occur between 50 and 300 Hz. Displacement sensitivity falls off steeply at frequencies below 100 Hz. However, acceleration sensitivity remains to frequencies at least as low as 0.1 Hz (Sand and Karlsen, 1986, 2000). Behavioral studies on infrasound detection and its mechanisms are discussed in more detail in Section 5.2.1.

It is difficult to explain the species variation in hearing mechanisms, sensitivity, and bandwidth. What are the adaptive advantages for sound pressure sensitivity and good hearing to several thousand hertz? Why have these adaptations developed among some species and families and not others? It is now reasonably clear that sound pressure sensitivity and a wide bandwidth of hearing are not necessarily coadaptations for detecting intraspecific communication sounds (e.g., Ladich, 2000). The correlation between communication sound production and auditory sensitivity is poor or nonexistent. Aside from the general advantage for obtaining more total information from the environment, one scenario for the development of sound pressure sensitivity is that animals evolving in shallow, relatively quiet underwater environments (possibly in freshwater) could gain fitness advantages by increasing sensitivity to the limits imposed by usual ambient noise levels and by increasing hearing bandwidth toward the higher frequencies that propagate more effectively in shallow water (Rogers and Cox, 1988). These and related issues are treated in more detail in Section 6.

5.2. Detection in the Infrasonic and Ultrasonic Range

Since the early 1990s, it has become apparent that at least some teleost species can detect infrasound (sound below 20 Hz) and ultrasound (sounds above 20 kHz). While sensitivity to infrasound may be a general feature of most fishes, the reported capability to detect ultrasound in some species suggests that fishes have been able to evolve highly specialized hearing capabilities in response to selective pressures, as also observed in other vertebrates, including mammals.

5.2.1. Infrasound Detection and Use

Fish detection of infrasound was not investigated until fairly recently since most laboratory sound sources were unable to produce undistorted tones below 20–30 Hz. In addition, most earlier fish audiograms indicated a steadily declining sensitivity toward lower frequencies (Fig. 1.4; Fay, 1988), and so infrasound detection in fishes seemed uninteresting. However, as we have already discussed (Sections 3 and 4), the unaided otolith organs of the inner ear are *not* sensitive to sound pressure but to linear accelerations. These organs may be modeled as critically damped, simple harmonic oscillators, and at frequencies below the natural frequency of the system, the deflection of the otolith relative to the sensory epithelium follows the acceleration of the organ (de Vries, 1950). The model indicates a working range of otolith organs reaching from 0 Hz to the upper frequency limit of hearing, and this range means that postural and locomotor activity provide "vestibular" stimulation that overlaps the frequency range for "auditory" stimulation.

A dramatic change in the shape of the audiogram becomes obvious when thresholds are related not to pressure but to particle acceleration, which is the more relevant stimulus parameter at very low frequencies, even in species possessing a swim bladder (Section 4.2; see also Fig. 1.6). The apparent drop in sensitivity toward low frequencies then disappears. It is therefore easy to reach erroneous conclusions when hearing capabilities and optimal frequency ranges in fishes are judged from the shape of sound pressure audiograms (Sand and Karlsen, 2000).

Infrasound sensitivity in fishes was first tested in the Atlantic cod using an acoustic tube and cardiac conditioning (see Fig. 1.6) (Sand and Karlsen, 1986). At 0.1 Hz the particle acceleration threshold was about 10^{-5} ms^{-2}. This represents a sensitivity to linear acceleration about 10,000 times higher than in humans, although a similar sensitivity to linear acceleration has been reported for the bullfrog saccule (Koyama et al., 1982). In the plaice (*Pleuronectes platessa*), a flatfish lacking a swim bladder, the threshold at 0.1 Hz is about $4 \cdot 10^{-5}$ ms^{-2} (Karlsen, 1992a), which corresponds to the particle motion thresholds previously determined for this species between 30 and 150 Hz (Chapman and Sand, 1974). Below the upper-frequency cutoff, the particle acceleration audiogram is virtually flat, as predicted by the model of de Vries (1950).

The infrasound sensitivity observed in these experiments depends on the otolith organs and not the lateral line. Unlike the dense otoliths, the mass of the lateral line cupulae is close to that of the surrounding water and no relative movements deflecting the sensory hair bundles will occur when the fish and the surrounding water are accelerated together in a sound field

(Section 7). Infrasound thresholds in the perch (*Perca fluviatilis*) were not affected by blocking the lateral line mechanosensitivity with Co^{2+} (Karlsen, 1992b), confirming that the otolith organs are the sensory system involved in the observed infrasound detection.

The acute sensitivity of at least some species of fishes to infrasound, or linear acceleration, may theoretically provide the animals with a wide range of information about the environment. An obvious potential use for this sensitivity is detection of moving objects in the surroundings, where infrasound could be important in, for instance, courtship and prey–predator interactions. The major acceleration components of the noise produced by swimming goldfish is in the infrasound range below 10–20 Hz (Kalmijn, 1989). Juvenile salmonids display strong avoidance reactions to infrasound (Knudsen et al., 1992, 1997), and it is reasonable to suggest that such behavior has evolved as a protection against predators. Infrasound has been used as an effective acoustic barrier for downstream migrating Atlantic salmon (*Salmo salar*) smolts (Knudsen et al., 1994), and it has recently been shown that downstream migrating European silver eels (*Anguilla anguilla*) are deflected by intense infrasound fields (Sand et al., 2000).

The ambient noise in the sea increases toward lower frequencies, and the spectral slope is particularly steep in the infrasound range (Urick, 1974). A highly speculative, and experimentally untested, hypothesis is that migratory fish may utilize infrasound patterns in the ocean for orientation and navigation (Sand and Karlsen, 1986). It may also be speculated that fishes could utilize their acute acceleration sensitivity for inertial navigation, detection of the relative speed and direction of layered ocean currents, and sensing of water movements associated with surface waves (Sand and Karlsen, 2000). The latter are distorted and refracted at shallow depths, providing potential cues for detecting underwater topography. The emerging picture is that fishes might be detecting a complex acoustic and hydrodynamic landscape, with distinct landmarks and information about distant structures, as well as the local environment of sounds and noise. In effect, the origin of the ear may have, at least in part, been involved with detection of this aspect of the auditory scene (see Sections 2 and 6). Moreover, the usefulness of this kind of detection may have predated fishes since sensitivity to infrasound, or low-frequency linear acceleration, has been found in several other aquatic animal groups including cephalopods (Packard et al., 1990) and crustaceans (Heuch and Karlsen, 1997).

5.2.2. Ultrasound Detection and Use

Just as we have recently learned that some fishes can detect very low frequencies, other data now show that certain species can detect sounds at very high frequencies. Despite the assumption that the only vertebrates with ultrasonic hearing are mammals, several reports in the early 1990s suggested that some fishes in the order Clupeiformes (herrings, shads, anchovies, and relatives) would swim away from echo sounders using sounds anywhere from 30 to 130 kHz (e.g., Nestler et al., 1992; Dunning et al., 1992; Ross et al., 1995, 1996). Several groups then went on to apply this finding to use ultrasound to repel species of shad, herring, and others from the water intakes of power plants (reviewed in Popper and Carlson, 1998).

Recent behavioral investigations demonstrated that American shad (*Alosa sapidissima*) are able to detect high-intensity sounds from below 100 Hz to over 180 kHz, while goldfish, used as controls, were insensitive to ultrasound (Mann et al., 1997, 1998, 2001; Popper, 2000). The hearing range of the American shad overlaps with the range of echolocation sounds used by dolphins, a major shad predator. Mann et al. (1998) showed that American shad will respond to echolocation-like sounds with rapid movement, and they further demonstrated that such sounds should be detectable by American shad at over 100 meters from the source. Thus, Mann et al. (1998) speculated that ultrasonic hearing evolved in some species for the detection of dolphin predators.

Ultrasound testing of other Clupeiformes showed that while the gulf menhaden (*Brevoortia patronus*) can detect ultrasound, several other species including the bay anchovy

(*Anchoa mitchilli*), scaled sardine (*Harengula jaguana*), and Spanish sardine (*Sardinella aurita*) can only detect sounds to about 4 kHz (Mann et al., 2001). While data are still needed on additional species, these results suggest that ultrasound detection may be limited to one subfamily of Clupeiformes, the Alosinae.

The most important question concerning ultrasound detection in Clupeiformes is how these sounds are detected. All known ultrasound detectors among both vertebrates and invertebrates are earlike structures, so the ear seems to be the likely site for detecting ultrasound in these fishes. Moreover, the utricle is very different in Clupeiformes than in any other vertebrate group, and so it has been suggested that this region has evolved for high-frequency, and in some cases ultrasound, detection (Mann et al., 1998). It remains unclear, however, why not all Clupeiformes are able to detect ultrasound. One possibility is that the specialized utricle in Clupeiformes evolved in shallow waters to enhance hearing (just as the Otophysi evolved Weberian ossicles) and that once they entered the oceans and encountered echolocating dolphins, some species evolved further modifications of the utricle for ultrasound detection, while other species did not.

Whether ultrasound detection is found in other fish groups remains to be seen. Mann et al. (1998) showed that goldfish could not detect ultrasound, while Astrup and Møhl (1993, 1998) found that the Atlantic cod is able to detect exceptionally loud sounds at about 39 kHz. However, they suggest that detection of 39 kHz in Atlantic cod may involve overstimulated skin receptors rather than the ear (Astrup and Møhl, 1998; Astrup, 1999).

5.3. Sound Source Localization

Most researchers believe that sound source localization is not possible by fishes unless the otolith organs receive input directly from the kinetic component of underwater sound as particle motion, since particle motion is a vector quantity that can be encoded by directionally sensitive hair cells (Sand, 1974; Fay, 1984; Fay and Edds-Walton, 1997a; Lu and Popper, 1997, 2001; Edds-Walton et al., 1999). The sound pressure component, on the other hand, is mediated by the swim bladder, or other gas chamber, that converts pressure fluctuations to motions detectable by the ears, and these motions then come equally to the two ears and always from the direction of this re-radiating source (van Bergeijk, 1964). Under the usual conditions, particle motion processing is most likely at low frequencies (below 300 Hz), at high sound levels (above 80 dB re: 1 μPa), and close to the sound source (within a wavelength or so).

Two kinds of experiments quantitatively measure localization acuity. One observes animals moving to choose (e.g., approach or orient toward) a particular source in a choice experiment. The other measures the smallest angle between two sources that still permits the animal to discriminate between them (the minimum audible angle—MAA). There are few quantitative studies on fishes using the first kind of experiment (e.g., Popper et al., 1973; Schuijf and Seimelink, 1974; Schuijf, 1975). However, there are some quantitative data from the second kind of experiment demonstrating that cod can discriminate between sources separated by 10°–20° in azimuth and elevation if the signal level or signal-to-noise ratio is sufficiently high (Chapman and Johnstone, 1974; Hawkins and Sand, 1977; Buwalda, 1981).

It has also been demonstrated that cod can discriminate between sound sources from opposing directions (180° apart) in the horizontal (Schuijf and Buwalda, 1975) and vertical (Buwalda et al., 1983) planes. In both experiments, it was shown that the phase relation between acoustic particle velocity and sound pressure provides information necessary for this discrimination. These results support the "phase model" of directional hearing by fishes developed by Schuijf (1975). This model proposes that the axis of acoustic particle motion is represented by the ensemble of primary afferents responding in a directional manner, and a decision as to which end of the axis points to the source is based on computations using the phase relations between particle motion and sound pressure.

Although it has been demonstrated that cod can discriminate between sources in different locations, including at different distances (Schuijf and Hawkins, 1983) at the same azimuth and elevation, it is still not clear whether, and to what extent, fishes can absolutely determine the actual location of sound sources. Furthermore, all of the foregoing considerations implicitly assume that the axis of particle motion possibly resolved by the auditory system makes a line that points to the source. As Kalmijn (1989) pointed out, this is true in the near-field only for monopole (e.g., pulsating sphere) sound sources, and most sources of likely biological significance are better modeled as dipole (e.g., translating sphere) or more complex sources in the near-field. In this case, the lines along which particles move are not generally parallel to the line from the source to the receiver. Thus, it is still not clear that resolving the axis of particle motion solves a fundamental problem of sound source localization for most biologically significant sound sources in the near-field. Clearly, more sophisticated behavioral experiments are required to determine whether fishes do, in fact, acquire useful information about the location of different types of sound sources.

An alternative view of sound source localization is that fishes may not be able to perceive the absolute location of sound sources but may be able to accurately approach or flee sources only by maintaining a particular body orientation with respect to the axes of particle motion received (Kalmijn, 1997). This strategy would allow effective guided approach or avoidance, even for dipole and more complex sources for which a sample of the axis of particle motion at one point in space may be insufficient to determine the source's actual location.

Recent experimental work on sound source localization in fishes investigated the peripheral neural codes that underlie the determination of the axis angle of acoustic particle motion. Responses from the saccular nerve of toadfish (*Opsanus tau*) and the unrelated sleeper goby (*Dormitator latifrons*) were obtained to 100-Hz translatory whole-body movements at a variety of angular axes in the horizontal and midsagittal planes (Fay and Edds-Walton, 1997a; Lu and Popper, 1998; Edds-Walton et al., 1999). Most saccular afferents had a directional response pattern resembling a cosine function, or the pattern expected from a hair cell's intrinsic directional response (Hudspeth and Corey, 1977).

Directional response functions (DRF) show the common cosinusoidal form of directional response patterns as plotted in polar coordinates (Fig. 1.7). In the horizontal plane (left side of Fig. 1.7), the best axes are near an azimuth of –30° to –60°. This angle roughly corresponds to the orientation of the saccular epithelium in the head as it hangs nearly vertically but diverging with respect to the fish's midline. Thus, the narrow range of best azimuths observed seems to be determined by the orientation of the receptor organ itself relative to the midline. In the midsagittal plane (right side of Fig. 1.7), a wide diversity of best elevations for the same five afferents show that this variation in directionality is likely from the diverse orientations in the vertical plane of saccular hair cells that the afferents innervate. Thus, hair cell orientation patterns on the saccular macula largely determine the range of best elevations observed in afferent response patterns. Distributions of best azimuth and elevation for over 400 such afferents recorded in toadfish saccules are consistent with the illustrative data of Figure 1.7 (Fay and Edds-Walton, 1997a; Edds-Walton et al., 1999).

For the toadfish and most other species investigated, the left saccule is angled to the left of the fish's midline while the right saccule is oriented an equal angle to the right. Thus, regardless of the overall directional orientation of the hair cells on the macula, stimulation in the horizontal plane and the resulting afferent responses will tend to be greatest when the relative otolith movement is parallel with the epithelial surface and along the general orientation axis of the organ in the head. Since the paired saccules are oriented differently in azimuth, there will be azimuth-dependent interaural differences in response magnitude for the two saccules, even though there are minimal interaural intensity differences reach-

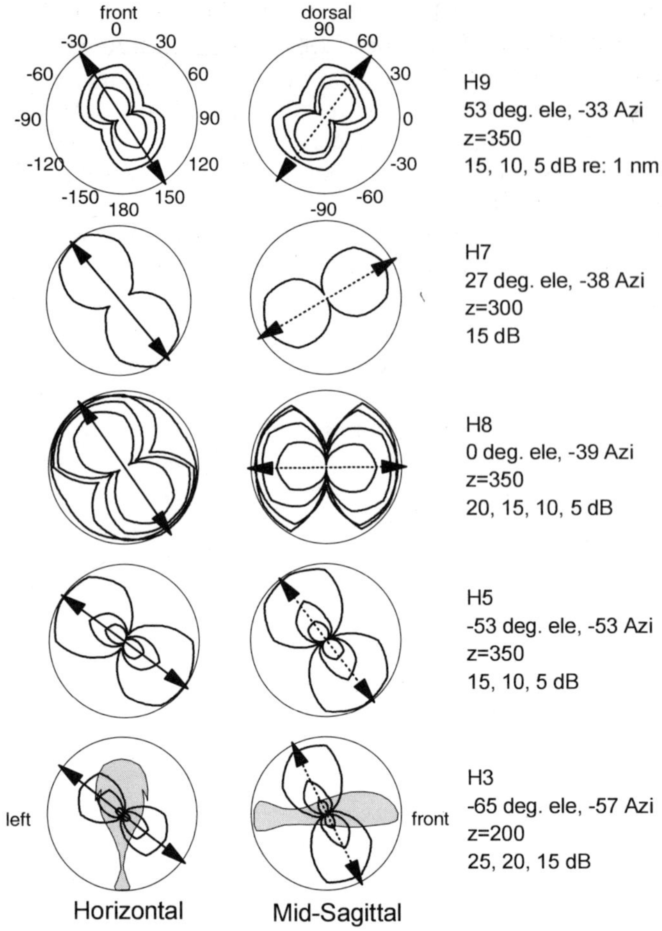

FIGURE 1.7. Directional response functions (DRF) for five afferents from the left ear of one toadfish (*Opsanus tau*). Response magnitude is plotted as a function of directional axis angle in polar coordinates. *Right column*: DRFs in the horizontal plane (azimuth). *Left column*: DRFs in the midsagittal plane (elevation). For most afferents, DRFs were determined at several displacement levels. On the right are afferent designation, best azimuth (degrees with respect to straight ahead), best elevation (degrees with respect to horizontal plane), maximum response magnitude (z) indicated by the circular scale, and displacement levels used in dB re: 1 nanometer, root mean square. (Unpublished data from Fay and Edds-Walton.)

ing the ears. These interaural response differences may be used to compute stimulus azimuth (Sand, 1974; Edds-Walton et al., 1999), so fishes would be conceptually like most other vertebrates studied in using interaural response differences as the basis for the computation of azimuth (Yost and Gourevitch, 1987), and hair cell orientation patterns over the surface of the otolithic epithelium may be unimportant for azimuthal sound source localization.

For altitude localization, it is important that the saccular epithelium is approximately a vertically oriented plane. Therefore, each differently oriented hair cell will respond best to motional stimuli having an axis elevation corresponding to its orientation on the sensory epithelium. An animal's determination of source elevation could be possibly made based on the profile of activity over a population of orientation-labeled saccular afferents (as originally conceived by Schuijf, 1975, for both elevation and azimuth).

5.4. Effects of Noise on Signal Detection

Studies of sound detection are typically carried out in unusually quiet environments because it has been shown that background sounds (noise) can influence detection thresholds for fishes (Tavolga, 1974; see reviews by Fay, 1988, 1992a). If a threshold is changed by background noise, it is called a "masked" threshold because another sound ("noise") partly hides ("masks") the sound of interest (the "signal") and raises its threshold for detection. In general, masking can be understood by assuming that signal detection is based on a decision about whether a detection channel is activated by noise alone or by a signal plus noise. It is assumed that a signal tone is detected at threshold by monitoring the output of the auditory filter centered on the signal frequency. Thus, only the masking noise falling within this filter's passband would be effective in interfering with tone detection.

The signal-to-noise (S/N) ratio at masked threshold is called the critical masking ratio (CR). The CR may be used to calculate an effective masking bandwidth in hertz (BW) using the equation BW = $10^{(CR/10)}$ (Fletcher, 1940). This estimate is known as the critical masking ratio bandwidth (CRB). In goldfish, the CR and the CRB increase with center frequency (Fay, 1974a). This upwardly sloping function means that the bandwidths of detection channels are less frequency selective (i.e., increase in width) and allow a greater noise power through them toward the higher frequencies. This sort of function for the CR has been observed in all vertebrates investigated (reviewed in Fay, 1992a) and is approximately independent of signal level. Filters of this sort would be expected to produce the phenomenon of "critical bandwidth," or the directly measured bandwidth within which a noise is integrated in its masking effect. Critical bandwidths have been demonstrated in goldfish (Enger, 1973; Tavolga, 1974) and cod (Hawkins and Chapman, 1975) as well as in all other avian and mammalian species investigated. Masking experiments on a variety of fish species are in accord in demonstrating that all fishes investigated have auditory filters analogous to those measured in humans and other vertebrates (reviewed in Fay, 1988, 1992a). By having a bank of auditory filters, fishes use a strategy that appears to be a widely shared characteristic of vertebrate auditory systems. Auditory filters appear to be most useful for restricting the frequency range for which noise is an effective masker of narrowband signals. In natural environments, most sounds to be detected are usually masked to some degree by other environmental noise, and auditory filters effectively increase S/N under normal listening conditions. Filters also probably play a role in resolving the shape of a signal's spectrum for sound source identification and auditory scene analysis.

5.5. Effects of Signal Duration

The threshold for detecting a sound generally improves as sound duration increases in most vertebrates studied (Fay, 1992a), including goldfish (Fay and Coombs, 1983) and cod (Hawkins, 1981). To a first approximation, threshold functions of duration for goldfish (Fay and Coombs, 1983) are power functions with an exponent (slope) of about −1.0, meaning that a 10-fold increase in duration produces about a 10 dB reduction in threshold. At long durations, these functions are limited by the temporal limitations of a hypothetical integrator that combines neural activity, or the decisions based on this activity, over time. For goldfish, durations greater than about 400 ms do not lead to further lowering of thresholds.

5.6. Sound Feature Discrimination

The acuity with which animals are able to distinguish between different sounds determines the amount of information about sources that is obtainable using the auditory system. Differences tested are usually for levels of intensity and for frequency.

Level discrimination thresholds (LDT) are measured by determining the smallest discernable difference between sounds differing only in intensity or level. This ability has obvious survival value (e.g., its role in the perception of source distance, changes in distance, and source level). LDTs for goldfish demonstrate that for stimulus durations between 10 and 200–300 ms, LDTs for both tones and noise decline nearly linearly with log duration (about −3 dB per 10-fold increase in duration), reaching an asymptote of 2–3 dB at durations greater than about 200 ms (Fay, 1985, 1989a). For long-duration tones, LDTs are independent of frequency and remain constant with increasing overall level, consistent with Weber's Law (Fay, 1989a). LDT values and the effects of overall level, duration, and frequency are similar in goldfish to those of other vertebrates studied, including humans (reviewed in Fay, 1988, 1992a), so it seems unlikely that there are significant species differences in level discrimination capabilities among fishes.

Frequency discrimination by fishes has been of great interest since the early studies of von Frisch and his colleagues (von Frisch, 1936; Wohlfahrt, 1939; Dijkgraaf and Verheijen, 1950). Otolith organs lack an analogue of the cochlear's basilar membrane, and it appears unlikely that a traveling wave occurs along the sensory epithelium. The movements of the otoliths may be frequency dependent (Sand and Michelsen, 1978), and some frequency-dependent regional damage of the sensory epithelium by intense sounds has been claimed (Enger, 1981). However, a macromechanical place mechanism of frequency analysis (von Békésy, 1960) seems not to occur, and any frequency analytic capacities would have to be explained mainly by other mechanisms, such as hair cell tuning (Crawford and Fettiplace, 1981), hair cell micromechanics (Fay, 1997; Fay and Edds-Walton, 1997b), or time-domain processing (Wever, 1949; Fay et al., 1978).

The smallest difference in frequency (*d*f) required for reliable discrimination is the frequency discrimination threshold (FDT). FDTs have been determined as a function of frequency for several hearing specialists and nonspecialists. In general, FDTs increase monotonically with frequency (f), maintaining a *d*f/f (Weber ratio) of about 0.04 for goldfish. The hearing specialists studied are more sensitive to frequency changes than are the hearing nonspecialists (*d*f/f at or above 0.1). FDTs tend to remain constant with changes in overall level for goldfish (Fay, 1989b). These features of the FDT in fishes are essentially similar to those of all vertebrates tested (Fay, 1992a), and the values of *d*f for goldfish and other hearing specialists are within the upper range of values determined for mammals and birds (reviewed in Fay, 1974b, 1988, 1992a).

The question of the mechanisms underlying frequency discrimination in fishes has not been completely resolved. The early (e.g., Fay, 1970a) suggestion that tone frequency is represented in patterns of interspike intervals in auditory nerve fibers has received support from measurements showing that the temporal error with which primary saccular afferents phase-lock to tones is approximately equal to the behaviorally measured *d*f expressed as a period discrimination threshold throughout the frequency range of hearing (Fay et al., 1978). On the other hand, primary afferents are frequency selective to some degree (Furukawa and Ishii, 1967; Fay and Ream, 1986; Fay, 1997), and this selectivity is enhanced at or below the auditory midbrain (Lu and Fay, 1993), apparently through lateral inhibition (Lu and Fay, 1996). Thus, in fishes as in all other vertebrates investigated, an across-cell spike probability representation of frequency exists in addition to a temporal representation at frequencies below about 3 kHz, and it cannot be ruled out that a code based on an across-neuron profile of activity plays a role in behavioral frequency analysis by fishes (cf. Enger, 1981).

6. The Sense of Hearing in Fishes

Psychophysical studies on detection and discrimination thresholds described above measure the limits of hearing sensitivity and acuity and thus help define the sense of hearing in fishes in a way that is comparable with psychoacoustic studies on human and other terrestrial vertebrate listeners. However, it is clear that the human sense of hearing is far richer and more complex than is usually revealed in psychoacoustic studies. The most general functions of the human auditory system may be to analyze the complex mixture of sounds reaching the ears from multiple sources into component frequencies and temporal events, group those components that arise from each source, and then synthesize a representation of the auditory scene (Bregman, 1990) in which the source of each sound is perceived as an object or entity. The human sense of hearing thus permits individuals to draw the right conclusions about the objects and events in the local world so that they may act accordingly. This general function would seem to be of adaptive value to all species, including fishes.

These important hearing functions have been studied quantitatively in nonhuman animals, including goldfish, using a paradigm known as "stimulus generalization" (e.g., Fay, 1970b, 1995,

2000; Fay et al., 1996). This method reveals what stimulus dimensions are salient or what the animals normally "pay attention to" and presents a way to measure an animal's perception of sound similarity. A gradient of response magnitude along a stimulus dimension such as frequency has been interpreted as revealing a parallel perceptual dimension (Guttman, 1963), perhaps similar to pitch or timbre. Generalization experiments have shown that goldfish behave as if they have perceptual dimensions similar to pure-tone pitch, complex pitch, roughness, and timbre as defined in studies on human listeners. In addition, stimulus generalization methods have been used to demonstrate analytic listening and auditory scene analysis in goldfish (Fay, 1992b, 1998a,b, 2000).

In a simple example of stimulus generalization, goldfish were first conditioned to a pure tone and then tested for response to novel tone frequencies that flanked the conditioning frequency (Fay, 1969). Responses to novel frequencies above and below the conditioning frequency fell to chance levels along substantially symmetrical, monotonic gradients over tone frequency (e.g., Fay, 1970b). Using the stimulus generalization method, the ability to "hear out" or independently analyze the individual frequency components in a multicomponent complex sound (analytic listening) was studied in goldfish (Fay, 1992b). Goldfish were shown to acquire independent information about the frequencies of two tones presented simultaneously and can be said to listen analytically. This demonstration of simultaneous frequency analysis (as called for years ago by van Bergeijk, 1964) suggests that this fundamental aspect of hearing is a primitive character shared among humans (Hartman, 1988), fishes, and perhaps all vertebrates. This capability plays an important role in permitting listeners to determine the individual simultaneous sources making up an auditory scene (Bregman, 1990).

Generalization experiments using periodic click trains at different rates suggest that goldfish have a perceptual dimension that is continuous and monotonic with pulse repetition rate and that this dimension has at least some of the properties of periodicity pitch or roughness perception as defined in experiments on human listeners. In addition, goldfish behave in stimulus generalization experiments as if they are able to distinguish complex periodic stimuli on the basis of both pitch (pulse repetition rate) and timbre (spectral and temporal envelope of the pulse) (Fay, 1995).

These kinds of effects suggest that goldfish probably perceive the pitch and timbre of complex sounds simultaneously and independently. This behavior is analogous to the kinds of knowledge humans have about pitches and sources. Imagine listening to a musical instrument (e.g., a flute), playing a given note (e.g., A). Humans are able to recognize the source and also recognize a pitch change when another note is played. Now if a saxophone plays the same notes, humans recognize that the pitch may be the same but that the instrument is different. Recognizing and comparing notes is a pitch judgment, and identifying the source instrument is a timbre judgment. In general, human listeners share some of these perceptual phenomena with goldfish.

Some experiments on human hearing have been designed and interpreted on the assumption that an important, and perhaps primary, function of the sense of hearing is the perceptual formation and analysis of the various sound sources normally making up an auditory scene (Bregman, 1990). One of the fundamental concepts of auditory scene analysis is that a perceptual correlate of a sound source or event (an "auditory stream") is formed by analyzing sound features from the complex mixture and then combining them in ways that lead to likely hypotheses regarding the identity of the sources. Determining one source from a mixture of sources is an example of what is termed auditory stream segregation in human listeners. Since most environments contain multiple, independent, and simultaneous sound sources, appropriate behavior with respect to these sources would seem to require perceptual processes similar to stream segregation and scene analysis. Experiments using classical conditioning in a stimulus generalization paradigm were carried out on goldfish to investigate behaviors that might be consistent with auditory stream segregation in a fish (Fay, 1998b, 2000). Results of this sort of experiment indicate that two concurrent pulse trains making up a conditioning stimulus were "heard out" or analyzed independently. This pattern of results is consistent with the hypothesis that goldfish are capable of auditory stream segregation as it is generally defined for human listeners (Bregman, 1990). These perceptual behaviors are thus shared among humans, starlings (Hulse et al., 1997), and goldfish.

The results of these and other generalization experiments have led to the conclusion that goldfish acquire a lot of information about the characteristics of sounds. So far, these experiments have been unable to demonstrate qualitative differences between fishes and human (or any other vertebrate) listeners in sound source perception and in the sense of hearing. Goldfish behave as if they have perceptual dimensions corresponding to spectral and complex pitch, timbre, and roughness as observed in human listeners and are able to listen analytically to mixtures of tones and complex sounds. These results suggest that these aspects of auditory perception are primitive in the sense of being shared among vertebrates. In addition to these sound feature analyses that seem shared with other vertebrates, fishes may even be capable of a greater capacity for processing information from the sound field. For example, through the simultaneous reception of acoustic particle motion and sound pressure, fishes may also be able to perceive acoustic intensity, or the direction and magnitude of acoustic energy flow (Fay, 1984). This, combined with infrasound processing and the lateral line's capacities for imaging local hydrodynamic flow (Coombs et al., 1996; Chapter 7 and Section 7, below), may give fishes an image of local objects and events that may exceed in complexity that of most terrestrial animals.

7. Division of Labor Between the Ear and Lateral Line

Within their linear range, the response of hair cells is proportional to the displacement of the ciliary bundle (Hudspeth and Corey, 1977). For the lateral line, a free (or "superficial") neuromast responds to the velocity of the water relative to its cupula, while the enclosed canal neuromasts respond to acceleration of the water relative to the fish (Kalmijn, 1989, Kroese and Schellart, 1992). As noted in previous sections, the unaided otolith organs respond to whole-body acceleration of the fish. Both the lateral line canal system and the inner ear otolith organs are thus acceleration detectors. However, the lateral line canal system responds to the spatial differences in the acceleration of the water along the length of the sensory arrays, whereas the otolith organs detect the acceleration of the water averaged over the volume occupied by the fish (Platt et al., 1989). Close to a moving source, the water acceleration vectors vary greatly in strength and direction over short distances and thus provide an excellent stimulus for the lateral line (see Kalmijn, 1989, for a detailed discussion). As the distance to the source increases, the field becomes more nearly uniform. As a consequence, the lack of local differential motions no longer provides a stimulus for the lateral line since the overlying cupula has a density that is nearly identical to the surrounding fluid. In contrast, the dense otoliths make the otolith organs suited to detect the whole-body motions of the almost neutrally buoyant fishes. This difference in detection range between the lateral line and the inner ear was emphasized in the classic review of lateral line function by Dijkgraaf (1963). However, in spite of the physical evidence, it has been a long-lived notion that the lateral line may be sensitive to propagated sound at considerable distances from the source (for review, see Sand, 1981, 1984).

The relative insensitivity of the teleost lateral line to homogeneous water motion was demonstrated experimentally by Sand (1981) by recording from the trunk lateral line nerves in a cyprinid, the roach (*Rutilus rutilus*). Responses to local water movements produced by a vibrating sphere close to the organ were compared with gross vibrations of the fish and surrounding water. The latter stimulus was generated by suspending the fish freely in the center of a water-filled acoustic tube in order to simulate the situation a fish will encounter at some distance from a sound source. Figure 1.8

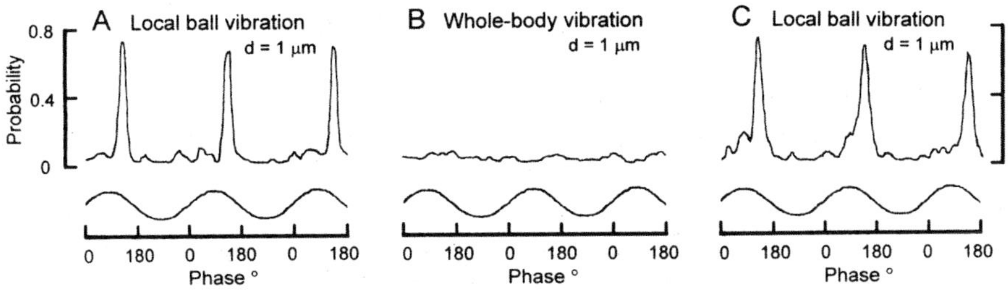

FIGURE 1.8. Recordings from a trunk lateral line nerve fiber in *Rutilus* showing the synchronization between afferent activity and applied vibrations. The synchronization is expressed as the probability of spike occurrence as a function of stimulus phase. (A) Initial response to vibrating ball stimulation close to the sensitive canal pore. (B) Response to vibration of the fish and surrounding water in an acoustic tube. (C) Control recording of the response to the vibrating ball obtained after the whole-body vibration in the acoustic tube. The afferent activity was recorded by an implanted electrode, and the fish was freely suspended in the tube. The vibration frequency was 50 Hz, and the estimated particle displacement at the canal pore was 1 µm in A and C. The measured particle displacement centrally in the tube was 1 µm in B. (Based on data from Sand, 1981.)

shows the response to 50 Hz, 1 µm, water displacements at the canal pore caused by the vibrating sphere compared to vibrations at the same frequency and amplitude of the fish and surrounding water in the tube. The lateral line was clearly insensitive to the latter stimulus except at excessively high sound pressures that were far above natural conditions. Direct measurements of the relative movements between the fish and surrounding water have indicated that the lateral line is only stimulated by moving objects within a distance of less than a few body lengths (Denton and Gray, 1982, 1989).

In order to elucidate the different tasks of the inner ear and the lateral line in the detection of moving objects, Enger et al. (1989) studied the feeding behavior of the predatory bluegill (*Lepomis macrochirus*) under infrared illumination. When presented with live goldfish or moving artificial prey in total darkness, bluegills with intact lateral lines perform sudden attacks when they are within about 2 cm from the target. Before the final attack, the bluegills frequently approached the prey from the tank periphery in smooth, apparently deliberate, moves. Directed approaches initiated by moving artificial prey were observed only if the generated water accelerations were below 10 Hz. When the function of the lateral line was blocked by cobalt ions (Karlsen and Sand, 1987), the bluegills never attacked the goldfish or the simulated prey, whereas the deliberate, directed approaches were intact. The authors concluded that the deliberate approaches from a distance are mediated by the inner ear in complete darkness, while a successful final attack requires the lateral line for its elicitation.

Because the neuromasts of the lateral line are arranged in extended, linear arrays, a fish can simultaneously sample the distribution of the water acceleration relative to the fish along the body. From the shape of the field, the animal may estimate the exact position of the target. In the final strike, the predator sucks in the prey if the mouth is positioned accurately relative to the prey. Even a small deviation may make a strike unsuccessful, explaining why the superior close-range locating power of the lateral line, compared to the inner ear, is essential for the final strike in darkness. The less accurate, but longer ranging, directional hearing linked to the inner ear may be essential for guiding the predator into the vicinity of the prey. Comparable arguments may of course be given to explain the roles of the inner ear and lateral line for the predator avoidance behavior in prey fishes. A similar experimental approach has been adopted by Coombs and coworkers in studies of the prey-catching behavior of the mottled sculpin (*Cottus bairdi*) (see Chapter 7 for a review).

The insensitivity of the lateral line organs to water movements generated by distant sources precludes masking by the major fraction of the hydrodynamic background noise. The lateral line organs are thus well suited to detect the minute, local flow fields caused by moving objects at close range in an inherently noisy environment (Denton and Gray, 1982; Sand, 1984). Furthermore, the superficial neuromasts have low-pass filtering properties, responding to the velocity of water flow relative to the fish, while the enclosed canal organs possess high-pass filtering characteristics and are sensitive to water acceleration (Denton and Gray, 1983; Kalmijn, 1989; Kroese and Schellart, 1992). The free neuromasts are thus permanently stimulated in running water and in swimming fishes, making them useful for rheotaxis (Montgomery et al., 1997). However, the masking effects of such stimuli render the free neuromasts unable to discriminate moving objects at close range in these situations. On the other hand, the filtering properties of the canal organs allow them to detect the hydrodynamic stimuli caused by close-range moving objects even in the presence of constant background water flow (Engelmann et al., 2000). The lateral line system is also sensitive to the turbulent wake of moving objects (see Sand, 1984 for a review), and a necklace of trailing vortices may persist after a swimming fish has left the scene, thus forming a track that may be detected by the lateral line of other fishes entering the wake (Plachta, 2000; and see Chapter 6).

In conclusion, the sensory equipment of teleosts for detection of acoustic and hydrodynamic stimuli is certainly impressive. The otolith organs are sensitive to the whole-body acceleration caused by the kinetic sound component but may also be sensitive to the sound pressure component in species possessing a swim bladder or a gas bubble lying in close proximity to the ear. The superficial neuromasts and the canal organs of the lateral line system detect water velocity and acceleration, respectively, relative to the fish, and in some species a third subdivision of the lateral line system is closely connected to gas compartments and is sensitive to sound pressure. The division of labor between these systems in extracting information about the environment and about local and distant sources of acoustic and hydrodynamic stimuli is a challenging field of research that is still just in the starting phase of exploration. The integration in the central nervous system of information from this spectrum of sensory subdivisions of the octavolateralis system has hardly been addressed.

8. Unresolved Problems and Suggestions for Future Work

In 1993, two of the authors of this paper did a brief review of fish hearing and developed a list of unresolved problems and issues that they suggested were the most critical for the next decades' work in the field (Popper and Fay, 1993). In reviewing that list, we find that while many of the questions remain open, additional questions now arise due to an increased level of understanding of auditory mechanisms in fishes, and our understanding of how fishes fit into the overall scheme of vertebrate hearing (see Fay and Popper 2000). In this section, we once again consider outstanding and important issues in fish hearing, with the hope that these may help set the agenda for the work that will be done between now and the next conference on sensory biology of aquatic animals. This list is purposefully kept short and it does not represent all of the outstanding questions. Instead, these are the questions that have the most interest to the authors of this paper, and we invite readers to share their personal lists with us. Since the four authors are most interested in the periphery and in acoustic perception, we acknowledge that there is an additional body of questions that need to be dealt with regarding acoustic processing in the CNS. Questions about the CNS might include the role of binaural interactions, the presence of maps of auditory space, detailed pathways for sound stimuli and potential mapping of the receptors, interactions between different end organs and the ear and lateral line (and perhaps visual and electrosensory systems, where present), and the role of the efferent system in auditory processing.

1. The inner ear and the lateral line detect both different and overlapping aspects of hydrodynamic and acoustic stimuli generated by moving or vibrating objects as well as self-motion. The functional interactions between these sensory systems, in order to obtain the most useful information about the surroundings, are not completely understood. While we have not covered the CNS in this chapter, it is important to understand the ways in which information from the ear and lateral line is integrated in the CNS.
2. What are the roles of the various structural components of the inner ear, and how do they interact? While we know a reasonable amount about the function of the saccule in hearing and the utricle as a vestibular receptor, almost nothing is known of the function of the lagena and, when present in bony fishes, the macula neglecta. Are these indeed multimodal end organs and, if they are, what are their functions? What are their specific contributions to different inner ear senses and at what level of the CNS do they interact to provide the fish with a general acoustic (or vestibular) view of the world?
3. What is the functional significance of the size and shape of the various otoliths? These vary considerably across species and particularly in the saccule. What functional significance do these differences imply, or are there any differences in function even when the otoliths vary considerably in shape? What are the movement patterns of the otoliths, and how do they interact with the sensory epithelium? While there has been limited discussion of these issues, the first comprehensive analysis has just appeared (Lychakov and Rebane, 2000).
4. What is the functional significance of having sensory hair cells oriented in different directions within and between maculae of the inner ear? There is considerable variation in the hair cell orientation patterns, especially when comparing hearing specialists and nonspecialists. Some patterns are correlated with connections between the inner ear and the swim bladder. What is the functional significance of these patterns, and are the patterns correlated with sound source localization?
5. Do hearing nonspecialists use pressure signals for sound detection? The nature and physical properties of the transmission channel between the swim bladder and the ear in hearing nonspecialists is presently unknown. Is there a contribution from the backbone?
6. Why are some fish species specialized for sound pressure reception, while others are not? It is becoming clear that the use of vocalizations or other deliberate communication sounds cannot explain this diversity.
7. What are the relationships between auditory sensitivity and water depth in species possessing a swim bladder? Since the swim bladder responds to rapid changes of depth by changing volume, does this impact hearing?
8. What is the relationship between size of the otolith organs and the swim bladder and hearing sensitivity within specific species? Related to this are questions of the impact of the constantly increasing number of hair cells in the ear on hearing ability. Do hearing capabilities change as the number of hair cells increase?
9. How does the coupling between gas-filled structures and the lateral line in clupeids and chaetodontids (and possibly in other species where such connections have yet to be described) contribute to the function of the organ? This coupling adds a new dimension to the analysis of complex hydrodynamic and acoustic stimuli. The possible existence of similar arrangements in other groups of teleosts should be explored.
10. What are the behavioral implications of infrasound and ultrasound detection? Several hypotheses regarding possible use of hearing infrasound and ultrasound have been put forward, but experimental evidence supporting these suggestions is essentially lacking. Future experiments should clarify the functional implications of infrasound and ultrasound detection in fishes.
11. How do fishes determine sound direction? Most theories of directional hearing in fishes are based on vectorial analysis of the particle motions of the incident sound. However, the most common natural underwater sound

sources are dipoles, and vectorial analysis is thus inadequate to determine direction to the source. The recent hypothesis by Kalmijn (1988, 1989, 1997) suggesting that fishes do not actually determine the direction to the source, but rather may be guided to approach the source by keeping a constant angle between the body axis and the vectorial sound component, should be experimentally tested.

12. How do fishes determine distance from the sound source? Might they do this by distinguishing between sounds with different ratios between the pressure and kinetic components, as suggested by a few studies?

13. Why did hearing evolve in fishes (and vertebrates)? The general uses of sound reception by fishes are not understood, except in the cases of vocal communication in some species. One hypothesis developed in this paper is that fishes use ambient sounds to obtain information about the state and structure of the environment (perhaps helping to image the environmental scene of objects and events, along with the other senses). In this way, the sense of hearing in fishes may be quite similar to those analyzed for other vertebrates, including humans. Mechanisms of encoding and processing auditory information in fishes may be representative of vertebrates generally. Adaptations specially developed for the specific aquatic behaviors of fishes have thus been highly successful also for the terrestrial way of life.

14. How well do bony fishes other than teleosts, such as sturgeons, gars, bowfin, and lungfish, hear? Would an understanding of hearing, the ear, and the auditory CNS in these species help our understanding of the evolution of vertebrate hearing?

Acknowledgments. The authors dedicate this chapter to our friend and colleague Professor Per Enger. In a career spanning more than five decades, Per has made numerous pivotal contributions that have helped to define marine bioacoustics. His numerous studies have asked critical questions, and his results stand today as some of the major contributions to our field. We take pleasure in being able to dedicate this chapter to Per, a gentlemen and scholar in every way.

References

Alexander, R. McN. (1959a). The physical properties of the swim bladder in intact Cypriniformes. *J. Exp. Biol.* 36:315–332.

Alexander, R. McN. (1959b). The physical properties of the swim bladders of fish other than Cypriniformes. *J. Exp. Biol.* 36:347–355.

Alexander, R. McN. (1966). Physical aspects of swim bladder function. *Biol. Rev.* 41:141–176.

Allen, J. (1997). OHCs shift the excitation pattern via BM tension. In: *Diversity in Auditory Mechanics* (Lewis, E.R., Long, G.R., Lyon, R.F., Narins, P.M., Steele, C.R., and Hecht-Poinar, E., eds.), pp. 167–175. Singapore: World Scientific Publishers.

Astrup, J. (1999). Ultrasound detection in fish: A parallel to the sonar-mediated detection of bats by ultrasound-sensitive insects? *Comp. Biochem. Physiol. A.* 124:19–27.

Astrup, J., and Møhl, B. (1993). Detection of intense ultrasound by the cod *Gadus morhua*. *J. Exp. Biol.* 182:71–80.

Astrup, J., and Møhl, B. (1998). Discrimination between high and low repetition rates of ultrasonic pulses by the cod. *J. Fish Biol.* 52:205–208.

Atema, A., Fay, R.R., Popper, A.N., and Tavolga, W.N. (eds.) (1988). *Sensory Biology of Aquatic Animals*. New York: Springer-Verlag.

Banner, A. (1967). Evidence of sensitivity to acoustic displacements in the lemon shark, *Negaprion brevirostris* (Poey). In: *Lateral Line Detectors* (Cahn, P.H., ed.), pp. 265–273. Bloomington, In: Indiana University Press.

Blaxter, J.H.S., Denton, E.J., and Gray, J.A.B. (1979). The herring swim bladder as a gas reservoir for the acoustico-lateralis system. *J. Mar. Biol. Assoc. UK* 59:1–10.

Blaxter, J.H.S., Denton, E.J., and Gray, J.A.B. (1981). Acousticolateralis system in clupeid fishes. In: *Hearing and Sound Communication in Fishes* (Tavolga, W.N., Popper, A.N., and Fay, R.R., eds.), pp. 39–53. New York: Springer-Verlag.

Boyle, R., Carey, J.P., and Highstein, S.M. (1991). Morphological correlates of response dynamics and efferent stimulation in horizontal semicircular calyx afferents of the toadfish, *Opsanus tau*. *J. Neurophysiol.* 66:1504–1521.

Bregman, A.S. (1990). *Auditory Scene Analysis: The Perceptual Organization of Sound*. Cambridge, MA: MIT Press.

Buerkle, U. (1969). Auditory masking and the critical band in Atlantic cod (*Gadus morhua*). *J. Fish. Res. Board Canada* 26:1113–1119.

Buwalda, R.J.A. (1981). Segregation of directional and nondirectional acoustic information in the cod. In: *Hearing and Sound Communication in Fishes* (Tavolga, W.N., Popper, A.N., and Fay, R.R., eds.), pp. 139–172. New York: Springer-Verlag.

Buwalda, R.J.A., Schuijf, A., and Hawkins, A.D. (1983). Discrimination by the cod of sound from opposing directions. *J. Comp. Physiol.* 150:175–184.

Carlström, D. (1963). A crystallographic study of vertebrate otoliths. *Biol. Bull.* 125:441–463.

Chang, J.S.Y., Popper, A.N., and Saidel, W.M. (1992). Heterogeneity of sensory hair cells in a fish ear. *J. Comp. Neurol.* 324:621–640.

Chapman, C.J., and Hawkins, A. (1973). A field study of hearing in the cod, *Gadus morhua L. J. Comp. Physiol.* 85:147–167.

Chapman, C.J., and Johnstone, A.D.F. (1974). Some auditory discrimination experiments on marine fish. *J. Exp. Biol.* 61:521–528.

Chapman, C.J., and Sand, O. (1974). Field studies of hearing in two species of flatfish, *Pleuronectes platessa* (L.) and *Limanda limanda* (L.) (Family Pleuronectidae). *Comp. Biochem. Physiol.* 47:371–385.

Clack, J.A. (1996). Otoliths in fossil coelacanths. *J. Vertebr. Paleontol.* 16:168–171.

Coates, M.L. (1999). Endocranial preservation of a Carboniferous actinopterygian from Lancashire, UK, and the interrelationships of primitive actinopterygians. *Philos. Trans. R. Soc. Lond. B* 354:435–462.

Coombs, S., Hastings, M., and Finneran, J. (1996). Measuring and modeling lateral line excitation patterns to changing dipole source locations. *J. Comp. Physiol.* 178A:359–371.

Coombs, S., and Popper, A.N. (1979). Hearing differences among Hawaiian squirrelfishes (family Holocentridae) related to differences in peripheral auditory system. *J. Comp. Physiol.* 132:203–207.

Cortopassi, K.A., and Lewis, E.R. (1998). A comparison of the linear tuning properties of two classes of axons in the bullfrog lagena. *Brain Behav. Evol.* 51:331–348.

Corwin, J.T. (1981). Audition in elasmobranchs. In: *Hearing and Sound Communication in Fishes* (Tavolga, W.N., Popper, A.N., and Fay, R.R., eds.), pp. 81–105. New York: Springer-Verlag.

Crawford, A.C., and Fettiplace, R. (1981). An electrical tuning mechanism in turtle cochlear hair cells. *J. Physiol. (Lond.)* 312:377–412.

Crawford, J.D., Cook, A.P., and Heberlein, A.S. (1997). Bioacoustic behavior of African fishes (Mormyridae): Potential cues for species and individual recognition in *Pollimyrus*. *J. Acoust. Soc. Am.* 102:1–13.

Demski, L., Gerald, G.W., and Popper, A. (1973). Central and peripheral mechanisms in teleost sound production. *Am. Zool.* 13:1141–1167.

Denton, E.J., and Blaxter, J.H.S. (1976). The mechanical relationships between the clupeid swim bladder, inner ear and lateral line. *J. Mar. Biol. Assoc. UK* 56:787–807.

Denton, E.J., and Gray, J.A.B. (1982). The rigidity of fish and patterns of lateral line stimulation. *Nature* 297:679–681.

Denton, E.J., and Gray, J.A.B. (1983). Mechanical factors in the excitation of clupeid lateral lines. *Proc. R. Soc. Lond. B.* 218:1–26.

Denton, E.J., and Gray, J.A.B. (1989). Some observations on the forces acting on neuromasts in fish lateral line canals. In: *The Mechanosensory Lateral Line: Neurobiology and Evolution* (Coombs, S., Görner, P., and Münz, M., eds.), pp. 229–246. Berlin: Springer-Verlag.

de Vries, H.I. de. (1950). The mechanics of the labyrinth otoliths. *Acta Oto-Laryngol.* 38:262–273.

Dijkgraaf, S. (1932). Untersuchungen über die Funktion der Seitenorgane an Fischen. *Z. Vergl. Physiol.* 20:162–214.

Dijkgraaf, S. (1947). Ein Töne erzeugender Fisch im Neapler Aquarium. *Experientia* 3:493–494.

Dijkgraaf, S. (1952). über die Schallwahrnehmung bei Meeresfischen. *Z. Vergl. Physiol.* 34:104–122.

Dijkgraaf, S. (1963). The functioning and significance of the lateral-line organs. *Biol. Rev.* 38: 51–105.

Dijkgraaf, S., and Verheijen, F. (1950). Neue Versuchen über das Tonunterscheidungsverm gen der Elritze. *Z. Vergl. Physiol.* 32:248–265.

Døving, K.B., Westerberg, H., and Johnsen, P.B. (1985). Role of olfaction in the behavioral and neuronal responses of Atlantic Salmon, *Salmo salar*, to hydrographic stratification. *Can. J. Fish. Aquat. Sci.* 42:1658–1667.

Dunning, D.J., Ross, Q.E., Geoghegan, P., Reichle, J.J., Menezes, J.K., and Watson, J.K. (1992). Alewives in a cage avoid high-frequency sound. *N. Am. J. Fish. Manage.* 12:407–416.

Edds-Walton, P.L., Fay, R.R., and Highstein, S. (1999). Dendritic arbors and central projections of physiologically characterized auditory fibers from the saccule of the toadfish, *Opsanus tau*. *J. Comp. Neurol.* 411:212–238.

Engelmann, J., Hanke, W., Mogdans, J., and Bleckmann, H. (2000). Hydrodynamic stimuli and the fish lateral line. *Nature* 408:51–52.

Enger, P.S. (1967). Hearing in herring. *Comp. Biochem. Physiol.* 22:527–538.

Enger, P.S. (1973). Masking of auditory responses in the medulla oblongata of goldfish. *J. Exp. Biol.* 59:415–424.

Enger, P.S. (1981). Frequency discrimination in teleosts: Central or peripheral? In: *Hearing and Sound Communication in Fishes* (Tavolga, W.N., Popper, A.N., and Fay, R.R., eds.), pp. 243–255. New York: Springer-Verlag.

Enger, P.S., Kalmijn, A.J., and Sand, O. (1989). Behavioral investigations on the functions of the lateral line and inner ear in predation. In: *The Mechanosensory Lateral Line: Neurobiology and Evolution* (Coombs, S., Görner, P., and Münz, M., eds.), pp. 575–587. Berlin: Springer-Verlag.

Fay, R.R. (1969). Auditory sensitivity of the goldfish within the near acoustic field. U.S. Naval Submarine Medical Center, Submarine Base, Groton, Connecticut, Report No. 605:1–11.

Fay, R.R. (1970a). Auditory frequency generalization in the goldfish (*Carassius auratus*). *J. Exp. Anal. Behav.* 14:353–360.

Fay, R.R. (1970b). Auditory frequency discrimination in the goldfish (*Carassius auratus*). *J. Comp. Physiol. Psychol.* 73:175–180.

Fay, R.R. (1974a). Masking of tones by noise for the goldfish (*Carassius auratus*). *J. Comp. Physiol. Psychol.* 87:708–716.

Fay, R.R. (1974b). Auditory frequency discrimination in vertebrates. *J. Acoust. Soc. Am.* 56:206–209.

Fay, R.R. (1984). The goldfish ear codes the axis of acoustic particle motion in three dimensions. *Science* 225:951–954.

Fay, R.R. (1985). Sound intensity processing by the goldfish. *J. Acoust. Soc. Am.* 78:1296–1309.

Fay, R.R. (1988). *Hearing in Vertebrates: A Psychophysics Databook*. Winnetka, IL: Hill-Fay Associates.

Fay, R.R. (1989a). Intensity discrimination of pulsed tones by the goldfish (*Carassius auratus*). *J. Acoust. Soc. Am.* 85:500–502.

Fay, R.R. (1989b). Frequency discrimination in the goldfish (*Carassius auratus*): Effects of roving intensity, sensation level, and the direction of frequency change. *J. Acoust. Soc. Am.* 85:503–505.

Fay, R.R. (1992a). Structure and function in sound discrimination among vertebrates. In: *The Evolutionary Biology of Hearing* (Webster, D.B., Fay, R.R., and Popper, A.N., eds.), pp. 229–263. New York: Springer-Verlag.

Fay, R.R. (1992b). Analytic listening by the goldfish. *Hear. Res.* 59:101–107.

Fay, R.R. (1995). Perception of spectrally and temporally complex sounds by the goldfish (*Carassius auratus*). *Hear. Res.* 89:146–154.

Fay, R.R. (1997). Frequency selectivity of saccular afferents of the goldfish revealed by revcor analysis. In: *Diversity in Auditory Mechanics* (Lewis, E.R., Long, G.R., Lyon, R.F., Narins, P.M., Steele, C.R., and Hecht-Poinar, E., eds.), pp. 69–75. Singapore: World Scientific Publishers.

Fay, R.R. (1998a). Perception of two-tone complexes by goldfish (*Carassius auratus*). *Hear. Res.* 120:17–24.

Fay, R.R. (1998b). Auditory stream segregation in goldfish (*Carassius auratus*). *Hear. Res.* 120:69–76.

Fay, R.R. (2000). Frequency contrasts underlying auditory stream segregation in goldfish. *J. Assoc. Res. Otolaryngol.* 1:120–128.

Fay, R.R., and Coombs, S. (1983). Neural mechanisms in sound detection and temporal summation. *Hear. Res.* 10:69–92.

Fay, R.R., and Edds-Walton, P.L. (1997a). Directional response properties of saccular afferents of the toadfish, *Opsanus tau*. *Hear. Res.* 111:1–21.

Fay, R.R., and Edds-Walton, P.L. (1997b). Diversity in frequency response properties of saccular afferents of the toadfish (*Opsanus tau*). *Hear. Res.* 111:235–246.

Fay, R.R., and Popper, A.N. (1975). Modes of stimulation of the teleost ear. *J. Exp. Biol.* 62: 379–387.

Fay, R.R., and Popper, A.N. (2000). Evolution of hearing in vertebrates: The inner ears and processing. *Hear. Res.* 149:1–10.

Fay, R.R., and Ream, T.J. (1986). Acoustic response and tuning in saccular nerve fibers of the goldfish (*Carassius auratus*). *J. Acoust. Soc. Am.* 79:1883–1895.

Fay, R.R., Ahroon, W.A., and Orawski, A.A. (1978). Auditory masking patterns in the goldfish (*Carassius auratus*): Psychophysical tuning curves. *J. Exp. Biol.* 74:83–100.

Fay, R.R., Chronopoulos, M., and Patterson, R.D. (1996). The sound of a sinusoid: Perception and neural representations in the goldfish (*Carassius auratus*). *Aud. Neurosci.* 2:377–392.

Finneran, J.J., and Hastings, M.C. (2000). A mathematical analysis of the peripheral auditory system mechanics in the goldfish (*Carassius auratus*). *J. Acoust. Soc. Am.* 108:1308–1321.

Fish, M.P., and Mowbray, W.H. (1970). *Sounds of Western North Atlantic Fishes*. Baltimore, MD: Johns Hopkins Press.

Fish, J.F., and Offutt, G.C. (1972). Hearing thresholds from toadfish, *Opsanus tau*, measured in the laboratory and field. *J. Acoust. Soc. Am.* 51:1318–1321.

Fletcher, H. (1940). Auditory patterns. *Rev. Mod. Phys.* 12:47–65.

Fletcher, L.B., and Crawford, J.D. (2001). Acoustic detection by sound-producing fishes (Mormyridae): The role of gas-filled tympanic bladders. *J. Exp. Biol.* 204:175–183.

Flock, Å. (1964). Structure of the macula utriculi with special reference to directional interplay of sensory responses as revealed by morphological polarization. *J. Cell Biol.* 22:413–431.

Furukawa, T., and Ishii, Y. (1967). Neurophysiological studies on hearing in the goldfish. *J. Neurophysiol.* 30:1337–1403.

Gauldie, R.W., Mulligan, K., and Thompson, R.K. (1987). The otoliths of a chimaera, the New Zealand elephant fish *Callorhynchus milii*. *N. Z. J. Mar. Freshwater Res.* 21:275–280.

Guttman, N. (1963). Laws of behavior and facts of perception. In: *Psychology: A Study of a Science, Vol. 5* (Koch, S., ed.), pp. 114–178. New York: McGraw-Hill.

Hartmann, W.M. (1988). Pitch perception and the segregation and integration of auditory entities. In: *Auditory Function: Neurological Bases of Hearing* (Edelman, G.M., Gall, W.E., and Cowan, W.M., eds.), pp. 623–645. New York: Wiley.

Hawkins, A.D. (1981). The hearing abilities of fish. In: *Hearing and Sound Communication in Fishes* (Tavolga, W.N., Popper, A.N., and Fay, R.R., eds.), pp. 109–137. New York: Springer-Verlag.

Hawkins, A.D., and Chapman, C.J. (1975) Masked auditory thresholds in the cod, *Gadus morhua* L. *J. Comp. Physiol. A.* 103:209–226.

Hawkins, A.D., and Johnstone, A.D.F. (1978). The hearing of the Atlantic salmon, *Salmo salar*. *J. Fish Biol.* 13:655–673.

Hawkins, A.D., and Myrberg, A.A. Jr. (1983). Hearing and sound communication under water. In: *Bioacoustics: A Comparative Approach* (Lewis, B., ed.), pp. 347–405. London: Academic Press.

Hawkins, A.D., and Sand, O. (1977). Directional hearing in the median vertical plane by the cod. *J. Comp. Physiol. A.* 122:1–8.

Heuch, P.A., and Karlsen, H.E. (1997). Detection of infrasonic water oscillations by copepods of *Lepeophtheirus salmonis* (Copepoda: Caligida). *J. Plankton Res.* 19:735–746.

Hudspeth, A.J. (1989). How the ear's works work. *Nature* 341:397–404.

Hudspeth, A.J., and Corey, D.P. (1977). Sensitivity, polarity and conductance change in the response of vertebrate hair cells to controlled mechanical stimuli. *Proc. Natl. Acad. Sci. USA* 74:2407–2411.

Hulse, S.H., MacDougall-Shackelton, S.A., and Wisniewski, B. (1997). Auditory scene analysis by songbirds: Stream segregation of birdsong by European starlings (*Sturnus vulgaris*). *J. Comp. Psychol.* 111:3–13.

Jacobs, D.W., and Tavolga, W.N. (1967). Acoustic intensity limens in the goldfish. *Anim. Behav.* 15:324–335.

Jerkø, H., Turunen-Rise, I., Enger, P.S., and Sand, O. (1989). Hearing in the eel (*Anguilla*). *J. Comp. Physiol.* 165A:455–459.

Jones, F.R.H., and Marshall, N.B. (1953). The structure and function of the teleostean swim bladder. *Biol. Rev.* 28:16–83.

Kalmijn, A.J. (1988). Hydrodynamic and acoustic field detection. In: *Sensory Biology of Aquatic Animals* (Atema, A., Fay, R.R., Popper, A.N., and Tavolga, W.N., eds.), pp. 83–130. New York: Springer-Verlag.

Kalmijn, A.J. (1989). Functional evolution of lateral line and inner ear sensory systems. In: *The Mechanosensory Lateral Line: Neurobiology and Evolution* (Coombs, S., Görner, P., and Münz, M., eds.), pp. 187–215. Berlin: Springer-Verlag.

Kalmijn, A.J. (1997). Electric and near-field acoustic detection, a comparative study. *Acta Physiol. Scand.* 161:Suppl 638, 25–38.

Karlsen, H.E. (1992a). Infrasound sensitivity in the plaice (*Pleuronectes platessa*). *J. Exp. Biol.* 171: 173–187.

Karlsen, H.E. (1992b). The inner ear is responsible for detection of infrasound in the perch (*Perca fluviatilis*). *J. Exp. Biol.* 171:163–172.

Karlsen, H.E., and Sand, O. (1987). Selective and reversible blocking of the lateral line in freshwater fish. *J. Exp. Biol.* 133:249–267.

Knudsen, F.R., Enger, P.S., and Sand, O. (1992). Awareness reactions and avoidance responses to sound in juvenile Atlantic salmon, *Salmo salar* L. *J. Fish Biol.* 40:523–534.

Knudsen, F.R., Enger, P.S., and Sand, O. (1994). Avoidance responses to low frequency sound in downstream migrating Atlantic salmon smolt, *Salmo salar* L. *J. Fish Biol.* 45:227–233.

Knudsen, F.R., Schreck, C.B., Knapp, S.M., Enger, P.S., and Sand, O. (1997). Infrasound produces flight and avoidance responses in Pacific juvenile salmonids. *J. Fish Biol.* 51:824–829.

Koslowsk, J., and Crawford, J. (2000). Transformations of an auditory temporal code in the medulla of a sound-producing fish. *J. Neurosci.* 20:2400–2408.

Koyama, H., Lewis, E.R., Leverenz, E.L., and Baird, R.A. (1982). Acute seismic sensitivity of the bullfrog ear. *Brain Res.* 250:168–172.

Kroese, A.B.A., and Schellart, N.A.M. (1992). Velocity- and acceleration-sensitive units in the trunk lateral line of the trout. *J. Neurophysiol.* 68:2212–2221.

Ladich, F. (2000). Acoustic communication and the evolution of hearing in fishes. *Philos. Trans. R. Soc. Lond.* 355:1285–1288.

Lanford, P.J., Platt, C., and Popper, A.N. (2000). Structure and function in the saccule of the goldfish (*Carassius auratus*): A model of diversity in the non-amniote ear. *Hear. Res.* 143:1–13.

Lewis, E.R., Everenz, E.L., and Bialek, W.S. (1985). *The Vertebrate Ear.* Boca Raton, FL: CRC Press.

Lowenstein, O. (1971). The labyrinth. In: *Physiology of Fishes, Vol. 5* (Hoar, W.S., and Randall, D.J., eds.), pp. 207–240. New York: Academic Press.

Lowenstein, O., and Roberts, T.D.M. (1950). The equilibrium function of the otolith organs of the thornback ray (*Raja clavata*). *J. Physiol. (Lond.)* 110:392–415.

Lowenstein, O., and Roberts, T.D.M. (1951). The localization and analysis of the responses to vibration from the isolated elasmobranch labyrinth: A contribution to the problem of the evolution of hearing in vertebrates. *J. Physiol. (Lond.)* 114: 471–489.

Lowenstein, O., Osborne, M.P., and Wersäll, J. (1964). Structure and innervation of the sensory epithelia of the labyrinth in the thornback ray (*Raja clavata*). *Proc. R. Soc. Lond. B.* 160:1–12.

Lu, Z., and Fay, R.R. (1993). Acoustic response properties of single units in the torus semicircularis of the goldfish, *Carassius auratus*. *J. Comp. Physiol.* 173:33–48.

Lu, Z., and Fay, R.R. (1996). Two-tone interaction in auditory nerve fibers and midbrain neurons of the goldfish, *Carassius auratus*. *Aud. Neurosci.* 2:57–273.

Lu, Z., and Popper, A.N. (1997). Encoding of acoustic particle motion by saccular ganglion cells of a fish: Intracellular recording and tracing. *Soc. Neurosci. Abstr.* 23:180.

Lu, Z., and Popper, A.N. (2001). Neural response directionality correlates of hair cell orientation in a teleost fish. *J. Comp. Physiol. A.* 187:453–465.

Lu, Z., Popper, A.N., and Fay, R.R. (1996). Behavioral detection of acoustic particle motion by a teleost fish, *Astronotus ocellatus*: Sensitivity and directionality. *J. Comp. Physiol. A.* 179:227–233.

Lu, Z., Song, J., and Popper, A.N. (1998). Encoding of acoustic directional information by saccular afferents of the sleeper goby, *Dormitator latifrons*. *J. Comp. Physiol. A* 182:805–815.

Lychakov, D.V., and Rebane, Y.T. (2000). Otolith regularities. *Hear. Res.* 143:83–102.

Maisey, J.G. (1988). Reply to Schulze. *Copeia* 1988:259–260.

Mann, D.A., Lu, Z., and Popper, A.N. (1997). Ultrasound detection by a teleost fish. *Nature* 389:341.

Mann, D.A., Lu, Z., Hastings, M.C., and Popper, A.N. (1998). Detection of ultrasonic tones and simulated dolphin echolocation clicks by a teleost fish, the American shad (*Alosa sapidissima*). *J. Acoust. Soc. Am.* 104:562–568.

Mann, D.A., Higgs, D.M., Tavolga, W.N., Souza, M.J., and Popper, A.N. (2001). Ultrasound detection by clupeiform fishes. *J. Acoust. Soc. Am.* 109:3048–3054.

Minnaert, F.M. (1933). On musical air-bubbles and the sounds of running water. *Phil. Mag.* 16:235–248.

McCormick, C., and Popper, A.N. (1984). Auditory sensitivity and psychophysical tuning curves in the elephant nose fish, *Gnathonemus petersii*. *J. Comp. Physiol. A.* 155:753–761.

Montgomery, J., Baker, C.F., and Carton, A.G. (1997). The lateral line can mediate rheotaxis in fish. *Nature* 389:960–963.

Moorman, S.J., Burress, C., Cordova, R., and Slater, J. (1999). Stimulus dependence of the development of the zebrafish (*Danio rerio*) vestibular system. *J. Neurobiol.* 38:247–258.

Moulton, J.M. (1963). Acoustic behaviour of fish. In: *Acoustic Behavior of Animals* (Busnel, R.G., ed.), pp. 655–687. Amsterdam: Elsevier.

Myrberg, A.A. Jr. (1980). Sensory mediation of social recognition in fishes. In: *Fish Behavior and its Use in the Capture and Culture of Fishes* (Bardach, J.E., Magnuson, J.J., and May, R.C., eds.), pp. 146–178. Manila: ICLARM.

Myrberg, A.A. Jr. (1981). Sound communication and interception in fishes. In: *Hearing and Sound Communication in Fishes* (Tavolga, W.N., Popper, A.N., and Fay, R.R., eds.), pp. 395–426. New York: Springer-Verlag.

Myrberg, A.A. Jr., and Riggio, R.J. (1985). Acoustically mediated individual recognition by a coral reef fish (*Pomacentrus partitus*). *Anim. Behav.* 33:411–416.

Myrberg, A.A. Jr., and Spires, J.Y. (1980). Hearing in damselfishes: An analysis of signal detection among closely related species. *J. Comp. Physiol.* 140:135–144.

Nestler, J.M., Ploskey, G.R., Pickens, J., Menezes, J., and Schilt, C. (1992). Responses of blueback herring to high-frequency sound and implications for reducing entrainment at hydropower dams. *N. Am. J. Fish. Manage.* 12:667–683.

Offutt, G. (1973). Structures for detection of acoustic stimuli in the Atlantic codfish *Gadus morhua*. *J. Acoust. Soc. Am.* 56:665–671.

Oman, C.M., Marcus, E.N., and Curthoys, I.S. (1987). The influence of semicircular canal morphology on endolymph flow dynamics: An anatomically descriptive mathematical model. *Acta Otoaryngol.* 103:1–13.

Packard, A., Karlsen, H.E., and Sand, O. (1990). Low-frequency hearing in cephalopods. *J. Comp. Physiol.* 166A:501–505.

Parker, G.H. (1902). Hearing and allied senses in fishes. *Bull. U.S. Fish. Comm.* 22:45–64.

Parker, G.H. (1903). The sense of hearing in fishes. *Am. Nat.* 37:185–203.

Parker, G.H. (1904). The function of the lateral line organs in fishes. *Bull. U.S. Bur. Fish.* 24:185–207.

Parker, G.H. (1909). The sense of hearing in the dogfish. *Science* 29:428.

Parvulescu, A. (1964). Problems of propagation and processing. In: *Marine Bioacoustics* (Tavolga, W.N., ed.), pp. 87–100. Oxford, UK: Pergamon Press.

Parvulescu, A. (1967). The acoustics of small tanks. In: *Marine Bioacoustics II* (Tavolga, W.N., ed.), pp. 7–14. Oxford, UK: Pergamon Press.

Plachta, D. (2000). Responses of toral lateral line units of the goldfish, *Carassius auratus*, to dipole and complex water wave stimuli. Doctoral dissertation, Univ. Bonn.

Platt, C. (1973). Central control of postural orientation in flatfish. I. Postural change dependence on central neural changes. *J. Exp. Biol.* 59:491–521.

Platt, C. (1977). Hair cell distribution and orientation in goldfish otolith organs. *J. Comp. Neurol.* 172: 283–297.

Platt, C. (1983). The peripheral vestibular system of fishes. In: *Fish Neurobiology. I. Brain Stem and Sense Organs* (Northcutt, R.G., and Davis, R.E., eds.), pp. 89–124. Ann Arbor, MI: U. Michigan Press.

Platt, C. (1988). Equilibrium in the vertebrates: Signals senses, and steering underwater. In: *Sensory Biology of Aquatic Animals* (Atema, A., Fay, R.R., Popper, A.N., and Tavolga, W.N., eds.), pp. 783–809. New York: Springer-Verlag.

Platt, C. (1993). Zebrafish inner ear sensory surfaces are similar to those in goldfish. *Hear. Res.* 65:133–140.

Platt, C., and Popper, A.N. (1981). Structure and function in the ear. In: *Hearing and Sound Communication in Fishes* (Tavolga, W.N., Popper, A.N., and Fay, R.R., eds.), pp. 3–38. New York: Springer-Verlag.

Platt, C., Popper, A.N., and Fay, R.R. (1989). The ear as part of the octavolateralis system. In: *The Mechanosensory Lateral Line: Neurobiology and Evolution* (Coombs, S., Görner, P., and Münz, M., eds.), pp. 633–651. Berlin: Springer-Verlag.

Poggendorf, D. (1952). Die absoluten Hörschwellen des Zwergwelses (*Amiurus nebulosus*) und Beiträge zur Physik des Weberschen Apparates der Ostariophysen. *Z. Vergl. Physiol.* 34:222–257.

Popper, A.N. (1970). Auditory capacities of the Mexican blind cave fish (*Astyanax jordani*) and its eyed ancestor (*Astyanax mexicanus*). *Anim. Behav.* 18:552–562.

Popper, A.N. (1978). Scanning electron microscopic study of the otolithic organs in the bichir (*Polypterus bichir*) and shovel-nose sturgeon (*Scaphyrhynchus platyrynchus*). *J. Comp. Neurol.* 18:117–128.

Popper, A.N. (1981). Comparative scanning electron microscopic investigations of the sensory epithelia in the teleost sacculus and lagena. *J. Comp. Neurol.* 200:357–374.

Popper, A.N. (1983). Organization of the inner ear and processing of acoustic information. In: *Fish Neurobiology. I. Brain Stem and Sense Organs* (Northcutt, R.G., and Davis, R.E., eds.), pp. 125–178. Ann Arbor, MI: U. Michigan Press.

Popper, A.N. (2000). Hair cell heterogeneity and ultrasonic hearing: Recent advances in understanding fish hearing. *Philos. Trans. R. Soc. Lond.* 355:1277–1280.

Popper, A.N., and Carlson, T.J. (1998). Application of the use of sound to control fish behavior. *Trans. Am. Fish. Soc.* 127:673–707.

Popper, A.N., and Coombs, S. (1982). The morphology and evolution of the ear in Actinopterygian fishes. *Am. Zool.* 22:311–328.

Popper, A.N., and Fay, R.R. (1993). Sound detection and processing by fish: Critical review and major research questions. *Brain Behav. Evol.* 41:14–38.

Popper, A.N., and Fay, R.R. (1997). Evolution of the ear and hearing: Issues and questions. *Brain Behav. Evol.* 50:213–221.

Popper, A.N., and Fay, R.R. (1999). The auditory periphery in fishes. In: *Comparative Hearing: Fish and Amphibians* (Fay, R.R., and Popper, A.N., eds.), pp. 43–100. New York: Springer-Verlag.

Popper, A.N., and Lu, Z. (2000). Structure-function relationships in fish otolith organs. *Fish. Res.* 46:15–25.

Popper, A.N., and Northcutt, R.G. (1983). Structure and innervation of the inner ear of the bowfin, *Amia calva*. *J. Comp. Neurol.* 213:279–286.

Popper, A.N., and Platt, C. (1983). Sensory surface of the saccule and lagena in the ears of ostariophysan fishes. *J. Morphol.* 176:121–129.

Popper, A.N., and Platt, C. (1993). Inner ear and lateral line of bony fishes. In: *The Physiology of Fishes* (Evans, D.H., ed.), pp. 99–136. Boca Raton, FL: CRC Press.

Popper, A.N., and Tavolga, W.N. (1981). Structure and function of the ear of the marine catfish, *Arius felis*. *J. Comp. Physiol.* 144:27–34.

Popper, A.N., Saidel, W.M., and Chang, J.S.Y. (1993). Two types of sensory hair cell in the saccule of a teleost fish. *Hear. Res.* 66:211–216.

Popper, A.N., Salmon, M., and Parvulescu, A. (1973). Sound localization by two species of Hawaiian squirrelfish, *Myripristis berndti* and *M. argyromus*. *Anim. Behav.* 21:86–97.

Rabbitt, R.D., Boyle, R., and Highstein, S.M. (1994). Sensory transduction of head velocity and acceleration in the toadfish horizontal semicircular canal. *J. Neurophysiol.* 72:1041–1048.

Retzius, G. (1881). *Das Gehörorgan der Wirbelthiere, Vol. I.* Stockholm: Samson and Wallin.

Rogers, P.H., and Cox, M. (1988). Underwater sound as a biological stimulus. In: *Sensory Biology of Aquatic Animals* (Atema, A., Fay, R.R., Popper, A.N., and Tavolga, W.N., eds.), pp. 131–149. New York: Springer-Verlag.

Ross, Q.E., Dunning, D.J., Menezes, J.K., Kenna, M.J., and Tiller, G. (1995). Reducing impingement of alewives with high-frequency sound at a power plant on Lake Ontario. *N. Am. J. Fish. Manage.* 15:378–388.

Ross, Q.E., Dunning, D.J., Thorne, R., Menezes, J.K., Tiller, G.W., and Watson, J.K. (1996). Response of alewives to high-frequency sound at a power plant intake on Lake Ontario. *N. Am. J. Fish. Manage.* 16:548–559.

Saidel, W.M., and Popper, A.N. (1987). Sound reception in two anabantid fishes. *Comp. Biochem. Physiol.* 88A:37–44.

Saidel, W.M., Lanford, P.J., Yan, H.Y., and Popper, A.N. (1995). Hair cell heterogeneity in the goldfish saccule. *Brain Behav. Evol.* 46:362–370.

Sand, O. (1974). Directional sensitivity of microphonic potentials from the perch ear. *J. Exp. Biol.* 60:881–899.

Sand, O. (1981). The lateral-line and sound reception. In: *Hearing and Sound Communication in Fishes* (Tavolga, W.N., Popper, A.N., and Fay, R.R., eds.), pp. 459–480. New York: Springer-Verlag.

Sand, O. (1984). Lateral line systems. In: *Comparative Physiology of Sensory Systems* (Bolis, L., Keynes, R.D., and Maddrell, S.H.P., eds.), pp. 3–32. Cambridge, UK: Cambridge University Press.

Sand, O., and Enger, P.S. (1973). Evidence for an auditory function of the swim bladder in the cod. *J. Exp. Biol.* 59:405–414.

Sand, O., and Hawkins, A.D. (1973). Acoustic properties of the cod swim bladder. *J. Exp. Biol.* 58:797–820.

Sand, O., and Hawkins, A.D. (1974). Measurements of swim bladder volume and pressure in the cod. *Norw. J. Zool.* 22:31–34.

Sand, O., and Karlsen, H.E. (1986). Detection of infrasound by the Atlantic cod. *J. Exp. Biol.* 125:197–204.

Sand, O., and Karlsen, H.E. (2000). Detection of infrasound and linear acceleration in fish. *Philos. Trans. R. Soc. Lond. B* 355:1295–1298.

Sand, O., and Michelsen, A. (1978). Vibration measurements of the perch saccular otolith. *J. Comp. Physiol.* 123:85–89.

Sand, O., Enger, P.S., Karlsen, H.E., Knudsen, F.R., and Kvernstuen, T. (2000). Avoidance responses to infrasound in downstream migrating European silver eels, *Anguilla anguilla*. *Environ. Biol. Fishes* 57:327–336.

Schellart, N.A.M., and Popper, A.N. (1992). Functional aspects of the evolution of the auditory system of actinopterygian fish. In: *The Evolutionary Biology of Hearing* (Webster, D.B., Fay, R.R., and Popper, A.N., eds.), pp. 295–322. New York: Springer-Verlag.

Schöne, H. (1964). über die Arbeitsweise der Statolithenapparate bei Plattfischen. *Biol. Jahresh.* 4:135–156.

Schuijf, A. (1975). Directional hearing of cod (*Gadus morhua*) under approximate free field conditions. *J. Comp. Physiol. A.* 98:307–332.

Schuijf, A., and Buwalda, R.J.A. (1975). On the mechanism of directional hearing in cod (*Gadus morhua*). *J. Comp. Physiol. A.* 98:333–344.

Schuijf, A., and Hawkins, A.D. (1983). Acoustic distance discrimination by the cod. *Nature* 302: 143–144.

Schuijf, A., and Siemelink, M. (1974). The ability of cod (*Gadus morhua*) to orient towards a sound source. *Experientia* 30:773–774.

Schuijf, A., Baretta, J.W., and Wildschut, J.T. (1972). A field investigation on the discrimination of sound direction in *Labrus berggylta* (Pisces: Perciformes). *Neth. J. Zool.* 22:81–104.

Schultze, H.-P. (1990). A new acanthodian from the Pennsylvanian of Utah, USA, and the distribution of otoliths in gnathostomes. *J. Vertebr. Paleontol.* 10:49–58.

Sento, S., and Furukawa, T. (1987). Intra-axonal labeling of saccular afferents in the goldfish *Carassius auratus*: Correlations between morphological and physiological characteristics. *J. Comp. Neurol.* 258:352–367.

Spanier, E. (1979). Aspects of species recognition by sound in four species of damselfishes, genus Eupomacentrus (Pisces: Pomacentridae). *Z. Tierpsychol.* 51:301–316.

Steen, J.B. (1971). The swim bladder as a hydrostatic organ. In: *Fish Physiology, Vol. IV* (Hoar, W.S., and Randall, D.J., eds.), pp. 413–443. New York: Academic Press.

Steinacker, A., and Rojas, L. (1988). Acetylcholine modulated potassium channel in the hair cell of the toadfish saccule. *Hear. Res.* 155:265–269

Steinacker, A., and Romero, A. (1991). Characterization of the voltage gated potassium current in toadfish saccule hair cell. *Brain Res.* 556:22–32.

Steinacker, A., and Romero, A. (1992). Voltage-gated potassium current resonance in the toadfish saccular hair cell. *Brain Res.* 574:229–236.

Stipetć, E. (1939). Über das Gehörorgan der Mormyriden. *Z. Vergl. Physiol.* 26:740–752.

Sundnes, G., and Gytre, T. (1972). Swim bladder gas pressure in cod in relation to hydrostatic pressure. *J. Cons. Perm. Int. Explor. Mer.* 34:529–532.

Sundnes, G., and Sand, O. (1975). Studies of a physostome swim bladder by resonance frequency analyses. *J. Cons. Perm. Int. Explor. Mer.* 36:176–182.

Tavolga, W.N. (1956). Visual, chemical and sound stimuli as cues in the sex discriminatory behavior of the gobiid fish, *Bathygobius soporator*. *Zoologica* 41:49–64.

Tavolga, W.N. (1958). The significance of underwater sounds produced by males of the gobiid fish, *Bathygobius soporator*. *Physiol. Zool.* 31:259–271.

Tavolga, W.N. (1971a). Sound production and detection. In: *Fish Physiology, Vol. V* (Hoar, W.S., and Randall, D.J., eds.), pp. 135–205. New York: Academic Press.

Tavolga, W.N. (1971b). Acoustic orientation in the sea catfish, *Galeichthys felis*. *Ann. NY Acad. Sci.* 188:80–97.

Tavolga, W.N. (1974). Signal/noise ratio and the critical band in fishes. *J. Acoust. Soc. Am.* 55:1323–1333.

Tavolga, W.N. (1976a). Acoustic obstacle detection in the sea catfish (*Arius felis*). In: *Sound Reception in Fish* (Schuijf, A., and Hawkins, A.D., eds.) pp. 185–204. Amsterdam: Elsevier.

Tavolga, W.N. (ed.). (1976b). *Sound Reception in Fishes: Benchmark Papers in Animal Behavior, Vol. 7*. Stroudsburg, PA: Dowden, Hutchinson & Ross.

Tavolga, W.N. (ed.). (1977). *Sound Production in Fishes: Benchmark Papers in Animal Behavior, Vol. 9*. Stroudsburg, PA: Dowden, Hutchinson & Ross.

Tavolga, W.N. (1982). Auditory acuity in the sea catfish (*Arius felis*). *J. Exp. Biol.* 96:367–376.

Tavolga, W.N., and Wodinsky, J. (1963). Auditory capacities in fishes: Pure tone thresholds in nine species of marine teleosts. *Bull. Am. Mus. Nat. Hist.* 126:177–240.

Urick, R.J. (1974). Sea-bed motion as a source of the ambient noise background of the sea. *J. Acoust. Soc. Am.* 56:1010–1011.

van Bergeijk, W.A. (1964). Directional and nondirectional hearing in fish. In: *Marine Bioacoustics* (Tavolga, W.N., ed.), pp. 281–299. Oxford, UK: Pergamon Press.

van Bergeijk, W.A. (1967). The evolution of vertebrate hearing. *Contrib. Sens. Physiol.* 2:1–49.

von Békésy, G. (1960). *Experiments in Hearing* (Wever, E.G., ed.). New York: McGraw-Hill.

von Frisch, K. (1923). Ein Zwergwels der kommt, wenn man ihm pfeift. *Biol. Zentralbl. Leipzig* 43:439–446.

von Frisch, K. (1936). Über den Gehörsinn der Fische. *Biol. Rev.* 11:210–246.

von Frisch, K. (1938a). The sense of hearing in fish. *Nature* 141:8–11.

von Frisch, K. (1938b). Über die Bedeutung des Sacculus und der Lagena für den Gehörsinn der Fische. *Z. Vergl. Physiol.* 25:703–747.

von Frisch, K., and Dijkgraaf, S. (1935). Können Fische die Schallrichtung Wahrnehmen? *Z. Vergl. Physiol.* 22:641–655.

von Holst, E. (1950). Die Arbeitsweise des Statolithenapparates bei Fischen. *Z. Vergl. Physiol.* 32:60–120.

Webb, J.F. (1998). The laterophysic connection: a unique link between the swim bladder and the lateral-line system in *Chaetodon* (Perciformes: Chaetodontidae). *Copeia* 1032–1036.

Webb, J.F., and Smith, W.L. (2000). The laterophysic connection in chaetodontid butterflyfish: Morphological variation and speculation on sensory function. *Philos. Trans. R. Soc. Lond. B.* 355:1125–1129.

Weber, E.H. (1820). *De aure et auditu hominis et animalium. Pars I. De aure animalium aquatilium.* Leipzig., Gehard Fleischer.

Wersäll, J. (1961). Vestibular receptor cells in fish and mammals. *Acta Oto-Laryngol. Suppl.* 163:25–29.

Wever, E.G. (1949). *Theory of Hearing*. New York: Wiley.

Winn, H.E. (1964). The biological significance of fish sounds. In: *Marine Bioacoustics* (Tavolga, W.N., ed.), pp. 213–231. New York: Pergamon Press.

Wohlfahrt, T.A. (1939). Untersuchungen über das Tonunterscheidungsverm gen der Elritze. *Z. Vergl. Physiol.* 26:570–604.

Yan, H.Y., and Curtsinger, W.S. (2000). The otic gasbladder as an ancillary auditory structure in a mormyrid fish. *J. Comp. Physiol. A.* 186:595–602.

Yan, H.Y., Fine, M.L., Horn, N.S., and Colón, W.E. (2000). Variability in the role of the gasbladder in fish audition. *J. Comp. Physiol. A.* 186:435–445.

Yost, W.A., and Gourevitch, G. (eds.) (1987). *Directional Hearing*. New York: Springer-Verlag.

Zelick, R., Mann, D., and Popper, A.N. (1999). Acoustic communication in fishes and frogs. In: *Comparative Hearing: Fish and Amphibians* (Fay, R.R., and Popper, A.N., eds.), pp. 363–411. New York: Springer-Verlag.

2
Trails in Open Waters: Sensory Cues in Salmon Migration

Kjell B. Døving and Ole B. Stabell

Abstract

Salmon display layer-dependent swimming patterns in open waters. Analysis of the behavior shows that they move within certain microstructure layers of the water. However, anosmic salmon do not demonstrate this behavior, either in fresh- or in saltwater. One explanation for such layer preferences is that salmon detect characteristic odorants from the home river that are localized in a particular layer. The olfactory system of salmon is necessary for home stream detection, and is also very sensitive to odorants that emanate from conspecific fishes. These features combined suggest an important role of the olfactory system and also a possible involvement of kin recognition in homing behavior. A change in the magnetic field surrounding the free-swimming salmon does not influence the characteristic layer preferences of sensory intact animals. However, recent studies have shown that salmon are particularly sensitive to low-frequency stimuli in the range of 0.1 to 10 Hz (i.e., infrasound). This fact implies that a salmon may detect linear acceleration to which it is exposed when moving from one layer to an adjacent one. Based on this knowledge, the following scenario is proposed for homing orientation: In open waters salmon swim with small-scale oscillations from one layer with a particular smell to neighboring layers, and thereby gain information about the chemical differences between layers. The salmon may further detect the relative movements of the layers by means of their auditory system. Thus, the salmon can sense the heading of a particular layer and may orient in a countercurrent direction of that layer. If this layer contains characteristic odors of the home stream, such rheotactic behavior will eventually lead the fish to the sources of the attractive chemical signals.

1. Introduction

How can a salmon off the coast find its way back to the river it left two to five years earlier? The migration of the salmon from its home river as a juvenile fish, out into the ocean to grow, and back to the native stream as an adult, has fascinated humankind for centuries. The attempts to explain this homing orientation (i.e., how a salmon finds its way through open water and back to the outlet of its home river) have called for a great variety of proposals regarding sensory mechanisms.

In 1599, Norwegian clergyman Peder Claussøn Friis suggested that salmon were guided in the sea by the kingfish (opah), which possessed supernatural powers (Storm, 1881). Buckland (1880) proposed that instinct, combined with "the power of smell," was used by the salmon for finding their way in the ocean. Much later it was shown that olfaction is mandatory for selecting the correct tributary of a river (Wisby and Hasler, 1954), but only recently has the importance of the olfactory sense in the ocean phase of migration been fully acknowledged (reviewed by Stabell, 1992). Hasler (1966) suggested that orientation of fishes in open waters took place by a sun-compass mechanism, coordinated with a biological chronometer. However, blinded fishes have been found to maintain their homing ability in coastal waters (Hiyama et al., 1967; Toft, 1975), and therefore, vision is probably not involved in the ocean phase of migration. Also navigation by the use of a magnetic sense has been proposed by several investigators (Yano and Nakamura, 1992; Walker et al., 1997; Yano et al., 1997), but the use of a magnetic sense by fishes remains controversial. Recently, the layered structure of lakes and sea has been found to be of importance for migrating fishes (Westerberg, 1982a,b). This microstructure of water, combined with its content of chemical signals, constitutes a basis for a particular behavior of fishes. The layer-dependent behavior, and the use of sensory systems to detect these layers, calls for a critical and updated evaluation on how the salmon find their way.

In this review, we evaluate options for navigation versus orientation by fishes in the open sea. The layered properties of the water will receive particular attention and the different sensory systems that could be essential for salmon during its migration will also be discussed. A layer-dependent behavior and its coupling to an intact olfactory sense calls for an explanation of the orientation mechanisms underlying the directed movements. Since the auditory system has recently been found to be particularly sensitive to infrasound (Sand and Karlsen, 1986, 2000), the possibility that the ear can be used for detection of linear acceleration between water layers will be discussed. Basic properties of the olfactory system will be described, and the possibility that salmon migration in the sea should be viewed as an integral part of the home range and kin recognition concepts is finally considered.

2. Orientation in Open Waters

A bicoordinate system must be present for directional navigation in open waters. This concerns fishes (Neave, 1964) as well as humans (Brown, 1956). Accurate longitude determination was not achieved before the first exact chronometer was invented and constructed by John Harrison during the eighteenth century (Brown, 1956; Sobel, 1995). In fishes, a biological chronometer of this kind has never been demonstrated. Such a chronometer should both be able to tell the fishes about local time of day as well as "home time" throughout changing seasons (Stabell, 1984). Accordingly, if the fishes don't know their correct point of residence, a correct compass course toward home cannot be found.

Salmon follow ocean currents during their migration to and from feeding areas. In addition, they generally travel downstream during their ocean life, following currents associated with gyres in clearly defined areas (Royce et al., 1968). According to Neave, the view that currents may provide guiding clues fails to explain how orientation to a current can be affected in the absence of fixed reference points

(Neave, 1964). An answer to this question may be found in the proposal of Royce et al. (1968) that salmon are capable of detecting the interfaces between moving bodies of water. If the fishes can detect the heading direction of a particular water layer, then the claim for a fixed reference point is unwarranted. Such water layers may then be followed in both an upstream as well as a downstream direction, in a manner similar to the within-stream movements of descending smolt and ascending mature fishes. Thus, we shall advocate the idea that salmon do not navigate but orient themselves in relation to specific features of their near environment.

3. Water, the Layered Environment

Movements in the ocean take place in a three-dimensional system. In open waters, there are few, if any, visual cues, and the absence of fixed points makes orientation a challenge to sensory organs and brain structures that perceive the environment. Water contains compounds that may serve as chemical signals and is also an environment where chemical cues function in a different way compared to air. The mixing processes, given by turbulence and molecular diffusion, are basically the same in air as in water, but they have dramatically different scales. Odorous substances released in water are spread in all directions by diffusion, but if the source is moving or the surrounding medium moves, a chemical trail will form by advection of the substance. In water, low turbulence will also permit a trail to remain for a long time because diffusion in water is a slow process. In the absence of photochemical breakdown and under quiet conditions, such odor trails can linger for days.

Migrating salmon move in the ocean during both day and night, on sunny days and on days with overcast sky. They move in depths covering mainly the upper layers of the sea (Døving et al., 1985; Ogura and Ishida, 1995; Westerberg et al., 1999). Properties of the immediate environment must therefore be of vital interest for the salmon. It is with this in mind that the small-scale structure of water comes into focus.

3.1. Origin of the Layers

In oceanographic terms, vertical microstructure means the small-scale features given by horizontal layers of water with thickness less than one meter (Cooper, 1967). Continuous recordings of salinity and temperature profiles have revealed such types of distinct stratification in all natural bodies of water. Variations in the vertical distribution can be obtained as a result of four different processes, depending on the source of energy for mixing (Munk, 1981). First, there is surface mixing, where heating, precipitation, and runoff to the surface layer cause anomalies in local temperatures and salinities. Mechanical stirring by wind will cause well-mixed volumes of water that spread vertically according to their appropriate density levels. Second, there is bottom-boundary mixing, where the stratified fluid of the interior is mixed vertically along the bottom topography, whereby the mixed volumes separate and spread by gravity and advection into the interior. Third, there will be internal mixing from local shear-induced turbulence, or overturning of internal waves. Fourth and last, there is double diffusive mixing, where stable stratification given by density can become convecting if the vertical gradient of either salt or temperature is destabilized in its contribution to density (salt fingers).

3.2. Properties of Layers

Certain features of the horizontal layers in water may be of particular significance for the migrating fishes. Evidently, each layer has a distinct origin and carries information about that source by its dissolved content of chemicals. The layers can be extensive in size and cover large areas. The horizontal distribution of the layers is much larger than their thickness; the ratio is in the order of between 100:1 and 1,000:1. For example, intrusion of water at the Antarctic shelf spreads with a thickness of about 200 m at a depth of 2,000 m, and extends horizontally in an area of about 100 times 800 km (Carmack and Killworth, 1978). However, dimensions of about 10 m in thickness and 10 km in extension are more common. The lifetime of an intrusion is also dependent on the thickness of the layer. A layer 20–50 m thick may last for several months, and layers with a thickness of 5–10 m can last for a week. Finally, the layers are not mutually stationary objects but move in relation to each other. The velocity differences between layers are typically 0.01 to 0.1 $m \cdot s^{-1}$, measurable across an intersection with thickness between 10 and 100 cm.

An issue of particular importance for orientation purposes is that the river water is mixed with a limited amount of seawater in the estuary, and that this mixture forms an intrusion spreading at a depth according to its density. Thus, dissolved substances in the river water will be diluted in a limited volume of water, and not to an extent corresponding to, for instance, the total masses of water in a fjord. The dilution will also vary with local conditions and changes in temperature, wind, and tide, as given by the mixing types just described. Such variations suggest that the concentration of dissolved substances within a particular layer and carrying information about a particular river will be unpredictable. Thus, the presence, and not the concentration gradients, of a chemical signal should be regarded as the important issue for orientation purposes (Powers, 1941).

It should also be emphasized that a particular body of water will eventually move away from the river outlet. Thus, a body of water from a river, forming a layer in the ocean, will carry two important tags of information, the chemical signals telling about its origin, and the directional heading of its movement. These particular but obvious aspects seem to have been ignored in many reviews and discussions evaluating salmon migration.

An important question then arises: Can salmon detect these types of information? The importance of the olfactory organ in relation to salmon migration has been repeatedly stressed (Stabell, 1984, 1992). Therefore, a more appropriate question to ask seems to be, What additional type of sensory system can the salmon make use of? Recent developments in our understanding of the auditory system and infrasound detection in fishes may indicate a possible solution to these problems (Sand and Karlsen, 1986, 2000). An evaluation of the sensory systems used by salmon in navigational purpose therefore seems a natural next step.

4. Involvement of Sensory Systems in the Migratory Behavior of Salmon

4.1. Salmon with an Intact Sense of Smell

Buckland suggested a possible role of the olfactory sense in the homing behavior of salmon (Buckland, 1880). Not until several decades later, however, was the involvement of olfaction in coastal orientation tested experimentally (Craigie, 1926; Hiyama et al., 1967; Bertmar and Toft, 1969; Toft, 1975). In total, the results of these studies strongly suggest that olfaction is crucial for all species of salmon, of both Pacific and Atlantic origin, in maintaining the ability to orient in open waters (Stabell, 1984, for review). In the final phases of homing in rivers, olfaction is also mandatory for all salmonid species tested (Wisby and Hasler, 1954; Groves et al., 1968; Stabell, 1992). In this context, it should be noted that a substantial amount of literature has accumulated on the importance of "kin recognition" and "kin preference behavior" in the homing of salmonid fishes (Olsén, 1992).

The behavioral mechanisms underlying homing performance in open waters were first dealt with by Westerberg (1982a,b), who investigated the movements of salmon in relation to their immediate environment. Until then, scientists had not realized that salmon move in close association with the distinct layering of the water. Westerberg (1982a,b), in particular, studied the vertical swimming behavior of the salmon. He conducted his experiments with Atlantic salmon (*Salmo salar*) in the Baltic Sea, in a Swedish lake system and in a Norwegian fjord. Techniques were developed to study swimming depth of fishes and made continuous

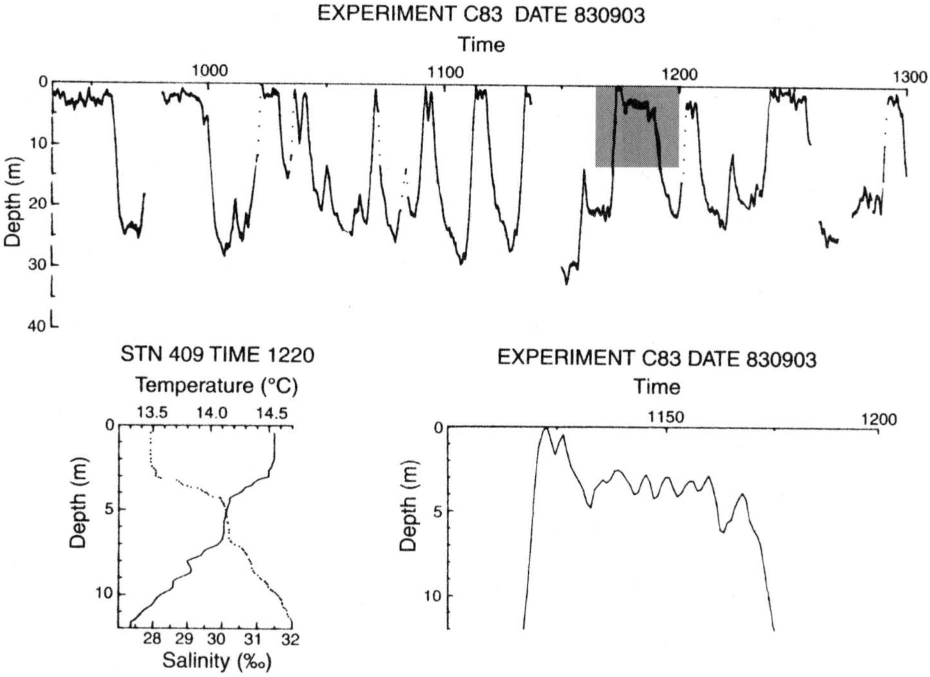

FIGURE 2.1. Upper trace, swimming depth recording of a salmon with intact olfactory organ in Høgsfjorden, Norway. The shaded area is enlarged in the lower trace (*right*) together with an adjacent temperature and salinity profile (*left*). (From Døving et al., 1985.)

tracking within a precision of ±5 cm possible using a transmitter with maximum recording depth of 20 m.

Such technical advances allowed Westerberg (1982a,b) to conclude that the salmon tended to follow the fine-structure gradients in the quasi-mixed surface layer or in the thermocline. In between these periods of swimming at a certain depth, the salmon made rapid excursions either down to the thermocline or up to the mixed layer at the surface (Fig. 2.1). The observed excursions were interpreted as exploratory searches for the vertical distributed home stream odor. The downward dives were made with a vertical speed of 0.1–0.2 $m \cdot s^{-1}$ at an angle of approximately 10°, while the swimming angle for the fish in the upward phase was approximately 25°. The frequency of these exploratory dives was about one per hour. Westerberg concluded that the olfactory signal is a probable sign stimulus for orientation in relation to water currents. He also suggested that the salmon is able to detect the acceleration resulting from the relative movement of layers, which may be detected when crossing from one layer to another.

Studies in the Norwegian Fjord system also demonstrated that the free-swimming salmon follows a certain layer in the water for prolonged periods (Fig. 2.2). The tracking data revealed that, during a surveillance period of one hour, the intact salmon preferred a specific temperature (Fig. 2.3, top), making few, if any, excursions to regions with other temperatures. Studies of the small-scale preference behavior of Pacific salmon have confirmed the same type of vertical movements for the species studied as previously observed for Atlantic salmon (Quinn and terHart, 1987; Ogura and Ishida, 1995; Yano et al., 1997; Tanaka et al., 2000).

4.2. Anosmic Salmon

If the hypothesis for homeward orientation suggested by Westerberg (1982a,b) is correct, then salmon deprived of their olfactory organ

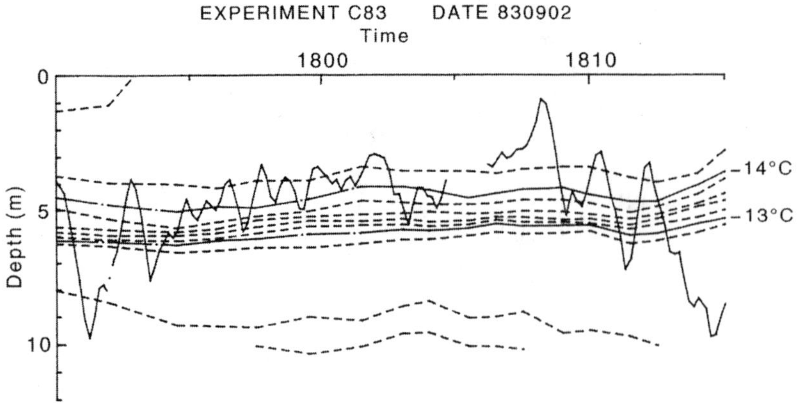

FIGURE 2.2. Detail of the swimming depth of a salmon with an intact olfactory organ in relation to the thermal structure along the track. The temperature soundings were made in approximately 2-min intervals, and the isotherms are drawn with a spacing of 0.2°C. (From Døving et al., 1985.)

should behave differently from salmon with the olfactory sense intact. This has been confirmed several times. As can be seen in Figure 2.3, bottom, anosmic salmon in Lake Vänern displayed a dramatically different behavior compared to the control fish. The anosmic salmon did not prefer a specific layer, but made use of almost the total available temperature range during the recording session. Such anomalous behavior was also confirmed with anosmic salmon in the Norwegian fjord system (Fig. 2.4).

Critics of experiments with anosmic salmon have frequently stated that due to the trauma resulting from impairment of the olfactory sense, the fishes will not perform normally. One salmon tagged and followed in the Norwegian fjord system was made anosmic on one side only. This salmon behaved similarly to intact fishes in the tracking session, and was later captured in the River Imsa (i.e., its home river). Thus, ablation of one olfactory nerve did not cause a loss of normal behavior, nor did it influence the ability to track the home river (Døving et al., 1985).

4.3. Influence of the Earth's Magnetic Field

The idea that the salmon can navigate in the ocean by detecting the Earth's magnetic field has attracted many advocates. Several investigators have searched for magnetic material in fishes. In one study, magnetic material was found within all species of the main teleost groups, but the data revealed a diffuse localization of the magnetite particles, generally in areas in close connection to the bone tissue. There were no significant differences in the amount of magnetic material between migrating versus more stationary species (Hanson and Westerberg, 1987). However, magnetic material has been described in the olfactory organ of rainbow trout (Walker et al., 1997). These authors traced nervous connections to the brain, and specifically suggested that a magnetite-based magnetic sense may make an important contribution to long-distance orientation by animals. Recordings of nervous activity from a trigeminal branch were also made, but this branch was not cut to ascertain that the change in activity was related to a peripheral input.

Walker et al. (1997) propose that the animals are able to form a "magnetic map." Before accepting that salmon make use of an assumed magnetic sense to form a hypothetical magnetic map, one should acknowledge the variations, or noise, in the Earth's magnetic field of several tens of nanoTeslas (nT) at any location, changing with a time scale of hours or days. These variations with time are in the same order of magnitude observed when moving 10 km in a north–south direction, or as the anomalies caused by natural variations of magnetic minerals in the bedrock (Dobrin and Savit, 1988). In addition, both temporary variations due to magnetic storms (±200 nT), as well as the fixed magnetic anomalies (±200 nT) caused by the magnetic minerals in the oceanic crust, may cast doubt on the prospects of forming an applicable magnetic map. Elaborate corrections using modern computers are always carried out on magnetic survey data before magnetic maps can be produced. It is not likely that a fish has this capacity. It should be noted also, that such a map should not only be formed, but also memorized by the migrating animal. Capabilities of this kind have yet to be demonstrated in fishes (see Chapter 3, however, for a review of

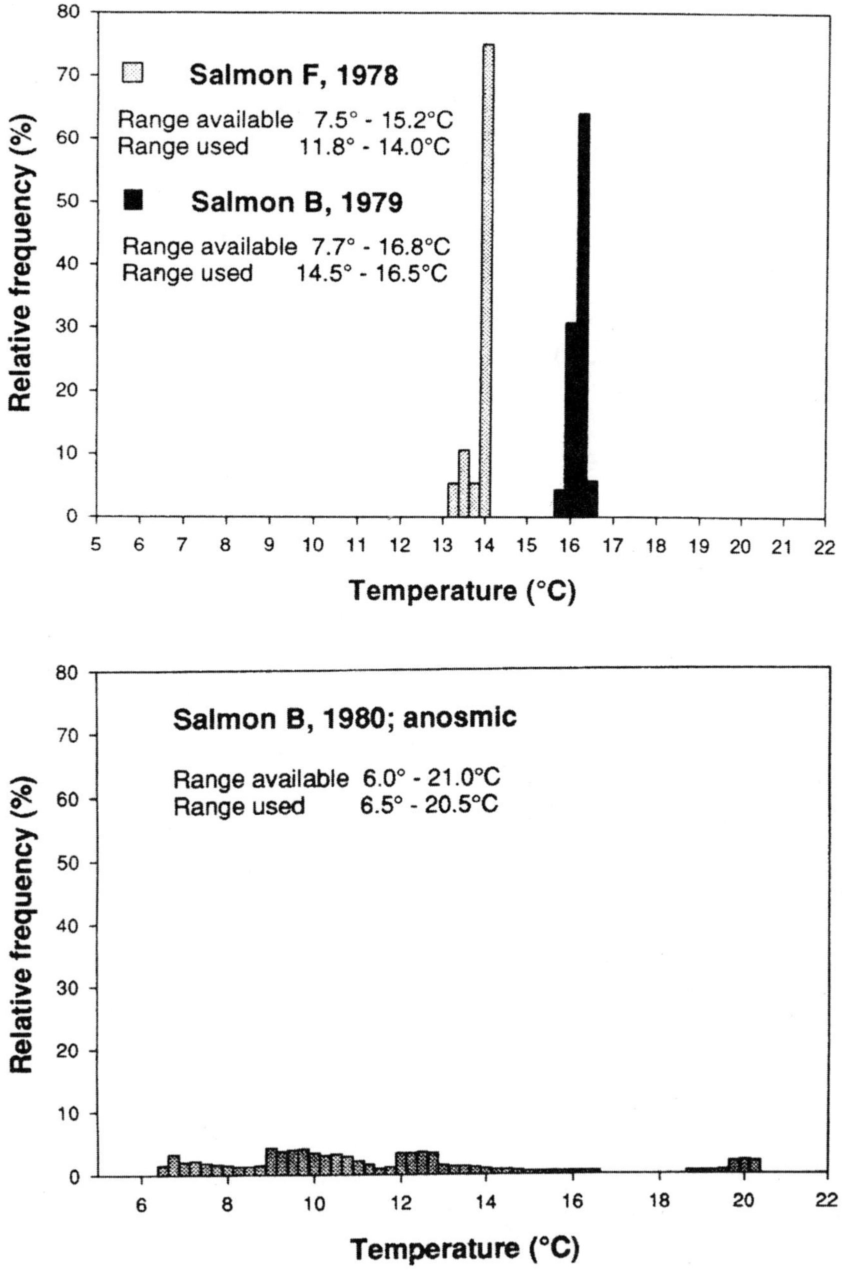

FIGURE 2.3. Relative frequency of time spent at different temperatures (0.25°C intervals) of free-swimming salmon in Lake Vänern, Sweden. The salmon F 1978 and B 1979 had intact olfactory organs. The salmon B 1980 was anosmic. (Redrawn from Westerberg, 1982b.)

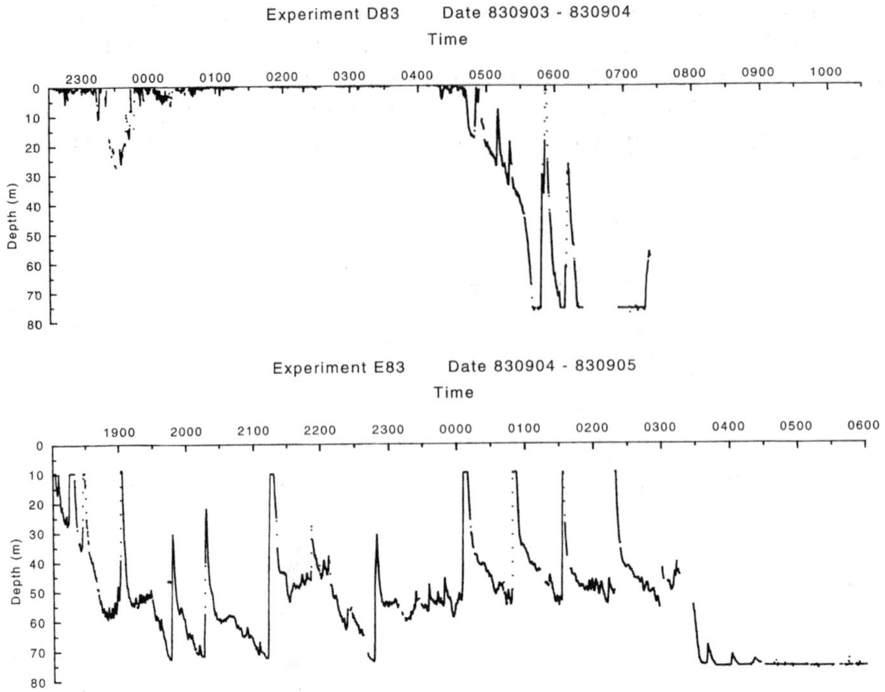

FIGURE 2.4. Swimming depth profiles of two anosmic salmon. (From Døving et al., 1985.)

the magnetic sense). Finally, induced magnetic disturbances have not been found to affect the vertical swimming behavior of salmon (Yano et al., 1997), suggesting that magnetic detection may at best be a remote tool in orientation ability of migrating fishes.

4.4. The Auditory System and Detection of Infrasound

Westerberg proposed that salmon use the current shear within layer gradients to detect flow directions, and thereby the heading toward the source of the home stream odors (Westerberg, 1982a). It seems plausible that salmon can detect such relative movements of layers by means of their auditory system. This proposition results from the discovery that the fish auditory system is very sensitive to infrasound (Sand and Karlsen, 1986, 2000), and not only sensitive to frequencies in the range between 50 and 300 Hz as previously believed (Popper and Fay, 1993). The detection threshold of infrasound is in the order of 10^{-4} m·s^{-2} (Sand and Karlsen, 2000). Thus, if it takes the salmon one second to swim from one layer to the next it can detect differences in water currents as small as 0.1 mm·s^{-1}, which is well below the typical differences in relative velocities between neighboring layers. It seems conceivable that the salmon can detect the direction of movements in addition to the scalar size of the movements. The small-scale motion from one layer to an adjacent one can then be interpreted as behavior to gain information about the relative movements of the layers. However, detection of the relative movements by two neighboring layers may not be sufficient to take a correct course. Evidently, if the layer of interest is the slower moving one, "upstream" heading will lead the wrong way. Accordingly, the regular deep-dives observed by fishes may, in addition, be necessary to gain an overall view of the general heading of several water layers combined.

Another interesting aspect of auditory sensitivity to linear acceleration is the possibility for fishes to detect deviation from a straight line during motion in the horizontal plane. If the salmon deviate from a straight path during forward movement it will be exposed to acceleration. The magnitude of this centrifugal acceleration **a** is given by the formula:

$$\mathbf{a} = \mathbf{v}^2 \cdot \mathbf{R}^{-1}$$

where **v** is the swimming velocity and **R** is the radius of the curvature. If **v** is $1\,\text{m}\cdot\text{s}^{-1}$ and **a** equals $10^{-4}\,\text{m}\cdot\text{s}^{-2}$, **R** will be 10 km. The monitoring of small-scale vertical movements of salmon has revealed a frequency of about one per minute, suggesting that the fish is also moving horizontally for prolonged periods of time. If a 1-m-long fish moves forward with a velocity of 1 body length$\cdot\text{s}^{-1}$, it will move 60 m during one minute. If it swims at a curvature with radius **R** = 10 km, it will deviate $(60 \cdot (2\pi\mathbf{R})^{-1}) \cdot 360° = 0.34°$ from a straight line (Sand and Karlsen, 2000). This equals 36 cm off the bull's-eye at a distance of 60 m, suggesting an additional "sense of inertia" (Harden Jones, 1984) to support accurate orientation in the open sea.

5. The Olfactory System: A Key to Path Finding

The olfactory system shows great similarities both in anatomical layout and in functional properties throughout the vertebrate phylum. A large number of primary sensory neurons converge and make synapses with a small number of relay neurons. Three different morphological types of sensory neurons have been described in teleosts (Thommesen, 1983; Hansen and Finger, 2000). The sensory neurons project to specific loci in the olfactory bulb (Morita and Finger, 1998). There is evidence that sensory neurons with cilia participate in pheremone detection of alarm substance in crucian carp, and make synaptic contacts with relay neurons that send axons to the brain via the medial part of the medial olfactory tract (Hamdani et al., 2000; Hamdani and Døving, 2002). The sensory neurons with microvilli participate in feeding behavior and converge to relay neurons that reach the brain via the lateral olfactory tract (Hamdani et al., 2001a, b).

The functional specificity in the projection of sensory neurons has also been established by electrophysiological recording in salmonids. For instance, bile salts induce responses mainly in the medial part of the olfactory bulb while amino acids induce activity in the lateral part (Døving et al., 1980). Spatial distribution in the connection of sensory neurons implies that fishes can discriminate between classes of chemical compounds, and that separate parts of the brain are used in each case for handling of information.

5.1. Sensitivity

The olfactory system in salmonid fishes is very sensitive to substances emanating from conspecifics, in particular, bile-salt-like compounds (Døving et al., 1980; Thommesen, 1983). Recordings from the salmon olfactory bulb on stimulation of the olfactory epithelium with water from a salmon river have revealed that responses are obtained even when the river water is diluted 1,000 times (K.B. Døving, unpublished). For sulphotaurolithocholic acid, a bile salt derivative, the threshold concentration estimated from electrophysiological recordings has been found in the order of 100 nM. Behavioral thresholds, however, are generally found down to 100 times below those obtained by electrophysiological recordings, suggesting that compounds of this kind are detected by the fish at concentrations of at least 1 nM. In this context, a recent study shows that larvae of the sea lamprey (*Petromyzon marinus*) release specific bile acids that attract migratory adults to spawning rivers at concentrations of 0.1 nM (Bjerselius et al., 2000; Polkinghorne et al., 2001).

5.2. Discrimination

Recordings from single neurones in the olfactory bulb have also demonstrated two important aspects of olfactory communication. First, recordings from olfactory neurons of Arctic charr have shown that the responses to water from different charr populations evoke responses that were independent from one another (Døving et al., 1974). In other words, the olfactory system of charr has the capacity

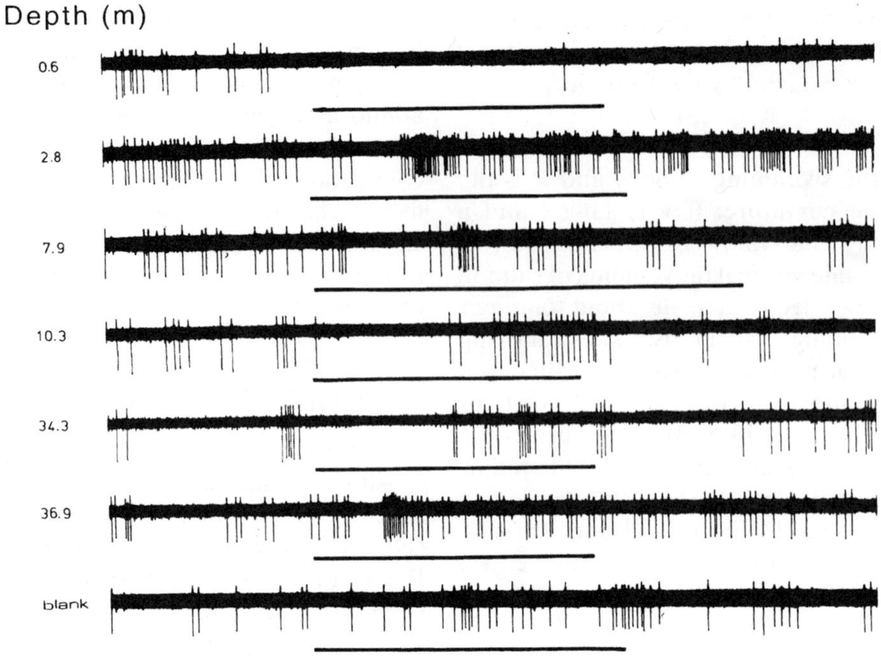

FIGURE 2.5. Responses from a single neuron in the olfactory bulb of salmon. Water samples were taken from the depths indicated to the left of each trace. A line underneath each trace indicates the stimulation period. Each trace is 20s. (From Døving et al., 1985.)

to discriminate between odorants emanating from various charr populations.

Second, the bulbar neurons of salmon respond differentially when the olfactory epithelium of salmon is exposed to water samples from different depths in a Norwegian fjord (Døving et al., 1985) (Fig. 2.5). The results of these experiments indicate a powerful discriminating capacity of the olfactory system. In addition, they support the conclusions reached from experiments with free-swimming salmon, suggesting an important impact of the olfactory sense on the normal migratory behavior.

5.3. Adaptation

It may be argued that the movements of salmon in and out of particular layers, as observed when following fishes with acoustic tags, is a way of avoiding adaptation to a continuous olfactory stimulus. However, a great number of primary receptors converge to a small number of secondary neurons in the olfactory system (Trotier and Døving, 1996), and an extremely slow adaptation of the olfactory system has been previously described as a general functional property (Ottoson, 1956). Recordings of the nervous activity from single neurons in the olfactory tract of cod have also demonstrated that secondary neurons do not adapt to a continuous stimulus. Thus, a continuous stimulation of the olfactory organ with a 1.0 μM solution of methionine caused an elevated impulse activity during 30 min of monitoring (Kjøstolfsen, 1983). Properties of this kind should be expected also in the secondary neurons of salmonids.

6. Orientation in Streams and the Concept of Home Range

The location in which a stream-dwelling fish spends most of its life is called its "home range," and within that range it normally establishes a more restricted "home area" or territory (Gerking, 1953, 1959; Gunning, 1959). When displaced outside their home range, the majority of fishes will return back to that location whether displaced upstream or downstream. This kind of behavior has been observed for several species of fishes (Stott, 1967; Hill and Grossman, 1987), including brown trout (Halvorsen and Stabell, 1990). Removal of the olfactory organ will drastically lower the return rate of the brown trout (Halvorsen and Stabell, 1990). This finding supports the suggestion made by Gunning (Gunning, 1959), that homing ability of longear sunfish *Lepomis megalotis megalotis* is mediated by the olfactory sense. Conspecific chemosensory cues have been shown important for lake trout *Salvelinus namaycus* in the detection of spawning sites (Foster, 1985), and such chemical cues may also be important for the detection of home area by brown trout (Arnesen and Stabell, 1992).

Adult Arctic charr, as well as Atlantic salmon parr, can recognize and are attracted to the odor of their conspecific strain compared to that of a strain from another river (Selset and Døving, 1980; Stabell, 1982). In fact, salmonid fishes are even able to discriminate between the odor of sibling groups from within a single river system (Quinn and Busack, 1985; Olsén, 1992). The attractive odor is found in the intestinal content of fishes (Selset and Døving, 1980; Olsén, 1987; Stabell, 1987), and seems to be deposited on the substrate by the stationary fishes (Stabell, 1987). This last observation, in combination with the findings by Foster (1985), may establish an important connection between kin recognition, homing behavior, and home area detection. Experiments made by Olsén at al. (1998) indicate that the major histocompatability complex (MHC) has a significant influence on the odors used for kin recognition and discrimination in juvenile Arctic charr.

7. A Unified Model of Orientation

Olfaction is important for all stages of homing, where salmon seem able to detect flow directions in streams as well as in open waters. Accordingly, homeward orientation of salmon may consist of a single type of behavior (i.e., a positive rheotactic response released by chemical signals) (Stabell, 1992). Homing of anadromous salmonids in open waters has previously been proposed as an integral part of a uniform homing process, consisting of innate responses to conspecific chemical signals (Nordeng, 1971, 1977). In fact, the entire migration cycle of salmon may be based on rheotactic responses in the presence of conspecific odors. The controlling factor for these responses should then be the motivation status of the fish, specifically related to the chemical signals in question. Such a mechanism would further depend on the developmental stages of the animal.

For all species of anadromous salmonids, a simple set of physiological "switches" could be postulated, each releasing a specific change in behavior in response to a unique set of chemical signals. For instance, during the endocrine process of smolt transformation the fish may undergo a shift from positive to negative rheotaxis in the presence of chemical signals from juvenile fishes of its local population. In contrast, sexual maturation in the fish may induce a shift from negative to positive rheotaxis in response to the same set of chemical signals. In this way, odor trails may be followed to and from feeding areas in the open sea during a full migration cycle, relying on behavioral mechanisms and sensory cues that are the same all along the route.

Acknowledgments. We are grateful to Johan B. Steen and Odleiv Olesen for comments and suggestions relating to an earlier version of this manuscript.

References

Arnesen, A.M., and Stabell, O.B. (1992). Behaviour of stream-dwelling brown trout towards odours present in home stream water. *Chemoecol.* 3:94–100.

Bertmar, G., and Toft, R. (1969). Sensory mechanisms of homing in salmonid fish. I. Introductory experiments on the olfactory sense in grilse of Baltic salmon (*Salmo salar*). *Behav*. 35:234–241.

Bjerselius, R., Li, W., Teeter, J.H., Seelye, J.G., Johnsen, P.B., Maniak, P.J., Grant, G.C., Polkinghorne, C.N., and Sorenson, P.W. (2000).

Direct behavioural evidence that unique bile acids released by larval sea lamprey (*Petromyzon marinus*) function as a migratory pheromone. *Can. J. Fish. Aquat. Sci.* 57:557–569.

Brown, L.A. (1956). The longitude. In: *The World of Mathematics.* (Newman, J.R., ed.), pp. 780–819. New York: Simon & Schuster.

Buckland, F. (1880). *Natural History of British Fishes.* London: Unwin.

Carmack, E.C., and Killworth, P.D. (1978). Formation and interleaving of abyssal water masses off Wilkes Land, Antarctica. *Deep-Sea Res.* 25:357– 370.

Cooper, L.H.N. (1967). Stratification in the deep ocean. *Sci. Progress* 55:73–90.

Craigie, E.H. (1926). A preliminary experiment upon the relation of the olfactory sense to the migration of the sockeye salmon (*Oncorhynchus nerka*, Walbaum). *Trans. Roy. Soc. Can.* 5:215–224.

Dobrin, M.B., and Savit, C.H. (1988). *Introduction to Geophysical Prospecting*, 4th ed. New York: McGraw-Hill.

Døving, K.B., Nordeng, H., and Oakley, B. (1974). Single unit discrimination of fish odours released by char (*Salmo alpinus* L.) populations. *Comp. Biochem. Physiol. A.* 47:1051–1063.

Døving, K.B., Selset, R., and Thommesen, G. (1980). Olfactory sensitivity to bile acids in salmonid fishes. *Acta. Physiol. Scand.* 108:123–131.

Døving, K.B., Westerberg, H., and Johnsen, P.B. (1985). Role of olfaction in the behavioral and neural responses of Atlantic salmon, *Salmo salar*, to hydrographic stratification. *Can. J. Fish. Aquat. Sci.* 42:1658–1667.

Foster, N.R. (1985). Lake trout reproductive behaviour: Influence of chemosensory cues from young-of-the-year by-products. *Trans. Amer. Fish. Soc.* 114:794–803.

Gerking, S.D. (1953). Evidence for the concepts of home range and territory in stream fishes. *Ecology* 34:347–365.

Gerking, S.D. (1959). The restricted movements of fish populations. *Biol. Rev.* 34:221–242.

Groves, A.B., Collins, G.B., and Trefethen, P.S. (1968). Roles of olfaction and vision in choice of spawning site of homing adult chinook salmon (*O. tshawytscha*). *J. Fish. Res. Bd. Can.* 25:867–876.

Gunning, G.E. (1959). The sensory basis for homing in the longear sunfish, *Lepomis megalotis megalotis* (Rafinesque). *Invest. Ind. Lakes Streams* 5:103–130.

Halvorsen, M., and Stabell, O.B. (1990). Homing behavior of displaced stream-dwelling brown trout. *Anim. Behav.* 39:1089–1097.

Hamdani, E.H., Alexander, G., and Døving, K.B. (2001b). Projection of sensory neurones with microvilli to the lateral olfactory tract indicates their participation in feeding behaviour in crucian carp. *Chem. Senses* 26:1139–1144.

Hamdani, E.H. and Døving, K.B. (2002). Alarm reaction in the crucian carp is mediated by olfactory neurones with long dendrites. *Chem. Senses* 27:395–398.

Hamdani, E.H., Kasumyan, A., and Døving, K.B. (2001a). Is feeding behaviour in the crucian carp mediated by the lateral olfactory tract? *Chem. Senses* 26:1133–1138.

Hamdani, E.H., Stabell, O.B., Alexander, G., and Døving, K.B. (2000). Alarm reaction in the crucian carp is mediated by the medial part of the medial olfactory tract. *Chem. Senses* 25:103–109.

Hansen, A., and Finger, T.E. (2000). Phylogenetic distribution of crypt-type olfactory receptor neurons in fishes. *Brain Behav. Evol.* 55:100–110.

Hanson, M., and Westerberg, H. (1987). Occurrence of magnetic material in teleosts. *Comp. Biochem. Physiol.* 86A:169–172.

Harden Jones, F.R. (1984). Could fish use inertial clues when on migration? In: *Mechanisms of Migration in Fishes* (McCleave, J.D., Arnalod, G.P., Dodson, J.J., and Neill, W.H., eds.), pp. 67–78. New York: Plenum.

Hasler, A.D. (1966). *Underwater Guideposts: Homing of Salmon.* Madison, Milwaukee, and London: University of Wisconsin Press.

Hill, J., and Grossman, G.D. (1987). Home range estimates for three North American stream fishes. *Copeia* 1987:376–380.

Hiyama, Y., Taniuchi, T., Suyama, K., Ishioka, K., Sato, R., Kajihara, T., and Maiwa, R. (1967). A preliminary experiment on the return of tagged chum salmon to the Otsuchi River, Japan. *Bull. Jap. Soc. Sci. Fish.* 33:18–19.

Kjøstolfsen, I. (1983). Adaptasjon i lukteorganet hos torsk (*Gadus morhua* L.). In: *Department of Biology*, p. 54. Oslo: University of Oslo.

Morita, Y., and Finger, T.E. (1998). Differential projections of ciliated and microvillous olfactory receptor cells in the catfish, *Ictalurus punctatus. J. Comp. Neurol.* 398:539–550.

Munk, W. (1981). Internal waves and small-scale processes. In: *Evolution of Physical Oceanography* (Warren, B.A., and Wunsch, C., eds.), pp. 264–291. Cambridge, MA and London, England: MIT Press.

Neave, F. (1964). Ocean migrations of Pacific salmon. *J. Fish. Res. Bd. Can.* 21:1227–1244.

Nordeng, H. (1971). Is the local orientation of anadromous fishes determined by pheromones? *Nature (Lond.)* 233:411–413.

Nordeng. H. (1977). A pheromone hypothesis for homeward migration in anadromous salmonids. *Oikos* 28:155–159.

Ogura, M., and Ishida, Y. (1995). Homing behaviour and vertical movements of four species of Pacific salmon (*Oncorhynchus* spp.) in the central Bering Sea. *Can. J. Fish. Aquat. Sci.* 52:532–540.

Olsén, H.K. (1992). Kin recognition in fish mediated by chemical cues. In: *Fish Chemoreception* (Hara, T.J., ed.), pp. 229–248. London: Chapman & Hall.

Olsén, K.H. (1987). Chemoattraction of juvenile Arctic charr (*Salvelinus alpinus* L.) to water scented by intestinal content and urine. *Comp. Biochem. Physiol. A.* 87:641–643.

Olsén, K.H., Grahn, M., Lohm, J., and Langefors, A. (1998). MHC and kin discrimination in juvenile Arctic charr, *Salvelinus alpinus* (L.). *Anim. Behav.* 56:319–327.

Ottoson, D. (1956). Analysis of the electrical activity of the olfactory epithelium. *Acta Physiol. Scand.* 35:1–83.

Polkinghorne, C.N., Olson, J.M., Gallaher, D.G., and Sorensen, P.W. (2001). Larval sea lamprey release two unique bile acids to the water at a rate sufficient to produce detectable riverine pheromone plumes. *Fish Physiol. Biochem.* 24:15–30.

Popper, A.N., and Fay, R.R. (1993). Sound detection and processing by fish: Critical review and major research questions. *Brain. Behav. Evol.* 41:14–38.

Powers, E.B. (1941). Physico-chemical behaviors of waters as factors in the "homing" of the salmon. *Ecology* 22:1–16.

Quinn, T.P., and Busack, C.A. (1985). Chemosensory recognition of siblings in juvenile coho salmon (*Oncorhynchus kisutch*). *Anim. Behav.* 33:51–56.

Quinn, T.P., and terHart, B.A. (1987). Movements of adult sockeye salmon (*Oncorhynchus nerka*) in British Columbia coastal waters in relation to temperature and salinity stratification: Ultrasonic telemetry results. In: *Sockeye Salmon* Oncorhynchus nerka, Population Biology *and Future Management*. (Smith, H.D., Margolis, L., and Wood, C.C., eds.), pp. 61–77. Ottawa: Canadian Government Publishing Centre, Dept. of Fisheries and Oceans.

Royce, W.F., Smith, L.S., and Hartt, A.C. (1968). Models of oceanic migrations of Pacific salmon and comments on guidance mechanisms. *Fish Bull. Wash.* 66:443–462.

Sand, O., and Karlsen, H.E. (1986). Detection of infrasound in the Atlantic cod. *J. Exp. Biol.* 125:197–204.

Sand, O., and Karlsen, H.E. (2000). Detection of infrasound and linear acceleration in fishes. *Phil. Trans. R. Soc. Lond. B.* 355:1295–1298.

Selset, R., and Døving, K.B. (1980). Behaviour of mature anadromous charr (*Salmo alpinus* L.) towards odorants produced by smolts of their own population. *Acta Physiol. Scand.* 108:113–122.

Sobel, D. (1995). *Longitude: The True Story of a Lone Genius Who Solved the Greatest Scientific Problem of His Time.* New York: Walker and Co.

Stabell, O.B. (1982). Detection of natural odorants by Atlantic salmon parr using positive rheotaxis olfactometry. In: *Proceedings of the Salmon and Trout Migratory Behavior Symposium, June 1981* (Brannon, E.L., and Salo, E.O., eds.), pp. 71–78. Seattle: University of Washington.

Stabell, O.B. (1984). Homing and olfaction in salmonids: A critical review with special reference to the Atlantic salmon. *Biol. Rev. Camb. Philos. Soc.* 59:333–388.

Stabell, O.B. (1987). Intraspecific pheromone discrimination and substrate marking by Atlantic salmon parr. *J. Chem. Ecol.* 13:1625–1644.

Stabell, O.B. (1992). Olfactory control of homing behaviour in salmonids. In: *Fish Chemoreception* (Hara, T.J., ed.), pp. 249–270. London: Chapman & Hall.

Storm, G. (1881). The collected writings by Peder Claussøn Friis. In: *Samlede Skrifter af Peder Claussøn Friis*, pp. 111–118. Christiania: Brögger Forlag.

Stott, B. (1967). The movements and population densities of roach (*Rutilus rutilus* L.) and gudgeon (*Gobio gobio* L.) in the river Mole. *J. Anim. Ecol.* 36:407–423.

Tanaka, H., Takagi, Y., and Naito, Y. (2000). Behavioural thermoregulation of chum salmon during homing migration in coastal waters. *J. Exp. Biol.* 203:1825–1833.

Thommesen, G. (1983). Morphology, distribution, and specificity of olfactory receptor cells in salmonid fishes. *Acta Physiol. Scand.* 117:241–249.

Toft, R. (1975). The significance of the olfactory and visual sense in the behaviour of spawning migration in Baltic salmon. *Swed. Salmon Res. Inst. Rep.* 10:1–75.

Trotier, D., and Døving, K.B. (1996). Functional role of receptor neurons in encoding olfactory information. *J. Neurobiol.* 30:58–66.

Walker, M.M., Diebel, C.E., Haugh, C.V., Pankhurst, P.M., Montgomery, J.C., and Green, C.R. (1997). Structure and function of the vertebrate magnetic sense. *Nature* 390:371–376.

Westerberg, H. (1982a). Ultrasonic tracking of Atlantic salmon (*Salmo salar* L.). I. Swimming depth and temperature stratification. *Rep. Inst. Freshwat. Res. Drottningholm.* 60:102–115.

Westerberg, H. (1982b). Ultrasonic tracking of Atlantic salmon (*Salmo salar* L.). II. Movements in coastal regions. *Rep. Inst. Freshwat. Res. Drottningholm.* 60:81–101.

Westerberg, H., Sturlaugson, J., Ikonen, E., and Karlsson, L. (1999). Data storage tag study of salmon (*Salmo salar*) migration in the Baltic: Behaviour and the migration route as reconstructed from SST data. International Council for the Exploration of the Sea CM 1999/AA:06:1–18.

Wisby, W.J., and Hasler, A.D. (1954). The effect of olfactory occlusion on migrating silver salmon (*O. kisutch*). *J. Fish. Res. Bd. Can.* 11:472–478.

Yano, A., Ogura. M., Sato, A., Sakaki, Y., Shimizu. Y., Baba, N., and Nagasawa, K. (1997). Effect of modified magnetic field on the ocean migration of maturing chum salmon, *Oncorhynchus keta. Mar. Biol.* 129:523–530.

Yano, K., and Nakamura, A. (1992). Observations on the effect of visual and olfactory ablation on the swimming behavior of migrating adult chum salmon, *Oncorhynchus keta. Jap. J. Ichthyol.* 39:67–84.

3
Detection and Use of the Earth's Magnetic Field by Aquatic Vertebrates

Michael M. Walker, Carol E. Diebel, and Joseph L. Kirschvink

Abstract

Although the hypothesis that animals use a magnetic sense to navigate over long distances in the sea is intuitively appealing, evidence that aquatic vertebrates respond to the magnetic field in nature has been difficult to obtain until recent years. Aquatic vertebrates have, however, been prominent in laboratory-based demonstration and analysis of the magnetic sense and its mechanism. The key conclusions of these studies have been that the magnetic sense exhibits fundamental properties found in other specialized sensory systems and that the magnetic senses of aquatic vertebrates and birds exhibit substantial similarities. In particular, the magnetic sense appears to be *selective* for the magnetic field stimulus; that is, it responds only to the magnetic field stimulus and does not extract magnetic field information from interactions of the magnetic field with the detector components in other specialized sensory systems. The magnetic sense of aquatic vertebrates is also likely to be *highly sensitive* to small changes in magnetic fields, with its detector cells operating at close to the limit set by background thermal energy. Finally, it seems likely that the magnetic senses of birds and aquatic vertebrates exhibit substantial similarities in their structure and function.

Laboratory experiments have demonstrated behavioral and neural responses to magnetic direction and intensity in species from four classes of aquatic vertebrates. Magnetic impairment experiments also strongly imply that magnetic field detection in both sea turtles and elasmobranchs is based on single-domain particles of magnetite. At the receptor level, an array of new imaging and microscopic techniques has identified magnetoreceptor cells that contain 1-μm-long chains of single-domain magnetite crystals within the olfactory lamellae of rainbow trout. These chains of magnetite crystals will respond only to magnetic fields and appear to have been selected for high sensitivity to small changes in magnetic field stimuli. Recent experiments have demonstrated that the magnetic sense of birds is also based on magnetite located in the nasal region and that the same nerve carries magnetic field information to the brain in both fishes and birds. It therefore seems likely that magnetite is the basis of magnetic field detection in a wide range of vertebrate groups. We conclude that, in the aquatic vertebrates, the magnetic sense can now be demonstrated and analyzed in the laboratory using experimental approaches developed for the study of other sensory modalities. Careful selection of experimental subjects will be required, however, to overcome the challenge of applying insights gained in the laboratory to experimental analysis of the use of the magnetic field in the aquatic environment.

1. Introduction

How animals navigate over long distances is one of the great, unsolved mysteries in biology today. Nowhere is this more true than in aquatic environments, where swimming animals are subject to passive displacement by water currents that may be very difficult to detect, particularly in deep water. There are, however, abundant examples that pelagic animals traveling in deep water (e.g., Holland et al., 1990; Klimley, 1993; Papi et al., 1997, 2000) know where they are and can travel direct routes between important locations in their environment even when traveling within major current systems. What such studies do not provide is answers to questions about the external stimuli used by animals to navigate over these long distances.

The hypothesis that animals navigate using the earth's magnetic field was first proposed in the nineteenth century (Viguier, 1882), and has an abiding intuitive appeal. This appeal serves only to add to the mystery of animal navigation, however, because the difficulty of achieving reproducible behavioral responses to magnetic field stimuli in the laboratory and the lack of an identifiable magnetic sense "organ" led instead to widespread skepticism about the existence of the magnetic sense (e.g., Griffin, 1982). It was not until the early 1970s that the first experimental evidence was obtained for detection of magnetic fields by birds (Keeton, 1971; Wiltschko, 1972) and it was some years before the first reproducible responses to magnetic fields by aquatic species were reported (Phillips, 1977; Quinn, 1980). These results were not sufficient to dispel skepticism entirely, however, because the locus and mechanism of magnetic field detection and the neural pathway transmitting magnetic field information to the brain remained unidentified.

In this chapter, we summarize the evidence from field studies suggesting that sharks and whales use the magnetic field to guide long-distance movements. We then focus on experimental demonstration and analysis of the magnetic sense and its mechanism. Our central thesis is that the magnetic sense will share key properties with other sensory systems. In particular, the cells that detect magnetic fields should be selective for and have high sensitivity to magnetic fields (Block, 1992). That is, the receptor cells should respond only to magnetic fields (their adequate stimulus) and their sensitivity to changes in magnetic fields should approach the limit set by the background thermal energy, kT. Experimental results suggest that the magnetic sense of aquatic vertebrates does indeed respond only to its adequate stimulus, but it remains to be demonstrated experimentally that the magnetic sense also shares the property of being highly sensitive to magnetic fields. We conclude that a coherent picture is emerging but that much more work is required to elucidate the structure, function, and use of the magnetic sense in aquatic vertebrates. Of particular importance will be demonstration of the links among the components of the magnetic sense and experimental testing of the use of the sense in nature.

2. The Magnetic Field as a Stimulus

2.1. Sources of the Observed Magnetic Field

By far the bulk of the magnetic field that can be observed within the biosphere is generated through heat convection currents flowing within the molten core of the Earth. These produce the well-known magnetic dipole (represented schematically in Fig. 3.1) that attracts the north-seeking pole of a hand-held compass. The magnetic dipole is responsible for systematic increases in the *intensity* (the force the magnetic field exerts on a unit dipole) and *inclination* (the angle formed between the magnetic field vector and the local horizontal) between the equator and the poles of the Earth's magnetic field. A mathematical model of the dipole and non-dipole components of the field produced in the Earth's core permits calculation of the systematic variation in the observed field. The model does not account for all of the fields due to crustal rocks, which constitute the residual field (sometimes termed magnetic anomalies). The *declination* of the Earth's field is defined as the angle between magnetic and geographic north and arises because the axes of the earth's rotation and its magnetic dipole are not aligned. Magnetic declination varies rapidly near the magnetic poles and relatively slowly near the magnetic equator (see Skiles, 1985, for a comprehensive review of the Earth's magnetic field relevant to living organisms).

In addition to the dipole field produced in the Earth's core, non-dipole components of the field produced in the core and crustal rocks produce magnetic fields (magnetic anomalies) that add to or subtract from the dipole field produced in the core. The fields due to crustal rocks are generally small (<5% of the total field) but can vary rapidly over short distances relative to the field produced in the core. The non-dipole components of the core field produce magnetic anomalies that vary more slowly and spread over much larger areas than those normally produced by crustal rocks. Figure 3.1 shows in schematic form a magnetic anomaly caused by the interaction between the Earth's magnetic dipole and a dipole magnetic source (such as a volcanic or iron deposit) in the Earth's crust. The inset graph in Figure 3.1 is a simplified illustration of the field that would

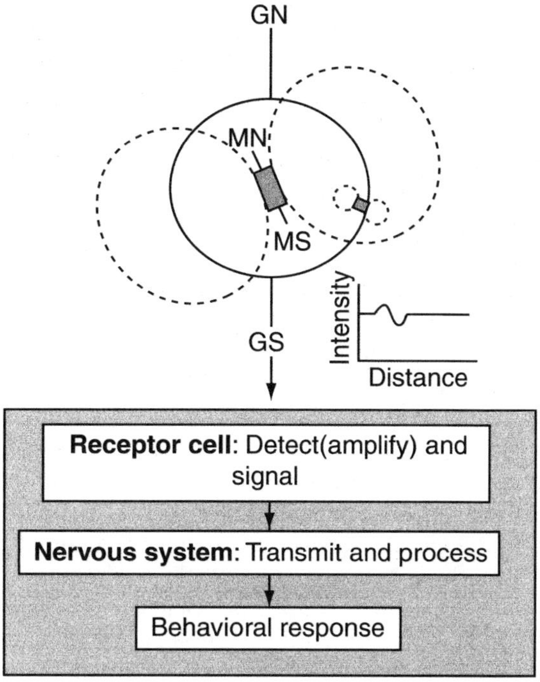

FIGURE 3.1. Schematic of the dipole magnetic field produced in the Earth's core, its interaction with a simple dipole source in the Earth's crust, and hypothesized locations and interactions with key components (boxes) of the magnetic sensory system. MN, MS: magnetic north and south poles; GN, GS: geographic north and south poles; solid circle: surface of the Earth; broken lines: magnetic field lines around magnetic dipole sources (filled bars) in the core and crust of the Earth. (Inset: schematic plot of intensity as a function of distance along a transect through a simple dipole anomaly arising from the interaction of a magnetic source in the crust with the dipole produced in the Earth's core.) The magnetic field stimulus enters the body (shaded box) unchanged where it interacts with the detector element in a receptor cell. The transduced magnetic signal may then be amplified before conversion into a change in the membrane potential of the receptor cell that transmits the transformed signal to the afferent nerve. The peripheral afferent nerve then transmits the signal to the brain where it is processed and a behavioral output is produced.

be recorded by a magnetometer along a transect across the source of the anomaly. The intensity measured by the magnetometer is constant outside the anomaly and represents the field produced in the Earth's core. The interaction between the dipole fields from the core and the source of the anomaly produces the sinusoidal section of the intensity plot in Figure 3.1.

The Earth's dipole magnetic field thus provides consistent information about direction and latitude throughout the biosphere. In particular, the polarity of the field provides information about absolute direction, whereas the inclination of the field identifies the Earth's magnetic axis together with the directions to the magnetic equator and nearest magnetic pole. Animals sensitive enough to magnetic fields to detect the systematic variations in the inclination and intensity of the field could obtain information about their location relative to the magnetic equator and pole (a magnetic latitude akin to geographic latitude). Note, however, that the information about latitude from the intensity of the Earth's field is embedded in considerable noise due to the nondipole fields produced in the core, fields produced by crustal rocks, and short-term variations in the field produced by the solar wind and solar flares (Skiles, 1985). Over longer time periods, secular variation can cause considerable variations in total intensity at any given site while the dipole field can reverse completely at time intervals of from tens of thousands to hundreds of thousands of years (Skiles, 1985; Courtillot et al., 1997).

2.2. Implications for Detection and Use of the Earth's Magnetic Field

Magnetic fields are relatively simple stimuli that have only the two dimensions of intensity and direction. The magnetic field observed at a point on the surface of the Earth includes stable components that vary systematically over very long distances (thousands of km) and randomly over much shorter distances (meters to tens of km; Fig. 3.1). Systematic variations in intensity due to the field in the Earth's core range from 2 to 5 nanoTesla (nT) per km between the magnetic equator and the magnetic poles, whereas the intensity variations due to crustal rocks range from 10^1 to 10^2 nT per km with extremes on the order of 10^3 nT per km.

If we consider only the above stable components of the Earth's magnetic field, we can see that animals will only experience change in the magnetic field they observe when they move. Because the Earth's magnetic field varies over different scales, the separate components of the observed field will change at different rates. As animals move around their environment, they will thus be exposed to field changes over a range of frequencies. The lowest frequencies will spring from the systematic variation due to the dipole produced in the Earth's core, whereas the highest frequencies will be those experienced when animals move through areas where there are strong magnetic anomalies.

Based on our central thesis above, we present in schematic form (Fig. 3.1) the entry of the magnetic field stimulus into the body of the animal and its interactions with the key components of the magnetic sensory system. Because tissues are transparent to magnetic fields, the magnetic field stimulus potentially can enter the receptor cell directly, where it will impinge on a detector element that responds only to magnetic fields. The transduced magnetic field signal is likely to be amplified after detection (Block, 1992) and to result in a change in the membrane potential of the receptor cell. The change in the membrane potential of the receptor cell is transmitted across the afferent synapse to the afferent nerve, which then transmits the encoded information about the magnetic field stimulus to the central nervous system. This information is then processed by the brain and a behavioral output specified if necessary.

The hypothesis that animals use the Earth's magnetic field for navigation predicts that animals should be differentially associated with particular features of the field. The fields produced in the Earth's core and in crustal rocks both potentially contain navigational information that animals might use to guide movement. It has proven difficult, however, to determine what might be the respective roles of the two components in navigation by animals. The systematic variation produced in the Earth's core provides information about location between the magnetic equator and pole (a latitude), but a second coordinate (a longitude) that might be used with the latitudinal information to determine position has not yet been clearly identified. The fields produced in the crustal rocks produce a magnetic topography that animals might use as landmarks or guides during homing and migration (e.g., Kirschvink et al., 1986). Results presented in the next section are consistent with the hypothesis that both sharks and whales respond to features of the magnetic environment associated with magnetic anomalies. Note, however, that there is as yet no direct experimental evidence outside the laboratory for response by aquatic animals to magnetic fields.

3. Evidence for Response to the Earth's Magnetic Field in the Aquatic Environment

Differential association with magnetic field parameters of the positions where whales strand themselves alive (Klinowska, 1985; Kirschvink et al., 1986), where fin whales are sighted at sea (Walker et al., 1992), and of the tracks of hammerhead sharks (Klimley, 1993) are consistent with the hypothesis that aquatic animals respond to the magnetic intensity topography produced by crustal rocks. Thus Klinowska (1985) hypothesized that whales that strand themselves alive have made a significant orientation error and that examination of geophysical variables at such stranding sites should give clues to the nature of the sensory information that was being used when the mistake was made. When she superimposed the locations of live stranding sites on magnetic anomaly maps of the United Kingdom, Klinowska (1985) observed an association between the locations of live stranding sites and areas where minima (or valleys) in the magnetic topography intersected the coast. This pattern was confirmed and extended for the coast of the eastern United States by Kirschvink et al. (1986), who used Monte Carlo simulations to show that the observed patterns of live strandings could not have occurred by chance. These results were consistent with the ability of the whales to detect geomagnetic topography (Kirschvink et al., 1986) as suggested by Klinowska (1985).

It must be acknowledged, however, that live strandings of whales are, first, rare events that in no way reflect the normal behavior of whales and, second, subject to significant sampling biases associated with unrelated phenomena such as human population density (Mead, 1979). These problems were overcome in an analysis of associations between the sighting positions at sea of fin whales collected by the Cetacean and Turtle Assessment Programme (CETAP) run by the Bureau of Land Management during the late 1970s and early 1980s. The CETAP surveys collected systematic sighting data to assess the abundance of large marine animals over the continental shelf between Cape Hatteras and the Gulf of Maine.

Sighting positions for fin whales from the CETAP data set were superimposed on geophysical data for the continental shelf obtained from the NOAA Geophysical Data Center (Fig. 3.2). Monte Carlo simulations showed that the sighting positions were preferentially associated with areas of low magnetic intensity and gradient at times when the whales were migrating (spring and fall but also with low intensity in winter), but not in summer when the whales were at their summer feeding areas in the Gulf of Maine (Table 3.1). These results were consistent with the hypothesis that the whales traveled in the magnetic valleys, which will be characterized by low values of intensity and gradient. Kirschvink et al. (1986) suggested that use of magnetic topography would permit whales, and perhaps other animals, to guide north–south migrations in the deep ocean using the marine magnetic lineations produced by seafloor spreading.

Differential associations with magnetic topography have also been reported for scalloped hammerhead sharks tracked during nocturnal homing movements between Las Animas Island and the Espiritu Santo seamount in the Gulf of California. Klimley (1993) found that the sharks were highly oriented. They swam in the same directions for extended periods while

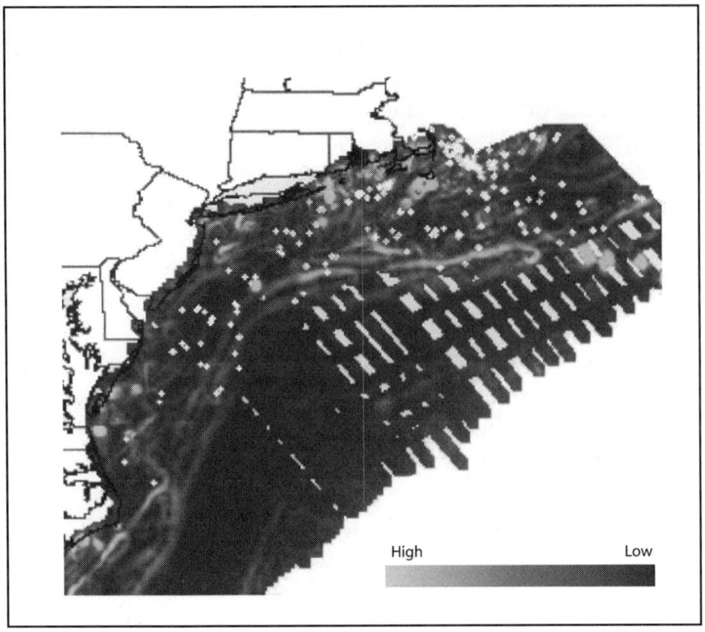

FIGURE 3.2. Sighting positions (all seasons) for fin whales superimposed on an image of magnetic field intensity gradients over the outer continental shelf off the northeastern United States. Magnetic data are from the Decade of North American Geology data set. Sighting positions from the CETAP dedicated aerial surveys are indicated by white crosses. Magnetic field gradients are indicated by shades of gray, with 256 steps between minimum (dark) and maximum (light) gradients. (Adapted from Walker et al., 1992.)

remaining at depths out of sight of both the surface and the seafloor as they traveled between the island and the seamount. The movements could not be correlated with bathymetric features but were associated with areas of high intensity slope (37 nT/km) in the Earth's magnetic field. On the basis of these results, Klimley (1993) proposed that the sharks navigate using geomagnetic topotaxis in which they actively track features of the magnetic topography, such as magnetic intensity ridges and valleys.

4. Behavioral Responses to Magnetic Fields in the Laboratory

Over the last two decades, a variety of experiments have provided experimental confirmation of the above evidence that aquatic vertebrates from different classes respond to magnetic fields in nature. Amphibians, salmonid fishes, and sea turtles have been

TABLE 3.1. Results of Monte Carlo simulations used in two-tailed tests of the hypothesis that mean values of geophysical parameters at positions where fin whales were sighted in different seasons were significantly different from the mean values at simulated sighting positions on the CETAP flight tracks.

	All	Spring	Summer	Fall	Winter
No. of sightings	82	31	29	7	15
Bottom depth	>.1	>.1	>.1	>.1	>.1
Bottom slope	>.1	>.1	>.1	>.1	>.1
Field intensity	>.1	>.1	>.1	>.1	.034
Field gradient	>.1	.024	>.1	.008	.038

Note: Animals observed feeding or engaged in behavior associated with feeding were excluded from the analysis on the grounds that their sighting positions would have been determined by the location of food. Cells in the table contain estimates of the probabilities that the mean values of the geophysical parameters for the simulated positions that are equal to or lower than the mean values for the parameters at sighting positions could be obtained by chance.

shown to respond to magnetic field direction in orientation experiments, whereas both teleost and elasmobranch fishes have been successfully conditioned to magnetic fields in the laboratory. In the paragraphs that follow, we examine key results from these experiments with amphibians, sea turtles, and fishes.

4.1. Orientation Responses to Magnetic Direction and Intensity

The critical assumption of orientation arena experiments is that the spontaneous directional choices made by animals placed in featureless orientation arenas match the directions they would choose in their normal environment (Emlen, 1975). Thus, during their migration seasons, many birds become active at night, and orient in the same directions when placed in a featureless arena as their migrating conspecifics are flying. In the laboratory setting, animals can be induced to establish an orientation direction to a key feature of their living environment such as a water flow direction or a shore. The animals are then tested for that orientation direction when placed in a featureless arena.

The first experimental evidence of magnetic orientation by aquatic vertebrates came in cave salamanders (Phillips, 1977) and in two salmon species (Quinn, 1980; Taylor, 1986, 1987). The cave salamanders were trained to move either parallel or perpendicular to the direction of the magnetic field present in training corridors (Phillips, 1977). When released in a cross-shaped testing assembly in which corridors were aligned parallel and perpendicular to the direction of the magnetic field, the salamanders were significantly oriented along the axes in which they had been trained. In ongoing work with amphibians by Phillips and his colleagues (e.g., Fischer et al., 2001; see also Deutschlander et al., 1999), eastern red spotted newts are trained to escape sudden temperature changes in their living tank by swimming toward an artificial shore. The newts subsequently swim in the training direction when they are placed in an orientation arena without a shore. Similarly, juvenile chinook salmon were allowed to establish an orientation facing into a current that carried their food and flowed from west to east in their living tank. When placed in a featureless arena where there was a radial current flow, the fishes oriented in the east–west axis (Taylor, 1986, 1987).

In contrast with the above examples, Quinn (1980) tested the orientation of sockeye salmon fry during their migration to the lakes in which they would disperse to live. Newly hatched sockeye salmon fry leave the gravel beds where they hatch and swim upstream to lakes where they live until their seaward migration (Quinn, 1980). Migrating fry were captured as they swam toward the lake in which they would live until they migrated downstream to the sea. The fishes were then placed in an orientation arena. The directions chosen by the fry in the arena were consistent with the hypothesis that the fishes were orienting to the axis of the lake in which they would live until they reached the smolt stage and began their migration to the sea.

More recently, Lohmann and colleagues (1996, 2000, 2001) have demonstrated orientation to both magnetic inclination and intensity by hatchling loggerhead turtles. When placed in an orientation arena, the hatchling turtles oriented in the offshore direction as indicated by the magnetic field to which they were exposed (Fig. 3.3A). When presented with fields of inclinations and intensities found at several different locations around the central North Atlantic Ocean, the hatchlings oriented in directions that would have caused them to move toward the center of the North Atlantic gyre (Lohmann and Lohmann, 1996; Lohmann et al., 2001). Such a pattern would be expected to keep the turtles entrained within the North Atlantic gyre and prevent them from being carried into colder waters to the north of the gyre (Lohmann and Lohmann, 1996; Lohmann et al., 2001).

4.2. Conditioned Responses to Magnetic Intensity

Although it is difficult to change magnetic intensity without also changing magnetic field direction, it appears that animals can discriminate changes in magnetic intensity in conditioning experiments subject to two constraints. These constraints are that (1) the fields to be discriminated are spatially distinctive and (2) the subjects must be moving. The simplest pair of spatially distinctive fields is the case where the animal discriminates the presence and absence of a magnetic intensity anomaly induced by an electromagnetic coil. Because the intensity of the Earth's magnetic field is constant within an experimental arena, the animal is thus asked to discriminate the presence and absence of intensity variations due to the coil. The animal must then move in order to gain exposure to the presence or absence of intensity variations in the experimental situation.

Yellowfin tuna have been trained to discriminate the presence and absence of a nonuniform magnetic field in experimental tanks (Walker, 1984). Nonuniform fields (produced by passing direct current through vertically oriented coils) added localized fields of varying intensities to the uniform Earth's field within

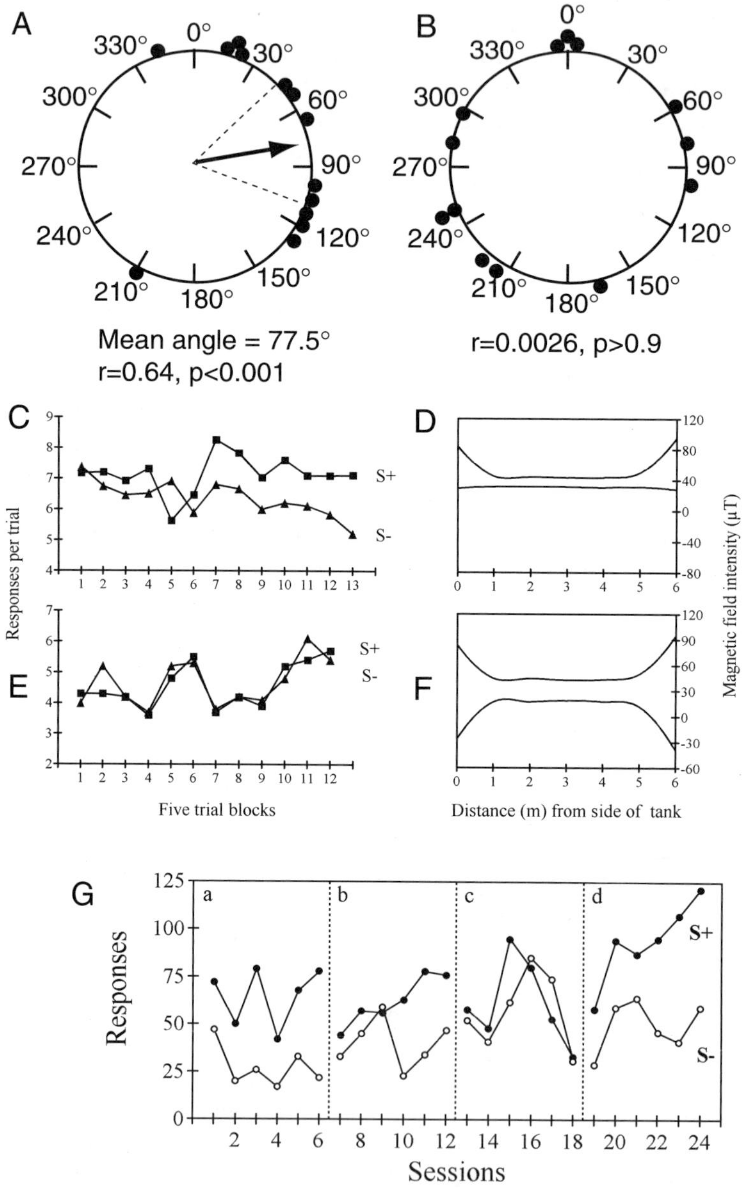

the tanks in which fishes were trained. Reversing the polarity of the current to the coils caused the nonuniform field to be added to or subtracted from the Earth's magnetic field in the tank. Individually-trained yellowfin tuna swam repeatedly through a hoop lowered into an experimental tank for a 30-second trial period. At the end of each trial and depending on the presence or absence of the magnetic field produced by the coil, the fishes were rewarded or not rewarded with food for swimming through the hoop. Discrimination learning was then detected as a change over time in the rates of response during reinforced (S+) and nonreinforced (S–) trials.

The fishes readily learned to discriminate the presence and absence of the nonuniform field but not between the two nonuniform fields produced by reversing the polarity of the current to the coils. For fishes tested with the presence and absence of the nonuniform field due to the coil (Fig. 3.3E), response rates during both S+ and S– trials remained similar over the first 6 five-trial blocks (Fig. 3.3C). After 6 five-trial blocks, however, response rates were consistently higher in the presence of S+ than in the presence of S– (Fig. 3.3C). For fishes trained with two nonuniform fields produced by reversing the polarity of the current to the coils (Fig. 3.3F), there was no separation of response rates to S+ or S– at any stage of the experiment (Fig. 3.3D).

The basic result with the tuna has now been replicated in two other fish species, a teleost and an elasmobranch. Rainbow trout (*Oncorhynchus mykiss*; Walker et al., 1997) and the short-tailed stingray (*Dasyatis brevicaudata*; Hodson, 2000) discriminated between the presence and absence of magnetic anomalies superimposed on the background field in experimental tanks. The pattern of discrimination learning by these two species and the yellowfin tuna was remarkably similar despite variations among the three species in the number of trials required for discrimination to appear. In the case of the trout, reversal learning, a well-known learning phenomenon, has also been demonstrated (Haugh and Walker, 1998). Taken together, these results demonstrate that the magnetic sense can be analyzed using conditioning approaches in the same manner as better-known sensory systems.

FIGURE 3.3. Behavioral responses to magnetic fields (A, B) Distributions of mean bearings for hatchling loggerhead turtles in an orientation arena. The animals wore a nylon lycra harness into which brass weights (controls) or stirring bar magnets (experimentals) were placed. Distributions of bearings in A and B are for control and experimental animals respectively. (C, D) Magnetic discrimination learning in individually trained yellowfin tuna (n = 7 in C; n = 2 in D). Each point is the mean of five trials in which responding was reinforced with food (S+; filled squares) given at the end of each trial or five trials in which responding was not reinforced no matter how often the fish responded. (E, F) Variations in magnetic intensity (in microTesla; μT) with distance from the edge of the experimental tanks used in the experiments plotted in C and E. In D, the background field of the Earth is a uniform 37 μT, whereas the field produced by a coil wrapped around the tank wall adds between 10 μT and 60 μT respectively to the Earth's field at the center and edge of the tank. In F, the field produced by the coil is added to (upper trace) or subtracted from (lower trace) the background field of the Earth in the tank. The fields shown in D and F were used in the experiments presented in C and E respectively. (G) Impairment of learned magnetic discrimination by short-tailed stingrays. The experimental procedure differed little from the procedure used for the experiments with the tuna in C and D. Each point represents the mean number of responses per session made by the experimental animals in the presence of the reinforced stimulus (S+; filled circles) and the nonreinforced stimulus (S–; open circles). Panels A and B show the discrimination performance before and after the insertion of brass weights into the nasal cavities of the animals. Panel C shows impairment of discrimination by replacement of the brass weights by neodymium-iron-boron magnets of the same size as the brass weights. Panel D shows the recovery of discrimination after removal of the magnets. (A, B: redrawn from Irwin and Lohmann, 2000; C–F: adapted from Walker, 1984; G: Hodson, 2000.)

5. Implications for Magnetoreceptor Mechanisms

These behavioral experiments have wider implications for the mechanism of magnetic field detection in other aquatic vertebrates. Hypotheses concerning the magnetoreceptor mechanism have proposed that the magnetic field signal is either (1) extracted from interactions of the magnetic field with the detector components in other specialized sensory systems (Leask, 1977; Kalmijn, 1978); or (2) detected directly using magnetite that is linked somehow to the nervous system (Kirschvink and Gould, 1981). Three hypotheses have been proposed. The "light-dependent" (also known as the "optical pumping") hypothesis proposes that electrons from visual pigments that have been excited by light will interact with the external magnetic field to produce a signal that could be detected by the visual system (Leask, 1977; Deutschlander et al., 1999). The electrical induction hypothesis proposes that the electroreceptor systems of the elasmobranchs (sharks and rays) detect electric current flows induced as the animals, and/or the water mass in which they are swimming, move through the Earth's magnetic field (Kalmijn, 1978, 1981, 1982). In contrast, the magnetite hypothesis proposes that the motion of single-domain crystals of magnetite, a magnetic mineral, converts the force exerted on the crystals by an external magnetic field into a mechanical signal that can be detected by the nervous system. Kirschvink and Gould (1981) suggested several mechanisms that might be used to convert the magnetic signal from the movement of the magnetite into an electrical signal at the membrane of a receptor cell.

The competing hypotheses of electrical induction-based and magnetite-based magnetoreception can be distinguished on the basis of the predictions they make concerning the effect of attached magnets on magnetic field detection. Because attached magnets impose a constant field relative to the body, they will make no contribution to the electrical signal induced by an elasmobranch fish as it swims. Magnets should therefore not affect magnetic field detection using electroreceptors but they should impair magnetite-based magnetoreception provided they are placed close enough to the magnetite to interfere with its use in magnetic field detection.

We have tested the magnetite-based magnetoreception hypothesis in an elasmobranch, the short-tailed stingray, *Dasyatis brevicaudata*. We assumed that, if magnetite located in the nose of the rainbow trout (Diebel et al., 2000; see below) is to form the basis of a general mechanism of magnetic field detection in the vertebrates, then magnetite-based magnetoreceptors are likely to occur in similar locations in representatives of major vertebrate taxa. We therefore sought to impair magnetite-based magnetoreceptors by attaching magnets over the noses of stingrays that had been trained to discriminate the presence and absence of a magnetic intensity anomaly in an experimental tank (Fig. 3.3G(a)). When the rays carried brass weights (3 mm × 2 mm cylinders) implanted in the nasal cavity, they were still able to discriminate the presence and absence of the anomaly (Fig. 3.3G(b)). The rays could no longer discriminate the anomaly, however, when the brass weights were replaced with rare-earth (neodymium-iron-boron) magnets of the same dimensions (Fig. 3.3G(c)). The rays were able to make the discrimination again immediately after the magnets were removed (Fig. 3.3G(d)). Although use of the ampullary electroreceptors to detect magnetic fields is not excluded by this finding, it seems likely that magnetite-based magnetoreception is at least the primary means of magnetic field detection in the stingray and perhaps also in other elasmobranchs.

A similar experiment has been carried out with sea turtles. In orientation experiments, hatchling loggerhead turtles wore a harness to which a stirring bar magnet or a brass weight of equivalent size and weight could be attached (Irwin and Lohmann, 2000). Figure 3.3A shows that turtles with a brass bar attached to the harness were significantly oriented with a mean heading of 77.5° ($r = 0.64$, $n = 15$, $p < 0.001$, Rayleigh test). The 95% confidence interval for the mean bearing included the expected orientation direction (90°) for the hatchlings. Turtles with a bar magnet attached to the harness were not significantly oriented as a group ($r = 0.0026$, $n = 13$, $p > 0.9$, Rayleigh test; Fig. 3.3B). Experimental data consistent with the magnetite-based magnetoreception have thus been obtained in aquatic species from three vertebrate classes.

6. Neural Transmission

The discovery of magnetite suitable for use in magnetoreception in the front of the head in a variety of teleost fishes (Walker et al., 1984; Kirschvink et al., 1985; Mann et al., 1988; Diebel et al., 2000) provided a focus for the search for the sensory nerve that might transmit magnetic field information to the brain. The olfactory (ON), trigeminal (TN), and anterior lateral line (ALLN) nerves are sensory nerves that innervate the front of the head and that could each potentially carry magnetic field information to the brain. The ON is the major source of afferent innervation for the olfactory mucosa. The TN is a mixed nerve that, inter alia, carries afferent signals from mechanoreceptor cells and that, in rats, is known to innervate the olfactory epithelium (Finger et al., 1990). The ALLN innervates the highly sensitive mechanoreceptors of the lateral line and, in the elasmobranchs, innervates mechanoreceptors that have been adapted for electroreception.

Responses to magnetic field stimuli were found to occur in the superficial ophthalmic branch (SO) of the TN of the trout (Walker et al., 1997), the same branch of the TN system that responded to magnetic field stimuli in birds (Beason and Semm, 1987; Semm and Beason, 1990). The responsive units in the trout showed regular firing patterns except during transient responses to a trebling of magnetic intensity presented as square waves at frequencies of 0.5 and 1 Hz (Fig. 3.4A–D). Both excitatory and inhibitory responses were observed but the units responded only to either the onset or the offset of a stimulus (Fig. 3.4D). Surprisingly, no unit responded when magnetic field direction was reversed without a simultaneous change in intensity (Fig. 3.4B). The response of the units could also be modulated by varying the presentation rate of a change in magnetic intensity. The latency and time-course (the first point after the stimulus step and the period during which the firing rate was more than two standard deviations above the mean for each unit) of the responses by the two units exposed to both stimulation frequencies were similar but the peak amplitudes of the responses decreased and increased, respectively, when the rate at which intensity changed was presented increased from 0.5 to 1 Hz (Fig. 3.4C).

The neural responses to magnetic fields in the trout have not been localized to any branch of the TN, shown to depend on magnetite such as that found in the cells in the nose, nor to underpin behavioral responses to magnetic fields by the trout. The responses to changes in magnetic intensity found in the TN are, however, consistent with detection of magnetic fields in the front of the head of the trout and led to a search for detector cells associated with the TN.

7. The Search for the Site of Magnetic Field Detection

The behavioral and electrophysiological experiments led us to search for candidate magnetite-based magnetoreceptor cells in the rainbow trout. This search was complicated by the transparency of tissues to magnetic field stimuli, the nature of the Earth's magnetic field as a stimulus, and the extremely small size of the magnetite crystals themselves. New techniques, and combinations of techniques, have had to be developed to overcome these obstacles.

7.1. The Magnetoreceptor Cells

We have used the crystal and magnetic properties of single-domain magnetite to identify magnetoreceptor cells in the nose of the rainbow trout despite the small size (<50 nm) and extreme rarity (<5 p.p.b. by volume) of the crystals. We first used reflection mode confocal laser scanning microscopy (CLSM) to demonstrate detection of the chains of magnetite crystals present in magnetotactic bacteria (Walker et al., 1997). We then searched for similar

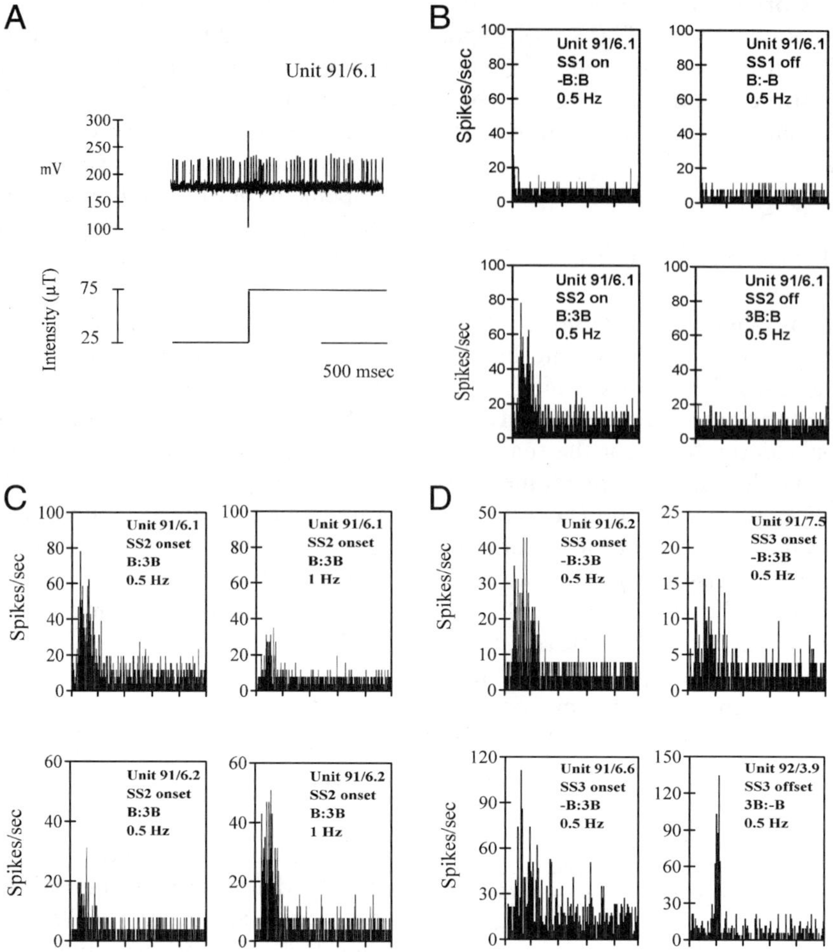

FIGURE 3.4. Neural responses to magnetic field stimuli in rainbow trout (adapted from Walker et al., 1997). (A) Peristimulus activity of a single unit (91/6.1) in the SO branch of the TN in the rainbow trout. The onset of a standard search stimulus (labeled SS2; bottom trace) that trebled the intensity without changing the direction of the magnetic field in the experimental situation is aligned with the associated stimulus artifacts (clipped for clarity) in the top trace. To the left of the artifact, the unit is spontaneously active in the background magnetic field. To the right of the artifact, the trace shows the activity of the unit for 1 s after the onset of SS2. Acceleration of the firing rate of the unit is evident for the first 100 ms after the stimulus step. (B) Poststimulus time histograms (PSTHs) of responses by the unit shown in A to SS1 (a search stimulus that reversed the direction without changing the intensity of the field in the experimental situation) and SS2 presented 128 times at 0.5 Hz (on for 1 s then off for 1 s). (C) PSTHs of responses by two spontaneously active units to the onsets of SS2 presented 128 times at 0.5 and 1 Hz. (D) PSTHs of responses by four spontaneously active units to SS3 (a search stimulus whose onsets reversed the direction and trebled the intensity of the magnetic field in the experimental situation and whose offsets reversed the effect of the onsets). The stimuli were presented 128 times in each case. Each plot in B–D begins at the step change in the field and is of duration 500 ms. The magnetic field remained constant throughout the period. Unit identification number, search stimulus number, stimulus step, and presentation rates are listed. Sampling bin widths are 2 ms in B, C, and in the upper-left panel of D, and 4 ms in the remaining panels of D. Tick marks on the abscissae are at 100 ms intervals.

reflections in heads of rainbow trout that had been embedded in plastic (Fig. 3.5A,B). Mapping the reflections in three dimensions then permitted us to image single crystals in thin sections in the transmission electron microscope (Fig. 3.5C,D) and to identify the crystals uniquely as magnetite using atomic and magnetic force microscopy (Fig. 3.5E; Walker et al., 1997; Diebel et al., 2000).

The cells containing the magnetite particles are 10–12 μm in length, have a distinctive multilobed shape, and are consistently located near the basal lamina of the olfactory epithelium (Fig. 3.5A,B). The cells are relatively rare and were found only near the tips of the olfactory lamellae (distal to the cells of the olfactory sensory epithelium). The cells each have several processes that extend out to and are surrounded by tubular-shaped fibroblastic cells (with two processes) that help delineate the basal layer (Fig. 3.5B). The chain of magnetite crystals in each cell is about 1 μm long (range 0.5 μm–1.5 μm, $n = 4$; estimated from the CLSM; Diebel et al., 2000) and we estimate each chain will have a magnetic-to-thermal-energy ratio of about 4. The location of the chain of magnetite crystals within each cell suggests that a mechanical linkage of the chain to the cell could transduce the movement of the chain in response to the external magnetic field into changes in the membrane potential of the cell.

7.2. Neuroanatomy

In the first step toward testing the hypothesis that the magnetite-containing cells may be functionally linked to the TN, we sought to trace the superficial ophthalmic branch of the TN from the site where electrophysiological recordings of responses to magnetic field stimulation were made, to the endings of the individual nerve cells (Fig. 3.6). We used serial histological sections and DiI, a fluorescent lipophilic dye, placed on the cut ends of the TN to trace the nerve in both anterograde and retrograde directions. The dye migrated along both myelinated and unmyelinated fibers in the TN. Posterior to the orbit, the SO branch joined other branches of the TN and ended in cell bodies that make up part of the anterior ganglion (Fig. 3.6C). From the ganglion, the labeled nerve tracts entered the anterior dorsal area of the medulla oblongata. Anterior to the orbit, the SO branch has branches that innervate the skin, surround the olfactory nerve and olfactory capsule (processes 1–3 in Fig. 3.6C), and also penetrate the olfactory lamellae within the olfactory capsule itself (Fig. 3.6C). Fine branches of the TN penetrate the olfactory lamellae both from the top and from the base before terminating in finer processes within the olfactory lamellae, where the magnetite-containing cells are most often found (Fig. 3.6A).

Although we can propose a link from the candidate magnetoreceptor cells in the lamina propria of the olfactory lamellae through the SO branch of the TN to the brain, afferent synaptic contacts between the nerve endings and the magnetoreceptor cells have not yet been identified. Detection of both magnetite and the endings of stained nerves in the confocal microscope has not yet been achieved due to the different media required for best detection of the magnetite and the nerves. In addition, it has been impossible so far to recognize the chains of magnetite crystals in the transmission electron microscope, at least in part because there is a very low probability that more than one crystal in a chain will fit within one thin section. There is thus only indirect evidence from the magnetic impairment experiments and the magnetic-to-thermal-energy ratio of the magnetite chains that the magnetoreceptor cells are functionally linked to the nervous system.

8. Discussion

We conclude first that the mystery surrounding the magnetic sense is well on the way to being dispelled. Experimental results now demonstrate that the magnetic sense has key properties, in particular selectivity and sensitivity, in common with other senses (Block, 1992). Although behavioral and electrophysiological results from aquatic species bear directly only on the issue of selectivity, their similarity with results from terrestrial species is consistent with the widespread use of magnetite in magnetic field detection. Second, there is now some evidence to underpin the intuitive appeal of the hypothesis that aquatic animals use a magnetic sense to navigate over long distances. Much remains to be learned, however, because the results so far permit no more than an outline description of the structure and function of the magnetic sense in a single species and there is as yet no direct experimental evidence for use of the magnetic field for any purpose by aquatic species.

8.1. How Magnetic Fields Are Detected

Although we have focused on magnetite, it cannot yet be determined whether there is a single common mechanism or multiple, inde-

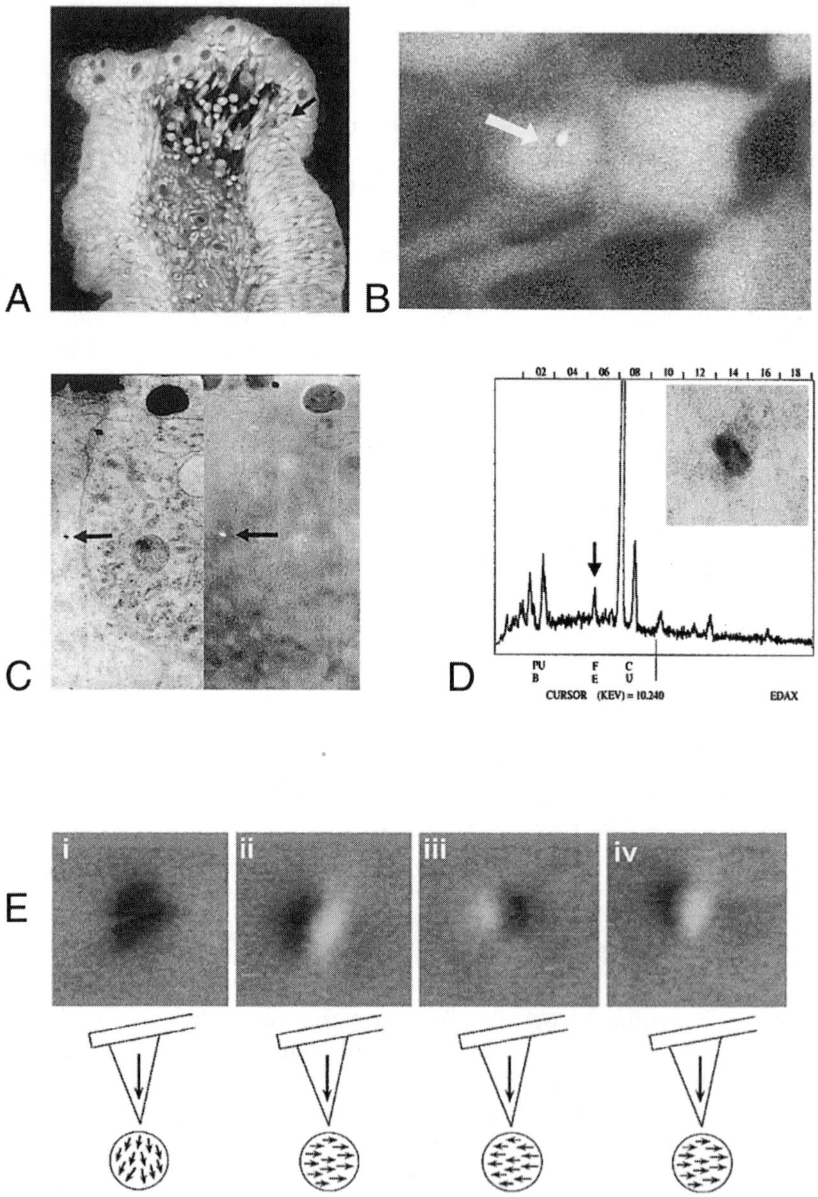

pendently derived mechanisms of magnetic field detection in aquatic vertebrates. Magnetic field detection has been proposed to occur in the visual system of amphibians (Deutschlander et al., 1999) and the electroreceptor system of elasmobranchs (Kalmijn, 1981) as well as in specialized cells that contain magnetite in the teleost fishes (Walker et al., 1984, 1997; Diebel et al., 2000). The number of species studied so far in each of these vertebrate classes is small and no more than two of the proposed mechanisms for magnetic field detection have been investigated in any of them. The small number of species and variation in experimental techniques used to date make it difficult to evaluate the detection hypotheses by comparing experimental results from among the different taxa that have been studied.

Block's (1992) argument that evolutionary pressures should produce highly specialized sensory systems does, however, provide a theoretical basis for evaluation of magnetic field detection hypotheses. The magnetite hypothesis assumes that magnetite will form the basis of a sensory system that specializes in detecting magnetic fields. If, on the other hand, magnetic fields are detected secondarily in other sensory systems, then subsets of receptor cells in those sensory systems should be specialized for magnetic field detection. In both cases, the detector cells should be both *selective for* and *highly sensitive to* magnetic field stimuli (Block, 1992).

The magnetite-based magnetoreceptor cells in the nose of rainbow trout will be clearly selective for magnetic field stimuli (Walker et al., 1997; Diebel et al., 2000). The magnetoreceptor cells in the trout could respond only to pervasive stimuli, such as magnetic fields, gravity, and temperature variations, that pass through tissue because the cells do not contact

FIGURE 3.5. Detection of intracellular magnetite. (A) Magnetite detected using a confocal laser scanning microscope (CLSM) in reflection mode shows as a spot (arrow) and has been overlaid onto an autofluorescence image of the olfactory lamella taken at the same depth and magnification (×190). (B) Autofluorescence image of a magnetite-containing cell viewed using transmission mode CLSM. The white spot (arrow) shows where the reflection due to magnetite has prevented light passing through the cell. (C) Bright-field (left) and dark-field (right) transmission electron micrograph (TEM) of a crystal associated with a reflectance in the trout olfactory lamellae. In bright-field TEM, both the crystal (arrow) and a much larger pigment granule (top center) are electron-dense. In dark-field TEM, the crystal (arrow) reflects the electron beam strongly whereas the large pigment granule (upper right) does not (magnification 12,500). (D) Energy dispersive analysis of X-ray emissions (EDAX) of the crystal in C. Inset shows the crystal (length 50 nm) at higher magnification. The copper (Cu) peak is due to the copper grid used, and lead (Pb) and uranium (U) peaks are from TEM stains. The peak from iron (Fe) present in the crystal is indicated by an arrow. This peak was absent in control regions of the same section. (E) Magnetic force microscopy (MFM) images that show the response of a putative single magnetic particle (within trout olfactory tissue) in the presence of an applied magnetic field. The magnetic field applied in the plane of the sample was +1.4, +150, −150 and +130 milliTesla (mT) for images A–D, respectively. MFM images (75-nm squares) are shown on top, with a representation of the MFM tip and magnetization of the particle underneath. The MFM tip (inverted triangle) is permanently magnetized with a coercivity of +500 mT at right angles (arrow in inverted triangle) to the applied field. The small arrows within each circle under the tip represent the alignment of the individual magnetic dipole moments that might act as the field source. (Ei.) Image shows a dark patch at the location of the particle. This dark patch indicates an attractive reaction between the tip and sample, consistent with the magnetic field from the MFM tip weakly magnetizing the particle and causing an attractive interaction. (Eii–iv.) MFM images show the nearly dipolar responses of the magnetic particle under a strong applied magnetic field. These are consistent with an MFM image of a single-domain particle magnetized along the direction of the applied field. Note that the reversal of the field and dipolar response in C are consistent with the particle magnetization flipping in the reversed applied field. In images B–D, the applied field was large enough to completely align the magnetic moment of the crystals within the field. (A, C, D: adapted from Walker et al., 1997; B, E: adapted from Diebel et al., 2000.)

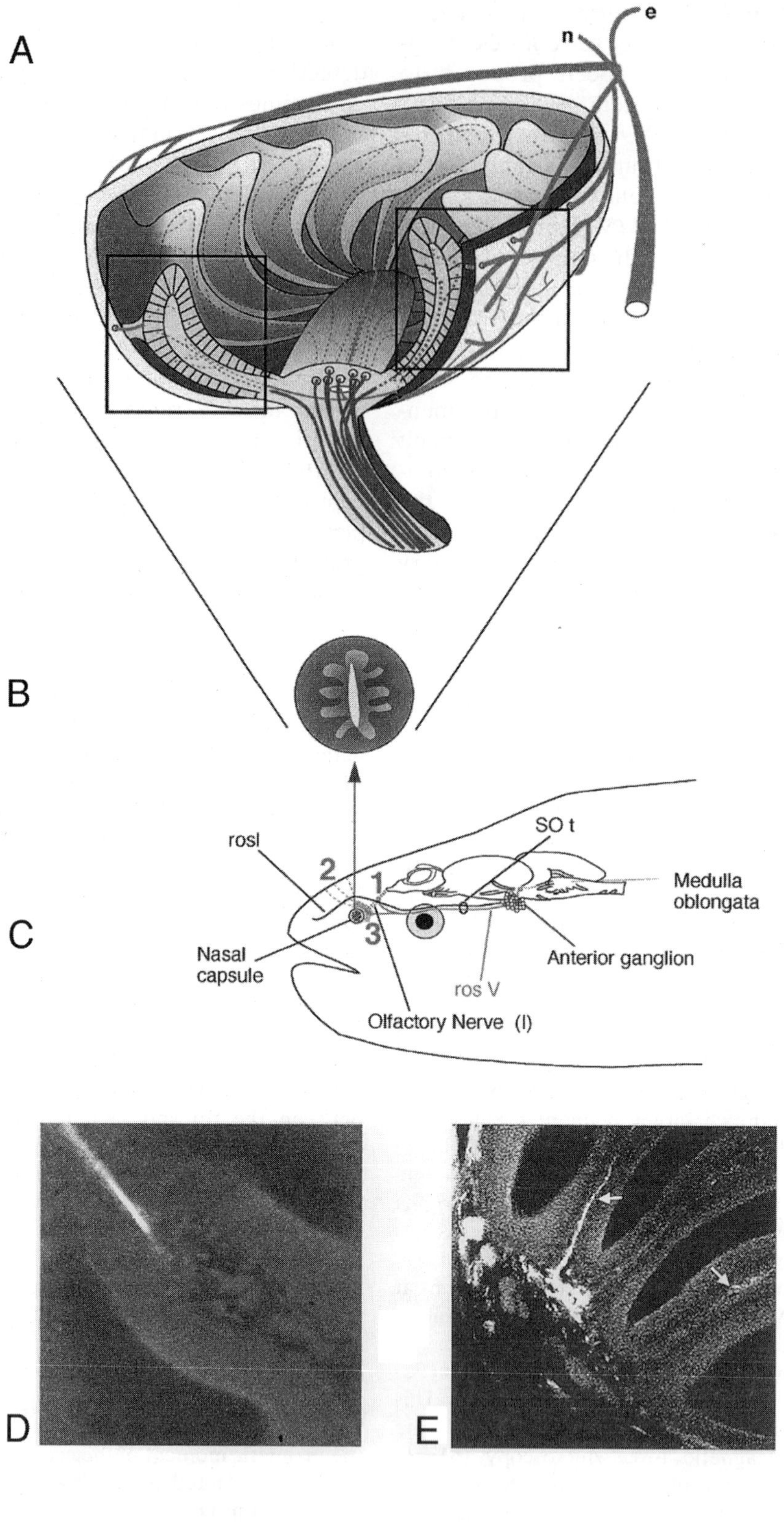
A
n
e
B
C
rosl
2
1
3
SO t
Medulla
oblongata
Nasal
capsule
Anterior ganglion
ros V
Olfactory Nerve (I)
D
E

the external environment directly. The magnetite crystals in the receptor cells are too small to be affected by gravity but their motion will be affected by external temperature variations (Kirschvink and Gould, 1981; Kirschvink and Walker, 1985). Poikilotherms such as fishes may therefore require processes that compensate for temperature effects on magnetic field detection.

The magnetite-based magnetoreceptor cells in the nose of rainbow trout are also likely to achieve high sensitivity to magnetic field stimuli (Diebel et al., 2000). Theoretical analyses (Kirschvink and Walker, 1985) predicted the existence of such arrays. These analyses also predicted that the energy of interaction of these arrays with the Earth's magnetic field would be two and six times the background thermal energy, kT, in arrays specialized for detecting the intensity and direction respectively of the Earth's magnetic field (Kirschvink and Walker, 1985). We have estimated the energy of interaction of the chains of magnetite in the trout with the Earth's magnetic field to be about 4 kT (Diebel et al., 2000).

Although we cannot yet explain why this magnetic interaction energy should be intermediate between the values predicted by Kirschvink and Walker (1985), we note that these magnetite chains will respond only to magnetic fields and, if present in sufficient numbers, will permit high sensitivity to changes in magnetic field stimuli.

In contrast with magnetite-based magnetoreceptors, there is as yet no evidence for specialized receptors that are selective for magnetic fields in either the visual or the electroreceptor systems respectively of amphibians and elasmobranchs. In the absence of specialized subsets of receptors, it also seems unlikely that these sensory systems could achieve the high sensitivity necessary to explain the close associations of hammerhead sharks with magnetic topography (Klimley, 1993) or the sensitivity to small changes in magnetic field inclination proposed for amphibians (Fischer et al., 2001). In the elasmobranchs, high sensitivity would require extremely long ampullary canals in the electroreceptor system (Kirschvink et al., 2001). Similarly, the eyes of amphibians would require either much greater numbers of receptor cells or receptor cells with much greater volumes of visual pigments than are necessary for vision (Kirschvink et al., 2001).

FIGURE 3.6. Innervation of the head region and nasal capsule of the trout by the SO branch of the TN (adapted from Walker et al., 1997). (A) A three-dimensional diagram of the innervation by the SO into olfactory lamellae in the nasal capsule of the trout. One process innervates the nasal membrane and flap (**n**) and the other (top right (**e**)) innervates the skin (process 2 in C). Others form a network of nerves that surround the nasal capsule (box at right; 3 in C). Within this network, the smaller branches have fine processes that pass through the nasal membrane lining the nasal capsule and innervate, at both the top and base, individual olfactory lamellae that form the olfactory rosette. The olfactory nerve (dark gray) is the combination of all axons of the olfactory sensory cells that are situated in the mucosa and send their axons to the olfactory bulb. The network of nerves surrounding the capsule generally lies in a fatty layer that is typically found between the neurocranium (not shown) and the outer membrane (stippled) that lines the nasal capsule. The pale area in the front two lamellae represents the folded layers of the olfactory epithelium that are separated internally by the lamina propria. New lamellae are formed in the area of the nasal capsule (not shown). The box at left outlines the areas where the TN enters the olfactory lamellae from the top (shown in D) and the bottom (shown in E). (B) Olfactory rosette within the trout nasal capsule (top view). The nasal flap that lies over the top of the olfactory rosette has been removed for clarity. (C) Schematic of the innervation of the SO (labeled SO t) in the head region of the trout. (D, E) Optical slices showing two different branching patterns of DiI-labeled nerve processes entering trout olfactory lamellae. In D, a labeled fine process from a branch of the SO ramus of the TN can be seen entering a single lamella through the top (magnification ×135). In E, fine processes can also be seen entering the lamina propria of several lamellae (arrows) from their bases (magnification ×55). These processes originate from a different branch of the SO than the one that innervates the top area in D.

Such elaboration of the sensory cells and associated structures in these sensory systems has yet to be reported.

In the two vertebrate groups where current experimental results bear on more than one of the magnetic field detection hypotheses, we suggest that the evidence favors the magnetite hypothesis. Hatchling sea turtles orient in complete darkness (Lohmann and Lohmann, 1996; Irwin and Lohmann, 2000; Lohmann et al., 2001) when there will be no photons to excite the electrons in visual pigments as required by the optical pumping hypothesis (Leask, 1977). They cannot orient, however, when they are carrying magnets in a harness on their backs. Similarly, magnetic field discrimination in short-tailed stingrays is abolished by magnets attached over the likely location of magnetite used in magnetoreception. Because the electroreceptor system would still have been stimulated in these latter experiments, the failure to respond in the presence of the magnets suggests that the stingrays do not normally attend to signals in the electroreceptor system that are derived from the external magnetic field. The similarity of these experimental results in the elasmobranchs and the turtles, combined with the results from the teleost fishes, suggests that a single common mechanism is more likely than multiple, independently derived mechanisms of magnetic field detection in the aquatic vertebrates.

8.2. Use of the Magnetic Sense in the Aquatic Environment

Our developing ability to study the structure and function of the magnetic sense in aquatic vertebrates using orthodox approaches to the study of sensory systems is not matched by our ability to test experimentally for use of the Earth's magnetic field by aquatic animals. Travel paths and positions of sharks and whales can be correlated with minute variations in magnetic intensity (Kirschvink et al., 1986; Walker et al., 1992; Klimley, 1993). These results are consistent with the hypothesis that sharks and whales may navigate using the magnetic topography produced by magnetic anomalies but are contradicted by the apparent disorienting effect of magnetic anomalies on homing pigeons (Walcott, 1977).

Significant issues of scale and experimental control explain why the few attempts to test experimentally for use of the Earth's magnetic field in navigation by aquatic animals have been inconclusive (e.g., Papi et al., 1997, 2000; Yano et al., 1997). First, sensory systems operate over short time scales (milliseconds to minutes) but long-distance movements can take from hours to months. Detection of magnetic effects on behavior are therefore likely to require that activity be monitored on the temporal and spatial scales over which the magnetic sense will operate. Second, because free-living animals will frequently carry out other normal functions such as feeding, resting, and avoiding predators during long-distance journeys, it will be difficult to predict the direction an animal will travel on short time scales and so to predict the effects of experimental manipulations on the direction of travel. Greater experimental control than has been achieved so far is likely to come through careful selection of subjects for which a direction of travel can be predicted in advance. We suggest seabirds that roost on land at night and feed well away from land during the day could be studied experimentally in much the same way as homing pigeons. An exciting new development that would support such experiments is the development of global positioning devices small enough to be carried by birds as small as homing pigeons (Steiner et al., 2000).

8.3. Comparison with Results from Birds

The far greater volume of research that has been done on the magnetic sense in birds has produced a number of similarities and differences from the results obtained so far with aquatic vertebrates. There is experimental evidence for both selectivity and high sensitivity of the magnetic sense in birds to go with the evidence for selectivity obtained from the aquatic vertebrates. Recordings from the superficial ophthalmic branch of the trigeminal nerve and the trigeminal ganglion in the bobolink, *Dolichonyx oryzivorus*, demonstrated sensitivity to changes in magnetic intensity of 200 nT (Semm and Beason, 1990), whereas studies of the effects of magnetic anomalies on the initial orientation of homing pigeons suggested that the behavioral threshold for changes in magnetic intensity may be as low as 10 nT (Gould, 1982). These results are consistent with the sensitivities to intensity changes estimated for whales (Kirschvink et al., 1986; Walker et al., 1992). Recent impairment experiments have demonstrated the likely dependence of magnetic field detection on magnetite located in the front of the head in homing pigeons (Haugh et al., 2001). Although there are still major gaps in our knowledge, the similarities in the results from laboratory studies of the magnetic sense in birds and aquatic vertebrates give us confidence in the hypothesis that magnetite provides the basis for a general mechanism of magnetic field detection in the vertebrates.

Although use of the Earth's magnetic field by birds has been difficult to demonstrate reliably and the results of experiments have sometimes been difficult to interpret (Walcott, 1992), several consistent results now permit a sketch of how birds such as homing pigeons may use the magnetic field to navigate. First, magnetic coils and attached magnets disrupt the initial orientation of homing pigeons on cloudy but not on sunny days, apparently by preventing the birds from using their magnetic compass on cloudy days (Keeton, 1972; Walcott and Green, 1974). Recent experiments using rare-earth magnets attached over the olfactory cavity of pigeons have demonstrated a highly reproducible effect on the initial orientation of homing pigeons on sunny days (Haugh et al., 2001). This result is consistent with the hypothesis that the magnets interfered with determination by the birds of their position rather than of direction because the birds use the sun compass in preference to the magnetic compass when the sun is available to them (Keeton, 1971). Finally, pigeon orientation errors associated with magnetic storms and anomalies have been interpreted to be errors in position determination (the "map step" of Kramer, 1953; Gould, 1982). This interpretation is consistent with the hypothesis that magnetic anomalies affect pigeon orientation by disrupting the ability of the birds to determine position using systematic variations in the intensity of the Earth's magnetic field (Gould, 1982).

8.4. Future Research

A key conclusion from our work is that the structure and function of the magnetic sense can now be studied in the laboratory using orthodox approaches for the study of sensory systems. Skepticism that the magnetic sense exists was reasonable in the absence of a clearly identified detector system and afferent nerves. In the last five years, it has been possible to identify candidate detector cells that meet the criterion of selectivity for the magnetic field stimulus and permit high sensitivity of the magnetic sense (Diebel et al., 2000). Psychophysical studies using electrophysiological and conditioning techniques have confirmed the high sensitivity of the magnetic sense and its dependence on magnetite in birds (Semm and Beason, 1990), bees (Walker and Bitterman, 1988, 1989), and fishes (Walker, 1984; Walker et al., 1997; Hodson, 2000). The magnetic sense thus shares key properties of all specialized sensory systems (Block, 1992).

What the above studies of the magnetic sense have not achieved has been demonstration that the separate components of the sense are functionally linked. Thus, there is no ultrastructural evidence that the magnetic chains in the trout are linked to the candidate magnetoreceptor cells in a way that will produce changes in the membrane potential of the cells due to movements of the chains in response to the external magnetic field. Nor has any histological or cytological evidence been obtained that demonstrates the existence of afferent synaptic contacts between the magnetite-containing cells and the trigeminal nerve. There is also a complete lack of knowledge of both the central projections of the magnetic sensory nerves and how the stimulus is processed in the brain.

There are even greater challenges to be had in the study of the use of the magnetic sense by animals in nature. The case for use of magnetic compasses by homing pigeons is reasonably clear but there is no experimental evidence yet that clearly identifies how animals might determine their position using the Earth's magnetic field. We suggest that experimental designs for field experiments could be usefully informed by the results of sensory studies that have been published over the last two decades or so. We suggest also that field experiments will benefit from careful selection of experimental subjects and from the availability of new tracking technologies that permit reconstruction with high resolution of the paths traveled by animals. In summary, there are many exciting opportunities for experimental study of the magnetic sense in both the laboratory and the field and we eagerly await the results of research over the years to come.

References

Beason, R.C., and Semm, P. (1987). Magnetic responses of the trigeminal nerve system of the bobolink (*Dolichonyx oryzivorus*). *Neurosci. Lett.* 80:229–234.

Block, S.M. (1992). Biophysical principles of sensory transduction. In: *Sensory Transduction* (Corey, D.P., and Roper, S.D., eds.), pp. 1–17. Society of General Physiologists 45th Annual Symposium, Rockefeller University Press.

Courtillot, V., Hulot, G., Alexandrescu, M., le Mouel, J.-L., and Kirschvink, J.L. (1997). Sensitivity and evolution of sea-turtle magnetoreception: Observations, modelling and constraints from geomagnetic secular variation. *Terra Nova* 9:203–207.

Deutschlander, M.E., Phillips, J.B., and Borland, S.C. (1999). The case for light-dependent magnetic orientation in animals. *J. Exp. Biol.* 202:891–908.

Diebel, C.E., Proksch, R., Green, C.R., Neilson. P., and Walker, M.M. (2000). Magnetite defines a magnetoreceptor. *Nature* 406:299–302.

Emlen, S.T. (1975). Migration: Orientation and navigation. In: *Avian Biology* (Farner, D.S., and King, J.R. eds.), Vol. 5, pp. 129–219. New York: Academic Press.

Finger, T.E., St. Jeor, V.L., Kinnamon, J.C., and Silver, W.L. (1990). Ultrastructure of substance P- and CGRP-immunoreactive nerve fibers in the nasal epithelium of rodents. *J. Comp. Neurol.* 294: 293–305.

Fischer, J.H., Freake, M.J., Borland, S.C., and Phillips, J.B. (2001). Evidence for the use of magnetic map information by an amphibian. *Anim. Behav.* 62: 1–10.

Gould, J.L. (1982). The map sense of pigeons. *Nature* 296:205–211.

Griffin, D.R. (1982). Ecology of migration: Is magnetic orientation a reality? *Quart. Rev. Biol.* 57:293–295.

Haugh, C.V., and Walker, M.M. (1998). Magnetic discrimination learning in rainbow trout (*Oncorhynchus mykiss*). *J. Navigation* 51:35–45.

Haugh, C.V., Wiltschko, R., Wiltschko, W., and Walker, M.M. (2001). P-GPS (Pigeon Geomagnetic Positioning System): II. Consistent effect of attached magnets on initial orientation of homing pigeons (*Columba livia*). *Royal Institute of Navigation Conference on Animal Navigation*, Oxford University, April 2001.

Hodson, R.B. (2000). Magnetoreception in the short-tailed stingray, *Dasyatis brevicaudata*. MSc thesis, University of Auckland, New Zealand.

Holland, K.N., Brill, R.W., and Chang, R.K.C. (1990). Horizontal and vertical movements of yellowfin and bigeye tuna associated with fish aggregating devices. *Fish. Bull. U.S.* 88:493–507.

Irwin, W.P., and Lohmann, K.J. (2000). Orientation behavior of sea turtle hatchlings: Disruption by magnets. Abstract, *Annual Meeting, Society for Integrative and Comparative Biology. Amer. Zoologist* 39:5.

Kalmijn, A.J. (1978). Experimental evidence of geomagnetic orientation in elasmobranch fishes. In: *Animal Migration, Navigation and Homing* (Schmidt-Koenig, K., and Keeton, W.T., eds.), pp. 347–353. New York: Springer-Verlag.

Kalmijn, A.J. (1981). Biophysics of geomagnetic field detection. *IEEE Trans. Mag.* 17:1113–1124.

Kalmijn, A.J. (1982). Electric and magnetic field detection in elasmobranch fishes. *Science* 218:916–918.

Keeton, W.T. (1971). Magnets interfere with pigeon homing. *Proc. Nat. Acad. Sci. USA* 68:102–106.

Keeton, W.T. (1972). Effects of magnets on pigeon homing. In: *Animal Orientation and Navigation* (Galler, S.R., Schmidt-Koenig, K., Jacobs, G.J., and Belleville, R.E. eds.) pp. 579–594. Washington DC: U.S. Government Printing Office.

Keeton, W.T., Larkin, T.S., Walcott, C., and Windsor, D.M. (1974). Normal fluctuations in the earth's field influence pigeon orientation. *J. Comp. Physiol.* 95:95–103.

Kirschvink, J.L., and Gould, J.L. (1981). Biogenic magnetite as a basis for magnetic field detection in animals. *Biosystems* 13:181–201.

Kirschvink, J.L., and Walker, M.M. (1985). Particle-size considerations for magnetite-based magnetoreceptors. In: *Magnetite Biomineralization and*

Magnetoreception by Living Organisms: A New Biomagnetism (Kirschvink, J.L., Jones, D.S., and MacFadden, B.J. eds.), pp. 243–254. New York: Plenum.

Kirschvink, J.L., Dizon, A.E., and Westphal, J.A. (1986). Evidence from strandings for geomagnetic sensitivity in cetaceans. *J. Exp. Biol.* 120:1–24.

Kirschvink, J.L., Walker, M.M., and Diebel, C.E. (2001). Magnetite-based magnetoreception. *Current Opinion in Neurobiology* 11:462–467.

Kirschvink, J.L., Walker, M.M., Chang, S.-B., Dizon, A.E., and Peterson, K.A. (1985). Chains of single-domain magnetite particles in the chinook salmon, *Oncorhynchus tshawytscha. J. Comp. Physiol. A.* 157:375–381.

Klimley, A.P. (1993). Highly directional swimming by scalloped hammerhead sharks, *Sphyrna lewini*, and substrate irradiance, temperature, bathymetry and geomagnetic field. *Mar. Biol.* 117:1–22.

Klinowska, M. (1985). Cetacean live stranding sites relate to geomagnetic topography. *Aquatic Mammals* 1:27–32.

Kramer, G. (1953). Wird die Sonnenhöhe bei der Heimfindeorientierung verwertet? *J. Ornithol.* 4:201–219.

Leask, M.J.M. (1977). A physicochemical mechanism for magnetic field detection by migratory birds and homing pigeons. *Nature* 267:144–145.

Lohmann, K.J., and Lohmann, C.M.F. (1996). Orientation and open-sea navigation in sea turtles. *J. Exp. Biol.* 199:73–81.

Lohmann, K.J., Cain, S.D., Dodge, S.A., and Lohmann, C.M.F. (2001). Regional magnetic fields as navigational markers for sea turtles. *Science* 294:364–366.

Mann, S., Sparks, N.H.C., Walker, M.M., and Kirschvink, J.L. (1988). Ultrastructure, morphology and organization of biogenic magnetite from sockeye salmon, *Oncorhynchus nerka*: Implications for magnetoreception. *J. Exp. Biol.* 140: 35–49.

Mead, J.G. (1979). An analysis of cetacean strandings along the eastern coast of the United States. In: *Biology of Marine Mammals: Insights Through Strandings* (Geraci, J.B., and St. Aubin, D.J. eds.), pp. 54–68. U.S. Marine Mammal Commission Report MMC-77/13.

Papi, F., Luschi, P., Crosio, E., and Hughes, G.R. (1997). Satellite-tracking experiments on the navigational ability and migratory behaviour of the loggerhead turtle *Caretta caretta. Mar. Biol.* 129:215–220.

Papi, F., Luschi, P., Åkesson, S., Capogrossi, S., and Hays, G.C. (2000). Open-sea migration of magnetically disturbed sea turtles. *J. Exp. Biol.* 203:3435–3443.

Phillips, J.B. (1977). Use of earth's magnetic field by orienting cave salamanders (*Eurycea lucifuga*). *J. Comp. Physiol. A.* 121:273–288.

Quinn, T.P. (1980). Evidence for celestial and magnetic compass orienation in lake-migrating sockeye salmon fry. *J. Comp. Physiol. A.* 137: 243–248.

Semm, P., and Beason, R.C. (1990). Responses to small magnetic field variations by the trigeminal system of the bobolink. *Brain Res. Bull.* 25: 735–740.

Skiles, D.D. (1985). The geomagnetic fields: Its nature, history and biological relevance. In: *Magnetite Biomineralization and Magnetoreception by Living Organisms: A New Biomagnetism* (Kirschvink, J.L., Jones, D.S., and MacFadden, B.J. eds.), pp. 43–102. New York: Plenum.

Steiner, I., Bürgi, C., Werffel, S., Dell'Omo, G., Valenti, P., Tröster, G., Wolfer, D.P., and Lipp, H.-P. (2000). A GPS logger and software for analysis of homing in pigeons and small mammals. *Physiol. Behav.* 71:589–596.

Taylor, P.B. (1986). Experimental evidence for geomagnetic orientation in juvenile salmon, *Oncorhynchus tshawyscha Walbaum. J. Fish Biol.* 28:607–623.

Taylor, P.B. (1987). Experimental evidence for juvenile Chinook salmon, *Oncorhynchus tshawytscha* Walbaum orientation at night and in sunlight after a 7° change in latitude. *J. Fish Biol.* 31:89–111.

Viguier, C. (1882). Le sens d'orientation et ses organs chez les animaus et chez l'homme. *Rev. Philosophique de la France et de l'Étrangere* 14:1–36.

Walcott, C. (1977). Anomalies in the earth's magnetic field increase the scatter of pigeons' vanishing bearings. In: *Animal Migration, Navigation and Homing* (Schmidt-Koenig, K., and Keeton, W.T. eds.), pp. 143–151. New York: Springer-Verlag.

Walcott, C. (1992). Pigeons at magnetic anomalies: The effect of loft location. *J. Exp. Biol.* 170: 127–141.

Walcott, C., and Green, R.P. (1974). Orientation of homing pigeons altered by a change in the direction of an applied magnetic field. *Science* 184:180–182.

Walker, M.M. (1984). Learned magnetic field discrimination in the yellowfin tuna, *Thunnus albacares. J. Comp. Physiol. A.* 155:673–679.

Walker, M.M., and Bitterman, M.E. (1988). Attached magnets disrupt magnetic field discrimination by honeybees. *J. Exp. Biol.* 141:447–451.

Walker, M.M., and Bitterman, M.E. (1989). Honeybees can be trained to respond to very small changes in geomagnetic field intensity. *J. Exp. Biol.* 145:489–494.

Walker, M.M., Kirschvink, J.L., Ahmed, G., and Dizon, A.E. (1992). Fin whales (*Balaenoptera physalus*) avoid geomagnetic gradients during migration. *J. Exp. Biol.* 171:67–78.

Walker, M.M., Kirschvink, J.L., Chang, S.-B.R., and Dizon, A.E. (1984). A candidate magnetic sense organ in the yellowfin tuna, *Thunnus albacares*. *Science* 224:751–753.

Walker, M.M., Diebel, C.E., Haugh, C.V., Pankhurst, P.M., Montgomery, J.C., and Green, C.R. (1997). Structure and function of the vertebrate magnetic sense. *Nature* 390:371–376.

Wiltschko, W. (1972). The influence of magnetic total intensity and inclination on directions chosen by migrating European robins. In: *Animal Orientation and Navigation* (Galler, S.R., Schmidt-Koenig, K., Jacobs, G.J., and Belleville, R.E. eds.), pp. 569–578. Washington DC: US Government Printing Office.

Yano, A., Ogura, M., Sato, A., Sakaki, Y., Shimizu, Y., Baba, N., and Nagasawa, K. (1997). Effect of modified magnetic field on the ocean migration of maturing chum salmon. *Mar. Biol.* 129:523–530.

Part 2
Finding Food and Other Localized Sources

R. Glenn Northcutt

Finding food and shelter, avoiding predators, and finding a mate are universal activities of primary importance among metazoans, and a wide variety of strategies and sensory systems are used to accomplish these goals. This section focuses on two of those sensory systems—the octavolateral system and the visual system—and explores how these systems process biologically relevant stimuli in fishes.

In Chapter 4, Kalmijn discusses the physical features of the aquatic environment that sharks and rays can detect with their octavolateral system and how different aspects of these physical features are utilized for prey detection and orientation. In the next chapter, von der Emde and Bell explore how those fishes that possess an electric organ use active electrolocation to generate "electrical images" of their surroundings, and they summarize what is currently known regarding the central neural processing of active electrolocational information. In the two subsequent chapters Bleckmann and colleagues discuss how the ordinary mechanoreceptive lateral line system in fishes processes complex hydrodynamic stimuli, and Coombs and Braun review areas of research that have recently yielded significant insights into how fishes process mechanoreceptive lateral line information. In a final chapter, Collin and Shand summarize the diverse range of retinal specializations in fishes and the selective pressures responsible for generating these specializations.

Although three of the chapters focus on information that is processed by either the electroreceptors or mechanoreceptors of the lateral line system, Kalmijn's analysis of the physical features of the aquatic world stresses the remarkable similarity in the electric and near-field "acoustic" fields and in so doing provides important new information on the initial functions of the inner ear and its subsequent evolution.

Ten years ago, it was possible to sequentially characterize the electrical properties of both electro- and mechanoreceptors of the lateral line system. At that time, however, very little was known about the kinds of information processed by these receptors and their complex central circuits. The chapters on how information is processed in active electroreception (von der Emde and Bell) and mechanoreceptors (Bleckmann and colleagues) demonstrate the remarkable progress that has taken place in this area. In this context, one of the most exciting discoveries is the realization that superficial and canal neuromasts appear to form two parallel sensory channels, one sensitive to water-velocity (superficial neuromasts) and the other sensitive to pressure gradients (canal neuromasts), with both subserving very different classes of behavior. This work is nicely summarized by Coombs and Braun.

In contrast to the amazing progress that has been made in understanding how different receptor classes and their associated lower brain centers process information, we are reminded in these chapters of how little is still known about the location of higher brain centers related to the lateral line system and the types of operations performed by these centers. If progress continues at the same pace for the next ten years, however, far more information about these centers will soon be available.

Collin and Shand's chapter on retinal specializations and their ecological correlations is a wonderful counterpoint to the other chapters on the lateral line system. There is no question that we know far more about the parameters of visual stimuli than we do about either biologically relevant acoustic or lateral line stimuli, and of course it is easier to manipulate these parameters. For this reason the final chapter is a tour de force, explaining how retinal organization has been altered for prey detection under essentially every underwater environmental condition.

4 Physical Principles of Electric, Magnetic, and Near-Field Acoustic Orientation

Ad. J. Kalmijn

Abstract

To elucidate the weakly electric and earth's magnetic sensory capabilities of sharks and rays and to introduce the concept of inertial hearing in the acoustic near field, this chapter discusses the physical features of the underwater world, the particular receptivities of the sense organs, and the spatial information the animals seek to orient at sea and to arrive at their prey. The objective is to establish the logical connections between the directional cues in the natural environment and the orientational responses they elicit from the animals. The features of the underwater world are presented in their most simple form, just as they are detected, processed, and perceived by the animals. In this chapter, we observe early fishes proficiently practice the laws of physics, purely from eons of experience. The remarkable similarity in the electric and low-frequency acoustic fields of underwater objects is emphasized to reveal the close relationship between the two sensory modalities.

1. Introduction

Animals shrive in their environment by recognizing features and events enabling them to predict and control subsequent developments. Their ability to respond appropriately is founded on experience gathered through the process of evolution in the form of reflexes, instinct, and intuition. In higher animals, particularly in humans, the faculties of thought and reason play a major role as well. Yet, the sensory performances of lower vertebrates are certainly not less impressive, as they involve the most advanced physical principles we know.

Hence, after establishing the astounding sensitivity of sharks and rays to DC and low-frequency electric fields, I felt compelled to study the oceans' weak electric fields and the information they offer before accrediting these fishes with a true electric sense. This search led to (1) the discovery of the common DC bioelectric fields of aquatic animals, guiding sharks and rays to their prey, and (2) the idea that sharks and rays electrically sense their magnetic compass heading and their drift in ocean streams (Kalmijn, 1966, 1971, 1974).

My foray into the acoustic near field started when I realized (1) that the lateral line of fishes responds to a higher-order fractional derivative of fluid motion than thought at the time and (2) the inner ear entered into the detection of acoustic pressure at a later stage, *after* evolving as a system of linear and angular inertial sensors. Seeking the relevant near-field features, I also noticed a close analogy between a prey's acoustic near field and bioelectric field, providing predators with similar directional cues (Kalmijn, 1988b, 1989, 1997).

Sensory biology is well served by the present analysis of the fishes' abilities to execute biologically meaningful behavioral responses to electric, magnetic, and low-frequency acoustic fields. Nevertheless, it remains intriguing how sharks learned from experience to apply the very principles that Faraday (1832) discovered and Maxwell (1891) expressed mathematically, yet were not correctly understood until Einstein (1905) introduced the new concepts of space and time, clarifying the "electrodynamics of moving bodies" (Kalmijn, 1984, 1988a, 1997).

2. Biologically Relevant Features of Ambient Electric and Magnetic Fields

2.1. Kinds of Electric Fields Sharks and Rays Detect in Ocean Waters

In the ocean, sharks and rays are exposed to at least three kinds of electric fields: (1) the fields resulting from their motion through the water in the presence of the earth's magnetic field, (2) the fields associated with ocean streams and ionospheric circulations, and (3) the fields of biological organisms in general (Kalmijn, 1974). To discuss the detection of these fields and the particular kinds of sensory information they present, I will first treat them separately, under idealized electric and magnetic environmental conditions. Then, I will render the situation more realistic and contemplate how sharks and rays may sort out the information and sense (1) their magnetic compass heading, (2) their orientation in ocean streams and coastal waters, and (3) the presence of prey or other local sources of electric fields. To evoke well-oriented behavioral responses, DC and low-frequency electric fields of less than 8 Hz suffice, even at voltage gradients of a few nanovolts per centimeter (Kalmijn, 1966, 1982, 1997). For the sake of brevity, I will often speak only of sharks, regarding them as representing all cartilaginous fishes, including the skates and the other families of rays, as well as the more distantly related chimaeras or ratfish.

2.2. Electric Fields Sharks Detect in the Earth's Magnetic Field, Without an Ambient Electric Field

Imagine a shark traveling in a horizontal magnetic field in equatorial waters devoid of ambient electric fields. Then, in the shark's frame of reference, the *proper* frame in which the ampullary sense organs are at rest, the animal is exposed to a virtually uniform, purely electric field (Einstein, 1905). This field, an observer in the earth frame would explain, is normal to the ambient magnetic field and to the shark's velocity with respect to the water. Consequently, when a shark, traveling in a horizontal plane, bears in an easterly direction, it receives a *ventro-dorsal* electric field, and when it bears in a westerly direction, a *dorso-ventral* electric field. The observer would consider the strength of this field to be given by $\mathbf{v} \times \mathbf{B}$, that is, proportional to (1) the velocity of the animal $\mathbf{v}$, (2) the magnetic induction $\mathbf{B}$ of the earth's magnetic field, and (3) the sine of the angle between the shark's velocity and the magnetic field, or equivalently, the cosine of the complementary angle of the shark's deviation from due east, the direction in which the ventro-dorsal field is strongest.

Actually, the shark's electroreceptors, the ampullae of Lorenzini, are not true DC input devices, but very low-frequency AC devices with an adaptation time constant on the order of 3–5 seconds. Hence, when a shark moves for a while at a constant velocity in a uniform magnetic field, its sense organs do not register the DC electric field it receives, since they only detect the AC *changes* in the field at the electroreceptive skin surface. The shark must explore the situation by purposely varying either, or both, its speed and direction of travel, preferably within a period of time short compared to the adaptation time constant of the sense organs. For this, the elasmobranchs' normal thrust-varying, swerving mode of travel may suffice. Expressed operationally, the transient AC electric field a shark detects is the low-frequency AC component of the DC electric field obtained by vectorially subtracting the ambient DC electric field that the animal received just before changing its velocity from the ambient DC electric field it receives just after changing its velocity. As commonly holds true in sensory biology, the animal's exploratory behavior constitutes an obligatory part of the detection mechanism. Thus, I will speak of a "DC electric field that a shark *receives*" before the field is transduced by the sense organs into receptor potentials and of an "AC electric field that a shark *detects*" after the field has been transduced into receptor potentials. The receptor potentials drive the process of synaptic transmission and, thus, control the afferent nerve-potential rate, conveying the information to the central nervous system.

From the observer's $\mathbf{v} \times \mathbf{B}$ expression for the electric field the shark receives, filtered by the differentiating process of sensory adaptation, one may infer the electric field $\mathbf{E}$ that the animal actually detects. Thus, in a horizontal magnetic field, the transient AC electric field a shark detects after increasing its speed of travel along a straight path, is strongest ventro-dorsally when the animal is heading due east, and strongest dorso-ventrally when it is heading due west. After decreasing its speed, the fields the animal detects are opposite in direction. The AC electric field the animal detects after changing its speed, without turning, is nil when it is heading due north or south. Conversely, the transient AC electric field the shark detects after veering right, without changing its speed, is strongest ventro-dorsally when the animal is heading due north, and strongest dorso-ventrally when it is heading due south. After veering left, the AC electric fields the animal detects are opposite in direction. The AC electric fields a shark detects after veering right or left, without changing its speed, are nil when the animal is heading due east or west. Strictly speaking, we should say "when the shark is heading due east at the moment it has turned through half the angle of veering right or left." To be sure, even though the electric field a shark receives during the turn *about* its aft–fore body axis has nonzero circulation, its consequences are negligible owing to the inherent weakness of the circulation and to the diverging anatomical arrangement of the electroreceptor canals, causing the circulation to remain all but undetected (Kalmijn, 1984).

In temperate waters, where the earth's magnetic field is not horizontal, but inclined to the vertical, we may analyze the situation by vectorially adding the electric field the shark receives from interaction with the vertical component of the magnetic field to the electric field it receives from interaction with the horizontal component of the magnetic field. The question then arises as to what extent the vertical component of the earth's magnetic field complicates matters. When in the southern hemisphere, where the magnetic field slopes up, a shark increases its speed, with or without changing its heading, it detects a left–right AC electric field owing to its interaction with the vertical component of the magnetic field, in addition to the ventro-dorsal AC field it detects owing to its interaction with the horizontal component of the magnetic field. When, in the northern hemisphere, where the magnetic field slopes down, the shark increases its speed or, in the southern hemisphere decreases its speed, the AC electric fields it detects are right–left, instead of left–right. Remember, for later reference, that any changes in the direction of the shark's horizontal velocity, without changes in speed, do not result in left–right or right–left AC electric fields from interaction with the vertical component of the earth's magnetic field.

To summarize, when a shark increases its speed of travel, without changing its heading, the ventro-dorsal AC field it detects is proportional to the component of the magnetic field along its *right–left* body axis and to its *change* in speed. When a shark changes its heading, without changing its speed, the ventro-dorsal AC field it detects is proportional not only to the component of the magnetic field along its *aft–fore* body axis and to the angle of turning right or left, if sufficiently small, but also to the magnitude of its *constant* forward speed. When a shark increases its speed, whether or not it also changes its heading, the concurrent left–right AC field it detects is proportional to the component of the magnetic field along its *ventro-dorsal* body axis and to its *change* in speed. To make the AC electric fields stand out most clearly, the animal might, prior to changing its speed or heading, travel at a constant velocity for several seconds to adapt its AC sense organs to the DC level of the electric field it receives. This shows how much the animals' behavior is part of the sensory mechanism. Below, I will discuss how a shark may distinguish between the AC electric fields it detects due to interaction with the earth's magnetic field and the AC electric fields it detects due to the ubiquitous presence of the oceans' virtually uniform DC ambient electric fields and the nonuniform DC bioelectric fields of prey and other local sources of electricity.

2.3. Electric Fields Sharks Detect in Ocean Streams, Without the Earth's Magnetic Field

Imagine a shark traveling in a horizontal, virtually uniform DC electric field in a body of seawater in which the earth's magnetic field has been canceled instrumentally. The animal must again probe the field to detect it, as the electroreceptors are not DC, but very low-frequency AC devices. However, in this instance, the animal detects an AC electric field only when it changes its heading, not when it merely changes its speed. This observation, though perhaps not immediately apparent, is of fundamental importance, as it provides the key to understanding how conveniently simple it is for a shark to discriminate between the AC electric field it detects due to its interaction with the earth's magnetic field and the AC electric field it detects due to the presence of an ambient DC electric field. When the shark does not change its heading, whether or not it changes its speed, it successfully fails to be affected by the ambient DC electric field by virtue of the AC nature of its sense organs. If they were true DC devices, the animal would not only forgo the benefit of *not* being affected by the ambient DC electric field when merely changing its speed, but also have to deal with its own DC bioelectric field and the internal DC offsets of its sensory system.

When turning, the direction of the AC electric field a shark detects is, shortly after the animal has started to change its heading, nearly normal to the direction of the ambient DC electric field, being the vector difference between the ambient DC electric field it receives directly before and shortly after the start of the turn. Actually, the angle between the AC field the animal detects and the DC field it receives deviates from 90° by half the angle of turning. That is, the AC electric field the animal detects rotates at half the rate the animal turns, but does so in the opposite direction. For small angles of turning, the strength of the AC electric field the animal detects is nearly proportional to the size of the angle, but is progressively less than proportional for larger angles. For a shark to detect the AC electric field at full strength, it must execute the turn within a period of time short compared to the 3–5-second adaptation time constant of the sense organs. To render the AC electric fields it detects most prominent, the shark may expressly refrain from turning for several seconds prior to veering right or left to probe the horizontal ambient DC electric field.

Optionally, one may visualize the situation by the conceptual aid of a "parting line," as I will call the imaginary line dividing the electroreceptive surface into (1) a region where the sense organs detect *negative* transient AC *voltages* at their skin pores and (2) a region where they detect *positive* transient AC *voltages* at their skin pores. When, at the beginning of a turn, the parting line first appears, it is practically normal to the AC electric field the shark detects and, therefore, parallel to the ambient DC electric field it receives, regardless of the direction of the ambient DC field. Moreover, from the previous paragraph it follows that, when the animal turns right or left, the parting line rotates over the electroreceptive skin surface in the opposite direction at half the rate and over half the angle that the shark turns. Moreover, the larger the angle of turning, the stronger is the AC voltage detected. Note that (1) the terms "positive" and "negative" refer to the electrical potentials, V, that are defined by the line integrals of the electric field $\mathbf{E}$ from the respective ampullary pores to the parting line, (2) the electric field $\mathbf{E}$ and the voltage drop $-\nabla V$ point "down-hill" from + to −, and (3) by convention, the voltage gradient ∇V points "up-hill" from − to +.

The strength of the transient AC electric field that a shark detects when veering right or left is proportional to the strength of the ambient DC electric field and, for small angles of turning, to the animal's angular change in heading, but *independent* of the animal's speed and direction of travel. As an aid to visualizing the situation, we may give the parting line a directional sense by furnishing it with an arrow in such a fashion that, when looking in the direction of the arrow, the AC voltages the animal detects at its electroreceptor pores are positive to the right and negative to the left when it turns right, or positive to the left and negative to the right when it turns left. Thus defined, the parting line indicates the direction of the ambient DC electric field at all times. Furthermore, the animal may interpret the strength of the AC voltages it detects at the electroreceptor pores, taking into account the angle over which it turns, in terms of the strength of the ambient DC electric field. Hence, sharks and rays receive unambiguous information about the weak electric fields induced by ocean streams and, particularly in coastal waters, by the diurnal, solar-driven, ionospheric circulations. The electrical return currents of ocean streams cause significant electric fields in adjacent waters as well, so that there is hardly any place in the world's oceans where the ambient DC electric field escapes detection by sharks and rays (Kalmijn, 1974).

2.4. Electric Fields Sharks Detect Both from the Earth's Magnetic Field and from Ocean Streams

In order to consider the two kinds of fields a shark receives conjointly, imagine the animal traveling in the inclined earth's magnetic field of the southern hemisphere in the presence of a uniform ambient DC electric field. When the animal increases its speed, but does not change its heading, it detects a ventro-dorsal AC electric field proportional to the component of the earth's magnetic field along its *right–left* axis as well as a left–right AC electric field proportional to the component of the earth's magnetic field along its *ventro-dorsal* axis. Yet, it does not detect an AC electric field from the virtually uniform ambient DC electric field. On the other hand, when the animal changes its heading, veering to the right, without changing its speed, it detects a ventro-dorsal AC electric field proportional to the component of the earth's magnetic field along its *aft–fore* axis as well as a horizontal AC electric field from the ambient DC electric field. When the animal decelerates without turning, or veers to the left without changing its speed, the AC electric fields it detects are opposite in direction. When traveling in a horizontal plane, a shark will not detect an AC electric field from the presence of the vertical component of an ambient DC electric field, if at all present despite the proximity of the sea surface. Hence, by changing its speed but not its heading and by changing its heading but not its speed, a shark receives unambiguous information about the earth's magnetic and about the ambient electric field. When the animal changes its speed as well as its heading, it will in some fashion have to tell the signals apart.

3. Detection and Processing of Ambient Electric and Magnetic Fields

3.1. Detection of Magnetic Compass Heading and Orientation to Ambient Electric Fields

To determine its *magnetic* compass heading, a shark may explore the earth's magnetic field by varying its speed of travel in a reproducible manner successively in two different directions, say 45° apart. From the ratio between the strengths of the ventro-dorsal or dorso-ventral

AC electric fields the animal successively detects and from their polarities, it may sense magnetic north, an unambiguous task equivalent to determining the horizontal component of the earth's magnetic field from its parallel projections onto two oblique axes normal to the two successive directions of travel. Furthermore, from the ratio between the strengths of the horizontal component of the earth's magnetic field thus determined and the vertical component of the earth's magnetic field available from the left–right or right–left AC electric field the shark concurrently detects when varying its speed in either of the two successive directions, the animal may sense the inclination of the magnetic field. Since the horizontal AC signal upon a speed increase is left–right in the southern, and right–left in the northern hemisphere, it provides complete information of the magnetic latitude. From our analysis, we also learn that the AC electric fields a shark detects when *steadily* swerving left and right are not sufficient for the animal to establish its magnetic compass heading, unless it either varies its forward speed or maintains a set speed at a set angle of swerving and compares the strength of the field it detects with the maximal strength attained, that is, when heading due north at the same velocity and the same angle of swerving. Nonetheless, there remains an ambiguity between easterly and westerly bearings, which the animal may remove, for example, by executing its swerving mode of travel consecutively in different compass directions (Kalmijn, 1997). Note that the swerving method requires the animal to set or measure its forward speed *per se*, rather than detect its *change* in forward speed.

To sense the direction of an ambient, virtually uniform, horizontal DC *electric* field, a shark may explore the situation by veering either right or left, preferably without changing its speed of travel during the turn to avoid having to correct for the electric field resulting from interaction with the vertical component of the earth's magnetic field. The horizontal component of the earth's magnetic field gives rise only to a dorso-ventral or ventro-dorsal AC electric field. In fact, just turning the head, without otherwise moving with respect to the water, suffices for the animal to explore the ambient DC electric field. The strength and direction of the horizontal AC electric field the shark detects on completion of a given turn indicates the strength and direction of the ambient DC electric field. The parting line and the voltage distribution over the electroreceptive skin region offer similar information. As mentioned above, the AC electric field the animal detects *at the onset* of a turn is normal to the ambient DC electric field, whereas the parting line is parallel to it. To explore a DC electric field, the animal may equally well swim steadily in a meandering fashion, swaying the head alternately left and right, without the need of varying its forward speed.

3.2. Pictorial Summary of Electric Fields a Shark Detects Along an Octagonal Path

To envision the AC electric fields a shark detects at sea (see Fig. 4.2), we will follow the animal as it travels an imaginary octagonal circuit, in a body of water near the surface, flowing uniformly with respect to the ocean floor, (i) maintaining in the initial part of each straight leg a constant speed sufficiently long to let its electroreceptors adapt to any uniform DC electric field it might receive, (ii) swimming at a set, elevated speed for a period of time short compared to the time of adaptation, (iii) resuming its original, constant speed sufficiently long to have the sense organs once again adapt, (iv) turning through a set angle, 45° in the diagram, within a period of time short compared to the time of adaptation, and so forth. From the diagram, it may be learned that a shark receives unambiguous information about the earth's magnetic field and an existing uniform ambient DC electric field merely by traveling two consecutive straight courses, connected by one change in heading (Kalmijn, 1997).

3.3. Behavioral Proof of Orientation to Earth's Magnetic and Ambient Electric Fields

In behavioral experiments conducted to prove that small stingrays are able to orient in uniform magnetic and DC electric fields of natural strengths, the animals were free to swim over the sandy bottom along the perimeter of their shallow, cylindrical tank. The stingrays were conditioned to obtain food by entering one, though passing the other, of two feeding stations located on opposite sides of the tank (Fig. 4.1). Which station to enter or to pass depended on the direction of the test field, chosen in a random order to be either "normal" or "reversed." In circling the test tank, the stingrays moved, turned, sped up, and slowed down in a fashion similar to traveling the octagonal circuit of Figure 4.2. The results were evaluated by the statistical method of sequential analysis (Kalmijn, 1982). Presently, I will "explain" the animals' behavior following the line of reasoning outlined above.

In the magnetic-field experiments, the animals were conditioned to enter the station in the magnetic east and to pass the one in the magnetic west. To obtain food, a stingray had to enter the station at which the vertical AC electric field it detected on increasing its speed went through zero, changing from ventro-dorsal to dorso-ventral, and to pass the station at which the field went through zero from dorso-ventral to ventro-dorsal. Or, in other words, it had to enter the station at which the AC voltages at the receptor pores went through null, changing from ventral-positive, dorsal-negative to ventral-negative, dorsal-positive, and to pass the station at which they went through null, from ventral-negative, dorsal-positive to ventral-positive, dorsal-negative, independent of whether the animal was circling the tank clockwise or counterclockwise (Kalmijn, 1978).

In the electric-field tests, the stingrays were conditioned to enter the station on the left side of the tank, looking into the direction of the field, and to pass the station on the right side of the tank. Thus, a stingray, whether continuously veering to the right when circling the tank clockwise or veering to the left when circling the tank counterclockwise, had to enter the feeding station when the AC electric field it detected pointed to its left, and to pass the station when it pointed to its right. Or, equivalently, it had to enter the station when the parting line was parallel to its aft–fore body axis and the AC voltages at the skin pores were negative on its left side and positive on its right side, and to pass the station when they were positive on its left side and negative on its right side, again independent of the direction in which the animal was circling the tank (Kalmijn, 1982).

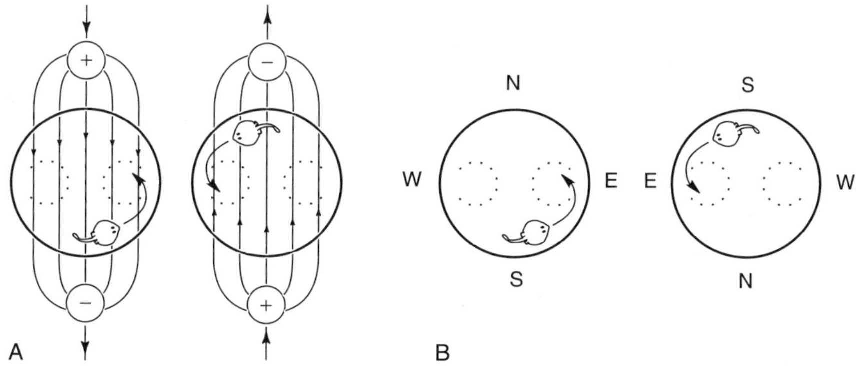

FIGURE 4.1. Proof of orientation in ambient electric and magnetic fields. (A) Stingray enters a feeding station to the left but avoids a similar station to the right with respect to an ambient DC electric field of 5 nV/cm, simulating the electric field of an ocean stream. (After Kalmijn, 1982.) (B) Stingray enters a feeding station in the magnetic east but passes a similar station in the magnetic west, in an ambient magnetic field simulating the conditions in northern-hemisphere and equatorial waters. (After Kalmijn, 1981, 1984.)

3.4. Active and Passive Orientation in Earth's Magnetic and Ambient Electric Fields

Sharks may concurrently orient with respect to the oceans' DC electric and the earth's magnetic field. In view of the difference in origin between the two kinds of electric fields sharks detect, one may speak, in engineering terms, of *active* electric orientation when the animal relies on electric fields energetically ensuing from its own motional activity in the presence of the earth's magnetic field, and of *passive* electric orientation when it relies on electric fields energetically ensuing from processes taking place in its surroundings. Hence, a shark senses its magnetic compass heading in the active mode, but it senses ocean streams and

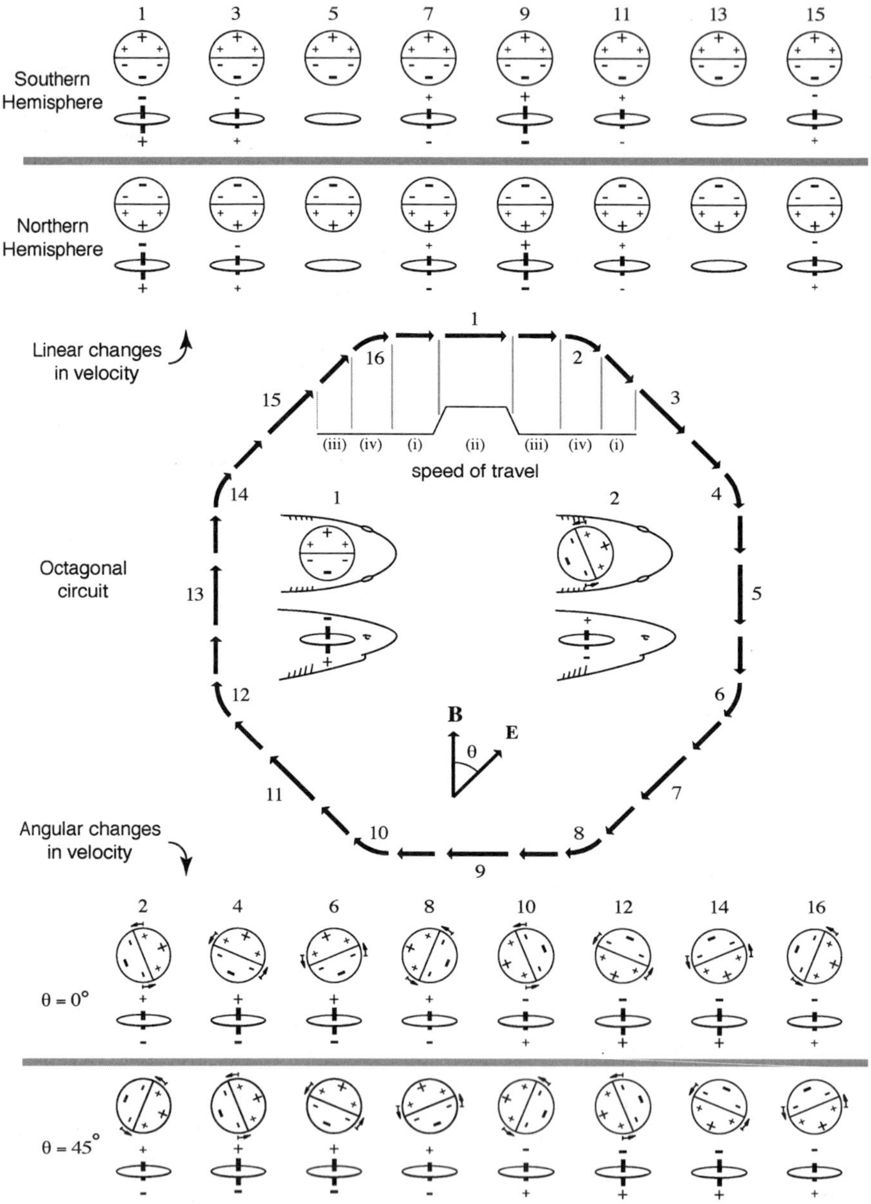

prey in the passive mode. On the basis of these definitions, the designation of "passive-electroreceptive fishes" is a misnomer for sharks and rays (Kalmijn, 1974). Note also that an animal's probing of the ambient DC electric field does not render the mode of operation active, as it does not energetically produce the field, but only changes it from purely DC to low-frequency AC, thus bringing it into the frequency range of the sense organs.

My lifetime fascination with the "secret" electric and magnetic sensory world of sharks and rays must not give the erroneous impression that elasmobranch fishes exclusively rely on their keen electric sense. My electromagnetic theory requires the animals to execute specific locomotory behaviors. To do so, they must monitor the efficacy of their muscular efforts. Hence, one may expect sharks and rays to use the otolith organs and the semicircular canals of the inner ear to gauge their linear and angular accelerations, use the hydrodynamic sensors of the lateral line to gauge the velocity or a higher, fractional derivative of the water rushing past the skin, and use the eyes and the organs of olfaction and taste where applicable. The integration of the various sensory signals undoubtedly has made great demands on the processing capabilities of the central nervous system.

4. Orientation in Bioelectric Fields of Prey

4.1. Two Crucial Hypotheses Concerning Bioelectric and Ambient-Electric Fields

I will now formulate two closely related, vitally important hypotheses, assumed only tacitly in earlier work. In traveling the ocean, a shark continuously monitors the ambient electric as well as the earth's magnetic field for two good reasons: (1) Where and when the ambient fields appear to be steady and uniform, the shark uses them to set and maintain its compass heading by subtracting the electrical signals it expects from the electrical signals it detects and correcting its course of travel so as to null the difference. (2) Where and when the electrical signals it detects differ from the electrical signals it expects more than it can attribute to a mere straying off-course, the shark interprets the difference between the two as resulting from a nearby source. Such a source could be an object generating an electric field of its own or, conceivably, an object distorting an existing ambient electric field. In orienting to the ambient and local electric fields, the animals practice the same cybernetic principles of correcting for the angular deviations from a set norm. Note that, in sensing

FIGURE 4.2. Interpretation of electric fields a shark detects in ocean waters. The lengths of the arrows, forming the octagonal path, indicate the shark's speed of travel along the pertinent stretches of the circuit. The odd numbers refer to the straight parts the animal travels at a set, elevated speed, the even numbers to the curved parts along which it changes its heading without changing its speed (top insert). For each of the numbered parts, the AC electric field the animal detects in a horizontal plane is indicated by the circular rose (top view) and the AC electric field it detects along a vertical axis by the oval rose (side view, slightly from above). The horizontal component of the earth's magnetic field of induction **B** points to the top of the page. The horizontal ambient DC electric field **E** makes an angle θ with the horizontal component of the magnetic field **B**. Top row miniatures and situation #1 inside the octagonal: AC electric fields the shark detects along the straight parts of the circuit, in the southern and northern hemispheres. Bottom row miniatures and situation #2 inside the octagonal: AC electric fields the shark detects at the end of each 45° turn, for two directions of the ambient DC electric field, $\theta = 0°$ and $\theta = 45°$, referred to north. The small, curved arrows indicate the apparent rotation of the parting line between the AC positive and negative regions during each of the turns. The miniatures are depicted for a shark facing right. The circular and oval roses are considered fixed to the head and thus rotate with the animal. Note that the parting line rotates with respect to the head through half the angle the animal turns with respect to the water, though in the opposite direction.

its prey, a shark not only relies on the vector difference between the electric field it actually detects and the far-from-trivial fields it expects as a result of its swimming effort, but also makes excellent use of the intrinsic geometric properties of the prey's field to complement the otherwise utterly incomplete sensory information it receives. However difficult it may seem for sharks and rays to sort out the information they receive from the earth's magnetic and ambient electric fields, for certain, the bioelectric fields of prey guide them to their target, as shown in numerous experiments conducted in the laboratory and at sea.

Consequently, in approaching prey, a shark must take into account the electrical signals resulting from its own deliberate changes in speed and its sharp turns in the presence of the ambient electric and earth's magnetic field, not a trivial task. However, by analogy with the performances of other sensory systems, including our own, I consider this assumption quite reasonable, as it signifies a common principle in sensory processing. Therefore, I have thus far treated the bioelectric fields a shark encounters as the only fields present. As to how the animal, in approaching prey, deals with the concurrent sensory signals it receives from the ambient electric and earth's magnetic field, theoretical evidence suggests that the cybernetic approach algorithm, which I believe the animal relies on, by its very nature accommodates at least the ambient DC electric field. Whether the algorithm also accommodates the signals resulting from interaction with the earth's magnetic field remains to be determined, though simplifying conditions prevail. When the approach takes place in a horizontal plane, as in bottom-feeding sharks and rays, the vertical component of the earth's magnetic field gives rise to a left–right electric field in the southern hemisphere, or a right–left field in the northern hemisphere, the strength of which depends on the magnetic latitude, but *not* on the direction of approach. Moreover, the horizontal component of the earth's magnetic field gives rise only to a vertical electric field, which does not affect the horizontal component of the prey's field.

4.2. Anecdotal Behavioral Evidence in Support of Sensory Hypotheses Formulated

In addition to the original behavioral experiments on the function of the electric sense in the detection of prey, I will present some (largely unpublished) anecdotal evidence in support of the hypotheses advanced above. (1) Captive stingrays were readily trained to orient with respect to uniform DC electric fields in the presence of the earth's magnetic field, but failed to be trained in the same ambient electric field after the earth's magnetic field had been canceled. It seems as if the mere absence of the electrical signals expected from their interaction with the earth's magnetic field (i.e., nothing to correct for) thoroughly confused the animals. (2) When, during behavioral tests in a shallow bay, by mishap a current electrode cable came loose, thereby abruptly changing the direction of the ambient electric field, two steadily traveling rays immediately dove to the bottom and displayed their characteristic feeding behavior of vigorously digging in the sand. Such an unexpected change in the direction of the ambient electric field normally occurs only near a local source, usually an animal that may well serve as a meal. (3) The feeding responses to electrically simulated prey often seemed to be more lively and to take place from larger distances when an ambient DC electric field was present, both in captivity and in the wild, in shallow bays and in open water. (4) One last piece of anecdotal evidence pertains to the sensitivity of sharks to ambient DC electric fields, despite the low-frequency AC nature of the ampullary sense organs. Years ago, as a graduate student, I applied a uniform electric field of slowly, smoothly increasing strength to a small shark resting on the bottom of a vinyl observation pool. The animal did not respond to the field, not even when it reached a strength of several microvolts-per-centimeter, until it made a hardly noticeable sideways movement of the head, upon which it suddenly darted away as in fright.

4.3. Approach Algorithm: A Realistic Model of Processing Directional Information

When a shark enters upon a local electric field suggesting prey, I envision it to correct its course of swimming in such a fashion as to keep the angle between its aft–fore body axis and the electric field it detects at its electroreceptive skin surface constant (Kalmijn, 1988c; Fig. 4.3). As a consequence, the shark arrives almost invariantly at the source of the field and bites it to ascertain its edibility. The approach algorithm has proven to be extremely robust, in that it is insensitive to the angle of approach, to the polarity of the prey's field, and to changes in the strength and the direction of the field and in the position of its source. The algorithm uses the sensory apparatus to its utmost, allowing the animal to detect fields of mere nanovolts per centimeter at its electroreceptive skin surface by averaging over all its ampullae of Lorenzini. It thus arrives at the best possible estimate of the direction of the field in view of the inevitable environmental and receptor-system noise. The *direction* of the field at the shark's electroreceptive skin surface, which the proposed approach algorithm is based on, I believe constitutes the most significant feature of the prey's bioelectric field, certainly at striking distance, where the field is barely detectable. However, near the source, the predator may use the spatial differences in the field *strength* as well. The approach paths predicted are consistent with all behavioral observations made so far and, I suggest, shall in the near future be verified by the use of phantom prey.

5. On Inertial Hearing in the Low-Frequency Acoustic Near Field

5.1. Similarity Between Bioelectric and Low-Frequency Acoustic Fields

Near moving underwater objects, but outside the viscous hydrodynamic boundary layer and outside the narrow trailing wake, the acoustic near field is predominantly governed by Laplace's equation, which the acoustic wave equation reduces to where the medium behaves as if it were incompressible and free of vorticity. Expressed mathematically, $\nabla^2\Phi = 0$, or the divergence of the gradient of the velocity potential Φ is nil. The Laplace region extends to distances from the source much shorter than the wavelength of the sound. Yet, given the 4.5 times higher speed of sound in water (almost 1500 m/s) than in air, the incompressible condition prevails up to distances of 10 m or more from the source at the most prominent frequencies, of 10 Hz and less, produced by quietly moving underwater objects. Hence, the Laplace region is so extensive as to comprise most, if not all, of the prey's low-frequency acoustic near field that a predatory fish can possibly detect against the background of the ever-present environmental and receptor noise (Enger et al., 1989; Kalmijn, 1989).

Even though at higher source frequencies the Laplace region is smaller, a predator still must negotiate the inner, practically incompressible part of the acoustic near field to reach its target. The great, only recently recognized, importance of these considerations is evident from the remarkable mathematical similarity between the practically incompressible region of the prey's acoustic near field and its bioelectric field. Since mathematically the tasks of orienting in the Laplace region of the acoustic near field and in the bioelectric field are so similar, the approach algorithm I have advanced in connection with the electric sense may apply to the old question of "directional hearing" in the acoustic near field just as well. Hence, at low source frequencies, the term "acoustic guidance" might be more appropriate, since the predator's acceleration-sensitive lateral-line and inner-ear sense organs, in general, do not detect the straight-line direction to the source, but the direction of the curved multipole field lines at the position of the head and along the sides of the body (Fig. 4.3).

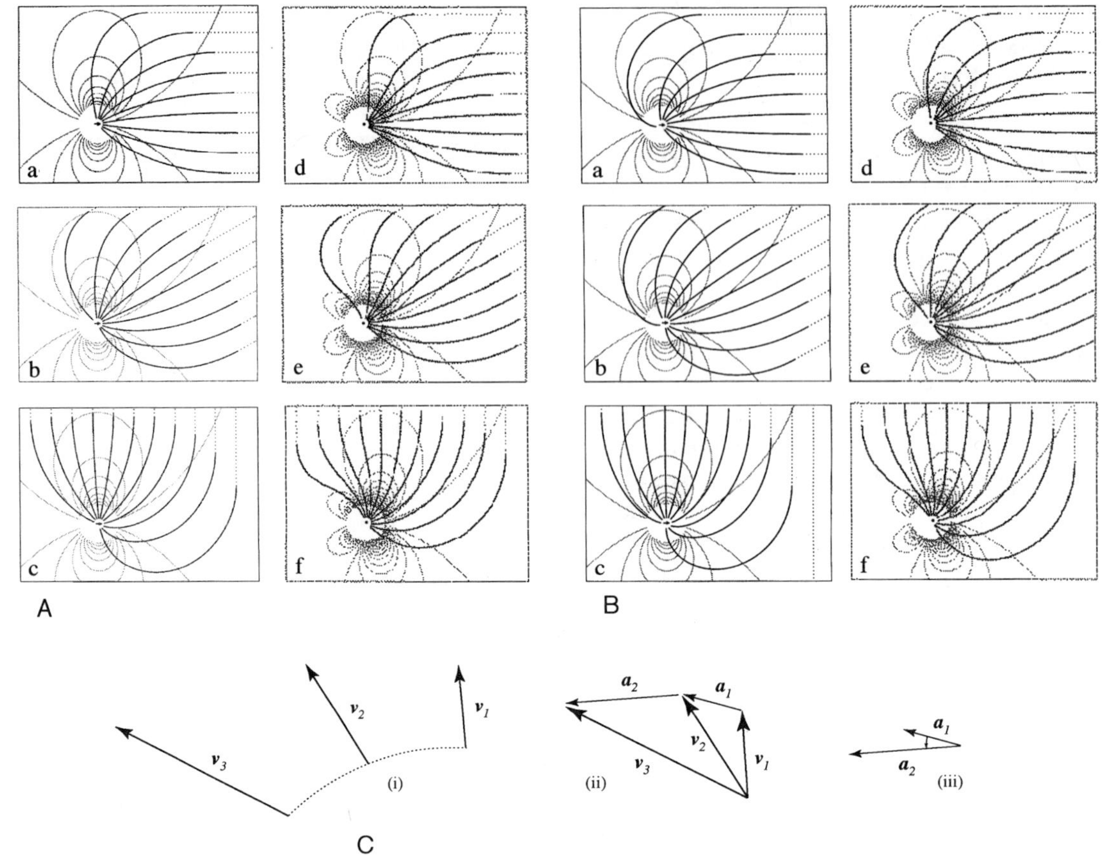

FIGURE 4.3. Electric and near-field acoustic multipole approach algorithm. (A) Guided approach in electric or acceleration field. The gray dipole (*a–c*) and multipole (*d–f*) field lines represent the bioelectric fields of stationary prey, in the electrical case; or the acceleration fields of moving prey, in the acoustical case. The predator enters the fields from three different directions along the paths indicated by the dotted lines, viewed in the frame of the prey. When the electrical stimuli received by the electroreceptors, or the local acceleration stimuli received by the inertial sense organs of the inner ear, are sufficiently strong, the predator begins its guided approach. The solid lines indicate the approach paths along which the predator maintains a constant angle between the electric field or the local accelerations it receives and its body axes, respectively. (B) Guided approach in velocity field. The gray dipole (*a–c*) and multipole (*d–f*) field lines represent the velocity fields of quietly moving prey. The predator enters the fields from three different directions along the paths indicated by the dotted lines, viewed in the frame of the prey. When the vective acceleration stimuli received by the inertial sense organs of the inner ear are sufficiently strong, the predator begins its guided approach. The solid lines indicate the approach paths along which the predator maintains a constant angle between the vective accelerations it receives and its body axes. (C) Vector diagrams illustrating the vective derivative of a prey's velocity field. The fluid velocities and vective accelerations to which the predator is subjected are depicted for three consecutive positions along the approach path. (i) Velocity vectors, $\mathbf{v}_1$, $\mathbf{v}_2$, and $\mathbf{v}_3$, sampled one unit of time apart. (ii) Time-rate of change vectors, $\mathbf{a}_1$, and $\mathbf{a}_2$, associated with velocity vectors (i.e., the vective accelerations experienced by the predator). (iii) Angular velocity (i.e., change in direction of vective acceleration per unit time), indicated by small, curved arrow. (Adapted from Kalmijn, 1997, 2000.)

5.2. Physically Adequate Stimuli of Lateral-Line and Inner-Ear Sense Organs

The vertebrate inner ear originally evolved as a three-axis complement of inertial sense organs responding to the linear and angular accelerations imparted to the animals' cranium. The earliest function of hearing, I believe, consisted in the inertial detection of the water perturbations created by moving underwater objects. Detection of acoustic pressure became possible only at a later date, by the aid of gas-filled sacs, such as the swim bladder, serving as pressure-to-motion converters. The directionally sensitive otolith organs of the inner ear indeed appear exquisitely suited to serve as the inertial sensors for the approach algorithm. The kindred lateral-line system features similar receptor cells (the stereotyped hair cells) as the inner ear, but operates differently in that it detects the *relative* motion between the water and the recipient fish, rather than the motion of the recipient fish *per se*. By virtue of the frequency-dependent properties of the hydrodynamic boundary layer, the lateral-line sense organs respond to a fractional derivative of the relative velocity: the free neuromasts to a derivative typically halfway between the velocity and the acceleration and the canal organs to the acceleration per se or to an even higher-order fractional derivative (Kalmijn, 1997). Mathematical identification of the relevant features of the low-frequency acoustic near field certainly has greatly advanced our appreciation of these sensory systems.

5.3. Inertial Detection of Local and Vective Derivatives of Velocity Field

In an endeavor to conceive of the accelerations a predatory fish experiences in the near field of moving prey, I realized that they are of a twofold origin: (1) At a given position in the field of a swimming or hovering prey, the water velocity may change over time owing to the accelerated motions of the source. At this position, a neutrally buoyant predator is subject to the first kind of acceleration, the *local* derivative of the velocity field. (2) When a predator moves relative to the source of a velocity field, it will, by transecting the field, from place to place be exposed to different velocities, even when the local velocities do not change over time. The resulting *vective* derivative (Kalmijn, 1997) of the velocity is again experienced by the predator as an acceleration, that of the second kind. This implies that a predator may use the *inertial* sense organs of the inner ear to detect a steady *velocity* field by appreciating the differences in the velocity experienced by moving relative to the source of the field. A nonuniform velocity field in the frame of the prey is perceived by a predator as an acceleration field, even when the two are moving relative to one another at a constant, nonzero velocity. No prey can swim, or even glide, stealthily enough through the water to escape detection by the inertial sense organs of a predator's inner ear.

In any case, whether the approach algorithm is applied to the local derivative or to the vective derivative of the velocity field, or to both, it efficiently guides the predator to its prey (Fig. 4.3). Since the vective derivative falls off more steeply with increasing distance from the source than the local derivative, far from the source the local derivative always prevails unless both, the local and the vective derivative, are already completely buried in the noise. Hence, inertial hearing in the far field, wherever it occurs, is more likely to be of the local, than the vective variety. We humans actually are thoroughly familiar with local and vective accelerations, but, as terrestrial animals, we receive them not by the route of the surrounding medium, but via the substrate. In that respect, surfers present the missing link. For the detection of sound in air, aquatic animals have acquired sophisticated hearing aids (first pioneered by fishes with air sacs) that prevented them from going deaf due to a catastrophic mismatch in acoustic impedance when they crawled on land.

5.4. Inertial Origin of Vertebrate Hearing in the Acoustic Near Field

Whether or not the reader concurs in calling the inertial detection of fluid acceleration in the acoustic near field of prey an early form of hearing depends on one's interpretation of the term *hearing*. Remember, though, the acceleration fields a predatory fish encounters in the vicinity of moving objects (1) present solutions of the acoustic wave equation and (2) are detected by the sense organs of the inner ear. The contrast in appearance between the acoustic near field and far field is merely a matter of boundary conditions. The inner part of the near field is bordered by the moving surface of the source; the far field is wide open to the water volume. The transition from near-field to far-field inertial hearing is gradual. The detection of acoustic pressure, however, is a secondary development made possible by the advent of gas sacs and ingenious structural elements, mechanically coupling the pressure-sensitive sacs to the motion-sensitive inner-ear organs. In modern bony fishes, the detection of acoustic pressure does not override, but complements the original inertial function of hearing. The conversion from pressure to motion plays a major role in the detection of the high-frequency sounds fishes produce for the purpose of signaling and communication. For this highly specialized form of hearing, I gladly refer the reader to my expert colleagues in the acoustic far field.

6. Conclusion

The orientation theories discussed in this chapter are founded on the physical laws that, to my knowledge, the animals practice. Whether or not sharks behave as I envisioned here is not an issue. I shall be satisfied if they indeed act in conformity with the *principles* outlined in this chapter, however different and endlessly more elegant their implementations may be.

Dijkgraaf (1963) taught us to bear in mind the biological significance of the stimulus. A theory can be physically correct, yet biologically irrelevant. Only the animals can tell in properly designed behavioral experiments, conducted under biologically valid conditions, where possible, in the natural habitat (Kalmijn, 1982, 1988a).

Ernst Mach aptly suggested that, from the study of sensory biology, one may learn a better physics. I fully agree, and like to add that, from the study of physics, we may learn a better sensory biology. Certainly, the senses of fishes present delightful examples. Life arose from a barren, physical world and has evolved into the most fascinating part of it.

Acknowledgments. I am greatly indebted to Michael C. McClune and Ivan F. Gonzalez for their invaluable advice and critique, to Vera Kalmijn for her artistic contributions and moral support, and to Drs. S.P. Collin and N.J. Marshall for inviting me to write this article. I accomplished this difficult, yet enjoyable project in partial fulfillment of my obligations to the National Science Foundation, the Office of Naval Research, NAVSEA, Schlumberger, and Scripps Institution of Oceanography.

References

Dijkgraaf, S. (1963). The functioning and significance of the lateral-line organs. *Biol. Rev.* 38:51–105.

Einstein, A. (1905). Zur Elektrodynamik bewegter Körper. *Annalen der Physik* 17:891–921.

Einstein, A. (1952). On the electrodynamics of moving bodies. In: *The Principle of Relativity* (Lorentz, H.A., Einstein, A., Minkowski, H., and Weyl, H., eds.), pp. 35–65. Republication of original Methuen 1923 translation. New York: Dover Publications.

Enger, P.S., Kalmijn, A.J., and Sand, O. (1989). Behavioral investigations on the functions of the lateral line and inner ear in predation. In: *The Mechanosensory Lateral Line* (Coombs, S., Görner, P., and Münz, H., eds.), pp. 187–215. New York: Springer-Verlag.

Faraday, M. (1832). Experimental researches in electricity. *Phil. Trans. R. Soc. (Lond.)* 122(1):125–194.

Kalmijn, A.J. (1966). Electro-perception in sharks and rays. *Nature (Lond.)* 212:1232–1233.

Kalmijn, A.J. (1971). The electric sense of sharks and rays. *J. Exp. Biol.* 55:371–383.

Kalmijn, A.J. (1974). The detection of electric fields from inanimate and animate sources other than electric organs. In: *Handbook of Sensory Physiology, Vol. III/3* (Fessard, A., Autrum, H., Jung, R., Loewenstein, W.R., McKay, D.M., and Teuber, H.L., eds.), pp. 147–200. New York: Springer Verlag.

Kalmijn, A.J. (1978). Electric and magnetic sensory world of sharks, skates, and rays. In: *Sensory Biology of Sharks, Skates, and Rays* (Hodgson, E.S., and Mathewson, R.F., eds.), pp. 507–528. Washington DC: U.S. Government Printing Office.

Kalmijn, A.J. (1981). Biophysics of geomagnetic field detection. *IEEE Trans. Mag.* 17:1113–1124.

Kalmijn, A.J. (1982). Electric and magnetic field detection in elasmobranch fishes. *Science* 218: 916–918.

Kalmijn, A.J. (1984). Theory of electromagnetic orientation: a further analysis. In: *Comparative Physiology of Sensory Systems* (Bolis, L., Keynes, R.D., and Maddrell, S.H.P., eds.), pp. 525–560. Cambridge: Cambridge University Press.

Kalmijn, A.J. (1988a). Electromagnetic orientation: a relativistic approach. In: Electromagnetic Fields and Neurobehavioral Function (O'Connor, M.E., and Lovely, R.H., eds.), pp. 23–45. New York: Alan R. Liss.

Kalmijn, A.J. (1988b). Hydrodynamic and acoustic field detection. In: *Sensory Biology of Aquatic Animals* (Atema, J., Fay, R.R., Popper, A.N., and Tavolga, W.N., eds.), pp. 83–130. New York: Springer-Verlag.

Kalmijn, A.J. (1988c). Detection of weak electric fields. In: *Sensory Biology of Aquatic Animals* (Atema, J., Fay, R.R., Popper, A.N., and Tavolga, W.N., eds.), pp. 154–189. New York: Springer Verlag.

Kalmijn, A.J. (1989). Functional evolution of lateral line and inner ear sensory systems. In: *The Mechanosensory Lateral Line* (Coombs, S., Görner, P., and Münz, H., eds.), pp. 187–215. New York: Springer-Verlag.

Kalmijn, A.J. (1997). Electric and near-field acoustic detection, a comparative study. *Acta Physiol. Scand. Stockholm* 161, suppl 638: 25–38.

Kalmijn, A.J. (2000). Detection and processing of electromagnetic and near-field acoustic signals in elasmobranch fishes. *Phil. Trans. R. Soc. (Lond.) B.* 355:1135–1141.

Maxwell, J.C. (1891). *A Treatise on Electricity And Magnetism*, 3rd ed. Republication of original Clarendon Press edition, 1954, in two volumes. New York: Dover Publications.

5
Active Electrolocation and Its Neural Processing in Mormyrid Electric Fishes

Gerhard von der Emde and Curtis C. Bell

Abstract

Weakly electric fishes use active electrolocation to orient in their environment at night. They emit an electric organ discharge (EOD) with a specialized organ in their tail and sense this signal with epidermal electroreceptors. Objects in the vicinity of the fish locally alter the transepidermal current flow evoked by the EOD and thereby project an "electrical image" on the fish's skin. By analyzing this image, the fish can detect and three-dimensionally localize objects and also can determine some of their properties, such as the object's electrical impedance, its shape, and maybe also its size. During electrolocation, peak amplitude and waveform of the local EOD inform the fish about the object's complex impedance. Object distance is determined by measuring the slope of the electric image, while the determination of object shape and size requires some complex neural calculation of spatial-temporal image parameters.

Mormyrids possess a special type of electroreceptor organs for active electrolocation, called mormyromasts. Each of these organs contains two types of receptor cells, both of which respond to signal amplitude changes, and one also responds to signal waveform. Primary receptor afferents project to the electrosensory lateral line lobe (ELL) forming two somatotopic maps of the body surface. The ELL also receives input originating from the command nucleus in the medulla, which initiates each discharge of the electric organ. The dynamic interaction of this electric organ corollary discharge with the input provided by the primary afferents is essential for the extraction of information about the electric image and thus about the object. During initial processing in ELL, basic features of the electric image are extracted by the projection of a plastic copy of the peripheral electric image onto the ELL maps. The spatial-temporal relationships of the voltage distributions on the fish's skin are analyzed in relation to the changes that are occurring constantly during active electrolocation. Little is known about the physiology of higher-order electrosensory structures beyond ELL. A major feature of the electrosensory pathway is extensive feedback from higher to lower centers, which might be responsible for many of the dynamic neural response patterns observed.

1. Introduction

The sensory systems of animals have been markedly shaped by evolutionary processes because the ability to detect and identify environmental events is crucial for survival and reproduction (Dusenbery, 1992). Evolution has shaped both peripheral sense organs and central nervous system processing to meet the animal's needs in its environment. Special sensory abilities have sometimes appeared and developed because they have opened up particular environmental niches for various groups of animals. The weakly electric fishes of Africa (Mormyriformes) and South America (Gymnotiformes) are examples of this. These freshwater fishes can live in completely dark habitats because of their electric sense and ability to produce weak electric signals. They do not need visible light during their active period but can rely exclusively on electrolocation and electrocommunication. Thus, most weakly electric fishes are either nocturnal or live in dark or turbid habitats that are devoid of light during the day (Hopkins, 1988; Moller, 1995).

2. Passive and Active Electrolocation

Weakly electric fishes use both passive and active electrolocation. During *passive electrolocation*, fishes "listen" to electrical signals generated in the external environment (Kalmijn, 1974; Peters et al., 1999). Such signals are usually DC or low frequency and may be either biotic or abiotic in origin.

During *active electrolocation*, weakly electric fishes produce electric signals known as EODs that create three-dimensional electric fields in the water around the fishes (Fig. 5.1; see color plate). This article focuses on mormyrid fishes from Africa in which the EODs are brief electric pulses, with durations between 0.2 and several milliseconds that are different in different species. Intervals between EODs can vary from 9 ms during territorial displays to several minutes when the fish is "hiding" (Hopkins, 1988; Kramer, 1990). The electric organ of these animals consists of four rows of regularly oriented electrocytes (Hopkins, 1999). The electrocytes are activated synchronously by motoneurons in the spinal cord, and the voltages of the individual electrocytes add up to generate an EOD with an open circuit amplitude of up to 10 volts (Bell et al., 1976).

During active electrolocation the fishes sense their own EOD with cutaneous electroreceptors (Lissmann and Machin, 1958) that are distributed over most of the body surface (Harder, 1968). The electroreceptors that mormyrids use for active electrolocation are called mormyromasts (Szabo and Hagiwara, 1967). Each mormyromast senses the local transepidermal voltage caused by the EOD at a particular location on the skin surface. The responses of all the mormyromasts together signal the pattern of EOD induced voltages over the entire body surface. The mormyrid EOD is an all-or-none event of essentially constant amplitude, with the result that the voltage distribution during each EOD also remains constant, as long as the fish swims in the open water. However, objects near the fish, with different electrical properties than those of the surrounding water, change the electrical field and alter the EOD-induced voltage pattern on the skin surface (Fig. 5.1). The fish senses nearby objects by analyzing this EOD-induced voltage pattern on its skin.

3. Electric Images of Objects

The effect of a nearby object on the EOD-induced voltage pattern on the skin surface is known as the *electric image* of the object

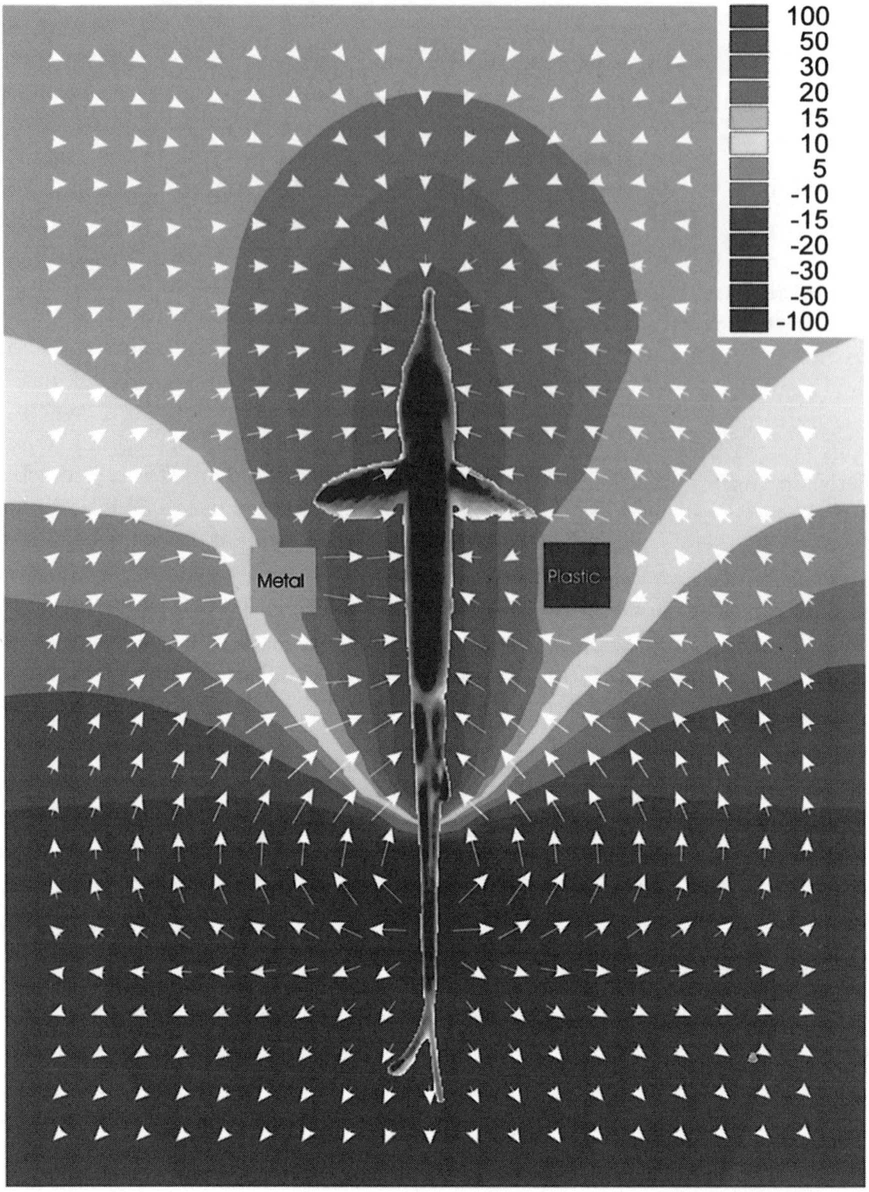

FIGURE 5.1. The electric field produced by *G. petersii* during the first peak of an electric organ discharge. Equipotential lines are color coded with positive voltages shown in red and negative voltages in blue. The scale in the upper-right corner depicts the voltages in mV. The arrows represent local field vectors, which were recorded at the corresponding locations. To the left and the right of the fish seen from above, objects are positioned. The left object is a metal cube (good electrical conductor), which locally increases field vector lengths in a region between the object and the fish's skin. This leads to an increase in the voltages perceived by the electroreceptors located at the affected skin region. The object on the right is a plastic cube, and because it is an isolator it leads to a decrease in local field vector lengths and thus to a decrease in the perceived local voltages. (This figure was kindly provided by S. Schwarz.) (See color plate)

(Rasnow, 1996; Caputi et al., 1998; von der Emde et al., 1998; Assad et al., 1999). The electric image of an object is the *modulation* that it causes in the pattern of EOD-induced transepidermal voltages, that is, the difference between the voltage pattern in the presence of the object and the voltage pattern in its absence.

The different properties of the object affect the electric image in the following ways: (1) *Location:* The location of the object relative to the fish determines the position of the center of the image on the skin surface, with the center of the image usually being located at the point on the skin that is closest to the object. (2) *Ohmic resistance:* Purely resistive objects modulate transcutaneous voltage amplitude but not waveform. The general form of the image is that of a "Mexican hat" (Caputi et al., 1998). If the object is more conductive than the water, more current flows through the center of the image, and amplitudes are increased there. If the object is less conductive than the water, less current flows, and amplitudes are decreased in the center. In both cases, there is a narrow rim around the larger center in which voltage changes are in the opposite direction to changes in the center. (3) *Capacitance:* Capacitive objects alter both the amplitude and the waveform of local EOD-induced voltages. The amplitude modulations caused by a capacitance depend on the value of the impedance in Ohms, as with Ohmic resistance, but waveform distortions occur only within a certain range of capacitive values (von der Emde and Ronacher, 1994). The distortion is maximum at some value within this range and decreases toward both higher and lower capacitances. (4) *Object size:* The width of the electric image of an object is positively related to the object's size (i.e., larger objects project larger images). (5) *Distance:* The width of an object's image increases as the object moves away from the fish. Maximal amplitude and waveform modulations also decrease with distance. (6) *Shape:* The influence of object shape has not been systematically analyzed so far. It is known that spherical or ellipsoid objects can have an influence on the "slope" of the electrical image (see below), which in extreme cases can cause "electrical illusions" (i.e., an object can appear to be further away than it actually is) (von der Emde et al., 1998).

Thus, the different properties of an object in the environment are systematically reflected in the electrical image that the object "projects" onto the skin surface. The electrical image is in turn measured by mormyromast electroreceptors that relay this information to the nervous system via primary afferent fibers. The task of the nervous system and of the fish is to analyze the information from receptors in order to accurately locate and identify objects in the environment. The ability of the fish to carry out these perceptual tasks has been measured in behavioral experiments.

4. Behavioral Measurements of Active Electrolocation

4.1. Resistance Detection

The ability of weakly electric fishes to detect objects with resistivities that are different from that of water was established behaviorally by Lissmann in 1958 (Lissmann and Machin, 1958). Most stones are isolators, whereas organic "objects" are conductors. The detection of such objects is based on the measurement of small changes in the amplitude of the local EOD, as described in the previous section, and behavioral experiments have shown that mormyrids can reliably detect at least a 1% change in EOD amplitude (Hall et al., 1995; von der Emde and Zelick, 1995).

4.2. Capacitance Detection

Until recently it was assumed that electric fishes can measure only the single quality of object resistance and would therefore perceive only a "black-and-white electric picture" of their surroundings. But many objects within the aquatic environment have capacitive as well as resistive properties (Heiligenberg, 1973). This is especially true for living objects such as water plants, fishes, and insect larvae, which possess a complex impedance consisting of resistive and capacitive components (Schwan, 1963). Behavioral experiments have revealed that mormyrids as well as gymnotids can detect the capacitive properties of objects (Fig. 5.3A) (Meyer, 1982; von der Emde and Ringer, 1992; von der Emde, 1998). For capacitance detection the animals measure waveform distortions (mormyrids) or phase shifts (gymnotids) in the local EOD that are caused by the capacitive object. Behavioral experiments have also shown that mormyrids can measure extremely small EOD waveform distortions independently from amplitude modulations (von der Emde and Ronacher, 1994). Capacitance perception adds "color" to the otherwise black-and-white electric picture of the environment and allows much more information to be gathered. During food search, the electrically colored food items with their capacitive properties will pop out of an electrically gray background and will thus be more easily detected (von der Emde and Bleckmann, 1998).

4.3. Distance Measurement

Mormyrids can localize objects three-dimensionally in space using active electrolocation. *Gnathonemus petersii* can learn to discriminate between objects based on their distance from the fish, independently of the size or electrical properties of the object (Fig. 5.3B) (von der Emde et al., 1998; Schwarz and von der Emde, 2001). Thus, they possess a true sense of depth perception. Their depth perception is based on a newly found sensory mechanism in which they determine object distance by measuring the normalized maximal slope of the image (i.e., the transition from the rim area of the Mexican hat to its center), which depends only on distance and not on object size or properties. This mechanism of depth perception is unique because it requires only a single two-dimensional receptive surface, which does not have to be moved (von der Emde et al., 1998).

Recognition of an object's size independently of its distance is an important perceptual ability known as *size constancy* (Douglas et al., 1988). This is a difficult task for both active electrolocation and vision, because changes in image size in both systems can be caused by either a change in distance or a change in object size. In active electrolocation, the image grows larger with object distance (Fig. 5.2; see color plate) (Caputi et al., 1998), whereas in vision the retinal image decreases with distance. Size constancy therefore requires independent determination of object distance in both active electrolocation and vision. Thus, mormyrid fishes possess an ability that is essential for size constancy, although a capacity for size constancy itself has not yet been investigated. The ability to measure object distance is necessary, of course, for many other perceptual tasks besides size constancy.

4.4. Shape Determination

Recent experiments have shown that *G. petersii* can determine the three-dimensional shapes of objects using active electrolocation and that they spontaneously categorize objects according to their shapes. In a computer-controlled observation tank with an infrared video camera mounted above, individual fishes were rewarded by playback of social signals for swimming close to a certain object but not to a differently shaped alternative object. During the following nonrewarded observation period, fishes continued to stay close to the previously positive object even if it was relocated to a different position in the arena (Fig. 5.3C). When the material or the size of the object was varied, many fishes continued to prefer it as long as its shape remained the same (Schwarz and von der Emde, 1999). The cues which the fish uses for determining shape are not yet known.

4.5. Foraging Behavior

Gnathonemus petersii employs active electrolocation in finding small insect larvae, its main prey (von der Emde and Bleckmann, 1998). Active electrolocation is especially helpful in a dark and complex environment where prey identification as well as prey detection is required. However, the animals never use just one sense to find their food, but also employ all of the other senses that provide information about their prey. These other senses include passive electrolocation, the mechanosensory lateral line system (see Chapter 6), olfaction, touch, and, if available, vision (von der Emde

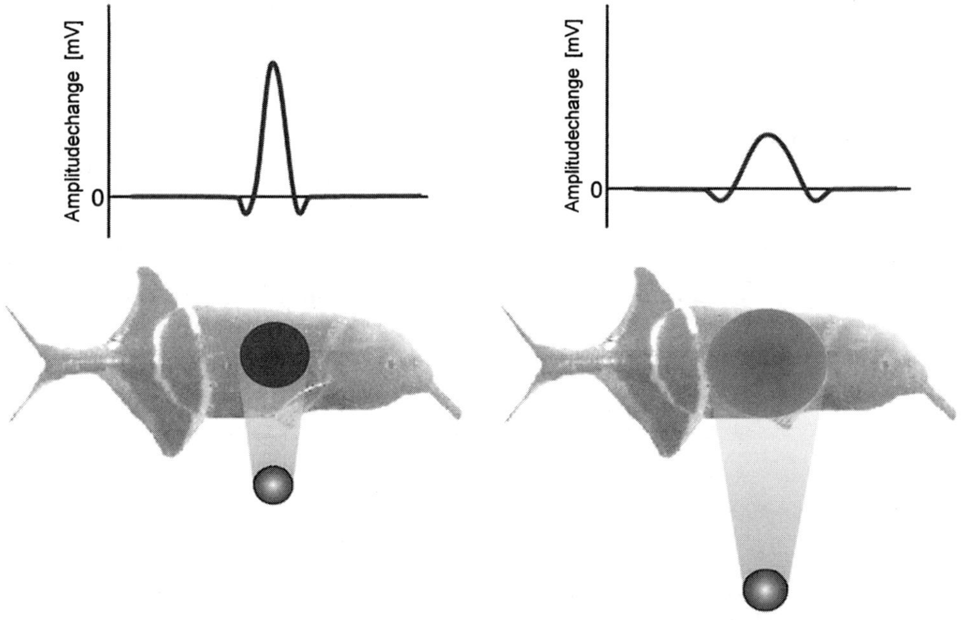

FIGURE 5.2. Color-coded representation of the electric images of a sphere projected onto the skin surface of a *G. petersii*. On the left, the sphere is at a closer distance to the fish than at the right. Red color illustrates an increase in amplitude caused by the presence of the sphere, blue depicts an amplitude decrease. The upper row shows a one-dimensional measurement of the electric images by a pair of electrodes moved along the midline from the tail toward the mouth of the fish. The ordinates give the change in amplitude by the presence of the sphere compared to the situation without it. Note that an increase in distance causes an increase in size of the electric image, and a decrease in the maximal amplitude change. (See color plate)

and Bleck mann, 1998). Active electrolocation, when added to these other senses, gives weakly electric fishes a crucial advantage over their nocturnal competitors.

5. Electroreceptors

Mormyromast electroreceptors are responsible for active electrolocation and are the most numerous type of electroreceptor in mormyrids. The other two types of electroreceptors are: Knollenorgane, which have a role in sensing the discharges of other fishes; and ampullary organs, which sense low-frequency voltage sources in the environment and are responsible for passive electrolocation (Bell and Szabo, 1986; Zakon, 1987; Hopkins, 1988).

All three types of electroreceptors respond to the fish's own EOD, but responsibility for active electrolocation can nevertheless be assigned to mormyromasts, for two main reasons: (1) The responses of mormyromast afferents are exquisitely sensitive to small changes in EOD amplitude as is required for active electrolocation. The responses of Knollenorgan afferents are insensitive to such changes and the responses of ampullary afferents are only minimally sensitive (Szabo and Hagiwara, 1967; Bell and Russell, 1978; Bell, 1990b). (2) The central effects of the reafferent (i.e., self-induced) (von Holst and Mittelstaedt, 1950) responses of mormyromast afferents to the fish's own EOD are selectively enhanced at the first stage of central processing by a corollary discharge signal associated with the EOD motor command (see below). Such enhancement accords with a role for these receptors in active electrolocation. In contrast, the reafferent responses of Knollenorgan afferents are completely blocked by the corollary discharge signal and the responses of ampullary afferents are minimized. Such suppression of reafferent responses from the latter types of electroreceptors is in agreement with their roles in sensing external signals rather than the fish's own EOD.

Mormyromast electroreceptors have two distinct types of sensory cells that are referred to as *A* and *B* (Szabo and Wersäll, 1970). The two types are separately innervated, and their primary afferents project to separate regions of the electrosensory lobe (ELL) in the medulla (Bell, 1990b). Physiologically, both types of fiber are silent in the absence of electrosensory stimulation and both types respond to a brief EOD-like pulse of current with one or more spikes. Stimulation near threshold evokes a single spike at a latency of about 10 ms. As stimulus intensity is increased, the latency of the first spike shows a smooth decrease to a minimum of about 2 ms, and additional spikes are added. The smooth decrease in latency of the first spike suggested to Szabo and Hagiwara (1967) that mormyromast afferents may use

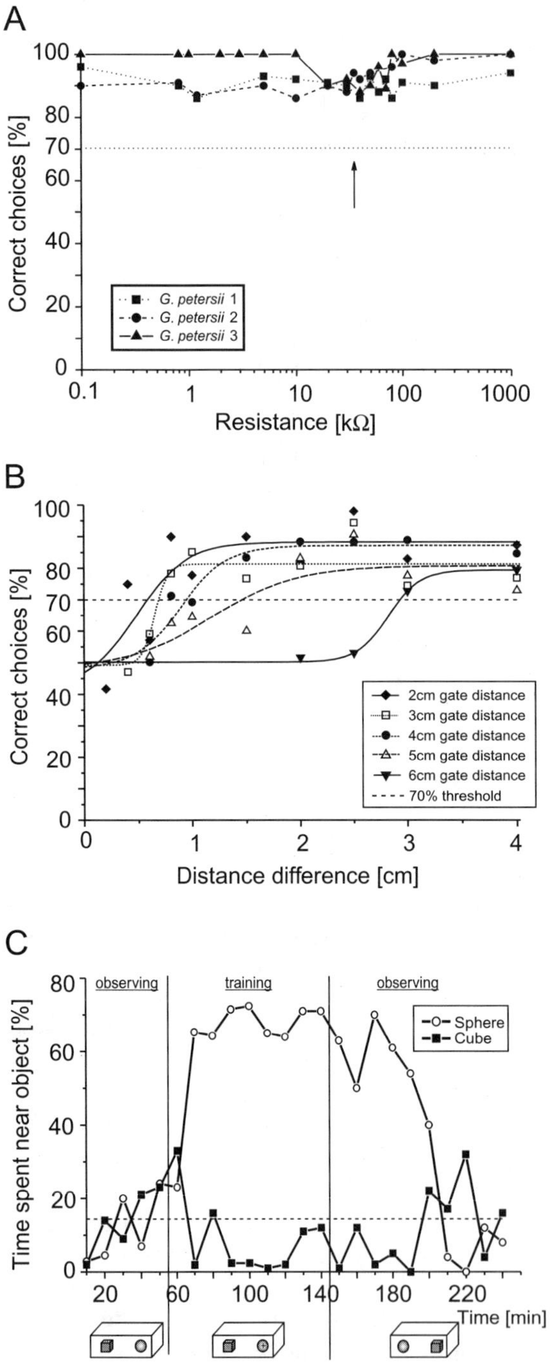

FIGURE 5.3. Behavioral performance of *G. petersii* during active electrolocation. (A) Capacitance detection. Performance of three *G. petersii* discriminating between a fixed capacitive object (1 nF) and various resistive objects whose values are given on the abscissa. The arrow marks the impedance value of the 1 nF object. (Modified after von der Emde, 1990.) (B) Distance discrimination. A *G. petersii* discriminated between the distances of two objects, each located behind a gate. The distance of the closer object from its gate was kept constant (gate distance), while the other object was moved from larger distances toward the closer object. A choice by the fish of the object located further away from its gate was rewarded. The abscissa gives the distance differences of the two objects from their respective gates. Each curve shows the performance of the fish at a given gate distance. (Modified after von der Emde et al., 1998.) (C) Object shape detection. A *G. petersii* was placed in a tank with two objects (a sphere and a cube of equal size), and the time it spent close to each of the objects was observed (ordinate). During a training period (training), the fish was rewarded when staying close to the sphere by playback of social signals, which attracted the fish to stay close to the sphere most of the time. In the following observation period, the fish continued to prefer the sphere, despite the fact that the locations of the two objects was exchanged.

latency as a code for stimulus intensity, and several physiological and behavioral observations support this hypothesis (Hall et al., 1995).

Afferent fibers from A and B sensory receptors are physiologically distinct in a number of ways (Bell, 1990b). The most important difference is probably in their sensitivity to distortions of the EOD waveform, such as those that are caused by capacitive objects (von der Emde and Bleckmann, 1997). Fibers from type B cells are exquisitely sensitive to such distortions, whereas those from type A cells are not. Both fibers are similarly sensitive to changes in EOD amplitude. These findings suggest that the fish may sense the capacitive properties of objects, independently of the resistive properties, by centrally comparing the EOD responses of A and B fibers. If the responses of both fiber types change in the same direction and by a similar amount in the presence of a new object, then the object is resistive. If, on the other hand, the responses of B fibers are more strongly affected than those of A fibers, then the object has capacitive properties.

6. Generation of the Electric Organ Discharge

The motor command that is responsible for the EOD is initiated in the command nucleus of the brainstem (Grant et al., 1986). The EOD is initiated by a synchronized two-spike volley in the command nucleus that evokes a similar volley in the nearby medullary relay nucleus (Bennett et al., 1963). The motor command signal is then conveyed by the axons of the medullary relay nucleus to the caudal tip of the spinal cord where it activates electromotoneurons. The electromotoneurons in turn evoke an EOD from the electric organ. The command nucleus is influenced, but not driven, by afferent input from a precommand nucleus and by other cell groups in the mesencephalon (Bell et al., 1983; Grant et al., 1986; Niso et al., 1989; von der Emde et al., 2000).

Axons of cells in the command nucleus are branched. One branch projects to the medullary relay nucleus and is responsible for evoking the EOD, whereas the other branch projects to the bulbar command associated nucleus (BCA). BCA is the starting point for the *corollary discharge* pathway, which runs from the command nucleus via BCA and several other command-associated nuclei to the ELL and other electrosensory areas of the mormyrid brain (Bell et al., 1983, 1995; Bell and von der Emde, 1995). The electric organ corollary discharge (EOCD) provides information about the timing of the EOD and has a strong effect on the processing of reafferent sensory input evoked by the EOD, as described more fully in the next section.

7. Central Processing of Mormyromast Information in the Electrosensory Lobe

7.1. ELL Circuitry

Afferent fibers from mormyromast electroreceptors terminate in the ELL, a cerebellum-like cortical structure in the medulla (Fig. 5.4). Fibers from A-type sensory cells terminate in the medial zone (MZ), fibers from B-type sensory cells terminate in the dorsolateral zone (DLZ), and fibers from ampullary receptors terminate in the ventrolateral zone (VLZ). The electroreceptive surface is somatotopically mapped within each of these zones.

Mormyromast afferent fibers terminate with mixed chemical-electrical synapses on small granular cells in the deeper layers of ELL (Fig. 5.5). Axons of these granular cells terminate in turn on the basilar dendrites of larger cells located more superficially. Some of the granular cell effects on more superficial cells are excitatory and others inhibitory. The larger, more superficial cells have apical dendrites as well as basilar dendrites. The apical dendrites extend throughout the molecular layer of ELL where they are contacted by parallel fibers that originate from an external granular cell mass known as the eminentia granularis posterior (EGp). The larger cells include two types of efferent

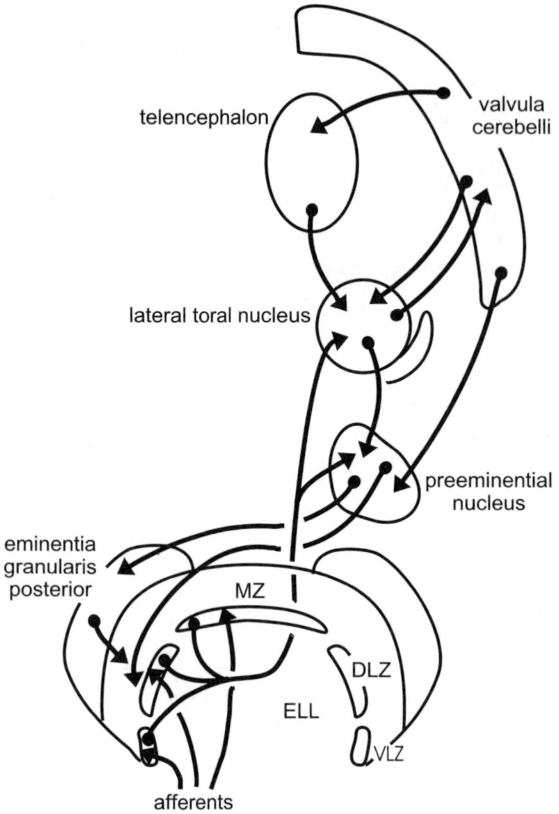

FIGURE 5.4. Schematic drawing of major electrosensory structures and their connections in mormyrid fish. Note the extensive descending connections allowing for the results of higher-level processing to influence lower-level processing. MZ—medial zone of ELL; DLZ—dorsolateral zone of ELL; VLZ—ventrolateral zone of ELL.

cells, large ganglion (LG) and large fusiform (LF), that relay electrosensory information to higher levels of the system, as well as two types of Purkinje-like inhibitory interneurons known as medium ganglion cells (MG1 and MG2). MG1 cells appear to inhibit LF cells preferentially and MG2 cells appear to inhibit LG cells preferentially (Han et al., 1999). LG cells are inhibited by electrosensory stimuli in the center of their receptive fields, whereas LF cells are excited by such stimuli (Bell et al., 1997b). Recent findings suggest that MG1 cells are inhibited by electrosensory stimuli in the center of the receptive field whereas MG2 cells are excited (C. Mohr, G. von der Emde, and C.C. Bell, unpublished).

These anatomical and physiological findings have been put together with the additional hypothesis of mutual inhibition between MG1 and MG2 cells to suggest the functional circuit shown in Figure 5.5. The circuit includes LF cells, which are efferent "on cells" conveying increases in transcutaneous voltage to higher centers, and LG cells, which are efferent "off cells" conveying decreases in transcutaneous voltage to higher centers. The hypothesized mutual inhibition between MG1 and MG2 cells could mediate contrast enhancement and other functions, as discussed elsewhere (Han et al., 1999).

Bidirectional topographically organized projections are present between the two mormyromast regions of ELL (Bell et al., 1981). This suggests that the comparison between the responses of A- and B-type mormyromast afferents, necessary for distinguishing the capacitive and resistive properties of objects, could occur in ELL. However, in a study directed at this question (von der Emde and Bell, 1994), cells in the medial zone of ELL, for example, behaved just like the A-type afferents that terminate there in being unaffected by small waveform distortions, distortions that had marked effects in the dorsolateral zone where B type afferents terminate. No functional interaction between the two zones was observed. Thus, the comparison of information from A and B types seems to occur at a higher level of the system.

7.2. The Effect of the Electric Organ Corollary Discharge

The mormyromast regions of ELL receive two different kinds of input with each EOD. EOD-evoked reafferent input from the periphery and corollary discharge signals associated with the EOD motor command. The corollary discharge signals are conveyed to ELL from various central structures. The EOCD effects are examined by blocking the EOD with curare. The EOD motor command signal continues to be emitted under these conditions, but without the normally consequent EOD. The effects of the EOCD can thus be examined in isolation and in combination with electrosensory stimuli controlled by the experimenter.

Interaction between EOD-evoked reafferent input and EOCD signals begins at the very first stage of central processing. Recordings of synaptic potentials show that the EOCD evokes a prominent excitatory postsynaptic potential (EPSP) in granular cells and arrives at these cells at the same time as EOD-evoked reafferent input. The prominent EOCD-driven EPSP appears to serve two different functions. The first is to selectively enhance transmission of the reafferent input that is evoked in mormyromast afferents by the fish's own EOD. Such reafferent input is the only input that can be used for active electrolocation. Input due to other voltage sources such as the discharges of nearby fishes is noise. The disruptive effects of such external noise stimuli are reduced by the EOCD enhancement of responses arriving just

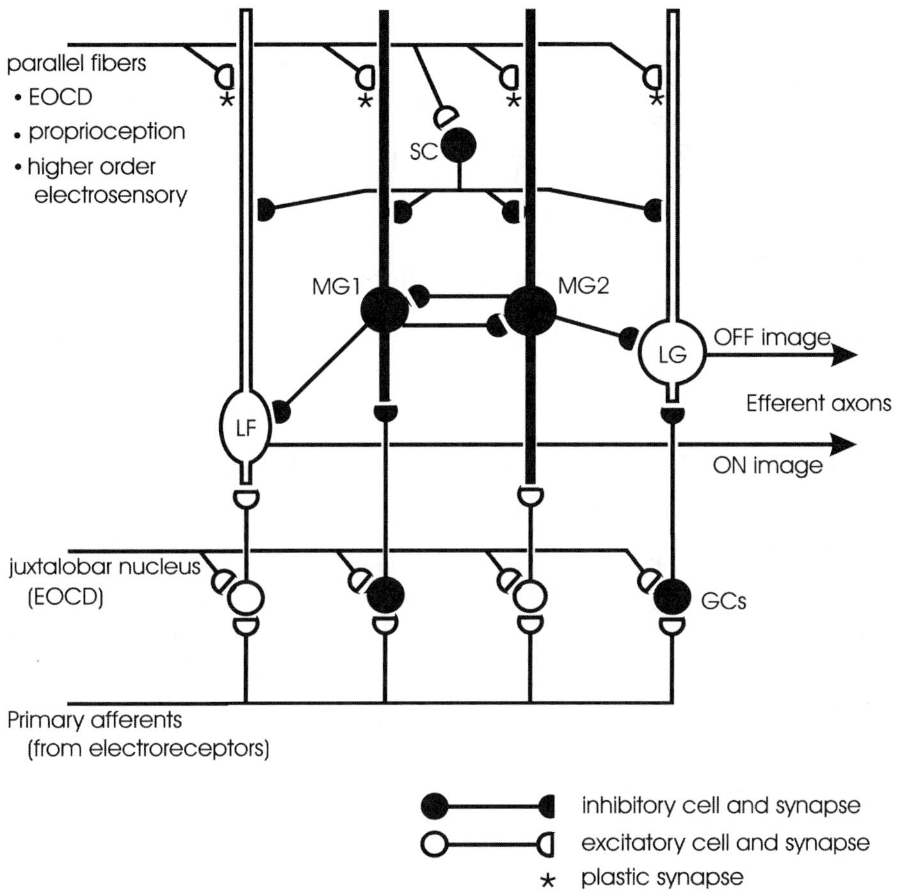

FIGURE 5.5. Schematic drawing of circuitry module for mormyromast region of mormyrid ELL. Primary afferent fibers excite inhibitory and excitatory granular cells (GCs), which are also excited and gated by an electric organ corollary discharge (EOCD) signal from the juxtalobar nucleus. The granular cells excite or inhibit more superficial-lying cells that include two types of efferent cells (large fusiform, LF, and large ganglion, LG) as well as two types of Purkinje-like medium ganglion cells (MG1 and MG2). The apical dendrites of these four types of superficial cells are excited by parallel fibers that convey EOCD, proprioceptive, and higher-order electrosensory signals. The parallel fiber synapses are plastic as indicated by an *. Parallel fibers also excite inhibitory stellate cells (SCs).

after the EOD. This gating of reafferent input by the EOCD has been demonstrated behaviorally (Meyer and Bell, 1983; Hall et al., 1995). The EOCD-driven enhancement of reafferent input from mormyromast afferents is one of the arguments for assigning these receptors their central role in active electrolocation, particularly since re-afferent input in the other two classes of afferent fibers from the Knollenorgane and from ampullary organs is suppressed by the EOCD.

The second function of the EOCD-driven EPSP in granular cells is to provide a timing signal for deriving stimulus intensity from post EOD latency. The granular cell receives both EOCD-evoked and the afferent-evoked EPSPs. The sum of these two EPSPs within the granular cell grows larger as the latency of the afferent EPSP gets smaller, reaching a maximum when the peaks of the two EPSPs coincide at the minimal afferent latency. Thus, depolarization and activation of the granular cell by EOD-evoked reafferent responses depend on the latency of the spikes in the afferent fibers.

The EOCD EPSP in granular cells is evoked by a single EOCD-driven spike in fibers from the juxtalobar nucleus (Fig. 5.5), one of the command-associated nuclei that is linked to the command nucleus through a series of three other command-associated nuclei (Bell and von der Emde, 1995; Bell et al., 1995). The timing of the EOD motor command is conveyed with great accuracy over this pathway that includes four synapses. A crude measurement showed that the time delay between the command signal in the command nucleus and a spike in a juxtalobar cell varies by less than 50 μs. This remarkable preservation of timing information argues in itself for the importance of accurate latency measurement in the mormyromast system.

Granular cells are strongly inhibited by electrosensory stimuli outside the excitatory centers of their receptive fields (Bell, 1990a). This inhibition appears to be mediated by a remarkable inhibitory interneuron in the deeper layers of ELL known as the large multipolar interneuron (LMI; Meek et al., 1996). Both the dendrites and axons of these cells are myelinated and the cells appear to receive very little excitatory synaptic input. Both axonal and dendritic processes give rise to large inhibitory terminals that cover $\frac{1}{4}$ to $\frac{1}{2}$ of the surface area of granular cells. The lack of excitatory synaptic input to these cells, as well as physiological findings (Han et al., 2000), suggest that they are activated by a rapid nonsynaptic mechanism (e.g., ephaptically) through their large terminals on granular cells following primary afferent-induced depolarization of the granular cells. These anatomical and physiological specializations indicate a very rapidly acting and powerful lateral inhibition that could serve to enhance small differences in latency between neighboring groups of mormyromast afferents and thereby sharpen the electric image of an object.

7.3. EOCD-Associated Neural Plasticity

Electric organ corollary discharge (EOCD) effects on higher-order cells in ELL are also prominent, and many of these effects are plastic in that they depend on the reafferent input to the cell that has occurred in the recent past (Bell et al., 1992; Bell et al., 1997a). This plasticity is in contrast to the lack of plasticity in granular cells where the EOCD-evoked EPSP is not affected by previous pairing with afferent input. The plasticity is usually studied by first examining the effect of the EOCD alone on a cell and then delivering an electrosensory stimulus at a fixed delay following each EOD motor command signal. The chosen delay is usually the delay at which the EOD would normally occur in the absence of curare. After maintaining the pairing for a few seconds to several minutes, the sensory stimulus is turned off and the EOCD effect is examined again (Fig. 5.6).

Pairing the EOCD with a sensory stimulus leads to a marked change in the EOCD effect. The EOCD changes in such a way as to oppose the effect of the paired sensory stimulus. If the sensory stimulus excites the cell, then the EOCD change is one of increased inhibition. If the sensory stimulus inhibits the cell then the EOCD change is one of increased excitation. Thus, in the normal life of the animal, the plasticity results in an EOCD-evoked negative

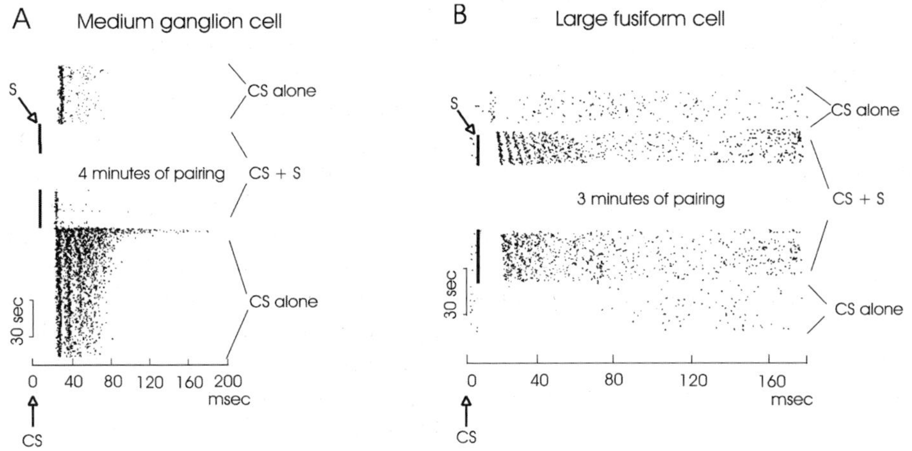

FIGURE 5.6. Examples of EOCD plasticity following pairing with an electrosensory stimulus. Recordings were extracellular with identification of cell type based on similarity of response properties to those of other cells recorded intracellularly and labeled with biocytin. Each sweep of the rasters is initiated by the fish's EOD motor command signal (CS) and the dots in the rasters are action potentials. In each case the initial response to the CS is shown first (CS alone), followed by the response to the CS plus a sensory stimulus (CS + S). The sensory stimulus (S) was delivered at the EOD delay as indicated by a black vertical bar. After several minutes of pairing CS with S, the sensory stimulus was turned off and the effect of CS alone was examined again. (A) This medium ganglion cell was inhibited by the sensory stimulus. Note that several minutes of pairing with such a stimulus results in a strongly excitatory response to CS alone (i.e., a negative image of the previously paired inhibition). (B) This large fusiform cell was excited by the sensory stimulus. Note that several minutes of pairing result in a strong inhibition to CS alone (i.e., a negative image of the previously paired excitation).

image of the pattern of reafferent responses that followed the EOD in the recent past. Addition of this negative image to the actual image of current EOD responses removes the expected component. This allows for novel or unexpected features in the EOD-evoked afferent input to stand out more clearly. In the context of active electrolocation, EOCD plasticity and the accompanying subtraction of a previous pattern of sensory responses generates the electric image of a suddenly appearing object. Recall that the electric image of an object is the *difference* between the pattern of EOD-evoked transcutaneous voltages in the presence of an object and the pattern in the object's absence.

The plasticity of EOCD responses in ELL appears to be due to synaptic plasticity at synapses between EOCD-conveying parallel fibers and the apical dendrites of MG, LF, and LG cells (synapses marked with * in Fig. 5.5). Synaptic plasticity has been identified at this synapse in in vitro slice preparations (Bell et al., 1997c), and computer modeling has shown that the plasticity found at this synapse is just the type that is required to generate EOCD-driven negative images of expected sensory input (Roberts and Bell, 1999, 2000).

The neurons of EGp that give rise to parallel fibers receive not only EOCD-driven input but also signals from other sources, including proprioceptive signals conveying information about body and fin positions, and electrosensory signals descending from higher levels of the electrosensory system beyond ELL (Fig. 5.4). Changes in the latter two types of signals will often be associated with changes in primary afferent input to the deeper layers of ELL. Thus, these signals could serve as predictors of the electrosensory patterns with which they have been recently associated, just like the EOCD. The generation of negative images of predicted sensory input following associations between peripheral electrosensory input and proprioceptive signals or descending electrosensory signals conveyed by parallel fibers has been clearly demonstrated in the cerebellum-like sensory structures of both gymnotid electric fishes (Bastian, 1996) and elasmobranch fishes (Montgomery and Bodznick, 1994), although such generation has not yet been tested in mormyrid fishes.

The utility of generating negative images based on past associations with proprioceptive or descending electrosensory signals is readily apparent. When the fish bends its body, for example, the electric organ in the tail is moved either closer to or further away from electroreceptors on one side of the body. The changes in transcutaneous voltage induced by such bending convey no useful information and can be predicted based on proprioceptive activity. Subtracting out the bending-induced changes in ELL allows other unpredictable, information-bearing sensory inputs to stand out more clearly. Similarly, previous electrosensory input, as signaled perhaps by descending signals from higher centers, will often permit the prediction of current electrosensory input, and subtracting such predictions would again allow for information-bearing new sensory input to stand out.

In summary, we have described in this section some of the sensory processes in ELL that are important for active electrolocation. These processes include: selective enhancement of EOD-evoked reafferent input; decoding of afferent latency to determine stimulus intensity; contrast enhancement by rapidly acting inhibitory interneurons; generation of separate systems to signal increases (on) and decreases (off) in stimulus amplitude; possible contrast enhancement by mutual inhibition between the on and off components of the circuit; and the enhancement of information-bearing sensory signals by subtracting out the predictable features of the sensory inflow.

8. Higher Stages of the Electrosensory System

The ELL is only one of several major structures in the mormyrid brain that are concerned with the electrosensory system (Fig. 5.4). At the top of the hierarchy is the valvula cerebelli, which is also known as the mormyrid gigantocerebellum because it is extraordinarily large and covers all the rest of the brain in most species (Nieuwenhuys and Nicholson, 1969). One-third to one-half of the valvula surface is devoted to the electrosensory system. The valvula receives input from and projects back to the lateral and preeminential nuclei. The lateral nucleus receives input from ELL and projects back to the preeminential nucleus as well as to other structures. The preeminential nucleus projects back to ELL both directly to the deep molecular layer and indirectly via EGp and the parallel fibers (Bell et al., 1981; Finger et al., 1981; G. von der Emde, G. Kirchberg, J. Meek, and K. Grant, unpublished).

These anatomical findings demonstrate the presence of feedback from higher to lower stages of the mormyrid electrosensory system. Such feedback allows for the results of higher-level processing, or for memories that may be stored at the higher levels, to be returned to lower levels for interaction with ascending information from the periphery. Extensive feedback from higher to lower centers is also a major feature of many other sensory systems, including the mammalian visual and auditory systems (Huffman and Henson, 1990; Van Essen and Gallant, 1994), but the roles of such feedback are not understood.

Very little is known about the physiology of higher-order electrosensory structures beyond ELL. Electrosensory responses have been recorded in the preeminential nucleus (von der Emde and Bell, 1995), the lateral nucleus (Mohr and von der Emde, 1998), parts of the valvula cerebelli (Russell and Bell, 1978), and in the telencephalon (Prechtl et al., 1998). Independent EOCD effects that cannot be explained by prior interaction in ELL have been observed in the preeminential nucleus and valvula cerebelli, and EOCD plasticity that appears to be different from the EOCD plasticity in ELL has been observed in preeminential nucleus. Thus, only a small beginning has been made in understanding the higher levels of the mormyrid electrosensory system and much remains to be learned about processing at each level and about the interactions between levels.

References

Assad, C., Rasnow, B., and Stoddard. P.K. (1999). Electric organ discharges and electric images during electrolocation. *J. Exp. Biol.* 202:1185–1193.

Bastian, J. (1996). Plasticity in an electrosensory system. I. General features of a dynamic filter. *J. Neurophysiol.* 76:2483–2496.

Bell, C., Bodznick, D., Montgomery, J., and Bastian, J. (1997a). The generation and subtraction of sensory expectations within cerebellum-like structures. *Brain Behav. Evol.* 50:17–31.

Bell, C.C. (1990a). Mormyromast electroreceptor organs and their afferent fibers in mormyrid fish. II. Intra-axonal recordings show initial stages of central processing. *J. Neurophysiol.* 63:303–318.

Bell, C.C. (1990b). Mormyromast electroreceptor organs and their afferent fibers in mormyrid fish. III. Physiological differences between two morphological types of fibers. *J. Neurophysiol.* 63:319–332.

Bell, C.C., and Russell, C.J. (1978). Effect of electric organ discharge on ampullary receptors in a mormyrid. *Brain Res.* 145:85–96.

Bell, C.C., and Szabo, T. (1986). Electroreception in Mormyrid fish. Central Anatomy. In: *Electroreception* (Heiligenberg, W., ed.), pp. 375–421. New York: Wiley.

Bell, C.C., and von der Emde, G. (1995). Electric organ corollary discharge pathways in mormyrid fish. II. The medial juxtalobar nucleus. *J. Comp. Physiol. A.* 177:463–479.

Bell, C.C., Bradbury, J., and Russell, C.J. (1976). The electric organ of a mormyrid as a current and voltage source. *J. Comp. Physiol. A.* 110:65–88.

Bell, C.C., Caputi, A., and Grant, K. (1997b). Physiology and plasticity of morphologically identified cells in the mormyrid electrosensory lobe. *J. Neurosci.* 17:6409–6423.

Bell, C.C., Finger, T.E., and Russell, C.J. (1981). Central connections of the posterior lateral line lobe in mormyrid fish. *Exp. Brain Res.* 42:9–22.

Bell, C.C., Grant, K., and Serrier, J. (1992). Sensory processing and corollary discharge effects in the mormyromast regions of the mormyrid electrosensory lobe. I. Field potentials. *J. Neurophysiol.* 68:843–858.

Bell, C.C., Libouban, S., and Szabo, T. (1983). Pathways of the electric organ discharge command and its corollary discharges in mormyrid fish. *J. Comp. Neurol.* 216:327–338.

Bell, C.C., Dunn, K., Hall, C., and Caputi, A. (1995). Electric organ corollary discharge pathways in mormyrid fish. I. The mesencephalic command associated nucleus. *J. Comp. Physiol. A.* 177:449–462.

Bell, C.C., Han, V.Z., Sugawara, Y., and Grant, K. (1997c). Synaptic plasticity in a cerebellum-like structure depends on a temporal order. *Nature* 387:278–281.

Bennett, M.V.L., Aljure, E., Nakajima, Y., and Pappas, G.D. (1963). Electrotonic junctions between teleost spinal neurons: Electrophysiology and ultrastructure. *Science* 141:262–264.

Caputi, A.A., Budelli, R., Grant, K., and Bell, C.C. (1998). The electric image in weakly electric fish: Physical images of resistive objects in *Gnathonemus petersii*. *J. Exp. Biol.* 201:2115–2128.

Douglas, R.H., Eva, J., and Guttridge, N. (1988). Size constancy in goldfish (*Carassius auratus*). *Behav. Brain Res.* 30:37–42.

Dusenbery, D.B. (1992). *Sensory Ecology: How Organisms Acquire and Respond to Information.* New York: W.H. Freeman.

Finger, T.E., Bell, C.C., and Russell, C.J. (1981). Electrosensory pathways to the valvula cerebelli in mormyrid fish. *Exp. Brain Res.* 42:22–33.

Grant, K., Bell, C.C., Clausse, S., and Ravaille, M. (1986). Morphology and physiology of the brainstem nuclei controlling the electric organ discharge in mormyrid fish. *J. Comp. Neurol.* 245: 514–530.

Hall, C., Bell, C., and Zelick, R. (1995). Behavioral evidence of a latency code for stimulus intensity in mormyrid electric fish. *J. Comp. Physiol. A.* 177: 29–39.

Han, V.Z., Grant, K., and Bell, C.C. (2000). Rapid activation of GABAergic interneurons and possible calcium independent GABA release in the mormyrid electrosensory lobe. *J. Neurophysiol.* 83:1592–1604.

Han, V.Z., Bell, C.C., Grant, K., and Sugawara, Y. (1999). The mormyrid electrosensory lobe in vitro: Morphology of cells and circuits. *J. Comp. Neurol.* 404:359–374.

Harder, W. (1968). Die Beziehungen zwischen Elektrorezeptoren, elektrischen Organen, Seitenlinienorganen und Nervensystem bei den Mormyridae (Teleostei, Pisces). *Zeit. Vergl. Physiol.* 59:272–318.

Heiligenberg, W. (1973). Electrolocation of objects in the electric fish *Eigenmannia* (Rhamphichthyidae, Gymnotoidei). *J. Comp. Physiol.* 87:137–164.

Hopkins, C.D. (1988). Neuroethology of electric communication. *Ann. Rev. Neurosci.* 11:497–535.

Hopkins, C.D. (1999). Design features for electric communication. *J. Exp. Biol.* 202:1217–1228.

Huffman, R.F., and Henson, O.W. (1990). The descending auditory pathway and acousticomotor systems: Connections with the inferior colliculus. *Brain Res. Rev.* 15:295–323.

Kalmijn, A.J. (1974). The detection of electric fields from inanimate and animate sources other than electric organs. In: *Handbook of Sensory Physiology* (Fessard, A., ed.), pp. 148–200. Berlin: Springer-Verlag.

Kramer, B. (1990). *Electrocommunication in Teleost Fishes: Behavior and Experiments.* Berlin: Springer-Verlag.

Lissmann, H.W., and Machin, K.E. (1958). The mechanism of object location in *Gymnarchus niloticus* and similar fish. *J. Exp. Biol.* 35:451–486.

Meek, J., Grant, K., Sugawara, Y., Hafmans, T.G.M., Veron, M., and Denizot, J.P. (1996). Interneurons of the ganglionic layer in the mormyrid electrosensory lateral line lobe: Morphology, immunohistochemistry, and synaptology. *J. Comp. Neurol.* 375:43–65.

Meyer, J.H. (1982). Behavioral responses of weakly electric fish to complex impedances. *J. Comp. Physiol.* 145:459–470.

Meyer, J.H., and Bell, C.C. (1983). Sensory gating by a corollary discharge mechanism. *J. Comp. Physiol. A.* 151:401–406.

Mohr, C., and von der Emde, G. (1998). Mapping of the nucleus lateralis (torus semicircularis) of the electrosensory system of *Gnathonemus petersii*. In: New Neuroethology on the Move (Wehner, R., ed.), *Proc. 26th Göttingen Neurobiology Conference 1998*, p. 54. Thieme, Stuttgart, New York.

Moller, P. (1995). *Electric Fishes: History and Behavior*. London: Chapman & Hall.

Montgomery, J.C., and Bodznick, D. (1994). An adaptive filter that cancels self-induced noise in the electrosensory and lateral line mechanosensory systems of fish. *Neurosci. Lett.* 174:145–148.

Nieuwenhuys, R., and Nicholson, C. (1969). A survey of the general morphology, the fiber connections, the possible functional significance of the gigantocerebellum of mormyrid fish. In: *Neurobiology of Cerebellar Evolution and Development* (Llinas, R., ed.), pp. 107–134. Chicago: American Medical Association.

Niso, R., Serrier, J., and Grant, K. (1989). Mesencephalic control of the bulbar electromotor network in the mormyrid *Gnathonemus petersii. Europ. J. Neurosci. Suppl.* 2:176.

Peters, R.C., Loos, W.J.G., Bretschneider, F., and Baretta, A.B. (1999). Electroreception in catfish: Patterns from motion. *Belgian J. Zool.* 129:263–268.

Prechtl, J.C., von der Emde, G., Wolfart, J., Karamürsel, S., Akoev, G.N., Andrianov, Y.N., and Bullock, T.H. (1998). Sensory processing in the pallium of a teleost fish, *Gnathonemus petersii. J. Neurosci.* 18:7381–7393.

Rasnow, B. (1996). The effects of simple objects on the electric field of *Apteronotus. J. Comp. Physiol. A.* 178:397–411.

Roberts, P.D., and Bell, C.C. (1999). Computational consequences of temporally asymmetric learning rules: II. Sensory image cancellation. *J. Comput. Neurosci.* 1–15.

Roberts, P.D., and Bell, C.C. (2000). Computational consequences of temporally asymetric learning rules: II. Sensory image cancelation. *J. Comput. Neurosci.* (in press).

Russell, C.J., and Bell, C.C. (1978). Neuronal responses to electrosensory input in mormyrid valvula cerebelli. *J. Neurophysiol.* 41:1495–1510.

Schwan, H.P. (1963). Determination of biological impedances. In: *Physical Techniques in Biological Research* (Nastuk, W.L., ed.), pp. 323–407. New York: Academic Press.

Schwarz, S., and von der Emde, G. (1999). Object classification by the weakly electric fish, *Gnathonemus petersii.* In: (Eysel, U., ed.), *Göttingen Neurobiology Report, 1999, 27th Göttingen Neurobiology Conference*, p. 332. Thieme, Göttingen.

Schwarz, S., and von der Emde, G. (2001). Distance discrimination during active electrolocation in the weakly electric fish *Gnathonemus petersii. J. Comp. Physiol. A.* 186:1185–1197.

Szabo, T., and Hagiwara, S. (1967). A latency change mechanism involved in sensory coding of electric fish (mormyrids). *Physiol. Behav.* 2:331–335.

Szabo, T., and Wersäll, J. (1970). Ultrastructure of an electroreceptor (Mormyromast) in a mormyrid fish, *Gnathonemus petersii.* II. *J. Ultrastructure Res.* 30:473–490.

Van Essen, D.C., and Gallant, J.L. (1994). Neural mechanisms of form and motion processing in the primate visual system. *Neuron* 13:1–10.

von der Emde, G. (1990). Discrimination of objects through electrolocation in the weakly electric fish, *Gnathonemus petersii. J. Comp. Physiol. A.* 167: 413–421.

von der Emde, G. (1998). Capacitance detection in the wave-type electric fish *Eigenmannia* during active electrolocation. *J. Comp. Physiol. A.* 182: 217–224.

von der Emde, G., and Bell, C.C. (1994). Responses of cells in the mormyrid electrosensory lobe to EODs with distorted waveforms: implications for

capacitance detection. *J. Comp. Physiol. A.* 175: 83–93.
von der Emde, G., and Bell, C.C. (1995). The nucleus prae-eminentialis of mormyrid electric fish: Field potentials, somatotopy and single unit activity. *25th Annual Meeting, Society for Neuroscience*, p. 184. San Diego, CA.
von der Emde, G., and Bleckmann, H. (1997). Waveform tuning of electroreceptor cells in the weakly electric fish, *Gnathonemus petersii. J. Comp. Physiol. A.* 181:511–524.
von der Emde, G., and Bleckmann, H. (1998). Finding food: Senses involved in foraging for insect larvae in the electric fish, *Gnathonemus petersii. J. Exp. Biol.* 201:969–980.
von der Emde, G., and Ringer, T. (1992). Electrolocation of capacitive objects in four species of pulse-type weakly electric fish. I. Discrimination performance. *Ethology* 91:326–338.
von der Emde, G., and Ronacher, B. (1994). Perception of electric properties of objects in electrolocating weakly electric fish: Two-dimensional similarity scaling reveals a City-Block metric. *J. Comp. Physiol. A.* 175:801–812.
von der Emde, G., and Zelick, R. (1995). Behavioral detection of electric signal waveform distortion in the weakly electric fish, *Gnathonemus petersii. J. Comp. Physiol. A.* 177:493–501.
von der Emde, G., Gomez Sena, L., Niso, R., and Grant, K. (2000). The midbrain pre-command nucleus of the mormyrid electromotor network. *J. Neurosci.* 20:5483–5495.
von der Emde, G., Schwarz, S., Gomez, L., Budelli, R., and Grant, K. (1998). Electric fish measure distance in the dark. *Naturwissenschaffen* 395: 890–894.
von Holst, E., and Mittelstaedt, H. (1950). Das Reafferenzprinzip. *Naturwissenschaffen* 37:464–476.
Zakon, H.H. (1987). The electroreceptors: diversity in structure and function. In: *Sensory Biology of Aquatic Animals* (Tavolga, W.N., ed.), pp. 813–850. Berlin, Heidelberg, New York: Springer-Verlag.

6
Processing of Dipole and More Complex Hydrodynamic Stimuli Under Still- and Running-Water Conditions

Horst Bleckmann, Joachim Mogdans, and Guido Dehnhardt

Abstract

As an adaptation to their environment, aquatic animals have developed sophisticated hydrodynamic receptor systems for the detection of water motions. The hydrodynamic receptor system of fishes is the mechanosensory lateral line. The sensory units of the lateral line are the neuromasts that are dispersed over the body surface. Superficial neuromasts are freestanding on the surface of the skin and are sensitive to water velocity. Canal neuromasts are embedded in lateral line canals and respond to pressure gradients between canal pores. The peripheral lateral line responds strongly to sinusoidal water motions generated by a stationary vibrating sphere. In running water, superficial neuromast responses to hydrodynamic stimuli are masked, whereas trunk canal neuromast responses are hardly affected, indicating a clear form-function relationship of the peripheral lateral line. Neurons in the fish brainstem and midbrain are less sensitive to sine wave stimuli but show a variety of responses to moving object stimuli. In the brainstem, a functional subdivision can be found similar to that in the lateral line periphery. These findings show that natural stimulus conditions, for example, moving sources and background noise, are necessary to reveal the functional limitations and evolutionary adaptations of the lateral line system.

1. Introduction

Hydrodynamic sensory systems are widespread in aquatic animals (see review by Bleckmann, 1994). Fishes, for instance, use the lateral line system for the detection of water motions caused by conspecifics, predators, or prey (Bleckmann, 1994). To study the behavioral capabilities and the physiology of the lateral line system, researchers successfully used single-frequency water motions generated with a stationary vibrating sphere (dipole), introduced as a hydrodynamic stimulus by Harris and van Bergeikj (1962) 40 years ago. Dipole stimuli of constant amplitude and frequency can be easily generated and are well defined, both in time and space (Coombs et al., 1989a). Unfortunately natural animal-generated water motions often have ill-defined temporal and spatial distributions and usually show amplitude and frequency fluctuations. For instance, the hydrodynamic trails caused by subundulatory swimming fishes during steady locomotion are irregular in time course and consist of a chain of slightly deformed vortex rings (e.g., Blickhan et al., 1992) that may last for more than one minute (Fig. 6.1 [see color plate] and Hanke et al., 2000). Since animal-induced water motions contain biologically relevant information (Hanke et al., 2000), it is conceivable that hydrodynamic sensory systems like the lateral line are adapted to sense and analyze such water motions.

This article reviews recent data that show that, after four decades of lateral line research using stationary vibrating sphere stimuli in still water, it is now time to study the lateral line with more complex hydrodynamic stimuli or with vibrating sphere stimuli applied in noisy environments. We believe that only if we include more complex stimulus regimes we will be able to fully understand the information processing capabilities of the lateral line and other hydrodynamic sensory systems. To cover the whole potential range of hydrodynamic reception we include recent data about hydrodynamic sensation in seals.

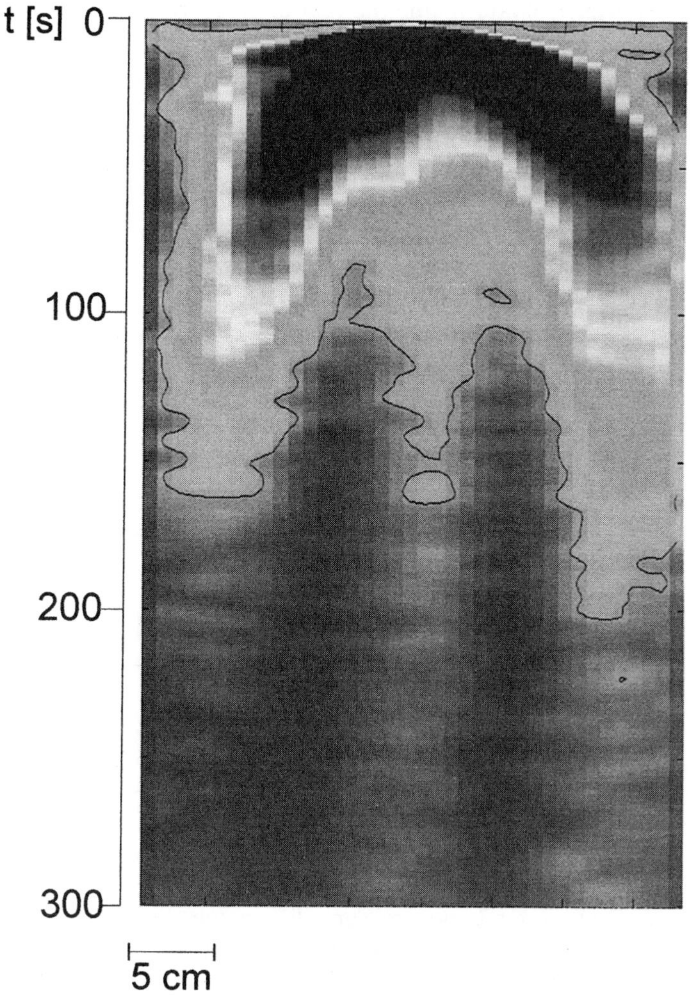

FIGURE 6.1. Color-coded water velocity behind a swimming goldfish averaged over the columns of the camera field. Dark red indicates high water velocities, dark blue indicates water velocities below 0.1 mm/s. The axis of ordinates shows the width of the camera field, the abscissa the time that has passed since the fish entered the camera field. The blue line indicates a water velocity of 0.2 mm/s (for further details see Hanke et al., 2000). (See color plate)

2. Morphology of the Peripheral Lateral Line

The peripheral lateral line of fishes consists of neuromasts that occur freestanding on the skin or in subepidermal canals. A fish may have more than 1,000 neuromasts that are distributed across its entire surface. Neuromasts contain two populations of sensory hair-cells with opposite orientation of their hair bundles. Afferent fibers may innervate hair cells in more than one neuromast (Münz, 1979). However, most likely single fibers innervate only hair-cells of identical orientation (Görner, 1963). The ciliary bundles of lateral line hair cells project into a cupula that connects the hair cells with the water surrounding the fish or with the canal fluid. For a more detailed description of the peripheral lateral line the reader is referred to Coombs et al. (1989b).

3. Physiology of the Lateral Line Periphery

Hair cells are displacement detectors. Displacement of the ciliary bundle toward the kinocilium causes a depolarization, displacement in the opposite direction of hyperpolarization (Kroese and van Netten, 1988). The degree of the response amplitude varies with stimulus angle in a cosine fashion. Due to the orientation of hair cells in a neuromast (see above) and due to the separate innervation of unidirectionally polarized hair cells, neuromasts provide directional information (Kroese and van Netten, 1988).

Under natural conditions, water motions cause the cupula to move, which in turn moves the ciliary bundles of the underlying hair cells. Due to inertial and frictional forces, neuromasts respond within their operating range in proportion to the velocity of the water surrounding the cupula. Thus superficial neuromasts (operating range DC to about 100 Hz) are velocity detectors. Within lateral line canals, fluid flow depends on pressure differences between neighboring canal pores. As a consequence, canal neuromasts respond within their operating range (>DC to about 150 Hz) approximately in proportion to outside water acceleration (Kalmijn, 1988).

3.1. Responses to Water Motions Caused by a Vibrating Sphere

3.1.1. Still-Water Conditions

Primary lateral line afferents show ongoing activity. They respond with high sensitivity and with a sustained and phase-coupled discharge to a stationary sinusoidally vibrating sphere (Fig. 6.2A,B). In a low-noise environment two types of fibers can be distinguished: Type I afferents function as velocity detectors; that is, they have a gain of approximately 6 dB/octave and thus most likely innervate superficial neuromasts. Type II afferents are more sensitive to water acceleration; that is, their gain is about 12 dB/octave (Fig. 6.2C). Type II afferents most likely innervate canal neuromasts (see above). Both types of afferents encode the amplitude of a sinusoidal hydrodynamic stimulus by the degree of phase-coupling and by firing rate (Fig. 6.2D). Tests with amplitude-modulated stimuli have demonstrated that primary lateral line afferents phase lock to both the carrier frequency and the amplitude modulation frequency (Mogdans and Bleckmann, 1999). When stimulated near threshold with a small vibrating sphere, RF of type I and type II lateral line afferents are punctate. This indicates that primary lateral line afferents innervate only a single or, at best, a few neighboring neuromasts.

3.1.2. Running-Water Conditions

If exposed to running water, type I afferents respond with a burst-like increase of ongoing activity. Consequently the responses of type I afferents to a vibrating sphere stimulus are masked (Fig. 6.3, top). In contrast, ongoing activity of type II afferents barely changes if the fish is exposed to unidirectional water flow. Consequently, type II responses are not masked in running water (Fig. 6.3, bottom, and Engelmann et al., 2000). In running water, DC flow immediately drives superficial neuromasts of still-water fishes (*Carassius auratus*) into

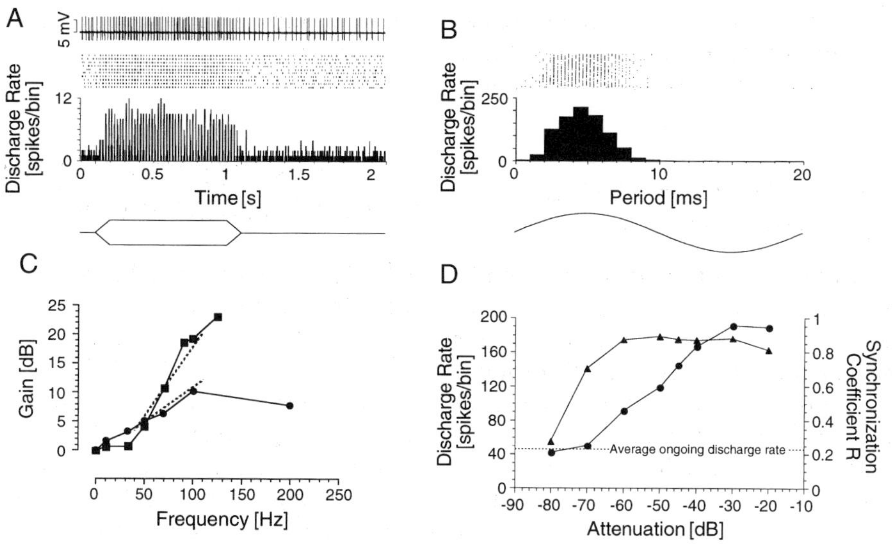

FIGURE 6.2. Characterization of the responses of goldfish posterior lateral line (PLLN) fibers to the water motions caused by a stationary vibrating sphere. (A) Raster diagram and peri stimulus time histogram (binwidth 2 ms) of the responses of a PLLN fiber to 10 repetitions of a 50-Hz stimulus (vibration amplitude 4 µm, sphere diameter 8 mm). *Top trace*: original recording; *bottom trace*: stimulus. (B) Raster diagram of the distribution of spikes within each cycle of the 50-Hz stimulus and the corresponding period histogram. Graph was derived from the data shown in A. (C) Gain of two PLLN fibers plotted as a function of frequency. One fiber (circles) had a gain of about 6 dB/octave; the other fiber (squares) a gain of about 12 dB/octave. Dotted lines indicate slopes of 6 and 12 dB, respectively. (D) Input-output function of a PLLN fiber. Discharge rates (line connecting circles re: left-hand axis) and synchronization coefficients R as a measure of phase coupling (lines connecting triangles re: right-hand axis) are plotted as function of level (in rel. dB). An attenuation of –20 dB corresponds to a peak-to-peak vibration amplitude of 425 µm.

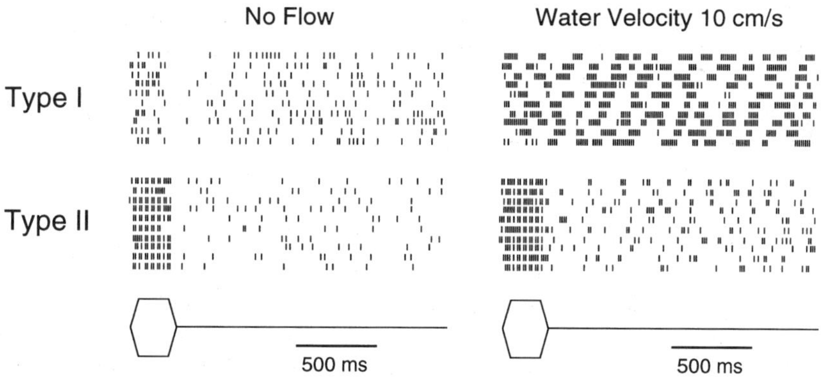

FIGURE 6.3. Raster diagrams of responses of type I and type II fibers in the goldfish posterior lateral line nerve to 10 repetitions of a 50-Hz constant-amplitude sine wave stimulus (100 µm p-p displacement amplitude) generated by a vibrating sphere of 8 mm diameter. Distance between fish and sphere was 6 mm. *Left*: response to the sphere without background water flow. *Right*: response to the sphere in the presence of a 10-cm/s DC background flow. Flow direction was from anterior to posterior. Type I fibers are putatively innervating superficial neuromasts. Note that background flow masks the response to the vibrating sphere. Type II fibers are putatively innervating canal neuromasts. The response to the vibrating sphere is not masked by background flow.

saturation and therefore renders them useless for the detection of water motions like those generated, for example, by swimming zooplankton.

3.2. Responses to Water Motions Caused by Moving Objects

Moving aquatic animals often generate complicated flow patterns (see above). To stimulate the lateral line with more complex water motions, Bleckmann and Zelick (1993) used a small moving object as a stimulus source. A moving object causes changes in water velocity that consist of an initial predictable short transient followed by an ill-defined long-lasting wake (Mogdans and Bleckmann, 1998). Water motions are usually associated with changes in hydrodynamic pressure. However, the pressure changes caused by a moving object are prominent only during the initial transient and are small in the object's wake (Mogdans and Bleckmann, 1998). If the lateral line system is stimulated with an object that passes the fish laterally, again two types of afferents can be distinguished (Mogdans and Bleckmann, 1998): Type I afferents respond with a well-defined single peak of excitation followed by inhibition or vice versa. The response pattern inverses when object motion direction reverses (Fig. 6.4, top). Type I afferents in addition discharge numerous bursts of spikes after the initial response (i.e., after an object has passed the fish). These unpredictable bursts are caused by the wake of the moving object (Mogdans and Bleckmann, 1998). Since the wake contains high-velocity water motions (Mogdans and Bleckmann, 1998) and since superficial neuromasts respond in proportion to water velocity, type I afferents are believed to receive input from superficial neuromasts.

Type II afferents also respond to a moving object with a single peak of excitation followed by inhibition or vice versa. However, type II afferents barely respond after the object has passed the fish (Fig. 6.4, bottom). Since canal neuromasts respond to the spatial difference in the steepness of the pressure gradient along the lateral line canal and since pressure gradients in the wake caused by the moving object are small (Mogdans and Bleckmann, 1998), type II afferents most likely receive input from canal neuromasts.

4. Physiology of the Medial Octavolateralis Nucleus

4.1. Responses of MON Cells to Sine Wave Stimuli

4.1.1. Still-Water Conditions

In bony and cartilaginous fishes, the medial octavolateralis nucleus (MON) is the first site of sensory integration in the ascending lateral line pathway (e.g., McCormick and Braford, 1988). Among MON cells, only a proportion responds with a sustained discharge to single-frequency wave stimuli (Mogdans and Goenechea, 1999). Most cells respond in a phasic-tonic or purely phasic fashion (Coombs et al., 1998; Montgomery et al., 1996). Some MON cells respond with periods of intermittent excitation and inhibition or with a short burst of activity after stimulus offset. In other cells, responses take considerable time to reach a maximal level of discharge. A few cells are inhibited by sine wave (50 Hz) stimuli.

The degree of phase-coupling is highly variable among MON cells. Cells with primary-like responses exhibit a high degree of phase-coupling, whereas cells with responses unlike those of primary afferents may show only weak or no phase-coupling (Coombs et al., 1998). The excitatory RF of MON units range from primary-like to large, covering most of the surface of a fish (e.g., Coombs et al., 1998; Wubbels et al., 1993; S. Kröther, unpublished). This suggests that one population of cells receives input from a restricted portion of the lateral line periphery, whereas other cells get input from neuromasts that are widely distributed across the surface of the fish (Mogdans et al., 1999). The RF of other cells may include regions in which dipole stimulation leads to inhibition of ongoing activity. Anatomical (New et al., 1996), pharmacalogical (D. Fay, and J. New, unpublished), and modeling data (Coombs et al., 1996) are in agreement with the hypothesis that primary-like RF in the medulla are sharpened by neural mechanisms based on lateral inhibition.

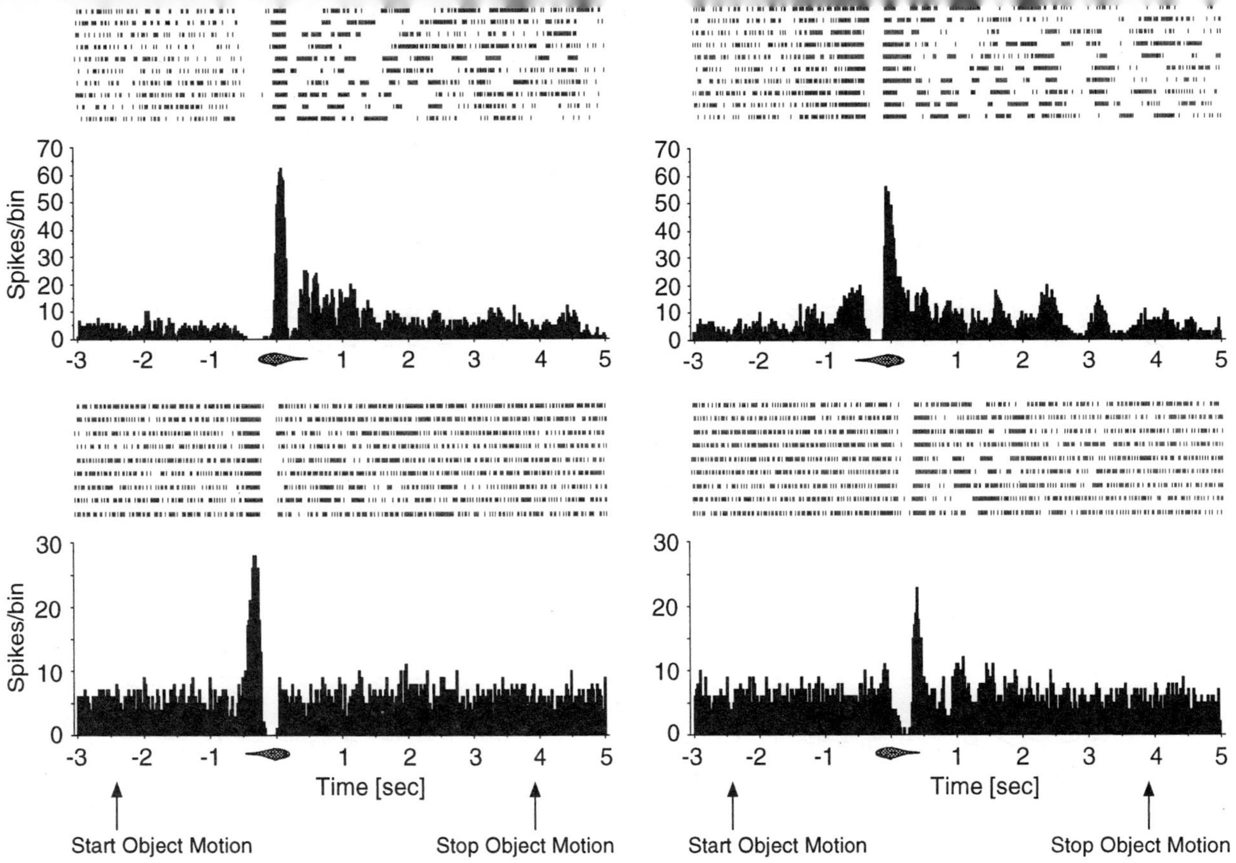

FIGURE 6.4. Responses of goldfish PLLN fibers to an object passing the fish laterally with a speed of 15 cm/sec. Raster diagrams and peri stimulus time (PST) histograms (binwidth 20 msec) of the responses to 10 stimulus presentations are shown. *Upper left and lower right*: motion direction from anterior to posterior. *Upper right and lower left*: motion direction from anterior to posterior. The fish symbol indicates size, location, and orientation of the fish relative to the orbit of the moving object. At zero time, the lateral distance between object and fish was minimal. Negative times refer to the object's approach of the fish. *Top*: example of a unit that responded with a triphasic main response pattern that consisted of inhibition followed by excitation and again inhibition (IEI pattern) in the AP direction. When object motion was in the opposite direction, the main response pattern was inverse; that is, the unit responded with excitation followed by inhibition and again excitation (EIE pattern). The unit continued to fire unpredictable bursts of spikes for a long time after the object had passed along the side of the fish. *Bottom*: example of a unit that responded with a biphasic discharge pattern that consisted of inhibition followed by excitation (IE pattern) in the AP direction. When object motion was in the opposite direction, the main response pattern was inverse; that is, the unit responded with excitation followed by inhibition (EI pattern). Compared to the unit shown above this unit barely responded after the object had passed along the side of the fish.

4.1.2. Running-Water Conditions

If MON units are stimulated with a vibrating sphere in the presence of background flow, at least three types can be distinguished. Type I MON units respond to running water with a change in discharge rate. In running water, the responses of these units to a dipole stimulus are altered such that either response rates or the degree of phase-coupling or both are decreased. Thus type I MON units most likely receive input from type I afferents (i.e., from superficial neuromasts). Type II MON units hardly respond to running water. Moreover, these units do not alter their responses to dipole stimuli under running-water conditions. Type II MON units most likely receive input from canal neuromasts. Type III MON units also do not respond to running water but the response to a dipole presented in background flow is significantly decreased (S. Kröther, unpublished). Most likely excitatory input to type III MON units, mediated via type II primary affernts, is inhibited by input from type I primary afferents.

4.2. Responses of MON Units to Moving Object Stimuli

In goldfish, about 30% of the MON units are insensitive to a stationary dipole stimulus, even if tested with peak-to-peak vibration amplitudes substantially higher (800 μm) than those causing rate saturation in primary lateral line afferents (Mogdans and Goenechea, 1999). Many of these seemingly insensitive cells readily respond to the water motions generated by a moving sphere. The responses of MON cells to moving-object stimuli are highly variable. However, two consistent response types can be distinguished. Many MON units show responses to the passing object and to the water motions in the wake of the object. These cells probably get input from superficial neuromasts. Other MON units also respond to the passing object with excitation; however, these cells do not respond to the wake caused by the object. These cells probably get input from canal neuromasts. In contrast to primary afferents (cf. Fig. 6.4) the responses of both types of MON units are often independent of the direction of object motion both in terms of the temporal discharge pattern and in terms of discharge rate (Mogdans and Goenechea, 1999). Some MON units respond with inhibition to a passing object (Mogdans et al., 1997) (Fig. 6.5).

5. Responses of Toral Units

In bony and cartilaginous fishes most ascending MON efferents terminate in the contralateral torus semicircularis and in the dorsomedial nucleus, respectively, of the midbrain (e.g., Boord and Montgomery, 1989; McCormick, 1989). In cartilaginous fishes, a further midbrain center of lateral line information processing is the anterior nucleus (Bleckmann et al., 1987). Ongoing activity of midbrain lateral line units is low or absent (for review see Bleckmann and Bullock, 1989).

5.1. Responses to Vibrating Sphere Stimuli

Units in the torus semicircularis may be highly sensitive to a vibrating sphere stimulus, often with minimal displacement thresholds in the frequency range 75–150 Hz (Münz, 1989; Bleckmann and Münz, 1990). Toral lateral line units tuned to certain stimulus frequencies are rare but have been found (Müller et al., 1996; Plachta et al., 1999). The temporal responses of different midbrain lateral line units to a dipole stimulus can be fairly different. Units may discharge only a few spikes at stimulus onset or they may show, depending on stimulus amplitude and/or frequency, a phasic discharge followed by an off-response. Other units respond in a nonadapting fashion to constant-frequency stimuli or with a strong suppression of activity at certain frequencies (Bleckmann and Bullock, 1989; Plachta et al., 1999). Toral lateral line units barely phase-couple to sinusoidal water motions. Excitatory RF may be single-peaked and small or multipeaked and large (Plachta et al., 1999). Midbrain lateral line units of goldfish respond well to amplitude-modulated sinusoidal water motions. For carrier frequencies ≥50 Hz and a modulation depth

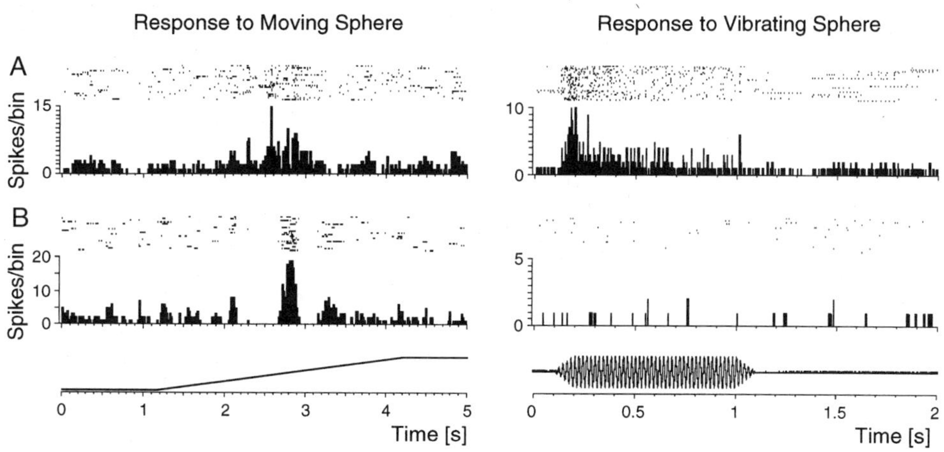

FIGURE 6.5. Responses of medullary lateral line units to a moving sphere (left) and to a stationary vibrating sphere (right). Spike activity over time is illustrated by raster diagrams and PST histograms (binwidths 20ms in left-handed figures and 2ms in right-handed figures). Stimulus traces are shown in the bottom graphs. *Left*: slope indicates time of object motion; *right*: voltage delivered to the vibrator. (A) Example of a unit that responded to both the moving sphere and the vibrating sphere stimulus. (B) Example of a unit that responded to the moving sphere but not to the vibrating sphere. Speed of moving sphere was 10cm/s; motion direction was from anterior to posterior. Vibration frequency of the stationary sphere was 50Hz; p-p displacement amplitudes were 330μm.

≥36%, responses are characterized by periodic decreases and increases in firing rate that correspond to the amplitude modulation frequencies. The large modulation depth necessary to elicit a response shows that toral lateral line units are not especially sensitive to amplitude modulations. Consequently, units tuned to particular amplitude modulation frequencies (test range 4Hz–14Hz) were not found (Plachta et al., 1999). Toral lateral line responses to a vibrating sphere stimulus have not yet been tested under flow-water conditions.

5.2. Responses to Moving Object Stimuli

In goldfish about 10% of all toral lateral line units do not respond to a stationary vibrating sphere. However, many of these units can be driven with water motions caused by a slowly moving sphere (Plachta et al., 1999). Thus, up to the level of the midbrain, there are at least two functionally seperated lateral line pathways, one of which processes the water motions caused by stationary vibrating objects, the other the water motions caused by moving objects. Like the periphery and the MON, the torus semicircularis also contains at least two types of units. Type I toral lateral line units respond with a prominent peak of excitation to a moving object and may continue to fire after the object has passed the fish (e.g., Fig. 6.6A). These units most likely receive input from superficial neuromasts. Type II toral units respond with a single short peak of excitation but not to the water motions in the wake of the moving object (e.g., Fig. 6.6C; i.e., these units most likely receive input from canal neuromasts). Type II toral units are organized systematically in the torus semicircularis; that is, units located in the anterior torus have RF in the anterior part of the fish, whereas successively more caudal units have successively more caudal RF. Some toral lateral line units are inhibited while an object passes the fish (Wojtenek et al., 1998; Plachta et al., 1999). In terms of spike number and temporal response pattern, toral units, like many MON units, are often not directionally sensitive (e.g., Fig. 6.6A). This again indicates that the two populations of hair cells in a neuromast converge onto these units. Other toral units respond only if the object moves in a certain

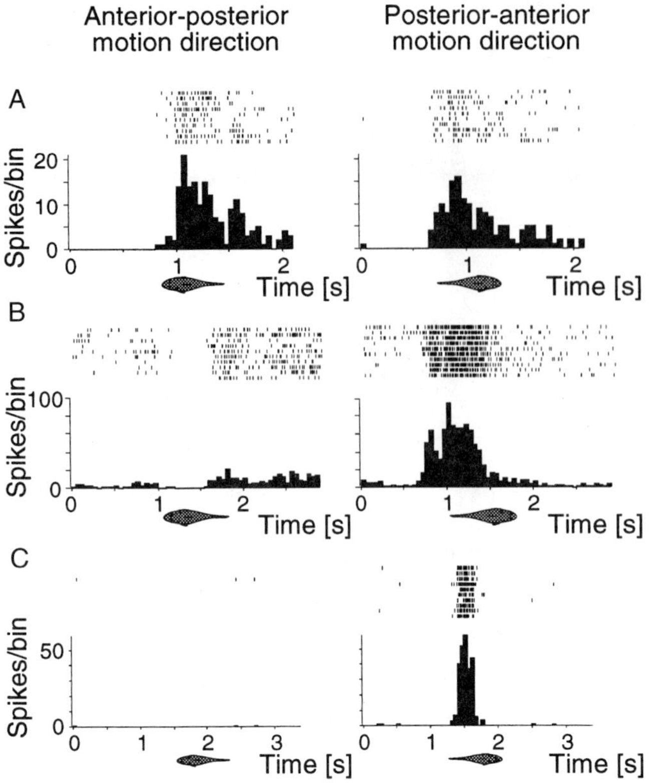

FIGURE 6.6. Effects of direction of object motion on the temporal discharge patterns of lateral line units in the midbrain of goldfish. (A) Example of a unit in which the temporal pattern of the response barely changed when the direction of object motion was reversed. (B) Data from a unit for which the temporal response pattern clearly inversed when motion direction was reversed. Note that the response to PA object motion was characterized by a broad excitatory peak. In contrast, the response to AP object motion shows a broad zone of suppressed neural activity that is followed by a period of increased neural activity. (C) Data from a unit that responded with a sharp peak of excitation to PA object motion but did not respond to AP object motion. Data in A–C were recorded from units in different animals but in response to almost identical stimuli (i.e., object speed was between 15 and 20 cm/s and minimal object distance was 1 cm).

direction (e.g., Fig. 6.6C). In some units, the response pattern inverses if the direction of object motion is reversed; that is, object motion in one direction causes inhibition of ongoing discharge rate (Fig. 6.6B, left) while object motion in the opposite direction causes excitation (Fig. 6.6B, right and Wojtenek et al., 1998).

Even though the differences between the responses of peripheral, medullary, and midbrain lateral line units to a moving object are striking, there is yet no indication that aspects of the source other than motion direction and the rostro-caudal location of a moving object are represented neuronally.

6. Behavior

Since their discovery we have learned a lot about the behavioral capacities of hydrodynamic sensory systems. Fishes, for instance, can be conditioned to respond to water surface waves or to the water motions caused by a stationary vibrating sphere (review Bleckmann, 1994). Using this approach it has been shown that fishes not only can localize a wave source but also can discriminate stimulus amplitude and frequency. Recent experiments show that fishes can also discriminate water motions caused by moving objects; that is, in a two-choice experiment fishes can use lateral line input to discriminate the direction of object motion, object speed, size, and shape (Vogel and Bleckmann, 2001).

Besides fishes, marine mammals may use hydrodynamic information for the detection and localization of prey. One sensory system capable of detecting minute water motions is the facial vibrissae, which, in seals, have a spectral sensitivity tuned to the frequency range of fish-generated water movements (Dehnhardt et al., 1998). Seals lack a sonar system. Nevertheless, even blind seals successfully hunt fish (Newby et al., 1970). Recent experiments (Dehnhardt et al., 2001) demonstrate that a blindfolded seal can detect and follow the hydrodynamic trail generated by a propeller-driven miniature submarine (Fig. 6.7). The seal

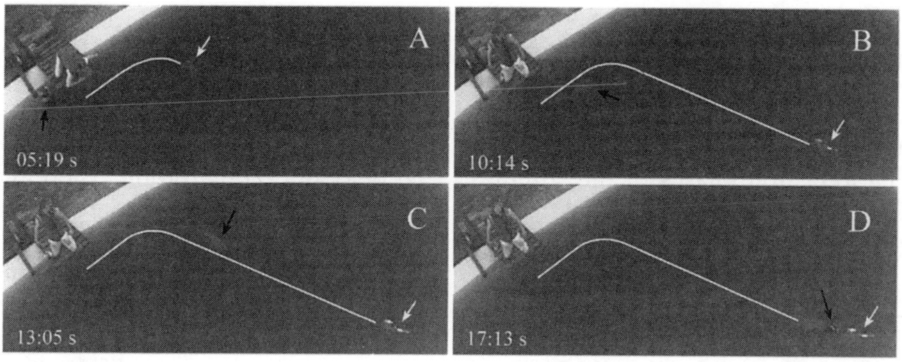

FIGURE 6.7. Example of hydrodynamic trail following by a seal. The figure shows a sequence of single frames taken from video recordings. A submarine was started while the seal's head, supplied with a visually opaque stocking mask and headphones for acoustical masking, was placed in a hoop stationed above the water surface. After the submarine's engine had turned off (running time 3 and 5 s), the headphones were removed and the seal started its search. White line indicates the hydrodynamic trail generated by the submarine. *White arrow*: position of the submarine; *black arrow*: position of harbor seal. In experiments during which whisker movements were impeded by a stocking mask that covered the seal's muzzle (opening of the mouth was still possible for potential gustatory sampling), the seal started its search as usual but always failed to detect the hydrodynamic trail.

followed the hydrodynamic trail and located the submarine in 78% (n = 326) of the trials, even if the submarine's course contained unpredictable changes.

There is evidence that some fishes can also track the hydrodynamic trail caused by swimming prey fishes (Hanke, 2001; Pohlmann et al., 2001). If this is true, the capabilities and the operating range of the lateral line system are much better and larger than originally thought. Further behavioral and physiologigal studies are needed to confirm this.

7. Discussion

In sensory systems other than the lateral line, we repeatedly find (1) a separation of sensory channels, (2) a systematic representation of stimulus parameters along central maps, and (3) the emergence of more and more selective responses to specific, biologically relevant stimuli as one ascends to higher brain levels. Data obtained over the last 15 years have provided some answers to the question to which degree the same holds true for the mechanosensory lateral line.

7.1. Separation of Sensory Channels

The peripheral lateral line consists of two main sensory channels, one of which processes superficial-, the other canal-neuromast information. While superficial neuromasts respond approximately in proportion to water velocity, canal neuromasts respond more in proportion to water acceleration. As one consequence of this functional separation, superficial neuromasts respond well to a vibrating sphere stimulus only under still-water conditions, whereas canal neuromasts respond well in both still and running water (Fig. 6.3 and Engelmann et al., 2000). This finding most likely explains why still-water fishes possess many superficial neuromasts distributed across the whole body surface whereas river fishes, fishes living in turbulent waters, and fast-swimming fishes have only a small number of superficial neuromasts (Dijkgraaf, 1963). Due to the directional sensitivity of hair cells and the fact that each afferent nerve fiber innervates only hair cells of identical orientation, neuromasts give rise to two populations of afferent fibers with opposite directional sensitivity. Thus the peripheral lateral line keeps not only canal and superficial neuromast information separated, but also the information about the direction of water flow with respect to the most sensitive axis of the neuromast innervated. If central lateral line units phase-lock to a sinusoidal stimulus, they usually respond—like primary afferents—only to the compression or rarefaction phases of the stimulus (Bleckmann and Bullock, 1989). This indicates that many central units receive input from hair cells that are aligned in the same direction. Some central lateral line neurons respond to both the compression and rarefaction phases of a stimulus (S. Kröther, unpublished). Units of this type must receive input from hair cells that are aligned in opposite directions.

Irrespective of whether a primary lateral line afferent innervates a superficial or a canal neuromast, it will respond to both the water motions caused by a stationary vibrating sphere and those caused by a moving object. This delineates central lateral line units, many of which separate the information provided by a stationary vibrating sphere from the information provided by a moving sphere (object).

Cells that respond to both a moving object and a stationary vibrating sphere may be part of a channel in which local hydrodynamic events like those generated by planktonic prey (Montgomery, 1989) are analyzed. This channel should be sensitive to punctuate stimuli and should have a high spatial resolution. Whether units that respond exclusively to moving objects are part of a channel in which complex water motions that stimulate large parts of the lateral line periphery are analyzed is not known.

7.2. Central Maps

In all bony fishes and in aquatic amphibians primary lateral line afferents distribute throughout the medulla where fibers from the anterior lateral line nerve are represented ventromedially and fibers from the posterior lateral line nerve are represented dorsolaterally (e.g., Fritzsch et al., 1984; McCormick, 1989; Song and Northcutt, 1991). The projections from the anterior and posterior lines run parallel but do not mix. Moreover, the position of neuromasts is represented in the MON in that the projection of anterior neuromasts lie ventrolateral to the projections of more posterior neuromasts. The same applies to the projections of the posterior lateral line of zebrafish (Alexandre and Ghysen, 1999). Thus, the primary projections of the lateral line of fishes can be doubly somatotopic. There may also be a crude somatotopy with the dorsolateral and ventrolateral surfaces of the trunk represented ventrally and dorsally, respectively, in the MON of some fishes (Song and Northcutt, 1991). There are no indications that superficial and canal neuromasts map differentially in the MON, nor that vertically and horizontally oriented neuromasts are mapped separately (Song and Northcutt, 1991). Whether the two hair cell populations in a neuromast map differentially in the medulla and/or midbrain is also not clear (Fritzsch, 1988).

Recently a midbrain lateral line map has been demonstrated in the torus semicircularis of goldfish. Cells in the anterior torus respond while an object passes the anterior part, while cells in the posterior torus respond while the object passes the posterior part of the fish (Plachta, 2000). Physiological studies in which water surface waves were used to stimulate the lateral line of amphibians (Zittlau et al., 1986; Bartels et al., 1990) have revealed computed lateral line maps that represent the direction of surface wave propagation. Presumably more lateral line maps will be uncovered if we use more natural, complex lateral line stimuli.

7.3. Responses Selective to Specific Stimuli

In the MON, but mainly in the torus semicircularis, cells can be found that respond only if an object moves in a particular direction (e.g., Müller et al., 1996; Wojtenek et al., 1998; Plachta et al., 2001). Other cells, which seem to integrate the water motions across the whole fish body, respond, depending on object motion direction, with inhibition or with excitation (Wojtenek et al., 1998). Other parameters than object direction may also be represented in the central lateral line system. For instance, a certain number of MON and toral units respond neither to a vibrating sphere nor to a moving sphere (Mogdans and Goenechea,

1999; Plachta, 2000). Some units of this type may require a distinct wave pattern in order to respond. This idea is in line with the observation that primary lateral line afferents terminate throughout the MON (e.g., Claas and Münz, 1981; Song and Northcutt, 1991; Alexandre and Ghysen, 1999); that is, many secondary cells probably receive input from primary afferents that innervate neuromasts of different orientation that may be distributed across large portions of the fish (Bleckmann, 1994). This argument is supported by the large RF of many secondary and higher-order lateral line cells (review Bleckmann and Bullock, 1989). It should be noted that a recent anatomical study indicates that lateral line fibers innervating a canal neuromast terminate in a single well-defined field in the MON (New et al., 2000). If this finding can be extended to more than one canal neuromast and to other species, this could mean that the canal neuromast system maintains a high spatial resolution throughout the ascending pathway and thus can be used for object localization. This is in line with the observation that type II toral lateral line neurons, which most likely receive input from canal neuromasts, encode the rostrocaudal location of a moving object (see above) and that canal neuromasts mediate orientation to prey (Coombs et al., 2001).

7.4. Detection Range of Hydrodynamic Receptor Systems

Seals can detect and track the hydrodynamic trail caused by a miniature submarine (Dehnhardt et al., 2001), suggesting that they also can track the wake caused by swimming fish. There are some indications that nocturnal piscivorous fishes also may use hydrodynamic cues for fish tracking (Hanke, 2001; Pohlmann et al., 2001). Succesfully tracking fish-generated wakes not only requires the detection and recognition of a track, but also the ability to determine the swimming direction and identity of the track generator. Information about the age of a track probably also is useful.

Up to now hydrodynamic sensory systems were viewed as systems for close range detection. However, the ability to track fish-generated wakes once again raises the question regarding the detection range of hydrodynamic sensory systems. Under still-water conditions even after 1 minute the hydrodynamic trail caused by a small swimming fish contains information about motion direction and time of passage (Hanke, 2001). Assume that prey fishes like herring swim with $1\,m\,s^{-1}$ (Videler, 1991); theoretically they are still detectable for a predator intercepting the wake after the fishes have covered a distance of about 60 m. This detection range may be even larger for a predator intercepting the wake caused by a large fish or by a school of fishes, which, for instance, can easily consist of 10^5 or 10^7 individuals (Pitcher, 1993). In any case it would surpass the detection range of the sonar system of bottlenose dolphins (Au and Snyder, 1980). More behavioral data are needed to determine the maximum detection range of hydrodynamic sensory systems.

References

Alexandre, D., and Ghysen, A. (1999). Somatotopy of the lateral line projection in larval zebrafish. *Proc. Natl. Acad. Sci. USA* 13:7758–7762.

Au, W.W.L., and Snyder, K.J. (1980). Long-range target detection in open waters by an echolocating Atlantic bottlenose dolphin (*Tursiops truncatus*). *J. Acoust. Soc. Am.* 68:1077–1084.

Bartels, M., Münz, H., and Claas, B. (1990). Representation of lateral line and electrosensory systems in the midbrain of the axolotl, *Ambystoma mexicanum*. *J. Comp. Physiol. A.* 167:347–356.

Bleckmann, H. (1994). Reception of hydrodynamic stimuli in aquatic and semiaquatic animals. In: *Progress in Zoology, Vol. 41* (Rathmayer, W., ed.), pp. 1–115. Stuttgart, Jena, New York: Gustav Fischer-Verlag.

Bleckmann, H., and Bullock, T.H. (1989). Central nervous physiology of the lateral line system, with special reference to cartilaginous fishes. In: *The Mechanosensory Lateral Line: Neurobiology and Evolution* (Coombs, S., Görner, P., and Münz, H., eds.), pp. 387–408. Berlin, Heidelberg, New York, Tokyo: Springer-Verlag.

Bleckmann, H., and Münz, H. (1990). Physiology of lateral-line mechanoreceptors in a teleost with highly branched, multiple lateral lines. *Brain Behav. Evol.* 35:240–250.

Bleckmann, H., and Zelick, R. (1993). The responses of peripheral and central mechanosensory lateral line units of weakly electric fish to moving objects. *J. Comp. Physiol. A.* 172:115–128.

Bleckmann, H., Bullock, T.H., and Jørgensen, J.M. (1987). The lateral line mechanoreceptive mesencephalic, diencephalic, and telencephalic regions in the thornback ray, *Platyrhinoidis triseriata* (Elasmobranchii). *J. Comp. Physiol. A.* 161:67–84.

Blickhan, R., Krick, C., Breithaupt, T., Zehren, D., and Nachtigall, W. (1992). Generation of a vortex-chain in the wake of a subundulatory swimmer. *Naturwissenschaften* 79:220–221.

Boord, R.L., and Montgomery, J.C. (1989). Central mechanosensory lateral line centers and pathways among the elasmobranchs. In: *The Mechanosensory Lateral Line* (Coombs, S., Görner, P., and Münz, H., eds.), pp. 323–340. Neurobiology and evolution. Berlin, Heidelberg, New York, Tokyo: Springer-Verlag.

Claas, B., and Münz, H. (1981). Projection of lateral line afferents in a teleost brain. *Neurosci. Lett.* 23:287–290.

Coombs, S., Braun, C.B., and Donovan, B. (2001). The orienting response of lake michigan mottled sculpin is mediated by canal neuromasts. *J. Exp. Biol.* 204:337–348.

Coombs, S., Fay, R.R., and Janssen, J. (1989a). Hot-film anemometry for measuring lateral line stimuli. *J. Acoust. Soc. Am.* 85:2185–2193.

Coombs, S., Hastings, M., and Finneran, J. (1996). Modeling and measuring lateral line excitation patterns to changing dipole source locations. *J. Comp. Physiol. A.* 178:359–371.

Coombs, S., Janssen, J., and Webb, J.F. (1989b). Diversity of lateral line systems: Evolutionary and functional considerations. In: *The Mechanosensory Lateral Line: Neurobiology and Evolution* (Coombs, S., Görner, P., and Münz, H., eds.), pp. 353–393. Berlin, Heidelberg, New York, Tokyo: Springer-Verlag.

Coombs, S., Mogdans, J., Halstead, M., and Montgomery, J. (1998). Transformation of peripheral inputs by the first-order lateral line brainstem nucleus. *J. Comp. Physiol. A.* 182:609–626.

Dehnhardt, G., Mauck, B., and Bleckmann, H. (1998). Seal whiskers detect water movements. *Nature* 394:235–236.

Dehnhardt, G., Mauck, B., Hanke, W., and Bleckmann, H. (2001). Hydrodynamic trail-following in harbor seals. *Science* 293:102–104.

Dijkgraaf, S. (1963). The functioning and significance of the lateral line organs. *Biol. Rev.* 38:51–106.

Engelmann, J., Hanke, W., Mogdans, J., and Bleckmann, H. (2000). Hydrodynamic stimuli and the fish lateral line. *Nature* 408:51–52.

Fritzsch, B. (1988). The lateral line and inner ear afferents in larval and adult urodeles. *Brain Behav. Evol.* 31:325–348.

Fritzsch, B., Nikundiwe, A.M., and Will, U. (1984). Projection patterns of lateral line afferents in anurans: A comparative HRP study. *J. Comp. Neurol.* 229:451–469.

Görner, P. (1963). Untersuchungen zur Morphologie und Elektrophysiologie des Seitenlinienorgans vom Krallenfrosch (*Xenopus laevis* Daudin). *J. Comp. Physiol. A.* 47:316–338.

Hanke, W. (2001). Hydrodynamische Spuren schwimmender Fische und ihre mögliche Bedeutung für das Jagdverhalten fischfressender Tiere. Doctoral thesis, University of Bonn.

Hanke, W., Brücker, C., and Bleckmann, H. (2000). The aging of water disturbances caused by swimming goldfish. *J. Exp. Biol.* 203:1193–1200.

Harris, G.G., and Bergeijk, W.A. van (1962). Evidence that the lateral line organ responds to near-field displacements of sound sources in water. *J. Acoust. Soc. Am.* 34:1831–1841.

Kalmijn, A.J. (1988). Hydrodynamic and acoustic field detection. In: *The Mechanosensory Lateral Line: Neurobiology and Evolution* (Coombs, S., Görner, P., and Münz, H., eds.), pp. 83–130. Berlin, Heidelberg, New York, Tokyo: Springer-Verlag.

Kroese, A.B.A., and Netten, S.M. van (1988). Sensory transduction in lateral line hair cells. In: *The Mechanosensory Lateral Line: Neurobiology and Evolution* (Coombs, S., Görner, P., and Münz, H., eds.), pp. 265–284. Berlin, Heidelberg, New York, Tokyo: Springer-Verlag.

McCormick, C.A. (1989). Central lateral line mechanosensory pathways in bony fish. In: *The Mechanosensory Lateral Line: Neurobiology and Evolution* (Coombs, S., Görner, P., and Münz, H., eds.), pp. 341–364. Berlin, Heidelberg, New York, Tokyo: Springer-Verlag.

McCormick, C.A., and Braford, M.R. Jr. (1988). Central connections of the octavolateralis system: Evolutionary considerations. In: *The Mechanosensory Lateral line: Neurobiology and Evolution* (Coombs, S., Görner, P., and Münz, H., eds.), pp. 733–756. Berlin, Heidelberg, New York, Tokyo: Springer-Verlag.

Mogdans, J., and Bleckmann, H. (1998). Responses of the goldfish trunk lateral line to moving object. *J. Comp. Physiol. A.* 182:659–676.

Mogdans, J., and Bleckmann, H. (1999). Peripheral lateral line responses to amplitude-modulated sinusoidal wave stimuli. *J. Comp. Physiol. A.* 185: 173–180.

Mogdans, J., and Goenechea, L. (1999). Responses of medullary lateral line units in the goldfish, *Carassius auratus*, to sinusoidal and complex wave stimuli. *Zoology* 102:227–237.

Mogdans, J., Bleckmann, H., and Menger, N. (1997). Sensitivity of central units in the goldfish, *Carassius auratus*, to transient hydrodynamic stimuli. *Brain Behav. Evol.* 50:261–283.

Mogdans, J., Wojtenek, W., and Hanke, W. (1999). The puzzle of hydrodynamic information processing: How are complex water motions analyzed by the lateral line? *European J. Morphol.* 37:195–199.

Montgomery, J.C. (1989). Lateral line detection of planktonic prey. In: *The Mechanosensory Lateral Line: Neurobiology and Evolution* (Coombs, S., Görner, P., and Münz, H., eds.), pp. 561–574. Berlin, Heidelberg, New York, Tokyo: Springer-Verlag.

Montgomery, J., Bodznick, D., and Halstead, M. (1996). Hindbrain signal processing in the lateral line system of the dwarf scorpionfish Scopeana papillosus. *J. Exp. Biol.* 199:893–899.

Müller, H.M., Fleck, A., and Bleckmann, H. (1996). The responses of central octavolateralis cells to moving sources. *J. Comp. Physiol. A.* 179:455–471.

Münz, H. (1979). Morphology and innervation of the lateral line system in *Sarotherodon niloticus* L. (Cichlidae, Teleostei). *Zoomorphology* 93:73–86.

Münz, H. (1989). Functional organization of the lateral line periphery. In: *The Mechanosensory Lateral Line: Neurobiology and Evolution* (Coombs, S., Görner, P., and Münz, H., eds.), pp. 285–298. Berlin, Heidelberg, New York, Tokyo: Springer-Verlag.

New, J., Braun, C.B., and Walter, K. (2000). Central projections of nerve fibers innervating individual canal neuromast organs in the muskellunge, *Esox masquinongy*. *Soc. Neurosci.* 26:146 (Abstract).

New, J.G., Coombs, S., McCormick, C.A., and Oshel, P.E. (1996). Cytoarchitecture of the medial octavolateralis nucleus in the goldfish, *Carassius auratus*. *J. Comp. Neurol.* 366:534–546.

Newby, T.C., Hart, F.M., and Arnold, R.A. (1970). Weight and blindness in harbour seals. *J. Mamm.* 51:152.

Pitcher, T.J. (1993). Functions of schoaling behaviour in teleosts. In: *Behaviour of Teleost Fishes* (Pitcher, T.J., ed.), pp. 363–440. London: Chapman and Hall.

Plachta, D., Mogdans, J., and Bleckmann, H. (1999). Responses of midbrain lateral line units of the goldfish, *Carassius auratus*, to constant-amplitude and amplitude modulated water wave stimuli. *J. Comp. Physiol. A.* 185:405–417.

Plachta, D. (2000). Responses of toral lateral line units of the goldfish, *Carassius auratus*, to dipole and complex water wave stimuli. Doctoral thesis, Bonn.

Pohlmann, K., Grasso, F.W., and Breithaupt, T. (2001). Tracking wakes: The nocturnal predatory strategy of piscivorous catfish. *PNAS* 98:7371–7374.

Song, J., and Northcutt, R.G. (1991). The primary projections of the lateral-line nerves of the Florida gar, *Lepisosteus platyrhincus*. *Brain Behav. Evol.* 37:38–63.

Videler, J.J., and Wardle, C.S. (1991). Fish swimming stride by stride: Speed limits and endurance. *Rev. Fish Biol. Fish* 1:23–40.

Vogel, D., and Bleckmann, H. (2001). Behavioral discrimination of water motions caused by moving objects. *J. Comp. Physiol. A.* 186:1107–1117.

Wojtenek, W., Mogdans, J., and Bleckmann, H. (1998). The responses of midbrain lateral line units of the goldfish *Carassius auratus* to moving objects. *Zoology* 101:69–82.

Wubbels, R.J., Kroese, A.B.A., and Schellart, N.A.M. (1993). Response properties of lateral line and auditory units in the medulla oblongata of the rainbow trout (*Oncorhynchus mykiss*). *J. Exp. Biol.* 179:77–92.

Zittlau, K.E., Claas, B., and Münz, H. (1986). Directional sensitivity of lateral line units in the clawed toad *Xenopus laevis* Daudin. *J. Comp. Physiol. A.* 158:469–477.

7 Information Processing by the Lateral Line System

Sheryl Coombs and Christopher B. Braun

Abstract

This review covers four areas of research that have fruitfully contributed to our understanding of lateral line function within the past 10 years. One striking aspect of the lateral line system is its tremendous diversity. Recent findings, however, indicate a functional constancy that may be maintained by relatively subtle morphological features. Other morphological variations have been shown to enhance sensitivity at particular frequency bandwidths. A second area of research has focused on hydrodynamic imaging and the peripheral patterns of receptor excitation that might encode stimulus features such as amplitude, distance, location, and direction of motion. A detailed model is described and provides several predictions for the types of information passed from the periphery to the central nervous system (CNS). The third topic covered is the mechanisms that enhance signal detection in noisy backgrounds. It is becoming clear that canals act as biomechanical filters to improve signal-to-noise ratios in the presence of low-frequency noises such as uniform, ambient water motions. Two central mechanisms, efferent modulation of receptor excitation and a central dynamic filter mechanism, have been shown to reduce reafference due to self-generated noise and may enhance signal detection in general. The second central mechanism is postulated to be similar to the anti-hebbian learning mechanism that has been well documented within the related electrosensory system. Finally, this review covers the recently documented roles of the lateral line system in natural behaviors, including courtship and prey capture. Some of these recent studies have led to the exciting conclusion that the lateral line may be composed of two distinct information channels, one served by canal and the other by superficial neuromasts, and that each may be dedicated to different behavioral tasks.

1. Introduction

There is probably no other sensory system as specialized for sensory processing in the aquatic environment as the lateral line system. It is, after all, a water current detector found exclusively in aquatic vertebrates (fishes and some amphibians). By its very nature, the lateral line system is generally a close-range system, capable of detecting current-generating sources (e.g., nearby swimming fishes) no more than one or two body lengths away. The lateral line system can also detect ambient water motions, such as those in a stream or ocean current, as well as distortions in ambient or self-generated motions due to the presence of stationary obstacles (e.g., rocks in a stream). As such, the lateral line system has been implicated in a number of different behaviors, including schooling (Partridge and Pitcher, 1980), prey capture (Hoekstra and Janssen, 1985; Bleckmann et al., 1989), courtship and spawning (Satou et al., 1994b), and rheotaxis (Montgomery et al., 1997). In a more general sense, the lateral line system is also undoubtedly used to form hydrodynamic images of the environment, enabling fishes to determine the size, shape, identity, and location of both animate and inanimate entities in their immediate vicinity. Thus, in many ways, the lateral line system shares with other sensory systems the ability to process information not only about external entities of interest in the environment, but also about the consequences of the animal's own movements through the environment.

In this chapter, we highlight four research areas where we think significant advances have been made in the last decade to increase our general understanding of information processing by the lateral line system. We examine (1) the functional consequences of structural diversity in the lateral line periphery, (2) the role of spatial excitation patterns in forming hydrodynamic images of nearby biotic and abiotic entities, (3) peripheral and central nervous system mechanisms for enhancing signal-to-noise ratios, and (4) newly discovered behavioral roles for the lateral line.

2. Structural Diversity and Information Processing: Comparative Approaches

The basic functional unit of the lateral line system is the neuromast, a conglomeration of sensory hair cells and one or more classes of supporting cell. The apical surface of each hair cell contains a directionally sensitive ciliary bundle. All the ciliary bundles within a neuromast are further ensconced in a gelatinous mass, the cupula. Neuromasts are located either superficially on the skin surface, in lines or clusters, or within dermal, fluid-filled canals that open to the surface via a series of pores or tubules (Fig. 7.1A). Any given species might possess from fewer than 100 to thousands of neuromasts, located in various, but stereotypical, positions on the head and trunk (for review of lateral line anatomy, see Coombs and Montgomery, 1999).

2.1. Structural Diversity: Functional Diversity?

One of the advantages of the lateral line system as a model of sensory processing is the diversity in its peripheral morphology across primitively aquatic vertebrates. Neuromasts themselves are reasonably conservative organs, showing variation in a few easily measured variables, such as size, shape, and hair cell number. These and other factors, including cupular size and shape, canal width and shape, and canal fluid viscosity, can be expected to affect neuromast response properties (Denton and Gray, 1988, 1989; van Netten and Kroese, 1989). The most dramatic aspect of lateral line variation, though, is the diversity of secondary structures (such as canals, pits, or mounds that flank each neuromast) that act to filter input to the cupula and underlying neuromast. Within gnathostome fishes, the primitive condition is the presence of both canal lines on the head and trunk, and lines of superficial neuromasts on the head and perhaps trunk. Within any given modern family, however, there is tremen-

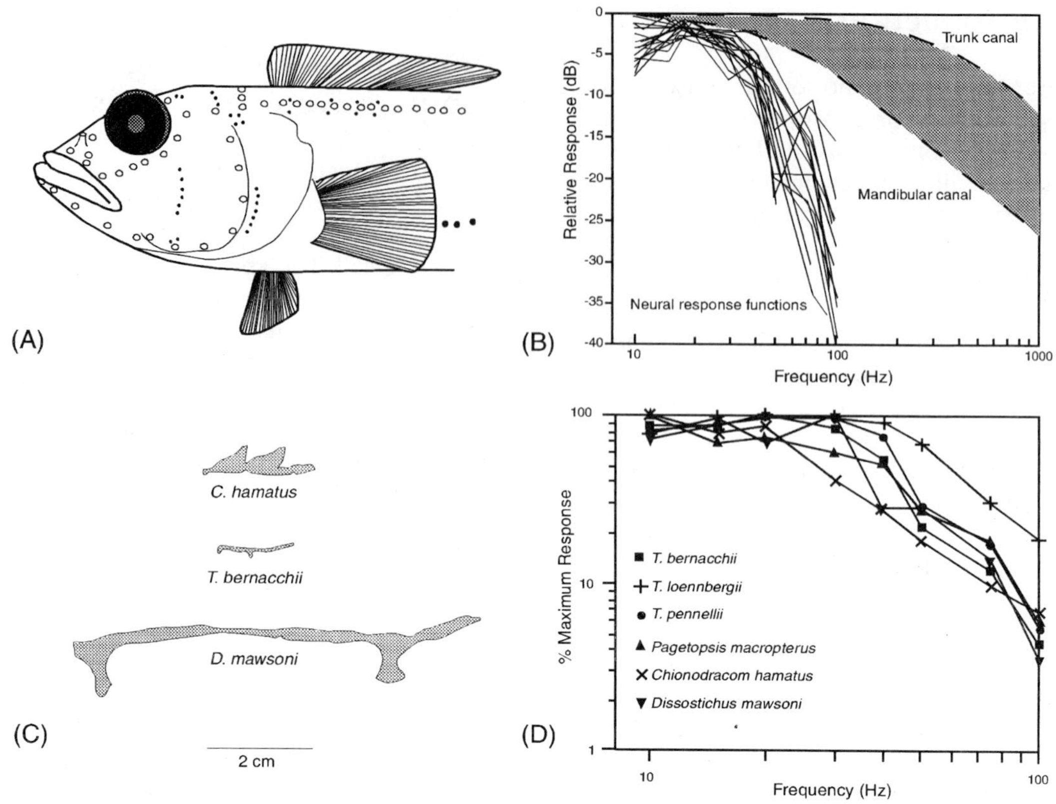

FIGURE 7.1. (A) Schematic illustration of the lateral line system of *Trematomus bernacchii*. Neuromasts are found superficially (solid circles) and within canals that open to the surface through pores (open circles). (B) Neural response functions (thin lines) of fibers innervating mandibular, preopercular, and trunk canal neuromasts. The modeled response properties for canals with dimensions equal to the mandibular or trunk canals are shown as dashed lines. (C) Canal morphology of three notothenioids. Note the thinned section of canal in *D. mawsoni*. (D) Comparison of canal neuromast response properties within the Notothenioidei. (All illustrations are redrawn from Coombs and Montgomery, 1994b.)

dous variation in the relative development of canal or superficial systems. Further, the lateral line can encompass the full range of variation within an individual animal, with a range of neuromast sizes and shapes at different locations, both canal and superficial organs, regional variation in canal and/or pore size, and so on. As predicted by biophysical models (Denton and Gray, 1983), canal and superficial neuromasts differ greatly in their response properties (Münz, 1989; Kroese and Schellart 1992) and the types of stimuli to which they respond. Accordingly, the diversity in lateral line morphology has long been assumed to reflect the information processing needs of individual species (Dijkgraaf, 1963).

In recent years workers have attempted to address the functional outcome of this variability in several ways, with mixed success. Unfortunately, most comparisons have been primarily correlative and hampered by a reliance on speculation. Still, modern comparative studies have at least begun to examine variability in a phylogenetic context and among closely related species, a tradition begun by Jakubowski (e.g., Jakubowski, 1966). Recent examples include Vischer's (1990) study of cottid fishes, Honkanen's (1993) study of sticklebacks, Braun's (1995) study of zoarcoids, Marshall's (1996) description of deep-sea fishes, and Macdowall's (1997) study of galaxiids. These studies all document examples of variability that suggest specializations for information processing or dealing with environmental noise (i.e., different levels of turbulence or types of ambient water motions), but none have progressed beyond speculation. Further work is needed to model functional outcomes of variables like branching patterns in canal tubules or neuromast spacing. These models would generate testable hypotheses of functional specializations that could then be addressed by behavioral comparisons. Studies that examine ecologically and anatomically diverse, but closely related species (e.g., Vischer, 1990; Braun, 1995) can suggest exemplary test cases and relevant comparisons. Only by integrating modern comparative approaches (e.g., Harvey and Pagel, 1991) with performance-based behavioral comparisons can we begin to understand the role of ecological adaptation in sensory information processing.

Two studies have directly examined behavioral performance in relation to morphological variation within two teleost families. In these studies, Janssen and colleagues found significant differences between prey detection ranges and predatory approach behaviors that could be correlated with differences in canal morphology in two percid (Janssen, 1997) and two cottid (Janssen et al., 1999) species. Those species with larger canal diameters and pore sizes detected prey at greater distances, perhaps as a consequence of a greater sensitivity to low frequencies provided by enlarged canals (see Section 4.1). Janssen and colleagues argue that both signal-sensitivity, as inferred from prey detection ranges, and self-generated noises, as inferred from different predatory search and approach behaviors (e.g., a slow, constant-velocity glide vs. a rapid, accelerative lunge), were each matched to the signal-to-noise processing capabilities of the different lateral line canal morphologies in these species.

2.2. Structural Diversity: Functional Constancy?

The relationship between morphology and the particular information processing needs of a given species is complex and many other factors might obscure or override the effect of any single factor (e.g., neuromast size) in the final measures of sensory performance. This complexity is clearly demonstrated by a series of papers on form-function relationships in a monophyletic lineage (suborder Notothenioidei) of antarctic fishes (Coombs and Montgomery, 1992, 1994a; Montgomery et al., 1994; reviewed in Coombs and Montgomery, 1994b). In these papers, several aspects of lateral line morphology were measured at different body locations and in several species of notothenioids. These measurements were used to estimate the relationship between external water motions and neuromast response, following the mathematical models of Denton and Grey (1983, 1988) for canal fluid motions and van Netten and Kroese (1987, 1989) and van Netten (1991) for cupular mechanics. These predictions were then compared to neurophysiological measures of the responses of the same neuromasts. The models based on the biophysical filtering properties of the canal and friction-coupling of the cupula were quite poor at predicting neuromast response properties above ~30 Hz (Fig. 7.1B).

In spite of large differences in canal morphologies and neuromast sizes, all canal neuromast fibers had essentially identical frequency response functions, regardless of which canal they innervated. Similar results were found for superficial neuromast fibers in the same species, despite large differences in the sizes of superficial neuromasts (Coombs and Montgomery, 1994a). These investigators concluded that other filtering mechanisms, such as temperature-dependent, membrane-kinetic properties that limit hair cell responsiveness at higher frequencies, are as important as neuromast size and canal morphology. Much of the differences in canal morphology or neuromast size must therefore be nonadaptive, at least with respect to frequency responsiveness.

A similar result was also found by Mohr and Bleckmann (1998) in a study of lateral line afferent physiology in the surface-feeding topminnow, *Aplocheilus lineatus*. In spite of the obvious importance of frequency analysis in localizing surface wave sources (see Bleckmann et al., 1989), the lateral line periphery in this species contained only two differently tuned channels. These two classes of afferents were found to innervate all the neuromasts examined, and there was no relationship found between neuromast size or shape and frequency responsiveness. While there may be important reasons for preserving functional constancy within a species (e.g., for accurate encoding of pressure-gradient patterns, see Section 3), there are fewer reasons to think that functional constancy will prevail between species when there may actually be differences in information processing needs. Continuing their research on antarctic fishes, however, Montgomery et al. (1994) showed that even in the most extreme case, frequency response curves of canal neuromast fibers in *Dissostichus mawsoni*, the species with the largest canal diameters, were very similar to those from *Tremtomus bernacchii*, a species with very narrow canals (Fig. 7.1C,D). These investigators argue that there may be a single conserved feature (i.e., narrow canal diameter in the vicinity of the neuromast in *D. mawsoni*) (Fig. 7.1C) that underlies the functional constancy. In this light, it is interesting to note that constrictions or ridges near the neuromasts have been reported in several species (e.g., as described for *Notopterus* by Coombs et al., 1988), and may be a mechanism of constraining canal function in light of other conflicting selective factors, such as reduced dermal bone mass to increase buoyancy.

These studies suggest that functional conclusions drawn from morphological data may be severely limited and that morphological diversity may not always reflect functional differences. Downstream filtering mechanisms may have an equal or greater influence on system responsiveness than canal or neuromast size alone. Further, small, perhaps less obvious morphological features, such as a narrow region or a constricting ridge within the canal, might also play a relatively large role in shaping lateral line responsiveness.

3. Hydrodynamic Imaging of Moving and Stationary Sources

There is now mounting evidence to suggest that while superficial neuromasts are important to behaviors like rheotaxis, canal neuromasts are important to behaviors like prey capture and obstacle avoidance, in which animals must determine the location of current-generating or current-distorting sources (see summary of evidence in Section 5.2). In this section, we review recent modeling studies that have provided us with new insights into how spatial excitation patterns along lateral line canals might encode different kinds of information (e.g., location) about both current-generating and current-distorting sources.

As a spatial array of sensors, the lateral line canal system is capable of encoding, in fine spatial resolution and at a single instant in time, the steep pressure gradients associated with incompressible flows in the near-field of a moving source. That is, the response of each canal neuromast to fluid motions inside the canal is proportional to the pressure difference between the two surrounding canal pores divided by the interpore distance (i.e., the net acceleration between the fish and the surrounding water at each canal segment). Although pressure-gradient patterns arising from a moving source with respect to a stationary fish differ from those produced by a moving fish with respect to a stationary source, the principal is the same. Relative movement between the fish and a nearby source will produce information-rich pressure-gradient patterns (Denton and Gray, 1983; Gray, 1984; Hassan, 1985, 1992a,b, 1993; Gray and Best, 1989; Coombs et al., 1996, 2000b).

3.1. Encoding Source Distance

Nearly all of what we know about hydrodynamic imaging of current-distorting sources comes from Hassan's very elegant models of expected flow patterns along the surface of a fish-shaped body as it glides past obstacles of different sizes and shapes (e.g., cylinder; Hassan, 1985). In the last decade, Hassan has refined his modeling approach to compute the distribution of pressure differences between pores of an idealized lateral line canal system (Hassan, 1992a,b, 1993), thereby approximating excitation patterns along the lateral line canal system. When these distributions are plotted as a function of distance between the fish and a planar surface for canals at different body locations (i.e., above and below the eye, along the lower jaw, down the check, and along the trunk) (Fig. 7.2), two things become quite clear. One is that the polarity of the pressure-difference distribution along the fish changes from positive pressure differences on the head to negative pressure differences along the trunk. In other words, canal fluids are flowing in one direction inside head canals and in the opposite direction inside trunk canals. This partly reflects the fact that as a fish moves through the water, pressure is built up in front of the advancing fish and is reduced in the rear. The other emergent principle is that as the fish gets closer to the planar surface, the maximum pressure difference gets larger and the slopes of these pressure-difference distributions over the sensory surface get steeper. Interestingly, current-distorting sources of a different kind (i.e., conductive and nonconductive sources in an electric field) produce similar, distance-dependent changes in the spatial patterns of potential (voltage) differences across the skin surface of weakly electric fishes (Rasnow and Bower, 1997; Assad et al., 1999). Moreover, pressure-difference patterns along the mechanosensory lateral line arising from sources that move relative to a stationary fish also show distance-dependent changes in maximum amplitudes and slopes; this principle holds for both vibrating sources at a single location (Coombs et al., 1996) (Fig. 7.3A) and nonvibrating sources that change their location by moving at a constant velocity (Montgomery and Coombs, 1998).

These striking similarities for different types of current-distorting and current-generating sources suggest that both lateral line and electrosensory systems may be using a novel mechanism for distance perception that depends on a single, two-dimensional array of sensors, using maximum amplitude and/or slope of the excitation pattern along the array. If only one of these cues were used, a high-amplitude source that is far away would not be distinguished from a low-amplitude source that is nearby (Fig. 7.3B). Apparently fishes are not confused by source amplitude, however, as blinded mottled sculpin are able to accurately determine the distance of a source randomly varying in vibration amplitude (Janssen and Corcoran, 1998). Von der Emde and colleagues suggested that weakly electric fish might solve similar ambiguity problems by relying on the slope-to-amplitude ratio—the only combination of cues found to provide unambiguous information about source distance (von der Emde et al., 1998; see also Chapter 5). They demonstrated

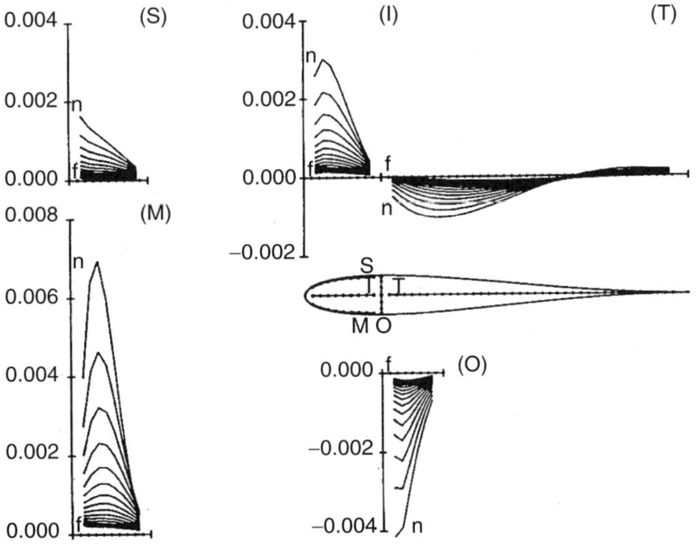

FIGURE 7.2. Expected pressure differences between adjacent pore pairs on the supraorbital (S), infraorbital (I), preopercular (O), mandibular (M), and trunk (T) canals as a fish-like body glides above a plane surface from near (n) and far (f) distances—in this case, from 0.01 body lengths (top curve) to 0.1 body lengths (bottom curve).

that weakly electric fishes actually appear to use this cue by showing that fishes erroneously judged spheres, which have smaller slope-to-amplitude ratios than most other objects, as being further away than cubes at the same distance. The slope-to-amplitude ratio also provides unambiguous information to the lateral line system about the distance of vibrating sources (Fig. 7.3C). Furthermore, this ratio predicts that (1) the precision with which fishes can determine source distance will decrease as source distance increases and that (2) the effective range over which source distance can be determined will be independent of source amplitude (Fig. 7.3C).

3.2. Encoding Source Location and Direction of Movement

Excitation or pressure-difference distributions contain information not only on the distance of the source but also on its direction of movement and location with respect to the long axis of the canal. Information about direction of movement is contained in the spatial distribution and/or time-course of pressure-difference polarities. As a vibrating source accelerates in one direction, pressure is increased above ambient levels in front of the advancing source and reduced in the rear (top panel, Fig. 7.4A). As a consequence, the pressure-difference distribution along the length of the fish and along a single idealized canal consists of a central positive peak corresponding to the rostro-caudal location of the source and two smaller, negative pressure-difference peaks on either side (bottom panel, Fig. 7.4A). This means that fluid flow inside the canal is in one direction in the center of the canal and in opposite directions on either side (see arrows in bottom panel of

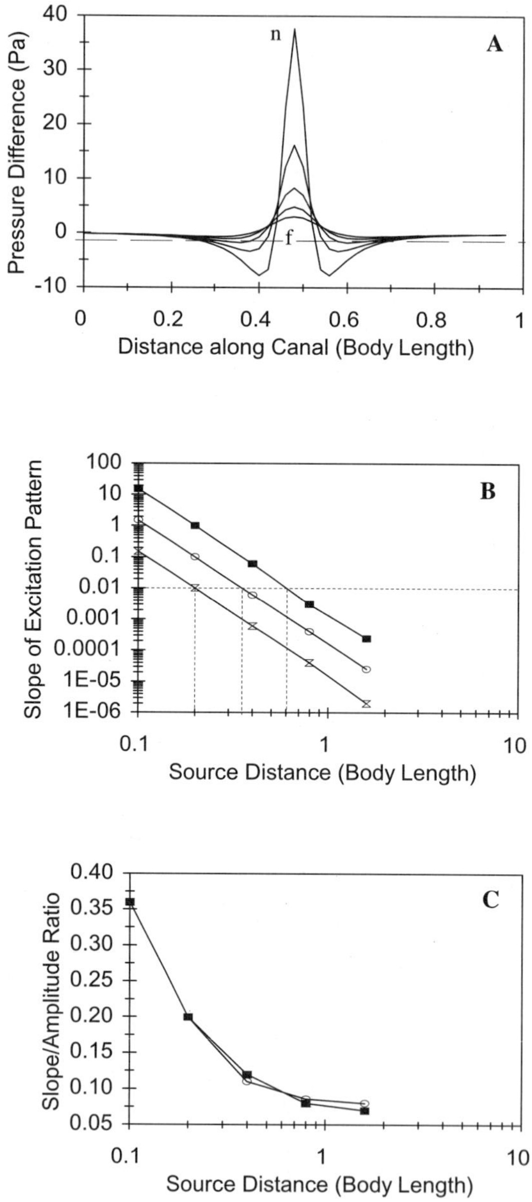

FIGURE 7.3 (A) Pressure-difference distributions along an idealized canal (as in Fig. 7.4A) for different source distances, from near (n) (0.06 body lengths) to far (f)(0.14 body lengths). (B) Maximum slope of pressure-difference distribution as a function of source distance and vibration amplitude: 0.1 m/sec (double triangles), 1 m/sec (open circles), and 10 m/sec (filled squares). The ambiguous nature of slope information is illustrated with the horizontal dashed line showing that a given slope of 0.01 corresponds to different source distances (vertical dashed lines), depending on source amplitude. (C) Slope/amplitude ratio as a function of source distance and vibration amplitude. Note that functions for different source amplitudes all lie on top of one another, illustrating that this ratio provides an unambiguous cue for source distance.

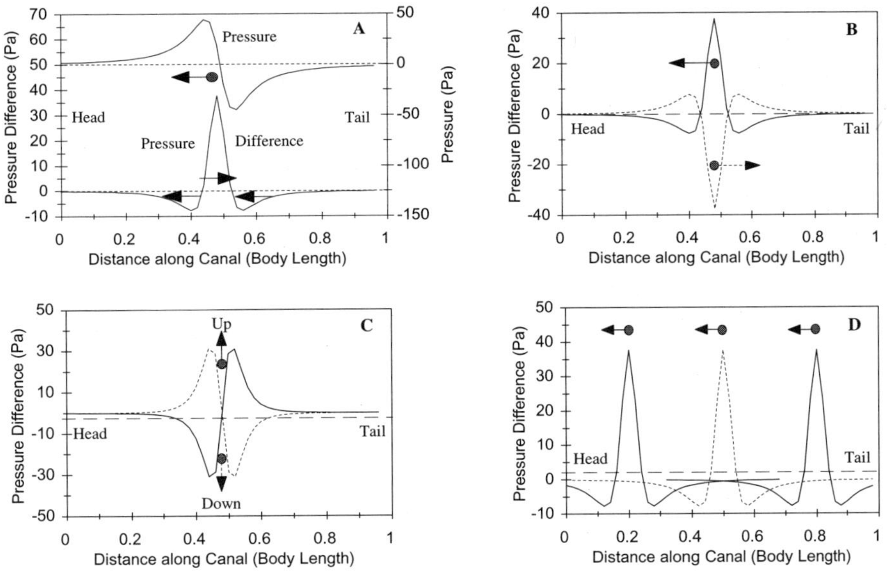

FIGURE 7.4. (A) Pressure (top, right-hand axis) and pressure-difference (bottom, left-hand axis) distributions along a single, idealized lateral line canal determined from dipole flow field equations for a small (radius = 0.03 body lengths) vibrating sphere that is 0.06 body lengths away from the canal. Sphere is centered on the canal and axis of sphere vibration is parallel to the long axis of the canal. Pressure-difference calculations are based on interpore distances of 0.02 body lengths. Actual pressures and pressure-differences are based on calculations for a fish 10 cm long and for a 50-Hz source velocity of 1 m/sec (pk-pk). (B) Pressure-difference distributions for a vibration axis that is parallel to the long axis of the canal. (C) Pressure-difference distributions for a vibration axis that is orthogonal to the long axis of the canal. Solid lines represent headward (B) and upward (C) directions of movement, whereas dashed lines represent tailward (B) and downward (C) directions of movement. (D) Pressure-difference distributions as a function of source location during a headward phase of sinusoidal vibration.

Fig. 7.4A). When the source moves in the opposite direction, it produces a negative pressure-difference peak in the center and positive pressure-difference peaks in the surround (dashed line, Fig. 7.4B). A vertically oriented axis of vibration produces a different set of alternating patterns (Fig. 7.4C).

Because information about source location is contained in the relative location of pressure-difference maxima along the length of the canal, changes in the relative location of these maxima over time might also provide information about the speed of movement and the overall direction in which a vibrating source may be moving (Fig. 7.4D). Nonvibrating sources moving at constant velocities also produce direction- and velocity-dependent effects in the spatiotemporal patterns of pressure-difference amplitudes and polarities (Montgomery and Coombs, 1998). Although the location of pressure-difference maxima (e.g., Fig. 7.4A,C) may encode information about the relative location of a vibrating source, this information may be confounded by the axis of vibration relative to the long axis of the canal. A vertically vibrating sphere, for example, will produce two pressure-difference maxima, alternating in polarity over time, along a rostro-caudally oriented canal (Fig. 7.4C). These maxima correspond to locations along the fish that are either at the headward or tailward side of the source, but not directly at the source. In this case, the steepest part of the pressure-difference distribution, or the pressure-difference minimum, corresponds to the location of the source. It may very well be that as with source distance, source location requires information about both the maximum amplitude and slope of the excitation pattern. It is also worth noting that most fishes possess lateral line canals that are oriented in three dimensions and that tend to converge near a single location at the back of the head: the rostro-caudally oriented trunk canal, the latero-medially directed supratemporal canal, and the dorso-ventrally oriented preopercular canal. The extent to which and the way in which pressure-difference distributions may be combined across these differently oriented canals to represent source location is unknown, but comparisons of excitation patterns along all three axes may very well help to resolve any ambiguities associated with vibration axis.

The picture that is beginning to emerge from these modeling studies is that spatial excitation patterns contain two bits of information in the form of pressure-difference amplitudes and pressure-difference directions or polarities. While it is quite clear from physiology experiments that peripheral lateral line fibers encode both pressure-difference amplitudes and directions associated with different types of source movement (Coombs et al., 1996; Montgomery and Coombs, 1998; see also Chapter 6), we do not yet know exactly how these two bits of information are combined and used by the central nervous system (CNS) to determine different source attributes. However, a comparison of spatial excitation patterns across fibers would make it theoretically possible for the system to extract nearly instantaneous information about many different source attributes, including distance, location, and direction of movement—a truly marvelous feat!

4. Extracting Signals from Noise

Extracting information from biologically relevant signals in the presence of background noises is a problem faced by all sensory systems. For the lateral line system, there are both biotic and abiotic sources of water movements that could be regarded as either signals or noises, depending on the behavioral context. For example, abiotic flows (e.g., streams, tidal currents) might function as signals in evoking rheotactic behavior or as noises that interfere with the fish's ability to detect other current-generating sources (e.g., prey). Likewise, biotic flows (e.g., the wake of a swimming fish) might represent biologically significant signals (e.g., potential prey, predators, or mates) to other fishes, while at the same time representing unwanted noise to the fish doing the swimming. The same self-generated flow that may be a contaminating noise in one context (e.g., interfering with prey detection) might also serve as part of a useful signal in another context (e.g., enabling the detection of current-distorting sources).

Given that the definition of signal or noise depends on behaviorally labile contexts, it is not surprising that there are a number of potential strategies and mechanisms that fishes might use to reduce or filter out different kinds of noises. These range from relatively simple biomechanical mechanisms at the level of the sensory organ to complex behaviors. One example of a simple static mechanism for reducing self-generated noise is the placement of lateral line end organs away from moving body parts (e.g., fins) (Dijkgraaf, 1963). At the other end of the continuum, fishes might employ behaviors (e.g., don't move, stop breathing) to eliminate the problem of self-generated noise. Indeed, many sit-and-wait predators are likely to improve their signal-to-noise ratios for detecting prey while at the same time decreasing the prey's signal-to-noise ratio for detecting them. Here we review new insights on the role of canals as biomechanical filters and two recently discovered dynamic filter mechanisms in the CNS for suppressing unwanted sensory reafference.

4.1. Canals as Biomechanical Filters

Although it has been known for quite some time that the amplitudes of fluid motions inside lateral line canals are altered relative to those outside the canal (Denton and Gray, 1988, 1989), the function and possible biological significance of canals as biomechanical filters in improving signal-to-noise ratios has been less well appreciated and understood. This is primarily because the filter properties of the canal (or rather our view of them) depends on whether the amplitude of water motion inside and outside the canal is expressed in terms of displacement, velocity, or acceleration. There is substantial theoretical and empirical evidence in support of viscous (velocity) forces as the primary stimulus to superficial neuromasts and inertial (acceleration) forces as the effective stimulus to canal neuromasts (Denton and Gray 1983, 1989; Kroese and Schellart, 1992; Montgomery et al., 1994), suggesting that the appropriate frame of reference should depend on the type of sense organ in question (Kalmijn, 1988, 1989). With respect to acceleration, canals operate as low-pass filters, with fluid velocities inside the canal (the effective stimulus to the neuromast) being equal for frequencies up to around 100 Hz (Fig. 7.5A). While the acceleration frame of reference gives us clear insight into the types of signals most likely to activate canal neuromasts (e.g., rapid movements), it may also obscure the function of canals as filters to reduce unwanted low-frequency noises (Montgomery et al., 1994). That is, with respect to velocity, canals operate as high-pass filters to reduce the effects of low-frequency motions, such as those likely to be generated from slow ventilatory movements (~1–3 Hz) or slow, ambient water motions (Fig. 7.5B). Indeed, recent behavioral experiments have demonstrated that mottled sculpin exhibit canal-mediated orienting responses to a 50-Hz vibration amplitude as low as 0.0001 cm/sec in the presence of unidirectional flows up to 8 cm/sec (Kanter and Coombs, 2000). In light of these behavioral results and recent physiology results in which responses of superficial, but not canal, neuromast fibers to a 50-Hz vibrating sphere were diminished in the presence of a 10–15 cm/sec ongoing flow (Engelmann et al., 2000; see also Chapter 6), these results show a clear advantage of canal neuromasts over superficial neuromasts for detecting high-frequency AC signals in the presence of low-frequency ambient water motions.

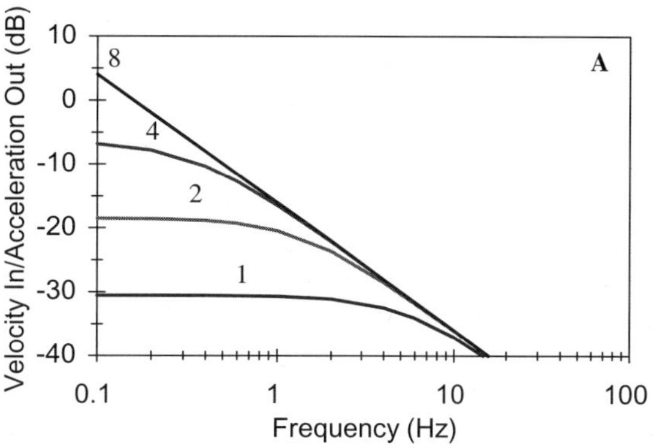

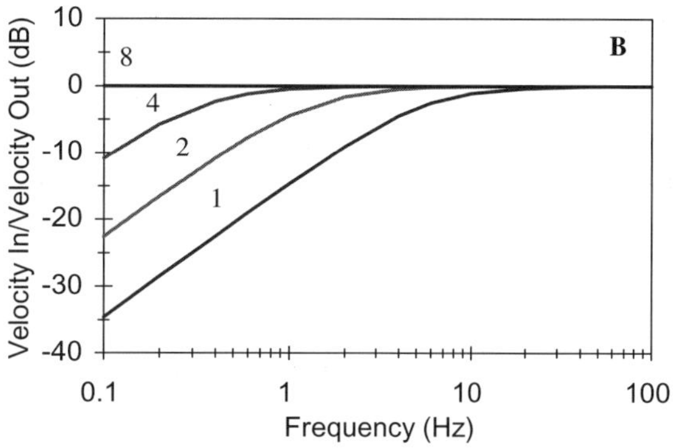

FIGURE 7.5. Fluid velocity inside the canal relative to fluid acceleration (A) and velocity (B) outside the canal. Functions modeled after the equations of Denton and Gray (1988, 1989) for rigid-walled tubes 1, 2, 4, and 8 mm in diameter. Fluid motion inside the canal is expressed as velocity because neuromast (cupula) responses are driven by viscous forces and are thus proportional to the velocity of fluid motion inside the canal.

4.2. Dynamic Filters in the CNS

In recent years, two new and very different types of dynamic neural mechanisms for suppressing unwanted sensory reafference (self-generated noise) have emerged in the CNS (see also Chapter 20). One involves the octavolateralis efferent nucleus (OEN) located in the brainstem and its inhibitory modulation of peripheral activity in hair cells and their afferent fibers by descending efferent fibers (Tricas and Highstein, 1991). The second mechanism involves modulation of CNS activity in second-order cells located in the medial octavolateralis nucleus (MON), the first-order nucleus in the brainstem (Montgomery and Bodznick, 1994; Bodznick et al., 1999).

In an elegant, but difficult set of experiments, Tricas and Highstein (1991) demonstrated that the activation of the octavolateralis efferent system by the visual system could be best understood by the arousing properties of novel stimuli and their ability to activate feedforward mechanisms for reducing responses to different kinds of noises, including those caused by ambient substrate vibrations and the fish's own breathing. These investigators recorded the activity of primary lateral line afferent and efferent fibers in chronically implanted free-swimming toadfish. Among other things, they showed that afferent activity was robustly modulated by respiratory gill movements and that this modulated activity was significantly, but only transiently reduced when the efferent system was activated by stroboscopic illumination. Similar reductions in afferent activity occurred when the toadfish was presented with live prey. So in this instance, at least, activation of the efferent system appears to be closely associated with arousal behaviors and the onset of novel stimuli, rather than protection from overstimulation (as was the prevailing view, e.g., Roberts and Meredith, 1989). Its main function may be to enhance the signal-to-noise processing capabilities of the lateral line system when arousing signal sources, like predators or prey, are at hand.

As the work of Tricas and Highstein (1991) clearly showed, afferent activity in primary lateral line afferents can be rather robustly modulated by ventilatory movements in toadfish. Spontaneous activity in lateral line afferents of the dwarf scorpionfish is also strongly modulated by ventilation, but the activity of second-order neurons (crest cells) in the MON of the same species is not (Montgomery et al., 1996). Given that lateral line afferents provide the primary input to the MON and crest cells provide the primary output, the question is, How is the ventilatory-modulated afferent input to the MON canceled at the level of the crest cells? The answer is that there is a dynamic filter mechanism in the molecular layer of the medial nucleus that enables crest cells to learn how to selectively ignore expected signals (Montgomery and Bodznick, 1994; Bodznick et al., 1999). Unlike the transient and immediate modulation of sensitivity by the efferent system caused by unexpected events in the context of arousal, this filter takes time (several minutes) to build and functions in the context of expected events. What is particularly intriguing and remarkable about this filter mechanism, apart from the fact that it is adaptive, is that it appears to be a generalized form of the modifiable efference copy mechanism first discovered in the electrosensory lateral line lobe (ELL) of weakly electric mormyrid fishes and now also recognized in first-order electrosensory nuclei of elasmobranchs and weakly electric gymnotids (see review by Bell et al., 1997). Moreover, the circuitry that underlies this mechanism may be generalized to other sensory systems with cerebellar-like structures (e.g., the dorsal cochlear nucleus (DCN), of mammals) (Montgomery et al., 1995; Bell et al., 1997).

5. Lateral Line Information Processing and Its Role in Natural Behaviors

The preceding sections of this chapter discuss information processing and its mechanisms in the lateral line system. It is instructive, then, to conclude by examining exactly what sorts of information are normally processed by animals in their habitat, and determining how that information is used to improve an animal's chance of survival or reproduction. In other words, how does the lateral line system impact the lives of the animals that possess it? Can it be used to find prey, avoid predators, navigate, communicate with conspecifics, or contribute to an overall perception of environmental features? Like many aspects of lateral line biology, these questions were well addressed by Dijkgraaf, nearly 40 years ago (Dijkgraaf, 1963). More recent advances in lateral line ethology have been covered in recent reviews (Bleckmann et al., 1989; Coombs and Montgomery, 1999). In the present context, we will confine ourselves to findings that relate to information processing, and how lateral line information contributes to behaviors. In this light, the most significant new findings can be divided into two categories: (1)

studies investigating the importance of lateral line information and its integration with other modalities in behaviors that depend on multiple sensory systems and (2) new investigations into the functional distinctions between canal and superficial neuromasts.

5.1. Contributions of the Lateral Line System to Multisensory Tasks

At first glance, it is reasonable to assume that animals use as many information channels as possible to gather information from the environment. Thus animals may be using lateral line information channels in concert with auditory, visual, and chemosensory cues. The majority of lateral line behavioral studies have concentrated on prey capture and feeding behaviors in lateral line specialists. More recently, however, workers have begun to consider the role of the lateral line system as a single channel among many that contribute to feeding behaviors, especially in more generalized fish species.

For instance, the Chinese perch (*Siniperca chuatsi*) studied by Liang et al. (1998) is a diurnal predator, but after monitoring prey capture success in these fishes when all sensory systems were intact and after selective inactivation of different sensory systems one at a time, these investigators found that fishes consumed statistically similar numbers of prey under intact conditions, after removal of the eyes, or with their lateral lines inactivated. Although loss of both vision and lateral line senses severely impaired hunting success, loss of one system alone did not have an overall significant effect. In another diurnal predator, green sunfish (*Lepomis cyanellus*), Janssen and Corcoran (1993) showed that attack trajectories to capture a visible prey item were dependent on the position of a nearby water jet, indicating that lateral line information actually overrides vision in the animal's final estimation of prey location. In a similar, but more detailed and controlled set of experiments, New and his coworkers (New and Kang, 2000; New et al., 2001) showed that muskellunge (*Esox musquinongy*) have a two-phase predatory sequence, consisting of an initial, visually mediated stalking phase from far distances, followed by lateral line- and visually mediated strike behaviors at short distances. These interactions between vision and the lateral line system are especially interesting in light of Tricas and Highstein's (1991) findings that visual activation of the lateral line efferent system by prey causes transient inhibition of neuromast afferent activity and may help to improve signal-to-noise processing capabilities. (See Section 4). The exact functional outcome(s) of visually mediated modulation of lateral line activity is still unclear, but it is obvious that many behaviors will have multisensory interactions.

It has generally been well appreciated that courtship and mating behaviors rely on multisensory information, particularly olfaction, vision, and hearing. It has also been well known that fishes often incorporate courtship dances and vibratory motions in their mating and spawning rituals (Tinbergen, 1951), yet the water motions created by these vibrations and movements have not received much attention as an information channel served by the lateral line system in coordinating reproduction (Sargent et al., 1998). Satou and his coworkers have provided the first demonstration that vibratory motions are part of an information channel used to coordinate spawning in salmon (Satou, 1987; Satou et al., 1987, 1991), and that this information channel is served by the lateral line system (Satou et al., 1994b). Further, successful coordination of spawning behaviors requires both visual and lateral line inputs (Takeuchi et al., 1987; Satou et al., 1994a). These studies are the first to implicate the lateral line in communicative behaviors, but it is likely that this is a taxonomically more widespread function of the lateral line system and deserves further study.

5.2. Do Canal and Superficial Neuromasts Comprise Distinct Information Channels?

Perhaps the most exciting development in recent lateral line research has been the emerging awareness of the relative importance of canal and superficial neuromasts to different behaviors. A distinct contribution of canal neuromasts to prey capture behavior, for example, has been supported by studies in which the accuracy (Janssen, 1996) or frequency (Coombs et al., 2000a) of the initial prey-orienting response was diminished after mechanical or pharmacological blocking of lateral line canal neuromasts, but not after physical destruction of superficial neuromasts. Even 10 Hz stimuli, well within the bandwidth of superficial neuromasts, produced similar results (Coombs et al., 2000a). Thus, although stimuli produced by moving prey may have greatest energy at low frequencies (e.g., Montgomery and MacDonald, 1987) and would appear to be more effective in stimulating superficial neuromasts, all evidence indicates that canal rather than superficial neuromasts mediate orienting behaviors. Canal obstruction also severely limits spatial discrimination performance by blind cave fish (Abdel-Latif et al., 1990). Thus, it is quite likely that canal neuromasts contribute to several behaviors in addition to orienting responses associated with prey capture behavior.

If superficial neuromasts do not contribute to orienting behaviors in the context of feeding, what then is their behavioral role? According to the recent findings of Montgomery et al. (1997), superficial neuromasts play a crucial role in rheotaxis to slow currents. Since the early work of Lyon (1904), it has been widely assumed that the sensory basis of rheotaxis is primarily visual and tactile (Arnold, 1974). Animals use the shift in the visual scene to determine their direction and speed of motion, or in extreme cases, their tactile senses can be called into play if their displacement brings them into contact with the substrate or other objects. Benthic fishes, however, are not displaced by low-energy currents, so are deprived of visual cues to current direction and speed. Yet many benthic fishes, as well as blind cave fishes, show strong tendencies to orient to currents, including those too weak to displace the fishes. Using ablation of either the entire lateral line system or selective destruction of canal and superficial neuromasts, Montgomery and colleagues have shown that, in blind or benthic (at least in the context of their experiments) fishes, superficial neuromasts are essential for rheotactic behaviors but destruction of canal neuromasts alone does not affect rheotactic thresholds. (Montgomery et al., 1997; Baker and Montgomery, 1999a,b).

It is interesting to note, particularly in light of the multisensory interactions described above, that addition of a chemical stimulus decreased the lateral line mediated threshold for rheotaxis in blind cave fishes (Baker and Montgomery, 1999a). Whether this increase in sensitivity is mediated centrally or by peripheral mechanisms is unknown, but chemical senses and hydrodynamic senses are inexorably linked. Diffuse chemical cues have a low information content on their own but as they are dispersed by hydrodynamic structures, information on source strength and direction is added (see Weissburg, 2000). Our understanding of information processing by both the lateral line and chemosensory systems is thus likely to be greatly enhanced by paying closer attention to hydrodynamic structure and the interactions between these sensory systems.

It is widely appreciated that canal and superficial neuromasts respond to stimuli quite differently, primarily due to the biophysics of canals (see Section 4.1). It is possible that these biophysical distinctions alone are enough to account for the behavioral distinctions mentioned above, but other intrinsic features of these two neuromast types might also contribute to delimiting their information processing roles and distinct behavioral contributions. One feature that differs between canal and superficial neuromasts is their pattern of afferent innervation. In at least some species, canal neuromasts each receive a unique set of afferents (each afferent fiber innervates only a single neuromast), but afferents that contact superficial neuromasts have divergent branching patterns (each afferent fiber innervates multiple neuromasts; Münz, 1989). On this basis alone, one might expect the canal neuromast system to be better suited to behavioral tasks that require conservation of spatial information. Such tasks might include source localization, hydrodynamic imaging, and perhaps communicative behaviors. The superficial neuromast system, on the other hand, would be better suited to behaviors requiring more spatial integration across the body surface, such as orienting to bulk water currents (rheotaxis; cf. Montgomery et al., 1997).

6. Conclusion

As this selective review demonstrates, the past 10 years of lateral line research have provided new insights on how information is processed by the lateral line system. We are beginning to understand how the lateral line contributes to an animal's perceptions of its environment and to the initiation and coordination of different behaviors. Several questions still remain, of course, and there are several areas where we think future studies will add to this understanding. For instance, we still need more information on how secondary structures, like canals, papillae, or pits affect the response properties of neuromasts. Future modeling studies should examine how each of these structures in combination with associated variables (e.g., internal canal shape, canal fluid viscosity, tubule branching, pore size and/or number) alters the fluid motions in the vicinity of the neuromast and the pattern of fluid motions across neuromast arrays. Modeled expectations must also be confirmed by measuring distal water motions outside the structures with respect to proximal water motions in the vicinity of the neuromast and by measuring afferent fiber responses in species with different types of secondary structures. With these combined approaches, a number of different questions could be addressed. For example, how do these various structures, or regional variations in them, affect the spatial resolution or sensitivity of the system? Does the relative development of canal and superficial systems reflect a compromise between conserving spatial information and preserving a reasonable signal-to-noise ratio in a given hydrodynamic environment? We also need more information on the ambient noise characteristics of different habitats and on the hydrodynamic signals created by different biotic sources (e.g., Bleckmann et al., 1991; Hanke et al., 2000).

Gentamicin susceptibility studies (Song et al., 1995) have shown differences between canal and superficial neuromast hair cells. Is there a functional significance to this hair cell heterogeneity, and if so, how does it affect information processing? The lateral line may be an ideal system for approaching general questions in hair cell diversity and function in all octavolateralis systems. Given the potential richness of information in the excitation pattern of the lateral line system, we are still quite ignorant of the use to which that information is put, and how it is processed by the CNS. Can animals discriminate between sources that are moving in different directions, at different speeds, or that are at different distances? Difficult behavioral experiments are needed to determine exactly what information about a source is actually available to animals.

The finding that the lateral line is involved in intraspecies communication (see above) opens new vistas in lateral line research. Where else might vibratory motions be used as a communication channel between fishes? Substrate vibrations, generated with slapping motions by benthic fishes, might also be an important communication channel. We know that lateral line information is required to localize substrate sources in the context of prey capture behavior (Janssen, 1990), but might it also be used in the context of intraspecific communication?

Perhaps the most fruitful area for future research is the possibility that there are two functionally distinct subsystems within the lateral line: one that receives input from superficial neuromasts and the other that receives input from canal neuromasts. What is the relationship between these two information streams? Is information from these two subsystems integrated in the brain, and if so, where? Clearly much work remains to be done, but we expect future research into the functional submodalities within the lateral line system, and the other issues raised above, to greatly enhance our understanding of how the lateral line system provides animals with information about their biotic and abiotic surroundings and about the consequences of their own interactions with the environment.

References

Abdel-Latif, H., Hassan, E.S., and Campenhausen, C. von (1990). Sensory performance of blind cave fish after destruction of the canal neuromasts. *Naturwissenschaften* 77:237–239.

Arnold, G.P. (1974). Rheotropism in fishes. *Biol. Rev.* 49:515–576.

Assad, C., Rasnow, B., and Stoddard, P.K. (1999). Electric organ discharges and electric images during electrolocation. *J. Exp. Biol.* 202:1185–1193.

Baker, C.F., and Montgomery, J.C. (1999a). The sensory basis of rheotaxis in the blind Mexican cave fish, *Astyanax fasciatus*. *J. Comp. Physiol. A.* 184:519–527.

Baker, C.F., and Montgomery, J.C. (1999b). Lateral line mediated rheotaxis in the antarctic fish *Pagothenia borchgrevinki*. *Polar. Biol.* 21:305–309.

Bell, C.C., Bodznick, D., Montgomery, J.C., and Bastian, J. (1997). The generation and subtraction of sensory expectations within cerebellum-like structures. *Brain Behav. Evol.* 50:17–31.

Bleckmann, H., Tittel, G., and Blübaum-Gronau, E. (1989). Lateral line system of surface-feeding fish: Anatomy, physiology and behavior. In: *The Mechanosensory Lateral Line: Neurobiology and Evolution.* (Coombs, S., Görner, P., and Münz, H., eds.), pp. 501–526. New York: Springer-Verlag.

Bleckmann, H., Breithaupt, T., Blickhan, R., and Tautz, J. (1991). The time course and frequency content of hydrodynamic events caused by moving fish, frogs, and crustaceans. *J. Comp. Physiol. A.* 168:749–757.

Bodznick, D., Montgomery, J.C., and Megan, C. (1999). Adaptive mechanisms in the elasmobranch hindbrain. *J. Exp. Bio.* 202:1357–1364.

Braun, C.B. (1995). Ecomorphological studies of lateral line systems: Phyletic versus ecological effects. *Am. Zool.* 35:16A.

Coombs, S., and Montgomery, J.C. (1992). Fibers innervating different parts of the lateral line system of an antararctic notothenioid, *Trematomus bernacchii*, have similar frequency responses, despite large variation in the peripheral morphology. *Brain Behav. Evol.* 40:217–233.

Coombs, S., and Montgomery, J.C. (1994a). Function and evolution of superficial neuromasts in an antarctic notothenioid fish. *Brain Behav. Evol.* 44:287–298.

Coombs, S., and Montgomery, J.C. (1994b). Structural diversity in the lateral line system of antarctic fish: Adaptive or non-adaptive? *Sensornye Sistemy* (Sensory Systems) 8:42–52.

Coombs, S., and Montgomery, J.C. (1999). The enigmatic lateral line system. In: *Comparative Hearing: Fishes and Amphibians* (Popper, A.N., Fay, R.R. eds.), pp. 319–362. New York: Springer-Verlag.

Coombs, S., Braun, C.B., and Donovan, B. (2000a). Orienting response of Lake Michigan mottled sculpin is mediated by canal neuromasts. *J. Exp. Biol.* (in press).

Coombs, S., Finneran, J., and Conley, R. (2000b). Hydrodynamic imaging by the lateral line system of the Lake Michigan mottled sculpin. *Phil. Trans. Roy. Soc. (Lond.)* 355:1111–1114.

Coombs, S., Hastings, M., and Finneran, J. (1996). Modeling and measuring lateral line excitation patterns to changing dipole source locations. *J. Comp. Physiol.* 178:359–371.

Coombs, S., Janssen, J., and Webb, J.C. (1988). Diversity of lateral line systems: Evolutionary and functional considerations. In: *Sensory Biology of Aquatic Animals* (Atema, J., Fay, R.R., Popper, A.N., and Tavolga, W.N., eds.), pp. 553–594. New York: Springer-Verlag.

Denton, E.J., and Gray, J.A.B. (1983). Mechanical factors in the excitation of clupeid lateral lines. *Proc. Roy. Soc. Lond. B.* 218:1–26.

Denton, E.J., and Gray, J.A.B. (1988). Mechanical factors in the excitation of the lateral line of fishes. In: *Sensory Biology of Aquatic Animals* (Atema, J., Fay, R.R., Popper, A.N., and Tavolga, W.N., eds.), pp. 595–618. New York: Springer-Verlag.

Denton, E.J., and Gray, J.A.B. (1989). Some observations on the forces acting on neuromasts in fish lateral line canals. In: *The Mechanosensory Lateral Line: Neurobiology and Evolution* (Coombs, S., Görner, P., and Münz, H., eds.), pp. 229–246. New York: Springer-Verlag.

Dijkgraaf, S. (1963). The functioning and significance of the lateral-line organs. *Biol. Rev.* 38:51–105.

Englemann, J., Hanke, W., Mogdans, J., and Bleckmann, H. (2000). Hydroynamic stimuli and the fish lateral line. *Nature* 408:51–52.

Gray, J. (1984). Interaction of sound pressure and particle acceleration in the excitation of the lateral-line neuromasts of sprats. *Proc. R. Soc. Lond. B.* 220:299–325.

Gray, J.A.B., and Best, A.C.G. (1989). Patterns of excitation of the lateral line of the ruffe. *J. Mar. Biol. Assn. UK* 69:289–306.

Hanke, W., Brücker, C., and Bleckmann, H. (2000). The ageing of the low-frequency water disturbances caused by swimming goldfish and its possible relevance to prey detection. *J. Exp. Bio.* 203: 1193–1200.

Hassan, E.-S. (1985). Mathematical analysis of the stimulus for the lateral line organ. *Biol. Cybern.* 52:23–36.

Hassan, E.-S. (1992a). Mathematical description of the stimuli to the lateral line system of fish derived from a three-dimensional flow field analysis. I. The case of moving in open water and of gliding towards a plane surface. *Biol. Cybern.* 66:443–452.

Hassan, E.-S. (1992b). Mathematical description of the stimuli to the lateral line system of fish derived from a three-dimensional flow field analysis. II. The case of gliding alongside or above a plane surface. *Biol. Cybern.* 66:453–461.

Hassan, E.-S. (1993). Mathematical description of the stimuli to the lateral line system of fish, derived from a 3-dimensional flow field analysis. III. The case of an oscillating sphere near the fish. *Biol. Cybern.* 69:525–538.

Harvey, P.H., and Pagel, M.D. (1991). *The Comparative Method in Evolutionary Biology.* Oxford: Oxford University Press.

Hoekstra, D., and Janssen, J. (1985). Non-visual feeding behavior of the mottled sculpin, *Cottus bairdi*, in Lake Michigan. *Environ. Biol. Fishes* 12:111–117.

Honkanen, T. (1993). Comparative study of the lateral-line system of the three-spined stickleback (*Gasterosteus aculeatus*) and the nine-spined stickleback (*Pungitius pungitius*). *Acta Zool. (Stockholm)* 74:331–336.

Jakubowski, M. (1966). Cutaneous sense organs of fishes. III. The lateral line organs in some Cobitidae. *Acta Biol. (Krakow)* 9:71–84.

Janssen, J. (1990). Localization of substrate vibrations by the mottled sculpin (*Cottus bairdi*). *Copeia* 1990:349–355.

Janssen, J. (1996). Use of the lateral line and tactile senses in feeding in four antarctic nototheniid fishes. *Environ. Biol. Fishes* 47:51–64.

Janssen, J. (1997). Comparison of response distance to prey via the lateral line in the ruffe and yellow perch. *J. Fish Biol.* 51:921–930.

Janssen, J., and Corcoran, J. (1993). Lateral line stimuli can override vision to determine sunfish strike trajectory. *J. Exp. Biol.* 176:299–305.

Janssen, J., and Corcoran, J. (1998). Distance determination via the lateral line in the mottled sculpin. *Copeia* 1998:657–662.

Janssen, J., Sideleva, V., and Biga, H. (1999). Use of the lateral line for feeding in two Lake Baikal sculpins. *J. Fish Biol.* 54:404–416.

Kalmijn, A.J. (1988). Hydrodynamic and acoustic field detection. In: *Sensory Biology of Aquatic Animals* (Atema, J., Fay, R.R, Popper, A.N., and Tavolga, W.N., eds.), pp. 83–130. New York: Springer-Verlag.

Kalmijn, A.J. (1989). Functional evolution of lateral line and inner ear sensory systems. In: *The Mechanosensory Lateral Line: Neurobiology and Evolution* (Coombs, S., Görner, P., and Münz, H., eds.), pp. 187–216. New York: Springer-Verlag.

Kanter, M., and Coombs, S. (2000). Lateral-line mediated detection of artificial prey in the presence of background flow by Lake Michigan mottled sculpin (*Cottus bairdi*). *Neurosci. Abst.* 26:146.

Kroese, A.B.A., and Schellart, N.A.M. (1992). Velocity- and acceleration-sensitive units in the trunk lateral line of the trout. *J. Neurophysiol.* 68:2212–2221.

Liang, X.F., Liu, J.K., and Huang, B.Y. (1998). The role of sense organs in the feeding behavior of Chinese perch. *J. Fish Biol.* 52:1058–1067.

Lyon, E.P. (1904). On rheotropism: I. Rheotropism in fishes. *Am. J. Physiol.* 12:149–161.

MacDowall, R.M. (1997). An accessory lateral line in some New Zealand and Australian galaxiids (Teleostei: Galaxiidae). *Ecol. Freshw. Fish.* 6:217–224.

Marshall, N.J. (1996). The lateral line systems of three deep-sea fish. *J. Fish Biol.* 49 (supplement A): 239–258.

Mohr, C., and Bleckmann, H. (1998). Electrophysiology of the cephalic lateral line of the surface-feeding fish *Aplocheilus lineatus*. *Comp. Biochem. Physiol.* 119A: 807–815.

Montgomery, J.C., and Bodznick, D. (1994). An adaptive filter that cancels self-induced noise in the electrosensory and lateral line mechanosensory systems of fish. *Neurosci. Lett.* 174:145–148.

Montgomery, J.C., and Coombs, S. (1998). Peripheral encoding of moving sources by the lateral line system of a sit-and-wait predator. *J. Exp. Biol.* 201(1):91–102.

Montgomery, J.C., and Macdonald, J.A. (1987). Sensory tuning of lateral line receptors in Antarctic fish to the movements of planktonic prey. *Science* 235:195–196.

Montgomery, J.C., Baker, C.F., and Carton, A.G. (1997). The lateral line can mediate rheotaxis in fish. *Nature* 389:960–963.

Montgomery, J., Bodznick, D., and Halstead, M. (1996). Hindbrain signal processing in the lateral line system of the dwarf scorpionfish *Scorpaena papillosus*. *J. Exp. Biol.* 199:893–899.

Montgomery, J.C., Coombs, S., and Janssen, J. (1994). Form and function relationships in lateral line systems: Comparative data from six species of antarctic notothenioid fish. *Brain. Behav. Evol.* 44:299–306.

Montgomery, J.C., Coombs, S., Conley, R.A., and Bodznick, D. (1995). Hindbrain sensory processing in lateral line, electrosensory and auditory systems: A comparative overview of anatomical and functional similarities. *Audit. Neurosci.* 1:207–231.

Münz, H. (1989). Functional organization of the lateral line periphery. In: *The Mechanosensory Lateral Line: Neurobiology and Evolution* (Coombs, S., Görner, P., and Münz, H., eds.), pp. 285–298. New York: Springer-Verlag.

New, J.G., and Kang, P.Y. (2000). Multimodal sensory integration in the strike-feeding behavior of predatory fishes. *Phil. Trans. Roy. Soc. (Lond.) B.* 355:1321–1324.

New, J.G., Fewkes, L.A., and Khan, A.N. (2001). Strike-feeding behavior in the muskellunge *Esox masquinongy*: Contributions of lateral line and visual sensory systems. *J. Exp. Biol.* 204:1207–1221.

Partridge, B., and Pitcher, T.J. (1980). The sensory basis of fish schools: Relative roles of lateral line and vision. *J. Comp. Physiol.* 135:315–325.

Rasnow, B., and Bower, J.M. (1997). Imaging with Electricity: How weakly electric fish might perceive objects. In: Computational Neuroscience: Trends in Research (Bower, J.M., ed.), Plenum Press, N.Y. pp. 795–800.

Roberts, B.L., and Meredith, G.E. (1989). The efferent system. In: *The Mechanosensory Lateral Line: Neurobiology and Evolution* (Coombs, S., Gorner, P., and Munz, H., eds.), pp. 445–459. New York, Berlin: Springer-Verlag.

Sargent, R.C., Rush, V.N., Wisenden, B.D., and Yan, H.Y. (1998). Courtship and mate choice in fishes: Integrating behavioral and sensory ecology. *Am. Zool.* 38:82–96.

Satou, M. (1987). A neuroethological study of reproductive behavior in the salmon. *Third International Symposium on the Reproductive Physiology of Fish*, pp. 154–159. St. Johns, Newfoundland, Canada.

Satou, M., Takeuchi, H.A., Takei, K., Hasegawa, T., Matsushima, T., and Okumoto, N. (1994a). Characterization of vibrational and visual signals which elicit spawning behavior in the male himé salmon (landlocked red salmon, *Oncorhynchus nerka*). *J. Comp. Physiol.* 174:527–537.

Satou, M., Takeuchi, H., Takei, K., Hasegawa, T., Okumoto, N., and Ueda, K. (1987). Involvement of vibrational and visual cues in eliciting spawning behavior in male himé salmon (landlocked red salmon, *Oncorhynchus nerka*). *Anim. Behav.* 35: 1556–1584.

Satou, M., Takeuchi, H.A., Nishii, J., Tanabe, M., Kitamura, S., Kudo, Y., and Okumoto, N. (1991). Intersexual vibrational communication during spawning behavior in the himé salmon (landlocked red salmon, *Oncorhynchus nerka*). *Fourth International Symposium on the Reproductive Physiology of Fish*, pp. 185–187. University of East Anglia, Norwich, U.K.

Satou, M., Takeuchi, H.A., Nishii, J., Tanabe, M., Kitamura, S., Okumoto, N., and Iwata, M. (1994b). Behavioral and electrophysiological evidences that the lateral line is involved in the inter-sexual vibrational communication of the himé salmon (landlocked red salmon, *Oncorhynchus nerka*). *J. Comp. Physiol.* 174:539–549.

Song, J., Yan, H.Y., and Popper, A.N. (1995). Damage and recovery of hair cells in fish canal (but not superficial) neuromasts after gentamicin exposure. *Hear. Res.* 91:63–71.

Takeuchi, H., Takei, K., Satou, M., Matsushima, T., Okumoto, N., and Ueda, K. (1987). Visual cues as key stimuli for courtship behavior in male himé salmon (landlocked red salmon, *Oncorhynchus nerka*). *Anim. Behav.* 35:936–939.

Tinbergen, N. (1951). *The Study of Instinct*. Oxford: Clarendon Press.

Tricas, T.C., and Highstein, S.M. (1991). Action of the octavolateralis efferent system upon the lateral line of free-swimming toadfish, *Opsanus tau*. *J. Comp. Physiol.* 169:25–37.

van Netten, S.M. (1991). Hydrodynamics of the excitation of the cupula in the fish canal lateral line. *J. Acoust. Soc. Am.* 89:310–319.

van Netten, S.M., Kroese, A.B.A. (1987). Laser interferometric measurements on the dynamic behavior of the cupula in the fish lateral line. *Hear. Res.* 29:55–61.

van Netten, S.M., and Kroese, A.B.A. (1989). Dynamical behavior and micromechanical properties of the cupula. In: *The Mechanosensory Lateral Line: Neurobiology and Evolution* (Coombs, S., Görner, P., and Münz, H., eds.), pp. 247–264. New York: Springer-Verlag.

Vischer, H.A. (1990). The morphology of the lateral line system in three species of Pacific cottid fishes occupying disparate habitats. *Experientia* 46: 244–250.

von der Emde, G.S., Schwarz, L., Gomez, L., Budelli, R., and Grant, K. (1998). Electric fish measure distance in the dark. *Nature* 395:890–894.

Weissburg, M.J. (2000). The fluid dynamical context of chemosensory behavior. *Biol. Bull.* 198: 188–202.

8
Retinal Sampling and the Visual Field in Fishes

Shaun P. Collin and Julia Shand

Abstract

The retina of aquatic vertebrates comprises a complex array of sampling elements, each subtending a specific region of the visual field. The transformation of light energy (an optical image) into electrical energy (a neural image) across the retina is nonuniform and reflects the complexity and diversity of the visual field. The identification of localized differences in the function, arrangement, and distribution of retinal neurons in fishes has revealed a level of plasticity unparalleled in vertebrates. Specialized retinal regions (areae centrales, horizontal streaks, and foveae) are examined with respect to both photoreceptor and ganglion cell sampling and the optimization of spatial resolving power and sensitivity. Selective sampling of the binocular visual field, specific asymmetries in the sampling of the dorsal and ventral hemifields, and the number of specializations that comprise only subpopulations of neurons also indicate that the mechanisms controlling the spacing, density, and regularity of retinal neurons are highly complex. Both photoreceptor and ganglion cell arrays change during development, and are affected by changes in the spectral composition, intensity, and symmetry of the photic environment. These arrays typically alter during a transition from one feeding strategy to another. The location of specialized retinal regions subtending the binocular visual field can even alter during development due to the continual growth of the retina throughout life and a changing visual environment.

1. Introduction

Bony and cartilaginous fishes occupy a large diversity of natural environments including the light-limited deep-sea, the turbid water of estuaries and rivers, and brightly lit coral reefs. Their activity patterns also vary according to the ambient light levels, where many species are characterized as diurnal, crepuscular, or nocturnal, although many may cross these behavioral boundaries. The eyes of these aquatic vertebrates must therefore provide an accurate representation of the visual world within an enormous number of photic environments. The large directional differences in intensity and spectral composition of the underwater light, the amount of particulate matter suspended in the water, turbidity, and the need to find food, predator, and mate have all been the driving forces in the evolution of a diverse range of retinal specializations. Although most aquatic animals are active during the day, many are crepuscular or even nocturnal with some species changing their diurnal activity and habitat during development. In all cases, the photic environment differs, which places selective pressure on the eye and its ability to effectively sample its visual field. Although the various components of the ocular media (cornea, lens, and vitreous) may be adapted to differentially filter different parts of the electromagnetic spectrum, it is the retina that ultimately samples the visual environment, transforming light energy or an "optical image" into electrical energy or a "neural image." The process of transduction is performed by the photoreceptors, which line the back of the retina, with each receptor subtending a minute part of the visual field. The transfer of information to the visual centers of the brain is mediated via a series of interneurons. Signals reaching the output (ganglion) cells, lining the inside of the retina, are carried to the central nervous system via ganglion cell axons, which comprise the optic nerve. Every point in each species' visual field is subtended by a corresponding point on the neural retina, which, in turn, is retinotopically mapped onto the optic tectum.

The distribution of both photoreceptors and ganglion cells across the retina is known to be nonuniform in many aquatic vertebrates, and topographic analyses have recently become a powerful tool in predicting the importance each species places on observing objects in specific regions of their visual field (Collin and Pettigrew, 1988a,b; Collin, 1999; Bozzano and Collin, 2000; Shand et al., 2000a). Localized increases in retinal cell density determine the spatial resolving power of the eye, where the distribution of retinal neurons in each species is unique and closely reflects each species' perceived world (Collin and Pettigrew, 1988a,b; 1989). Optical constraints such as the size of the eye, the size and shape of the pupil, the location of the eyes in the head, the degree of eye movement, the size of the visual field, and the degree of binocular overlap all contribute to the optimal placement of retinal regions for increased sampling (Fig. 8.1; see color plate). This chapter reviews the adaptive strategies employed by a range of cartilaginous and bony fishes from various lifestyles and habitats. These are examined with respect to the optimization of retinal sampling of an image in specific regions of the visual field. These retinal specializations will also be discussed in the context of changes in both the photic environment and the visual field during development and growth and the magnification of these specialized zones within the central nervous system.

2. The Retinal Elements of Visual Sampling

2.1. Photoreceptors

The photoreceptor array in fishes may be comprised of single, unequal double, equal double, triple, or even quadruple cones and rods, which are most commonly aligned as a single layer of receptors along the outer limiting membrane (OLM). However, possibly due to the need to reduce chromatic aberration produced by the lens, some subpopulations of receptors migrate out of this monolayer, with the short wavelength-sensitive photoreceptors lying in a more vitreal position than long-wavelength receptors (Eberle, 1967; Scholes, 1975; Shand, 1994). Photoreceptor types also undergo retinomotor movements in response to changing levels of light intensity and therefore may be arranged in staggered rows. Multiple rows or banks of photoreceptors (rods) occur in many deep-sea teleosts where, in some species, up to 28 banks of rods may form a fovea externa (see below). This is done in an effort to increase sensitivity (Locket, 1985; Fröhlich and Wagner, 1998) and provides the potential to mediate color vision (Denton and Locket, 1989).

The area of the visual environment sampled by a single photoreceptor may differ among

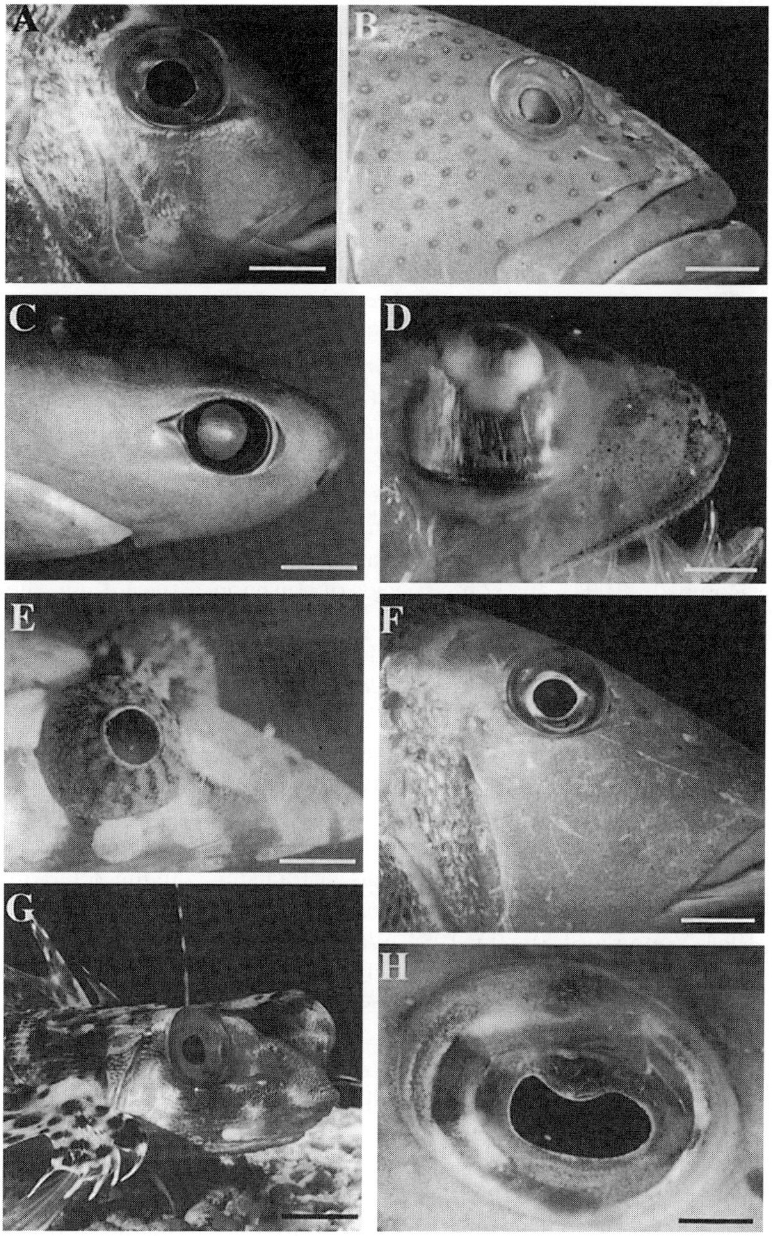

FIGURE 8.1. Diversity of eye size and shape in bony and cartilaginous fishes. (A) The collared sea bream *Gymnocranius bitorquatus* showing a binocular sighting groove etched into the snout. (B) The coral trout *Plectropoma leopardus* showing a colored corneal reflex and a rostrally tapered pupil. (C) The cookiecutter shark *Isistius brasiliensis* with large eyes and a blue tapetal reflex. (D) The deep-sea pearleye *Scopelarchus michaelsarsi* with its tubular eyes and large spherical lens. (Photograph kindly provided by N.J. Marshall.) (E) The sandlance *Limnichthyes fasciatus* showing its dorsal, protruding eyes. (F) The sweetlip, *Lethrinus chrysostomus* showing its binocular sighting groove. (G) The oriental sea robin *Dactyloptaena orientalis* perched on its modified pectoral fins and showing its highly positioned eyes. (H) A close up of the eye of the weever fish *Parapercis cylindrica* showing its crescent-shaped pupil. Scale bars, 10mm (A, B); 20mm (C); 2mm (D); 0.5mm (E); 15mm (F); 10mm (G); 1mm (H). (See color plate)

species and appears to be governed by the visual demands of the animal. Increased spatial resolving power is mediated by tightly packed arrays of small photoreceptors. However, the minimum diameter for a receptor outer segment, even in a small eye, is approximately 1μm. Below this diameter, the photoreceptors fail to act as waveguides and light propagates both inside and outside the receptor, causing "optical crosstalk" and degradation of the neural signal (Kirschfeld, 1976; Land, 1981). At the other end of the scale, some photoreceptor outer segments reach 8μm in diameter, specializing in maximizing photon capture to increase sensitivity. Teleosts inhabiting low-light conditions are also known to compromise resolution by grouping their photoreceptors into bundles, effectively sampling the visual field with large macroreceptors each comprised of up to 50 sampling elements (e.g., in the pearleye, *Scopelarchus michaelsarsi*, Collin et al., 1998 and the goldeye, *Hiodon alosoides*, Braekevelt, 1982).

A range of intracellular inclusions characterize many photoreceptors both morphologically and spectrally. Aggregations of glycogen located within the inner segment constitute a paraboloid in various species of primitive ray-finned fishes including the Florida garfish *Lepisosteus platyrhynchus* (Collin and Collin, 1993) and the bowfin *Amia calva* (Munk, 1968). These myoidal inclusions are thought to act as an energy store. Although rare in euteleosts, oil droplets have been described in various species of primitive fishes including the coelacanth (Walls, 1942), the Australian lungfish (Robinson, 1994), and some species of sturgeon (Munk, 1968; Govardovskii et al., 1992). As in reptiles (Ohtsuka, 1985; Kolb and Jones, 1987), birds (Bowmaker and Martin, 1985; Partridge, 1989; Hart et al., 1998), and nonplacental mammals (O'Day, 1938; Braekevelt, 1973; Arrese et al., 2002), the different types of oil droplets in these fishes (colored and colorless) are thought to either act as cutoff filters, especially for short wavelengths, or aid in focusing light onto the outer segments (Pedler and Tilly, 1964; Young and Martin, 1984; Wong, 1989). A similar function has been attributed to the large ellipsoidal inclusions (ellipsosomes) within the inner segment of cones in a small group of cyprinids (Borwein and Hollenberg, 1973; Anctil and Ali, 1976; MacNichol et al., 1978; Nag and Bhattacharjee, 1989, 1995; Nag, 1995; Novales-Flamarique and Hárosi, 2000). Described by Franz (1932) as false-oil droplets, these large globules, formed from mitochondria, contain a dense heme pigment (Avery and Bowmaker, 1982), are shortwave-absorbing cutoff filters, and tune the incoming light before it strikes the outer segment. The recent finding of ellipsosomes in lampreys (Collin et al., 1999; Collin and Potter, 2000) suggests that these inclusions may be the precursors to (gnathostomatous) vertebrate oil droplets.

Once light enters each photoreceptor cell, it is ultimately the visual pigments within the outer segment that initiate the process of visual transduction by the absorption of light in specific regions of the electromagnetic spectrum. Diurnal shallow-living fishes are known to possess up to four different cones types, providing sensitivity across the spectrum from ultraviolet (short wavelengths) to infrared (long wavelengths). Those living in turbid and low-light conditions, along with crepuscular species, tend to possess only two cone types. However, in most species thus far investigated, the spectral absorption of the long wavelength-sensitive double cones can be related to the spectral qualities of the body of water in which the fishes are found (see reviews by Lythgoe, 1984; Lythgoe and Partridge, 1989; Bowmaker, 1995; Partridge and Cummings, 1999; Chapter 17). The absorption of the single cones is offset to shorter wavelengths (blue and/or UV sensitive) than the ambient light and is thought to facilitate detection of contrast against the prevailing background light (Lythgoe and Partridge, 1991). The rod photoreceptors are predominantly sensitive to medium wavelengths (approximately 500 nm) and show little variation in spectral sensitivity, with the exception of deep-sea fishes, which mainly absorb shorter-wavelength light than their shallow-living relatives and are thought to be "tuned" to increase sensitivity and detect prey that use bioluminescence to camouflage their silhouettes against the downwelling light (Douglas et al., 1998b). The structure of visual pigments and the mechanisms by which they are spectrally tuned is discussed below (see Sections 4.2 and 5.3) and in Chapter 17.

2.2. Retinal Interneurons

In the vertebrate retina, light energy is transformed into graded potentials at the level of the photoreceptor cells and conveyed, via interneurons such as bipolar, horizontal, and amacrine cells, to the third-order neurons (ganglion cells) before being transmitted as impulses to the brain via the optic nerve. The bipolar cells transmit signals directly from the photoreceptors to the ganglion cells and a diverse range of cell shapes and sizes have been identified (about 10 different subtypes according to Naka and Carraway, 1975 and Scholes, 1975). Horizontal cells transfer information laterally across the retina and are capable of mediating chromatic interactions between different spectral cone types, generating the antagonistic surround of the bipolar cell receptive field and modulating spatial summation (Wagner, 1990). There is a large diversity of amacrine cell types in teleosts and the characterization of the range of cell types is beyond the scope of this chapter.

2.3. Ganglion Cells

Retinal ganglion cells are third-order neurons lining the inner retina that possess an axon that transmits the retinal output to the visual centers of the brain via the optic nerve. Ganglion cells predominantly lie within the ganglion cell layer but specific populations are also located within the inner plexiform and inner nuclear layers (Collin and Northcutt, 1993). The axons of ganglion cells are predominantly unmyelinated within the retina but become myelinated within the optic nerve. Ganglion cells receive input from bipolar and amacrine cells via both ribbon and chemical synapses, respectively, although Sakai et al. (1986) have established that the distal dendrites of some ganglion cells also possess presynaptic terminals to bipolar, amacrine, and other ganglion cell processes. The finding suggests that ganglion cells may be able to establish direct interactions with other ganglion cells and that they are able to shape their own response properties through microcircuits formed by their presynaptic dendrites (Sakai et al., 1986).

Various ganglion cell types have been identified in teleosts (Ito and Murakami, 1984; Hitchcock and Easter, 1986; Collin, 1989) and up to 11 different morphological types have been identified in the channel catfish, *Ictalurus punctatus* (Naka and Carraway, 1975; Dunn-Meynell and Sharma, 1986). Morphologically, the dendritic field size, stratification, and central projections are required for characterization, although all three criteria have not been analyzed concurrently in a single species of teleost. As in the human fovea, it is thought that subtypes of ganglion cells mediate high spatial resolving power and are located within defined retinal regions. In the coral trout *Plectropoma leopardus*, small Class 1 cells with a soma size of up to $16.8\,\mu m^2$ and a fan-shaped dendritic arbor are the only ganglion cell type to be found within the temporo-ventral area centralis (Collin, 1989).

Retrograde labeling from the optic nerve is necessary for definitive identification of the retinal ganglion cell population. However, analyses of the distribution and peak density of all neurons located within the ganglion cell layer closely mirrors that of retrogradely labeled ganglion cells, despite the inclusion of amacrine cells known to be present in the ganglion cell layer of fishes (Collin and Pettigrew, 1988c; Shand et al., 2000b). However, the proportions of ganglion to displaced amacrine cells can vary between species. The axons of the optic fiber layer abut the inner limiting membrane and traverse the retina as a continuous sheet or as fascicular bundles (Collin and Northcutt, 1993; Douglas et al., 2002). In lampreys, where a significant proportion of the retinal ganglion cells (75%) lie within the inner nuclear layer, axons exit the retina from the inner nuclear and inner plexiform layer boundary (Fritzsch and Collin, 1990).

As fish grow, new retinal cells, including ganglion cells, are continually being added to the retinal periphery throughout life (Johns, 1977). The new ganglion cells form annuli and, as new retinal tissue is added, subtend continually changing regions of the visual field as each cohort of cells becomes located more centrally. Since the visual world is retinotopically mapped onto the optic tectum, the synapses formed by the ganglion cell axons must also be continually changing.

3. The Visual Field in Fishes

3.1. Eye Position and the Visual Field

The size and shape of the visual field is crucial for the detection of mates, predators, and prey and, in conjunction with eye mobility, govern the development and location of retinal specializations. Despite the growing number of studies investigating the range of retinal sampling strategies in fishes, very few have concomitantly measured the extent of the visual field. Visual field size is species-specific and varies according to the position of the eyes within the head, the shape of the cranium, the depth of the eye socket, and eye mobility (Fig. 8.1). However, the visual field of most laterally placed eyes, when stationary, is also governed by the shape of the scleral eyecup, the extent to which the eye protrudes from the body contour, the distance of accommodatory lens movement, and the shape of the pupil. In most fishes, the pupil is not moveable, although some benthic species use pupillary movement for camouflage (Douglas et al., 1998a, 2002). In the Florida garfish, *Lepisosteus platyrhynchus*, which possesses laterally placed eyes, the monocular visual field is approximately 137° in the horizontal plane (Collin and Northcutt, 1995; Fig. 8.2), while in the gurnard, *Triglia*

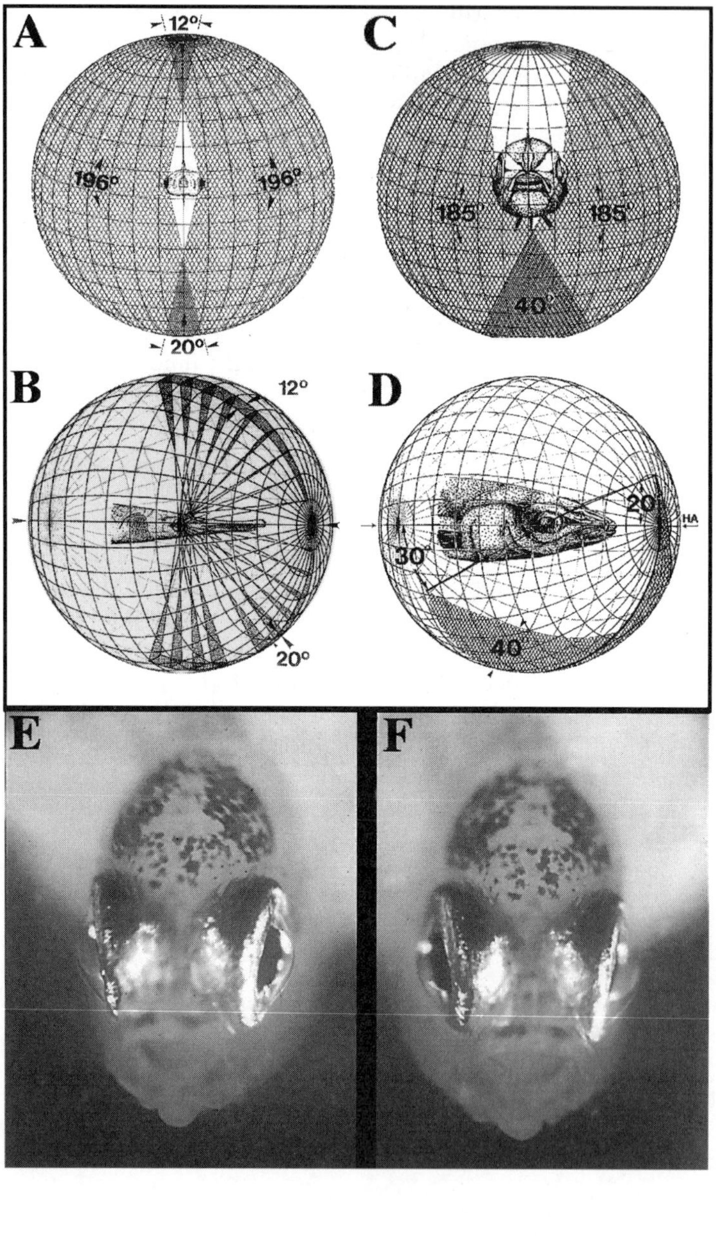

FIGURE 8.2. Visual fields in bony fishes. (A, B) The monocular visual field of the two eyes in the Florida garfish *Lepisosteus platyrhincus* measured ophthalmoscopically shown from the front (A) and from the side (B). Note the small degree of binocular overlap in the dorsal and ventral fields. (Adapted from Collin and Northcutt, 1993.) (C, D) The monocular visual field of the two eyes in the deep-sea smoothead *Conocara macroptera* from the front (C) and from the side (D) showing the large degree of binocular overlap in the ventral visual field. (S.P. Collin and, D.L. Lloyd, unpublished.) (E, F) Stereo pair of the head of the black bream *Acanthopagrus butcheri*. When these two images are fused, the part of the visual field subtended by both eyes can be better appreciated. HA, horizontal axis.

corax, the monocular visual field extends over 170° (Kahmann, 1934).

Although the refractive index of the cornea and the surrounding water effectively negate any refraction at the corneal interface in aquatic vertebrates, the radius of curvature of the cornea also determines the size of the visual field in a number of unusual fishes. The position of the eyes in the head and the ability to maintain the head so that the dorsal region of the eyes is partially exposed to the air governs the extent of the visual field in the four-eyed fish *Anableps anableps* (Sivak, 1976). Gliding just beneath the surface of the water, this species enjoys both aerial and aquatic visual fields simultaneously by possessing a lens (and overlying cornea) with different radii of curvature along the two axes. The unique cornea of the sandlance *Limnichthyes fasciatus* possesses a corneal lenticle with a refractive power of 200 dioptres (in water) and a radius of curvature of 0.24 mm (Pettigrew and Collin, 1995; Pettigrew et al., 2000). This compares to a radius of curvature of 3.2 mm for the nonrefractive cornea in the salamanderfish *Lepidogalaxias salamandroides* (Collin and Collin, 1988a, 1996). Combined with a flattened lens with a power of 550 dioptres, the small radius of curvature of the cornea in *L. fasciatus* increases the visual field. However, the independent and highly mobile eyes of this species extend its visual field to 180° horizontally and 90° vertically (Fritsches and Marshall, 1999) providing a substantial area of binocular overlap (Fig. 8.1E).

Ocular adaptations to increase the binocular field include a rostral tapering of the pupil leaving an aphakic gap that allows more light to strike the (temporal) retina, a binocular sighting groove etched into the snout (Fig. 8.1A,B,F), conjugated frontal eye rotation, and an invagination of the iris, for example, in some species of deep-sea fishes with tubular eyes (Collin et al., 1997; Fig. 8.1D). In the gurnard, *Triglia corax*, the binocular overlap is low (8°) while in various predatory serranids (*Plectropoma leopardus* 36°, *Serranus scriba* 40°, *Epinephalus fasciatus* 54°) the binocular overlap is high (Kahmann, 1934; S.P. Collin and J.D. Pettigrew, unpublished). These species all possess a temporal area centralis and accurately align the retinal specialization upon prey objects by conjugated saccadic eye movements before striking. In the Florida garfish *Lepisosteus platyrhincus*, a large (vertical) monocular field of 196° gives rise to a binocular overlap of 12° (dorsal) and 20° (ventral) (Fig. 8.2A,B). The deep-sea smootheads *Conocara* sp. possess a ventral binocular overlap of 40° (Fig. 8.2C,D) while the eyes of the larval black bream *Acanthopagrus butcheri* subtend a binocular overlap of 21° in the frontal visual field (Fig. 8.2E,F).

3.1. Eye Movements and the Visual Field

Although the extent of the visual field does not alter during eye movement, the degree of binocular overlap changes during ocular fixation. Governed structurally by both the development of the extraocular eye muscles and the relative size of the eye and its supporting socket, the degree of eye mobility in fishes is also closely related to habitat and visual demands. Most species possess some eye mobility but the extent may be masked to the observer by the movement of the globe beneath a secondary spectacle (or scleral cornea), which offers protection or streamlines the contour of the eye (Collin and Collin, 2001). Although head movements are relatively rare in fishes, when this presents a predatory threat (e.g., in the salamanderfish *Lepidogalaxias salamandroides*), a structural modification of the vertebrae actually enables this species to flex its "neck" in order to scan its visual field without any apparent eye movement (Berra and Allen, 1989; Collin and Collin, 1996).

Eye movements, independent of head or body movements, are likely to have evolved to stablize the image of the visual world on the retina. Fast eye shifts or saccades, smooth pursuit movements that follow moving objects, and vergence, which adjusts the eyes for different viewing distances, are correlated with the development of retinal specializations such as a fovea or area centralis (Walls, 1942) (see below). A range of eye movements has been characterized in fishes by Fritsches and Marshall (2002). These include conjugate (moving the eyes in the same direction), vergent (moving the eyes in opposite directions), and independent eye movements. Fritsches and Marshall (2002) analyzed the optomotor range of three foveate teleosts and found that the position of the fovea was correlated with dynamic eye movement. The sandlance *Limnichthyes fasciatus* can strike at prey within a strike zone of 130° using its centrally located fovea without any changes in body position. In contrast, the pipefish *Corythoichthyes* sp. and sandperch *Parapercis cylindrica*, possess asynchronous eye movements and both use body movements to orient their temporal foveae toward prey (Collin and Collin, 1999; Fritsches and Marshall, 2002).

Various species of fishes have evolved mechanisms to inhibit "retinal slip" during swimming and elicit compensatory eye movements in response to changes in orientation and the visual scene. This is achieved by fast gaze shifts (saccades), which render the visual system blind for a short time but effectively stabilize the visual field, thereby reducing retinal image motion. Most fishes adhere to this oculomotor strategy with the exception of one species, *L. fasciatus*, where the eyes drift up to a distance of 35° following a saccade (Fritsches and Marshall, 1999). However, the drifts are slow (3–4° degrees per second), which may not degrade image quality. These drifts may be important to realign the fovea within a "preferred" frontal visual field and/or may be a biproduct of a rather loose optokinetic stabilization mechanism due to a relatively featureless upwardly directed visual field (Fritsches and Marshall, 1999; Land, 1999).

4. Specialized Sampling of the Visual Field

4.1. Areae Centrales

Almost all aquatic vertebrates examined thus far possess a specialized retinal region where subpopulations of neurons are concentrated, thereby sampling a specific part of their visual field with increased resolving power. The term *area* was originally adopted by Chievitz (1891) to describe the macula lutea of humans, a term that was described later as an *area nasalis*, *area temporalis*, or *area centralis* in other vertebrates according to the region of the retina in which it lay. However, an area centralis is now defined as a concentric increase in retinal cell density in any region of the retina, most authors dispensing with the regional suffix. Sometimes associated with either a deep (convexiclivate) or shallow (concaviclivate) pit-like invagination in the retina termed a *fovea* (see below), an area centralis enables accurate fixation of the two eyes and often defines the main visual axis.

The degree of inhomogeneity across the fish retina has recently initiated detailed studies of the distribution and arrangement of subpopulations of photoreceptor and ganglion cells. Based on complementary criteria such as morphology (Reckel et al., 2001), cytochemistry (Hisatomi et al., 1997; Ishikawa et al., 1997), spectral sensitivity (Carleton et al., 2000), and, in some cases, physiology (Lasater, 1982), identification of these subpopulations has enabled more informed predictions of how each specialization may underlie behavior(s).

4.2. Photoreceptor Specializations

Based on morphology, photoreceptors have been characterized into a number of types including short and long single cones, double cones (equal and unequal), triple cones, quadruple cones, and rods. These different populations can be arranged in a variety of conformations in different species, for example, a row pattern (in the cutlips minnow, Collin et al., 1996), a twisted row pattern (in salmon, Ahlbert, 1976), a square pattern (in zebrafish, Cameron and Easter, 1995), a pentagonal pattern (in the garfish, Reckel et al., 2001), and a hexagonal pattern (in the deep-sea pearleye, Collin et al., 1998). However, all of these conformations can also be found in different regions of the same retina, for example, in the garfish *Belone belone* (Reckel et al., 2001; Fig. 8.3). Other less common arrangements include a triangular mosaic in the retina of the northern pike *Esox lucius* (Braekevelt, 1975) and four equal double cones bordering a rod in the salamanderfish *Lepidogalaxias salamandroides* (Collin and Collin, 1998). Despite this variation

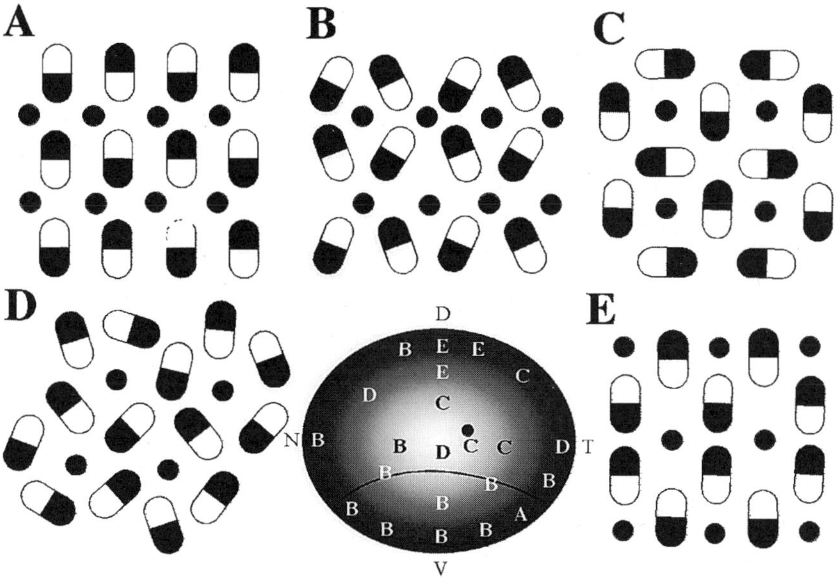

FIGURE 8.3. The diversity of retinal mosaics found throughout the retina in a single species (the garfish *Belone belone*). (A) *Pure row pattern*: Double and single cones form parallel rows. (B) *Twisted row pattern*: Double cones are twisted to form parallel rows. (C) *Square pattern*: Four double cones surround a single cone. (D) *Pentagonal pattern*: Five double cones surround a central single cone. (E) *Hexagonal pattern*: Six double cones surround a central single cone. The lower middle panel shows the retina and the location of each type of mosaic. D, dorsal; N, nasal; T, temporal; V, ventral. The optic nerve head is depicted as a black circle. The line separating ventral retina is an intraocular septum (see text). (Adapted from Reckel et al., 2001.)

in sampling, little is known of the role these regular arrays play in vision. However, of all the receptor patterns described, the square mosaic comprising four double cones surrounding a central single cone appears to be relatively common and has received particular attention. Although not definitively tested, a number of theories about the function of the square mosaic have been put forward, which include increasing both visual acuity (Engström, 1963a,b; Ahlbert, 1976) and contrast (Marc and Sperling, 1976; Meer, 1994), providing a more uniform spectral sampling (Bowmaker, 1990), allowing more detailed chromatic patterns to be resolved (Lythgoe, 1979), and enhancing the detection of polarized light (Cameron and Easter, 1993; Novales-Flamarique and Hawryshyn, 1998; Chapter 13). Although not mutually exclusive, the square mosaic may also aid in the analysis of movement in all directions in contrast to a row mosaic, which may be suited to the perception of movement in two directions (Engström, 1963a; Anctil, 1969; Bathelt, 1970). This correlation is confirmed for a number of predatory salmonids that strike at moving prey and perceive a more three-dimensional environment using a square mosaic (Lyall, 1957a; Ahlbert, 1976; Beaudet et al., 1997) compared to schooling salmonids, which rely less on accurate strikes for prey and perceive a more two-dimensional environment along the horizontal plane using a row mosaic. The nonschooling coral trout *Plectropoma leopardus* (Collin, 1989; Fig. 8.1B), the sand-lance *Limnichthyes fasciatus* (Collin and Collin, 1988b; Fig. 8.1E), the tuskfish *Pseudolabrus miles* (Fineran and Nicol, 1974), the weeverfish *Trachinus vipera* (Kunz et al., 1985), the archer-fish *Toxotes jaculatrix* (Braekevelt, 1985) and the pipefish *Corythoichthyes paxtoni* (Collin and Collin, 1999) all strike moving prey with precision in a three-dimensional environment and possess a regular square photoreceptor mosaic.

Although double cones often constitute the bulk of photoreceptors, triple and quadruple cones may also occur in large numbers in some species. Triple cones occur in two varieties, linear (e.g., in *Perca fluviatilis*) and triangular (e.g., in *Phoxinus laevis*), but their function is unknown (Lyall, 1956; Engstrom, 1960). Both quadruple and triple cones develop in central retina by the formation of subsurface cisternae along selected membrane borders of large numbers of neighboring single cones in developing black bream (Shand et al., 1999). These cone multiples fail to integrate into the regular square mosaic that predominates the retina in this species, and their central location within the retina suggests that these cone multiples are not necessarily aberrant but lie outside the influence of intercellular signaling cues that may be localized in the periphery and are responsible for the formation of the retinal mosaic during growth (Raymond, 1995; Shand et al., 1999). Interestingly, the lesioned retina of the green sunfish, *Lepomis cyanellus*, regenerates double and triple cones but is unable to reestablish the repeating cone mosaic following damage to the retina (Cameron and Easter, 1995).

Surrounded by either the retinal pigment epithelium or dense collections of tapetal material, both cones and rods may form aggregations, grouped together as macroreceptors to increase sensitivity in low light. Rods (e.g., in the weever fish *Trachinus vipera*, Kunz et al., 1985), or cones (e.g., in the pacific tarpon *Megalops cyprinoides*, McEwan, 1938), or both receptor types (e.g., in the mormyrid *Marcusenius longianalis*, Engström, 1963a, the goldeye *Hiodon alosoides*, Braekevelt, 1982, and the featherfin *Xenomystus nigri*, Ali and Anctil, 1976) are electrically and optically isolated into groups, but the sampling advantage this pattern provides is unknown. A detailed topographic study of both the photoreceptor and ganglion cell populations in the tubular-eyed deep-sea pearleye *Scopelarchus michaelsarsi* shows that groups of rods lie in the caudal region of the main retina, where electrotonic synapses may physiologically link each receptor in the group and summate input from a large area to increase sensitivity at the expense of spatial resolving power (Locket, 1970; Collin et al., 1998).

The large diversity in photoreceptor mosaics in different species of fishes and within different parts of the retina suggests that the environment and habitat play the most important roles in each species' sampling strategy rather than being controlled by any phylogenetic constraints. A detailed study by Reckel et al. (2001) in the garfish *Belone belone* is a case in point (Fig. 8.3). Three areae centrales are identified, a ventro-nasal (18×10^3 double cones per mm^2), a ventro-temporal (18×10^3 double cones per mm^2), and a dorso-temporal (14×10^3 double cones per mm^2) area, which mirrors the specialized zones of acute vision in the Florida garfish *Lepisosteus platyrhincus*, a species with a similar predatory lifestyle living close to the surface despite the two species being separated phylogenetically by at least 200 million years (Collin and Northcutt, 1993). The complexity of the visual environment being sampled is further emphasized by the large number of photoreceptor conformations within the one retina (e.g., in *B. belone*, Reckel et al., 2001).

The upper and lower visual fields of most species are very different but none more so than those species that exist near the surface of the water, such as the garfish *Belone belone*. In this unique environment, coping with high light intensities, avoiding predation and the refractive problems associated with seeing through the surface of the water (into Snell's window) must be overcome by the ventral retina, while the need to optimize sensitivity in often low light intensities must be accomplished by the dorsal retina. In *B. belone*, the upper and lower visual fields are sampled by a twisted row and a hexagonal array pattern of cone photoreceptors, respectively (Reckel et al., 2001; Fig. 8.3). As in the osteoglossid *Pantodon buchholzi* (Saidel, 1987), *B. belone* also possesses an intraocular septum that divides the dorsal and ventral retinal regions and inhibits the refractive disturbances associated with the borders of Snell's window (Schwartz, 1971; Fig. 8.3). In the pelagic blue marlin *Makaira nigricans*, the ventral retina contains a square mosaic, while the dorsal retina that peers into dim, monochromatic light possesses a row mosaic

(Fritsches et al., 2000). Marked differences in retinomotor movements (Reckel et al., 2001), tapetal material (Collin, 1988; Collin and Northcutt, 1993; Takei and Somiya, 2001), and optomotor control (Saidel and Fabiane, 1998; Saidel, 2000) between the dorsal and ventral retina reflect a divergence in spatial (finding a compromise between resolving power and sensitivity) and chromatic sampling.

Cone types characterized by the spectral sensitivity of their visual pigments govern chromatic sampling. The wavelength of maximum absorbance of a visual pigment (λ_{max}) is dependent on the amino acid sequence of the opsin protein and the interaction with its associated chromophore, which may be based on either rhodopsin (the aldehyde of vitamin A_1) or porphyropsin (the aldehyde of vitamin A_2). Concomitant with morphological variations, the opsin expressed within each photoreceptor type also varies and can be classified into one of five evolutionarily distinct groups of vertebrate visual pigments (Yokoyama, 1997). Changes in spectral sensitivity across species or within species during development (as discussed further below) are mediated by different ratios of the two chromophores (see review by Loew, 1995) and/or variations in the expression of genes coding the opsin protein (Carleton and Kocher, 2001; J. Shand and N. Thomas unpublished). Just as the density of specific cone types (based on morphology) varies across the retina and dictates changes in spatial resolving power, the spectral sensitivity of chromatic sampling units also varies, and can be related to the spectral transmission of light through the water from different regions of the visual field (Levine and MacNicol, 1979).

4.3. Ganglion Cell Specializations

In general, the topography of the ganglion cell population is closely aligned with the topography of the photoreceptor population, although few studies have compared both within the same species. Ganglion cell densities in reef teleosts range from a peak of 6×10^3 cells per mm^2 in the frogfish *Halophryne diemensis*, a lie-in-wait predator (Collin and Pettigrew, 1988a), to over 150×10^3 cells per mm^2 in the sandlance *Limnichthyes fasciatus*, a predator that launches its whole body to engulf its prey (Collin and Collin, 1988c; Pettigrew and Collin, 1995). The spacing of the ganglion cells dictates the spatial resolving power of the eye and has a close relationship to the behavioral acuity (Fig. 8.4). The spatial resolving power in reef fishes has been found to vary from 4 to 27 cycles per degree, respectively (Collin and Pettigrew, 1989). Although *H. diemensis* and *L. fasciatus* possess an area centralis in the middle of the retina, where the eyes subtend a large binocular overlap, in species with lateral eyes, the temporal retina is the most commonly specialized region for acute vision, subtending a part of the frontal visual field (Fig. 8.5). However, some fishes possess more than one acute zone, mediating increased spatial resolving power in different parts of the visual field (Collin and Pettigrew, 1989; Fig. 8.5). Some balistids possess both a temporal and a nasal area centralis for feeding and swimming backward, respectively (Ito and Murakami, 1984; Collin and Pettigrew, 1988b). Once thought to be devoid of any retinal specialization (Johns and Easter, 1977), the goldfish *Carassius auratus* has been found to possess a temporal area centralis with respect to both ganglion cell (Mednick and Springer, 1988) and photoreceptor (Mednick et al., 1988) densities, but many other species of cyprinids possess up to three areae (Zaunreiter et al., 1991; Collin and Ali, 1994).

In addition to photoreceptor arrays, other cell types including ganglion cells also form regular mosaics that effectively sample all parts of the eyes' visual field and may have specific functions. Recent studies have identified a subset of large ganglion cells with large soma (Fig. 8.4B), extensive dendritic fields, and terminal stratification within different levels of the inner plexiform layer (IPL). These have been identified in a range of fishes (Cook et al., 1999) and, although not unequivocally established, may have functional and evolutionary homologies to the large cells described for amphibians (Shamim et al., 1997) and mammals (Boycott and Wässle, 1974). On morphological criteria, these cells appear comparable to the alpha or Y class cells described in cats by Wässle et al.

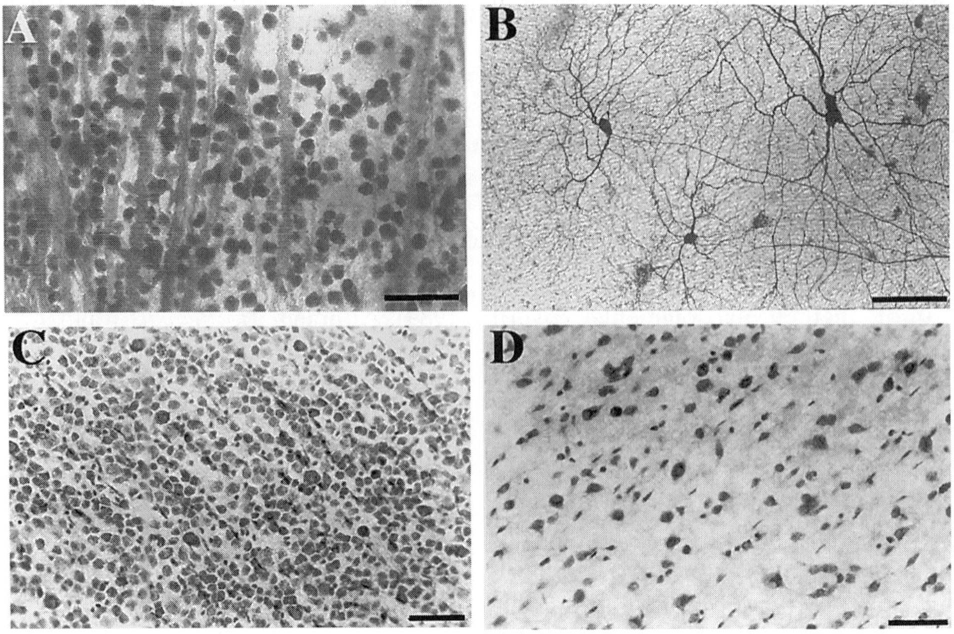

FIGURE 8.4. Ganglion cell sampling of the visual field. (A) Ganglion cells of the Florida garfish *Lepisosteus platyrhincus* retrogradely labeled with horseradish peroxidase from the optic nerve. Bundles of axons separate the cells into columns. (B) Giant ganglion cells retrogradely labeled with cobaltous lysine in the garfish showing the large dendritic fields of these alpha-like cells that form regular mosaics across the retina. (C, D) Ganglion cells of the black bream *Acanthopagrus butcheri* in the area centralis (C) and peripheral (D) regions of the retina. (Adapted from Shand et al., 2000a). Scale bars, 20 μm (A); 100 μm (B); 50 μm (C); 50 μm (D).

(1981) and therefore may be motion sensitive with brisk transient receptive fields. These large ganglion cells can be divided into subgroups: cells that stratify within the vitread (inner or ON cells), sclerad (outer or OFF cells) (Cook and Becker, 1991; Cook et al., 1992; Collin and Northcutt, 1993), middle (Cook and Sharma, 1995), and within both the inner and middle laminae of the IPL (Cook et al., 1999). Although these cells represent less than 1% of the ganglion cell population, each subtype forms a spatially independent mosaic providing full coverage of the retina and therefore the visual field (albeit at low density), despite having soma "displaced" into the inner nuclear layer. Regular arrays of biplexiform ganglion cells have also been discovered in both lampreys (Fritzsch and Collin, 1990) and bony fishes (Hitchcock and Easter, 1986; Cook et al., 1996), suggesting that, like the alpha cells, cellular mosaics may be a common feature of most vertebrate classes. The relationship between the mechanisms underlying the elaboration of dendritic arbors and the patterning of cell bodies still remains theoretical (Cook and Chalupa, 2000), however, the finding of regular arrays of cells confined to specific retinal regions (e.g., temporal retina, see below) suggests that, in some species, these cells may have specific functions relevant to their lifestyle.

4.4. The Area Giganto Cellularis

In contrast to the contiguous dendritic fields of large (alpha-like) ganglion cells that cover the entire retina, and therefore subtend all of the visual field in the shallow-water fishes described above, there are a group of deep-sea fishes that possess a regular array of large ganglion cells restricted to temporal retina. In the pearleye *Scopelarchus michaelsarsi* (Collin et al., 1998) and the daggertooth *Anotopterus pharao* (Uemura et al., 2000), the aptly named area giganto cellularis (AGC) is comprised of large soma with dendritic terminals stratifying within the outer and inner parts of the IPL, respectively. Due to their size and large dendritic field, these cells are thought to be motion sensitive to objects crossing the binocular visual field. The difference in dendritic stratification between the two species may be related to the background illumination, where the ON-type cells of *S. michaelsarsi* may respond to a bright point source of bioluminescent light against a dark background and the OFF-type cells of *A. pharao* may respond to a silhouetted prey item against the bright background of downwelling skylight (the daggertooth adopts a head-up posture within the water column, Collin et al., 1998; Uemura et al., 2000). Interestingly, an AGC comprising ON-type cells has also been described in the temporal retina of five species of procellariform sea birds, which feed on prey moving against a dark background (Hayes et al., 1991).

4.5. Horizontal Streaks

Chievitz (1889, 1891) and Slonaker (1897) were the first to describe horizontal band-shaped

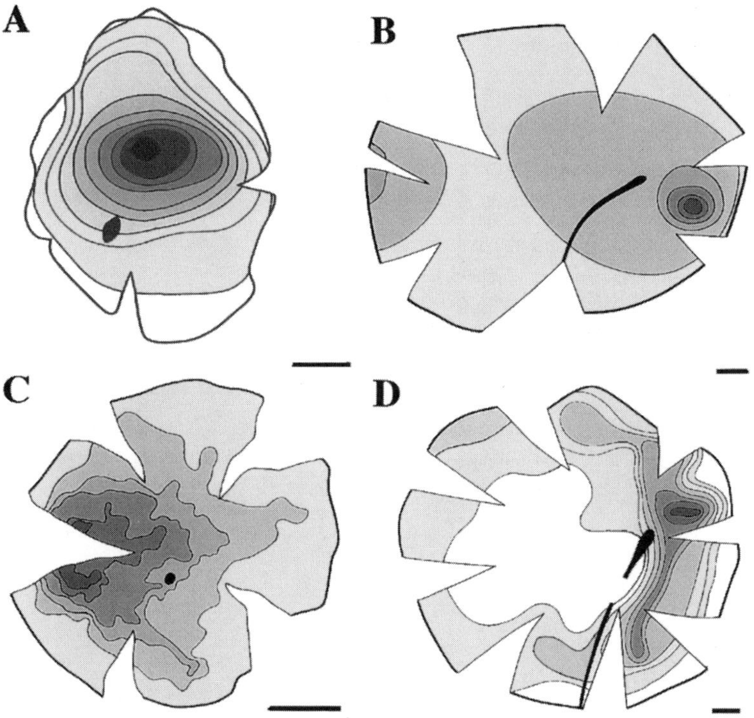

FIGURE 8.5. Ganglion cell specializations (areae centrales). (A) The sandlance *Limnichthyes fasciatus* (<50.0 to 150.0×10^3 ganglion cells per mm^2). (B) The coral cod *Cephalopholis miniatus* (<10.0 to 47.0×10^3 ganglion cells per mm^2). (C) The cookie-cutter shark *Isistius brasiliensis* (<0.8 to 1.6×10^3 ganglion cells per mm^2). (D) The deep-water bass *Howella sherborni* (<12.0 to 24.0×10^3 ganglion cells per mm^2). Note the concentric arrangement of iso-density contours with the exception of *H. sherborni*, which has a vertical streak. The optic nerve head and falciform process (where present) is depicted in black. The progressively darker shading represents increases in cell density. Dorsal is toward the top and the area centralis is located in temporal retina in each species with the exception of *L. fasciatus*. Scale bars, 1.0 mm (A, B); 5 mm (C); 1 mm (D). (Adapted from Collin and Collin, 1988a (A); Collin and Pettigrew, 1988a (B); Bozzano and Collin, 2000 (C); Collin, 1997 (D)).

retinal areae in bony fishes and these specializations or visual streaks can take two forms: either a band-shaped increase in thickness across the retinal meridian, as found in the cyprinodontid *Fundulus heteroclitus* (Butcher, 1938) and two species of mudskippers, *Boleophthalmus* and *Periophthalmus* (Munk, 1970), or simply a marked increase in ganglion and photoreceptor cell density as found in the balistids, *Navodon modestus* (Ito and Murakami, 1984) and *Balistoides conspicillum* (Collin and Pettigrew, 1988b) and the yellow-finned trevally, *Caranx ignobilis* (Collin, 1999; Fig. 8.6). The visual streak can be pronounced (e.g., a peak of over 90.0×10^3 ganglion cells per mm^2 in *C. ignobilis* and a centroperipheral gradient of 10:1, Collin, 1997) or weak (peak of 0.93×10^3 cells per mm^2 with a centroperipheral gradient of 1.8:1 in the velvet belly dogfish *Etmopterus spinax* (Bozzano and Collin, 2000; Fig. 8.6).

The area of the visual field subtended by a streak shows interspecific variation. Most are horizontal, maintaining a retinal seam of high spatial resolving power across the central meridian that allows a panoramic sampling of an elongated region of the lateral visual field. However, upwardly and downwardly directed

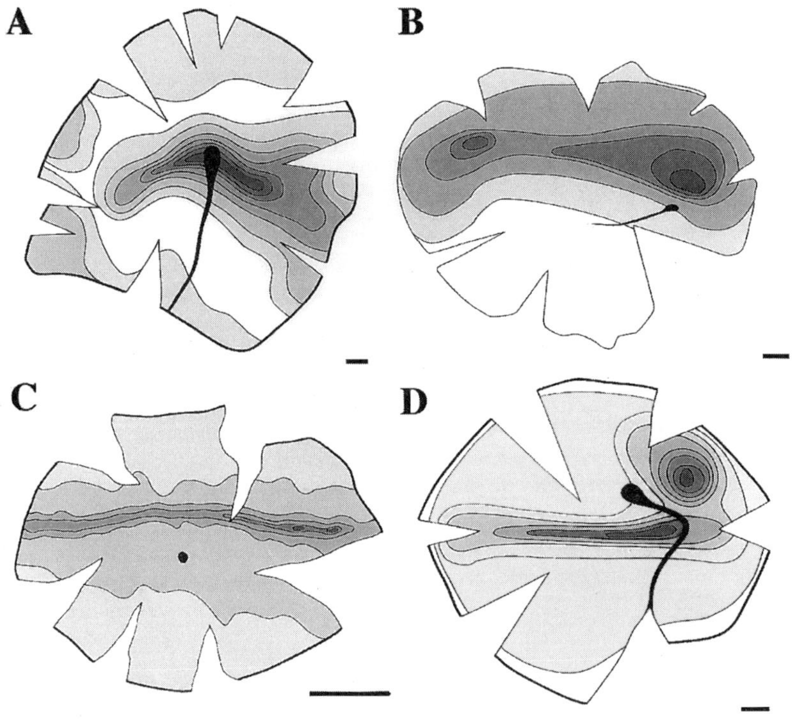

FIGURE 8.6. Ganglion cell specializations (horizontal streaks). (A) The trevally *Caranx ignobilis* (<10.0 to 90.0×10^3 ganglion cells per mm^2). (B) The weaverfish *Parapercis cylindrica* (<5.0 to 40.0×10^3 ganglion cells per mm^2). (C) The small-spotted dogfish *Scyliorhinus canicula* (<0.5 to 2.4×10^3 ganglion cells per mm^2). (D) The blue tuskfish *Choerodon albigena* (<10.0 to 83.0×10^3 ganglion cells per mm^2). Note the elongated iso-density contours across the retina, which may subtend both dorsal and ventral hemifields. Note that the streak is often associated with a second specialization located either within the visual streak or often in temporal retina. The optic nerve head and falciform process (where present) is depicted in black. The progressively darker shading represents increases in cell density. Dorsal is toward the top and the area centralis (where present) a located in temporal retina in each species. Scale bars, 0.25 mm (A); 1 mm (B); 5 mm (C); 1 mm (D). (Adapted from Collin, 1997 (A); Collin and Pettigrew, 1988a (B); Bozzano and Collin, 2000 (C); Collin and Pettigrew, 1988b (D)).

streaks are also found in ventral and dorsal retina, respectively (Fig. 8.6). The Florida garfish, *Lepisosteus platyrhincus*, possesses a visual streak (9.40×10^3 ganglion cells per mm^2) across the ventral meridian of the eye in conjunction with a temporal area centralis (6.25×10^3 cells per mm^2). The horizontal streak subtends the surface of the water, which it uses as a background to prey upon live fishes with its long snout, armed with needle-like teeth (Collin and Northcutt, 1993). A population of "displaced" ganglion cells situated in the inner nuclear layer is also concentrated into a ventral visual streak (6.25×10^3 cells per mm^2) in the garfish retina. This ectopic streak is inflected 20° from the horizontal and may provide a finer control for the stabilization of the retina during compensatory eye movements, while the temporal area centralis could provide an increase in visual acuity for prey observed in frontal visual space. Interspecific differences in the location of the streak also occur in cartilaginous fishes, where the position of the eyes within the head and the visual axis vary markedly among the three major groups: batoids, selachians, and the chimaerids (Bozzano and Collin, 2000; Fig. 8.6).

Many species of ray-finned fishes possess a retinal specialization in addition to a visual streak. In the blue tuskfish *Choerodon albigena* and the red-throated emperor *Lethrinus chrysostomus*, a temporal area centralis lies in conjunction with a horizontal streak (Collin and Pettigrew, 1988b; Fig. 8.6). The existence of two specializations subtending different regions of the visual field suggests that each may provide a specific sampling strategy, especially when each specialization is comprised of cells of different morphology and receptive field size (Collin, 1989). In *Aplocheilus lineatus* and *Epiplatys grahami*, which prey upon insects that lie trapped in the surface film of the water (Munk, 1970), each retina contains two band-shaped thickenings across the horizontal meridian. These central band-shaped areae, although both oriented parallel to the surface of the water, are separated by 40° and subtend different regions of the visual field. It is assumed that the upper band is able to view lateral visual field in search of food within the water column, while the lower (ventral) band is able to detect prey venturing into the edges of Snell's window (Munk, 1970).

The visual streak may have various functions. Thus far, the horizon (either the sand–water or air–water interface) has predominated the visual field of each species of ray-finned fishes found to possess a horizontal streak (Collin and Pettigrew, 1988a,b). A panoramic view of the visual field sampled with increased spatial resolving power negates the need for the saccadic eye movements necessary using an area centralis. Disturbances in the horizontal part of the visual field subtended by a streak will also lower the threshold for the perception of movement and signal either prey or predator. Where the symmetry of the visual world is particularly two-dimensional, but there is some need for higher spatial resolving power within a specific part of the visual field, a compromise is reached where a cell density peak lies within a visual streak (Yew et al., 1984; Bozzano and Collin, 2000; Fig. 8.6).

4.6. Foveae

Shallow foveae in the eyes of bony fishes were noted over 100 years ago (Slonaker, 1897). Later, Kahmann (1934) identified a large number of species with deep foveae located predominantly in temporal retina and mediating acute binocular vision. Foveae were always associated with fixating eye movements and, in a few species, foveal location varied with the position of the eyes in the head, sometimes mediating monocular vision.

More recently, the relationship between the presence of a fovea and increased eye mobility has been noted in a number of teleosts, including the kelp bass *Paralabrax clathratus* (Schwassmann, 1968), the sandlance *Limnichthyes fasciatus* (Pettigrew and Collin, 1995), the clingfish *Gobiesox strumosus* (Wagner et al., 1976), and the sandperch *Parapercis nebulosus* (Easter, 1992). Over 42 species of bony fishes that frequent shallow water have been found to possess foveae (reviewed in Collin and Collin, 1999). The teleost fovea is an indentation of the retina (Fig. 8.7) and, thus far, has always been associated with an increase in

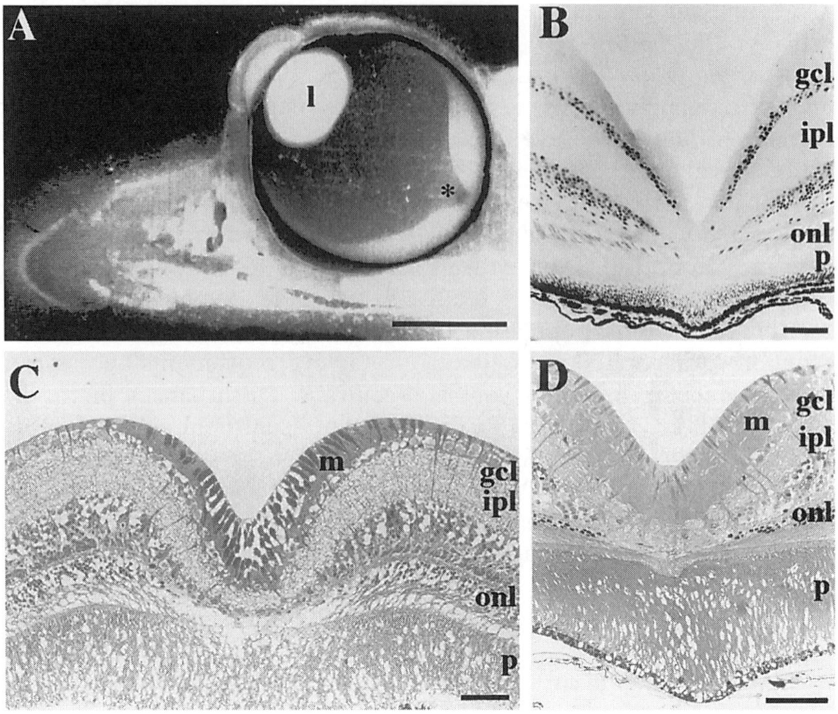

FIGURE 8.7. Fovea in bony fish. (A) A frozen section of the head of the sandlance *Limnichthyes fasciatus* in the sagittal plane showing the depth of the convexiclivate fovea (*) in the eye, the nonspherical lens (l), and the corneal lenticle, all aligned along the visual axis. (B) A resin section of the sandlance fovea showing the displacement of the inner retinal layers. (C, D) Transverse sections of the fovea in the deep-sea smoothheads *Conocara murrayi* (C) and, *C. macroptera* (D). Note the thick foveal lining of Müller cell processes (m) and the elongation of the photoreceptors beneath the foveal clivus. gcl, ganglion cell layer; ipl, inner plexiform layer; onl, outer nuclear layer; p, photoreceptor layer. Scale bars, 0.5 mm (A); 50 μm (B); 100 μm (C, D). (B is adapted from Collin and Collin, 1988c, and D is adapted from Collin et al., 2000.)

both photoreceptor and ganglion cell density (area centralis), although a clivus in the otherwise smooth retinal surface lining the eyecup predicts that the fovea also provides an optical advantage.

One optical advantage of a convexiclivate fovea (Munk, 1975; Collin and Collin, 1988b,c) may be image magnification resulting from a marked change in refractive index between the vitreous and the sloping sides of the fovea, thereby increasing visual resolution (Walls, 1937, 1940, 1942; Snyder and Miller, 1978; Locket, 1992; Collin et al., 1994). The fovea may play a major role in the detection and maintenance of fixation, providing an increased sensitivity to small angular movement as the image of a moving object is distorted by the curvature of the pit (Pumphrey, 1948). In conjunction with a high degree of independent eye mobility, a deep foveal pit may also act as a directional and monocular indicator of accommodative focus as has been found in the chameleon (Harkness and Bennett-Clarke, 1978) and in the marine sandlance (Pettigrew and Collin, 1995). Where the foveal axes fall within a pronounced binocular overlap, skewing of eccentric images may also provide a useful cue about range and possibly a method of breaking luminescent camouflage in a number of foveate teleosts, which have ventured into the deep-sea (Steenstrup and Munk, 1980; Locket, 1985, 1992; Wagner et al., 1998).

Based on morphological, ecological, and functional diversity, the teleost fovea has recently been characterized into at least four distinct types (Collin and Collin, 1999). Type I is exemplified by the syngnathid fovea, such as that in *Corythoichthyes paxtoni*, and is characterized by a steep-sided (convexiclivate) retinal pit without a lateral displacement of the inner retinal layers (Collin and Collin, 1999). A Type II fovea is also convexiclivate but the inner retinal layers are displaced laterally, leaving an unimpeded path for the incident light to strike the underlying photoreceptors. Examples of a Type II fovea are found in the sandlance, *Limnichthyes fasciatus* (Collin and Collin, 1988b,c; Fig. 8.7A,B) and the notosudid, *Scopelosaurus hoedti* (Munk, 1975). Type III foveae are similarly convexiclivate but possess a thick foveal lining of radial fiber processes putatively thought to be refractive. Examples of this foveal type are found in the deep-sea alepocephalids, *Conocara macroptera* (Collin et al., 1994; Wagner et al., 1998; Fig. 8.7C,D) and *Alepocephalus bairdii* (Locket, 1992). Although changes in refractive index within foveal and perifoveal retinal regions still need to be examined, indices of 1.3353 (vitreous) and 1.3494 (retina) measured in *Chondrostoma nasus* (Nicol, 1989) suggest that this gradation may produce refraction. Changes in foveal thickness, the displacement of the inner retinal layers, and variations in the shape of the foveal clivus may also produce different optical effects to satisfy specific ecological needs. Type III foveae may prove to be of particular interest given the widespread occurrence of the foveal lining in other vertebrates, such as birds (Locket, 1992), where the relative thickness of the dense radial fiber processes may comprise up to 40% of the foveal thickness (Locket, 1992; Wagner et al., 1998). Finally, the Type IV fovea is a shallow (concaviclivate) invagination of the retina where there is neither a lateral displacement of inner retinal layers nor a radial fiber lining. Examples of this type are found in the banded toado, *Sphaeroides pleurostictus* (Collin, 1987), and the deep-sea *Bathylagus benedicti* (Vilter, 1954).

5. Sampling a Changing Visual Environment

Many aquatic animals undergo dramatic changes in their visual environment during development, changes frequently associated with an alteration in feeding behavior. The transition of amphibians from an aquatic to a terrestrial habitat at metamorphosis imposes different optical demands, as does the migration of fishes between different bodies of water, resulting in the need to change their visual axis or spectral sampling. The reason that migration between different aquatic habitats results in pronounced changes in visual environment relates to the variable physical constraints that are imposed on the transmission of light in different bodies of water and at different depths. For example, as surface-living larval fishes metamorphose and move deeper in the water column, there will be a concomitant reduction in both the intensity and spectral composition of the ambient light. Similarly, anadromous migrations between the sea and freshwater for the purposes of maturation and breeding result in animals inhabiting an environment with very different spectral qualities. Where changes in feeding behavior take place, there may also be a need to change the main visual axis and retinal sampling array.

5.1. The Relationship Between Changing Visual Demands and the Visual Field

In general terms, prey species require a wide field of view to scan the visual scene, whereas predators often require a large overlap in the visual fields of both eyes to increase binocularity (Lythgoe, 1979). As tadpoles, frogs possess a wide monocular field of view in water but increase their binocular overlap to allow for depth perception and the detection of insectivorous prey on land (Sivak and Warburg, 1983). Migrations from aquatic to aerial habitats are also accompanied by optical changes to counteract the increased refractive power of the cornea in air (Sivak, 1988; Collin and Collin, 2001).

Despite avoiding the optical problems associated with a transition from an aquatic to an aerial lifestyle, developmental changes in feeding behavior of fishes can result in changes in the visual axis. The migration of the eyes in the head in flatfishes is perhaps the most dramatic. In the early stages of development, pelagic larvae feeding on plankton in the water column possess laterally placed eyes. During the transition to a benthic existence, instead of a dorso-ventral flattening of the body as seen in cartilaginous skates, flatfishes move the left (e.g., sole) or right (e.g., flounder) eye to the other side of the head. As adults, the fishes lie on their sides on the substrate with both eyes directed upward, subtending a large binocular overlap (Beaudet and Hawryshyn, 1999).

Metamorphosis in other pelagic larval fishes may be less radical, but it is likely there will be changes in the visual field associated with changes in feeding behavior following settlement. For example, during the transition from a pelagic juvenile stage, where they feed on plankton in the water column, goatfishes settle onto inter-reef substrate, where they begin using sensory chin barbels to disturb invertebrates from the sediment (McCormick, 1993). During this metamorphosis, the goatfish undergoes changes in head and body shape and the eyes move dorsally (J. Shand, unpublished). Similarly black bream, *Acanthopagrus butcheri*, alter the axis of maximal binocular overlap from 30° to 50° below the horizontal axis as they move out of the water column to become benthic feeders (S.P. Collin, K. Doving, and J. Shand, unpublished; Fig. 8.2E,F). The part of the visual field subtended by the two eyes changes from dorsal to ventral in the Australian lungfish, *Neoceratodus forsteri* (S.P. Collin and J. Joss, unpublished). In larval clown fish, *Premnas biaculeatus*, the functional visual field rapidly increases in the horizontal plane as shown by feeding experiments that record the position of the prey relative to the eye when the prey is first detected (Job and Bellwood, 1996). Eels are also known to increase the size of their eyes, and the size of their visual field, prior to migration into deeper waters for breeding (Pankhurst, 1982).

5.2. Changes in Photoreceptor Complement and the Cone Mosaic During Ontogeny

The retinal cone photoreceptors of fishes can be arranged in a number of different repeating arrays, which have been related to different sampling tasks (see Section 4.2). However, if visual demands change, it may be advantageous to alter the arrangement of the cone mosaic. Until recently, any changes to the mosaic have been attributed to either a loss or addition of cones rather than any reorganization. For example, during smoltification in salmonids, the single cones at the corner of the mosaic are lost (Lyall, 1957b; Ahlbert, 1976; Bowmaker and Kunz, 1987; Hawryshyn et al., 1989) or cease to be incorporated into the retina (Novales-Flamarique, 2001) between the parr and adult stages. However, in a recent study of the rainbow trout, changes in the cross-sectional shape of double cones have been found to alter the mosaic formation from a square to a row arrangement (Novales-Flamarique, 2001).

Many marine fishes hatch with only single cones arranged in a hexagonal (or row) mosaic, presumably to optimize packing and maximize acuity in small eyes (Blaxter, 1986; Evans and Fernald, 1993; Pankhurst et al., 1993; Shand et al., 1999). New cones are added during growth and double cones are formed at a later stage by the association of large numbers of neighboring single cones and the formation of subsurface cisternae along the common membranes (Shand et al., 1999, 2001). At the time of double cone formation, the hexagonal mosaic of single cones breaks down and, at least in black bream, a square mosaic comprising four double cones surrounding a single cone eventually results (Fig. 8.8).

Acuity in larval fishes rapidly increases as the eye and lens grow, producing an increase in the magnification of the image on the retina (Fernald, 1988; Shand, 1997). Sensitivity is improved as the photoreceptors increase in size, as new cones are differentiated and added at the retinal periphery and as rods are inserted throughout the retina (van der Meer, 1994; Poling and Fuiman, 1997; Fuiman and Delbos,

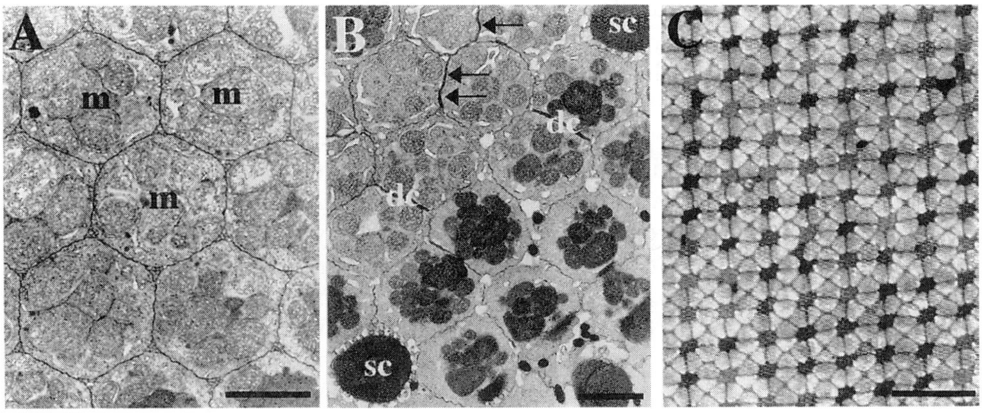

FIGURE 8.8. Development of a square cone mosaic. (A) Electron micrograph of the retina in a 3–5 mm standard length (SL) black bream *Acanthopagrus butcheri* showing the regular hexagonal array of a single morphological class of photoreceptor in tangential section. (B) Tangential section of the retina of a black bream of 8 mm SL showing cone multiples with subsurface cisternae along their apposing membranes (arrows) interrupted by single cones (sc). (C) The regular array of the square mosaic in a black bream of 30 mm SL where four double cones surround a central single cone (dark profiles). dc, double cone; m, mitochondria. Scale bars, 2 µm (A, B); 20 µm (C). (Adapted from Shand et al., 1999).

1998; Pankhurst and Hilder, 1998). The developmental changes in the complement and number of photoreceptors resulting in greater sensitivity provides the potential for migration to progressively lower light intensity habitats. For example, larval cardinal fishes (Apogonidae) are able to feed at greater depths than other coral reef fish larvae of a similar age due to the presence of larger double cones and a greater rate of addition of new rods (Shand, 1997; Job and Bellwood, 2000; Job and Shand, 2001). Similarly, dhufish *Glaucosoma hebraicum* rapidly increase both the length of their cones and their rod densities from an early stage of retinal development as they move into deeper water (Shand et al., 2001). Ontogenetic migration to a habitat with lower light intensity can also result in the reduction or loss of cones and an increase in rod density to enhance sensitivity. For example, in the surface-living juveniles of *Sebastes diploproa* (Boehlert, 1979) and *Gempylus serpens* (Munk, 1990) the cones are lost as the fishes mature and move into deep water. Another possible mechanism for increasing sensitivity during development is observed in the southern hemisphere lamprey *Geotria australis*, as it returns to the heavily tannin-stained rivers for spawning. The ellipsoid region of one of the cone photoreceptors develops a large transparent ellipsosome, which may gather and focus light onto the outer segment (Collin et al., 1999).

5.3. Spectral Sampling and Visual Pigment Changes

Particular combinations of visual pigments appear to be "suited" for the detection of contrast in different bodies of water (Lythgoe and Partridge, 1989, 1991; Chapter 17). Therefore, it is perhaps not surprising that during migration or development, concomitant changes in spectral sensitivity that match the changes in the photic environment have also been found in a number of species. The changes in spectral sensitivity can result from structural modifications in the retina or physiological changes to the visual pigments themselves.

The loss of single cones from the corners of the cone mosaic, mentioned above, has been correlated with the loss of ultraviolet (UV) sensitivity as fishes move into deeper water, switch from a planktivorous to a benthopelagic existence, or begin their migration to the ocean (Bowmaker and Kunz, 1987; Hawryshyn et al., 1989; Loew and Wahl, 1991). In the salmonid, *Oncorhynchus mykiss*, the loss of single cones appears to affect only the ventral retina. UV sensitivity is known to reappear when the fishes return to the rivers, however, it is not known if new cones are regenerated in ventral retina or whether the sensitivity results from the population of single cones that remains in dorsal retina (Hawryshyn et al., 1989; Browman and Hawryshyn, 1992; Beaudet et al., 1993; Deutschlander et al., 2001).

Changes in the physiology of the visual pigments can bring about changes in their absorption characteristics by two main mechanisms: either a change from one chromophore to another or a switch in opsin expression (Loew, 1995 for review). Opsins with the A_2 chromophore absorb at longer wavelengths than their A_1 analogue, and the difference is greatest at long wavelengths (Dartnall and Lythgoe, 1965; Whitmore and Bowmaker, 1989; Parry and Bowmaker, 2000). Hence species inhabiting long wavelength-transmitting estuarine or freshwater frequently possess A_2 (porphyropsin)-based visual pigments. Conversely, the majority of marine species possess A_1 (rhodopsin)-based visual pigments (Wald, 1939; Lythgoe, 1979; Bowmaker, 1995). Wald (1958), after examining lampreys and eels, came to the conclusion that species undergoing metamorphosis and moving between marine and freshwater switched from A_1 to A_2 and vice versa. Examples of species switching either their chromophores, or the proportion of A_1 to A_2, during migration include salmonids (Beatty, 1966), eels (Carlisle and Denton, 1959; Beatty, 1975), and lampreys (Crescitelli, 1956). Many species of cyprinids, whose habitats can be spectrally variable on a seasonal basis, possess a mixture of both chromophores, and the ratio of the two pigments can vary according to environmental conditions such as day length and temperature (see Beatty, 1984; Muntz and Mouat, 1984; Whitmore and Bowmaker, 1989).

Changes in rod opsin structure are also possible in combination with chromophore changes, as demonstrated in eels as they migrate either into deep marine water (Carlisle and Denton, 1959; Beatty, 1975) or into freshwater (Wood et al., 1992) habitats, where the peak absorption (λ_{max}) shifts to shorter and longer wavelengths, respectively. Opsin substitutions have also been monitored within individual rods in eels (Wood and Partridge, 1993) and inferred in the rods of the deep-sea pearleye, *Scoperlarchus analis* (Partridge et al., 1992).

In developing marine fishes, changes in cone opsin structure have been inferred in retinae containing purely A_1-based visual pigments. During the larval/juvenile transition in the winter flounder, *Pseudopleuronectes americanus*, only single cones with a λ_{max} at 520 nm are present, whereas in adults three different visual pigments occur with λ_{max} values at 457, 531, and 547 nm (Evans et al., 1993). In the pollack, *Pollachius pollachius*, the λ_{max} of the single cones shifts from violet (420 nm) to blue (460 nm) as the fishes migrate to deeper green-transmitting coastal water during growth (Shand et al., 1988). Similarly, in the goatfish, *Upeneus tragula*, double cones lose red sensitivity by shifting the λ_{max} of the double cones from a 487/580 nm pair to a 515/530 nm pair as the fishes leave the broad spectrum surface waters to take up a benthic habitat with reduced proportions of red light (Shand, 1993). In the estuarine black bream *Acanthopagrus butcheri*, the λ_{max} of both single and double cones changes as they migrate to deeper, predominantly red-transmitting, tannin-stained water and begin feeding from the substrate (Shand et al., 2002). In the bream, the short wavelength-sensitive single cones shift their λ_{max} from 420 to 480 nm and the medium wavelength-sensitive double cones from about 530 to 565 nm (Fig. 8.9). While all the above changes in visual pigments can be rationalized in terms of improving the efficiency of vision in differing spectral environments, the exact mechanisms that initiate and bring about the changes from one opsin to another are presently unknown.

Until recently, the spectral sensitivity of cones in early stage larval fishes has been unclear. The possession of only one morphological cone type (single cones) suggests the presence of only one visual pigment (Evans et al., 1993; Loew and Sillman, 1993). However, it

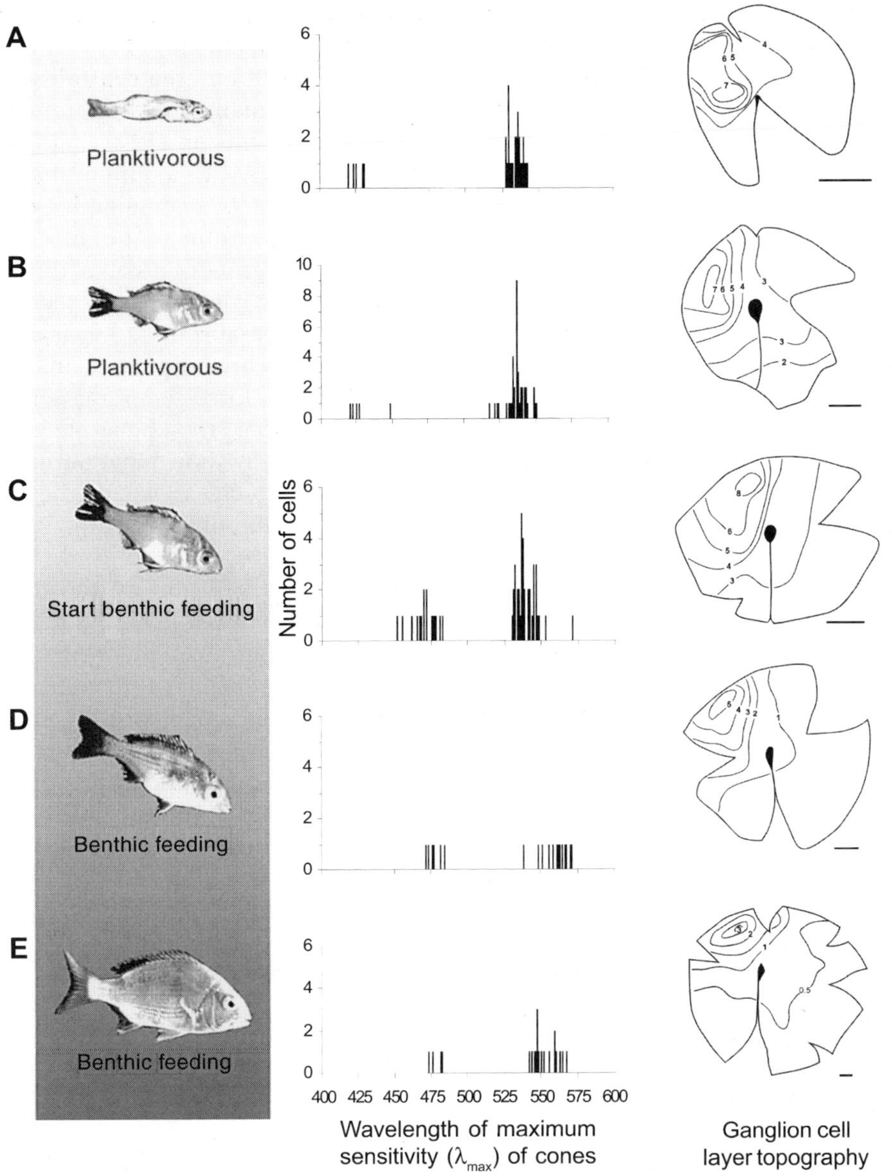

FIGURE 8.9. Summary of behavioral and visual changes in the black bream *Acanthopagrus butcheri*, during juvenile transition. The larvae and juveniles younger than 100 days (A, B) are primarily shallow-water plankton feeders, but gradually begin feeding from the substrate in deeper tannin-stained estuarine water (C). Concomitant with the behavioral changes, the visual pigments of the cones have been found to shift to longer wavelengths (frequency histograms of λ_{max} records from individual cones) as the position of the ganglion cell area centralis relocates from the temporal to the dorso-temporal region of the retina by 170 days of age (D) (retinal topography maps of right eyes) and become similar to adults (E). Scale bars, 1 mm. Cell densities $\times 10^4$ per mm^2. See text for further details.

is now apparent that at least two visual pigments are present in the retinae of many marine larval fishes when there is only one cone type present (Britt et al., 2001; Helvick et al., 2001; Shand et al., 2002). The majority of cones are medium wavelength (green) sensitive and would increase sensitivity to the predominant wavelengths transmitted in coastal waters. The second visual pigment appears to be short wavelength (either violet or UV) sensitive. The possession of a short wavelength-sensitive pigment may be an aid to planktivory in a shallow-water environment rich in UV wavelengths where zooplankton either reflect or absorb UV light (Bowmaker, 1991; Novales-Flamarique and Browman, 2001). Indeed, UV light has been found to facilitate plankton feeding (Loew et al., 1993; Browman et al., 1994) and several adult planktivorous fishes have also been found to possess UV-sensitive visual pigments (McFarland and Loew, 1994). From *in situ* opsin labeling in larval halibut, UV sensitivity is observed in only the ventral retina, which may be useful in feeding on plankton in the upper visual field (Helvick et al., 2001).

In the black bream, no changes in visual pigments are observed at the time the double cones are formed and the cone mosaic is reorganized (Shand et al., 2002; Figs. 8.8, 8.9). Changes in visual pigments occur only when behavioral and habitat transitions are underway (Shand et al., 2002; Fig. 8.9). However, individual variability in the timing of visual pigment changes was observed, where fishes reared in tannin-stained water or caught from tannin-stained estuaries undergo these changes at an earlier stage than those reared in clear water (J. Shand and N. Thomas unpublished). The individual variability may reflect the ability for species living in unstable environments, such as estuaries, to respond to changing conditions. However, previous light history has now also been shown to affect feeding ability of the larvae of striped trumpeter, *Latris lineata*, (Cobcroft and Pankhurst, 2001). It is therefore possible that visual pigment lability is not restricted to species that inhabit variable-light environments.

5.4. Relocation of the Area Centralis During Ontogeny

The high densities of ganglion cells in specific regions of the vertebrate retina, the area centralis or visual streak, have been shown to provide high resolving power along specific visual axes and be related to the behavior or habitat of the animal (Hughes, 1977, 1985). In fishes, the area centralis is commonly found in temporal or temporo-dorsal retina (Collin and Pettigrew, 1988a,b; Collin, 1999; Fig. 8.5). The development of an area centralis in fishes that have a continually growing retina has been addressed by Easter (1992) and, in the case of those species with a temporal area centralis, the visual axis is maintained by asymmetric growth of the retina. The mechanisms that control such growth involve differential cell addition (Cameron, 1995) and increased retinal stretching in nasal retina (Zygar et al., 1999).

Adult black bream possess an area centralis located in the dorso-temporal periphery when they feed from the substrate. However, it has been recently shown that during early larval and juvenile stages, when the fishes feed on plankton, the area centralis is located in a temporal position (Shand et al., 2000a,c). The change in the position of the area centralis takes place when the juveniles move to deeper water and begin feeding from the substrate (Fig. 8.9), although, in keeping with visual pigment changes, the timing can be variable (Shand et al., 2000c). The change from a temporal to a dorso-temporal location is mediated by cell death in central retina and differential cell addition at the retinal margins as the eye grows (Shand et al., 2000b). Such a change in the position of the area centralis and its maintenance in dorso-temporal retina is possible only because the retina is continuously growing.

6. Central Representation of Specialized Retinal Sampling

6.1. Tectal Magnification Factor

The high density of retinal ganglion cells associated with an area centralis, fovea, or horizontal streak may constitute a large proportion of the total population of ganglion cells in some species of fishes, despite occupying only a small proportion of the retinal area. Given the retinotopic organization of the optic tectum, does this specialized region of the retina occupy a disproportionately large region of the optic tectum? In the kelp bass *Paralabrax clathratus*, the foveal and perifoveal retinal regions contain a ganglion cell density peak of 33.44×10^3 cells per mm^2 and subtend only 10–15° of arc in the frontal visual field, but occupy a disproportionately large area (five times larger than a comparable region of the ventral or dorsal visual field) of the optic tectum (Schwassmann, 1968). Similarly, a magnified representation of a 25°-wide horizontal streak across the retinal meridian has been found in the optic tectum of the four-eyed fish, *Anableps microlepis* (Schwassmann and Kruger, 1965). In the lemon shark *Negaprion brevirostris*, the horizontal streak subtends 26% of the total visual field but the terminal fields of the axons from the ganglion cells comprising the streak occupy 52% of the tectal surface. The tectal magnification factor (number of microns on the tectal surface per degree of visual field coverage) in *N. brevirostris* is 100 µm/° within the visual streak and 33 µm/° outside the visual streak (Hueter, 1991). Qualitative comparisons between the optic tecta of species with comparably sized eyes with and without a visual streak also reveal that the optic tectum in species with a visual streak is appreciably larger (Ito and Murakami, 1984; Collin, 1987). Therefore, it has been suggested that the temporal area centralis projects to the thalamus and the visual streak projects to the optic tectum. Both anatomical and physiological investigations are still required to explore parallel processing in fishes.

6.2. Central Input of Binocular Information

In mammals, the partial decussation of retinal ganglion cell axons brings information from corresponding regions of the binocular visual fields into register. These axons then make accurate choices and synapse with target nuclei within the central nervous system. Both contra- and ipsilaterally projecting axons are essential for binocular vision where there is generally a direct relationship between the size of the binocular field and the region of the retina containing ipsilaterally projecting ganglion cells. Previous reports of bilateral visual projections in lampreys (De Miguel et al., 1990), cartilaginous fishes (Northcutt and Wathey, 1980; Repérant et al., 1986; Northcutt, 1991), and early life history stages in bony fishes (Collin and Northcutt, 1991, 1995) has recently initiated a developmental study in the black bream *Acanthopagrus butcheri*, where a temporal area centralis "relocates" to a dorsal position during development (see Section 5.4; Fig. 8.9). In this model, temporal ganglion cells undergo a period of cell death and a new dorsal area centralis develops by the differential addition of new ganglion cells at the retinal margin (Shand et al., 2000b). Retro- and anterograde labeling studies confirm three sites in the optic tectum that receive ipsilateral input in small fishes. However, input to these retino-recipient tectal targets is progressively lost as the fish grows and the area centralis subtends different regions of the dorso-temporal hemifield (S.P. Collin, J. Shand, L.B.G. Tee, and L.D. Beazley, unpublished). The loss of input from a subset of temporal ganglion cells subtending the binocular visual field in *A. butcheri* constitute a transient ipsilateral projection and a putative switch in binocular processing. However, this needs to be confirmed physiologically, in addition to the anatomical and physiological substrates for binocular vision in adult fishes.

Acknowledgments. SPC is supported by an ARC Discovery Grant (DP0209452). JS is supported by the NHMRC (Australia) (Program Grant No. 993219). Nicole Thomas kindly helped with the preparation of Figures 8.4, 8.8, and 8.9.

References

Ahlbert, I.-B. (1976). Organization of the cone cells in the retinae of salmon (*Salmo salar*) and trout (*Salmo trutta trutta*) in relation to their feeding habits. *Acta Zool.* 57:13–35.

Ali, M.A., and Anctil, M.A. (1976). *Retinas of Fishes: An Atlas*. Berlin: Springer-Verlag.

Anctil, M. (1969). Structure du rétine chez quelques téléosteens marin du plateau continental. *J. Fish. Res. Bd. Can.* 26:597–628.

Anctil, M., and Ali, M.A. (1976). Cone droplets of mitochondrial origin in the retina of *Fundulus heteroclitus* (Pisces, Cyprinidontidae). *Zoomorphologie* 84:103–111.

Arrese, C.A., Hart, N.S., Thomas, N., Beazley, L.D., and Shand, J. (2002). Trichromacy in Australian marsupials. *Curr. Biol.* 12:657–660.

Avery, J.A., and Bowmaker, J.K. (1982). Visual pigments in the four-eyed fish *Anableps anableps*. *Nature* 298:62–64.

Bathelt, D. (1970). Experimentelle und vergleichend morphologische Untersuchungen am visuellen System von Teleostiern. *Zool. Jb. Anat.* 87:402–470.

Beatty, D.D. (1966). A study of the succession of visual pigments in Pacific salmon (*Oncorhynchus*). *Can. J. Zool.* 44:429–455.

Beatty, D.D. (1975). Visual pigments of the American eel, *Anguilla rostrata*. *Vision Res.* 15:771–776.

Beatty, D.D. (1984). Visual pigments and the labile scotopic visual system of fish. *Vision Res.* 24: 1563–1573.

Beaudet, L., and Hawryshyn, C.W. (1999). Ecological aspects of vertebrate visual ontogeny. In: *Adaptive Mechanisms in the Ecology of Vision* (Archer, S.N., Djamgoz, M.B.A., Loew, E.R., Partridge, J.C., and Vallerga, S., eds.), pp. 383–412. Dordrecht: Kluwer.

Beaudet, L., Browman, H.I., and Hawryshyn, C.W. (1993). Optic nerve response and retinal structure in rainbow trout of different sizes. *Vision Res.* 33: 1739–1746.

Beaudet, L., Novales Flamarique, I., and Hawryshyn, C.W. (1997). Cone photoreceptor topography in the retina of sexually mature pacific salmonid fishes. *J. Comp. Neurol.* 383:49–59.

Berra, T.M., and Allen, G.R. (1989). Burrowing, emergence, behavior and functional morphology of the Australian salamanderfish, *Lepidogalaxias salamandroides*. Fisheries 14:2–10.

Blaxter, J.H.S. (1986). Development of sense organs and behaviour of teleost larvae with special reference to feeding and predator avoidance. *Trans. Am. Fish. Soc.* 115:98–114.

Boehlert, G.W. (1979). Retinal development in postlarval through juvenile *Sebastes diplopora*: Adaptation to a changing photic environment. *Rev. Can. Biol.* 38:265–280.

Borwein, B., and Hollenberg, M.J. (1973). The photoreceptors of the "four-eyed" fish *Anableps anableps*, L. *J. Morphol.* 140:405–441.

Bowmaker, J.K. (1990). Visual pigments of fishes. In: *The Visual System of Fish* (Douglas, R.H., and Djamgoz, M.B.A., eds.), pp. 81–107. London: Chapman & Hall.

Bowmaker, J.K. (1991). The evolution of vertebrate visual pigments and photoreceptors. In: *Vision and Visual Dysfunction, Vol. 2* (Cronly-Dillon, J.R., and Gregory, R.L., eds.), pp. 63–81. Boca Raton: CRC Press.

Bowmaker, J.K. (1995). The visual pigments of fish. *Prog. Retinal Eye Res.* 15:1–31.

Bowmaker, J.K., and Kunz, Y.W. (1987). Ultraviolet receptors, tetrachromatic colour vision and retinal mosaics in the brown trout (*Salmo trutta*): Age-dependent changes. *Vision Res.* 27:2101–2108.

Bowmaker, J.K., and Martin, G.R. (1985). Visual pigments and oil droplets in the penguin, *Spheniscus humboldti*. *J. Comp. Physiol. A.* 156:71–77.

Boycott, B.B., and Wässle, H. (1974). The morphological types of ganglion cells of the domestic cat's retina. *J. Physiol. (Lond.)* 240:397–419.

Bozzano, A., and Collin, S.P. (2000). Retinal ganglion cell topography in elasmobranches. *Brain Behav. Evol.* 55:191–208.

Braekevelt, C.R. (1973). Fine structure of the retinal pigment epithelium and photoreceptor cells of an Australian marsupial (*Setonix brachyurus*). *Can. J. Zool.* 51:1093–1100.

Braekevelt, C.R. (1975). Photoreceptor fine structure in the northern pike (*Esox lucius*). *J. Fish Res. Bd. Can.* 32:1711–1721.

Braekevelt, C.R. (1982). Photoreceptor fine structure in the goldeye (*Hiodon alosoides*) (Teleostei). *Anat. Embryol.* 165:177–192.

Braekevelt, C.R. (1985). Photoreceptor fine structure in the archerfish (*Toxotes jaculatrix*). *Amer. J. Anat.* 173:89–98.

Britt, L.L., Loew, E.R., and McFarland, W.N. (2001). Visual pigments in the early life history stages of Pacific northwest marine fishes. *J. Exp. Biol.* 204:2581–2587.

Browman, H.I., and Hawryshyn, C.W. (1992). Thyroxine induces a precocial loss of ultraviolet photosensitivity in rainbow trout (*Oncorhynchus mykiss*, Teleostei). *Vision Res.* 32:2303–2312.

Browman, H.I., Novales-Flamarique, I., and Hawryshyn, C.W. (1994). Ultraviolet photoreception contributes to prey search behavior in two species of zooplanktivorous fishes. *J. Exp. Biol.* 186:187–198.

Butcher, E.O. (1938). The structure of the retina of *Fundulus heteroclitus* and the regions of the retina associated with the different chromatophoric responses. *J. Exp. Zool.* 79:275–293.

Cameron, D.A. (1995). Asymmetric retinal growth in the adult teleost green sunfish (*Lepomis cyanellus*). *Visual Neurosci.* 12:95–102.

Cameron, D.A., and Easter, S.S. Jr. (1993). The cone photoreceptor mosaic of the green sunfish (*Lepomis cyanellus*). *Visual Neurosci.* 10:375–384.

Cameron, D.A., and Easter, S.S. (1995). Cone photoreceptor regeneration in adult fish retina: Phenotypic determination and mosaic pattern formation. *J. Neurosci.* 15:2255–2271.

Carleton, K.L., and Kocher T.D. (2001). Cone opsin genes of African cichlid fishes: Tuning spectral sensitivity by differential gene expression. *Mol. Biol. Evol.* 18:1540–1550.

Carleton, K.L., Hárosi, F.I., and Kocher, T.D. (2000). Visual pigments of African cichlid fishes: Evidence for ultraviolet vision from microspectrophotometry and DNA sequences. *Vision Res.* 40:879–890.

Carlisle, D.B., and Denton, E.J. (1959). On the metamorphosis of the visual pigments of *Anguilla anguilla* L. *J. Mar. Biol. Assoc. U.K.* 38:97–102.

Chievitz, J.H. (1889). Untersuchungen über die Area centralis retinae. *Arch. Anat. Physiol. Anat. Abt. Suppl.* 1889:139–194.

Chievitz, J.H. (1891). Ueber das Vorkommen der Area centralis retinae in den vier höheren Wirbelthierklassen. *Arch. Anat. Physiol. Anat. Abt.* 1891:311–333.

Cobcroft, J.M., Pankhurst, P.M., Hart, P.R., and Battaglene, S.C. (2001). The effects of light intensity and algae-induced turbidity on feeding behaviour of larval striped trumpeter. *J. Fish. Biol.* 59:1181–1197.

Collin, H.B., and Collin, S.P. (1988a). The cornea of the sandlance, *Limnichthyes fasciatus* (Creeiidae). *Cornea* 7(3):190–203.

Collin, H.B., and Collin, S.P. (1996). Fine structure of the cornea in the freshwater salamanderfish, *Lepidogalaxias salamandroides*. *Cornea* 15:414–426.

Collin, S.P. (1987). Retinal topography in reef teleosts. PhD thesis. University of Queensland, Australia.

Collin, S.P. (1988). The retina of the shovel-nosed ray, *Rhinobatos batillum* (Rhinobatidae): Morphology and quantitative analysis of the ganglion, amacrine and bipolar cell populations. *Exp. Biol.* 47:195–207.

Collin, S.P. (1989). Topography and morphology of retinal ganglion cells in the coral trout, *Plectropoma leopardus* (Serranidae): A retrograde cobaltous-lysine study. *J. Comp. Neurol.* 281:143–158.

Collin, S.P. (1997). Specialisations of the teleost visual system: adaptive diversity from shallow-water to deep-sea. *Acta. Physiol. Scand.* 161 (Supplement 638):5–28.

Collin, S.P. (1999). Behavioural ecology and retinal cell topography. In: *Adaptive Mechanisms in the Ecology of Vision* (Archer, S.N., Djamgoz, M.B.A., Loew, E.R., Partridge, J.C., and Vallerga, S., eds.), pp. 509–535. Dordrecht: Kluwer Academic Publishers.

Collin, S.P., and Ali, M.A. (1994). Multiple areas of acute vision in two freshwater teleosts, the creek chub, *Semotilus atromaculatus* (Mitchell) and the cutlips minnow, *Exoglossum maxillingua* (Lesueur). *Can. J. Zool.* 72:721–730.

Collin, S.P., and Collin, H.B. (1988b). The morphology of the retina and lens of the sandlance *Limnichthyes fasciatus* (Creeiidae). *Exp. Biol.* 47:208–218.

Collin, S.P., and Collin, H.B. (1988c). Topographic analysis of the retinal ganglion cell layer and the optic nerve in the sandlance, *Limnichthyes fasciatus* (Creeiidae, Perciformes). *J. Comp. Neurol.* 278:226–241.

Collin, S.P., and Collin, H.B. (1993). The visual system of the Florida garfish, *Lepisosteus platyrhincus* (Ginglymodi). I. Retina. *Brain Behav. Evol.* 42:77–97.

Collin, S.P., and Collin, H.B. (1998). Retinal and lenticular ultrastucture in the aestivating salamanderfish, *Lepidogalaxias salamandroides* (Galaxiidae, Teleostei) with special reference to a new type of photoreceptor mosaic. *Histol. Histopathol.* 13:1037–1048.

Collin, S.P., and Collin, H.B. (1999). The foveal photoreceptor mosaic in the pipefish, *Corythoichthyes paxtoni* (Syngnathidae, Teleostei). *Histol. Histopathol.* 14:369–382.

Collin, S.P., and Collin, H.B. (2001). The fish cornea: Adaptations for different aquatic environments. In: *Sensory Biology of Jawed Fishes: New Insights* (Kapoor, B.G., and Hara, T.J., eds.), pp. 57–96. Plymouth, UK: Science Publishers.

Collin, S.P., and Northcutt, R.G. (1991). The development and evolution of ipsilateral projections in the retinofugal pathway of ray-finned fishes. *Invest. Ophthalmol. Vis. Sci.* 32:1033.

Collin, S.P., and Northcutt, R.G. (1993). The visual system of the Florida garfish, *Lepisosteus platyrhincus* (Ginglymodi). III. Retinal ganglion cells. *Brain Behav. Evol.* 42:295–320.

Collin, S.P., and Northcutt, R.G. (1995). The visual system of the Florida garfish, *Lepisosteus platyrhincus* (Ginglymodi). IV. Bilateral projections and the binocular visual field. *Brain Behav. Evol.* 45:34–53.

Collin, S.P., and Pettigrew, J.D. (1988a). Retinal topography in reef teleosts. I. Some species with well-developed areae but poorly-developed streaks. *Brain Behav. Evol.* 31:269–282.

Collin, S.P., and Pettigrew, J.D. (1988b). Retinal topography in reef teleosts. II. Some species with prominent horizontal streaks and high density areae. *Brain Behav. Evol.* 31:283–295.

Collin, S.P., and Pettigrew, J.D. (1988c). Retinal ganglion cell topography in teleosts: A comparison between Nissl-stained material and retrograde labeling from the optic nerve. *J. Comp. Neurol.* 276:412–422.

Collin, S.P., and Pettigrew, J.D. (1989). Quantitative comparison of the limits on visual spatial resolution set by the ganglion cell layer in twelve species of reef teleosts. *Brain Behav. Evol.* 34:184–192.

Collin, S.P., and Potter, I.C. (2000). The ocular morphology of the southern hemisphere lamprey *Mordacia mordax* Richardson with special reference to a single class of photoreceptor and a retinal tapetum. *Brain Behav. Evol.* 55:120–138.

Collin, S.P., Collin, H.B., and Ali, M.A. (1996). Ultrastructure and organization of the retina and pigment epithelium in the cutlips minnow, *Exoglossum maxillingua* (Cyprinidae, Teleostei). *Histol. Histopathol.* 11:55–69.

Collin, S.P., Hoskins, R.V., and Partridge, J.C. (1997). Tubular eyes of deep-sea fishes: A comparative study of retinal topography. *Brain Behav. Evol.* 50:335–357.

Collin, S.P., Hoskins, R.V., and Partridge, J.C. (1998). Seven retinal specializations in the tubular eye of the deep-sea pearleye, *Scopelarchus michaelsarsi*: A case study in visual optimization. *Brain Behav. Evol.* 51:291–314.

Collin, S.P., Lloyd, D., and Wagner, H.-J. (2000). Visual and olfactory input to the CNS in foveate deep-sea teleosts: The relative importance of vision. *Phil. Trans. Roy. Soc. B.* 355:1315–1320.

Collin, S.P., Lloyd, D.J., and Partridge, J.C. (1994). Retinal ganglion cell topography in deep-sea fishes: Intrafamilial variation within the family Alepocephalidae. *Proc. Aust. Neurosci. Soc.* 4:205.

Collin, S.P., Potter, I.C., and Braekevelt, C.R. (1999). The ocular morphology of the Southern Hemisphere Lamprey *Geotria australis* Gray, with special reference to optical specializations and the characterization and phylogeny of photoreceptor types. *Brain Behav. Evol.* 54:96–118.

Cook, J.E., and Becker, D.L. (1991). Regular mosaics of large displaced and non-displaced ganglion cells in the retina of a cichlid fish. *J. Comp. Neurol.* 306:668–684.

Cook, J.E., and Chalupa, L.M. (2000). Retinal mosaics: New insights into an old concept. *TINS* 23:26–34.

Cook, J.E., and Sharma, S.C. (1995). Large retinal ganglion cells in the channel catfish (*Ictalurus punctatus*): Three types with distinct dendritic stratification patterns form similar but independent mosaics. *J. Comp. Neurol.* 362:331–349.

Cook, J.E., Becker, D.L., and Kapila, R. (1992). Independent mosaics of large inner- and outer-stratified ganglion cells in the goldfish retina. *J. Comp. Neurol.* 318:355–366.

Cook, J.E., Kondrashev, S.L., and Pudugolnikova, T.A. (1996). Biplexiform ganglion cells, characterized by dendrites in both outer and inner plexiform layers, are regular, mosaic forming elements of teleost fish retinae. *Visual Neurosci.* 13:517–528.

Cook, J.E., Pudugolnikova, T.A., and Kondrashev, S.L. (1999). Species-dependent variation in the dendritic stratification of apparently homologous retinal ganglion cell mosaics in two neoteleost fishes. *Vision Res.* 39:2615–2631.

Crescitelli, F. (1956). The nature of the lamprey visual pigment. *J. Gen. Physiol.* 39:423–435.

Dartnall, H.J.A., and Lythgoe, J.N. (1965). The spectral clustering of visual pigments. *Vision Res.* 5:81–100.

De Miguel, E., Rodicio, M.C., and Anadon, R. (1990). Organization of the visual system in larval lampreys: An HRP study. *J. Comp. Neurol.* 302:529–542.

Denton, E.J., and Locket, N.A. (1989). Possible wavelength discrimination by multibank retinae in deep-sea fishes. *J. Mar. Biol. Assoc. U.K.* 69:409–435.

Deutschlander, M.E., Greaves, D.K., Haimberger, T.J., and Hawryshyn, C.W. (2001). Functional mapping of ultraviolet photosensitivity during metamorphic transitions in a salmonid fish, *Oncorhynchus mykiss*. *J. Exp. Biol.* 204:2401–2413.

Douglas, R.H., Collin, S.P., and Corrigan, J. (2002). The eyes of suckermouth armoured catfish (Loricariidae, subfamily Hypostomus): pupil response, lenticular longitudinal spherical aberration and retinal topography. *J. Exp. Biol.* 205:3425–3433.

Douglas, R.H., Harper, R.D., and Case, J.F. (1998a). The pupil response of a teleost fish, *Porichthys notatus*: A description and comparison to other species. *Vision Res.* 38:2697–2710.

Douglas, R.H., Partridge, J.C., and Marshall, N.J. (1998b). The eyes of deep-sea fish. I. Lens pigmentation, tapeta and visual pigments. *Prog. Retinal Eye Res.* 17:587–636.

Dunn-Meynell, A.A., and Sharma, S.C. (1986). The visual system of the channel catfish *(Ictalurus punctatus)*. I. Retinal ganglion cell morphology. *J. Comp. Neurol.* 247:32–55.

Easter, S.S. Jr. (1992). Retinal growth in foveated teleosts: Nasotemporal asymmetry keeps the fovea in temporal retina. *J. Neurosci.* 12:2381– 2392.

Eberle, H. (1967). Cone length and chromatic aberration in *Lebistes reticulatus*. *Z. Verlg. Physiol.* 57:172–173

Engström, K. (1960). Cone types and cone arrangements in the retina of some cyprinids. *Acta Zool.* 41:277–295.

Engström, K. (1963a). Cone types and cone arrangements in teleost retinae. *Acta. Zool. (Stockh.)* 44:179–243.

Engström, K. (1963b). Structure, organization and ultrastructure of the visual cells in the teleost family Labridae. *Acta Zool. (Stockh.)* 44:1–41.

Evans, B.I., and Fernald, R.D. (1993). Retinal transformation at metamorphosis in the winter flounder (*Pseudopleuronectes americanus*). *Visual Neurosci.* 10:1055–1064.

Evans, B.I., Harosi, F.I., and Fernald, R.D. (1993). Photoreceptor spectral absorbance in larval and adult winter flounder (*Pseudopleuronectes americanus*). *Visual Neurosci.* 10:1065–1071.

Fernald, R.D. (1988). Aquatic adaptations in fish eyes. In: *Sensory Biology of Aquatic Animals* (Atema, J., Fay, R.R., Popper, A.N., and Tavolga, W.N., eds.), pp. 435–466. New York: Springer-Verlag.

Fineran, B.A., and Nicol, J.A.C. (1974). Studies on the eyes of New Zealand parrot-fishes (Labridae). *Proc. Roy. Soc. (Lond.) B.* 186:217–247.

Franz, V. (1932). Auge und Akkommodation von *Petromyzon* (*Lampetra*) *fluviatilis*, L. *Zool. Jb. (Zool.)* 52:118–178.

Fritsches, K.A., and Marshall, N.J. (2002). Independent conjugate eye movements during optokinesis in teleost fish. *J. Exp. Biol.* 205:1241–1252.

Fritsches, K.A., and Marshall, J. (1999). A new category of eye movements in a small fish. *Curr. Biol.* 9:R272–R273.

Fritsches, K.A., Partridge, J.C., Pettigrew, J.D., and Marshall, N.J. (2000). Colour vision in billfish. *Phil. Trans. R. Soc. (Lond.) B.* 355:1253–1256.

Fritzsch, B., and Collin, S.P. (1990). Dendritic distribution of two populations of ganglion cells and the retinopetal fibres in the retina of the silver lamprey (*Ichthyomyzon unicuspis*). *Visual Neurosci.* 4:533–545.

Fröhlich, E., and Wagner, H.-J. (1998). Development of multibank rod retinae in deep-sea fishes. *Visual Neurosci.* 15:1–7.

Fuiman, L.A., and Delbos, B.C. (1998). Developmental changes in visual sensitivity of red drum, *Sciaenops ocellatus*. *Copeia* 1998:936–943.

Govardovskii, V.I., Rohlich, P., Szel, A., and Zueva, L.V. (1992). Immunocytochemical reactivity of rod and cone visual pigments in the sturgeon retina. *Visual Neurosci.* 8:531–537.

Harkness, L., and Bennett-Clarke, H.C. (1978). The deep fovea as a focus indicator. *Nature Lond.* 272:814–816.

Hart, N.S., Partridge, J.C., and Cuthill, I.C. (1998). Visual pigments, oil droplets and cone photoreceptor distribution in the European starling (*Sturnus vulgaris*). *J. Exp. Biol.* 201:1433–1446.

Hawryshyn, C.W., Arnold, M.G., Chaisson, D.J., and Martin, P.C. (1989). The ontogeny of ultraviolet sensitivity in rainbow trout (*Salmo gairdneri*). *Visual Neurosci.* 2:247–254.

Hayes, B.P., Martin, G.R., and Brooke, M. de L. (1991). Novel area serving binocular vision in the retinae of procellariform seabirds. *Brain Behav. Evol.* 37:79–84.

Helvick, J.V., Drivenes, O., Harboe, T., and Seo, H.-C. (2001). Topography of different photoreceptor cell types in the larval retinae of Atlantic halibut (*Hippoglossus hippoglossus*). *J. Exp. Biol.* 204:2553–2559.

Hisatomi, O., Satoh, T., and Tokunaga, F. (1997). The primary structure and distribution of killifish visual pigments. *Vision Res.* 37:3089–3096.

Hitchcock, P.F., and Easter, S.S. Jr. (1986). Retinal ganglion cells in goldfish: A qualitative classification into four morphological types, and a quantitative study of the development of one of them. *J. Neurosci.* 6:1037–1050.

Hueter, R.E. (1991). Adaptations for spatial vision in sharks. *J. Exp. Zool. Suppl.* 5:130–141.

Hughes, A. (1977). The topography of vision in mammals of contrasting life style: Comparative optics and retinal organisation. In: *Handbook of Sensory Physiology, Vol. VII\5* (Crescitelli, F., ed.), pp. 613–756. Berlin: Springer Verlag.

Hughes, A. (1985). New perspectives in retinal organization. In: *Progress in Retinal Research, Vol. 4* (Osbourne, N.N., and Chader, G.J., eds.), pp. 243–313. Oxford: Pergamon.

Ishikawa, M., Hashimoto, Y., Tonosaki, A., and Sakuragi, S. (1997). Preference of peanut agglutinin labeling for long wavelength sensitive cone photoreceptors in the dace retina. *Vision Res.* 37:383–387.

Ito, H., and Murakami, T. (1984). Retinal ganglion cells in two teleost species, *Sebastiscus marmoratus* and *Navodon modestus*. *J. Comp. Neurol.* 229: 80–96.

Job, S.D., and Bellwood, D.R. (1996). Visual acuity and feeding in larval *Premnas biaculeatus*. *J. Fish Biol.* 48:952–963.

Job, S., and Bellwood, D. (2000). Light sensitivity in larval fishes: implications for vertical zonation in the pelagic zone. *Limnol. Oceanog.* 45:362–371.

Job, S., and Shand, J. (2001). Spectral sensitivity of larval and juvenile coral reef fishes: Implications for feeding in a variable light environment. *Marine Ecol. Prog. Ser.* 214:267–277.

Johns, P.A. (1977). Growth of the adult goldfish eye. III. Source of the new retinal cells. *J. Comp. Neurol.* 176:343–358.

Johns, P.A., and Easter, S.S. Jr. (1977). Growth of the adult goldfish eye. II. Increase in retinal cell number. *J. Comp. Neurol.* 176:331–342.

Kahmann, H. (1934). Über das Vorkommen einer Fovea centralis im Knochenfischauge. *Zool. Anz.* 106:49–55.

Kirschfeld, K. (1976). The resolution of lens and compound eyes. In: *Neural Principles in Vision* (Zettler, F., and Weiler, R., eds.), pp. 354–370. Berlin: Springer-Verlag.

Kolb, H., and Jones, J. (1987). The distinction by light and electron microscopy of two types of cone containing colourless oil droplets in the retina of the turtle. *Vision Res.* 27:1445–1458.

Kunz, Y.W., Shuilleabhain, M.N., and Callaghan, E. (1985). The eye of the venomous marine teleost *Trachinus vipera* with special reference to the structure and ultrastructure of visual cells and pigment epithelium. *Exp. Biol.* 43:161–178.

Land, M.F. (1981). Optics and vision in invertebrates. In: *Handbook of Sensory Physiology, Vol. VII/6B: Vision in Invertebrates* (Autrum, H., ed.), pp. 471–592. Berlin: Springer-Verlag.

Land, M.F. (1999). Visual optics: The sandlance eye breaks all the rules. *Curr. Biol.* 9:R286–R288.

Lasater, E.M. (1982). Spatial receptive fields of catfish retinal ganglion cells. *J. Neurophysiol.* 48:823–825.

Levine, J.S., and MacMichol, E.F. (1979). Visual pigments in teleost fishes: Effects of habitat, microhabitat and behaviour on visual system evolution. *Sensory Process.* 3:95–131.

Locket, N.A. (1970). Deep-sea fish retinas. *British Med. Bull.* 26:107–111.

Locket, N.A. (1985). The multiple bank fovea of *Bajacalifornia drakei*, an alepocephalid deep-sea teleost. *Proc. Roy. Soc. (Lond.) B.* 224:7–22.

Locket, N.A. (1992). Problems of deep foveas. *Aust. New Zeal. J. Ophthalmol.* 20:281–295.

Loew, E.R. (1995). Determinants of visual pigment spectral location and photoreceptor cell spectral sensitivity. In: *Neurobiology and Clinical Aspects of the Outer Retina* (Djamgoz, M.B.A., Archer, S.N., and Vallerga, S., eds.), pp. 57–77. London: Chapman & Hall.

Loew, E.R., and Sillman, A.J. (1993). Age-related changes in the visual pigments of the white sturgeon (*Acipenser transmontanus*). *Can. J. Zool.* 71:1552–1557.

Loew, E.R., and Wahl, C.M. (1991). A short-wavelength sensitive cone mechanism in juvenile yellow perch, *Perca flavescens*. *Vision Res.* 31: 353–360.

Loew, E.R., McFarland, W.N., Mills, E.L., and Hunter, D. (1993). A chromatic action spectrum for planktonic predation by juvenile yellow perch, *Perca flavescens*. *Can. J. Zool.* 71:384–386.

Lyall, A.H. (1956). Occurrence of triple and quadruple cones in the retina of the minnow (*Phoxinus laevis*). *Nature* 177:1086–1087.

Lyall, A.H. (1957a). Cone arrangements in teleost retinae. *Q. J. Microsc. Sci.* 98:189–201.

Lyall, A.H. (1957b). The growth of the trout retina. *Q. J. Micros. Sci.* 98:101–110.

Lythgoe, J.N. (1979). *The Ecology of Vision*. Oxford: Oxford University Press.

Lythgoe, J.N. (1984). Visual pigments and environmental light. *Vision Res.* 24:1539–1550.

Lythgoe, J.N., and Partridge, J.C. (1989). Visual pigments and the aquisition of visual information. *J. Exp. Biol.* 146:1–20.

Lythgoe, J.N., and Partridge, J.C. (1991). The modelling of optimal visual pigments of dichromatic teleosts in green coastal waters. *Vision Res.* 31:361–371.

MacNichol, E.F., Kunz, Y.W., Levine, J.S., Harosi, F.I., and Collins, B.A. (1978). Ellipsosomes: Organelles containing a cytochrome-like pigment in the retinal cones of certain fishes. *Science* 200:549–552.

Marc, R.E., and Sperling, H.G. (1976). The chromatic organization of the goldfish retina. *Vision Res.* 16:1211–1224.

McCormick, M.I. (1993). Development and changes at settlement in the barbel structure of the reef fish, *Upeneus tragula* (Family: Mullidae). *Environ. Biol. Fish.* 37:269–282.

McEwan, M.R. (1938). A comparison of the retina of the Mormyrids with that of various other teleosts. *Acta Zool.* 19:427–465.

McFarland, W.N., and Loew, E.R. (1994). Ultraviolet visual pigments in marine fishes of the Family Pomacentridae. *Vision Res.* 34:1393–1396.

Mednick, A.S., and Springer, A.D. (1988). Asymmetric distribution of retinal ganglion cells in goldfish. *J. Comp. Neurol.* 268:49–59.

Mednick, A.S., Berk, M.F., and Springer A.D. (1988). Asymmetric distribution of cells in the inner nuclear and cone mosaic layers of the goldfish retina. *Neurosci. Letts.* 94:241–246.

Munk, O. (1968). The eyes of *Amia* and *Lepisosteus* (Pisces, Holostei) compared with the brachiopterygian and teleostean eyes. *Vidensk Meddr. Dansk. Naturh. Foren.* 131:109–127.

Munk, O. (1970). On the occurrence and significance of horizontal band-shaped retinal areae in teleosts. *Vidensk Meddr. Dansk Naturh. Foren.* 133:85–120.

Munk, O. (1975). On the eyes of two foveate notosudid teleosts, *Scopelosaurus hoedti* and *Ahliesaurus berryi*. *Vidensk Meddr. Dansk Naturh. Foren.* 138:87–125.

Munk, O. (1990). Changes in the visual cell layer of the duplex retina during growth of the eye of a deep-sea teleost *Gempylus serpens* Cuvier 1829. *Acta Zool.* 71:89–95.

Muntz, W.R.A., and Mouat, G.S.V. (1984). Annual variations in the visual pigments of brown trout inhabiting lochs providing different light environments. *Vision Res.* 24:1575–1580.

Nag, T.C. (1995). Ultrastructure of ellipsosomes in the retina of *Garra lamta*. *J. Electron Micros.* 44:405–407.

Nag, T.C., and Bhattacharjee, J. (1989). Retinal organisation in a hill stream cyprinid, *Crossocheilus latius latius* Hamilton. *Exp. Biol.* 48:197–202.

Nag, T.C., and Bhattacharjee, J. (1995). Retinal ellipsosomes: Morphology, development, identification, and comparison with oil droplets. *Cell. Tiss. Res.* 279:633–637.

Naka, K.-I., and Carraway, N.R.G. (1975). Morphological and functional identification of catfish retinal neurons. I. Classical morphology. *J. Neurophysiol.* 38:53–71.

Nicol, J.A.C. (1989). *The Eyes of Fishes*. Oxford: Clarendon Press.

Northcutt, R.G. (1991). Visual pathways in elasmobranches: organisation and phylogenetic implications. *J. Exp. Zool. Suppl.* 5:97–107.

Northcutt, R.G., and Wathey, J.C. (1980). Guitarfish possess ipsilateral as well as contralateral retinofugal projections. *Neurosci. Letts.* 20:237–242.

Novales-Flamarique, I. (2001). Gradual and partial loss of corner cone-occupied area in the retina of rainbow trout. *Vision Res.* 41:3073–3082.

Novales-Flamarique, I., and Browman, H.I. (2001). Foraging and prey-search behaviour of small juvenile rainbow trout (*Oncorhunchus mykiss*) under polarised light. *J. Exp. Biol.* 204:2415–2422.

Novales-Flamarique, I., and Hárosi, F.I. (2000). Photoreceptors, visual pigments, and ellipsosomes in the killifish, *Fundulus heteroclitus*: A microspectrophotometric and histological study. *Visual Neurosci.* 17:403–420.

Novales-Flamarique, I., and Hawryshyn, C.W. (1998). Photoreceptor types and their relation to the spectral and polarization sensitivities of clupeid fishes. *J. Comp. Physiol. A.* 182:793–803.

O'Day, K. (1938). The visual cells of the platypus (*Ornithorhincus*). *Brit. J. Ophthamol.* 22:321–328.

Ohtsuka, T. (1985). Relation of spectral types to oil droplets in cones of turtle retina. *Science* 229:874–877.

Pankhurst, N.W. (1982). Relation of visual changes to the onset of sexual maturation in the European eel *Anguilla anguilla* (L.). *J. Fish. Biol.* 21:127–140.

Pankhurst, P.M., and Hilder, P.E. (1998). Effect of light intensity on feeding of striped trumpeter *Latris lineata* larvae. *Mar. Fresh. Res.* 49:363–368.

Pankhurst, P.M., Pankhurst, N.W., and Montgomery, J.C. (1993). Comparison of behavioural and morphological measures of visual acuity during ontogeny in a teleost fish, *Forsterygion varium*, Tripterygiidae (Forster, 1801). *Brain Behav. Evol.* 42:178–88.

Parry, J.W.L., and Bowmaker, J.K. (2000). Visual pigment reconstitution in intact goldfish retina using synthetic retinaldehyde isomers. *Vision Res.* 40:2241–2247.

Partridge, J.C. (1989). The visual ecology of avian cone oil droplets. *J. Comp. Physiol. A.* 165:415–426.

Partridge, J.C., and Cummings, M.E. (1999). Adaptation of visual pigments to the aquatic environment. In: *Adaptive Mechanisms in the Ecology of Vision* (Archer, S.N., Djamgoz, M.B.A., Loew, E.R., Partridge, J.C., and Vallerga, S., eds.), pp. 251–283. Dordrecht: Kluwer Academic Publishers.

Partridge, J.C., Archer, S.N., and van Oostrum, J. (1992). Single and multiple visual pigments in deep-sea fishes. *J. Mar. Biol. Assoc. U.K.* 72: 113–130.

Pedler, C., and Tilly, R. (1964). The nature of the gecko visual cells: A light and electron microscopic study. *Vision Res.* 4:499–510.

Pettigrew, J.D., and Collin, S.P. (1995). Terrestrial optics in an aquatic eye: The sandlance, *Limnichthyes fasciatus* (Creediidae, Teleostei). *J. Comp. Physiol. A.* 177:397–408.

Pettigrew, J.D., Collin, S.P., and Fritsches, K. (2000). Prey capture and accommodation in the sandlance, *Limnichthyes fasciatus* (Creediidae, Teleostei). *J. Comp. Physiol. A.* 186:247–260.

Poling, K.R., and Fuiman, L.A. (1997). Sensory development and concurrent behavioural changes in Atlantic croaker larvae. *J. Fish Biol.* 51:402–421.

Pumphrey, R.J. (1948). The theory of the fovea. *J. Exp. Biol.* 25:299–312.

Raymond, P.A. (1995). Development and organization of photoreceptors. In: *Neurobiology and Clinical Aspects of the Outer Retina* (Djamgoz, M.B.A., Archer, S.N., and Vallerga, S., eds.), pp. 1–23. London: Chapman and Hall.

Reckel, F., Melzer, R.R., and Smola, U. (2001). Outer retinal fine structure of the garfish *Belone belone* (L.) (Belonidae, Teleostei) during light and dark adaptation: Photoreceptors, cone patterns and densities. *Acta Zool. (Stockh.)* 82:89–105.

Repérant, J., Miceli, D., Rio, J.P., Peyrichoux, J., Pierre, J., and Kipitchnikova, E. (1986). The anatomical organization of retinal projections in the shark *Scyliorhinus canicula* with special reference to the evolution of the selachian primary visual system. *Brain Res. Rev.* 11:227–248.

Robinson, S.R. (1994). Early vertebrate colour vision. *Nature* 367:121.

Saidel, W.M. (1987). An usual optic fiber pattern in the retina of the primitive fish *Pantodon buchholzi*, Peters. *Vision Res.* 27:511–516.

Saidel, W.M. (2000). Coherence in nervous system design: the visual system of *Pantodon buchholzi*. *Phil. Trans. Roy. Soc. (Lond.) B.* 355:1177–1181.

Saidel, W.M., and Fabiane, R.S. (1998). Optomotor response of *Anableps anableps* depends on the field of view. *Vision Res.* 38:2001–2008.

Sakai, H.M., Naka K.-I., and Dowling, J.E. (1986). Ganglion cell dendrites are presynaptic in catfish retina. *Nature* 319:495–497.

Scholes, J.H. (1975). Colour receptors and the synaptic connexions in the retina of a cyprinid fish. *Phil. Trans. Roy. Soc. (Lond.) B.* 270:61–118.

Schwartz, E. (1971). Ein septum papillaris im Auge von Pantodon buchholzi Pet. (Teleostei, Osteoglossiformes). *Zeitsch. für Morph. der Tiere* 70:119–127.

Schwassmann, H.O. (1968). Visual projection upon the optic tectum in foveate marine teleosts. *Vision Res.* 8:1337–1348.

Schwassmann, H.O., and Kruger, L. (1965). Experimental analysis of the visual system of the four-eyed fish *Anableps microlepis*. *Vision Res.* 5:269–281.

Shamim, K.M., Tóth, P., and Cook, J.E. (1997). Large retinal ganglion cells in the pipid frog *Xenopus laevis* form independent, regular mosaics resembling those of teleost fish. *Vis. Neurosci.* 14: 811–826.

Shand, J. (1993). Changes in the spectral absorption of cone visual pigments during settlement of the goatfish *Upeneus tragula*: The loss of red sensitivity as a benthic existence begins. *J. Comp. Physiol. A.* 173:115–121.

Shand, J. (1994). Changes in retinal structure during development and settlement of the goatfish *Upeneus tragula*. *Brain Behav. Evol.* 43:51–60.

Shand, J. (1997). Ontogenetic changes in retinal structure and visual acuity: A comparative study of coral-reef teleosts with differing post-settlement lifestyles. *Environ. Biol. Fish.* 49:307–322.

Shand, J., Archer, M.A., and Collin, S.P. (1999). Ontogenetic changes in the retinal photoreceptor mosaic in a fish, the black bream, *Acanthopagrus butcheri*. *J. Comp. Neurol.* 412:203–217.

Shand, J., Archer, M.A., Thomas, N., and Cleary, J. (2001a). Retinal development of West Australian dhufish, *Glaucosoma hebraicum*. *Visual Neurosci.* 18:711–724.

Shand, J., Chin, S.M., Harman, A.M., and Collin, S.P. (2000a). The relationship between the position of the area centralis and feeding behaviour in juvenile black bream *Acanthopagrus butcheri* (Sparidae: Teleostei). *Phil. Trans. Roy. Soc. (Lond.) B.* 355:1183–1186.

Shand, J., Chin, S.M., Harman, A.M., and Collin, S.P. (2000b). Mechanisms for changing the position of the area centralis in a retina that undergoes continual growth. *Proc. Aust. Neurosci. Soc.* 11:100.

Shand, J., Chin, S.M., Harman, A.M., Moore, S., and Collin, S.P. (2000c). Variability in the location of the retinal ganglion cell area centralis is correlated with ontogenetic changes in feeding behaviour in the black bream, *Acanthopagrus butcheri* (Sparidae, Teleostei). *Brain Behav. Evol.* 55:176–190.

Shand, J., Hart, N.S., Thomas, N., and Partridge, J.C. (2002). Developmental changes in the cone visual pigments of black bream *Acanthopagrus butcheri* *J. Exp. Biol.* (in press).

Shand, J., Partridge, J.C., Archer, S.N., Potts, G.W., and Lythgoe, J.N. (1988). Spectral absorbance changes in the violet/blue sensitive cones of the juvenile pollack, *Pollachius pollachius*. *J. Comp. Physiol. A.* 163:699–703.

Sivak, J.G. (1976). Optics of the eye of the "four-eyed" fish (*Anableps anableps*). *Vision Res.* 16: 531–534.

Sivak, J.G. (1988). Optics of amphibious eyes in vertebrates. In: *Sensory Biology of Aquatic Animals* (Atema, J., Fay, R.R., Popper, A.N., and Tavolga, W.N., eds.). pp. 467–485. New York: Springer-Verlag.

Sivak, J.G., and Warburg, M.R. (1983). Changes in the optical properties of the eye during metamorphosis of an anuran, *Pleobates syriacus*. *J. Comp. Physiol.* 150:329–332.

Slonaker, J.R. (1897). A comparative study of the area of acute vision in vertebrates. *J. Morphol.* 13:445–492.

Snyder, A.W., and Miller, W.H. (1978). Telephoto lens system of falconiform eyes. *Nature Lond.* 275:127–129.

Steenstrup, S., and Munk, O. (1980). Optical function of the convexiclivate fovea with special regard to notosudid deep-sea teleosts. *Optica Acta* 27: 949–964.

Takei, S., and Somiya, H. (2001). Guanine-type retinal tapetum and ganglion cell topography in the retina of a carangid fish, *Kaiwarinus equula*. *Proc. Roy. Soc. (Lond.) B.* 269:75–82.

Uemura, M., Somiya, H., Moku, M., and Kawaguchi, K. (2000). Temporal and mosaic distribution of large cells in the retina of a daggertooth aulopiform deep-sea fish (*Anotopterus pharao*). *Phil. Trans. Roy. Soc. (Lond.) B.* 355:1161–1166.

van der Meer, H.J. (1994). Ontogenetic change of visual thresholds in the cichlid fish *Haplochromis sauvagei*. *Brain Behav. Evol.* 44:40–49.

Vilter, V. (1954). Différenciation fovéale dans l'appareil visuel d'un poisson abyssal, le Bathylagus benedicti. *Société de Biol.* 148:59–63.

Wagner, H.-J. (1990). Retinal structure of fishes. In: *The Visual System of Fish* (Douglas, R.H., and Djamgoz, M.B.A., eds.), pp. 109–157. London: Chapman and Hall.

Wagner, H.-J., Fröhlich, E., Negishi, K., and Collin, S.P. (1998). The eyes of deep-sea fishes. II. Functional morphology of the retina. *Prog. Retinal Eye Res.* 17:637–685.

Wagner, H.-J., Menezes, N.A., and Ali, M.A. (1976). Retinal adaptations in some Brazilian tide pools fishes (Teleostei). *Zoomorphol.* 83:209–226.

Wald, G. (1939). On the distribution of vitamin A_1 and A_2. *J. Gen. Physiol.* 22:391–415.

Wald, G. (1958). The significance of vertebrate metamorphosis. *Science* 128:1481–1490.

Walls, G.L. (1937). Significance of the foveal depression. *Arch. Ophthalmol.* 18:912–919.

Walls, G.L. (1940). Postscript on image expansion by the foveal clivus. *Arch. Ophthalmol.* 23:831–832.

Walls, G.L. (1942). *The Vertebrate Eye and Its Adaptive Radiation*. Michigan: Cranbrook Institute of Science.

Wässle, H., Peichl, L., and Boycott, B.B. (1981). Morphology and topography of on- and off-alpha cells in the cat retina. *Proc. Roy. Soc. (Lond.) B.* 212:157–175.

Whitmore, A.V., and Bowmaker, J.K. (1989). Seasonal variation in cone sensitivity and short-wave absorbing visual pigments in the rudd *Scardinius erythrophthalmus*. *J. Comp. Physiol. A.* 166: 103–115.

Wong, R.O.L. (1989). Morphology and distribution of neurons in the retina of the American garter snake *Thamnophis sirtalis*. *J. Comp. Neurol.* 283: 587–601.

Wood, P., and Partridge, J.C. (1993). Opsin substitution induced in retinal rods of the eel (*Anguilla anguilla* (L.): A model for G-protein-linked receptors. *Proc. Roy. Soc. (Lond.) B.* 254:227–232.

Wood, P., Partridge, J.C., and De Grip, W.J. (1992). Rod visual pigment changes in the elver of the eel *Anguilla anguilla*, L. measured by microspectrophotometry. *J. Fish Biol.* 41:601–611.

Yew, D.T., Chan, Y.W., Lee, M., and Lam, S. (1984). A biophysical, morphological and morphometrical survey of the eye of the small shark (*Hemiscyllium plagiosum*). *Anat. Anz.* 155:355–363.

Yokoyama, S. (1997). Molecular genetic basis of adaptive selection: Examples from color vision in vertebrates. *Ann. Rev. Genet.* 31:315–316.

Young, R.W., and Martin, G.R. (1984). Optics of retinal oil droplets: a model of light collection and polarization detection in the avian retina. *Vision Res.* 24:129–137.

Zaunreiter, M., Junger, H., and Kotrschal, K. (1991). Retinal morphology of cyprinid fishes: A quantitative histological study of ontogenetic changes and interspecific variation. *Vision Res.* 31:383–394.

Zygar, C.A., Lee, M.J., and Fernald R.D. (1999). Nasotemporal asymmetry during teleost retinal growth: Preserving an area of specialization. *J. Neurobiol.* 41:435–442.

Part 3
The Coevolution of Signal and Sense

Jelle Atema

The first day God said "let there be light. . . ."
The second day God said "let the water under the sky be gathered to one place . . . ," which he called "seas."
The fourth day God said "let the water teem with living creatures. . . ."

Genesis 1

Life has always evolved against a backdrop of physics: electromagnetic fields, photons, gravity, movement of fluid and solid media, and chemistry. These physical forces are a rich and dynamic source of information about the external environment, from which signals for specific life-enhancing consequences can be extracted. Increasingly, life itself began to influence the physics and particularly the chemistry on Earth, thereby generating further riches in the signal environment, particularly in the realm of communication.

The aquatic and terrestrial environments present few fundamentally different sensing opportunities. The ionic content of water allows for electric sensing of electric and magnetic fields. Soluble chemistry allows for an external taste sense. Most differences, however, are quantitative and spectral. For example, near-field acoustics in the denser aquatic medium has led to a unique sense organ, the lateral line. Light scattering plus suspended and soluble matter has led to unique opportunities for camouflage, requiring visual receptors and signal processing to break it.

Ever-increasing complexity of life forms is accompanied by ever-greater sensory performance and specialization, both underwater and above. Consider the coevolution of locomotion speed and sensory processing. As competition drives the need for speed, animals must sense further ahead in time and space to alert them about food, mates, predators, and impending crashes. This leads to the development of a head where most sensory systems are located and to a brain in the head where the senses mix to get a consensus. That brain also coordinates locomotive action in response to, or in search of, important signals. Sensing and moving are inextricably connected Umwelts, as recognized a century ago by Von Uexkull.

Theoretically, multisensory integration, both simultaneous and sequential, would facilitate the detection of coincident signals. In some steering and navigation tasks, the sensory challenge is on detection of beacons and spatial/temporal gradients, both embedded in noisy backgrounds. Wakes and odor plumes, for example, are composed of partially coincident chemical and hydrodynamic eddy gradients. It would not be surprising to find this physical signal coincidence reflected in sensory coincidence to enhance gradient detection. Similarly, visual and acoustic/hydrodynamic signal coincidence generated during ordinary locomotion and amplified in special courtship displays would suggest that predators and partners develop close temporal integration of these senses. Finally, integration of different senses while hunting could be described as hunting not for the prey itself, but hunting for a sequence of prey signals, from smell to sound to vision to taste. While a few such sequences are known, the processes that rule the sensory transitions are not.

A complex subset of an animal's stimulus world is provided by communication signals, where signal design and sensing capabilities are under similar and often competing selection pressures: to be conspicuously identifiable to potential mates and competitors and to be cryptic to predators. Offspring, parasites, and other players can be added to the competitive mixture. Initially, a communication signal needs to be tuned at least generally to the sensory capabilities of the receiver. Once a sensory system is in operation it becomes a substrate for signal exploitation by mates, predators, and so on. And then the strategic arms race between signal and sense selection is on, leading to such outstanding examples of signal design as, for example, mimicry and camouflage. Sensory systems respond with increasing powers of spectral, temporal, and dynamic contrast detection such as the use of color, UV, and polarization vision in the periphery and the use of search images in central processing.

For communication signals it is useful to consider both the purpose and the structure of the signal. These two signal attributes have been called, respectively, the "strategic design" and the "tactical design." Strategic aspects drive the evolution of sensory systems. Form follows function. For intraspecific communication signals an important strategic question is signal "honesty" and the ultimate cost of cheating. For instance, while it may be relatively easy to cheat by overproducing a dominance pheromone signal, it may be more difficult to hide one's real physiological state when that same pheromone message is delivered in the urine, which contains metabolic "truth." Tactical aspects determine the efficacy of a signal, for example, the dynamic, spectral, and temporal contrasts of the signal that allow it to be tuned to the intended receiver's sensory system and to stand out over the noise.

Perhaps not surprising then, most chapters in this coevolution section deal with communication and most come from the visual world where we humans have some insight. Two chapters consider chemical signals and one deals with acoustic signals. Mostly unexplored remain the synergistic and at times perhaps conflicting information extracted by different sensory systems in an animal's behavioral tasks. Catfish and lobsters, for example, can repeatedly accept food with one taste sense and reject it with the next taste sense. Behavioral ambivalence results when there is no consensus among senses. Such conflict situations can be illuminating when trying to understand the role of multiple senses.

9 Underwater Sound Generation and Acoustic Reception in Fishes with Some Notes on Frogs

Friedrich Ladich and Andrew H. Bass

Abstract

Fishes have evolved diverse mechanisms to generate sound. These include rubbing of bony elements against each other (stridulation), vibrating swim bladders or pectoral girdles via rapidly contracting muscles, and plucking enhanced tendons of pectoral fins. While stridulatory or plucking mechanisms produce wideband pulsed sounds with frequencies extending up to several kHz, vibration of the swim bladder results in low-frequency (<1 kHz) tonal, often harmonic, signals. In shallow waters, where most of the vocalizing fishes (and frogs) live, sound propagation is very much limited below a certain cut-off frequency. This implies that the communication range is restricted to no more than a few meters, quite different therefore FROM open ocean marine mammals, whose sounds travel hundreds of kilometers.

Fishes have also evolved a large diversity of hearing abilities. While most species known as hearing generalists are restricted to detecting low frequencies below 500 Hz and the particle motion component of sound, other species have morphological structures such as Weberian ossicles that effectively couple the inner ear to air-filled vibrating chambers within the body, thus allowing sound pressure detection. These accessory hearing structures enable such hearing specialists (otophysines, mormyrids, and anabantoids) to extend their hearing range up to several kHz. A correlation between auditory sensitivity and sound spectra is found in numerous species, although the distribution of sonic organs and hearing abilities among otophysans and other teleosts does not always support this pattern.

Hearing specializations are often characteristic of whole taxa, whereas sonic organs appear in a limited number of species within these taxa. Therefore, it is assumed that hearing specializations evolved much earlier. The major selective pressure influencing the evolution of hearing specializations among teleosts is most likely predator avoidance and/or prey detection in quiet freshwater habitats and to a lesser degree the optimization of acoustic communication.

A small number of frogs call and communicate acoustically underwater. The pipid *Xenopus* produces click-like signals that resemble those of numerous fish species. Interestingly, these signals are produced by a modified larynx. Similar to fishes, auditory sensitivity is apparently enhanced by coupling the ear to air-filled spaces such as the lung and the middle-ear cavity. Thus, it seems that fishes and frogs evolved similar mechanisms for sound production and detection in response to the physical limitations of their environment.

1. Introduction: Sound-Generating Mechanisms

Teleost fishes have evolved diverse mechanisms to generate sound, although they do not possess a sound-generating (sonic) organ like that found among frogs, birds, or mammals (i.e., larynx or syrinx). There exists no generally accepted classification scheme for sound production among teleosts because of the large variety of sonic organs (Schneider, 1961, 1967; Tavolga, 1964, 1971; Zelick et al., 1999; Fine and Ladich, in press). Nevertheless, the majority of known sonic mechanisms fall into two major categories—stridulatory and swim bladder vibrating. Stridulatory sounds are produced, for example, by rubbing bony elements such as teeth or fin rays against each other; this sonic mechanism is very similar to that found among most insects (Aiken, 1985). By contrast, swim bladder mechanisms consist primarily of very fast contracting muscles that rapidly compress or extend the swim bladder, resulting in the emission of low-frequency sound. While our discussion focuses on these mechanisms, we also note that sonic organs are unknown in several well-known sound producers such as gobies, pomacentrids, and loaches (Valinsky and Rigley, 1981). Incidental sounds have also been described during rapid air release among species where the swim bladder maintains a connection with the oesophagus (i.e., in physostomes), during motion of fishes through the water (hydrodynamic sounds) or during feeding (Dufosse, 1874; Dijkgraaf, 1941; Tavolga, 1971). It remains questionable if any of the latter sounds have any function in social communication, although they may be used in either coordination of swimming movements among members of fish schools or in predator and prey detection.

1.1. Stridulatory Mechanisms

Stridulatory mechanisms are found among species that rub pharyngeal teeth against each other in connection with nonfeeding activities such as alarm reactions and defending territories. The best-known sound producers of this group are members of the family Haemulidae (grunts) (Tavolga, 1971). Pharyngeal teeth stridulation is attributed to several additional fish families such as centrarchids or cichlids, which produce burst-like sounds, although the supporting evidence remains sparse (Lanzing, 1974; Ballantyne and Colgan, 1978). Perhaps the best-studied stridulatory organs are those found in numerous catfish families; these organs consist primarily of enhanced pectoral spines with a series of ridges on their proximal end (Schachner and Schaller, 1981; Fine et al., 1999; Fine and Ladich, in press). Rubbing the ridges, which are located on a dorsal process at the base of the spine or against a slightly concave groove within the fused pectoral girdle (cleithrum, coracoid), results in a series of short pulses (Pfeiffer and Eisenberg, 1965; Ladich, 1997; Heyd and Pfeiffer, 2000). A sound-producing apparatus of the dorsal fin has been described in the sisorid catfish (Mahajan, 1963) and also in the triggerfish (Schneider, 1961).

1.2. Swim Bladder Vibration Mechanisms

Swim bladder vibrations typically involve the so-called sonic or "drumming muscles" that are either intrinsic or extrinsic. Intrinsic muscles have fibers that attach at both ends of the swim bladder, such as in toadfishes, triglids, and some gadids (Hawkins and Myrberg, 1983; Bass and Baker, 1991; Hawkins, 1993) (Fig. 9.1) and sexual dimorphisms in sonic muscles have been observed among a number of species (Bass and Marchaterre, 1989; Fine et al., 1990; Walsh et al., 1995). One of the more interesting cases of sexual dimorphism is found in a batrachoidid, the plainfin midshipman (*Porichthys notatus*). A large suite of dimorphisms in behavioral, somatic, neurobiological, and endocrinological traits exist between two male reproductive morphs (type I and type II) of the plainfin midshipman (Bass, 1996). Type I male midshipman acoustically court females from nests that they build under rocky shelters in the intertidal zone, whereas type II males sneak spawn and neither build nests nor acoustically court females. Differences in vocal behavior are paralleled by dimorphisms in the sonic muscle (Bass and Marchaterre, 1989; Brantley et al., 1993). On average, the sonic muscle of type I males is six times greater in relative mass (sonic muscle mass/body mass) and almost 25 times larger in absolute terms than the muscle of either females or type II males. Type I male sonic muscles further diverge from those of the similar type II male and female phenotypes in a number of other muscle traits, including fiber size and number, myofibril ultrastructure (Z-line width, mitochondrial density, extent of sarcoplasmic reticulum), and the sizes of neuromuscular junctions.

Extrinsic swim bladder muscles originate on structures, such as the skull or a vertebral process, and insert on the swim bladder directly or indirectly via a structure attached to the swim bladder. In the pimelodid catfish, the sonic muscles originate at the transverse process of the fourth vertebra and insert on the rostral and ventral surface of the swim bladder, thus covering it entirely (Schachner and Schaller, 1981; Ladich and Bass, 1998). In the tiger fish (family Teraponidae), a pair of short muscles originate on the occipital region of the skull and insert on the anterior-dorsal surface of the swim bladder (Schneider, 1964). Sonic muscles have a more indirect attachment to the swim bladder among a number of distantly related groups of teleosts. For example, ariid, doradid, and mochokid catfishes have drum-

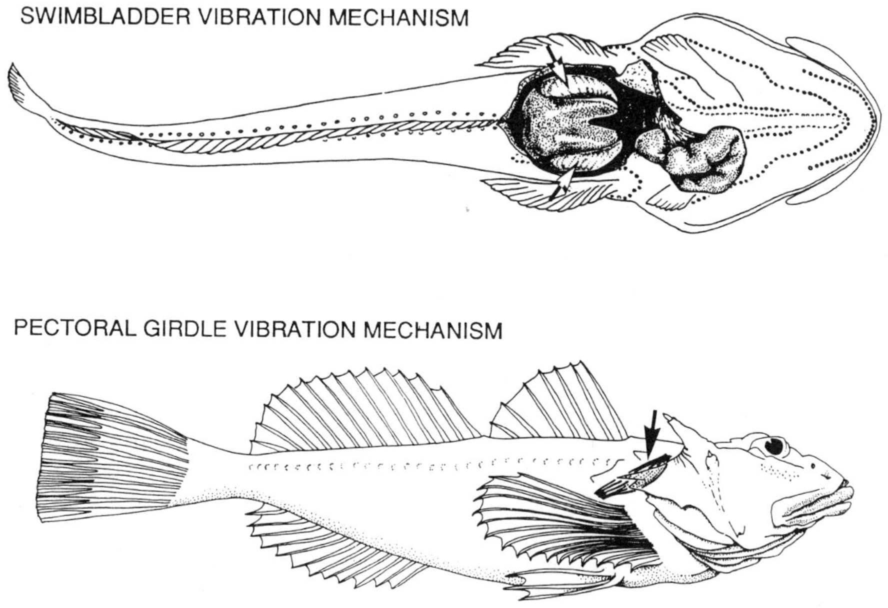

FIGURE 9.1. Swim bladder vibration mechanism of the midshipman fish *Porichthys notatus* (upper illustration) and pectoral girdle vibration mechanism of the longhorn sculpin *Myoxocephalus scorpius* (lower illustration). In the midshipman, sonic muscles appose the lateral walls of the bilobed swim bladder (arrows), whereas in sculpins, which lack swim bladders, sonic muscles (arrow) originate on the skull and insert on the pectoral girdle. (From Bass and Baker, 1991. Reprinted by permission of Karger, Basel.)

ming muscles that first insert on a thin bony plate (elastic spring), which is then attached to the swim bladder (Tavolga, 1962; Abu-Gideiri and Nasr, 1973; Kastberger, 1977; Ladich and Bass, 1996; 1998; Fine and Ladich, in press). This elastic spring mechanism ("Springfederapparat," according to Müller, 1857) varies in the origin of the sonic mucles as well as the shape of the elastic spring. Rapid contractions of the sonic muscles set the elastic spring and swim bladder wall into vibration. In squirrelfish (family Holocentridae), a bilateral pair of extrinsic muscles attach to the skull near the auditory bulla and then extend across the upper flattened part of the first two ventral ribs, which are firmly attached to the swim bladder, and end in ribbon-like tendons just in front of the third rib (Winn and Marshall, 1963). In yet other taxa, the swim bladder is vibrated by flat tendons surrounding it either dorsally or ventrally. For example, among drumfishes (family Sciaenidae), the sonic muscle fibers originate from the left and right side of the abdominal musculature and are attached to a broad central tendon, which extends between the muscles and dorsally crosses the swim bladder (Schneider and Hasler, 1960; Ono and Poss, 1982). In piranhas (family Characidae), sonic muscles originate on vertebral processes and insert in a tendon, which surrounds the bladder ventrally (Markl, 1971).

Numerous other sonic mechanisms have been documented for teleosts, none of which can be classified as using either a stridulatory or swim bladder vibrating mechanism. Croaking gouramis generate pulsed sounds by snapping two enhanced tendons of the pectoral fins over bony elevations of the fin rays during rapid pectoral fin beating (Kratochvil, 1978). This mechanism is similar to the plucking of guitar strings and has evolved in just three species of labyrinth fishes (anabantoids) (family Belontiidae; Kratochvil, 1980) (Fig. 9.2). Sculpins (family Cottidae) lack swim bladders but produce a series of knocking sounds or growls by vibrating the entire pectoral girdle. This is mediated via sonic muscles (musculus cephaloclavicularis, according to Barber and Mowbray, 1956), which originate on the skull, insert on the dorsal element of the pectoral girdle, and rapidly pull the girdle toward the skull (Barber and Mowbray, 1956; Ladich, 1989; Bass and Baker, 1991) (Fig. 9.1).

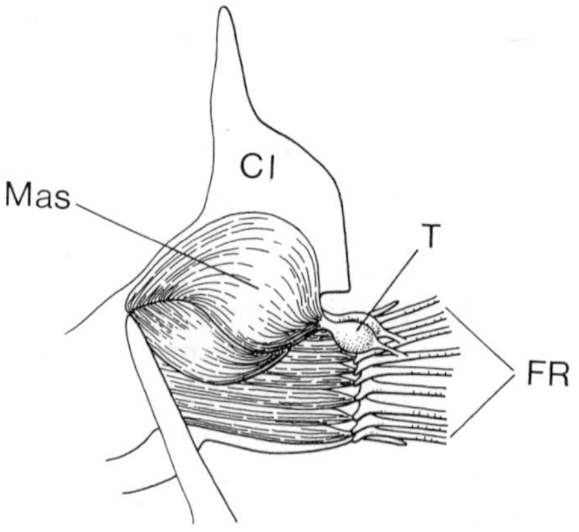

FIGURE 9.2. Pectoral sound generating mechanism of the croaking gourami *Trichopsis vittata*. (Redrawn from Kratochvil, 1978.) Abbreviations: Cl—cleithrum; FR—fin rays; Mas—superficial adductor muscle; T—enhanced tendons of the 4th and 5th fin ray.

The diversity of mechanisms and the apparent lack of phylogenetic patterns suggest that sound production has evolved independently in several groups, though it is likely that many of the sound-producing structures share embryological origins and thus might be considered homologues. For example, detailed ontogenetic and neuroanatomical studies within the order Scorpaeniformes reveal that apparently diverse sonic mechanisms such as intrinsic swim bladder muscles in the searobin *Trigla* and pectoral girdle vibration muscles in the sculpin *Cottus* seem to be homologous. Data suggest that the ancestral condition in scorpaeniforms is the presence of sonic muscles that originate on the cranium, passing posteriorly with attachment to the pectoral girdle, and finally inserting, by tendons, to ribs and vertebrae. Degeneration of some attachment sites (searobins, sculpins) probably represents a derived state (Rauther, 1945; Ladich and Bass, 1998).

2. Sound Characteristics

The diversity of sound-generating mechanisms results in sounds of widely varying spectral and temporal content. As the analysis of the spectral content of sound depends on the acoustic properties of the field or laboratory environment in which recordings take place (e.g., background noise, resonances, reflections), comparison of sounds may be partially affected by the differences in recording conditions (Tavolga, 1971). In general, fish sounds are short, percussive (<1 sec), highly repetitious acoustic signals with only a few exceptions such as the mating call of the midshipman, which may last for several minutes to over an hour (Ibara et al., 1983; Bass et al., 1999).

The spectral content of sound depends on the sound-generating apparatus and differs widely between species. Stridulatory sounds produced by the grating of pharyngeal teeth or rubbing of fin rays are essentially wideband sounds with no harmonic structure. In this case, sound energy either is mainly concentrated at a few hundred Hz, with some components extending continuously to several kHz, or has predominant frequencies between 1 and 4 kHz. The latter, in particular, is the case for sounds produced during abduction and adduction of the pectoral spines in catfishes. Ictalurid, pimelodid, mochokid, doradid, and callichthyid catfishes produce pulsed sounds and each pulse appears to be generated by the collision of an individual ridge at the base of the spine with the rubbing surface within the pectoral groove (Pfeiffer and Eisenberg, 1965; Ladich, 1997; Pruzsinszky and Ladich, 1998; Fine et al., 1999; Kaatz, 1999; Ladich 1999; Heyd and Pfeiffer, 2000). The peak frequencies are concentrated at or above 1 kHz and no clear dependency on body size has been observed. The duration of pectoral sounds varies from 30 to 100 ms and there is a trend for increased sound duration with increased length of the pectoral spine (Ladich, 1997). In cichlids, the main energies of stridulatory sounds uttered during sexual or aggressive behavior are concentrated below 500 Hz (Myrberg et al., 1965; Rowland, 1978), although the energies of some sound types extend up to 10 kHz (Nelissen, 1978). However, it is unclear if these latter sound types function in social communication or are merely byproducts of feeding and swimming. The stridulatory sounds of 14 species of grunts (family Haemulidae) show predominant frequencies between 100 and 500 Hz (Fish and Mowbray, 1970), while the main energies of the pharyngeal sounds of centrarchids extend to 12 kHz (Ballantyne and Colgan, 1978).

The sounds produced by the vibration of swim bladders are usually tonal and characterized by their harmonic content. The fundamental frequency of drumming sounds varies from 80 to over 200 Hz and sound energy is restricted to the low-frequency band with most energy below 1 kHz (some of these species differences will depend on ambient temperature; see Bass and Baker, 1991). The predominant frequency usually corresponds to the fundamental frequency of sound, but could be also found within the second or third harmonic (Ladich, 1997, 1999; Bass et al., 1999). As the drumming sounds are produced by fast-contracting muscles, the fundamental frequency or harmonic interval usually reflects the muscle's contraction rate or pulse repetition rate. In general, there are no obvious differences in the spectral or temporal properties of the sounds produced by intrinsic or extrinsic, directly or indirectly, vibrating swim bladder muscles. Similarly, acoustic differences cannot be attributed to the morphological diversity of swim bladder vibrating mechanisms. There is, however, one exception to these generalizations. The ultrastructural traits of the intrinsic sonic muscles of the plainfin midshipman fish are clearly structural adaptations associated with their ability to generate sound over unusually long periods of time (Bass and Marchaterre, 1989).

The pulse repetition rate can differ in closely related families. In the pimelodid catfish it is higher (165–177 Hz) than in the doradid catfish (96–114 Hz) (Ladich, 1997). Drumming sounds in both families are also frequency modulated. Although common in birds and many mammals, frequency modulation is relatively unusual in fishes (Fine et al., 1977). Long spawning calls of the male haddock *Melanogrammus aeglifinus* exhibit distinct frequency modulation (Hawkins and Rasmussen, 1978), the mormyrid *Pollimyrus isidori* modulates the frequency of "moans" in either an upward or downward direction (Crawford et al., 1986), while the "growls" of male midshipman *Porichthys notatus* show decreasing frequency as the call progresses (Bass et al., 1999).

Changes in fundamental (contraction) frequency are perhaps due to muscle fatigue, different levels of excitement, and water temperature (Fine, 1978; Hawkins, 1993; Brantley and Bass, 1994).

Several species are able to modify their drumming sounds. The midshipman *P. notatus* produces long-duration hums (>1 min), short-duration grunts (ms time scale), and growls (ms to min), which differ in their fundamental frequency and are obviously used in different behavioral contexts (Bass et al., 1999) (Fig. 9.3). The mormyrid *P. isidori* emits five different types of sounds (moans, hums, growls, hoots, and pops) that are also used in different contexts (Crawford et al., 1986; Crawford and Huang, 1999). In piranhas, two types of drumming sounds can be distinguished when held by hand, namely barks with pulse rates of 80–150 Hz and honks with much lower pulse rates (5–20 Hz) (Kastberger, 1981a). The maximum duration of calls is 200 ms and vocalization ceases after 1 to 2 minutes.

It is not fully understood if the swim bladder is responsible for the appearance of harmonics. Tavolga (1962), studying the role of the swim bladder in sound production in the sea catfish *Galeichthys felis*, argues that the harmonic content of the sound is not a function of the bladder but rather is dependent on environmental conditions. The swim bladder primarily amplifies the sound because severe damage or removal of air causes a large decrease in the amplitude of sounds produced by the elastic spring mechanisms (Tavolga, 1962). Kastberger (1981b), on the other hand, states that, in piranhas, the resonance effects of the swim bladder expand the more or less sinusoidal signal of the muscle response to a harmonic sound.

While muscle contraction rate determines the fundamental frequency and harmonic content of sound in catfishes, characids, and toadfishes, this is obviously not the case in species possessing other swim bladder vibrating mechanisms (we note that the muscle contraction rate is determined, in turn, in these species by a central pattern generator in the hindbrain; see Bass, 1996 and Bass and Baker, 1991). The drumming sounds of sciaenids show a broad spectrum with frequencies between 150 and 1,000 Hz lacking any harmonics; largest relative amplitudes are in the range of 250–600 Hz (Schneider and Hasler, 1960; Connaughton et al., 2000). Thus, the predominant frequencies do not follow the muscle contraction rate, which is approximately 25–30 Hz. Juvenile tiger fishes (*Terapon jarbua*) emit two types of sounds—drumming sounds and threatening sounds of higher intensity (Schneider, 1964). The spectrum does not show harmonics and intensity changes during growth.

A size dependency for swim bladder sound characteristics varies across species (Fine and Ladich, in press). No correlation between body size and mean fundamental frequency exists in doradid and pimelodid catfishes (Ladich, 1997). In contrast, the dominant frequencies of sounds decrease with body size in mormyrids, tiger fishes, and drums (Schneider, 1964; Crawford et al., 1997; Connaughton et al., 2000). A clear relationship between the dominant sound frequencies and body size has been found in species producing pulsed sounds with neither stridulatory nor swim bladder vibration mechanisms. Thus, in all three species of croaking gouramies (*Trichopsis*), the dominant frequency of sounds produced by pectoral fin tendons is inversely correlated to the body mass (Ladich et al., 1992). Mean frequencies of croaking sounds vary from 1.3 to 2.5 kHz, with the highest frequency found in the smallest pygmy gourami (Fig. 9.4). Furthermore, Henglmüller and Ladich (1999) and Wysocki and Ladich (2001) show that the dominant frequency of croaks decreases during ontogenetic development in *T. vittata*. The damselfish (family Pomacentridae) produces pulsed sounds by an unknown mechanism with main energies at about 700 Hz. Myrberg et al. (1993) provide evidence that the chirping sound of the bicolor damselfish *Pomacentrus partitus* reflects a clear inverse relationship to body size, which should be governed by swim bladder resonance.

The sound characteristics of fishes possessing yet other sonic mechanisms reveal a further diversity of vocalizations. Polypterids, sculpins, and gobies produce low-frequency sounds with peak frequencies at about 100 Hz (Ladich and Tadler, 1988; Ladich and Kratochvil, 1989; Lugli

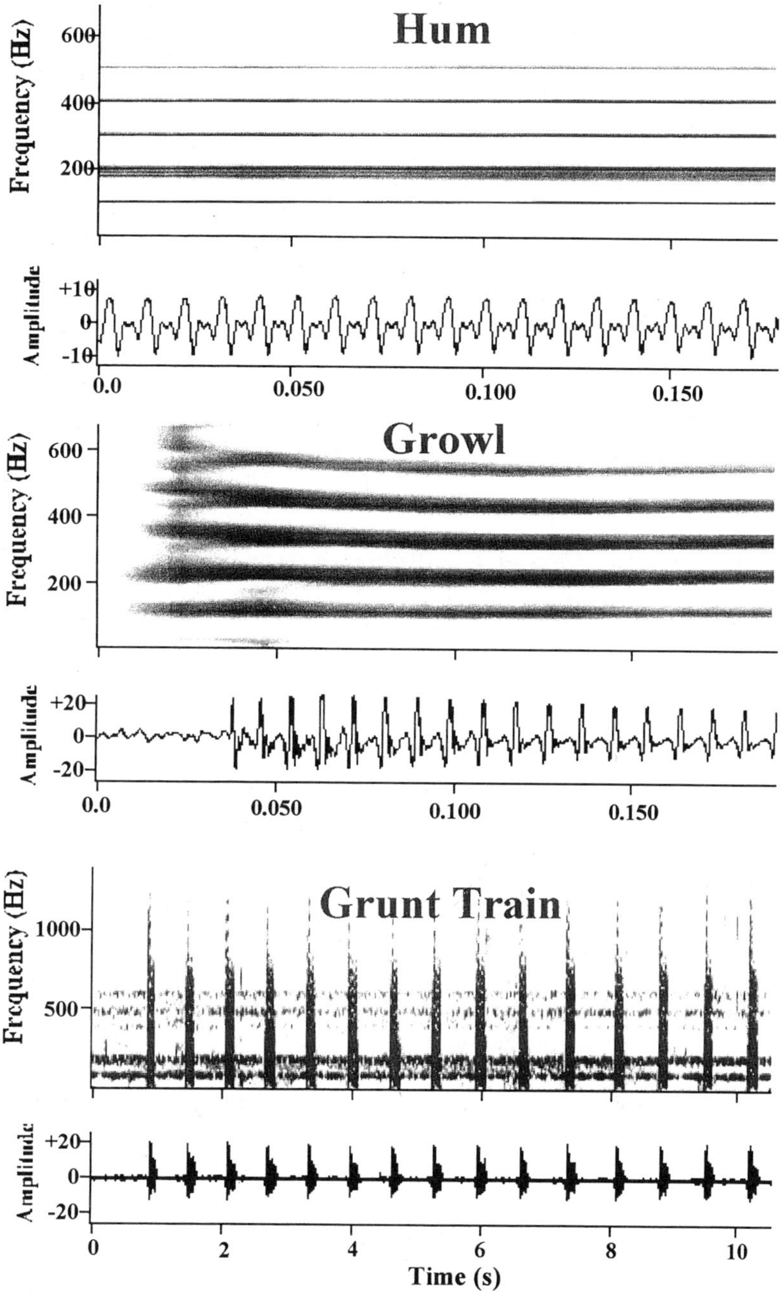

FIGURE 9.3. Sonograms and oscillograms of acoustic signals of the plainfin midshipman fish *Porichthys notatus*. Short segments of a hum, a growl, and a grunt train showing consecutive grunts from the nest of the same type I male. (From Bass and Clark, 2002, with permission)

et al., 1995). Knocking sounds of loaches are broadband but main energies are concentrated between 100 and 500 Hz (Ladich, 1999).

Temporal patterning of sound appears to be an important carrier of information in this vertebrate class (Winn, 1964; Fine et al., 1977; Myrberg et al., 1978). Distinct differences in pulse number, pulse period, and sound duration are found in representatives of several nonrelated genera such as *Lepomis* (family Centrarchidae; Gerald, 1971), *Pomacentrus* (family Pomacentridae; Spanier, 1979), *Pollimyrus* (family Mormyridae; Crawford et al., 1997), *Trichopsis* (family Belontiidae; Ladich et al., 1992), and the closely related toadfish *Opsanus* and midshipman *Porichthys* (family

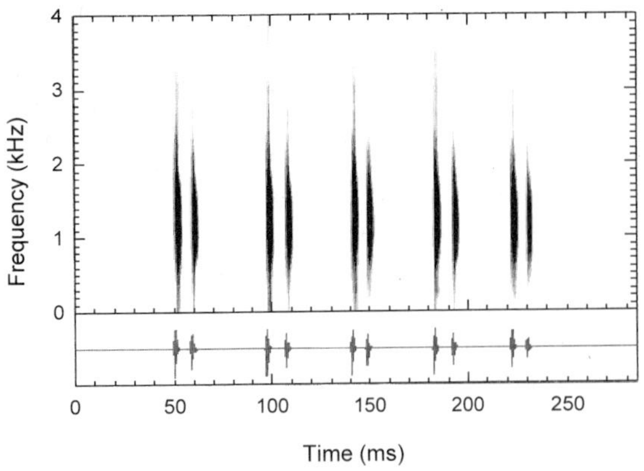

FIGURE 9.4. Sonogram and oscillogram of the croaking sound of *Trichopsis vittata*, consisting of a series of double pulses.

Batrachoididae; Fine et al., 1977; Bass et al., 1999). Spanier (1979) demonstrated that playbacks to conspecific sounds elicited the best response in four sympatric damselfish species. An intraspecific variety of sounds, which are used in different behavioral contexts, is mainly based on differences in temporal patterning. The squirrelfish *H. rufus* makes three types of sounds: single grunts, a series of variable-interval grunts, and equal-interval staccato calls (Winn et al., 1964). The cyprinid *Notropis analostanus* produces single knocks and series of knocks during agonistic interactions as well as purrs during courting (Stout, 1963). The repertoire of the male gadid *Melanogrammus aeglefinus* includes sounds such as a short series of slowly repeated knocks, a long series of rapidly repeated knocks, and fast, continuous hums (Hawkins and Amorim, 2000). Type I male midshipman (*P. notatus*) produce long-duration (>1 min) hums for attracting females to their nest, but much briefer (msec scale) grunts that are generated consecutively more than 100 times as grunt trains during agonistic encounters (Brantley and Bass, 1994; Bass et al., 1999). Two choice playback experiments show that female midshipman can distinguish sounds on the basis of their fundamental frequency, intensity, duration, gaps between successive signals, and the degree of amplitude modulation (McKibben and Bass, 1998, 2001). Duration and temporal patterning also changes during ontogeny; the number of pulses within a croak and pulse period increases during growth in the croaking gourami *T. vittata* (Henglmüller and Ladich, 1999).

3. Sound Propagation

Aquatic animals encounter strikingly diverse conditions for sound communication in different habitats (Bass and Clark, 2002). Open-water organisms, such as whales, can communicate over kilometers due to the low absorption (down to 0.001 dB/km) and increased velocity (five times higher) of sound in water in comparison to terrestrial habitats. In contrast, fishes living in shallow water are unable to communicate over long distances since low-frequency sound does not propagate efficiently. In deep water, sound can propagate by refracted pathways without ever contacting the surface or the bottom for up to 10,000 miles (Rogers and Cox, 1988). On the other hand, in shallow streams, ponds, tidal pools, or coral reefs, sound propagation is limited to frequencies above the cutoff frequency, which depends on water depth and bottom sediment. In one or two meters of water, the usable frequency range is above 100–2,000 Hz especially for slow (sandy) bottoms (Rogers and Cox, 1988). Interestingly, most vocalizing fish species described so far are substrate breeders living in shallow waters (Ladich, 1997). Pelagic open-ocean species are not known to be vocal (Marshall, 1967). In addition, numerous fish sounds are of low frequencies and therefore physical acoustics constrain long-distance underwater sonic communication.

Toadfishes commonly call at a depth of one meter and the fundamental frequency of their boatwhistle call is 200 Hz. Fine and Lenhardt (1983) broadcasted low-frequency acoustic signals (tones, noise, courtship calls) in 1-m-deep water over medium-to-fine sandy bottoms and observed that for pure tones transmission loss was greatest within the first 3 m from the transducer. For boatwhistles, the fundamental frequency was absorbed more quickly than its second harmonic, which is unlikely to be heard in this species. At a distance of 5 m, the signals were no longer above the background noise levels. According to Fine and Lenhardt (1983), low ambient noise did not exert a strong selection pressure on the frequency spectrum of the boatwhistle. Crawford et al. (1997) investigated sound production in the weakly electric fish *Pollimyrus isidori* in a freshwater flood plain of 2 to 3 meters in depth with a dense clay-like bottom in Mali, West Africa. The authors estimated that the lowest-frequency components in the sounds are near the cutoff frequency (estimated to be 330 Hz at a depth of 2 m) and are not likely to propagate well. Thus, the moan-sound produced may be a relatively short-range signal. Forrest et al. (1993) investigated sound propagation and transmission loss in a sloping pond, 2 m deep with a soft silt and leaf-litter bottom using sinusoidal sweeps from 100 to 20,000 Hz. They observed a sharp cutoff of 30–60 dB difference between propagating and nonpropagating frequencies. The frequency-dependent propagation was related to the depth of water at the shallowest transducer. At 6 cm depth, the cutoff frequency was 8 kHz, while at 50 cm, frequencies below 2 kHz did not propagate. The sound pressure levels of the calls of midshipman *P. notatus* diminished at 6 dB per distance doubling, which is close to that predicted for spherical spreading. Unlike previous studies in the closely related toadfish (see above), sounds were recorded at greater depths (~5 m) and over a hard rocky-gravel substrate (Bass and Clark, 2002), which likely explains the closer adherence to theoretical predictions. Kastberger (1977) found that the sound pressure levels of drumming sounds of the catfish *Doras* sp. also decreased by about 6 dB for every doubling of distance. High-frequency components (>500 Hz) disappeared totally within 1 m when the fish was held 10 cm below the water surface. Finally, the pulsed sounds of the damselfish attenuated with distance such that the signal-to-noise ratio decreased from 17 to 25 dB at 1–2 m to 5 to 10 dB at 11 m (Mann and Lobel, 1997). This study was performed at a depth of 7 m over a shell-and-coral substrate and it is unlikely that fishes can detect sounds at 11 m from the sound source. Pulse period was least affected by propagation when compared to peak frequency, pulse duration, interpulse interval, and variation of pulse amplitude within a call. These results suggest that damselfish sounds are more effective over a short distance and that the pulse period provides the most reliable basis for signal identification (Mann and Lobel, 1997).

The rapid attenuation of sound intensity and degradation of the signal quality over short distances in shallow waters has several implications for acoustic communication in fishes, especially for the majority, which maintain territories and breed on the substrate (also see Bass and Clark, in press). It is unlikely that these species can communicate acoustically over distances of more than 10 m. This is in agreement with numerous behavioral studies, which clearly show that fishes start vocalizing after the conspecific has been detected visually (for reviews see Myrberg, 1981; Ladich, 1997). Croaking gouramies, sculpins, gobies, toadfishes, and catfishes vocalize during agonistic encounters at distances of 1–5 cm (Ladich, 1989; Ladich et al., 1992; Brantley and Bass, 1994; Pruzsinszky and Ladich, 1998). At these distances, the acoustics of sound propagation are most likely less dependent on habitat. At distances of several meters, fishes might adapt to higher-frequency bands for sound communication and hearing. Although several groups such as pomacentrids, gouramies, and catfishes produce pulsed sound with main energies above 500 Hz or even 1,000 Hz (Myrberg and Spires, 1980; Ladich, 1997; Ladich and Yan, 1998), many fishes utilize swim bladder vibrating mechanisms, which usually result in low-frequency sounds (see above). As far as is known, the latter group is not exclusively restricted to deep water or vice versa. Fine and Lenhardt (1983) mentioned that male toadfishes occasionally call in depths of less than one foot. The freshwater goby *Proterorhinus marmoratus*, which utters low-frequency moans (90 Hz) during territorial defense and courtship, maintains territories in depths of 5–10 cm in the backwaters of the Danube River (F. Ladich, unpublished). This implies that selection pressures have apparently not acted to extend the communication range in fishes to more than one or a few meters, contrary to many marine mammals (Edds-Walton, 1997) and terrestrial insects and birds (Römer, 1993).

4. Sound Detection

Fishes have evolved numerous sound-generating mechanisms, which result in the production of different types of sounds. In addition, fishes possess diverse hearing abilities due to the evolution of numerous morphological structures that are utilized for the enhancement of hearing (see also Chapter 1). Teleost fishes can be subdivided into three major groups according to their hearing sensitivies (Hawkins and Myrberg, 1983): hearing nonspecialists (or generalists), hearing specialists, and a group that falls in between. Generalists detect sound just by way of their inner ears and no other peripheral, morphological structures are involved in the hearing process. The saccule is likely to be the end organ for sound detection in most fish species, yet the involvement of other end organs such as the utricle or lagena cannot be excluded (Blaxter et al., 1981; Popper and Tavolga, 1981; Popper and Fay, 1999; Chapter 1). Nonspecialists detect the kinetic component of the sound source in the acoustic field and are limited to sensitivity at low frequencies of about 100 to 300 Hz (Fay, 1988; Hawkins, 1993). Hearing specialists, on the other hand, are characterized by morphological specializations that form a close connection between an air-filled cavity within the body and the inner ear. Rapid volume changes of the air in the sound field, in response to pressure fluctuations, result in oscillations of the wall of the air-filled cavity, which are then transmitted directly to the inner ear. In this way, specialists can detect the sound pressure component of the acoustic field and are sensitive up to several kilohertz (von Frisch, 1936; Ladich, 1999).

Anabantoids (labyrinth fishes or gouramis) possess a suprabranchial chamber (labyrinth) located dorsal of the gills, which is utilized for air-breathing (Bader, 1937). As these labyrinths lie laterally of the inner ears and are only separated by a thin bony sheet or epithelium, they enhance their hearing sensitivity considerably. Labyrinth fishes can detect sound frequencies up to 5 kHz (Schneider, 1941; Ladich and Yan, 1998). Mormyrids possess gas-filled auditory bullae, directly attached to the saccule, that are cranial derivatives of the swim bladder and improve the hearing sensitivity of these weakly electric fishes up to 3.1 kHz (Stipetić, 1939; McCormick and Popper, 1984). In marine holocentrids, the swim bladder is connected via an elongated anterior extension to the ear, which enhances their hearing capacities in various degrees (Coombs and Popper, 1979; Hawkins, 1993).

Otophysans (cypriniforms, characiformes, siluriformes, and gymnotiformes) are the largest groups possessing accessory hearing structures (about 6,500 species; Nelson, 1994; also see Ladich and Bass, in press). They are characterized by having 1–4 Weberian ossicles (tripus, intercalarium, scaphium, claustrum) and modifications of the anterior vertebrae, which enhance their hearing abilities by directly transmitting swim bladder oscillations to the inner ear (Chranilov, 1927; von Frisch and Stettner, 1932). The functions of the swim bladder and Weberian ossicles of otophysans are similar to the tympanum and middle ear bones in birds and mammals. Poggendorf (1952) compared the sound amplitudes necessary to elicit a positive response in bullhead catfish, song bird (the bull finch), and humans. Similar sensitivities were observed at the best frequency (catfish: 800 Hz; bird and humans: 3.2 kHz). Major differences were, however, found in the optimum frequency range. The auditory sensitivity of the catfish was higher at low frequencies (e.g., 60 Hz) and lower at high frequencies (e.g., 5 kHz) compared to birds and humans. Poggendorf (1952) also found that the oscillation amplitude of the swim bladder and the tympana of the bull finch and humans were quite similar (catfish: 7×10^{-12} m; bull finch: 1×10^{-12} m; human 6×10^{-12} m). Extirpation of the malleus, the most caudal Weberian ossicle, resulted in a decrease in the catfish's auditory sensitivity by about 30–40 dB, but elicited no change in the shape of the hearing curve, hearing range, or the dependency of the response on the pressure component of sounds in the catfish.

Between nonspecialists and specialists falls a third group of fishes that are sound-pressure sensitive, possess enhanced sensitivities in terms of thresholds and frequency range, but do not possess morphological structures for hearing enhancement. This includes, for example, cods and damselfishes, which are sensitive to sound pressure apparently because of the placement of the swim bladder close to the inner ears. However, both groups are less sensitive and have more restricted frequency ranges than otophysans (Myrberg and Spires, 1980; Hawkins, 1993).

Differences in hearing sensitivity exist both between specialists and nonspecialists as well as between closely related species within one family. Among holocentrids, hearing ability depends on the degree of the association between the swim bladder and the ear (Coombs and Popper, 1979; Hawkins, 1993). *Myripristis kuntee* exhibits up to 40 dB lower sound pressure thresholds and responds over a wider frequency range than other holocentrids. The greater sensitivity of this species is associated with a strong contact between the gas-filled swim bladder and a thin membrane in the wall of the auditory bulla, lateral to the sacculus of the ear. In *Adioryx* and *Holocentrus*, the anterior wall of the swim bladder lies close to the skull, which results in a lower auditory sensitivity.

Comparative studies in other hearing specialists such as anabantoids and otophysans again reveal differences between closely related species. Among labyrinth fishes of the family Belontiidae, auditory sensitivity differs, especially in the range between 400 and 1,000 Hz. The blue gourami *Trichogaster trichopterus* differed from four other species by having absolute auditory thresholds 22–25 dB below those of the croaking gouramis of the genus *Trichopsis* and the dwarf gourami *Colisa lalia* (Ladich and Yan, 1998). Morphological differences, which might explain this diversity, are unknown. It is thought, however, that the size and shape of the air-breathing cavities lateral of the inner ears might result in different resonance frequencies and thus hearing abilities (Ladich and Yan, 1998).

In representatives of seven otophysan families from four orders, audiograms revealed major differences between and within families and orders (Fig. 9.5A). Hearing tuning curves were U-shaped or almost flat with maximum sensitivities between 400 and 1,500 Hz. While the hearing thresholds differed maximally by 30 dB from 100 to 1,000 Hz, this difference increased rapidly to more than 50 dB at 4 kHz (Ladich, 1999). However, no clear difference among otophysan taxa could be observed. Differences in auditory thresholds were found among representatives of different orders (*Carassius* (Cypriniformes) and *Eigenmannia* (Gymnotiformes) and *Platydoras*, *Pimelodus* (Siluriformes)), among families within one order (*Pimelodus* (Pimelodide) and *Corydoras* (Callichthyidae)), as well as within one family (Doradidae: *Platydoras* and *Agamyxis*). The large difference in hearing sensitivity, especially at higher frequencies (>1 kHz), seems to be related to the structural diversity of the auditory periphery in otophysans. Although a detailed investigation is lacking, it is assumed that the particularly low sensitivity in the callichthyid catfish is due to both an encapsulation of their small paired swim bladder sacs and a reduction in their number of Weberian ossicles (Alexander, 1964; Coburn and Grubach, 1998). Obviously a reduction in size and thickening of the wall reduces the ability of the swim bladder to oscillate (Ladich and Bass, in press).

5. Correlation Between Sound Generation and Perception

Are differences in auditory sensitivity between hearing generalists and specialists related to their ability to produce sounds? To what degree do hearing abilities, and in particular the appearance of accessory hearing structures, correlate with the evolution of sound-generating mechanisms in fishes? Is acoustic communication a driving force in the evolution of hearing specialization? If the major constraint in the evolution of hearing and sonic organs was the maximization of the effectiveness of intraspecific communication, natural selection would favor the evolution of hearing specializations in vocalizing species, whereby the main sound energy would be generated within the optimal hearing range of a particular species (Ladich, 2000).

Cohen and Winn (1967) observed a correlation between the fundamental frequency of drumming sounds and the saccular microphonic response at 150 Hz in the midshipman *Porichthys notatus*, although a slight mismatch was observed in the close relative, the oyster toadfish *Opsanus tau* (Fine, 1981). In the damselfish *Eupomacentrus partitus*, the peak energy

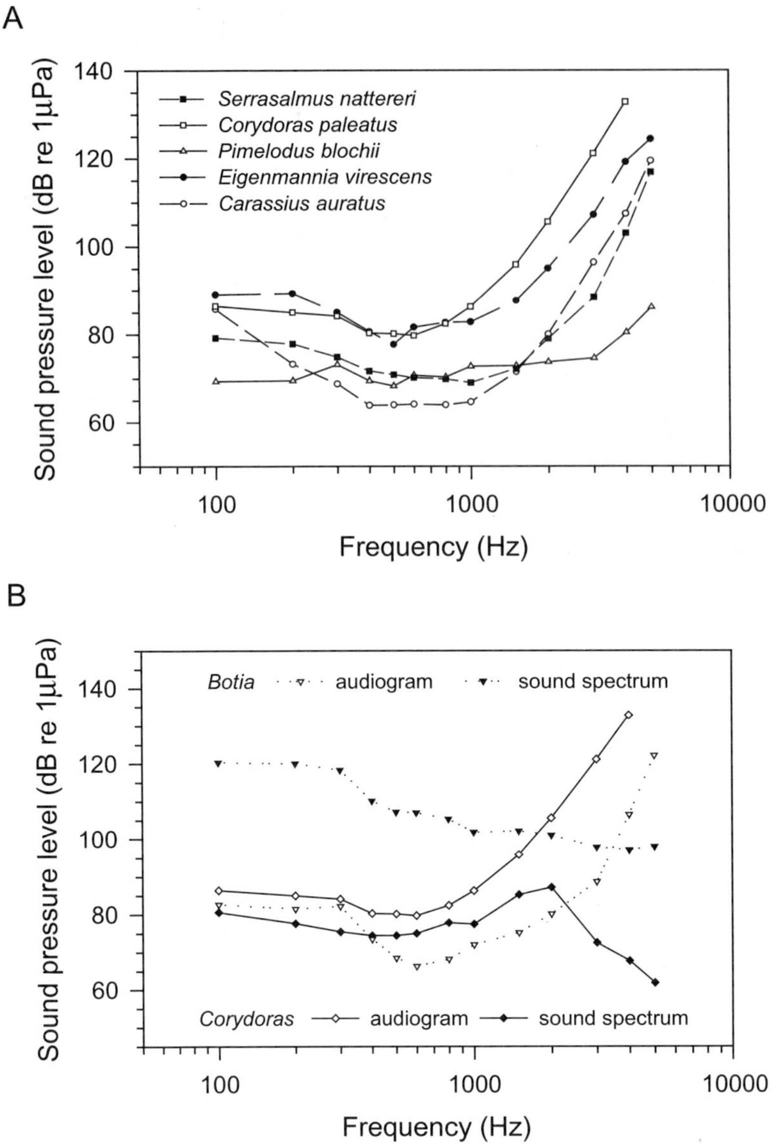

FIGURE 9.5 (A) Comparison of audiograms of representatives of all four otophysan orders: Cypriniformes — *Carassius auratus*; Characiformes—*Serrasalmus nattereri*; Siluriformes—*Corydoras paleatus*, *Pimelodus blochii*; Gymnotiformes — *Eigenmannia virescens*. (B) Audiograms of the loach *Botia modesta* and the catfish *Corydoras paleatus* in relation to spectral and intensity characteristics of sounds. (Modified after Ladich, 2000.)

of sounds matches the audiogram in the region of greatest sensitivity between 500 and 600 Hz (Myrberg and Spires, 1980). Stabentheiner (1988) found that the frequency spectrum of typical drumming sounds (barks) covers the range of best hearing (100–600 Hz) in the piranha *Serrasalmus nattereri*. Schellart and Popper (1992) analyzed 15 species of mostly marine teleosts and found a weak correlation between the best frequencies of hearing and dominant frequencies of sounds. Auditory neurons in the midbrain of the weakly electric fish *Pollimyrus isidori* and the midshipman *Porichthys notatus* had best sensitivities within the range of frequencies and levels of natural communication sounds (Crawford, 1993; Bodnar and Bass, 1997; also see McKibben and Bass, 1999 for primary afferents in midshipman). Ladich and Yan (1998) demonstrated that such a correlation also exists in a species producing high-pitched sounds, the anabantoid *T. vittata* (1–2 kHz). Midbrain auditory neurons in midshipman fishes also encode the difference frequency between the fundamental frequencies of concurrent hums (the calls of neighboring males often overlap during the breeding season; Bodnar and Bass, 1997, 1999, 2001; Bass et al., 1999).

The above correlations suggest that sound-producing organs evolved in tandem with hearing abilities and specializations in fishes. However, several points contradict this assumption. Morphologically similar sonic organs, such as swim bladder muscles and subsequently low-frequency drumming sounds, evolved in hearing specialists (catfishes, characids, mormyrids), nonspecialists (toadfishes, triglids, drums), and the group possessing intermediate hearing abilities (cods) (Hawkins and Rasmussen, 1978; Ono and Poss, 1982; Ladich and Bass, 1998; Crawford and Huang, 1999). In addition, the hearing sensitivities of nonvocal species have to be compared to closely related vocal species. Broadband high-frequency sounds (>500 Hz), on the other hand, are mainly known from groups having enhanced hearing abilities such as mormyrids (Rigley and Marshall, 1973), cyprinids, and catfishes (Ladich, 1988; Ladich, 1997), although they have also been described in hearing generalists, for example, in centrarchids and cichlids (Lanzing, 1974; Ballantyne and Colgan, 1978).

Recently, two comparative studies analyzed the correlated evolution of sound-generating and sound-detecting organs in more detail in two groups of hearing specialists that possess a variety of sound-generating mechanisms: otophysans and anabantoid fishes (Ladich and Yan, 1998; Ladich, 1999). In otophysans, several representatives emit low-frequency drumming as well as knocking sounds (100–400 Hz) or broadband stridulatory sounds (0.8–3 kHz). Representatives of some catfish families even possess two sonic organs (doradids, pimelodids). Sound spectra of pectoral stridulatory sound in catfishes match their flat hearing curves, except for *Corydoras*. In this callichthyid species, a clear mismatch was observed between its poor hearing ability above 1 kHz and the main energy of sounds, which were concentrated between 1 and 2 kHz (Fig. 9.5B). None of the species producing acoustic signals with dominant frequencies below 400 Hz (*Botia*, *Serrasalmus*, *Platydoras*, *Pimelodus*) possessed a pronounced low-frequency sensitivity maximum (Fig. 9.5, Table 9.1). Fishes emitting both low- and high-frequency sounds, such as pimelodid and doradid catfishes, did not possess two corresponding sensitivity maxima (Ladich, 1999). In anabantoid fishes, on the other hand, the main energies of the high-pitched sounds correspond with the best-hearing bandwidth in *T. vittata* and *Colisa lalia* (800–1,000 Hz). In the pygmy gourami *T. pumila*, dominant frequencies of sounds were found above 1.5 kHz and did not match the lowest threshold, which was below 1.5 kHz (Table 9.1) (Ladich and Yan, 1998).

The lack of a correlation between the main energies of sounds and best-hearing sensitivities in many otophysans and anabantoids contradicts data shown in other vocal teleosts. However, all other studies were limited to one species or representatives of one genus and seldom included nonvocal forms. Nonvocal species such as the goldfish and the blue gourami were found among the most sensitive species in both groups of hearing specialists (Ladich, 2000). The lack of a difference in auditory sensitivity between closely related vocal and nonvocal species is paralleled by a convergence in the fine structure of their inner ears (Ladich and Popper, 2001). In addition, poor hearing abilities occur in vocal and nonvocal forms (*Corydoras*, *Eigenmannia*). In summary, when comparing hearing curves of vocalizing and nonvocalizing otophysine species emitting drumming sounds to those producing stridulatory or knocking sounds, no clear relationship can be found between sound power spectra and auditory sensitivity (Ladich, 2000).

As in some of the abovementioned groups, sound-generating mechanisms and acoustic communication are not common features of all members of other taxa. Hawkins and Myrberg (1983) mentioned mute species among gadids and pomacentrids, groups of fishes that are known to contain large numbers of sound-producing species. Also within mormyrids, only a small number of species seem to be vocal, but hearing abilities of all species resemble each other. Among mormyrids, Crawford (1993) found a remarkable similarity in the best sensitivity in the sound-producing weakly electric fish *P. isidori* and the nonvocal *Brienomyrus niger*. Comparison of whole audiograms and total sound power spectra in vocalizing members of other taxa produces contradictory find-

TABLE 9.1. Frequency ranges containing the lowest hearing thresholds (audiogram) and the main energies of sounds (sound spectra). Minimum and maximum frequency were determined as those frequencies used during experiments where hearing sensitivity decreased by 10dB in relationship to the frequency of the maximum hearing sensitivity and where amplitudes within sound spectra dropped by 10dB relative to the dominant frequency of sounds. Two sound spectra ranges where given in species producing two types of sounds.

Order–Family–Species	Frequency range (Hz) Audiogram	Frequency range (Hz) Sound spectra
O: Cypriniformes		
F: Cyprinidae		
Carassius auratus	200–2000	—
Botia modesta	300–2000	100–400
O: Characiformes		
F: Characidae		
Serrasalmus nattereri	100–2000	100–600
O: Siluriformes		
F: Doradidae		
Platydoras costatus	100–4000	100–300, 100–4000
Agamyxis pectinifrons	100–3000	100–400, 100–5000
F: Pimelodidae		
Pimelodus blochii	100–4000	100–800, 100–4000
Pimelodus pictus	300–4000	100–800, 800–5000
F: Callichthyidae		
Corydoras paleatus	100–1500	600–3000
O: Gymnotiformes		
F: Sternopygidae		
Eigenmannia virescens	200–1500	—
O: Perciformes		
SO: Anabantoidei		
F: Belontiidae		
Trichopsis vittata	600–2500	800–2500
Trichopsis pumila	100–2500	1000–4000
Colisa lalia	100–2500	400–1600
Trichogaster trichopterus	300–2000	—

Source: After Ladich 2000.

ings as well. Crawford (1993) showed that midbrain neurons in *P. isidori* elicit their best excitatory responses to frequencies of 300Hz and below, while the audiogram indicated best sensitivities at around 500Hz and good sensitivity up to well over 1kHz. Thus the correlation between the auditory sensitivity and sound power spectra at the midbrain level is not paralleled at the level of the auditory periphery. Similarly, high-frequency components of sounds (>500Hz) are certainly not detectable by many perciform nonspecialists such as cichlids and centrarchids.

In conclusion, differences in auditory sensitivity between hearing generalists and specialists do not seem to be related to their ability to produce sounds, as vocalizing forms are found among fishes representing all types of hearing abilities. Therefore, vocal communication does not seem to be the major force for the evolution of hearing specializations. Since most studies lack a detailed analysis of the whole sound spectrum underwater, it is unclear to what degree sound production matches hearing sensitivity in other vocal species. Based on our current knowledge, a correlation has to be assumed in some species, although the distribution of sonic organs and hearing abilities among otophysans and other teleosts is not generally supported.

6. Behavior and Evolution

The lack of a correlation between the frequency composition of vocalizations and the frequency sensitivity of the auditory system in certain species does not necessarily imply that fishes cannot detect conspecific sounds. Hearing thresholds and sound pressure levels must also be taken into consideration. Acoustic signals can be easily recognized in species emitting high-amplitude sounds such as the loach *Botia horae* (40 dB above threshold; Ladich, 1999) despite the fact that the dominant frequency of sounds is outside the best auditory sensitivity. On the other hand, in the doradid *Agamyxis pectinifrons*, communication distances are limited because sound pressure levels are maximally 8–15 dB above hearing thresholds. Such comparisons cannot provide any information about which sound characteristics (spectral, temporal, intensity) are exploited by fishes (Myrberg et al., 1978).

Differences in sound intensity cannot be sufficiently explained by communication distance. During agonistic interactions, for example, fishes often emit very loud sounds despite the fact that they are very close to each other, typically 0–5 cm, and sound transmission is not hindered by the environment at this distance. Fighting sounds of croaking gouramis have sound pressure level of up to 123 dB (re: 1 μPa) and fishes can easily be heard outside the tanks despite their small body size (0.3–2 g) (Ladich et al., 1992). Differences in loudness might be explained at a number of different levels ranging from divergence in the efficiencies of sound-generating organs to different communicative functions of acoustic signals (e.g., mate attraction vs. nest defense) and predation risk. Territorial advertisement calls such as boatwhistles and hums in toadfishes, moans in mormyrids, or chirps in damselfishes (Crawford et al., 1997; Mann and Lobel, 1997; Brantley and Bass, 1994), which are used in mate attraction over several meters, need to be loud. On the other hand, courtship sounds uttered at the nest site after females have approached males should be of low intensity (in order to avoid predation), as observed in gobies, damselfish, and gouramis (Ladich and Kratochvil, 1989; Kenyon, 1994), and gouramis (F. Ladich, unpublished).

In summary, the occurrence of certain hearing abilities is independent of the physical attributes of sounds. This suggests that different selective pressures favored the evolution of hearing and hearing specializations on the one hand, and sound-generating mechanisms and vocalizations on the other. As hearing specializations are often characteristics of whole taxa, whereas sonic organs appear in a limited number of species within these taxa, it is assumed that the former evolved much earlier. This idea is supported by a phylogenetic analysis of the sound-producing organs within otophysans. Cyprini-formes is the most primitive group among otophysans with Characiphysi (Characiformes and Siluriphysi) being its sister group (Fink and Fink, 1996). Interestingly, only a few representatives of the large order of Cypriniformes are known to be vocal and in no case is there a sonic organ described. This indicates that sound-generating structures are less specialized within Cypriniformes. Highly specialized sonic mechanisms such as swim bladder drumming muscles only evolved in its sister group, the Characiphysi. Many characids as well as numerous catfish families possess intrinsic swim bladder muscles (Ladich and Bass, 1998). Phylogenetic analysis revealed that the Weberian apparatus is an important feature of all otophysans, while sound production evolved occasionally in Cypriniformes and on a regular basis in Characiphysi (Ladich, 1999, 2000).

Which environmental constraints have caused the ancestors of certain fishes to improve their hearing abilities? It appears that hearing specializations mostly occur in quiet environments such as lakes, slowly flowing waters, and the deep-sea (otophysans, mormyrids, anabantoids) and only occasionally in turbulent and noisy habitats such as coasts and reefs (holocentrids). Additionally, most sound energy that propagates in shallow freshwater habitats is of higher frequency (Rogers and Cox, 1988). Lowering the auditory threshold and extending the frequency range would allow fishes to analyze the auditory scene that surrounds them more precisely. Hearing out or determining a specific source from a mixture of sources is a fundamental feature shared by vertebrate animals (auditory stream segregation; Bregman, 1990). Nonvocal goldfish are able to separately analyze two pulse trains differing in both repetition rate and the spectral profile of pulses (Fay, 1998), indicating that complex perceptual processes evolved independently of acoustic communication. Such auditory abilities would clearly improve the chance of survival during attacks by predators and/or enable better prey detection (Canfield and Eaton, 1990). Markl (1972) observed that piranhas attacked prey items accompanied by splashing noise more often than those presented silently. Predator avoidance, in part through the development of hearing abilities, may largely explain the evolution of ultrasonic hearing in numerous nocturnally flying insects (Hoy, 1992) and perhaps some fishes as well. For example, clupeids are sensitive to ultrasound and respond to echolocating pulses of dolphins in playback experiments by startle behavior (Mann et al., 1997; A.N. Popper, unpublished). Together, these results suggest that the major selective pressures influencing the evolution of hearing specializations among teleosts could be predator avoidance and/or prey detection in quiet freshwater habitats and to a lesser degree the optimization of acoustic communication.

7. Sound Production and Underwater Hearing in Frogs

Numerous frog species have adopted, as adults, an aquatic rather than a terrestrial lifestyle but only a small number call and communicate acoustically underwater (Zelick et al., 1999). One representative of the leopard frog *Rana subaquavocalis* produces sounds from a depth of more than 1 meter, making it completely inaudible in air. Mating calls are pulsed snore-like sounds, which differ only slightly in sound characteristics (e.g., pulse number and duration) from their terrestrially calling relatives (Platz, 1993). Representatives of the pipid genus *Xenopus* are totally aquatic throughout their lifetime. They emit various trains of click-like signals during reproductive and agonistic behavior (Yager, 1992a; Kelley and Tobias, 1999). In *Xenopus borealis*, clicks are about 2–5 ms long and most sound energy is concentrated at 2.6 kHz. Interclick intervals vary from 40 ms in agonistic calls to 300–600 ms in advertisement calls (Yager, 1992a). These sounds are produced by a highly modified larynx, which lacks vocal chords, and without any moving air column. Rapid separation of calcified disc-like components of the arytenoid cartilages result in an air implosion and production of impulsive clicks in *X. borealis* (Yager, 1992b). While the call repertoire and behavioral contexts match those of many terrestrial anurans, the nature of the fundamental components of its calls (the click) set it apart from nonpipids. Its impulsive qualities and high repetition rates (25 to several hundreds per second) suggest that a major adaptation to underwater signaling involves the sound-producing mechanism itself. Interestingly, these series of short, broadband pulses resemble those produced by numerous fish species (gouramis, catfishes) and might be an adaptation to a shallow-water habitat where the transmission channel acts as a high-pass filter with the cutoff frequency related to the wavelength of the sound compared with the water depth (Rogers and Cox, 1988). Boatright-Horowitz et al. (1999) showed that the spectral information is mostly lost at distances of more than 1 meter. Therefore, temporal patterning and broadband pulsed sounds might have evolved for the transmission of acoustic information over longer distances (Myrberg et al., 1978; Wysocki and Ladich, 2002).

Given that numerous anuran species are aquatic and the species adapted for hearing in either air or water hear rather poorly when in the alternative medium, it would be interesting to know if frogs have adaptations for underwater, as well as aerial, hearing and how sound is transmitted from the medium to the inner ear. The middle ear of anurans is generally viewed as an adaptation for aerial hearing because of its tympanic membrane and middle-ear bone, and its similarity to that of amniotes. Like other anurans (see Capranica, 1976), the inner ear contains two separate sensory neuroepithelia: the amphibian papilla utilized for the detection of low-frequency sound and the basilar papilla for higher frequencies. Using chronic electrode implants in the midbrain auditory region (torus semicircularis), Lombard et al. (1981) compared auditory sound intensity threshold curves for *Rana catesbeiana* in air and underwater. Sound intensity threshold curves were lower in water than in air below 200 Hz, but similar above 400 Hz. The authors concluded that the amphibian papilla, responsible for low-frequency sound detection, is adapted for hearing underwater. Stimulation of *Xenopus laevis* with underwater sound, while measuring the tympanic disc vibration, showed two peaks in the range of 0.6–1.1 kHz and 1.6–2.2 kHz (Christensen-Dalsgaard et al., 1995). The first peak correlated with the sound-induced vibration of the lung, and the second with the pulsations of the air-bladder in the middle-ear cavity. Thus, sound sensitivity is apparently enhanced by coupling the ear to air-filled spaces. Again, this is very similar to the improvement of auditory sensitivity in hearing specialists among fishes, where vibrations of the swim bladder or suprabranchial chambers are transmitted to the inner ears. In summary, the available studies suggest that fishes and aquatic frogs evolved similar mechanisms for sound production and detection in response to the physical characteristics of their environment.

Acknowledgments. Research was supported by the Austrian Science Fund (FWF grant no. 12411 to F.L.) and the U.S. National Institutes of Health (DC00092 to A.H.B.).

References

Abu-Gideiri, J.B., and Nasr, D.H. (1973). Sound production by *Synodontis schall* (Bloch-Schneider). *Hydrobiologia* 43:415–428.

Aiken, R.B. (1985). Sound production by aquatic insects. *Biol. Rev.* 65:163–211.

Alexander, R.M.N. (1964). The structure of the Weberian apparatus in the Siluri. *Proc. Zool. Soc. Lond.* 142:419–440.

Bader, R. (1937). Bau, Entwicklung und Funktion des akzessorischen Atmungsorgans der Labyrinthfische. *Z. Wiss. Zool.* 149:323–401.

Ballantyne, P.K., and Colgan, P.W. (1978). Sound production during agonistic and reproductive behaviour in the pumpkinseed (*Lepomis gibbosus*), the bluegill (*Lepomis macrochirus*) and their hybrid sunfish. I. Context. *Biol. Behav.* 3:113–135.

Barber, S.B., and Mowbray, W.H. (1956). Mechanism of sound production in the sculpin. *Science* 124:219–220.

Bass, A.H. (1996). Shaping brain sexuality. *Amer. Scientist* 84:352–363.

Bass, A.H., and Baker, R. (1991). Evolution of homologous vocal control traits. *Brain Behav. Evol.* 38:240–254.

Bass, A.H., and Clark, C.W. (2002). The physical acoustics of underwater sound communication. In: *Springer Handbook of Auditory Research* (Simmons, A.M., Popper, A.N., and Fay, R.R., eds.). New York: Springer.

Bass, A.H., and Marchaterre, M.A. (1989). Sound-generating (sonic) motor system in a teleost fish (*Porichthys notatus*): Sexual polymorphism in the ultrastructure of myofibrils. *J. Comp. Neurol.* 286:141–153.

Bass, A.H., Bodnar, D., and Marchaterre, M. (1999). Complementary explanation for existing phenotypes in an acoustic communication system. In: *The Design of Animal Communication* (Hauser, M.D., and Konishi, M., eds.), pp. 493–514. Cambridge: MIT Press.

Blaxter, J.H.S., Denton, E.J., and Gray, J.A.B. (1981). Acousticolateralis system in clupeid fishes. In: *Hearing and Sound Communication in Fishes* (Tavolga, W.N., Popper, A.N., and Fay, R.R., eds.), pp. 39–56. New York: Springer.

Boatright-Horowitz, S.S., Cheney, C.A., and Simmons, A.M. (1999). Atmospheric and underwater propagation of bullfrog vocalizations. *Bioacoustics* 9:257–280.

Bodnar, D.A., and Bass, A.H. (1997). Temporal coding of concurrent acoustic signals in auditory midbrain. *J. Neurosci.* 17:7553–7564.

Bodnar, D.A., and Bass, A.H. (1999). A midbrain combinatorial code for temporal and spectral information in concurrent acoustic signals. *J. Neurophysiol.* 81:552–563.

Bodnar, D.A., and Bass, A.H. (2001). The coding of concurrent signals by the auditory midbrain: Effects of stimulus intensity and depth of modulation. *J. Acoust. Soc. Am.* 109:809–825.

Brantley, R.K., and Bass, A.H. (1994). Alternative male spawning tactics and acoustic signals in the plainfin midshipman fish *Porichthys notatus* Girard (Teleostei, Batrachoididae). *Ethology* 96: 213–232.

Brantley, R.K., Marchaterre, M., and Bass, A.H. (1993). Androgen effects on vocal muscle structure in a teleost fish with inter- and intrasexual dimorphism, *J. Morphol.* 216:305–318.

Bregman, A.S. (1990). *Auditory Scene Analysis: The Perceptual Organisation of Sound.* Cambridge: MIT Press.

Canfield, J.G., and Eaton, R.C. (1990). Swim bladder acoustic pressure transduction initiates Mauthner-mediated escape. *Nature* 347:760–762.

Capranica, R.C. (1976). Morphology and physiology of the auditory system. In: *Frog Neurobiology* (llina, R., and Precht, W., eds.), pp. 551–575. Berlin, Germany: Springer-Verlag.

Chranilov, N.S. (1927). Beiträge zur Kenntnis des Weber'schen Apparates der Ostariophysi 1. Vergleichend-anatomische Übersicht der Knochenelemente des Weber'schen Apparates bei Cypriniformes. *Zool. Jb. Anat. Ontog.* 49:501–597.

Christensen-Dalsgaard, J., Breithaupt, T., and Elepfandt, A. (1995). Underwater hearing in the clawed frog *Xenopus leavis. Naturwissenschaften* 77:135–137.

Coburn, M.M., and Grubach, P.G. (1998). Ontogeny of the Weberian Apparatus in the armored catfish *Corydoras paleatus* (Siluriformes: Callichthyidae). *Copeia* 1998(2):301–311.

Cohen, M.J., and Winn, H.E. (1967). Electrophysiological observation on hearing and sound production in the fish *Porichthys notatus. J. Exp. Zool.* 165:355–370.

Connaughton, M.A., Taylor, M.H., and Fine, M.L. (2000). Effects of fish size and temperature on weakfish disturbance calls: implications for the mechanism of sound generation. *J. Exp. Biol.* 203:1503–1512.

Coombs, S., and Popper, A.N. (1979). Hearing differences among Hawaiian squirrelfish (family Holocentridae) related to differences in the peripheral auditory system. *J. Comp. Physiol.* 132:203–207.

Crawford, J.D. (1993). Central auditory neurophysiology of a sound-producing fish: The mesencephalon of *Pollimyrus isidori* (Mormyridae). *J. Comp. Physiol. A.* 172:139–152.

Crawford, J.D., and Huang, X. (1999). Communication signals and sound production mechanism of mormyrid electric fish. *J. Exp. Biol.* 202:1417–1426.

Crawford, J.D., Hagedorn, M., and Hopkins, C.D. (1986). Acoustic communication in an electric fish, *Pollimyrus isidori* (Mormyridae). *J. Comp. Physiol. A.* 159:297–310.

Crawford, J.D., Jacobs, P., and Benech, V. (1997). Sound production and reproductive ecology of strongly acoustic fish in Africa: *Pollimyrus isidori*, Mormyridae. *Behaviour* 134:1–49.

Dijkgraaf, S. (1941). Haben die Lautäußerungen der Elritze eine biologische Bedeutung? *Zool. Anz.* 136:103–106.

Dufossé, M. (1874). Recherches sur les bruits et les sons expressifs que font entendre les poissons d'Europe et sur les organes producteurs de ces phenomenes acoustiques ainsi que sur les appareils de l'audition de plusiuers de ces animaux. *Ann. Sc. Nat.* 19, Art. No. 5.

Edds-Walton, P.L. (1997). Acoustic communication signals of mysticete whales. *Bioacoustics* 8:47–60.

Fay, R.R. (1988). *Hearing in Vertebrates: A Psychophysics Databook*, Illinois Winnetka: Hill-Fay Associates.

Fay, R.R. (1998). Auditory stream segregation in goldfish (*Carassius auratus*). *Hear. Res.* 120:69–76.

Fine, M.L. (1978). Seasonal and geographical variation of mating call of the oyster toadfish, *Opsanus tau* L. *Oecologia* 36:45–57.

Fine, M.L. (1981). Mismatch between sound production and hearing in the oyster toadfish. In: *Hearing and Sound Communication in Fishes* (Tavolga, W.N., Popper, A.N., and Fay, R.R., eds.), pp. 257–263. New York: Springer.

Fine, M.L., and Ladich, F. (in press). Sound production, spine locking, and related adaptations. In: *Catfishes* (Kapoor, B.G., Arratia, G., Chardon, M., and Diogo, R., eds.). New Delhi, India: Oxford and IBH Publishing.

Fine, M.L., and Lenhardt, M.L. (1983). Shallow water propagation of the toadfish mating call. *Comp. Biochem. Physiol.* 76:225–231.

Fine, M.L., Burns, N.M., and Harris, T.M. (1990). Ontogeny and sexual dimorphism of sonic muscle in the oyster toadfish. *Can. J. Zool.* 8:1378–1381.

Fine, M.L., Winn, H.E., and Olla, B.L. (1977). Communication in fishes. In: *How Animals Communicate* (Sebeok, T.A., ed.), pp. 472–518. Bloomington, IN: Indiana Univ. Press.

Fine, M.L., King, C.B., Friel, J.P., Loesser, K.E., and Newton, S. (1999). Sound production and locking of the pectoral spine of the channel catfish. *Amer. Fisher. Soc. Symp.* 24:105–114.

Fink, S.V., and Fink, W.L. (1996). Interrelationships of ostariophysan fishes. In: *Interrelationships of*

Fishes (Stiassny, M.L.J., Pasenti, L.R., and Johnson, G.D., eds.), pp. 209–249. San Diego: Academic Press.

Fish, M.P., and Mowbray, W.H. (1970). Sounds of Western North Atlantic Fishes. Baltimore: Johns Hopkins Press.

Forrest, T.G., Miller, G.L., and Zagar, J.R. (1993). Sound propagation in shallow water: Implications for acoustic communication by aquatic animals. *Bioacoustics* 4:259–270.

Gerald, J.W. (1971). Sound production in six species of sunfish (Centrarchidae). *Evolution* 25:75–87.

Hawkins, A.D. (1993). Underwater sound and fish behaviour. In: *Behaviour of Teleost Fishes* (Pitcher, T.J., ed.), pp. 129–169. London: Chapman & Hall.

Hawkins, A.D., and Amorim, M.C.P. (2000). Spawning sounds of the male haddock, *Melanogrammus aeglefinus. Environ. Biol. Fishes* 59:29–41.

Hawkins, A.D., and Myrberg, A.A. (1983). Hearing and sound communication underwater. In: *Bioacoustics: A Comparative Approach* (Lewis, B., ed.), pp. 347–405. London: Academic Press.

Hawkins, A.D., and Rasmussen, K.J. (1978). The calls of gadoid fish. *J. Mar. Biol. Ass. U.K.* 58:891–911.

Henglmüller, S.M., and Ladich, F. (1999). Development of agonistic behaviour and vocalization in croaking gourami. *J. Fish Biol.* 54:380–395.

Heyd, A., and Pfeiffer, W. (2000). Über die Lauterzeugung der Welse (Siluroidei, Ostariophysi, Teleostei) und ihren Zusammenhang mit der Phylogenese und der Schreckreaktion. *Rev. Suisse Zool.* 107:165–211.

Hoy, R.R. (1992). The evolution of hearing in insects as an adaptation to predation from bats. In: *The Evolutionary Biology of Hearing* (Webster, D.B., Fay, R.R., and Popper, A.N., eds.), pp. 115–129. New York: Springer.

Ibara, R.M., Penny, L.T., Ebeling, A.W., Dykhuizen, Gv., and Cailliet, G. (1983). The mating call of the plainfin midshipman fish, *Porichthys notatus*. In: *Predators and Prey in Fishes* (Noakes, D.L.G., ed.), pp. 205–212. The Hague, Netherlands: Junk Publishers.

Kaatz, I.M. (1999). The behavioral and morphological diversity of acoustic communication systems in a clade of tropical catfishes (Pisces: Siluriformes). PhD. dissertation. State University of New York, Syracuse.

Kastberger, G. (1977). Der Trommelapparat der Doradiden (Siluriformes, Pisces). *Zool. Jb. Physiol.* 81:281–309.

Kastberger, G. (1981a). Economy of sound production in piranhas (Serrasalminae, Characidae). I. Functional properties of sonic muscles. *Zool. Jb. Physiol.* 85:113–125.

Kastberger, G. (1981b). Economy of sound production in piranhas (Serrasalminae, Characidae). II. Functional properties of sound emitter. *Zool. Jb. Physiol.* 85:393–411.

Kelley, D.B., and Tobias, M.L. (1999). Vocal communication in *Xenopus laevis*. In: *The Design of Animal Communication* (Hauser, M.D., and Konishi, M., eds.), pp. 9–35, Cambridge: MIT Press.

Kenyon, T.N. (1994). The significance of sound interception to males of the bicolor damselfish, *Pomacentrus partitus. Environ. Biol. Fishes* 40: 391–405.

Kratochvil, H. (1978). Der Bau des Lautapparates vom Knurrenden Gurami (*Trichopsis vittatus* Cuvier and Valenciennes) (Anabantidae, Belontiidae). *Zoomorphologie* 91:91–99.

Kratochvil, H. (1980). Geschlechtsdimorphismus beim Lautapparat des Knurrenden Zwerggurami *Trichopsis pumilus* Arnold (Anabantidae, Teleostei). *Zoomorphologie* 94:204–208.

Ladich, F. (1988). Sound production by the gudgeon, *Gobio gobio* L.: A common European freshwater fish (Cyprinidae, Teleostei). *J. Fish. Biol.* 32: 707–715.

Ladich, F. (1989). Sound production by the river bullhead *Cottus gobio* L. (Cottidae, Teleostei). *J. Fish Biol.* 35:531–538.

Ladich, F. (1997). Comparative analysis of swim bladder (drumming) and pectoral (stridulation) sounds in three families of catfishes. *Bioacoustics* 8:185–208.

Ladich, F. (1999). Did auditory sensitivity and vocalization evolve independently in otophysan fishes? *Brain Behav. Evol.* 53:288–304.

Ladich, F. (2000). Acoustic communication and the evolution of hearing in fishes. *Phil. Trans. R. Soc. Lond. B.* 355:1285–1288.

Ladich, F., and Bass, A.H. (1996). Sonic/vocal-acousticolateralis pathways in teleost fishes: A transneuronal biocytin study in mochokid catfish. *J. Comp. Neurol.* 374:493–505.

Ladich, F., and Bass, A.H. (1998). Sonic/vocal motor pathways in catfishes: Comparison with other teleosts. *Brain Behav. Evol.* 51:315–330.

Ladich, F., and Bass, A.H. (in press). Audition in catfishes. In: *Catfishes* (Kapoor, B.G., Arratia, G., Chardon, M., and Diogo, R., eds.). New Delhi, India: Oxford and IBH Publishing.

Ladich, F., and Kratochvil, H. (1989). Sound production by the marmoreal goby, *Protherorhinus marmoratus* (Pallas) (Gobiidae, Teleostei). *Zool. Jb. Physiol.* 93:501–504.

Ladich, F., and Popper, A.N. (2001). Comparison of the inner ear ultrastructure between teleost fishes using different channels for communication. *Hear. Res.* 154:62–72.

Ladich, F., and Tadler, A. (1988). Sound production in *Polypterus* (Osteichthyes: Polypteridae). *Copeia* 1988(4):1076–1077.

Ladich, F., and Yan, H.Y. (1998). Correlation between auditory sensitivity and vocalization in anabantoid fishes. *J. Comp. Physiol. A.* 182:737–746.

Ladich, F., Bischof, C., Schleinzer, G., and Fuchs, A. (1992). Intra- and interspecific differences in agonistic vocalization in croaking gouramis (Genus: *Trichopsis*, Anabantoidei, Teleostei). *Bioacoustics* 4:131–141.

Lanzing, W.S.R. (1974). Sound production in the cichlid *Tilapia mossambica* Peters. *J. Fish Biol.* 6:341–347.

Lombard, R.E., Fay, R.R., and Werner, Y.L. (1981). Underwater hearing in the frog *Rana catesbeiana*. *J. Exp. Biol.* 91:57–71.

Lugli, M., Pavan, G., Torricelli, P., and Bobbio, L. (1995). Spawning vocalizations in male freshwater gobiids (Pisces, Gobiidae). *Environ. Biol. Fishes* 43:219–231.

Mahajan, C.L. (1963). Sound producing apparatus in an Indian catfish *Sisor rhabdophorus* Hamilton. *J. Linn. Soc. Zool.* 43:721–724.

Mann, D.A., and Lobel, P.S. (1997). Propagation of damselfish (Pomacentridae) courtship sounds. *J. Acoust. Soc. Am.* 101:3783–3791.

Mann, D.A., Lu, Z., and Popper, A.N. (1997). A clupeid fish can detect ultrasound. *Nature* 389:341.

Markl, H. (1971). Schallerzeugung bei Piranhas (Serrasalminae, Characidae). *Z. Vergl. Physiol.* 74:39–56.

Markl, H. (1972). Aggression und Beuteverhalten bei Piranhas (Serrasalminae, Characidae). *Z. Tierpsychol.* 30:190–216.

Marshall, N.B. (1967). Sound-producing mechanisms and the biology of deep-sea fishes. In: *Marine Bio-Acoustics*, (Tavolga, W.N., ed.), pp. 123–133. Oxford: Pergamon Press.

McCormick, C.A., and Popper, A.N. (1984). Auditory sensitivity and psychophysical tuning curves in the elephant nose fish, *Gnathonemus petersii*. *J. Comp. Physiol. A.* 155:753–761.

McKibben, J.R., and Bass, A.H. (1998). Behavioral assessment of acoustic parameters relevant to signal recognition and preference in a vocal fish. *J. Acoust. Soc. Am.* 104:3520–3533.

McKibben, J.R., and Bass, A.H. (1999). Peripheral encoding of behaviorally relevant acoustic signals in a vocal fish: single tones. *J. Comp. Physiol. A.* 184:563–576.

McKibben, J.R., and Bass, A.H. (2001). Effect of temporal envelope modulation on acoustic signal recognition in a vocal fish, the plainfin midshipman. *J. Acoust. Soc. Am.* 109:2934–2943.

Müller, J. (1857). Ueber die Fische, welche Töne von sich geben und die Entstehung dieser Töne. *Arch. Ant. Physiol. Wiss. Med*: 249–279.

Myrberg, A.A. (1981). Sound communication and interception in fishes. In: *Hearing and Sound Communication in Fishes* (Tavolga, W.N., Popper, A.N., and Fay, R.R., eds.), pp. 395–426. New York: Springer.

Myrberg, A.A., and Spires, J.Y. (1980). Hearing in damselfishes: An analysis of signal detection among closely related species. *J. Comp. Physiol.* 140:135–144.

Myrberg, A.A., Ha, S.J., and Shamblott, H.S. (1993). The sounds of bicolor damselfish (*Pomacentrus partitus*): Predictors of body size and a spectral basis for individual recognition and assessment. *J. Acoust. Soc. Am.* 94:3067–3070.

Myrberg, A.A., Kramer, E., and Heinecke, P. (1965). Sound production by cichlid fishes. *Science* 149:555–558.

Myrberg, A.A., Spanier, E., and Ha, S.J. (1978). Temporal patterning in acoustical communication. In: *Contrasts in Behaviour* (Reese, E.S., and Lighter, F.J., eds.), pp. 137–179. New York: Wiley.

Nelissen, M.H.J. (1978). Sound production by some tanganyikan cichlid fishes and a hypothesis for the evolution of the R communication mechanisms. *Behaviour* 64:137–147.

Nelson, J.S. (1994). *Fishes of the World*, 3rd ed. New York: Wiley.

Ono, R.D., and Poss, S.G. (1982). Structure and innervation of the swim bladder musculature in the weakfish, *Cynoscion regalis* (Teleostei: Sciaenidae). *Can. J. Zool.* 60:1955–1967.

Pfeiffer, W., and Eisenberg, J.F. (1965). Die Lauterzeugung der Dornwelse (Doradidae). *Z. Morph. Ökol. Tiere.* 54:669–679.

Platz, J.E. (1993). *Rana catesbeiana subaquavocalis*: A remarkable new species of leopard frog (*Rana pipiens* complex) from Southeastern Arizona that calls underwater. *J. Herpetol.* 27:154–167.

Poggendorf, D. (1952). Die absolute Hörschwelle des Zwergwelses (*Ameiurus nebulosus*) und Beiträge zur Physik des Weberschen Apparates der Ostariophysen. *Z. Vergl. Physiol.* 34:222–257.

Popper, A.N., and Fay, R.R. (1999). The auditory periphery in fishes. In: *Comparative Hearing: Fish and Amphibians* (Fay, R.R., and Popper, A.N., eds.), pp. 43–100. New York: Springer.

Popper, A.N., and Tavolga, W.N. (1981). Structure and function of the ear in the marine catfish, *Arius felis*. *J. Comp. Physiol.* 144:27–34.

Pruzsinszky, I., and Ladich, F. (1998). Sound production and reproductive behaviour of armoured catfish *Corydoras paleatus* (Callichthyidae). *Environ. Biol. Fishes* 53:183–191.

Rauther, M. (1945). Über die Schwimmblase und die zu ihr in Beziehung tretenden somatischen Muskeln bei den Trigliden und anderen Scleroparei. *Zool. Jb. Anat.* 69:159–250.

Rigley, L., and Marshall, J.A. (1973). Sound production by the elephant-nose fish *Gnathonemus petersi* (Pisces, Mormyridae). *Copeia* 1973(1): 134–135.

Rogers, P.H., and Cox, H. (1988). Underwater sound as a biological stimulus. In: *Sensory Biology of Aquatic Animals* (Atema, J., Fay, R.R., Popper, A.N., and Tavolga, W.N., eds.), pp. 131–149. New York: Springer.

Römer, H. (1993). Environmental and biological constraints for the evolution of long-range signalling and hearing in acoustic insects. *Phil. Trans. R. Soc. Lond. B.* 340:179–185.

Rowland, W.L. (1978). Sound production and associated behaviour in the jewel fish, *Hemichromis bimaculatus*. *Behaviour* 78:125–136.

Schachner, G., and Schaller, F. (1981). Schallerzeugung und Schallreaktionen beim Antennenwels (Mandim) *Rhamdia sebae sebae. Val. Zool. Beitr.* 27:375–392.

Schellart, N.A.M., and Popper, A.N. (1992). Functional aspects of the evolution of the auditory system of actinopterygian fish. In: *The Evolutionary Biology of Hearing* (Webster, D.E., Fay, R.R., and Popper, A.N., eds.), pp. 295–322. New York: Springer.

Schneider, H. (1941). Die Bedeutung der Atemhöhle der Labyrinthfische für ihr Hörvermögen. *Z. Vergl. Physiol.* 29:172–194.

Schneider, H. (1961). Neuere Ergebnisse der Lautforschung bei Fischen. *Naturwissenschaften* 15:513–518.

Schneider, H. (1964). Physiologische und morphologische Untersuchungen zur Bioakustik der Tigerfische (Pisces, Theraponidae). *Z. Vergl. Physiol.* 47:493–558.

Schneider, H. (1967). Morphology and physiology of sound-producing mechanisms in teleost fishes. In: *Marine Bio-Acoustics*, *Vol 2* (Tavolga, W.N., ed.), pp. 135–158. Oxford: Pergamon Press.

Schneider, H., and Hasler, A.D. (1960). Laute und Lauterzeugung beim Süßwassertrommler *Aplodinotus gruniens* Rafinesque (Sciaenidae, Pisces). *Z. Vergl. Physiol.* 43:499–517.

Spanier, E. (1979). Aspects of species recognition by sound in four species of damselfish, genus *Eupomacentrus* (Pisces: Pomacentridae). *Z. Tierpsychol.* 51:301–316.

Stabentheiner, A. (1988). Correlations between hearing and sound production in piranhas. *J. Comp. Physiol. A.* 162:67–76.

Stipetić, E. (1939). Über das Gehörorgan der Mormyriden. *Z. Vergl. Physiol.* 26:740–752.

Stout, J.F. (1963). The significance of sound production during the reproductive behaviour of *Notropis analostanus* (Family Cyprinidae). *Anim. Behav.* 11:83–92.

Tavolga, W.N. (1962). Mechanisms of sound production in the ariid catfishes *Galeichthys* and *Bagre*. *Bull. Amer. Mus. Nat. Hist.* 24:1–30.

Tavolga, W.N. (1964). Sonic characteristics and mechanisms in marine fishes. In: *Marine Bio-Acoustics* (Tavolga, W.N., ed.), pp. 195–211. Oxford: Pergamon Press.

Tavolga, W.N. (1971). Sound production and detection. In: *Fish Physiology, Vol. 5: Sensory Systems and Electric Organs* (Hoar, W.S., and Randall, D.J., eds.), pp. 135–205. London: Academic Press.

Valinsky, W., and Rigley, L. (1981). Function of sound production by the skunk loach *Botia horae* (Pisces, Cobitidae). *Z. Tierpsychol.* 55:161–172.

von Frisch, K. (1936). Über den Gehörsinn der Fische. *Biol. Rev.* 11:210–246.

von Frisch, K., and Stettner, H. (1932). Untersuchungen über den Sitz des Gehörsinnes bei der Elritze. *Z. Vergl. Physiol.* 17:687–801.

Walsh, P.J., Mommsen, T.P., and Bass, A.H. (1995). Biochemical and molecular aspects of singing in batrachoidid fishes. In: *Biochemistry and Molecular Biology of Fishes, Vol. 4: Metabolic Biochemistry* (Hochachka, P.W., and Mommsen, T.P., eds.), pp. 279–289. The Hague, Netherlands: Elsevier.

Winn, H.E. (1964). The biological significance of fish sounds. In: *Marine Bio-Acoustics* (Tavolga, W.N., ed.), pp. 213–231, Oxford: Pergamon Press.

Winn, H.E., and Marshall, J.A. (1963). Sound-producing organ of the squirrelfish, *Holocentrus rufus*. *Physiol. Zool.* 36:36–44.

Winn, H.E., Marshall, J.A., and Hazlett, B. (1964). Behavior, diel activities, and stimuli that elicit sound production and reactions to sounds in the long spine squirrelfish. *Copeia* 1964(2):413–425.

Wysocki, L.E., and Ladich, F. (2001). The ontogenetic development of auditory sensitivity, vocalization and acoustic communication in the labyrinth fish *Trichopsis vittata*. *J. Comp. Physiol. A.* 187: 177–187.

Wysocki, L.E., and Ladich, F. (2002). Can fishes resolve temporal characteristics of sounds? New insights using auditory evoked responses. *Hear. Res.* 169:34–36.

Yager, D.D. (1992a). Underwater acoustic communication in the African pipid frog *Xenopus borealis*. *Bioacoustics* 4:1–24.

Yager, D.D. (1992b). A unique sound production mechanism in the pipid anuran *Xenopus borealis*. *Zool. J. Linn. Soc.* 104:351–375.

Zelick, R., Mann, D.A., and Popper, A.N. (1999). Acoustic communication in fishes and frogs. In: *Comparative Hearing: Fish and Amphibians* (Fay, R.R., and Popper, A.N., eds.), pp. 363–411. New York: Springer.

10
The Design of Color Signals and Color Vision in Fishes

N. Justin Marshall and Misha Vorobyev

Abstract

This chapter attempts three things. First, some of the possible functions of the astonishing colors of reef fishes are examined. Second, the spectral sensitivities and potential color vision capabilities of marine fishes are discussed in the light of old and new data. Finally, an integrated approach is used to model what fishes look like to fishes and where, theoretically, one might expect them to place spectral sensitivities. Factors combined in models include body colors, spectral sensitivities, visual resolution, light environment at the microhabitat level, and behavior.

General conclusions are as follows. Colors are almost always for camouflage and both the spectral and spatial resolving power of fish eyes play an important role in the success of camouflage strategies. Colors used in camouflage may also, simultaneously, be used for advertisement or communication, critically dependent on background, depth, and viewing distance. Disappointingly, most reef fishes are probably dichromats or at most trichromats; however, their spectral sensitivities are surprisingly varied when compared to some other animals. This variation is mainly due to differing microhabitat light environments and more of these have been recently described than previously noted. There is some correlation between the colors of reef fishes and their spectral sensitivities, possibly driven by the need to detect and distinguish fishes of different colors on the reef. Light environments account for the broad envelope within which spectral sensitivities of reef fish are placed, in particular those of double cones as previously noted. Where intervening water between target and observer is considered, its effect rapidly overshadows other factors. At long wavelengths (yellow, orange, red), the reef is probably less colorful to many reef fishes than it appears to us. At short wavelengths, ultraviolet (UV), violet, and blue, it may be more colorful. UV sensitivity is possible in about half of the reef fishes so far examined, and this visual capability is currently best correlated with feeding strategies but may also play a role in secret communication on the reef.

1. Introduction

The deep-sea is perhaps the only habitat on earth where the spectral sensitivities of animals and their colors are tightly coevolved. In the lightless depths below 1,000 m, the only light available for vision is the bioluminescent flashes of other animals (Denton, 1990). As might be expected, the often-monochromatic, rod-based vision of most inhabitants here is well matched to the spectra of their bioluminescent emissions (Lythgoe, 1980; Frank and Case, 1988a,b; Partridge et al., 1992; Douglas and Partridge, 1997; Frank and Widder, 1999; Douglas, 2001; Chapters 17 and 18). If deep oceans are a spectrally simple world, shallow-water reef environments lie at the other extreme. The multicolored inhabitants of the reef live in one of the most spectrally diverse habitats on Earth. Color vision here has many tasks, from the constant demands of color communication over territory, mates, and food, to simply avoiding obstacles and not being eaten (Lorenz, 1962; Marshall, 2000a). Possibly as a result of the these varied demands and the many different microhabitat light environments on the reef, new data discussed here (and see McFarland et al., in press and Marshall et al., in press, a,b) reveal many different potential color vision systems in reef fishes. Surprisingly, however, few of these go beyond dichromacy (see Section 4).

Our chief aim in this chapter is to describe the visual world of reef fishes with the ultimate aim of seeing their world through their eyes. For readers not familiar with the species described here, a copy of Randall, Allen, and Steen's excellent guide to Great Barrier Reef fishes (Randall et al., 1991) would help some of the splendour of reef fish colors to come alive. Where possible, these authors were careful to present photographs of fishes in their natural habitat and, as a result, many of the photographs of parrotfish and their allies are somewhat disappointing when compared to the freshly caught fish in the hand. The reason behind the difficulty in photographing some reef fishes forms a major theme of the chapter. Camouflage is, perhaps surprisingly, the major function of many of the striking patterns and colors of reef fishes and is the reason why many photographs of any reef fishes often have the fishes merging into the background, even at moderate distances. As explained in Section 2.2, the parrotfish and wrasse are masters of this while simultaneously being astonishingly brilliantly colored close up. In the right context, communication with color on the reef is equally important as blending in, the issues of territory, sex, and food being paramount. The background against which a fish is seen is one of the critical factors in determining the effectiveness of camouflage *or* communication underwater (Thayer, 1909; Longley, 1914, 1916a,b, 1918; Cott, 1940; Endler, 1981; Endler, 1984; Crook, 1997a; Barry and Hawryshyn, 1999a; Marshall, 2000a,b; Marshall et al. in press a,b). Other important aspects in piecing together the puzzle of how fishes appear to fishes include the microhabitat light environment, the visual system of the observer (*not* the signaler necessarily), the distance between observer and observed, the colors under observation (and their combination), and the behavioral aspect to color (e.g., When and where is a color or color pattern revealed or turned on.). All these aspects are discussed in this chapter.

The data building blocks of body color, light, intervening water, background color, visual systems, and behavior form the essential constituents in understanding the visual world of any animal (Poulton, 1890; Hailman, 1977a,b, 1979; Lythgoe, 1979; Endler, 1990). The visual system is often overlooked as humans are visual animals and naturally assume that the world appears to other animals as it does to us. In fact, as primates, our color vision is good at detecting bananas and other ripening fruit or leaves (Mollon, 1989, 1991; Osorio and Vorobyev, 1996; Dominy and Lucas, 2001) and not good at other regions of the spectrum, notably blue (Mollon, 1989; Mollon et al., 1990). For reasons mainly concerned with light availability, many marine animals specialize in the blue region of the spectrum. What we see on snorkeling over or diving around a coral reef is certainly nothing close to the way its inhabitants see it. We are likely to be long-wavelength biased and more attracted to the yellows, oranges, and reds (bananas and cherries) while overlooking differences in blues. All currently available evidence suggests that reef and other marine fishes possess short-wavelength-based visual systems (Section 4; Lythgoe, 1966, 1968, 1979, 1980; Loew and Lythgoe, 1985; Lythgoe, 1988; Partridge, 1990; Lythgoe et al., 1994; Loew et al., 1996a; McFarland et al., in press) and that yellow may in fact be an effective camouflage color on the reef (Marshall, 2000b).

In truth, we know almost nothing of color vision capabilities in marine fishes, as little is known of the way their cones are interconnected and few behavioral tests of color vision have been performed (Hawryshyn, 1982; Hawryshyn et al., 1988; Douglas and Hawryshyn, 1990; Neumeyer, 1991; Douglas, 2001). The goldfish and its tetrachromatic color vision system is the only fish whose color vision *is* well known (Hawryshyn and Beauchamp, 1985; Hawryshyn et al., 1985; Palacios et al., 1989; Douglas and Hawryshyn, 1990; Neumeyer, 1991; Neumeyer et al., 1991; Neumeyer, 1992; Dorr and Neumeyer, 1996; Neumeyer et al., 2002). An assumption we make here is that spectral regions containing overlapping cone sensitivities are likely to be those regions where wavelength discrimination is best. From previous evidence, this is not unreasonable (Arnold and Neumeyer 1987); however, some caution is needed as in a few instances in the animal kingdom the output of spectral sensitivities is directly linked to behaviors (Scherer and Kolb, 1987a,b; Smith and Macagno, 1990; Kelber, 1997; Kelber, 1999), without the intermediate step of comparing cone output.

Our descriptive journey in this chapter starts with the colors of fishes and their possible function and goes on to examine light environments on the reef and review old and new data on spectral sensitivity and spectral sensitivity tuning. Finally, using simple mathematical models, some inferences and suggestions are made as to the way in which fishes see each other and the world they live in.

2. Reef Fish Colors

The remarkable colors of coral reef fishes and their possible functions have received special comment from many biologists including Darwin, Wallace, and Lorenz (Darwin, 1859; Lorenz, 1962). All three of these learned gentlemen came to basically the same conclusions that, in the crowded throng of the reef, fish colors help fishes to distinguish each other, set up territories, and attract and keep mates. Around the turn of the twentieth century Longley spent many patient hours underwater both observing and photographing fishes and noted that fishes often basically match their backgrounds, thus concluding that their colors were mostly for camouflage (Longley, 1914, 1916a,b, 1918, 1919). Cott's wonderful book *Adaptive Colors of Animals* also championed the camouflage theme of colors in fishes and other animals and recognized that even bold patterning may act as disruptive camouflage when the animal is in its natural context (Cott, 1940). This is of particular significance in the high-contrast, high-chroma world of the reef. More recent attempts to ascribe function to reef fish color have often overemphasized one or another function (Marshall, 2000a). Unlike the deep-sea, however, in the multifaceted visual world of the reef, both colors and the visual systems used to detect them are the result of compromises between different tasks (Endler, 1991a,b; Marshall, 2000a,b).

2.1. The Spectral Nature of Fish Colors

Once one realizes that colors do not appear to us the way they do to other animals, describing them becomes difficult. Bee vision and flower color is a good example of this, as bees are trichromatic like humans but possess a short-wavelength-shifted set of spectral sensitivities that best detect ultraviolet (UV), blue, and green, as opposed to our nominal blue, green, and red detectors (Menzel and Backhaus, 1991; Menzel and Shmida, 1993). The colors of flowers must then be described relative to the bee visual system, and a brave attempt was made by Lars Chittka and colleagues (Chittka and Menzel, 1992; Chittka et al., 1994). The result is terms such as "bee white" and "bee purple," which are distinct to the sensation of white and purple seen by us and result from a combination of human subjective terminology, bee spectral sensitivity, and the shape of the color spectra of flowers.

We face the same problem with fishes but with the added complexity that their spectral sensitivities are varied and largely unknown (Section 4). As a result, a first attempt uses both human subjective terms and the shape of spectra measured by spectrophotometry (Marshall, 2000a). This method of analysis has currently identified around 21 color categories in three broad groups: low saturation, simple, and complex. Low-saturation spectra are those with no sharp changes and are generally brown or black to our eyes. Simple spectra come in two varieties; those with a single peak and those with a step-shaped spectrum (Fig. 10.1; see color plate). Complex spectra may have combinations of two or more peaks or both peaks and steps (Fig. 10.1). In an attempt to examine the distribution of different fish spectra, peak-shaped spectra can be classified according to where the peak lies and step-shaped colors identified by the 50% point on the rise in the step (Chittka and Menzel, 1992; Menzel and Shmida, 1993; Marshall, 2000a,b). In general, spectra can be described with three terms: hue, saturation (chroma), and brightness, which refer to the way they are perceived and their shape. An excellent discussion of this is provided by Endler (1990). Other, more mathematical approaches to describing spectra using principle components, independent components, and a subjective segment analysis also are possible (Wyszecki and Stiles, 1982; Maloney, 1986; Endler, 1990; Cuthill et al., 1999; Chiao et al., 2000b,c; Grill and Rush, 2000; Wachtler et al., 2001) and are useful for some analyses.

More instructive than simply classifying spectra is an attempt to view them with the visual systems of the observer. This is now becoming more frequently possible with fishes and other animals (Vorobyev and Brandt, 1997; Vorobyev et al., 1998; Chiao et al., 2000a,d; Marshall, 2000b; Vorobyev et al., 2001a,b) and is what we attempt toward the end of the chapter.

2.2. Colors for Simultaneous Camouflage and Crypsis

The socially complex environment of the reef has resulted in a series of compromises in color signal design as fishes try to attract mates, ward off rivals, and also hide to prevent themselves being eaten. A number of examples are now emerging of color patterns that allow simultaneous camouflage and crypsis (Endler, 1991a; Marshall, 2000a,b).

It has been recognized a number of times in the past that yellow and blue are colors frequently used by marine animals, accounting for over 30% of colors on the reef (Marshall, 2000a). One reason for their prevalence in marine color schemes is that these colors transmit well in marine waters (Poulton, 1890; Longley, 1914, 1916a,b; Cott, 1940; Lythgoe, 1968; Lythgoe, 1979; Loew and Lythgoe, 1985; Lythgoe and Partridge, 1991; Marshall, 2000a,b). In freshwater, green and red transmit best, the exact optima contingent on the constituents found within different types of water bodies (Lythgoe, 1968; Jerlov, 1976; Lythgoe, 1979; Mobley, 1994). It is instructive to find a guidebook to freshwater fishes and compare it to the ones on marine fishes (Randall et al., 1991), just to get an overall impression of the different underwater color-schemes. As well as being able to transmit over long distances in the marine environment, yellow and blue are also a good combination for an animal to use to signal with, as they are complementary. That is, one color reflects in the spectrum where the other does not and vice versa, providing a strong signal for any color vision systems sensitive to short and middle wavelengths (Fig. 10.1). The human color vision system, which compares the output of the blue (S) cone to the medium (M) and long (L) wavelength cones combined (S − (L + M)—Wyszecki and Stiles, 1982; Backhaus et al., 1998), certainly finds this combination attractive given that it is used extensively in clothes, advertising, and art. What

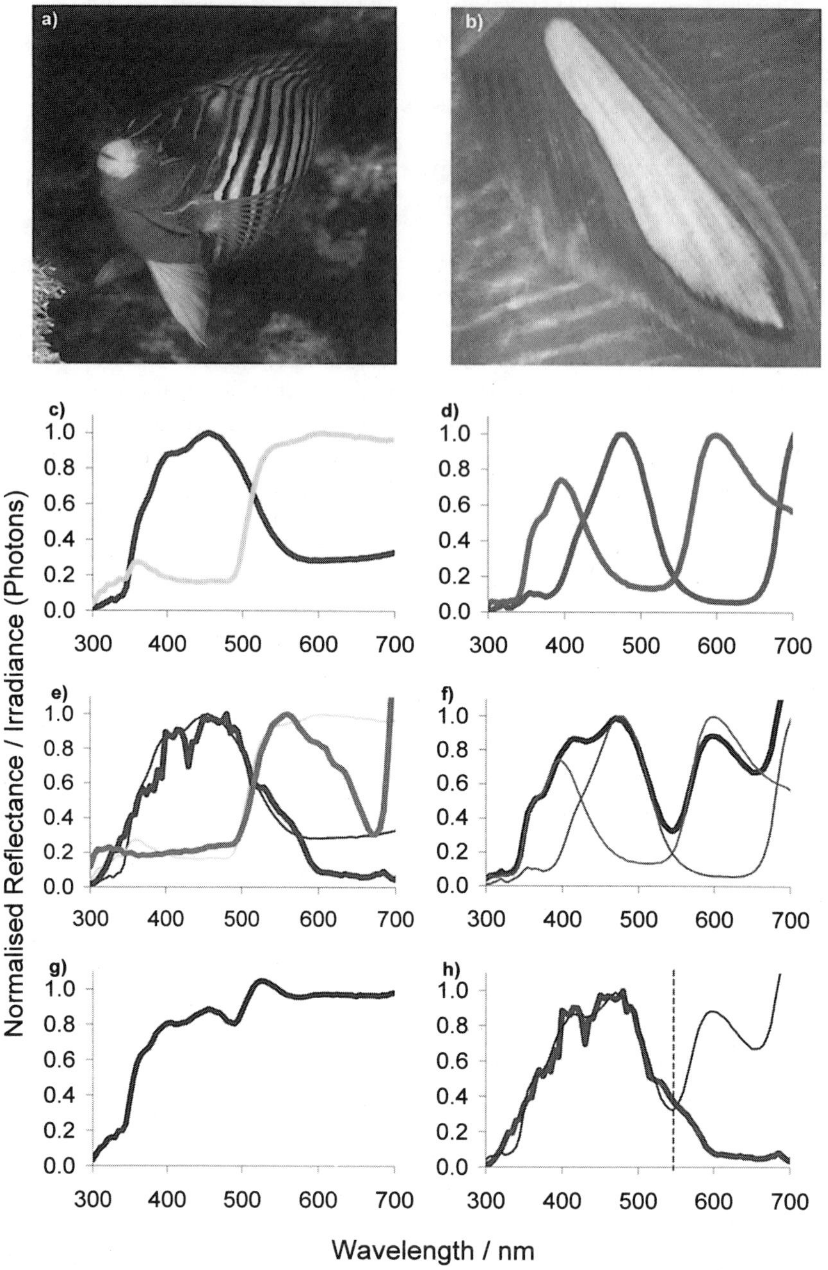

FIGURE 10.1. The colors of reef fish. (a) The regal angelfish *Pygoplites diacanthus*. Graphs in c, e and g relate to this fish. (b) The pectoral fin of the wrasse *Thalassoma lunare* to show fine stripe patterns. Graphs in d, f and h relate to this fish. (c) The reflectance of yellow and blue colors in *P. diacanthus*. The data have been normalized here and in subsequent graphs to emphasize chromatic difference. These are simple peak and step spectra. (d) The reflectance of pinkish-orange and green from *T. lunare*. These are complex multipeaked spectra. (e) Yellow and blue of *P. diacanthus* with the spectra of blue background water and shallow water coral (thick lines) plotted for comparison. (Data from Marshall, 2000b.) (f) The result of adding spectra in d together (black line). (g) The result of adding spectra in c together, gray from 400–700 nm. (h) The additive spectrum from f with background blue-water spectrum plotted for comparison. The vertical dotted line marks the long-wavelength limit of λ-max in reef fish cones known so far (Fig. 10.4). (See color plate)

we know of reef fish spectral sensitivities (Lythgoe, 1966; Lythgoe, 1972; Loew and Lythgoe, 1978; McFarland, 1991; Lythgoe et al., 1994; McFarland and Loew, 1994; Barry and Hawryshyn, 1999b; Marshall, 2000a,b; McFarland et al., in press; Marshall et al., in press, a,b) suggests that both yellow and blue will be a highly conspicuous combination to many reef fishes. Two other factors are especially important in determining the contrast of an object underwater: (1) the background it is viewed against and (2) the distance from which it is viewed.

If we examine factor 2 first, it is notable that many reef fishes, such as the angelfish (pomacanthids), damselfish (pomacentrids), rabbitfish (siganids), and butterflyfish (chaetodontids), all use rather fine patterns of yellow and blue spots or stripes (Fig. 10.1). Close up, these patterns are striking for reasons outlined above; however, as the fish swims into the distance, two factors will render these colors very effective for camouflage. First, the spatial resolving power of many fishes, including reef fishes is poor, at best around 10 times worse than our eyes (Lythgoe, 1979; Collin and Pettigrew, 1988a,b; Collin and Pettigrew, 1989; Marshall, 2000b), so fine patterns of two colors will tend to merge together. The resulting color, as the colors are complementary, is a dull gray (Fig. 10.1), an inconspicuous color on the reef (Kinney et al., 1967; Lythgoe, 1968). Thus, just considering the pattern of the colors within the body of the fish, they can be simultaneously conspicuous to close observers and well camouflaged to the eavesdropping eyes of potential predators at a greater distance (Marshall, 2000b). Second, as the distance between fishes increases, the veiling effect of the intervening water will decrease the visibility of the fish under examination (Lythgoe, 1968; Lythgoe, 1979; Muntz, 1990; Jagger and Muntz, 1993; Vorobyev et al., 2001b), and depending on the nature of the body of water, this effect may be wavelength or color dependent. We model this and other factors influencing fish visibility in Section 4.

Now considering factor 1 above, and moving outside the boundaries of the fish's body, the background against which a yellow and blue fish, or indeed a fish that is either all yellow or all blue (e.g., pomacentrids) is seen, determines whether a fish is camouflaged or conspicuous (Barry and Hawryshyn, 1999a,b). The effect can also be broken into two components, simple matching or contrast and disruptive matching (Poulton, 1890; Thayer, 1909; Cott, 1940). Marshall (2000a,b) argues that the yellow color of some fishes may be a good match to the color of coral. This is especially true in surface waters, rich in long wavelengths, and becomes less good with depth (see Section 4.3 and Fig. 10.5). The effectiveness of the match is also very dependent on the visual system examining it. Short-wavelength-biased visual systems such as in reef fishes (Section 4) are probably less good at discriminating or detecting yellow than ours. Remember that, as primates, humans are particularly interested in yellow as it is a fruit color. Yellow fishes may blend better with coral backgrounds to the eyes of their regular cohabitants than they appear to us (Fig. 10.1).

Marshall (2000a,b) and others (Hailman, 1977a; Lythgoe, 1979) have also argued that the blues of many fishes may be a good match to the backdrop of blue water on the reef (Fig. 10.1) and that many reef fishes contain blue as a component of body color for this reason. Although probably true for some blue fishes (Fig. 10.1), again the visual system of the observer and conditions of the visual interaction are critical (Section 4.3). Just as humans are good with bananas, reef fishes' short-wavelength-biased visual system may well be better at detecting and discriminating different blues, or indeed UVs and violets, than the visual system of a primate. Some behavioral evidence for this is now emerging (S. Job, unpublished; U. Siebeck and N.J. Marshall, unpublished; and S. Ostlund-Nilsson and N.J. Marshall, unpublished; and see Endler, 1991a and Smith et al., 2001 for freshwater comparisons); however, such ideas remain speculative.

Reef fishes such as the regal angelfish *Pygoplites diacanthus* (Fig. 10.1) possess both blue and yellow body colors painted on in fine stripes. Other species, such as the bicolor angelfish *Pomacanthus bicolour*, are more boldly colored and comprised of yellow and blue halves, or they may be either all yellow or all blue. Fish with these differing color schemes have a choice of how to be seen. A yellow fish viewed against a background of blue water will be conspicuous, while against a background of coral, well camouflaged. A blue fish viewed against a background coral will be conspicuous, and against blue water, well camouflaged (see Fig. 1 in Marshall, 2000b). Fish colored both yellow and blue may appear conspicuous against any one evenly colored background; however, reefs are often very disruptive, constructed with jutting coral, rubble, intervening water, and indeed the inhabitants of the reef themselves. Against such a backdrop, an apparently boldly colored fish may effectively disappear as the blocks of color from which it is made disrupt its body outline. This principle of disruptive camouflage is certainly used by many reef fishes, whether using yellow and blue or other color combinations (Fig. 10.1; Cott, 1940; Marshall, 2000a,b), but may depend for its success on the pattern size relative to the background. A 4-cm all-blue fish in a branching coral background with 4-cm holes between branches is well camouflaged through disruption, despite being only one color. Anyone who has seen a branching coral head full of brilliantly colored blue damselfish (often juvenile *Chromis viridis*) may think they are there just for mechanical protection but this only appears so looking down on the coral from above. A sideways view reveals a branched coral with gaps through to a background of blue space water that makes it hard to distinguish small blue fishes.

Yellow and blue are simple colors in the classification of Marshall (2000a). Wrasse (Labridae) and parrotfish (Scaridae), and a few other species outside of these families, often possess extraordinarily complex spectra with as many as three peaks or combinations of peaks and steps (Marshall, 2000a, Fig. 10.1). Possible adaptive reasons behind such complexity follow some of the same line of reasoning as just outlined for yellow and blue. At close proximity, the colors are likely to appear conspicuous as, in common with yellow and blue, they are often complementary. The purple/orange and green/blue colors of the wrasse *Thalassoma lunare* (as we see them) are a good example, the short-wavelength peak of the blue/green spectra neatly fitting in the trough left between the peaks of the purple/orange spectra (Fig. 10.1). As the two colors are in fine stripes, the colors will combine at a distance due to the poor spatial resolving power of fish eyes. Remarkably, the resulting combination is a very close match to the spectrum of blue background water, at least up to around 560 nm (Fig. 10.1). *T. lunare* is often to be found darting around the reef tops and down the front slope, and here it may be a close chromatic match to the background, rather than just graying out as yellow and blue-patterned fishes will (Fig. 10.1, Marshall, 2000b; Marshall et al., in press, a,b). As reef fish color vision is probably poor beyond 550 nm (McFarland et al., in press; Marshall et al., in press, a,b; and see Fig. 10.4 and Section 4), the combinatorial color of *T. lunare* viewed at a distance need not be well matched to this spectral region.

An aspect of fish color not mentioned here is their remarkable and often-overlooked ability to change color and pattern on a number of different timescales (Longley, 1918, 1919; Crook, 1997a,b; Marshall, 2000a). This is frequently in response to background changes (Longley, 1919; Crook, 1997b) and again emphasizes the importance of background to the way the fish expects it is being perceived. As this deserves a chapter unto itself, and because almost no recent data exist, we will postpone any discussion until future study by ourselves and hopefully others.

2.3. UV Colours: A Secret Communication Channel?

One aspect of animal coloration that has received much attention in recent years is UV colors and signals. This is partly due to the relatively recent realization that, in fact, many animals, unlike humans, both see and make use of UV (Hawryshyn and Beauchamp, 1985; Burkhardt, 1989; Burkhardt and Maier, 1989; Bowmaker, 1990; Bowmaker et al., 1991; Menzel and Backhaus, 1991; Fleishman et al., 1993; Jacobs, 1993; Frank and Widder, 1994a,b; Loew, 1994; Shashar, 1994; Douglas et al., 1995; Loew et al., 1996a; Andersson and Amundsen, 1997; Church et al., 1998; Losey et al., 1999;

Hart et al., 2000), and partly due to the availability of fairly cheap field spectroradiometers (Endler, 1990; Marshall, 1996; Cuthill et al., 1999; Cuthill et al., 2000). Since the human lens blocks UV wavelengths from reaching the retina (Backhaus et al., 1998), we do not see UV and therefore have no way of describing it. We are therefore inclined to treat this spectral region as somehow special. For animals with UV spectral sensitivity, however, UV will be just another color (Losey et al., 1999; Cuthill et al., 2000). In fact, very few UV-only colors are known (Burkhardt, 1989; Burkhardt and Finger, 1991; Finger et al., 1992; Losey et al., 1999; Marshall, 2000a), with many reflectances (usually peak-shapes) spilling over into the violet region and rendering the color pattern visible to us, but at a lower intensity or contrast. Nevertheless, for a visual system with dedicated UV sensitivity, UV patterns will obviously be more apparent (Fig. 10.2).

Interestingly, recent work has shown that about half a sample of 350 reef fishes have the capability to see UV (Losey et al., 1999; Siebeck and Marshall, 2001). In these studies, UV visual capability is usually based only on the transmission of UV by the ocular media. The UV sensitivity of underlying photoreceptors still needs to be tested. Most reef fishes possess a component of UV reflectance in their body

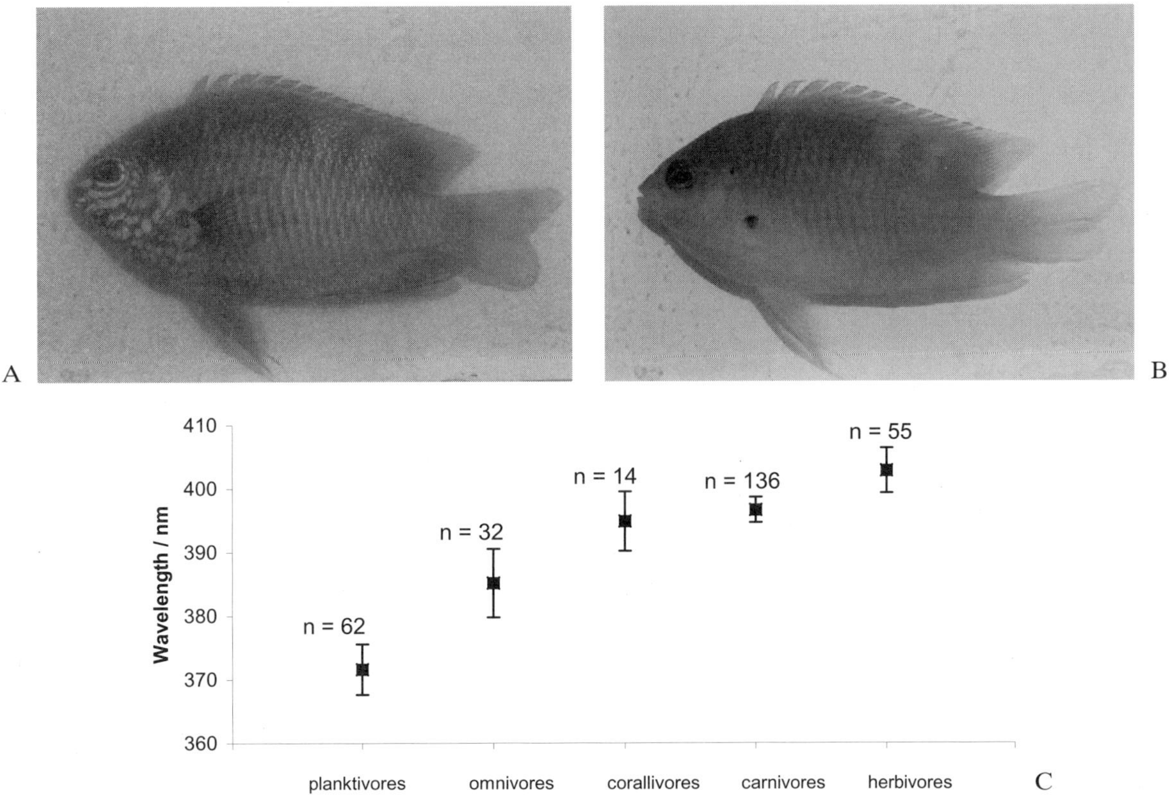

FIGURE 10.2. Ultraviolet (UV) colors of reef fish. The two top photographs show the UV markings in the damselfish *Pomacentrus moluccensis*. (A) Black and white print from a color video frame. (B) The same fish with the spectral window of the camera limited to 350–400 nm, the near UV (see Losey et al., 1999 for further methods). (C) Diet and ocular media (lens + cornea) transmission characteristics of 299 species of reef fishes. (Data from Siebeck and Marshall, 2001.) The *y*-axis plots the wavelength of 50% transmission of the ocular media. Planktivores possess significantly more UV transparent ocular media than carnivores and herbivores. (Siebeck and Marshall, 2001.)

colors (Marshall, 2000a), and to fishes possessing UV sensitivity, these color patterns might be a secret communication channel not open to UV-blind fishes (Siebeck and Marshall, 2001). As UV is rapidly attenuated by the scattering components in marine water (Jerlov, 1976; Lythgoe, 1979; Smith et al., 1979; Smith and Baker, 1981), UV colors may be most effective for close communication between fishes with the capability to see UV, while not being visible to predators lurking at a distance. Large predators also typically do not possess UV-transmitting ocular media (Fig. 10.2). UV patterns are often located on facial areas (Fig. 10.2) or fin edges, in positions only shown during fin display (N.J. Marshall and G.W. Losey, unpublished) and there is recent evidence that, as in birds (Cuthill et al., 1999), reef fishes use UV markings for sexual display (U. Siebeck, unpublished; S. Ostlund-Nilsson, unpublished; and see Smith et al., 2001 for a freshwater comparison).

UV visual capability is also correlated with diet (McFarland, 1991; McFarland and Loew, 1994; Loew et al., 1996a; Losey et al., 1999; Siebeck and Marshall, 2001), planktivorous fishes often possessing the capacity for UV vision (Fig. 10.2). It is thought that UV vision is particularly effective at detecting small particles against a background haze (Lythgoe, 1979; McFarland and Loew, 1994; Losey et al., 1999; Siebeck and Marshall, 2001). There are confounding phylogenetic factors in the relationship between diet and vision. For example, many, but by no means all pomacentrids are planktivorous. The pomacentrids of the genus *Stegastes*, for instance, are industrious algal farmers, while all but one species from the Pomacentridae so far examined transmit UV (Siebeck and Marshall, 2001). Conversely, a planktivorous wrasse from the Caribbean, *Clepticus parrai*, embedded in a family from which most species enthusiastically block UV, does have the potential to see wavelengths shorter than 400 nm (Kondrashev et al., 1986; Siebeck and Marshall, 2000). Careful studies of individual species, their colors, visual capabilities, and lifestyles are now required. Reef waters in the top 10–20 m are certainly rich in UV wavelengths (McFarland and Munz, 1975a; McFarland, 1991; Losey et al., 1999), making its use by the animals that live there a tantalizing, invisible mystery for us to investigate.

3. The Light Habitat of Reef Fish Reexamined

The use of modern spectrophotometers has revealed the presence of UV on reefs and in other waters (McFarland, 1991; Frank and Widder, 1994b; Frank and Widder, 1996). Along with the presence of UV, other large-scale characteristics of reef irradiance have been well characterized (Fig. 10.3; McFarland and Munz, 1975a; Jerlov, 1976; Levine and MacNichol, 1979; McFarland, 1986, 1991). An important and overlooked aspect of light on the reefs, however, is the light habitat at the scale of the inhabitants, the microhabitat light environment (Levine and MacNichol, 1979; Barry and Hawryshyn, 1999a,b; Marshall, 2000a; Cummings and Partridge, 2001). Many reef fish species are probably specialists in certain light regimes. The apogonids (cardinalfish), priacanthids (bigeyes), holocentrids (soldier and squirrelfish), and pempherids (sweepers), for example, spend the diurnal hours safe in a coral cave or crevice. They leave the protection of these hideaways at night to feed, but this does not mean that they are not visually active during the day. Many of the important sexual interactions of apogonids are conducted during the day (Thresher, 1984), where the light environment in their microhabitat is very long wavelength compared to those previously noted (Fig. 10.3). This is due to the influence of chlorophylls and other pigments in the algae, coral, and other organisms that coat and make up these daytime retreats. It is worth asking whether the visual systems of these species are specially adapted to this habitat. The retinae of holocentrids is roddominated, demonstrating the deep-sea origins of this family, and is clearly an adaptation to their nighttime foraging patterns (N.J. Marshall, unpublished). Priacanthid eyes have particularly bright tapeta and, in common with apogonids, a rod-dominated retina, both adaptations to vision at night (Ali and Anctil, 1976). Very little is known of the

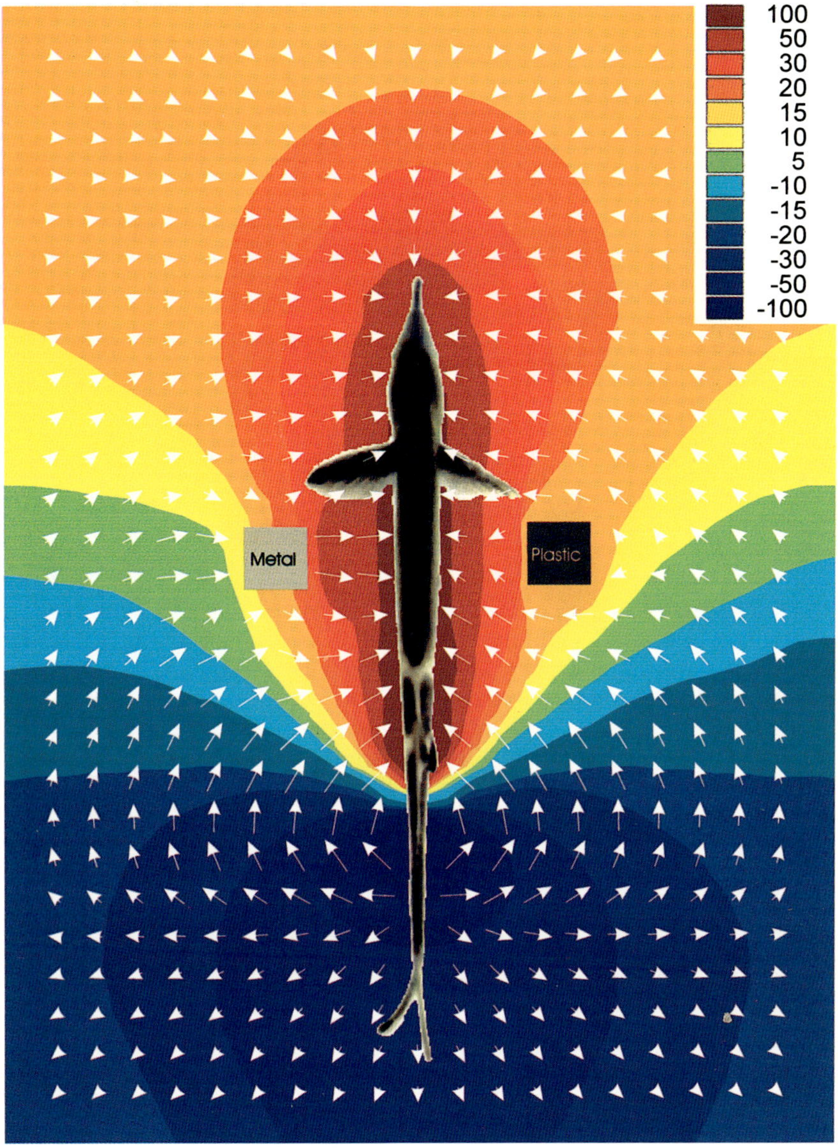

FIGURE 5.1. The electric field produced by *G. petersii* during the first peak of an electric organ discharge. Equipotential lines are color coded with positive voltages shown in red and negative voltages in blue. The scale in the upper-right corner depicts the voltages in mV. The arrows represent local field vectors, which were recorded at the corresponding locations. To the left and the right of the fish seen from above, objects are positioned. The left object is a metal cube (good electrical conductor), which locally increases field vector lengths in a region between the object and the fish's skin. This leads to an increase in the voltages perceived by the electroreceptors located at the affected skin region. The object on the right is a plastic cube, and because it is an isolator it leads to a decrease in local field vector lengths and thus to a decrease in the perceived local voltages. (This figure was kindly provided by S. Schwarz.)

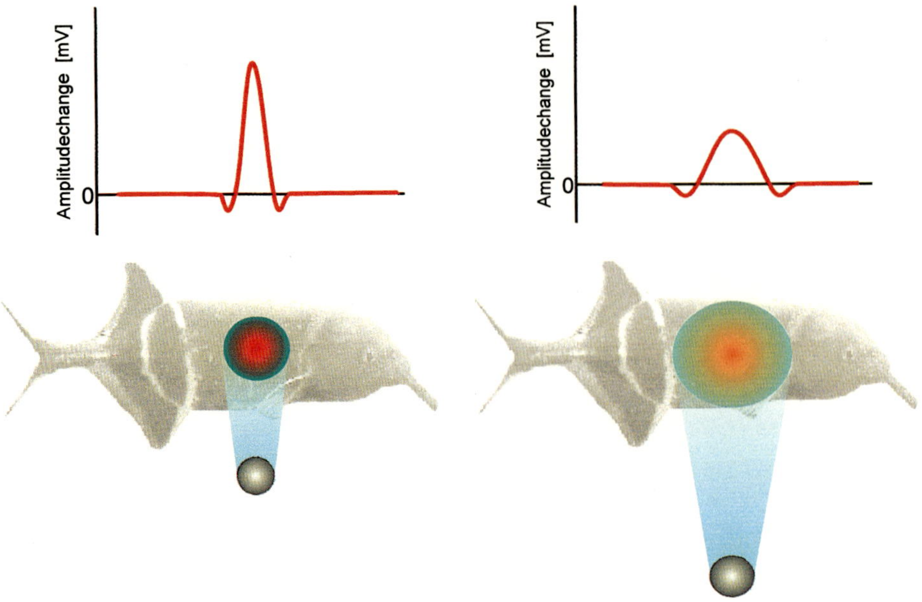

FIGURE 5.2. Color-coded representation of the electric images of a sphere projected onto the skin surface of a *G. petersii*. On the left, the sphere is at a closer distance to the fish than at the right. Red color illustrates an increase in amplitude caused by the presence of the sphere, blue depicts an amplitude decrease. The upper row shows a one-dimensional measurement of the electric images by a pair of electrodes moved along the midline from the tail toward the mouth of the fish. The ordinates give the change in amplitude by the presence of the sphere compared to the situation without it. Note that an increase in distance causes an increase in size of the electric image, and a decrease in the maximal amplitude change.

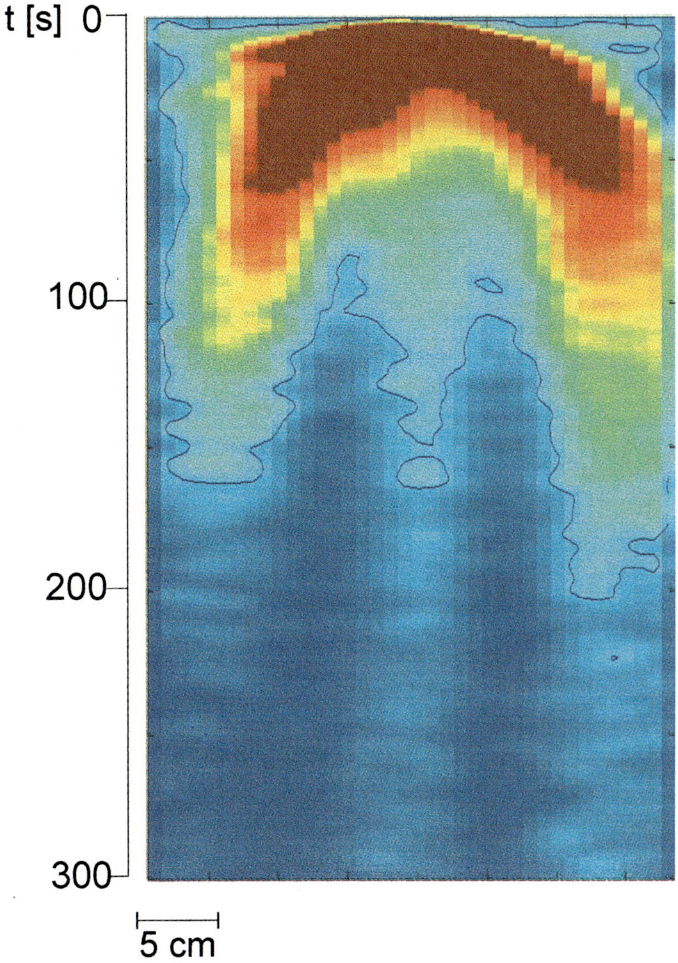

FIGURE 6.1. Color-coded water velocity behind a swimming goldfish averaged over the columns of the camera field. Dark red indicates high water velocities, dark blue indicates water velocities below 0.1 mm/s. The axis of ordinates shows the width of the camera field, the abscissa the time that has passed since the fish entered the camera field. The blue line indicates a water velocity of 0.2 mm/s (for further details see Hanke et al., 2000).

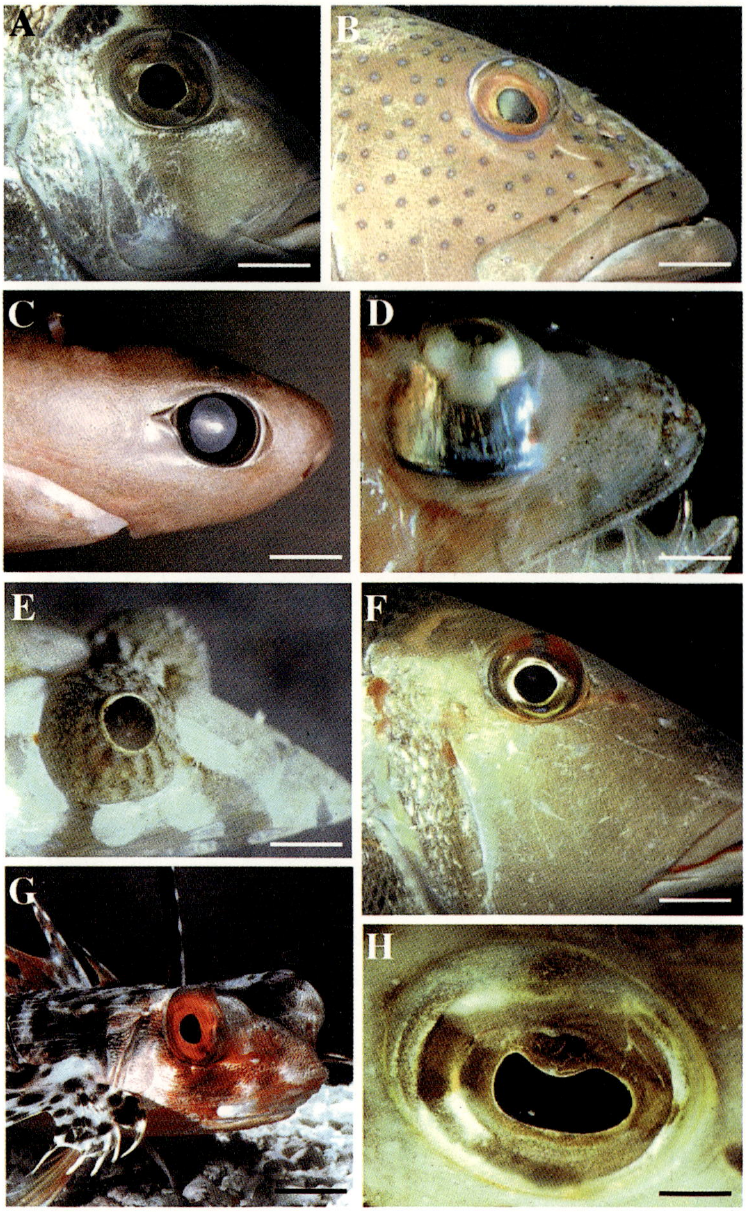

FIGURE 8.1. Diversity of eye size and shape in bony and cartilaginous fishes. (A) The collared sea bream *Gymnocranius bitorquatus* showing a binocular sighting groove etched into the snout. (B) The coral trout *Plectropoma leopardus* showing a colored corneal reflex and a rostrally tapered pupil. (C) The cookie-cutter shark *Isistius brasiliensis* with large eyes and a blue tapetal reflex. (D) The deep-sea pearleye *Scopelarchus michaelsarsi* with its tubular eyes and large spherical lens. (Photograph kindly provided by N.J. Marshall.) (E) The sandlance *Limnichthyes fasciatus* showing its dorsal, protruding eyes. (F) The sweetlip, *Lethrinus chrysostomus* showing its binocular sighting groove. (G) The oriental sea robin *Dactyloptaena orientalis* perched on its modified pectoral fins and showing its highly positioned eyes. (H) A close up of the eye of the weever fish *Parapercis cylindrica* showing its crescent-shaped pupil. Scale bars, 10 mm (A, B); 20 mm (C); 2 mm (D); 0.5 mm (E); 15 mm (F); 10 mm (G); 1 mm (H).

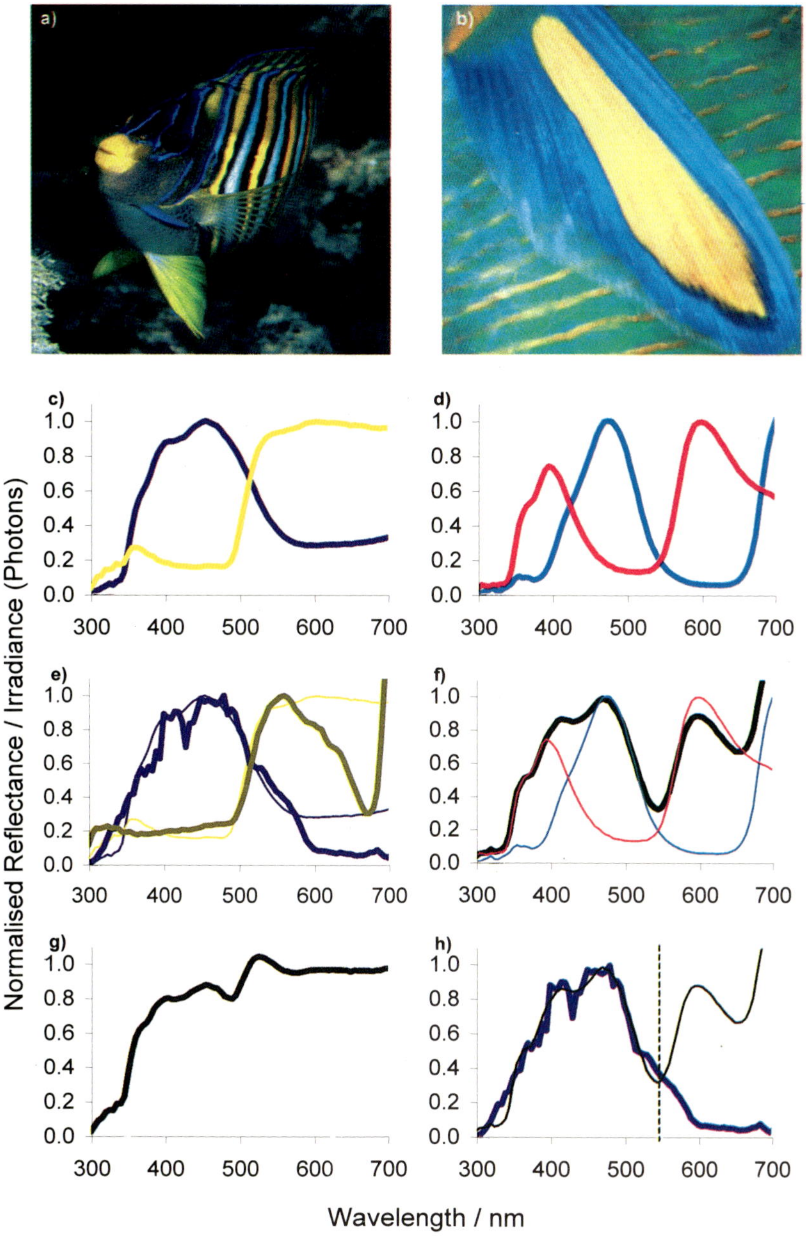

FIGURE 10.1. The colors of reef fish. (a) The regal angelfish *Pygoplites diacanthus*. Graphs in c, e and g relate to this fish. (b) The pectoral fin of the wrasse *Thalassoma lunare* to show fine stripe patterns. Graphs in d, f and h relate to this fish. (c) The reflectance of yellow and blue colors in *P. diacanthus*. The data have been normalized here and in subsequent graphs to emphasize chromatic difference. These are simple peak and step spectra. (d) The reflectance of pinkish-orange and green from *T. lunare*. These are complex multipeaked spectra. (e) Yellow and blue of *P. diacanthus* with the spectra of blue background water and shallow water coral (thick lines) plotted for comparison. (Data from Marshall, 2000b.) (f) The result of adding spectra in d together (black line). (g) The result of adding spectra in c together, gray from 400–700 nm. (h) The additive spectra from f with background blue-water spectra plotted for comparison. The vertical dotted line marks the long-wavelength limit of λ-max in reef fish cones known so far (Fig. 10.4).

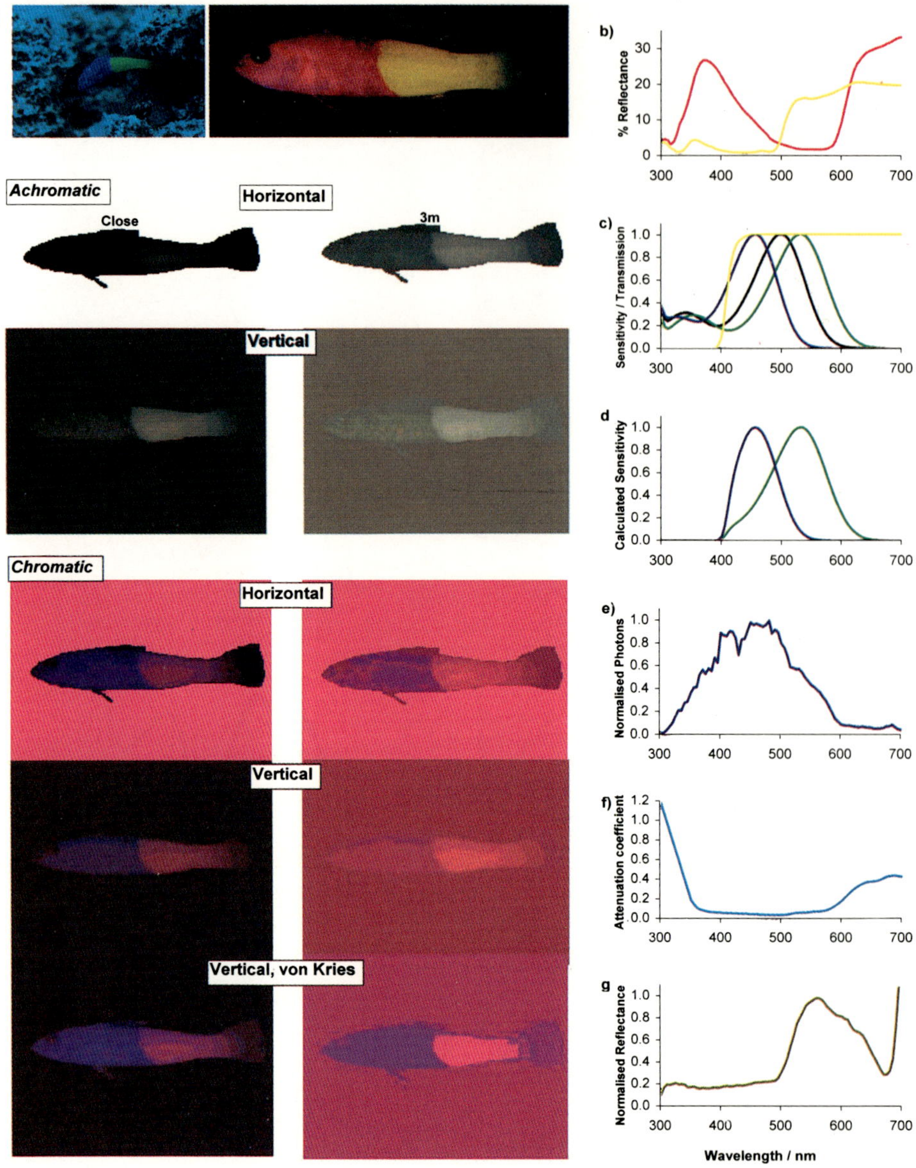

FIGURE 10.5 Looking at fish through fish eyes. (a) The royal dottyback, *Pseudochromis paccagnellae*. Top photographs as seen in the field at 10 m depth (note violet-blue color of head) and in the lab under broad field illumination. Subsequent fish images are representations of how the barracuda, *Sphyrena helleri*, might see the dottyback and are explained in Section 4.3. (b–g) Parameters used in model—Section 4.3. (b) The magenta and yellow colors of *P. paccagnellae* measured by spectrophotometer. (Marshall, 2000a.) (c) The visual system of *S. helleri*. *Yellow curve*: ocular media transmission. *Black curve*: rod sensitivity. *Blue curve*: single cone sensitivity. Green curve: twin cone sensitivity. (Data from McFarland et al. and Marshall et al., in press, a,b.) (d) Calculated sensitivities of *S. helleri* used for model in Section 4.3 of text. The rod is used in crepuscular vision and is not included; blue and green curves are the product of each of the single cone and twin cone sensitivities and the ocular media transmission from c. (e) Irradiance and blue-water background at around 7–10 m. (f) Attenuation of water intervening between fishes. (g) Coral background color. (Marshall, 2000b.)

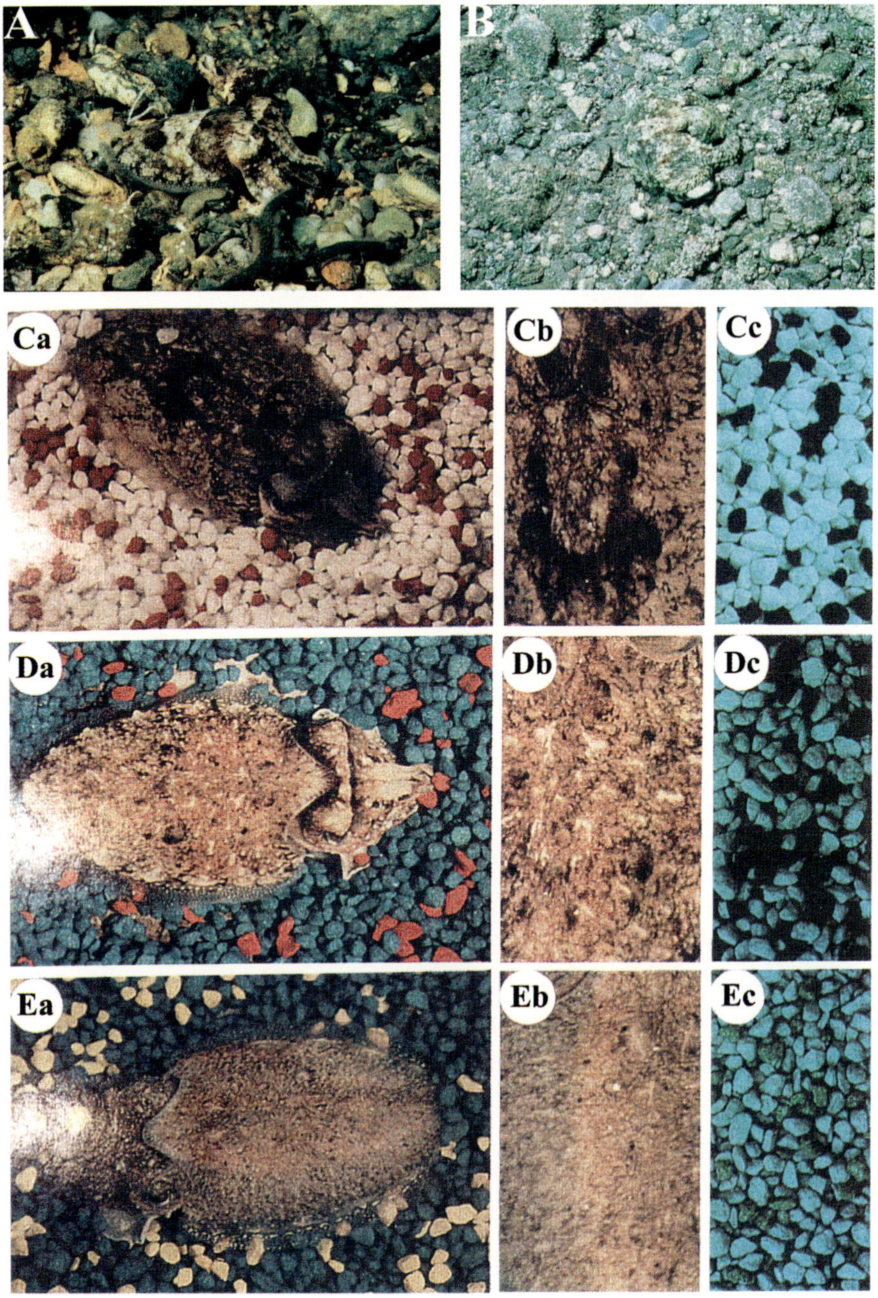

FIGURE 14.1. Camouflage by a color-blind cuttlefish. (A,B) Adult *Sepia officinalis* matching the color, brightness, texture, and pattern of the substrate in the Mediterranean Sea, and *Octopus bimaculatus* matching features of natural substrates off Catalina Island, California. (C–E) Subadult *S. officinalis* (100-mm mantle length) showing its color-blindness. Ca: photographed in daylight on a prepared background—red gravel on white; Cb: detail of mantle skin of the same animal; Cc: the same background photographed through a green interference filter. The D series is the same as C but with red gravel on blue; in E series, with yellow gravel on blue. The animals are not matching body color to that of the gravel background since they perceive it as uniform. (From Marshall and Messenger, 1996. Reprinted with permission from *Nature*. Copyright © 1996 Macmillan Magazines Limited.)

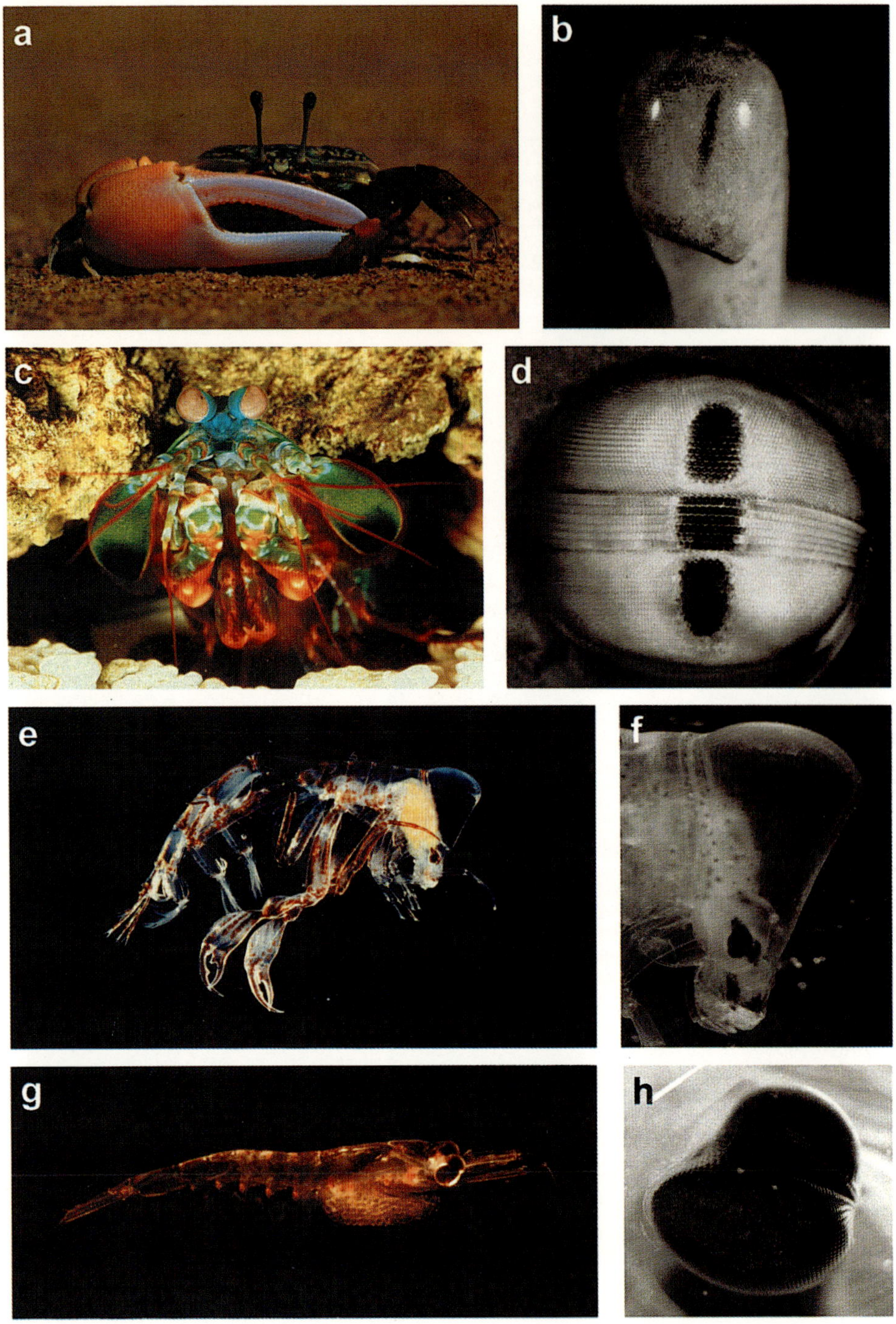

FIGURE 18.1. Crustaceans and their eyes. The fiddler crab *Uca polita* and closeup view of front of eye (a, b), (a—photograph by Jochen Zeil.) The stomatopod *Odontodactylus scyllarus* and closeup view of front of eye (c, d). The hyperiid amphipod *Phronima sedentaria* and closeup view of lateral aspect of head and eyes (e, f), (e—photograph by Mike Land.) The euphausiid *Nematobrachion megalops* and closeup view of lateral aspect of eye (g, h). Note the eye subdivisions in the last three examples.

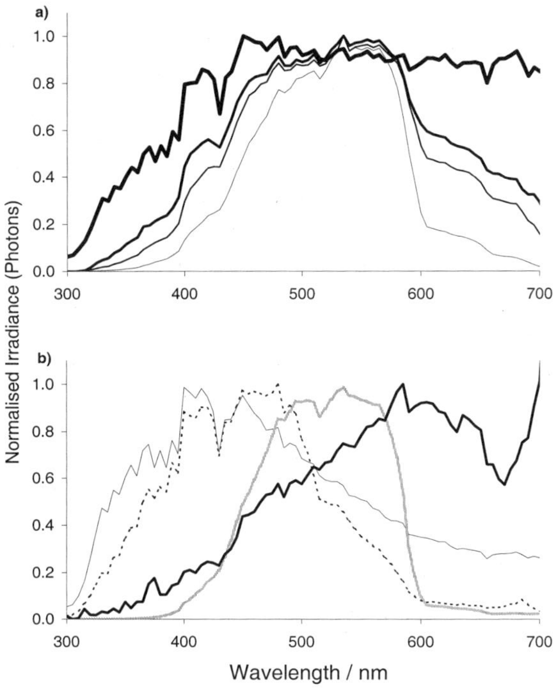

FIGURE 10.3. Light and microhabitat light environment on the reef. (a) Irradiance at different depths (0.1, 3.0, 6.0, and 10.0 m—progressively thinner lines) close to a reef in Kaneohe Bay, Oahu, Hawaii at a time close to midday under mixed overcast/blue sky. (b) Microhabitat light environments made with fiber-optic probe pointed at white Spectralon standard (a 99% white 300–800 nm, almost Lambertian surface) and placed in different reef areas. *Thick line*: at 5 m depth close to Lizard Island, Australia and 2–3 m into a reef cave/crack coated with algae, corals, and sponges. *Thin line*: 0.5 m in surface waters on the oceanic side of the Great Barrier Reef on a clear blue sky day, near midday. Gray line: 10.0 m close to the bottom of reef slope in relatively turbid reef water, Heron Island, Australia. *Dotted line*: 7.0 m on the oceanic side of the Great Barrier Reef on a clear blue sky day, near midday, probe angled to gather side-welling light. This simulates light striking a fish under a ledge.

color vision capabilities of these fishes (S. Job, unpublished). If their social interactions are conducted during the daytime, it is reasonable to think their cones (and all do possess cones—Ali and Anctill, 1976; N.J. Marshall, unpublished) may be red biased. Many of the fishes from these families possess red and orange body colors; however, the significance of this and their spectral sensitivities are speculations for future analysis (Marshall, 2000a; Marshall et al., in press, a,b; and see Section 4.4).

The red-light world of coral caves is entirely different from the UV/violet-dominant surface waters, the blue-green deep waters, and the broad-spectrum light environment many reef-top fishes experience (Fig. 10.3b). Nowhere else on earth can such a diverse set of light environments be found. Such disparate illuminants will have profound effects on the way in which colors are perceived due to failures in color constancy (Pokorny et al., 1991; Osorio et al., 1997). Even the relatively small chromatic differences in lighting in forest microhabitats have been implicated in changing signal efficiency on land (Endler, 1993; Endler and Thery, 1996; Fleishman et al., 1997), although the real influences on vision are more likely to be due to intensity differences in the illuminant rather than its color (Chiao et al., 2000c). On a reef, there are substantial differences in the chromatic content of light, largely due to the filtering effect of overlying water, but also, as we have just seen, due to the exact location of the fish's microhabitat. Future studies on social interactions and color signals in fishes should be careful to bear this in mind (Houde and Endler, 1990; Endler, 1991a; Marshall, 1996, 1998; Barry and Hawryshyn, 1999a). The effects illuminants have on color signals in reef animals are discussed in some detail elsewhere (Osorio et al., 1997; Chiao et al., 2000b; Marshall, 2000a; Vorobyev et al., 2001b; and Section 4.3). Useful comparative studies of kelp forest and freshwater microhabitats are provided by Endler (1991a), Partridge and Cummings (1999), and Cummings and Partridge (2001).

Finally, it is worth pointing out that some reef fishes such as the coral trout *Plectropomus leopardus*, the butterflyfish (chaetodontids), and angelfish (pomacanthids) may move frequently between microhabitats. This is either in search of food or in an effort to hide from predators. In any case, such fishes may be exposed to the full variety of microhabitat light environments and it is unclear which, if any, may be the most influential on their visual systems or if they are simply generalists.

4. Spectral Sensitivities and Color Vision in Reef Fish

4.1. A Brief Review of Past Ideas

A favorite exercise of fish ecologists has been to correlate the spectral sensitivities of different species to their chosen aquatic habitat (for a recent review see Douglas, 2001). Fish retinae are composed of two anatomically distinct types of cones: single cones and double/twin cones for diurnal vision and rods for scotopic or dim-light vision (Walls, 1942). The spectral sensitivity/light environment correlation is often expressed as the position of the peak absorbance of the rod or cone's visual pigment, the λ-max, versus the characteristics of the light environment (e.g., its wavelength of maximum transmission) (Lythgoe, 1966; Lythgoe, 1972; Lythgoe, 1979, 1984; Lythgoe et al., 1994).

Since rods operate predominantly in dim light, either in crepuscular periods or in deep water, it is reasonable to assume that their spectral sensitivity should be placed to maximize photon capture. Broadly speaking this is true; fish species inhabiting long-wavelength-dominated freshwater habitats possess more red-sensitive rod visual pigments than species living in green coastal waters or blue oceans (Munz and McFarland, 1973; Muntz, 1990; Partridge, 1990). Deep-sea fishes live in the narrowest of spectral worlds, centered close to 475 nm (Jerlov, 1976; Kirk, 1983), and their rod visual pigments possess λ-maxs close to 475 nm; they are often cited as the classic example of spectral sensitivity match to ambient illumination. This *Sensitivity Hypothesis* was first suggested more than 60 years ago (Bayliss et al., 1936; Clarke, 1936) and has also been used to explain the placement of double/twin cone (McFarland and Munz, 1975b; Loew and Lythgoe, 1978; Levine and MacNichol, 1979; Lythgoe, 1979; Bowmaker, 1990; Lythgoe and Partridge, 1991; Lythgoe et al., 1994; Douglas, 2001) and to a lesser extent single cone (McFarland, 1991; McFarland and Loew, 1994; Marshall, 2000a; Marshall et al., in press, a,b; McFarland et al., in press) sensitivities. In a comparison of freshwater and marine habitats, freshwater fishes tend to possess more long-wavelength-sensitive cones than marine fishes. Within the marine habitat, from coastal to offshore, as water changes color from green to blue, the cones of lutjanids (snappers) change from more green to more blue sensitive (Lythgoe et al., 1994). This is examined further in Section 4.2.

On close examination, at least for some fishes, the Sensitivity Hypothesis is clearly an oversimplification, both for rod and cone sensitivities. The scatter of λ-max values for deep-sea fish visual pigments (up to 100 nm either side of 475 nm; Bowmaker, 1990; Douglas, 2001) clearly indicates that visual pigments are important in other tasks as well as in maximizing photon capture. As suggested above, the final resting place of visual pigment sensitivities is likely to be a compromise, dependent on a variety of visual tasks and, due to its visual complexity, this is nowhere more likely than in coral reefs.

The simple world of the deep oceans becomes increasingly spectrally simple with depth and increasingly dominated by the light produced by the bioluminescent animals that live there. Although there is both a spread of spectral sensitivities and bioluminescent emissions, from blue/violet to red (Herring, 1977, 1982; Widder et al., 1983; Herring, 1985; Latz et al., 1988), a number of lines of evidence suggest that vision in the deep is tuned more specifically to bioluminescent light than to downwelling light (Douglas et al., 1998; J.C. Partridge, unpublished). Certainly for organisms below 1,000 m where no photons from the surface penetrate (see Chapter 16), this seems likely. (See Chapter 17 for the wonderful story of red bioluminescence and matching red sensitivity in the deep and Chapter 18 for a discussion of this relative to crustacean vision.)

Visual pigments offset from the maximum transmission of background space-light in surface waters may be better for detecting certain targets (Lythgoe, 1966, 1988; Partridge and Cummings, 1999), and this is known as the *Offset Hypothesis*. In this instance, it is the light reflected from the target relative to the background that is important and we will return to this theme when considering fish color vision in Section 4.3. The important point here is that visual systems have evolved to look at *things*, not just the available light spectrum (Partridge and Cummings, 1999). Cummings and Partridge (2001) recently demonstrated that in the variable illumination of the kelp forests, cones from different species of surf-perch (Embiotocidae) are well adapted to each species' different microhabitat light environment. Furthermore, cones are both matched to and offset from background illumination, an ideal combination for enhancing target contrast (Lythgoe, 1979). In common with reef fishes, all surf-perch species examined were also dichromatic, in this case probably through a combination of single and double cones (Cummings and Partridge, 2001). Barry and Hawryshyn (1999a,b) reached the same conclusions for *Thalassoma duperry*, the Hawaiian saddle wrasse.

A modification to the Sensitivity Hypothesis, the *Twilight Hypothesis*, suggests that visual pigments in fishes may be better tuned to the short-wavelength-shifted light environments of dawn and dusk (Hobson et al., 1981; McFarland, 1991). This suggests that spectral sensitivities are tuned to ambient light during the most important times of the day, in behavioral terms. There are arguably plenty of photons for vision over a broad spectral range during the day, releasing the restrictions of where to place spectral sensitivities. During crepuscular periods, as light becomes limited, it may be more critical for survival to match spectral sensitivity and available light. Particularly on coral reefs, dawn and dusk are times of increased activity (McFarland, 1986, 1991; McFarland and Wahl, 1996). Most predator strikes are recorded during these times and both predators and prey seem to know about possible compromises in visual function during crepuscular periods and have adjusted their spectral sensitivities accordingly (see Chapter 18 for further discussion).

Finally in this minireview of past work, as in our discussion of ocular media transmission (Section 2.3), a number of authors have pointed out that the placement of spectral sensitivity seems to be loosely related to phylogeny (Knowles and Dartnall, 1977; Partridge, 1990; Douglas, 2001; McFarland et al., in press). However, the degree to which this coexists or competes with adaptational explanations is unclear.

4.2. Some New Spectral Sensitivity Data

Until recently, the spectral sensitivities of reef fishes remained largely unknown or, for a variety of reasons, somewhat inaccurate (McFarland and Munz, 1975b; Lythgoe et al., 1994; Marshall, 2000a). The last few years have seen the number of species examined reach close to 50 (review in Marshall, 2000a, McFarland et al., in press, Marshall et al., in press, a,b). Here we examine these data, look for trends, and examine how fishes may look to fishes.

The λ-max values of the cones from 38 Hawaiian reef species measured with a microspectrophotometer (MSP) are plotted in Figure 10.4 (McFarland et al., in press; Marshall et al., in press, a,b). Before commenting on these, it is worth cautioning that, as this study was conducted to survey different fishes, rather than as an in-depth study of individual species, some spectral sensitivities in a few species may have been missed. However, the overall picture is likely to be relatively accurate (McFarland et al., in press). Four things are notable: (1) The spectral sensitivities of single cones occupy a broader and shorter wavelength-shifted region of the spectrum than those of the double/twin cones. (2) The single cone sensitivities fall into three distinct clusters centered around 360, 420, and 460 nm (UV, violet, and blue). The average double/twin cone sensitivity is around 500 nm (green). (3) Relative to the microhabitat light regimes, the double/twin cones seem better adapted to deeper or greener, more light-limited waters. (4) Single cones occupy a predominantly UV/blue region of the spectrum.

One of the problems with interpreting these data in any detail relative to the light microhabitat is that the depth ranges and habits of the fishes are, with a few exceptions, not well documented. The types of inferences that one would like to make are not currently possible. The data do support the idea that double/twin cones mediate luminance vision during the day

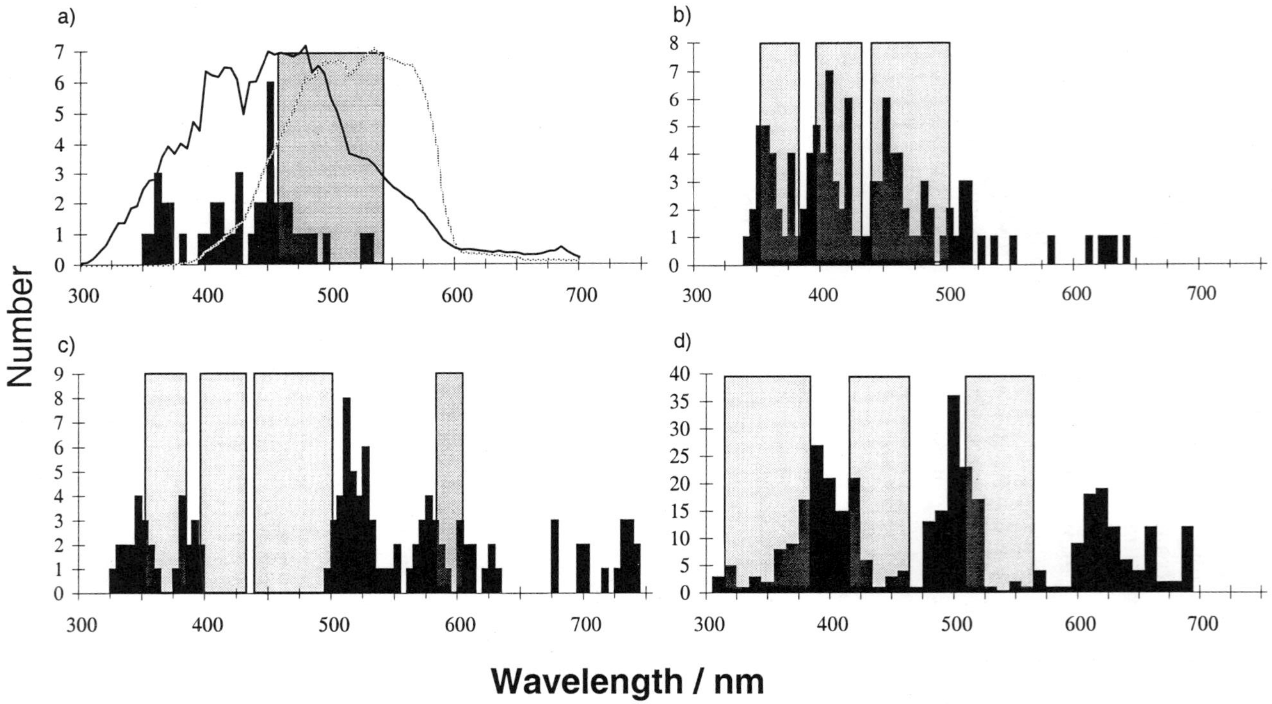

FIGURE 10.4. Spectral sensitivity and color characteristics of reef fishes. (a) Histogram of the λ-max values of single cone spectral sensitivities of 38 species of Hawaiian reef fishes. (McFarland et al., in press.) Shaded block overlay marks the λ-max range of double/twin cones of the same fish. The line overlays are microhabitat light environments from Figure 10.3. (b) Histogram of the spectral location of the peaks in peak-shaped colors of around 200 species of reef fishes. (Marshall, 2000a.) The shaded block overlays are the limits of single cone λ-max values plotted in A. (c) Histogram of the spectral location of color step-50% values for fish colors from the same species in B. Overlay same as B also but with addition of long-wavelength shaded block close to 600 nm, the result of modeling spectral sensitivities. See Section 4.5 for details. (d) Bee color vision and flower colors (data from Chittka and Menzel, 1992), for comparison with C. Histogram of flower step-shaped color 50% points from 180 flowers and shaded blocks demarks the λ-max positions of UV, blue, and green photoreceptos from 40 hymenopteran species.

(Smith et al., 1985; and see Section 4.4) as they sample a spectral region that would catch photons in any of the microhabitats and are almost ideal for deeper reef waters (Fig. 10.4). A fish that ventures into deeper zones of water may rely increasingly on this cone type. Interestingly, although double/twin cones generally account for the majority of photoreceptors in the retina of most reef fishes (Walls, 1942; Ali and Anctil, 1976), their exact functions are still not well known.

Two trends not obvious from these data but outlined by McFarland et al. (in press) and Marshall et al. (in press, a,b) are as follows. First, most of the fishes studied are likely to be di- or at most trichromatic. That is, most fishes possess only two of the spectral sensitivity types shown in Figure 10.4a. Interestingly, in two-thirds of these species, this pair of sensitivities is made up of one type of single cone and one type of double/twin cone. Only around one-third of the species examined possess two types of single cone. This is in marked contrast to freshwater fishes such as the goldfish, which has four types of single cone spectral sensitivities sampling from UV to far red (Palacios et al., 1989; and see Kusmic and Gualtieri, 2000 for a recent review of freshwater fish spectral sensitivities). As mentioned previously, it is possible that some cones have been overlooked in MSP preparations for this study, and due to their relatively small size, this is most likely for single cones. Second, λ-max positions for any one cone type of the UV, violet-blue, or double/twin are relatively widespread. That is, the spectral sensitivities of reef fishes are very varied. This interspecific variation is in marked contrast to birds, for example, whose tetrachromatic color vision is relatively conserved among species (Bowmaker, 1980; Bowmaker et al., 1997; Bowmaker, 1998; Hart et al., 2000; Hart, 2001).

The large differences in color vision systems of reef fishes may be driven by different microhabitat light environments (Fig. 10.3) or the result of other visual demands such as diet. An obvious alternative is that, if looking at things is so important to fishes, color vision in reef fishes may be correlated to the colors of individual species. This question is now examined in detail.

4.2.1. Are the Colours of Reef Fishes Correlated with Cone Sensitivities?

Comparing single cone sensitivities with the distribution of spectral peaks and 50% points (Fig. 10.4b,c) reveals some correlation, as has been previously noted with bee color vision and flower color (Chittka and Menzel, 1992, and Fig. 10.4d). Broadly speaking, spectral sensitivity λ-max values tend to lie either side of the step–50% points, and coincident with the reflectance peaks. The clusters apparent in color peak and color step–50% data are statistically different (Marshall et al., in press, a,b), but clearly there are colors between most clusters, meaning there are plenty of exceptions to this rule. It is worth noting the caveat provided by Vorobyev with relation to the bee data of Chittka and Menzel 1992 (but published in the next volume of the *Journal of Comparative Physiology*), that such placement of spectral sensitivities on either side of reflectance step–50% points may be more useful for detecting than discriminating colors. It is also more likely that colors have been constrained through evolution to existing spectral sensitivity positions, rather than the other way around, as again suggested recently in bees (Chittka, 1996, 1997).

The gap between 400 and 500 nm in step-shaped color 50% points is noteworthy and is in some respects an obvious rule of colors in nature. That is, colors which are UV, blue, or green, *must* be peak-shaped by definition to reflect only in midspectral locations. Step-shaped colors, which are UV/white, white, and then yellow, orange, and red (see nomenclature in Marshall, 2000a), lie on either side of these peaks as seen in a comparison of Figure 10.4b and c (Fox and Vevers, 1960; Marshall, 2000a). It is interesting that both in fishes and other animals, although theoretically possible, there are few peak-shaped yellow, orange, or red colors (Fig. 10.4; Vorobyev et al., 1998; and N.J. Marshall, unpublished).

What is the relevance of these relationships to color vision? As color vision in almost all systems works on differences between adjacent spectral sensitivities (see Marshall and Land, 1991, Neumeyer, 1991, and Chapter 11 for a possible exception to this), two spectral sensitivities placed either side of a step in reflectance will return a large difference in signal and be an effective detector of the color. For peak-shaped colors, the largest signal difference is from a pair of spectral sensitivities, one of which overlies the peak and one of which is placed to the side of the peak. Although a useful start, especially where not much information on spectral sensitivity exists, this is a somewhat simplistic view of color coding (and more detailed discussion of the relationship between colors and spectral sensitivities can be found in Chittka and Menzel, 1992; Osorio and Vorobyev, 1996; Chittka, 1997; Backhaus et al., 1998; Vorobyev and Osorio, 1998; Vorobyev et al., 1998; Chiao et al., 2000b; and Vorobyev et al., 2001b). Where species have to use a single cone and double/twin cones to construct color vision (see above) this line of reasoning seemingly breaks down as the double/twin cone sensitivities are placed over the top of yellow/orange step-shaped color 50% values (around 475–550 nm). At depth, however, the filtering effect of reef water will clip the long-wavelength side of yellow steps, rendering them effectively peak-shaped (see Fig. 10.5; see color plate), possibly rescuing this hypothesis (Marshall et al., in press, a,b).

In a comparison of fish colors to flower and bird color data-sets, fish color peaks and color step–50% points are more restricted between 350 and 550 (Marshall, 1998, 2000a; Marshall et al., in press, a,b; N.J. Marshall, unpublished; and compare Figs. 10.4b,c,d). The likely reason for this is the restricted light environment of the oceans at depths below a few meters (Fig. 10.3). That is, not only are spectral sensitivities tuned to the light environment; so too are colors. Fish colors do exist with peaks or step–50% points beyond 550 nm; there are just proportionally less of them than exist on land. As noted above, many of these long-wavelength colors belong to cave and crevice dwellers where the spectrum of ambient light is red biased.

4.3. What Does a Barracuda See Looking at a Small Meal over a Short Distance?

Armed with spectral sensitivity, color measurements, ambient light spectra, and some understanding of the behavior and habits of fishes, it is possible to model what fishes look like to fishes. This process necessarily has an element of subjectivity (how can a human represent UV, for example), but is worth doing in order to gain insight into camouflage and communication. The behavior we have chosen here is a predatory one: how a potential meal, the royal dottyback (*Pseudochromis paccagnellae*), looks to a barracuda (*Sphyraena helleri*).

An image of an object as seen through the eyes of a fish can be represented as a set of quantum catches corresponding to each point or pixel of the image. To show quantum catches we use the colors of a computer monitor (Vorobyev et al., 1998, 2001b; Chiao et al., 2000b,c). As we argue above, most reef fishes are cone dichromats and *S. helleri* possesses a blue-sensitive single cone with λ-max at 455 nm and twin cones with λ-max at 531 nm. For convenience, henceforth these are referred to as S and L cones. We code the quantum catch of the single cone with the "blue" and the twin cone with the "red" of a computer monitor and then generate color-weighted images (obviously purple in this case) based on the following model (Fig. 10.5). We also assume that fishes have achromatic vision (see Section 4.4), which is to some degree separated from chromatic vision. We show the quantum catch corresponding to the achromatic twin cone system on a separate panel using the brightness channel of a monitor (Fig. 10.5). How brightness and color are separated is a question of retinal wiring in fishes (Kamermans et al., 1998) and this is not known for the barracuda. It is important to note that we also do not know how fishes perceive colors—the code we use shows only the information from which the nervous system may construct color. Generally, the larger the changes in cone receptor signals, the larger are the changes in color appearance. Therefore, inspection of images, where quantum catches are coded with colors, allows us to judge whether an object can be easily detected against its background or whether the appearance of a fish changes strongly when viewing conditions change (Vorobyev et al., 2001b).

To calculate quantum catches, one needs to know the reflectance spectrum in each point of

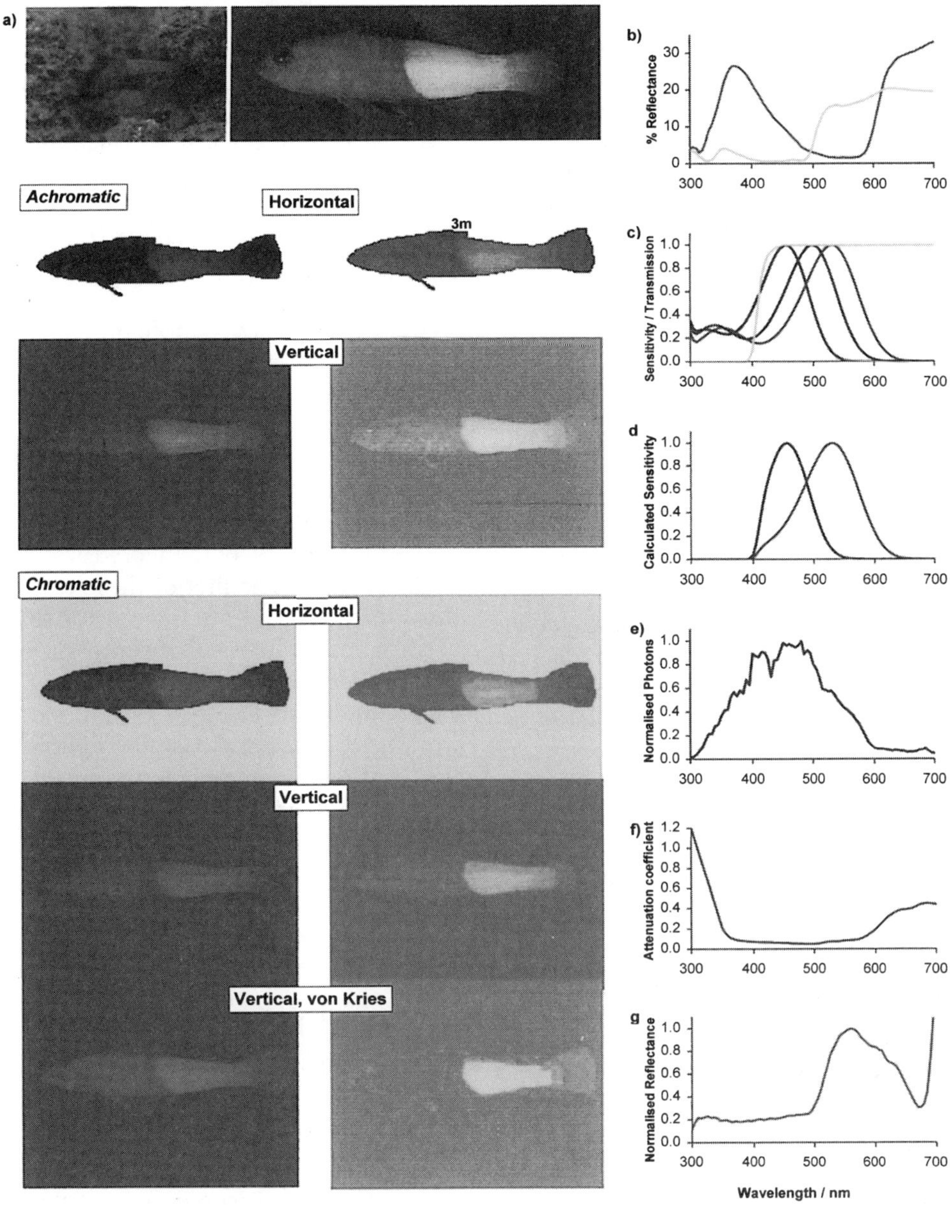

FIGURE 10.5. Looking at fish through fish eyes. (a) The royal dottyback, *Pseudochromis paccagnellae*. Top photographs as seen in the field at 10 m depth (note violet-blue color of head) and in the lab under broad field illumination. Subsequent fish images are representations of how the barracuda, *Sphyrena helleri*, might see the dottyback and are explaned in Section 4.3. (b–g) Parameters used in model—Section 4.3. (b) The magenta and yellow colors of *P. paccagnellae* measured by spectrophotometer. (Marshall 2000a.) (c) The visual system of *S. helleri*. *Yellow curve*: ocular media transmission. *Black curve*: rod sensitivity. *Blue curve*: single cone sensitivity. Green curve: twin cone sensitivity. (Data from McFarland et al. and Marshall et al., in press, a, b.) (d) Calculated sensitivities of *S. helleri* used for model in Section 4.3 of text. The rod is used in crepuscular vision and is not included; blue and green curves are the product of each of the single cone and twin cone sensitivities and the ocular media transmission from c. (e) Irradiance and blue-water background at around 7–10 m. (f) Attenuation of water intervening between fishes. (g) Coral background color. (Marshall 2000b.) (See color plate)

the image. In this case, we recorded images of fishes using a multispectral camera (Chiao et al., 2000d) rather than a spectrophotometer, the instrument used to generate the color data in Figures 10.1 and 10.4. The result is the same, regardless of method, and spectrophotometric measures of the colors of *P. paccagnellae* are plotted in Figure 10.5. The two methods have different advantages and disadvantages, and *P. paccagnellae* was partly chosen because of its simple body color pattern, magenta and yellow. The essence of the multispectral camera method is that the fish was filmed through a set of 40 narrow-band interference filters spanning the spectrum from 400 to 700 nm; as a result we had 40 frames, each of them corresponding to one particular wavelength band. Note that a limitation of this multispectral camera is its lack of UV sensitivity, and the short-wavelength peak of the magenta color of *P. paccagnellae* actually peaks in the UV. However, as the ocular media of *S. helleri* cut off UV wavelengths (Fig. 10.5), at least for this social interaction, the UV components of the color are irrelevant. For any point or pixel of the image, the set of 40 values for each wavelength band (scaled from 0 to 255, the dynamic range of the system), corresponding to different frames, gives the reflectance spectrum of this point.

As Section 3 outlines, absorption and scattering of light in water is wavelength dependent (Fig. 10.3). Therefore, the color appearance of objects in water changes rapidly with depth but also depends on whether an object is viewed from above or, at a distance, from the side (Lythgoe, 1979). The colors of objects change with distance due to the color of the intervening water. A formal description of this phenomenon is given by a dependence of quantum catches by each cone on the viewing distance. Let z be the distance from the observer to the object and $Q_i(z)$ be the quantum catch of a receptor of type i ($i = S, L$). The quantum catch then is given by:

$$Q_i(z) = k_i \int R_i(\lambda)\, G_S(\lambda, z) d\lambda \qquad (10.1)$$

where $R_i(\lambda)$ denotes the spectral sensitivity of a receptor i, and $G_S(\lambda,z)$ denotes the spectrum of the light entering the eyes. $G_S(\lambda,z)$ is dependent on depth color of and distance to the object. k_i are the scaling factors, which are explained shortly. $G_S(\lambda,z)$ also depends on the reflectance of a viewed surface, $S(\lambda)$, and on the distance to this surface, z. This dependence is given by:

$$G_S(\lambda, z) = I(\lambda)S(\lambda)\text{Exp}[-a_1(\lambda)z] + I^0(\lambda)(1-\text{Exp}[-a_2(\lambda)z]) \qquad (10.2)$$

where $I(\lambda)$ and $I^0(\lambda)$ denote the spectrum of the light illuminating the surface and the background respectively. $a_1(\lambda)$ and $a_2(\lambda)$ denote the narrow beam absorbance and scattering attenuation coefficients respectively (Fig. 10.5). These are the two factors that express how light traveling from object to observer is degraded or altered (Jerlov, 1976; Lythgoe, 1979; Kirk, 1983; Mobley, 1994). The first of these terms in Eq. 2 thus describes the attenuation of the light by absorbance as it travels from a surface to the eye. The second term describes one of the greatest problems of life in a turbid, particle-filled world such as a coral reef; the light that, due to the scatter, is added between a reflecting surface and the eye. The consequence of the light scatter is a veiling effect, which reduces contrast and changes color appearance in much the same way as fog on land (Lythgoe, 1979; Jagger and Muntz, 1993). When an object is viewed from a very close distance (i.e., $z = 0$), exponents in Eq. 10.1 are equal to unity, and Eq. 10.2 can be rewritten as $G_S(\lambda,z) = I(\lambda)S(\lambda)$. Then the quantum catch is given by

$$Q_i(z) = k_i \int R_i(\lambda)\, I(\lambda)\, S(\lambda) d\lambda \qquad (10.3)$$

an equation that is commonly used to calculate quantum catch in air and the one used in dichromatic models described in Section 4.5.

Visual systems have an ability to compensate for changes in illumination spectrum, a phenomenon known as color constancy (D'Zmura and Lennie, 1986; Maloney and Wandell, 1986; Pokorny et al., 1991; Osorio et al., 1997). One of the first proposed models of color constancy, a von Kries transformation, assumes that signals of photoreceptors are scaled so that the color of the illumination remains invariant. Such an algorithm is effectively implemented by known physiological receptor adaptation mechanisms. Although it is not known which algorithm of color constancy fishes use, the von Kries model yields predictions that agree nicely with results of behavioral experiments (Neumeyer, 1981; Dorr and Neumeyer, 1996). The mathematical formulation of von Kries color constancy is straightforward, using the scaling factors mentioned in Eq. 1. These depend on the illumination spectra

$$k_i = 1\Big/\int R_i(\lambda)I(\lambda)d\lambda \qquad (10.4)$$

From Eq. 4, it follows that the quantum catches for illumination alone are set to unity, and thus the color of the illumination remains invariant.

Figure 10.5a shows *P. paccagnellae* as we see it and as seen through the eyes of *S. helleri*. The left panels beneath the two "normal" photographs show a fish as it would be seen from a very close distance (i.e., Eq. 3 can be used). The right panels show the fish as it is seen from a distance of 3 meters (Eqs. 10.1, 10.2). The two attenuation coefficients, scatter and absorbance, are assumed to be the same, a good assumption over a short distance (Mobley, 1994). The spectral dependence of the attenuation coefficient, the spectrum of ambient light and background colors, colors of *P. paccagnellae* (measured spectrophotometrically; Marshall, 2000a), and spectral sensitivities of *S. helleri* are shown in Figures 10.5b–g. We consider two viewing conditions with both fish at a depth of approximately 10m. First, *P. paccagnellae* viewed from the side by *S. helleri* positioned against horizontal spacelight (i.e., blue water, Fig. 10.3), and illuminated by horizontal light (in this case assumed to be the same spectrum as spacelight, a fair assumption for water of this depth). Second, *S. helleri* looking down at *P. paccagnellae* against a background of coral, and illuminated by vertical light. *P. paccagnellae* spends much of its life in holes and cracks in the reef so this latter view is probably more common. A barracuda lurking close to the reef, as indeed they often do, may be lucky enough to see its prey against background spacelight, clearly an easier, higher-contrast target. Receptor quantum catches are scaled using von Kries' rule (Eq. 10.4).

P. paccagnellae has a yellow tail and magenta head. Interestingly, in depths of water over 5m a diver or camera perceives the color of the head of this fish as violet/blue (Fig. 10.5a). This is because the long-wavelength (red) component of the illumination is absorbed by water, and thus the red component of the head color (Fig. 10.5b) disappears. At the UV end of the spectrum the magenta color will also change markedly with depth and indeed distance viewed. In this instance, scatter of light rather than absorption is the important factor (Kirk, 1983; Mobley, 1994) and both these attenuating influences are clear in Figures 10.5e and f. As both barracuda and human ocular media prevent UV wavelengths below 400nm from reaching the retina (Fig. 10.5c), the depth and distance from which *P. paccagnellae* is viewed makes less difference at this end of the spectrum. However, for a fish with UV sensitivity, a large change in perception will result from these changing parameters. The biphasic nature of this color makes it one that will change substantially with microhabitat light type, and this may be no accident. Unfortunately, we currently know little about the visual system of *P. paccagnellae* or other fishes that may interact with it.

When viewed through barracuda eyes against horizontal spacelight, both head and tail contrast strongly to the background both for chromatic and achromatic vision. As the distance to the fish increases the tail becomes less visible. When viewed against a coral background, the head becomes more difficult to detect either close up or when viewed from 3m. The von Kries transformation neither helps to detect this fish better against the coral background, nor compensates for changes in color appearance caused by changes of the distance to this fish. This is consistent with earlier computations showing that von Kries color constancy fails underwater, because it cannot compensate for the veiling effect (Vorobyev et al., 2001b). In other words, the chromatic effects of increasing distances of intervening water are very strong, as best indicated by the steep-sloped nature of the attenuation coefficients at either end of the spectrum.

It seems likely that the head color has been adapted for camouflage as it is a good match to coral at a depth of 10 m, and for behavioral reasons. As previously mentioned, *P. paccagnellae* often hides in burrows among coral, with just its head but not its tail poking out. We speculate that the two colors of dottyback serve different functions. The head color is adapted for camouflage, but the tail color could be adapted for advertisement as it is a relatively strong contrast against coral or background water in both the chromatic and achromatic world (although less so at distance in achromatic space). John Lythgoe and others came to the same conclusion about yellow in ocean waters 35 years ago (Lythgoe, 1968; and see Section 2.2).

Also good for advertisement is the color contrast within the fish, ignoring the background, and this is clear in Figure 10.5a. Magenta and yellow are complementary colors in the same way as blue and yellow are (Section 2.2) and this contrast is enhanced by the red filtering effect of water, which removes the region of the spectrum in which magenta and yellow overlap. As a result, the head would contrast to the tail for almost any dichromatic visual system. Barry and Hawryshyn (1999a,b) quantified both within: fish and fish-to-background contrasts for several species of Hawaiian fishes as well as spectral sensitivity in *Thalassoma duperrey*, the Hawaiian saddle wrasse. They concluded that fish color patterns and their relative contrasts varied with habitat light environment and background (including within fishes) and that the two spectral sensitivities of *T. duperry* (460 and 550 nm) allowed good target detection through being matched and offset to target and background colors of conspecifics (Section 4.1).

The attentive reader may have noticed that the conclusions regarding camouflage and advertisement reached here are exactly the opposite to those discussed in Section 2.2. Previously, considering color space alone, we concluded that blue fishes may be best matched to blue background water and yellow fish may be well matched to the 500 nm reflectance step in coral reflectance (Fig. 10.1). As detailed shortly, reasons for this are threefold, but most importantly, this shows the importance of considering the whole system if at all possible. The reason for the surprising match of the blue head of *P. paccagnellae* (its head color at depth) to coral background and yellow tail to water in *achromatic* space is clear. Yellow is a relatively bright color, therefore matching the also-bright background spacelight. The magenta head, which at 10 m loses both high ends to its reflectance due to absorbance (the red end) and scatter (the UV end), becomes a dark violet-blue, better matched to the darker coral background at 10 m (see the first photograph in Fig. 10.5a). In the chromatic domain, it is also the depth of water and the resulting color of coral, especially for the 3 m viewing distance, which means that the dark violet-blue head matches the coral. In effect, color constancy is breaking down by being swamped by the blue light environment at depth.

4.4. Are Reef and Other Marine Fish Dichromats? The Functions of Single and Double Cones

The above model makes the assumption that barracuda possess a dichromatic color vision system. There are two parts to this assumption: first, that color vision exists at all as a result of two spectral sensitivities measured in the retina. Almost all amimals examined in sufficient behavioral detail with more than one spectral sensitivity possess color vision and fishes are not likely to be an exception. Second, we assume that the outputs of the twin and single cones are compared for color vision and that a luminance channel also exists, relying on the output of the twin cones alone. In humans, the daytime lumiance channel is a result of combining all cones, and color vision comes from comparisons between cones (Wyszecki and Stiles, 1982; Backhaus et al., 1998). Combining photoreceptor outputs in different ways for different functions is commonplace.

The two members of each double cone, known as principal and accessory (Walls, 1942), contain visual pigments, with a different λ-max, but in twin cones the visual pigment of both members is identical. Previous evidence suggests that the members of double or twin cones are both optically and electrically coupled (Smith et al., 1985). Therefore, we assume that the effective spectral sensitivity of double cones is equal to the combined sensitivities of their members and is necessarily broadened. This is not a good strategy for color discrimination, particularly in the spectrally different microhabitats of water (Osorio et al., 1997), but may be useful to boost sensitivity for luminance vision. The idea that chromatic vision is mediated by single cones only, and double cones are used for achromatic vision, has support from both behavioral and theoretical observations (Maier and Bowmaker, 1993; Vorobyev and Osorio, 1998). More behavioral evidence comes from Shearer and Neumeyer (1996), who found that the optomotor response in goldfish is driven by long-wavelength sensitive cones most likely to be double cones (Palacios et al., 1989). Remember the goldfish also has four single cone spectral sensitivities used in tetrachromatic vision (Neumeyer, 1992). Birds also possess double and single cone populations and the distinction between their four single cones for color vision and double cones for luminance vision is more clear-cut than in fishes (Hart, 2001). In the pigeon the spectral sensitivity of motion neurones is a good match to the spectral sensitivity of double cones and it is supposed that color vision plays little part in this process (Campenhausen and Kirschfeld, 1998).

In marine fishes current trends in spectral sensitivity data point to two distinct strategies: dichromacy by a combination of two single cones and dichromacy by combining double cones and a single class of single cones. We should emphasize that there is currently no behavioral proof behind this speculation. In species with two single cone types, double/twin cone accessory member spectral sensitivities often closely coincide with the longer-wavelength single cone sensitivity (McFarland et al., in press). This again points to a different function for double and single cones. We speculate that in species with two single cones, dichromacy is maintained without double/twin cone participation. Exceptions to this are the snapper *Lutjanus malabaricus* (Lythgoe et al., 1994), in which two single cones at 408 and 442 nm accompany double cones with members at 529 and 541 nm, and *Myripristis berndti*, a soldierfish with single cone sensitivities at 443 and 453 nm and double cone sensitivities at 514 and 506 nm. These species may be trichromats, the two single cone sensitivities combined with the additive double cone sensitivities (this addition an assumption in itself); however, no further evidence exists to support this idea at present.

Growing evidence in recent years suggests that double cones may be used in polarization sensitivity (PS) (Hawryshyn, 1992; Coughlin and Hawryshyn, 1994; Novales-Flamarique and Hawryshyn, 1998). As, ideally, polarization and color signals should be kept separate (Marshall et al., 1991), this could be seen as providing more evidence for a possible distinct noncolor vision function of double cones (but, see Chapter 13 for further discussion).

As noted above (Fig. 10.4) and elsewhere (Lythgoe, 1968; McFarland and Munz, 1975b; Bowmaker, 1990; Partridge, 1990; Cummings and Partridge, 2001), double cones sensitivities are often well matched to ambient microhabitat light while single cones, whether there be one or more of them, may be offset from this. This simple relationship led to the proposal of the Offset Hypothesis for visual pigment tuning (Section 4.1). In order to see objects, one visual pigment matched to the background and one visual pigment offset from the background, perhaps matched to a specific target, provides an ideal combination (Lythgoe, 1968, 1979). This is a simple form of color vision and seems to be favored by most marine fishes, including the reef fish species so far examined (Section 4.1, and McFarland et al., in press). The spectral sensitivity position of the offset single cone is more variable and certainly less tied to ambient illumination than double cones (Fig. 10.4; Lythgoe et al., 1994, and Douglas, 2001 for recent review). Its λ-max *may* relate somewhat to ambient light (e.g., no UV sensitive cones are found in deep-reef dwelling species). Alternatively, diet and color or contrast of objects may be of more importance (Endler, 1991; Barry and Hawryshyn, 1999a,b; Marshall, 2000a; Cummings and Partridge, 2001; Marshall et al., in press). Largely due to lack of data, this is still a relatively poorly understood area. As a result a summary of the ideas just discussed may be useful:

1. Marine fish are mostly dichromats.
2. Dichromacy is based on one single and one double cone or two single cones.
3. The spectral sensitivity window of double cones is closely correlated to ambient light at the microhabitat level.
4. Double cones may possess the following multiple functions within a single species and/or between species: luminance vision, polarization vision, and color vision.
5. Where two single cones are found in one species, the spectral sensitivity of one often coincides fairly closely with that of the double cone's accessory member. This suggests that the single cones have taken on the job of dichromatic color vision, perhaps freeing double cones for other tasks.
6. Spectral sensitivities of the dichromatic pair (however made up) partially support the Offset Hypothesis with one cone (either single or double) matched to ambient light and one cone offset toward shorter wavelengths (therefore either blue, violet, or UV sensitive).
7. The spectral position of the offset single cone may be related to the spectral envelope of available light, a planktivorous diet (Section 2.3), and the colors of objects of interest. In the latter case, it may be that colors have adapted to existing spectral sensitivities, but this is currently uncertain (Section 4.2.1).

Finally, in this section, it is worth recalling the *ghost cone* hypothesis of Lythgoe (1979). These virtual cone sensitivities result from the inhibitory center-surround interaction of two cone mechanisms at an unspecified level beyond the photoreceptors. A bimodal sensitivity results, with sensitivity peaks displaced to wavelengths shorter and longer than the spectral sensitivity peaks of the two cones. Although speculative, the idea has support from behavioral observation (see references in Lythgoe, 1979) and could work well in long-wavelength-biased habitats (such as the cave microhabitat detailed in Section 3), as it is dependent on stimulation of one cone more than the other. It is interesting in this context that reef fishes, in common with other marine species, are largely dichromats and that no spectral sensitivities beyond 540nm have been found. (Perhaps ghost cones are the reason and it would be worth reexamining this idea.)

4.5. Modeling Dichromatic Reef Fish Vision

As well as modeling what fishes see, with knowledge of background and object colors and the ambient light in different habitats, it is possible to ask questions about the theoretically ideal regions in the spectrum to place spectral sensitivities (Barlow, 1982; Govardovskii, 1983; Maloney, 1986; Lythgoe and Partridge, 1989; Endler, 1990; Lythgoe and Partridge, 1991; Vorobyev et al., 1998; Chiao et al., 2000b; Marshall et al., in press, a,b). The result of three such models are shown in Figure 10.6 and before going on to describe these, the model parameters are briefly discussed. (For detailed description see Marshall et al., in press, a,b.)

The aim of this model, like previous ones (Lythgoe and Partridge, 1989, 1991; Chiao et al., 2000a,d), is to ask which two visual pigments in a dichromatic color vision system are best suited to a variety of natural tasks. The tasks here are all non-noise limited, in that they are presumed to occur in bright daylight on the top of the reef, and photoreceptors are assumed to be adapted to the background in each task. Calculation starts at Eq. 3 above. That is, the quantum catch of photoreceptors looking at objects is calculated, from ambient light, reflectance spectra, and photoreceptor sensitivities. This second model then takes a slightly different approach.

As the task is presumed to be occurring in broad spectrum light, the von Kries scaling factors (k_i) are set to give equal photon catch for each photoreceptor. We define the number of photons q captured by cone type 1 within the range 300–800nm as:

$$q_1 = \sum^{300-800} \mathrm{I} \times \mathrm{S} \times \mathrm{R}_1 \quad (5)$$

As outlined above, rather than encoding total quantal catch, visual systems usually work on contrast signals so the response to a target

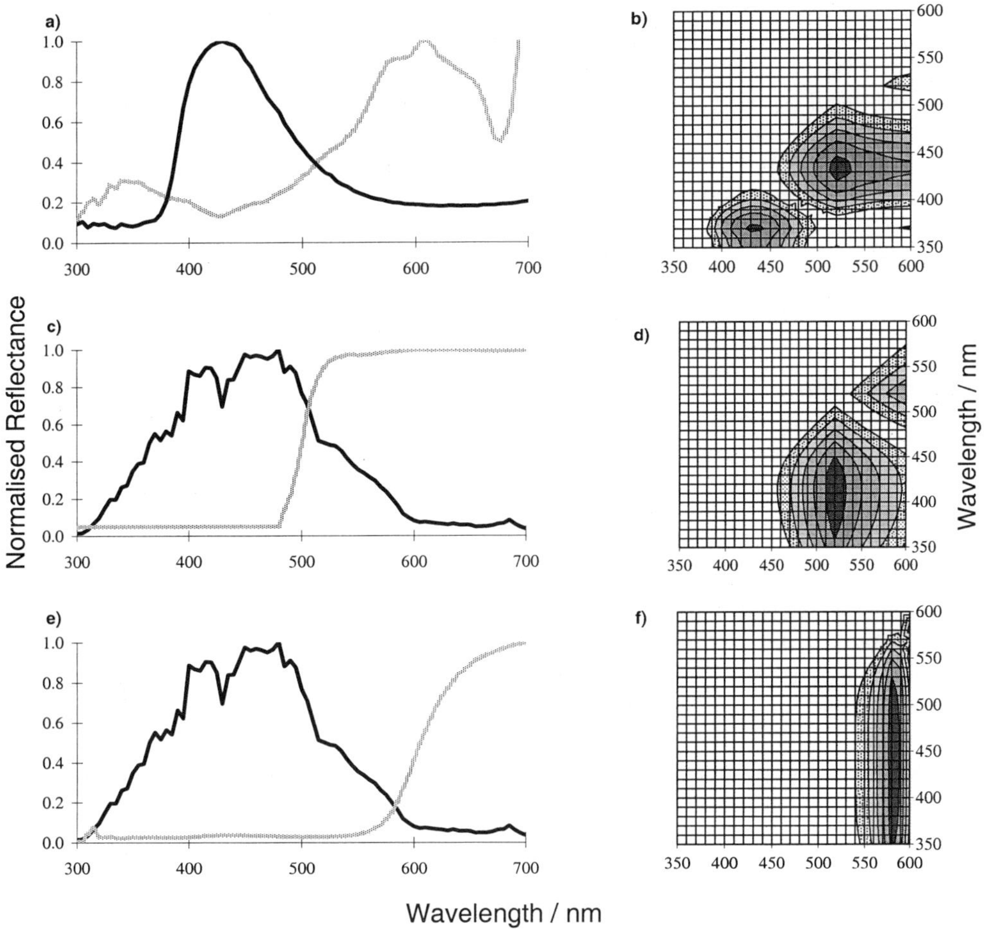

FIGURE 10.6. Models of theoretical dichromatic visual systems in reef fishes. Each of three possible tasks for reef fishes have same format. Left-hand graphs plot the colors of the tasks. The right-hand graph is the result of the model as a density plot, the darker the area the better the visual pigment pair (x and y axis—possible pairs from 350 to 600 nm) at the task. Further details in text. (a, b) Detection of a blue fish (black curve) against a coral/algae background (gray curve). (c, d) Detection of a yellow fish (gray curve) against a blue-water background (black curve). (e, f) Detection of a red fish (gray curve) against a blue-water background.

(t), or the color being examined, is relative to the background (b) against which it is viewed. The adapted response (r) of cone 1 is given by:

$$r_1 = \frac{q_{1(t)}}{q_{1(b)}} \tag{10.6}$$

The colors of objects are decoded by all known dichromatic systems as a chromatic signal (C), the difference between cone responses:

$$C = r_1 - r_2 \tag{7}$$

The larger the value of C, the stronger the chromatic signal or contrast between fish and background. Our second model then compares all possible pairs of visual pigments, in 5 nm steps, with peak values (λ-max) between 350 and 600 nm, and calculates C for each pair. Values of C are shown on a density plot, the axes being the λ-max values of the two hypothetical cone sensitivities (Fig. 10.6). The darker the plot, the better the cone pair at doing the chosen job.

The assumptions and differences from previous models are as follows: (1) Possible cone spectral sensitivities are calculated from the latest nomogram of Hart (1998). (2) Visual interactions occur in the top few meters of water where the full spectrum is abundantly available (we use the 0.1-m data from Fig. 10.3a), so the cone's response falls within its linear range (Wyszecki and Stiles, 1982). (3) The target is sited in clear reef water at a distance of only 1 m where the attenuation of intervening reef water is negligible. (4) Target and background are isoluminant, such that luminance differences between them are discounted (Osorio, 1997; Vorobyev and Osorio, 1998). That is, only the chromatic component of signals is considered, not brightness (Barry and Hawryshyn, 1999a,b).

The three natural tasks chosen (Fig. 6 from Marshall et al., in press, a) are (1) detecting a blue fish (*Chromis atripectoralis*) against an algae/coral background, (2) detecting a yellow fish (*Aulostomus chinensis*) against a blue-water background, and (3) detecting a red fish (*Priacanthus hamrur*) against a blue-water background. For (1) the best pair of visual pigments are at 525 and 435 nm, interestingly very close to those of the barracuda *S. helleri* (Fig. 16.5). For detecting yellow fish against a blue-water background, one pair of the pigments should lie anywhere from 390 to 439 nm and one at 520 nm. Again, these are similar to those of *S. helleri* and indeed almost all of the sensitivities from these and other natural situations investigated fall within ranges known for reef fishes (Marshall et al., in press, a,b; and Fig. 10.4). The last task predicts a range of possible pigments from 400 to 500 nm for one of the pair and long: wavelength sensitivity at 575 nm for the other. Intuitively, it is easy to see why, as the target in this case was a red fish. However, no reef fish has yet been found with a visual pigment λ-max this long, and it is interesting to consider why.

Douglas (2001) provides a useful summary of visual pigment positions in fishes, so we will not go into this in depth here. Clearly, one possible reason why no long-wavelength sensitivities have been found in reef fishes (the longest in the Hawaii data of McFarland et al., in press, being 538 nm) is that relatively few fishes have been examined. Simplistically it is also possible to say that as the marine habitat is a blue-water habitat, there is no ecological need for such a pigment. However, our model above, and the fact that many reef fishes live very close to surface waters where there is plenty of red light, argue against this. Furthermore, recall that one of the microhabitat light environments identified here (Fig. 10.3) is red biased, so these factors combined make it worth predicting that a red-sensitive receptor will be identified in the future. (It is interesting, but no surprise, to note that the predicted sensitivity range of such hypothetical red sensitivities, from this model and others in Marshall et al., in press, a,b, falls between two color step–50% color clusters—Figure 10.4). Alternatively, perhaps the long-wavelength ghost cones described in Section 4.4 are the answer; however, it would be much more efficient to possess spectral sensitivity near 600 nm as many freshwater fishes do (Lythgoe, 1979; Bowmaker, 1990; Partridge, 1990; Kusmic and Gaultieri, 2000).

5. Future Ideas and Directions

With more data now coming from coral reef fishes and their environment, some areas that predict would be worth a go in the future, are as follows:

1. Look for long-wavelength sensitivity especially on reef flat or cave/crevice habitat fishes and determine whether John Lythgoe's ghost cones (Lythgoe, 1979) are reality.
2. Examine diet and spectral sensitivities more closely. The chromatic demands of planktivory, carnivory, herbivory, and possible corallivory on the reef are likely to be very different.
3. Look with finer resolution at the depth and microhabitat distribution of reef fishes and see if this relates to body colors or spectral sensitivity.
4. Define the function of single and double cones in color and luminance vision on the reef.

5. Determine how fish color change is used in camouflage and display.

Acknowledgments. We would like to acknowledge inspiration from Eric Denton and John Lythgoe, and useful discussion with Julian Partridge, Mike Land, Tom Cronin, Danniel Osorio, Uli Siebeck, and Ron Douglas. We are particularly grateful to the following field stations for making this research possible: the Aquarius Underwater Habitat (UNCW/NOAA), Lizard Island Research Station, Heron Island Research Station. Financial support has come from ARC in Australia, NOAA and NSF in the United States and BBSRC and NERC in the U.K.

References

Ali, M.A., and Anctil, M. (1976). *Retinas of Fishes: An Atlas*. Berlin, Heidelberg, New York: Springer.

Arnold, K., and Neumeyer, C. (1987). Wavelength discrimination in the turtle. *Pseudemys scripa elegans. Vision Res*. 27:1501–1511.

Andersson, S., and Amundsen, T. (1997). Ultraviolet colour vision and ornamentation in bluethroats. *Proc. R. Soc. Lond. B.* 264:1587–1591.

Backhaus, W.G.K., Kliegl, R. and Werner, J.S. (1998). *Colour Vision*. Berlin, New York: Walter de Gruyter.

Barlow, H.B. (1982). What causes trichromacy? A theoretical analysis using comb-filtered spectra. *Vision Res*. 22:635–643.

Barry, K.L., and Hawryshyn, C.W. (1999a). Effects of incident light and background conditions on potential conspicuousness of Hawaiian coral reef fish. *J. Mar. Biol. Assoc. U.K.* 79:1–14.

Barry, K.L., and Hawryshyn, C.W. (1999b). Spectral sensitivity of the Hawaiian saddle wrasse, *Thalassoma duperrey*, and implications for visually mediated behaviour on coral reefs. *Env. Biol. Fishes*. 56:422–429.

Bayliss, L.E., Lythgoe J.N., and Tansley, K. (1936). Some forms of visual purple in sea fishes with a note on the visual cells of origin. *Proc. R. Soc. Lond. B.* 120:95–114.

Bowmaker, J.K. (1980). Colour vision in birds and the role of oil droplets. *TINS* 199:196–199.

Bowmaker, J.K. (1990). Visual pigments of fishes. In: *The Visual System of Fish* (Djamgoz, M.B.A., ed.), pp. 81–107. London: Chapman & Hall.

Bowmaker, J.K. (1998). Evolution of colour vision in vertebrates. *Eye* 12:541–547.

Bowmaker, J.K., Thorpe, A., and Douglas, R.H. (1991). Ultraviolet-sensitive cones in the goldfish. *Vision Res.* 31:349–352.

Bowmaker, J.K., Heath, L.A., Wilkie, S.E., and Hunt, D.M. (1997). Visual pigments and oil droplets from six classes of photoreceptor in the retinas of birds. *Vision Res.* 37:23–33.

Burkhardt, D. (1989). UV vision: A bird's eye view of feathers. *J. Comp. Physiol. A.* 164:787–796.

Burkhardt, D., and Finger, E. (1991). Black, white and UV: How birds see birds. *Naturwissenschaften* 78:279–280.

Burkhardt, D., Maier, E. (1989). The spectral sensitivity of a passerine bird is highest in the UV. *Naturwissenschaften* 76:82–83.

Chiao, C.-C., Cronin, T.W., and Marshall, N.J. (2000a). Eye design and colour signaling in a Stomatopod Crustacean *Gonodactylus smithii. Brain Behav. Evol.* 56:107–122.

Chiao, C.C., Cronin, T.W., and Osorio, D. (2000b). Color signals in natural scenes: characteristics of reflectance spectra and effects of natural illuminants. *J. Opt. Soc. Am. A.* 17:218–224.

Chiao, C.C., Osorio, D., Vorobyev, M., and Cronin, T.W. (2000c). Characterization of natural illuminants in forests and the use of digital video data to reconstruct illuminant spectra. *J. Opt. Soc. Am. A.* 17:1713–1721.

Chiao, C.-C., Vorobyev, M., Cronin, T.W., and Osorio, D. (2000d). Spectral tuning of dichromats to natural scenes. *Vision Res.* 40:3257–3271.

Chittka, L. (1992). The colour hexagon: A chromaticity diagram based on photoreceptor excitations as a generalised representation of colour opponency. *J. Comp. Physiol. A.* 170:533–543.

Chittka, L. (1996). Does bee colour vision predate the evolution of flower color? *Naturwissenschaften* 83:136–138.

Chittka, L. (1997). Bee color vision is optimal for coding flower color, but flower colors are not optimal for being coded. Why? *Israel J. Plant Sci.* 45:115–127.

Chittka, L., and Menzel, R. (1992). The evolutionary adaptation of flower colours and the insect pollinators' colour vision. *J. Comp. Physiol. A.* 171:171–181.

Chittka, L., Shmida, A., Troje, N., and Menzel, R. (1994). Ultraviolet as a component of flower reflections, and the colour perception of Hymenoptera. *Vision Res.* 34:1489–1508.

Church, S.C., Bennett, T.D., Cuthill, I.C., and Partridge, J.C. (1998). Ultraviolet cues affect the foraging behaviour of blue tits. *Proc. R. Soc. Lond. B.* 265:1509–1514.

Clarke, G.L. (1936). On the depth at which fishes can see. *Ecology* 17:452–456.

Collin, S.P., and Pettigrew, J.D. (1988a). Retinal topography in reef teleosts. I. Some species with well-developed areae but poorly-developed streaks. *Brain Behav. Evol.* 31:269–282.

Collin, S.P., and Pettigrew, J.D. (1988b). Retinal topography in reef teleosts. II. Some species with prominent horizontal streaks and high-density areae. *Brain Behav. Evol.* 31:283–295.

Collin, S.P., and Pettigrew, J.D. (1989). Quantitative comparison of the limits on visual spatial resolution set by the ganglion cell layer in twelve species of reef teleosts. *Brain. Behav. Evol.* 34:184–192.

Cott, H.B. (1940). *Adaptive Colouration in Animals.* London: Methuen.

Crook, A.C. (1997a). Determinants of the physiological colour patterns of juvenile parrotfish, *Chlorurus sordidus*. *Anim. Behav.* 53:1251–1261.

Crook, A.C. (1997b). Colour patterns in a coral reef fish Is background complexity important? *J. Exp. Mar. Biol. Ecol.* 217:237–252.

Coughlin, D.J., and Hawryshyn, C.W. (1994). A cellular basis for polarized light vision in rainbow trout. *J. Comp. Physiol. A.* 176:261–271.

Cummings, M.E., and Partridge, J.C. (2001). Visual pigments and optical habitats of Surfperch (Embiotocidae) in the California kelp forest. *J. Comp. Physiol. A.* 187:875–889.

Cuthill, I.C., Bennett, A.T.D., Partridge, J.C., and Maier, E.J. (1999). Plumage reflectance and the objective assessment of avian sexual dichromatism. *Am. Nat.* 160:183–200.

Cuthill, I.C., Partridge, J.C., Bennett, A.T.D., Church, S.C., Hart, N.S., and Hunt, S. (2000). Ultrviolet vision in birds. *Adv. Study Behav.* 29:159–214.

Darwin, C. (1859). *On the Origin of Species by Means of Natural Selection.* London: Murray.

Denton, E.J. (1990). Light and vision at depths greater than 200 metres. In: *Light and Life in the Sea* (Maddock, L., ed.), pp. 127–148. Cambridge: Cambridge University Press.

Dominy, N.J., and Lucas, P.W. (2001). Ecological importance of trichromatic vision to primates. *Nature* 410:363–366.

Dorr, S., and Neumeyer, C. (1996). The goldfish: A colour-constant animal. *Perception* 25:243–250.

Douglas, R.H. (2001). The Ecology of teleost fish visual pigments: A good example of sensory adaptation to the environment? In: *Ecology of Sensing* (Schmid, A., ed.), pp. 215–235. Berlin: Springer-Verlag.

Douglas, R.H., and Hawryshyn, C.W. (1990). Behavioral studies of fish vision: An analysis of visual capabilities. In: *The Visual System of Fish* (Djamgoz, M.B.A., ed.), pp. 373–418. London: Chapman & Hall.

Douglas, R.H., and Partridge, J.C. (1997). On the visual pigments of deep-sea fish. *J. Fish Biol.* 50: 68–85.

Douglas, R.H., Bowmaker, J.K., and Kunz, Y.W. (1995). Ultraviolet vision in fish. In: *Seeing Contour and Colour* (Murray, I.F., ed.), pp. 601–616. Oxford: Pergamon.

Douglas, R.H., Partridge, J.C., Dulai, K., Hunt, D., Mullineaux, C.W., Tauber, A.Y., and Hynninen, P.H. (1998). Dragon fish see using chlorophyll. *Nature* 393:423–424.

D'Zmura, M., and Lennie, P. (1986). Mechanisms of color constancy. *J. Opt. Soc. Am.* 3:1662–1672.

Endler, J.A. (1981). An overview of the relationships between mimicry and crypsis. *Biol. J. Linn. Soc.* 16:25–31.

Endler, J.A. (1984). Progressive background matching in moths, and a quantitive measure of crypsis. *Biol. J. Linn. Soc.* 22:187–231.

Endler, J.A. (1990). On the measurement and classification of colour in studies of animal colour patterns. *Biol. J. Linn. Soc.* 41:315–352.

Endler, J.A. (1991a). Variation in the appearance of guppy color patterns to guppies and their predators under different visual conditions. *Vision Res.* 31:587–608.

Endler, J.A. (1991b). Interactions between predators and prey. In: *Behavioral Ecology: An Evolutionary Approach* (Davies, N.B., ed.), pp. 169–196. Oxford: Blackwell Scientific Publications.

Endler, J.A. (1993). The color of light in forests and its implications. *Ecol. Monographs* 63:1–27.

Endler, J.A., and Thery, M. (1996). Interacting effects of lek placement, display behaviour, ambient light, and color patterns in three neotropical forest-dwelling birds. *Am. Nat.* 148:421–452.

Finger, E., Burkhardt, D., and Dyck, J. (1992). Avian plumage colors: Origin of UV reflection in a black parrot. *Naturwissenschaften* 79:187–188.

Fleishman, L.J., Loew, E.R., and Leal, M. (1993). Ultraviolet vision in lizards. *Nature* 365:397.

Fleishman, L.J., Bowman, M., Saunders, D., Miller, W.E., and Rury, M.J. (1997). The visual ecology of Puerto Rican anoline lizards: habitat light and spectral sensitivity. *J. Comp. Physiol. A.* 181: 446–460.

Fox, H.M., and Vevers, G. (1960). *The Nature of Animal Colours.* London: Sidgwick, Jackson.

Frank, T.M., and Case, J.F. (1988a). Visual spectral sensitivities of bioluminescent deep-sea crustaceans. *Biol. Bull.* 175:261–273.

Frank, T.M., and Case, J.F. (1988b). Visual spectral sensitivity of the bioluminescent deep-sea mysid, *Gnathophausia ingens*. *Biol. Bull.* 175:1–10.

Frank, T.M., and Widder, E.A. (1994a). Comparative study of behavioral-sensitivity thresholds to near-UV and blue-green light in deep-sea crustaceans. *Mar. Biol.* 121:229–235.

Frank, T.M., and Widder, E.A. (1994b). Evidence for behavioral sensitivity to near-UV light in the deep-sea crustacean *Systellaspis debilis*. *Mar. Biol.* 118:279–284.

Frank, T.M., and Widder, E.A. (1996). UV light in the deep-sea: *In situ* measurements of downwelling irradiance in relation to the visual threshold sensitivity of UV-sensitive crustaceans. *Mar. Fresh. Behav. Physiol.* 27:189–197.

Frank, T.M., and Widder, E.A. (1999). Comparative study of the spectral sensitivities of mesopelagic crustaceans. *J. Comp. Physiol. A.* 185:255–265.

Govardovskii, V.I. (1983). On the role of oil drops in colour vision. *Vision Res.* 23:1739–1740.

Grill, C.P., and Rush, V.N. (2000). Analysing spectral data: Comparison and application of two techniques. *Biol. J. Linn. Soc.* 69:121–138.

Hailman, J.P. (1977a). *Optic Signals: Animal Communication and Light.* Bloomington, London, Indiana University Press.

Hailman, J.P. (1977b). Communication by reflected light. In: *How Animals Communicate* (Sebeok, T.A., ed.), pp. 184–210. Bloomington, London: Indiana University Press.

Hailman, J.P. (1979). Environmental light and conspicuous colours. In: *The Behavioral Significance of Color* (Burtt, E.H.J., ed.), pp. 289–354. New York, London: Garland STMP Press.

Hart, N.S. (2001). Variations in cone photoreceptor abundance and the visual ecology of birds. *J. Comp. Physiol. A.* 187:685–697.

Hart, N.S., Partridge, J.C., and Cuthill, I.C. (1998). Visual pigments, oil droplets and cone photoreceptor distribution in the European starling (*Sturnus vulgaris*). *J. Exp. Biol.* 201:1433–1446.

Hart, N.S., Partridge, J.C., Cuthill, I.C., and Bennett, A.T.D. (2000). Visual pigments, oil droplets, ocular media and cone photoreceptor distribution in two species of passerine bird: The blue tit (*Parus caeruleus* L.) and the blackbird (*Turdus merula* L.). *J. Comp. Physiol. A.* 186:375–387.

Hawryshyn, C.W. (1982). Studies of aniaml color vision: Comments on some important theoretical considerations. *Can. J. Zool.* 60:2968–2970.

Hawryshyn, C.W. (1992). Polarization vision in fish. *Am. Sci.* 80:164–175.

Hawryshyn, C.W., and Beauchamp, R. (1985). Ultraviolet photosensitivity in goldfish: An independent UV retinal mechanism. *Vision Res.* 25:11–20.

Hawryshyn, C.W., Chou, B.R., and Beauchamp, R.D. (1985). Ultraviolet transmission by the ocular media of goldfish: implications for ultraviolet photosensitivity in fishes. *Can. J. Zool.* 63:1244–1251.

Hawryshyn, C.W., Arnold, M.G., McFarland, W.N., and Loew, E.R. (1988). Aspects of colour vision in the bluegill sunfish (*Lepomis macrochirus*): Ecological and evolutionary relevance. *J. Comp. Physiol. A.* 164:107–116.

Herring, P.J. (1977). Luminescence in cephalopods and fish. *Symp. Zool. Soc. Lond.* 38:127–159.

Herring, P.J. (1982). Aspects of the bioluminescence of fishes. *Oceanogr. Mar. Biol. Ann. Rev.* 20:415–470.

Herring, P.J. (1985). Bioluminescence in the crustacea. *J. Crust. Biol.* 5:557–573.

Hobson, E.S., McFarland, W.N., and Chess, J.R. (1981). Crepuscular and nocturnal activities of Californian nearshore fishes, with consideration of their scotopic visual pigments and the photic environment. *Fishery Bull.* 79:1–30.

Houde, A.E., and Endler, J.A. (1990). Correlated evolution of female mating preferences and male color patterns in the guppy *Poecilia reticulata*. *Science* 248:1405–1408.

Jacobs, G.H. (1993). The distribution and nature of colour vision among the mammals. *Biol. Rev.* 68:413–471.

Jagger, W.S., and Muntz, W.R.A. (1993). Aquatic vision and the modulation transfer properties of unlighted and diffusely lighted natural waters. *Vision Res.* 33:1755–1763.

Jerlov, N.G. (1976). *Marine Optics.* Amsterdam: Elsevier.

Kamermans, M., Kraaij, D.A., and Spekreijse, H. (1998). The cone/horizontal cell network: A possible site for color constancy. *Visual Neurosci.* 15:787–797.

Kelber, A. (1997). Innate preferences for flower features in the hawkmoth *Macroglossum stellatarum*. *J. Exp. Biol.* 200:827–836.

Kelber, A. (1999). Why "false" colours are seen by butterflies. *Nature* 402:251.

Kinney, J.A.S., Luria, S.M., and Weitzman, D.O. (1967). Visibility of colors underwater. *J. Opt. Soc. Am.* 57:802–809.

Kirk, J.T.O. (1983). *Light and Photosynthesis in Aquatic Ecosystems.* Cambridge, London, New York: Cambridge University Press.

Knowles, A., and Dartnall, H.J.A. (1977). Habitat and visual pigments. In: *The Eye, Vol. 2B: The Photobiology of Vision* (Davson, H., ed.), pp. 581–641. New York: Academic Press.

Kondrashev, S.L., Gamburtseva, A.G., Gnjubkina, V.P., Orlov, O.J., and My, P.T. (1986). Colouration of corneas in fish: A list of species. *Vision Res.* 26:287–290.

Kusmic, C., and Gualtieri, P. (2000). Morphology and spectral sensitivities of retinal and extraretinal photoreceptors in freshwater teleosts. *Micron.* 31:183–200.

Latz, M.I., Frank, T.M., and Case, J.F. (1988). Spectral composition of bioluminescence of epipelagic organisms from the Sargasso Sea. *Mar. Biol.* 98:441–446.

Levine, J.S., and MacNichol, E.F.J. (1979). Visual pigments in teleost fishes: Effects of habitat, microhabitat, and behaviour on visual system evolution. *Sensory Processes* 3:95–131.

Loew, E.R. (1994). A third, ultraviolet-sensitive, visual pigment in the Tokay gecko (*Gekko gekko*). *Vision Res.* 34:1427–1431.

Loew, E.R., and Lythgoe, J.N. (1978). The ecology of cone pigments in teleost fishes. *Vision Res.* 18:715–722.

Loew, E.R., and Lythgoe, J.N. (1985). The ecology of colour vision. *Endeavour* 14:170–174.

Loew, E.R., McAlery, F.A., and McFarland, W.N. (1996a). *Ultraviolet Sensitivity in the Larvae of Two Species of Marine Atherinid Fishes.* Australia: Gordon and Breach.

Loew, E.R., Govardovskii, V.I., Rohlich, P., and Szel, A. (1996b). Microspectrophotometric and immunocytochemical identification of ultraviolet photoreceptors in geckos. *Visual Neurosci.* 13: 247–256.

Longley, W.H. (1914). Report upon color of fishes of the Tortugas. *Year Book Carnegie Inst.* 13:207–208.

Longley, W.H. (1916a). Observations upon tropical fishes and inferences from their adaptive coloration. *Proc. Nat. Acad. Sci.* 2:733–737.

Longley, W.H. (1916b). The significance of the colors of tropical reef fishes. *Year Book Carnegie Inst.* 15:209–212.

Longley, W.H. (1918). Haunts and habits of tropical fishes. *Am. Museum J.* 18:79–88 plus 10 figures.

Longley, W.H. (1919). Report of additional observations and experiments upon problems of animal colouration. *Year Book Carnegie Inst.* 18:201–202.

Lorenz, K. (1962). The function of colour in coral reef fishes. *Proc. R. Inst. G.B.* 39:282–296.

Losey, G.S., Cronin, T., Goldsmith, T.H., Hyde, D., Marshall, N.J., and McFarland, W.N. (1999). The UV visual world of fishes: A review. *J. Fish Biol.* 54:921–943.

Lythgoe, J.N. (1966). Visual pigments and underwater vision. In: *Light as an Ecological Factor* (Rackham, O., ed.). Oxford: Blackwell.

Lythgoe, J.N. (1968). Red and yellow as conspicuous colours underwater. *Underwater Assoc. Rep.*: 1:51–53.

Lythgoe, J.N. (1972). The adaptation of visual pigments to the photic environment. In: *Handbook of Sensory Physiology* (Teuber, H.L., ed.). Berlin, Heidelberg, New York: Springer.

Lythgoe, J.N. (1979). *The Ecology of Vision.* Oxford: Clarendon Press.

Lythgoe, J.N. (1980). Vision in fishes: Ecological adaptations. In: *Environmental Physiology of Fishes* (Ali, M.A., ed.), pp. 431–445. London, New York: Plenum Publishing.

Lythgoe, J.N. (1984). Visual pigments and environmental light. *Vision Res.* 24:1539–1550.

Lythgoe, J.N. (1988). Light and vision in the aquatic environment. In: *Sensory Biology of Aquatic Animals* (Atema, J., Fay, R.R., Popper, A.N., Tarolga, W.N., eds.), pp. 57–82. New York: Springer Verlag.

Lythgoe, J.N., and Partridge, J.C. (1989). Visual pigments and the aquisition of visual information. *J. Exp. Biol.* 146:1–20.

Lythgoe, J.N., and Partridge, J.C. (1991). The modelling of optimal visual pigments of dichromatic teleosts in green coastal waters. *Vision Res.* 31: 361–371.

Lythgoe, J.N., Muntz, W.R.A., Partridge, J.C., Shand, J., and Williams, D.M. (1994). The ecology of the visual pigments of snappers (Lutjanidae) on the Great Barrier Reef. *J. Comp. Physiol.* 174:461–467.

Maier, E.J., Bowmaker, J.K. (1993). Colour vision in the passeriform bird, *Leiothrix lutea*: Correlation of visual pigment absorbance and oil droplet transmission with spectral sensitivity. *J. Comp. Physiol. A.* 172:295–301.

Maloney, L.T. (1986). Evaluation of linear models of surface spectral reflectance with small numbers of parameters. *J. Opt. Soc. Am. A.* 3:1673–1683.

Maloney, L.T., and Wandell, B.A. (1986). Color constancy: A method for recovering surface spectral reflectance. *J. Opt. Soc. Am. A.* 3:29–33.

Marshall, N.J. (1996). Measuring colours around a coral reef. *Biophotonics Int.* 3(4):52–56.

Marshall, N.J. (1998). Why are reef fish so colorful? *Sci. Am. Quarterly* 9(3):54–57.

Marshall, N.J. (2000a). The visual ecology of reef fish colours. In: *Animal Signals: Signalling and Signal*

Design in Animal Commumication (Espmark, Y., Amundsen, T., Rosenqvist, G., eds.), pp. 83–120. Trondheim, Norway: Tapir Academic Press.

Marshall, N.J. (2000b). Communication and camouflage with the same "bright" colours in reef fishes. *Phil. Trans. R. Soc. Lond. B.* 355:1243–1248.

Marshall, N.J., and Land, M.F. (1991). Vision in mantis shrimps (abstract). In: *Society for Experimental Biology Meeting*, p. 33. Birmingham.

Marshall, N.J., Jennings, K.J., Losey, G.W., and McFarland, W.N. (in press, a). Visual biology of Hawaiian coral reef fish. II. The colours of Hawaiian reef fish. *Copea.*

Marshall, N.J., Land, M.F., King, C.A., and Cronin, T.W. (1991). The compound eyes of mantis shrimps (Crustacea, Hoplocarida, Stomatopoda). I. Compound eye structure: The detection of polarised light. *Phil. Trans. R. Soc. Lond. B.* 334:33–56.

Marshall, N.J., Losey, G.W., McFarland, W.N., and Loew, E.R. (in press, b). Visual biology of Hawaiian coral reef fish. III. The ecology of reef fish vision. *Copea.*

McFarland, W.N. (1986). Light in the sea: Correlations with behaviours of fishes and invertebrates. *Amer. Zool.* 26:389–401.

McFarland, W.N. (1991). The visual world of coral reef fishes. In: *The Ecology of Fishes on Coral Reefs* (Sale, P.F., ed.), pp. 16–38. San Diego: Academic Press.

McFarland, W.N., and Loew, E.R. (1994). Ultraviolet visual pigments in marine fishes of the family Pomacentridae. *Vision Res.* 34:1395–1396.

McFarland, W.N., and Munz, F.W. (1975a). Part II: The photopic environment of clear tropical seas during the day. *Vision Res.* 15:1063–1070.

McFarland, W.N., and Munz, F.W. (1975b). Part III: The evolution of photopic visual pigments in fishes. *Vision Res.* 15:1071–1080.

McFarland, W.N., and Wahl, C.M. (1996). Visual constraints on migration behavior of juvenile French grunts. *Env. Biol. Fishes* 46:109–122.

McFarland, W.N., Loew, E.R., Marshall, N.J., and Losey, G.W., (in press). Visual biology of Hawaiian coral reef fish. I. Microspectrophotometry of retinal visual pigments. *Copea.*

Menzel, R., and Backhaus, W. (1991). Colour vision in insects. In: *The Perception of Colour* (Gouras, P., ed.), pp. 262–293. London: Macmillan Press.

Menzel, R., Shmida, A. (1993). The ecology of flower colours and the natural colour vision of insect pollinators: The Israeli flora as a study case. *Biol. Rev.* 68:81–120.

Mobley, C.D. (1994). *Light and Water Radiative Transfer in Natural Waters.* San Diego: Academic Press.

Mollon, J.D. (1989). "Tho' she kneel'd in that place where they grew . . .": The uses and origins of primate colour vision. *J. Exp. Biol.* 146:21–38.

Mollon, J.D. (1991). Uses and evolutionary origins of primate colour vision. In: *Vision and Visual Dysfunction: Evolution of the Eye and Visual System* (Gregory, R.L., ed.), pp. 306–319. London: Macmillan Press.

Mollon, J.D., Estevez, O., and Cavonius, C.R. (1990). The two subsystems of colour vision and their roles in wavelength discrimination. In: *Vision: Coding and Efficiency* (Blakemore, C., ed.), pp. 119–131. Cambridge: Cambridge University Press.

Muntz, W.R.A. (1990). Stimulus, environment and vision in fishes. In: *The Visual System of Fish* (Djamgoz, M.B.A., ed.), pp. 491–511. London: Chapman & Hall.

Munz, F.W., and McFarland, W.N. (1973). The significance of spectral position in the rhodopsins of tropical marine fishes. *Vision Res.* 13:1829–1874.

Neumeyer, C. (1981). Chromatic adaptation in the honeybee: Successive color contrast and color constancy. *J. Comp. Physiol. A.* 144:543–553.

Neumeyer, C. (1991). Evolution of colour vision. In: *Vision and Visual Dysfunction: Evolution of the Eye and Visual System* (Gregory, R.L., ed.), pp. 284–305. London: Macmillan Press.

Neumeyer, C. (1992). Tetrachromatic color vision in goldfish: Evidence from color mixture experiments. *J. Comp. Physiol. A.* 171:639–649.

Neumeyer, C., Wietsma, J.J., and Spekreijse, H. (1991). Separate processing of "colour" and "brightness" in goldfish. *Vision Res.* 31:537–549.

Neumeyer, C., Dorr, S., Fritsch, J, and Kardelky, C. (2002). Colour constancy in goldfish and man: Influence of surround size and lightness. *Perception* 31:171–187.

Novales-Flamarique, I., and Hawryshyn, C.W. (1998). Photoreceptor types and their relation to the spectral and polarization sensitivities of clupeid fishes. *J. Comp. Physiol. A.* 182:793–803.

Osorio, D. (1997). A functional view of cone pigments and colour vision. In: *John Dalton's Colour Vision Legacy* (Carden, D., ed.), pp. 483–489. London: Taylor and Francis.

Osorio, D., and Vorobyev, M. (1996). Colour vision as an adaptation to frugivory in primates. *Proc. R. Soc. Lond.* 263:593–599.

Osorio, D., Marshall, N.J., and Cronin, T.W. (1997). Stomatopod photoreceptor spectral tuning as an adaptation for colour constancy in water. *Vision Res.* 37:3299–3309.

Palacios, A.G., Varela, F.J., Srivastava, R., and Goldsmith, T.H. (1998). Spectral sensitivity of cones in the goldfish, *Carassius auratus. Vision Res.* 38: 2135–2146.

Partridge, J.C. (1990). The colour sensitivity and vision of fishes. In: *Light and Life in the Sea* (Maddock, L., ed.), pp. 167–184. Cambridge: Cambridge University Press.

Partridge, J.C., and Cummings, M.E. (1999). Adaptations of visual pigments to the aquatic environment. In: *Adaptive Mechanisms in the Ecology of Vision* (Archer, S.N., Djamgoz, E.R., Loew, E.R., Partridge, J.C., and Vallerga, S., eds.), pp. 251–284. Dordrecht, Boston, London: Kluwer.

Partridge, J.C., Archer, S.N., and van Oostrum, J. (1992). Single and multiple visual pigments in deep-sea fishes. *J. Mar. Biol. Assoc. UK.* 72:113–120.

Pokorny, J., Shevell, S.K., and Smith, V.C. (1991). Colour appearance and colour constancy. In: *Vision and Visual Dysfunction: The Perception of Colour* (Gouras, P., ed.), pp. 43–61. London: Macmillan Press.

Poulton, E.B. (1890). The colours of animals, their meaning and use. London: K., Paul, T., Trubner, Co. Ltd.

Randall, J.E., Allen, G.R., and Steene, R.C. (1991). *The Complete Diver's and Fishermen's Guide to Fishes of the Great Barrier Reef and Coral Sea*, 2nd Ed. Bathurst: Crawford House Publishing.

Scherer, C., and Kolb, G. (1987a). The influence of colour stimuli on visually controlled behaviour in *Aglais urticae* L. and *Parage aegeria* L. (Lepidoptera). *J. Comp. Physiol. A.* 161:891–898.

Scherer, C., and Kolb, G. (1987b). Behavioural experiments on the visual processing of colour stimuli in *Pieris brassicae* L. (Lepidoptera). *J. Comp. Physiol. A.* 160:645–656.

Shashar, N. (1994). UV vision by marine animals: Mainly questions. In: *Ultraviolet Radiation and Coral Reefs* (Jokiel, P.L., ed.), pp. 201–206. Hawai'i: Hawai'i Institute of Marine Biology.

Shearer, S., and Neumeyer, C. (1996). Motion detection in goldfish investigated with the optomotor response is "colour blind." *Vision Res.* 36:4025–4034.

Siebeck, U.E., and Marshall, N.J. (2000). Transmission of ocular media in labrid fishes. *Phil. Trans. R. Soc. Lond. B.* 355:1257–1262.

Siebeck, U.E., and Marshall, N.J. (2001). Ocular media transmission of coral reef fish: Can coral reef fish see ultraviolet light? *Vision Res.* 41:133–149.

Smith, E.J., Partridge, J.C., Parsons, K.N., White, E.M., Cuthill, I.C., Bennett, A.T.D., and Church, S.C. (2001). Ultraviolet vision and mate choice in the guppy (*Poecilia reticulata*). *Behav. Ecol.* 13:11–19.

Smith, K.C., and Macagno, E.R. (1990). UV photoreceptors in the compound eye of *Daphnia magna* (Crustacea, Branchiopoda): A fourth spectral class in a single ommatidia. *J. Comp. Physiol. A.* 166:597–606.

Smith, R.C., and Baker, K.S. (1981). Optical properties of the clearest natural waters (200–800 nm). *Applied Optics* 20:177–184.

Smith, R.C., Ensminger, R.L., Austin, R.W., Bailey, J.D., and Edwards, G.D. (1979). Ultraviolet submersible spectroradiometer. *Ocean Optics* 6:127–140.

Smith, R.L., Nishimura, Y., and Raviola, G. (1985). Interreceptor junction in the double cone of the chicken retina. *J. Submicrosc. Cytol.* 17:183–186.

Thayer, G.H. (1909). *Concealing-Colouration in the Animal Kingdom.* New York: Macmillan.

Thresher, R.E. (1984). *Reproduction in Reef Fishes.* T.F.H. Publication Inc. Ltd. Neptune City, USA.

Vorobyev, M., and Brandt, R. (1997). How do insect pollinators discriminate colors? *Israel. J. Plant Sci.* 45:103–113.

Vorobyev, M., and Osorio, D. (1998). Receptor noise as a determinant of color thresholds. *Proc. R. Soc. Lond. B.* 265:351–358.

Vorobyev, M., Brandt, R., Peitsch, D., Laughlin, S.B., and Menzel, R. (2001a). Colour thresholds and receptor noise: Behaviour and physiology campared. *Vision Res.* 41:639–653.

Vorobyev, M., Marshall, J., Osorio, D., Hempel de Ibarra, N., Menzel, R. (2001b). Colourful objects through animal eyes. *Color Res. and Appl.* S26: S214–S217.

Vorobyev, M., Osorio, D., Bennett, A.T.D., Cuthill, I.C., and Marshall, J. (1998). Tetrachromacy, oil droplets and bird plumage colours. *J. Comp. Physiol. A.* 183:621–633.

Wachtler, T., Lee, T.W., and Sejnowski, T.J. (2001). Chromatic structure of natural scenes. *J. Opt. Soc. Am. A.* 18:65–77.

Walls, G.L. (1942). *The Vertebrate Eye and its Adaptive Radiation.* Michigan: Cranbrook Press.

Widder, E.A., Latz, M.I., and Case, J.F. (1983). Marine bioluminescence spectra measured with an optical multichannel detection system. *Biol. Bull.* 165:791–810.

Wyszecki, G., and Stiles, W.S. (1982). *Colour Science: Concepts and Methods, Quantitative Data and Formulae*, 2nd Ed. New York: Wiley.

11
Color Vision in Fishes and Its Neural Basis

Christa Neumeyer

Abstract

The neurons in the retina of fishes possess chromatic response properties that are found in primates only in the visual cortex. The retinae mainly of cyprinid and cichlid fish species are being intensively studied using electrophysiological and neuroanatomical techniques. On the other hand, behavioral experiments are performed to establish the overall properties of color vision and other visual functions. Many details of color vision are known in goldfish, which is an ideal subject because it is especially suited for training experiments using food reward. To establish the neural basis of color vision and other visual functions a neuropharmacological approach in combination with behavioral experiments yields promising results. They indicate that there is a parallel processing of "color" and high visual acuity on the one hand, and "motion," "flicker," and "brightness" detection on the other hand, which is similar to the situation in the visual system of primates. In goldfish, as the best-investigated fish species, the following properties of color vision are described: wavelength discrimination, spectral sensitivity, color constancy, color contrast, and color perception. The shape of the $\Delta\lambda$-function gave the first hint that goldfish color vision may be tetrachromatic, being based on four cone types. This was proven to be the case in additive color mixture experiments. Goldfish color vision includes the ultraviolet spectral range and is, therefore, perhaps more complicated than human color vision.

1. Introduction

Color vision as a property of the entire visual system has to be shown and analyzed in behavioral experiments. However, not all behavioral techniques actually give insight into color vision—in fact most of them do not. Behaviors linked with food acquisition such as training experiments with food reward, and the innate prey-catching response elicited in amphibia by certain moving stimuli (Przyrembel et al., 1995), are the most informative. When using innate behaviors like the optomotor response and probably also the dorsal light reaction, goldfish and zebrafish (*Danio rerio*) behave as if they were entirely color-blind (Schaerer and Neumeyer, 1996; Krauss, 2001). Therefore, as known from the early experiments by Schlieper (1927), these methods cannot be used to investigate color vision. Even the training experiment using food reward may not necessarily indicate color vision (see below). Neumeyer et al. (1991) demonstrated that very different results were obtained depending on which of the two test fields the animal was trained on, a dark one or an illuminated one. Goldfish seem to decide on the basis of color only when they are trained on the dark test field. When trained on the illuminated field they are obviously using a brightness criterion.

Training experiments by Karl von Frisch (1913), in which fishes were rewarded with food, demonstrated color vision in animals for the first time. They showed that the minnow (*Phoxinus laevis*), a cyprinid fish, was able to discriminate the colors red, yellow, green, and blue from different shades of gray. This clearly indicated that minnows have a very good color vision probably similar to our own. In experiments with different species of cyprinid fishes (*Phoxinus laevis*, *Rhodeus amarus*, *Idus melanotus*, *Tinca vulgaris*), Burkamp (1923) found a discrimination ability between 20 different colors and shades of gray even under changes in the spectral composition of the illuminating light, and thus was able to demonstrate color constancy. Furthermore, in the first experiments using spectral colors, Schiemenz (1924) and Wolff (1925) showed that minnows can see ultraviolet light and that they possess three spectral ranges of best discrimination ability, and not only two as in human color vision. Both findings have been ignored for more than 50 years (for a more complete review of the early literature, see Douglas and Hawryshyn, 1990).

Beginning in the 1950s, color vision in fishes regained interest after the first intracellular recordings of "S-potentials" from the retina of teleost fishes by Gunnar Svaetichin. He found two types of graded responses, a "luminosity" and a "chromaticity" type. Whereas the first type revealed hyperpolarizing responses over the entire spectral range between 400 and 750 nm, the second type showed antagonistic response properties with depolarization in one part of the spectrum and hyperpolarization in another (Svaetichin, 1956; Svaetichin and MacNichol, 1958). Here, biphasic and triphasic response types were found. Originally, Svaetichin was convinced he had recorded from cones, and thought he had found the color-opponent cone types proposed by Ewald Hering to explain color opponency in human color vision. Later it became clear that he had recorded from horizontal cells. The first person who succeeded in recording intracellularly from single cones was Tsuneo Tomita (1963). In carp, he found three types with hyperpolarizing responses and maximal spectral sensitivities at about 460 nm, 530 nm, and 610 nm. Taken together with the first microspectrophotometric measurements of goldfish cone photopigments by Marks (1965), which yielded a similar result, it looked as if cyprinid fishes (i.e., goldfish and carp) had a trichromatic color vision system similar to humans. It took us more than 20 years to find out that this was not the case—their color vision is tetrachromatic and based on four types of cones (see below).

The cells in the retina of fishes and lower vertebrates in general are about 10 times as large as those in the retina of mammals and birds, and, therefore, can be more easily studied in electrophysiological investigations. Goldfish proved to be ideal subjects for investigations of color vision because they are highly suitable not only for electrophysiological, neuroanatomical, and microspectrophotometric investigation, but also for behavioral studies. Having been domesticated for about 2,000 years (Kuhn, 1935), the goldfish is rather tame and can be easily trained. Now, after 40 years of intensive investigation, the color vision system in goldfish is probably the best known of all vertebrates with the exception of primates. Therefore, it is goldfish that are considered in detail.

2. Goldfish: A Case Study of Color Vision

2.1. Wavelength Discrimination

For the characterization of a color vision system, the knowledge of wavelength discrimination is of particular importance. The shape of the so-called Δλ-function gives a hint about the number of the underlying cone types, and about the processing of hue. After the early experiments by Wolff (1925) on the minnow, wavelength discrimination was measured in goldfish by Yarczower and Bitterman (1965) in the spectral range between 450 nm and 650 nm. They reported two maxima of best discrimination ability, at 490 nm and at 610–630 nm. However, it is possible that discrimination in these experiments was based not only on hue but also on brightness, as the monochromatic light was not presented in equal fish-subjective brightness. Therefore, we remeasured the Δλ-function in goldfish by adjusting the intensity according to a spectral sensitivity function measured under the same experimental conditions (Neumeyer, 1984, 1986).

The Δλ-function (Fig. 11.1) reveals not two, but three ranges of best discrimination ability (at 400, 500, and 610 nm) similar to those found by Wolff (1925) in the minnow. The high discrimination at 400 nm (which is not found in human color vision) was very surprising. We first assumed that it is based on the β-band of the L-cone sensitivity, as only three cone types had previously been found in microspectrophotometric measurements (Marks, 1965; Hárosi, 1976). However, additive color mixture experiments and measurements of spectral sensitivity showed that the third range of discrimination at 400 nm is due to the existence of an ultraviolet sensitive cone type, in addition to three other cone types (Hawryshyn and Beauchamp, 1985; Neumeyer, 1985, 1992; Fratzer et al., 1994). Further additive color mixture experiments have proven that color vision in goldfish is tetrachromatic and indeed based on four cone types (Fig. 11.2). The first report of ultraviolet cones in cyprinid fishes was based on microspectrophotometric data from a Japanese species (*Tribolodon hakonensis*) by Hárosi and Hashimoto (1983) and from the roach (*Rutilus rutilus*) by Avery et al. (1983). The ultraviolet cone type in goldfish was identified microspectrophotometrically by Bowmaker et al. (1991), and more recently in patch clamp recordings by Palacios et al. (1998).

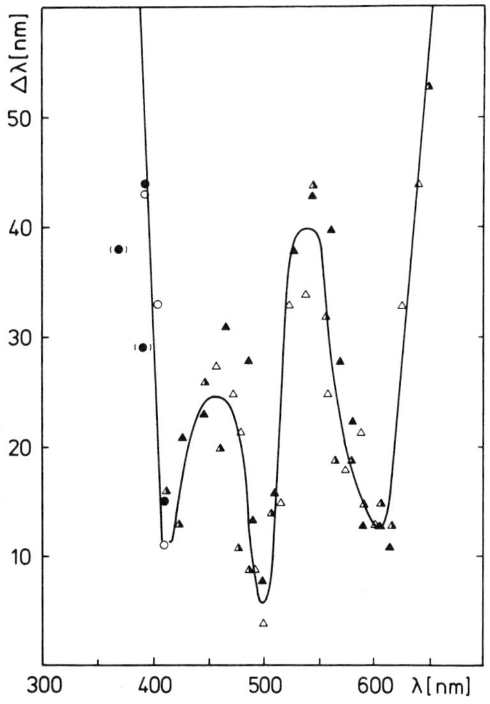

FIGURE 11.1. Δλ-function of goldfish. (From Fratzer et al., 1994.) Δλ represents the just-noticeable difference (with a threshold criterion set at 70% relative choice frequency) between a given training wavelength and an adjacent wavelength shown for comparison. Thus, small values of Δλ indicate high discrimination ability. The monochromatic light used in these experiments (Neumeyer, 1986) was adjusted to equal fish-subjective brightness according to the spectral sensitivity function measured under the same conditions (Neumeyer, 1984; shown as relative sensitivity in Fig. 11.3). Symbols stand for different fishes.

Complete measurements of wavelength discrimination (Δλ-) functions are very time consuming and were obtained in only a relatively small number of vertebrate species (for review see Neumeyer, 1991). It seems remarkable that (maybe with the exception of pigeons) the spectral ranges of best discrimination ability are

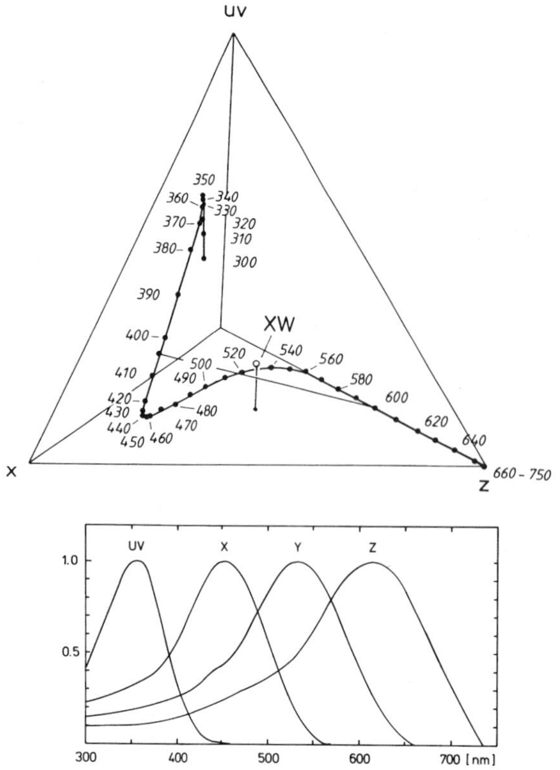

FIGURE 11.2. Color tetrahedron of the goldfish (from Neumeyer, 1992) calculated on the basis of the cone spectral sensitivity functions shown in the inset. (After Bowmaker et al., 1991.) Each point in this three-dimensional plot represents the relative excitation values of the four cone types (uv: UV-cone type; x: S-cone type; y: M-cone type, and z: L-cone type; with uv + x + y + z = 1). The tetrahedron is the equivalent of the color triangle for trichromatic color vision. It does not give information about the strength of excitation, and, therefore, represents only hue.

always found around 400, 500, and 600 nm, respectively, in creatures as different as honeybees, goldfish, turtles, squirrels, and humans. However, only goldfish and turtles are shown to have all three ranges of best discrimination ability (Neumeyer, 1986; Arnold and Neumeyer, 1987), with humans and honeybees possessing two (von Helversen, 1972), and squirrels and other nonprimate mammals possessing only one (Jacobs, 1993).

In many species of fishes the number of cone types is known from microspectrophotometric investigations, and we may assume that the dimensionality of color vision corresponds to this number. However, one has to be cautious. In goldfish and turtles the ultraviolet sensitive cone type has been overlooked in electrophysiological and microspectrophotometric studies for a long time because of its small size and low number. It is also theoretically possible that not all cone types found in such investigations contribute to color vision, resulting in a smaller dimensionality of color vision. However, so far there is no such example. Behavioral experiments always showed that all cone types play a role in color vision under photopic conditions. Under mesopic light conditions, however, we found that the goldfish L-cones do not contribute and color vision in this instance is trichromatic (Neumeyer and Arnold, 1989).

2.2. Spectral Sensitivity

There are several behavioral experiments investigating spectral sensitivity in goldfish with different methods (Cronly-Dillon and Muntz, 1965; Yager, 1967, 1969; Beauchamp and Rowe, 1977; Powers, 1978; Neumeyer, 1984; Neumeyer et al., 1991; Schaerer and Neumeyer, 1996). The results are very different and clearly show that it is not correct to speak about *the* spectral sensitivity function. Instead, there are several spectral sensitivity functions that have to be regarded as action spectra of different behavioral responses reflecting very different visual functions.

Using the optomotor response, a function was obtained with a single maximum in the long-wavelength range (Cronly-Dillon and Muntz, 1965; Schaerer and Neumeyer, 1996). It reflects the action spectrum of motion vision, which is dominated by the long-wavelength cone type (Fig. 11.3). In this behavioral context, motion vision is color blind, that is, there is no response to colored moving stripes when they do not modulate the L-cones. A similar function was obtained using the dorsal light reaction (Powers, 1978), and in training experiments measuring temporal resolution (Schaerer, 1990).

A spectral sensitivity function, which is obviously connected with the color vision system, was measured with a forced two-choice training technique in which goldfish were trained on a *dark* test field while the comparison test field was illuminated with monochromatic light of variable wavelength and intensity. Here a function with three very pronounced maxima and minima (between 400 and 720 nm) was obtained (Fig. 11.4). It indicates strong inhibitory interactions between the different spectrally adjacent cone types (Neumeyer, 1984). The function has a parallel in the increment threshold sensitivity of primate color vision (Sperling and Harwerth, 1971).

However, in the reversed training experiment under the same overall light conditions, in which the fishes were trained on the *illuminated* test field, a flat function was found with a sensitivity 1 to 1.5 log unit higher than in the other experiment in which we trained on the dark test field (Neumeyer et al., 1991). A similar function was reported by Yager (1967) in goldfish, and by Douglas (1986) in the roach (*Rutilus rutilus*) under similar training conditions. We assumed that this specific spectral sensitivity function resembles sensitivity for detection of brightness because two different wavelengths adjusted in intensity according to this function could not be discriminated.

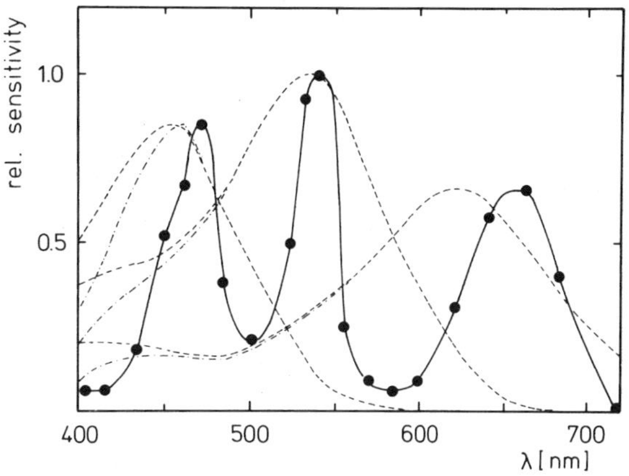

FIGURE 11.4. Relative spectral sensitivity function in goldfish in the range between 400 nm and 720 nm. (From Neumeyer, 1984.) The function was obtained in a forced two-choice training procedure, in which the goldfish were trained on the dark test field (D+), while the comparison test field was illuminated with monochromatic light of variable wavelength and intensity. Dashed lines: absorbance spectra of cone photopigments. (From Hárosi, 1976.)

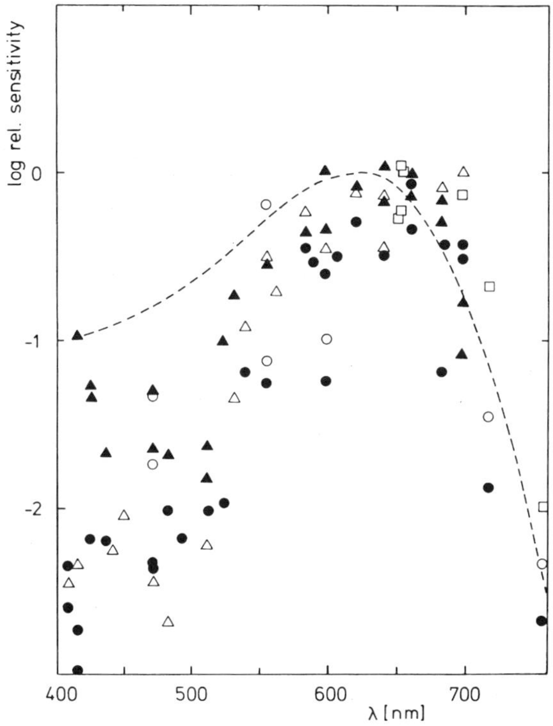

FIGURE 11.3. Action spectrum of the optomotor response in goldfish. (From Schaerer and Neumeyer, 1996.) Dashed line: L-cone spectral sensitivity. Symbols stand for different fishes tested.

2.3. Examining Color Vision with Retinal Neurons: A Neuropharmacological Approach

From all visual functions, color vision provides the unique opportunity to study the processing of color-specific information at different levels of the visual system. This can be done by comparing the spectral response properties of the photoreceptors with those of single neurons in the retina and brain, and with that of the entire system. However, the fact that a single neuron shows a "chromatic" response does not necessarily imply that this neuron is involved in color vision, because the photoreceptors always respond in a wavelength-specific way. Therefore, a simplistic comparison may be misleading. One way to assign a certain neuron type to color vision, or another visual function, seems to be a neuropharmacological approach in which the known effect of a drug at the cellular level is compared with the effect of the same drug in behavioral experiments (see below). A "direct" comparison can also be informative, as will be shown in the following section.

In the retina of cyprinid fishes, "color-opponent" responses are found at the level of horizontal, bipolar, and ganglion cells (see for review: Djamgoz and Yamada, 1990; Djamgoz et al., 1995; Kamermans and Spekreijse, 1999). It may not be the case that color-opponent neurons have a direct relationship to the perceptual property of "color opponency," but it may be assumed that they play an important role in color coding. Comparing the response maxima of biphasic (color-opponent) horizontal cells with the maxima of the behaviorally measured spectral sensitivity function (Fig. 11.3), one finds an interesting correspondence: In both cases the maxima in the long-wavelength range are shifted toward longer wavelengths in comparison to the maximum of the L-cone (from about 620nm to 660–680nm). This shift was assumed to be due to an unknown "far-red" photopigment by Naka and Rushton (1966). However, the shift can also be explained by an inhibitory action of the M-cone type on the L-cone type as shown in a theoretical paper by Sirovich and Abramov (1977). This inhibition has also been experimentally demonstrated in behavioral measurements of spectral sensitivity under chromatic adaptation to green light (Neumeyer, 1984).

A further hint that the spectral response characteristics of color-opponent neurons at a very peripheral level may have an influence on the properties of color vision was inferred from our experiments in a mesopic state of adaptation (Neumeyer and Arnold, 1989). Under a white-room illumination of 1 lx we found that the L-cone type does not contribute to color vision, the goldfish behaved as if red-green blind. This has a parallel at the cellular level of the retina: R/G-ganglion cells lose their opponent characteristics, and biphasic (H2) horizontal cells become monophasic at the transition from light- to dark-adaptation (Raynold et al., 1979; Weiler and Wagner, 1984). These functional changes are accompanied by morphological changes of the dendritic endings of horizontals cells, which indicate a remarkable plasticity of the wiring in the retina: Spinules are formed in the light-adapted state, and are redrawn during dark-adaptation (Wagner, 1980).

Spinule formation and horizontal cell response behavior seem to be controlled by dopamine released from interplexiform cells in the retina. The nonspecific dopamine antagonist haloperidol affects the horizontal cells in the same way as the transition from light to dark-adaptation: Biphasic horizontal cell responses become monophasic after application of the drug (Weiler et al., 1988). This fact was used in a neuropharmacological approach. In order to see whether dopamine has an effect on color vision, haloperidol was injected into the vitreous of the eyes of goldfish trained to discriminate colors in the mid- and long-wavelength range. It was shown that the fishes behaved as if they were red-green color-blind for about one hour (Mora-Ferrer and Neumeyer, 1996). The effect was specific for this spectral range. Wavelength discrimination in the blue-green and violet part of the spectrum was not affected after haloperidol treatment. Further investigations with specific dopamine antagonists indicated that red-green discrimination depends specifically on the D1-dopamine receptors (Mora-Ferrer and Neumeyer, 1996). Immunocytochemistry was used to localize the site of dopamine action in the retina. The stain was found in the outer as well as in the inner plexiform layers of the goldfish retina (Mora-Ferrer et al., 1999). Thus, it can be concluded that horizontal cells, but probably also amacrine and ganglion cells, are involved in red-green discrimination.

Furthermore, it could be shown that D1-receptors are important specifically in the context of color vision, but not in the context of motion vision measured in the optomotor response (Mora-Ferrer and Gangluff, 2000) and not in temporal resolution (Gangluff, 2000). However, these visual functions are sensitive to a blockade of D2-dopamine receptors. Sulpiride, a specific D2-receptor antagonist, did reduce sensitivity in the optomotor response, and lowered flicker fusion frequency. It did not affect wavelength discrimination. The location of D2-receptors in goldfish retina is presently under investigation.

Thus, the effect of dopamine in the neuropharmacological experiments indicate, at least for the L-cone contribution, two parallel ways of processing: color on the one hand, and motion and flicker detection on the other. The training experiments under mesopic light conditions (Neumeyer et al., 1991) and neuropharmacological investigations with the tuberculostatic drug ethambutol (Spekreijse et al., 1991) confirmed and extended this view. Color and spatial resolution are affected by mesopic light conditions, by ethambutol, and by dopamine via D1-receptors; whereas brightness, motion, and flicker detection are not. Electrophysiological recordings from ganglion cells under different states of adaptation let us assume that there are two types of color-opponent ganglion cells with L-cone input (Neumeyer et al., 1991). Parallel processing of color and spatial resolution on the one hand and brightness, motion, and temporal resolution on the other hand are remarkable parallels to the parvo- and magnocellular pathways in primate vision (Zeki, 1993).

2.4. Color Constancy and Color Contrast

The functional significance of a highly developed color vision system as found in goldfish, and most probably in other teleost fishes, is to recognize objects on the basis of color. However, the light reflected by an object is determined not only by the spectral absorbance of its surface but also by the spectral radiant flux of the illumination. As the spectral properties of natural daylight are continuously changing, depending on various factors such as the time of the day and part of the sky (Henderson, 1977), the spectral composition of light reflected by an object will change as well. Therefore, the cone types will be stimulated very differently by the same object under different illumination conditions and this will cause different excitation ratios. Without color constancy, a mechanism compensating for these changes, an object would appear in very different hues and could not be recognized on the basis of color, so that color vision would be useless. In fishes, the visual system has to cope not only with the spectral changes of daylight but also with changes due to water quality and depth (Kirk, 1983), and, therefore, color constancy should be especially important.

In fact, all animals in which the influence of illumination on color vision has been investigated behaved in a color-constant way (for review see Neumeyer, 1998). In fishes, as mentioned above, Burkamp (1923) has previously shown that different cyprinids are able to recognize their training color in tests under differently colored illuminations. Color constancy was also demonstrated in carp (*Cyprinus carpio*) by Dimentman et al. (1972). Most famous is an experiment in goldfish by Ingle (1985). Here goldfish were trained to select a yellow (or in one case a green) field among other gray and colored fields in a Mondrian configuration under white illumination. In the tests, the Mondrian was illuminated in such a way that the yellow field reflected the same amount of red, green, and blue light as the gray field in the training situation. Despite this fact, the goldfish selected the yellow field. However, from this experiment it can be concluded only that the fishes recognized their training color. It cannot be inferred that the yellow field was seen as exactly the same hue as in the training situation, or whether it was chosen just because it was the field that appeared to be most similar. Therefore, a more critical test for color constancy is to present a test field under colored illumination that stimulates the cone types in the same way as the training test field under white illumination. If the goldfish does *not* choose this test field but still chooses the training test field, color constancy in its strict sense is proven.

To investigate color constancy quantitatively and to show its limits, a method was applied that was used in earlier experiments with honeybees (Neumeyer, 1981). The trick in these experiments is to use a series of 10 to 15 test fields gradually ranging in hue (for example) from yellow through gray to blue. The training on a medium hue (gray for a human observer) is performed under white illumination. In the tests for color constancy, the illumination is changed to yellow or blue, respectively. Objectively, under the blue illumination, all test fields reflect more blue light, so that the training test field should appear more bluish. Therefore, without color constancy, the fish should *not* select the training test field but one of the yellowish test fields. Because of the small hue steps, there should always be a test field that reflects light in such a way that it stimulates the cone types in the same ratio as the training test field under white illumination. If the fish *does not* choose this test field but still prefers the training test field, the effectiveness of the color-constancy mechanism is shown. The choice of an intermediate test field indicates imperfect color constancy.

In the first of a series of experiments, color constancy was investigated for illumination colors between yellow and blue (Dörr and Neumeyer, 1996, 2000). These colors were chosen because they correspond to the main changes of natural daylight that are located on Planck's locus in color space representing color temperatures between 2,700 and 40,000°K. In the pretraining phase of the experiment, the goldfish had to learn to select the training test field among 12–14 slightly different small test fields, randomly distributed on a black background presented outside the tank and very close to its back wall. Correct choices were rewarded with a small amount of food paste given through a thin tube attached to a plexiglass stick. After a training period of several months under white illumination, the goldfish were tested under yellow and blue illuminations, each of different saturation. After each test, the colored illumination was exchanged with the white training illumination, and rewards were given after correct choices. The results indicate perfect color constancy under the yellow of medium saturation (yellow2) and the blue illumination of the lowest saturation (blue1). Color constancy was not perfect under more saturated illuminations (Fig. 11.5). Here, the choice frequency shifted from the training test field to neighboring test fields, indicating that other test fields appeared to the goldfish in the same hue as the training test field under white illumination. Using color metrics the effectiveness of the color-constancy mechanisms was inferred (Fig. 11.6).

Color constancy was also investigated with red/green illumination colors, located in color space perpendicular to Planck's locus (Fritsch, 1996; Neumeyer et al., 2002). Ten test fields were used, ranging in hue from red through gray to green. Color constancy was perfect only for the least-saturated illuminations when the test fields were presented on a *black* background. On a *gray* background, however, color constancy was also perfect under illumination colors of medium saturation. When using a *white* background, we found an interesting phenomenon: Illuminated with red light, the goldfish did not choose the training test field or one of the more green test fields as expected, but selected test fields that were more red! The effect was the same as in the case of test fields surrounded by a large red field, as found in experiments demonstrating simultaneous color contrast (Dörr and Neumeyer, 1997). Thus, the result with a white background seems to reflect color contrast and an overcompensation of the color-constancy mechanism. It indicates that color constancy requires a balance in lightness between test fields and surround, and that the lightness of surrounding areas may be important.

When varying the size of the background by surrounding each test field with a white annulus (presented on a black background) we found that a certain width was sufficient to yield the same effect as a large white background. Corresponding results were obtained with black annuli and a white background (Neumeyer et al., 2002). The ratio between surround width and test field diameter was about 1:1, and was, thus, very similar to the ratio found to produce maximal simultaneous color contrast (Dörr and Neumeyer, 1997). The effect of background size and lightness on color constancy in goldfish indicates the importance of lateral neural interactions, which may be one of the underlying mechanisms.

The overall effect of color constancy can be described by a von Kries mechanism of

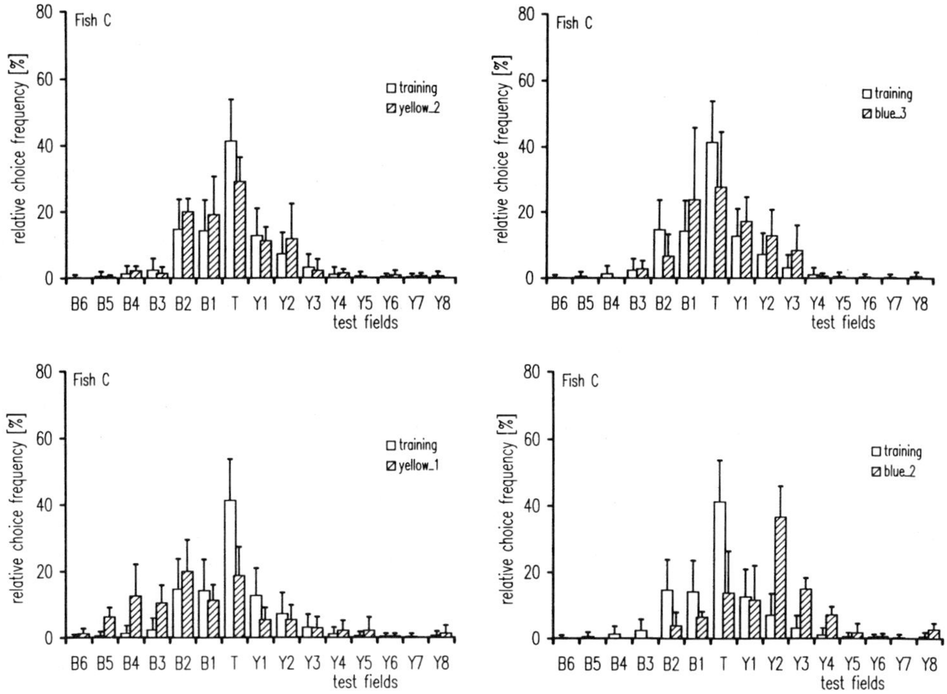

FIGURE 11.5. Choice behavior of one of two trained goldfish tested under colored illumination. (From Dörr and Neumeyer, 2000.) *Abscissa*: test field colors (T: training test field, gray for a human observer; B5–B1: blue test fields of decreasing saturation; Y1–Y8: yellow test fields of increasing saturation). *White columns*: training result under tungsten white illumination. *Hatched columns*: test results under colored illumination. *Left panel*: test results under yellow illumination (yellow1: most saturated illumination color); *right panel*: test results under blue illumination (blue1: slightest saturation; blue2: medium saturation).

selective chromatic adaptation (Dörr and Neumeyer, 2000, Fig. 11). However, it is not possible to assign the selective sensitivity reduction to a specific site in the visual pathway. At the most peripheral level, adaptation processes in each of the four cone types are certainly important. However, lateral interactions within the cone pedicles mediated by the horizontal cell network may be important in color constancy as well. This was shown by Kamermans et al. (1998) in model computations based on detailed studies of the horizontal cell responses (Kraaij et al., 1998; Kamermans and Spekreijse, 1999). "Double-opponent" cells found at the level of bipolar and ganglion cells in cyprinid retina can be also involved in color constancy and simultaneous color contrast as shown in earlier investigations (Daw, 1967; Maximova, 1977; Kaneko and Tachibana, 1983). In primates double-opponent cells, which are discussed in the context of color constancy, are found only in the visual cortex.

2.5. Color Perception

Color vision, as it is realized in goldfish and probably in other cyprinid fishes, is highly effective and in many respects similar to human color vision. It is more complicated as it is tetrachromatic and based on four different cone types. The consequences of tetrachromacy in color perception remain to be analyzed. For example, it is not known whether cyprinid fishes perceive light containing ultraviolet radiation, which stimulates all cone types in about equal strength, as "white" or "neutral." If so, it is an interesting question as to how light without

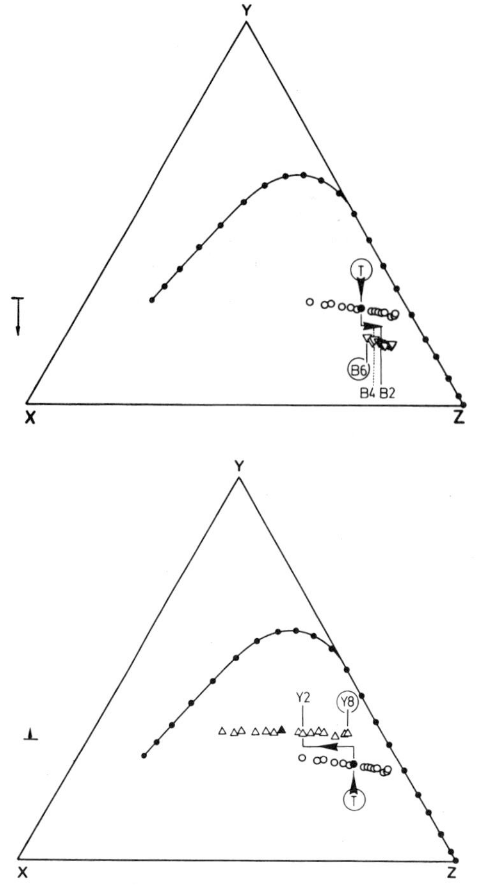

FIGURE 11.6. Color loci of the test field colors calculated for white training illumination (circles in both plots), for illumination yellow1 (above, triangles), and for illumination blue2 (below, triangles). The loci are shown in the basis triangle of the color tetrahedron (Fig. 11.2), as the colors used involved the ultraviolet range only slightly. The yellow illumination shifts all test field loci to the yz-side of the triangle. The test field located next to the training test field under white illumination is now test field B6. As shown in Figure 11.5, this test field is almost never chosen. Instead, test field B2 is preferred by this fish. Thus, the shift from T to B2 (arrow) is being compensated by the color-constancy mechanism. Analogously, under illumination blue2 (below), the shift from T to Y2 is also compensated. (From Dörr and Neumeyer 2000.)

ultraviolet will be perceived, which stimulates three of the four cone types equally strongly and which is "white" for a human observer. Is it a "ternary" color as proposed by Thompson et al. (1992), that is a perceptual compound of three colors? Such a color would be simultaneously "red," "green," and "blue"—unthinkable for humans. Such ternary colors should occur not only at the base triangle of the tetrahedron (Fig. 11.2) but also at the other three sides.

Another open question of color perception deals with the problem of color categorization. Does the fish brain organize the high number of discriminable colors in a small number of groups analogous to the human color categories? Is the grouping of colors independent of discrimination ability? The results of many training experiments and transfer tests indicate that there is a grouping of spectral colors, which is, however, not entirely independent of wavelength-discrimination ability (Kitschmann, 1999). It is not yet clear whether this is the consequence of a very prolonged training on a single wavelength, or whether color categories in a more strict sense do not exist. However, it is certain that generalization tests previously used to show color categories in goldfish (Goldman et al., 1991) give results that can be entirely explained with discrimination ability, and do not reflect categorization.

Thus, the open questions concerning fish color vision deal with color perception in its more strict sense. The properties of color vision to be investigated in threshold measurements, additive color mixture, and changes of surround and illumination are, at least for the goldfish, well established and in principle similar to the color vision systems of honeybees and humans.

Acknowledgments. I am very grateful to Susan Pinnells for correcting the English of the manuscript, and to the editors for improving its readability. The goldfish experimental work was supported by DFG (Ne 215/3 to 10) and the Human Frontier Science Program.

References

Arnold, K., and Neumeyer, C. (1987). Wavelength discrimination in the turtle *Pseudemys scripta elegans*. *Vision Res.* 27:1501–1511.

Avery, J.A., Bowmaker, J.K., Djamgoz, M.B.A., and Downing, J.E.G. (1983). Ultraviolet sensitive receptors in a freshwater fish. *J. Physiol. Lond.* 334:23.

Beauchamp, R.D., and Rowe, J.S. (1977). Goldfish spectral sensitivity: A conditioned heart rate measure in restrained or curarized fish. *Vision Res.* 17:617–624.

Bowmaker, J.K., Thorpe, A., and Douglas, R.H. (1991). Ultraviolet sensitive cones in the goldfish. *Vision Res.* 31:349–352.

Burkamp, W. (1923). Versuche über das Farbenwiedererkennen der Fische. *Zeitschrift für Sinnesphysiologie* 5:133–170.

Cronly-Dillon, J.R., and Muntz, W.R.A. (1965). The spectral sensitivity of the goldfish and the clawed tadpole under photopic conditions. *J. Exp. Biol.* 42:481–493.

Daw, N.W. (1967). Goldfish retina: Organization for simultaneous color contrast. *Science* 158:942–944.

Dimentman, A.M., Karas, A.Y., Maximov, V.V., and Orlov, O.Y. (1972). Constancy of object color perception in the carp (*Cyprinus carpio*). *Pavlov J. Higher Nervous Activity* 22/4:772–779 (in Russian; for a description of this experiment see Neumeyer (1998). pp. 332–335).

Djamgoz, M.B.A., and Yamada, M. (1990). Electrophysiological characteristics of retinal neurons: Synaptic interactions and functional outputs. In: *The Visual System of Fish.* (Douglas, R.H., and Djamgoz, M.B.A., eds.), Chapter 6, pp. 159–210. London: Chapman & Hall.

Djamgoz, M.B.A., Wagner, H.-J., and Witkovsky, P. (1995). Photoreceptor-horizontal cell connectivity, synaptic transmission and neuromodulation. In: *Neurobiology and Clinical Aspects of the Outer Retina* (Djamgoz, M.B.A., Archer, S.N., and Vallerga, S., eds.), Chapter 7, pp. 155–193. London: Chapman & Hall.

Dörr, S., and Neumeyer, C. (1996). The goldfish: A color-constant animal. *Perception* 25:243–250.

Dörr, S., and Neumeyer, C. (1997). Simultaneous color contrast in goldfish: A quantitative study. *Vision Res.* 37:1581–1593.

Dörr, S., and Neumeyer, C. (2000). Color-constancy in goldfish: the limits. *J. Comp. Physiol. A.* 186: 885–896.

Douglas, R.H. (1986). Photopic spectral sensitivity of a teleost fish, the roach (*Rutilus rutilus*), with special reference to its ultraviolet sensitivity. *J. Comp. Physiol. A.* 159:415–421.

Douglas, R.H., and Hawryshyn, C.W. (1990). Behavioural studies of fish vision: An analysis of visual capabilities. In: *The Visual System of Fish.* (Douglas, R.H., and Djamgoz, M.B.A., eds.), Chapter 11, pp. 373–418. London: Chapman & Hall.

Fratzer, C., Dörr, S., and Neumeyer, C. (1994). Wavelength discrimination of the goldfish in the ultraviolet spectral range. *Vision Res.* 34:1515–1520.

Frisch, K. von (1913). Weitere Untersuchungen über den Farbensinn der Fische. *Zool. Jahrb. Allg. Zool. Physiol.* 34:43–68.

Fritsch, J. (1996). Farbkonstanzuntersuchung im Grün-Purpur-Bereich am Goldfisch. Master-Thesis, Gutenberg-Universität, Mainz.

Gangluff, V. (2000). Die Bedeutung von Dopamin für das Bewegungssehen und das zeitliche und das räumliche Auflösungsvermögen des Goldfisches. Thesis, Gutenberg-Universität, Mainz.

Goldman, M., Lanson, R., and Rivera, G. (1991). Wavelength categorization by goldfish (*Carassius auratus*). *Intern. J. Comp. Psychol.* 4:195–206.

Hárosi, F.I. (1976). Spectral relations of cone photopigments in goldfish. *J. Gen. Physiol.* 68:65–80.

Hárosi, F.I., and Hashimoto, Y. (1983). Ultraviolet visual pigment in a vertebrate: A tetrachromatic cone system in the dace. *Science* 222:1021–1023.

Hawryshyn, C.W., and Beauchamp, R. (1985). Ultraviolet photosensitivity in goldfish: An independent u.v. retinal mechanism. *Vision Res.* 25:11–20.

Helversen, O. von (1972). Zur spektralen Unterschiedsempfindlichkeit der Honigbiene. *J. Comp. Physiol.* 80:439–472.

Henderson, S.T. (1977). *Daylight and its Spectrum*, 2nd ed. Bristol: Adam Hilger.

Ingle, D.J. (1985). The goldfish as retinex animal. *Science* 227:651–654.

Jacobs, G.H. (1993). The distribution and nature of colour vision among the mammals. *Biol. Rev.* 68:413–471.

Kamermans, M., and Spekreijse, H. (1999). The feedback pathway from horizontal cells to cones: A mini-review with a look ahead. *Vision Res.* 39:2449–2468.

Kamermans, M., Kraaij, D.A., and Spekreijse, H. (1998). The cone/horizontal cell system: A possible site for color constancy. *Vis. Neurosc.* 15:787–797.

Kaneko, A., and Tachibana, M. (1983). Double colour opponent receptive fields of carp bipolar cells. *Vision Res.* 23:381–388.

Kirk, J.T.O. (1983). *Light and Photosynthesis in Aquatic Ecosystems.* Cambridge: Cambridge University Press.

Kitschmann, M. (1999). Verhaltensphysiologische Untersuchung zur Kategorisierung von Spektralfarben beim Goldfisch (*Carassius auratus*) mit einer vergleichenden Studie am Farbensehen des Menschen. Thesis, Gutenberg-Universität, Mainz.

Kraaij, D.A., Kamermans, M., and Spekreijse, H. (1998). Spectral sensitivity of the feedback signal from horizontal cells to cones in goldfish retina. *Vis. Neurosci.* 15:799–808.

Krauss, A. (2001). Die Wellenlängenabhängigkeit des Bewegungssehens beim Zebrabärbling (*Danio rerio*) gemessen mit der optomotorischen Reaktion. Thesis, Joh. Gutenberg-Universität, Mainz.

Kuhn, F. (1935). *Der kleine Goldfischteich.* Leipzig: Insel-Verlag.

Marks, W.B. (1965). Visual pigments of single goldfish cones. *J. Physiol.* 178:14–32.

Maximova, E.M. (1977). Cellular mechanisms of color constancy. *Act. Nerv. Sup. (Praha)* 19:199–210.

Mora-Ferrer, C., and Gangluff, V. (2000). D2-dopamine receptor blockade impairs motion detection in goldfish. *Vis. Neurosci.* 17:177–186.

Mora-Ferrer, C., and Neumeyer, C. (1996). Reduction of red-green discrimination by dopamine D1 receptor antagonists and retinal dopamine depletion. *Vision Res.* 36:4035–4044.

Mora-Ferrer, C., Yazulla, S., Studholme, K.M., and Haak-Frendscho, M. (1999). Dopamine D1-receptor immunolocalization in goldfish retina. *J. Comp. Neurology* 4:705–714.

Naka, K.I., and Rushton, W.A.H. (1966). S-potentials from colour units in the retina of fish (Cyprinidae). *J. Physiol.* 185:536–555.

Neumeyer, C. (1981). Chromatic adaptation in the honeybee: Successive color contrast and color constancy. *J. Comp. Physiol.* 144:543–553.

Neumeyer, C. (1984). On spectral sensitivity in the goldfish: Evidence for neural interactions between different "cone mechanisms." *Vision Res.* 24:1223–1231.

Neumeyer, C. (1985). An ultraviolet receptor as a fourth receptor type in goldfish color vision. Naturwissenschaften 72:162–163.

Neumeyer, C. (1986). Wavelength discrimination in the goldfish. *J. Comp. Physiol. A.* 158:203–213.

Neumeyer, C. (1991). Evolution of colour vision. In: *Vision and Visual Dysfunction, Vol. 2.* (Cronly-Dillon, J., ed.), pp. 284–305. Houndmills and London: Macmillan Press.

Neumeyer, C. (1992). Tetrachromatic color vision in goldfish: Evidence by color mixture experiments. *J. Comp. Physiol. A.* 171:639–649.

Neumeyer, C. (1998). Comparative aspects of color constancy. In: *Perceptual Constancy: Why Things Look as They Do.* (Walsh, V., and Kulikowski, J., eds.), pp. 323-351. Cambridge: Cambridge University Press.

Neumeyer, C., and Arnold, K. (1989). The tetrachromatic color vision in the goldfish becomes trichromatic under white adaptation light of moderate intensity. *Vision Res.* 29:1719–1727.

Neumeyer, C., Wietsma, J.J., and Spekreijse, H. (1991). Separate processing of "color" and "brightness" in goldfish. *Vision Res.* 31:537–549.

Neumeyer, C., Dörr, S., Fritsch, J., and Kardelky, C. (2002). Color-constancy in goldfish and man: Influence of surround size and lightness. *Perception* 31:171–187.

Palacios, A., Varela, F.J., Srivastava, R., and Goldsmith, T.H. (1998). Spectral sensitivity of cones in the goldfish (*Carassius auratus*). *Vision Res.* 38:2135–2146.

Powers, M.K. (1978). Light-adapted spectral sensitivity of the goldfish: A reflex measure. *Vision Res.* 18:1131–1136.

Przyrembel, C., Keller, B., and Neumeyer, C. (1995). Trichromatic color vision in the salamander (*Salamandra salamandra*). *J. Comp. Physiol. A.* 176:575–586.

Raynold, J.P., Laviolette, J.R., and Wagner, H.-J. (1979). Goldfish retina: A correlate between cone activity and morphology of the horizontal cell in cone pedicles. *Science* 204:1436–1438.

Schaerer, S. (1990). Das zeitliche Auflösungsvermögen des Sehsystems beim Goldfisch und seine Abhängigkeit von der Wellenlänge. Master-Thesis, Gutenberg-Universität, Mainz.

Schaerer, S., and Neumeyer, C. (1996). Motion detection in goldfish investigated with the optomotor response is "color blind." *Vision Res.* 36:4025–4035.

Schiemenz, F. (1924). Über den Farbensinn der Fische. *Z. Vergl. Tierphysiol.* 1:175–220.

Schlieper, C. (1927). Farbensinn der Tiere und optomotorische Reaktion. *Z. Vergl. Tierphysiol.* 6:453–472.

Sirovich, L., and Abramov, I. (1977). Photopigments and pseudopigments. *Vision Res.* 17:5–16.

Spekreijse, H., Wietsma, J.J., and Neumeyer, C. (1991). Induced color blindness in goldfish: A behavioral and electrophysiological study. *Vision Res.* 31:551–562.

Sperling, H.G., and Harwerth, R.S. (1971). Red-green cone interactions in the increment threshold spectral sensitivity of primates. *Science* 172:180–184.

Svaetichin, G. (1956). Spectral response curves from single cones. *Acta Physiol. Scand.* 39 (Supp.) 134: 17–46.

Svaetichin, G., and MacNichol, E.F. (1958). Retinal mechanisms for chromatic and achromatic vision. *Ann. NY Acad. Sci.* 74:385–404.

Thompson, E., Palacios, A., and Varela, F.J. (1992). Ways of coloring. *Behav. Brain Sci.* 15:1–74.

Tomita, T. (1963). Electrical activity in the vertebrate retina. *J. Opt. Soc. Amer.* 53:49–57.

Wagner, H.-J. (1980). Light-dependent plasticity of the morphology of horizontal cell terminals in cone pedicles of fish retina. *J. Neurocyt.* 9:573–590.

Weiler, R., and Wagner, H.-J. (1984). Light-dependent change of cone-horizontal cell interactions in carp retina. *Brain Res.* 298:1–9.

Weiler, R., Kohler, K., Kolbinger, W., Wolburg, H., Kurz-Isler, G., and Wagner, H.-J. (1988). Dopaminergic neuromodulation in the retinas of lower vertebrates. *Neurosci. Res. Suppl.* 8: S183–S196.

Wolff, H. (1925). Das Farbenunterscheidungsvermögen der Elritze. *Z. Vergl. Tierphysiol.* 3:279–329.

Yager, D. (1967). Behavioral measures and theoretical analysis of spectral sensitivity and spectral saturation in the goldfish, *Carassius auratus*. *Vision Res.* 7:707–727.

Yager, D. (1969). Behavioral measures of spectral sensitivity in the goldfish following chromatic adaptation. *Vision Res.* 9:179–186.

Yarczower, M., and Bitterman, M.E. (1965). *Stimulus Generalization in the Goldfish.* In: *Stimulus generalization* (Mostovsky, D.J., ed.). Stanford: Stanford University Press.

Zeki, S. (1993). *A Vision of the Brain.* Oxford: Blackwell Scientific Publications.

12
Chemically Mediated Strategies to Counter Predation

Brian D. Wisenden

Abstract

Predator–prey interactions govern the evolution of many behavioral and morphological traits of aquatic animals. In aquatic environments, chemical cues reliably allow prey to assess and avoid predation risk. In this chapter, I review the classes of chemical cues involved in a predation event and ways in which these cues mediate predator–prey interactions. Predators release signature odors that prey use to detect risk of predation. Prey release several types of cues. Chemical deterrents are noxious substances that are either synthesized *de novo* or acquired from the diet. Disturbance cues are released by startled but noninjured prey. Alarm cues are chemicals released by damaged tissue injured by a predator's attack, or after passage through the digestive system of a predator (dietary alarm cues). Prey species are adept at detecting the odor of predators. Experience plays a role in allowing prey to learn to associate risk with visual and chemical correlates of predation risk. Prey respond to chemical cues of predation risk by adopting antipredator behaviors, shifting foraging and reproductive behaviors in a risk-sensitive fashion, producing morphological defences, and by shifting life-history traits. Future work should be directed at a more precise knowledge of the mechanisms of chemical communication (the chemical nature of the cues and the physiology of olfactory receptors) and better understanding of the role of chemical cues in shaping multispecies ecosystems, preferably from field experiments.

1. Introduction

Predation is the dominant agent of natural selection over evolutionary time (Lima and Dill, 1990). Consequently, extant organisms exhibit finely tuned mechanisms for the detection and avoidance of predation risk. Any source of information that confers on prey the smallest modicum of risk reduction will be promoted by natural selection. This chapter examines waterborne chemicals and their use as chemical cues.

2. Classes of Chemical Cues

Chemical cues are the most ancient and ubiquitous source of environmental information. Water, being the universal solvent, is ideal for the solution and dispersal of chemical information. Chemical cues persist longer than other modes of information, reveal the presence of predators concealed by structure, camouflage, turbidity, or darkness, and reveal the diet of novel predators. Published examples of chemically mediated predator–prey interactions involve virtually every major taxon found in marine and freshwater environments (Table 12.1).

Predation events proceed from detection to attack, capture, and finally, ingestion (Lima and Dill, 1990; Smith, 1992) and chemical cues are released at each stage of this predation sequence (Wisenden, 2000). The following sections detail five general classes of chemicals released by prey during a predatory event (Fig. 12.1).

2.1. Chemical Deterrents

Prey can discourage predatory attack directly with noxious chemicals. This strategy is best studied in marine species: sponges (e.g., Wilson et al., 1999; Waddell and Pawlik, 2000), gorgonian cnidarians (e.g., Harvell et al., 1996; Epifanio et al., 1999), and various marine macroalgae (e.g., Bolser and Hay, 1996; Powlik et al., 1997; van Alstyne et al., 1999; Stachowicz and Hay, 1999). Nudibranchs that forage on sponges overcome these chemicals and sequester them in their epidermis to deter their own predators (e.g., Marin et al., 1998; Ebel et al., 1999). I know of only one recent example of chemical deterrents in fishes (Tachibana and Gruber, 1988). Sole (*Pardachirus* spp.) produce a peptidic ichthyotoxin, coined Pardaxin, and additional lipophilic steroid glycosides that effectively repel sharks. Examples of behavioral acquisition of chemical deterrents are known for hermit crabs that place anemones on their back (e.g., Brooks, 1989) and for amphipods that carry noxious sea butterflies (Mollusca: Pteropoda) (McClintock and Janssen, 1990). In freshwater systems, larval amphibians (Kats et al., 1988; d'Heursel and Haddad, 1999; Hanifin et al., 1999) and insects (Scrimshaw and Kerfoot, 1987; White, 1989) deter predation by synthesizing noxious chemicals.

2.2. Disturbance Cues

Disturbance cues are released before prey injury at the detection stage of the predation sequence, that is, once the prey knows it is in trouble. Early warning leads to heightened vigilance and presumably reduced attack success by removing the element of surprise and by informing the predator that it has been detected. Disturbance cues have been demonstrated in crustaceans (Hazlett, 1985, 1990a; Zulandt Schneider and Moore, 2000), fishes such as darters (Wisenden et al., 1995a), sculpins (Bryer et al., 2001), catfish (Jordão and Volpato, 2000), frog tadpoles (Kiesecker et al., 1999), and salamanders (Hedberg, 1981; Mathis and Lancaster, 1998). Hazlett (1990b) provided experimental evidence that urinary ammonia is a likely candidate for the disturbance cue in crayfish. This hypothesis has since received supporting evidence in crayfish (Zulandt et al., 2000), darters (B.D. Wisenden, unpublished), and frog tadpoles (Kiesecker et al., 1999).

2.3. Injury-Released Chemical Alarm Cues

Following a successful attack, predators use teeth, claws, and other mechanisms to grasp prey and prevent escape. Aquatic taxa from protozoa to amphibians exhibit antipredator behavior in response to injury-released cues from conspecifics (Chivers and Smith, 1998), reflecting the historic and pervasive value of injury-released cues in risk assessment. This is clearly an important source of environmental information.

Avoidance of traps baited with alarm cues of conspecifics has been shown for field populations of fathead minnows (Mathis and Smith, 1992), dace (Chivers and Smith, 1994a), brook stickleback (Chivers and Smith, 1994a), guppies

TABLE 12.1. Cross tabulation by predator (column headings) and prey taxa (row headings) in chemically mediated interactions in freshwater (Fr) and marine (Mar) environments published between 1985 and 1999 inclusive.

Prey Taxon	Hab.	Amphibia	Fish	Crustacea	Insecta	Gastropoda	Cephalopod	Echinoderm	Cnidaria	Protozoa	Snake	Mammal	Bird	Misc.	Total
Amphibia	Fr	15	22	0	7	0	0	0	0	0	6	0	1	0	51
	Mar	0	0	0	0	0	0	0	0	0	0	0	0	0	0
Fish	Fr	0	37	0	1	0	0	0	0	0	2	0	1	0	41
	Mar	0	3	0	0	0	0	0	0	0	0	0	0	0	3
Crustacea	Fr	1	43	1	26	0	0	0	0	0	0	0	0	2	73
	Mar	0	10	5	0	0	3	0	1	0	0	0	0	0	19
Insecta	Fr	1	22	0	8	0	0	0	0	0	0	1	1	0	33
	Mar	0	0	0	0	0	0	0	0	0	0	0	0	0	0
Gastropoda	Fr	0	5	4	0	0	0	0	0	0	0	0	0	0	9
	Mar	0	18	8	0	3	0	11	2	0	0	0	0	0	42
Echinoderm	Fr	0	0	0	0	0	0	0	0	0	0	0	0	0	0
	Mar	0	3	1	0	0	0	0	0	0	0	0	0	0	4
Porifera	Fr	0	0	0	0	0	0	0	0	0	0	0	0	0	0
	Mar	0	18	0	0	4	0	6	0	0	0	0	0	0	28
Cnidaria	Fr	0	0	0	0	0	0	0	0	0	0	0	0	0	0
	Mar	0	21	0	0	0	2	3	0	0	0	0	0	0	26
Algae &	Fr	0	0	4	0	0	0	0	0	1	0	0	0	0	5
Vasc. plant	Mar	0	13	8	0	1	0	6	1	2	0	1	1	0	33
Protozoa	Fr	0	0	1	0	0	0	0	0	5	0	0	0	4	10
	Mar	0	0	0	0	0	0	0	0	0	0	0	0	0	0
Bivalvia	Fr	0	0	0	0	0	0	0	0	0	0	0	0	0	0
	Mar	0	2	1	0	2	0	2	0	0	0	1	1	0	9
Bacteria	Fr	0	0	0	0	0	0	0	0	0	0	0	0	0	0
	Mar	0	1	0	0	0	0	0	0	1	0	0	0	0	2
Bryozoa	Fr	0	0	0	0	0	0	0	0	0	0	0	0	0	0
	Mar	0	1	0	0	0	0	1	0	0	0	0	0	0	2
Ascidian	Fr	0	0	0	0	0	0	0	0	0	0	0	0	0	0
	Mar	0	2	0	0	1	0	0	1	0	0	0	0	0	4
Misc.	Fr	0	0	0	0	0	0	0	0	0	0	0	0	0	0
	Mar	0	2	0	0	0	0	0	0	0	0	0	0	0	2
Total	Fr	17	129	10	42	0	0	0	0	6	8	1	3	6	222
	Mar	0	94	23	0	11	5	29	5	3	0	2	2	0	174
	Total	17	223	33	42	11	5	29	5	9	8	3	5	6	396

Note: n = 403 articles retrieved from the Cambridge Scientific Abstracts database using the search criteria CHEMICAL and PREDATOR or PREY or PREDATION. Areas of emphasis differ slightly for marine and freshwater studies. The types of chemical cues studied in marine environments are represented by 69% deterrents, 10% alarm cues, and 21% kairomones, whereas those in freshwater environments are represented by 5% deterrents, 28% alarm cues, and 66% kairomones. Responses to chemical cues studied in marine environments were 96% behavioral, 3% morphological, and 1% life history shifts. In freshwater environments, only 76% of the studies (in the phrase 76% of the studies) were on behavioral responses, and 14% and 10% were on morphological and life historical responses, respectively.

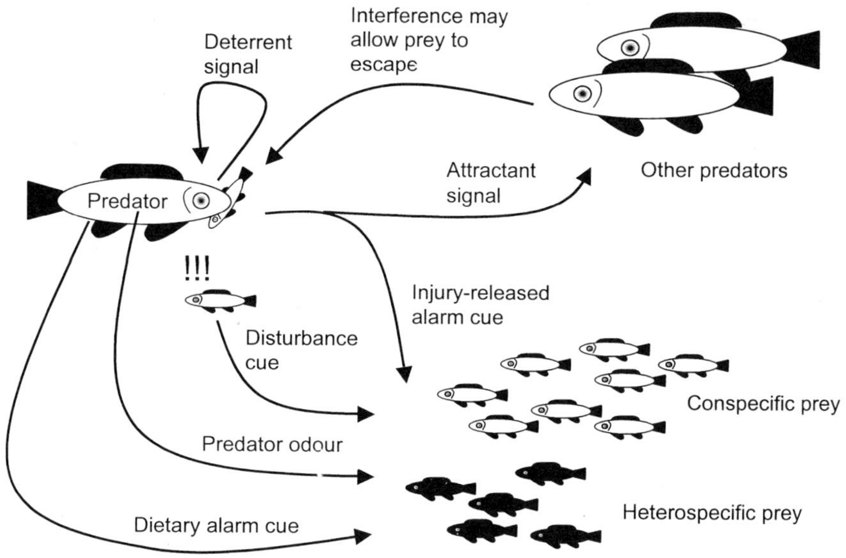

FIGURE 12.1. Schematic diagram of the classes of chemical cues that mediate predator–prey interactions in aquatic environments. Disturbance cues are released when prey are startled or frightened; alarm cues are released on injury by a predator. Alarm cues provide information about predation to conspecific and heterospecific prey with similar ecology (i.e., in the same prey guild) and to other predators looking to pirate a meal. Alarm cues are released at the time of attack and capture, and later when egested from the predator's digestive tract.

(Brown and Godin, 1999a), amphipod crustacea (Wisenden et al., 2001), and frog tadpoles (Adams and Claeson, 1998).

Avoidance of alarm cues of heterospecifics occurs among syntopic species that share predators; therefore cues from either species indicate danger to both. Cross-species reactions are known for marine mud snails (Snyder, 1967; Rahman et al., 2000), insects (Huryn and Chivers, 1999), insects and minnows (Wisenden et al., 1997), stickleback and minnows (Mathis and Smith, 1993a; Chivers and Smith, 1994a; Wisenden et al., 1995b), among stickleback species (Brown and Godin, 1997), among darter species (Smith, 1982; Commens and Mathis, 1999), minnows and darters (Chivers et al., 1995), among salmonid species (Mirza and Chivers, 2001), tadpoles (Adams and Claeson, 1998), salamanders (Lutterschmidt et al., 1994; Chivers et al., 1997), and newts (Marvin and Hutchison, 1995).

For the most part, injury-released chemical alarm cues are released inadvertently and are not specialized pheromones per se. An apparent exception is the alarm substance cell–alarm reaction system of ostariophysan fishes (minnows, suckers, catfish, characins). These species are generally small in size, occupy highly structured habitats, and form shoals with typical interindividual distances of several body lengths. These factors favor the reliance on chemical cues for risk assessment. Specialized epidermal club cells of ostariophysan fishes contain an alarm substance (Pfeiffer, 1977). The epidermis is the first tissue to encounter the grasping mechanisms of a predator. Ruptured cells release alarm substance into the surrounding waters. The release of a detectable substance into the water is detected by conspecific prey and predators alike. Alarm substance in minnow skin attracts predators to a predation event in progress (Mathis et al., 1995) and attempts of kleptoparasitism and/or cannibalism by late-arriving predators give the prey opportunity to escape (Chivers et al., 1996a) (Fig. 12.1). Thus, a selfish benefit accrues to the sender to offset the metabolic cost of these cells (Wisenden and Smith, 1997, 1998; Fig. 12.2).

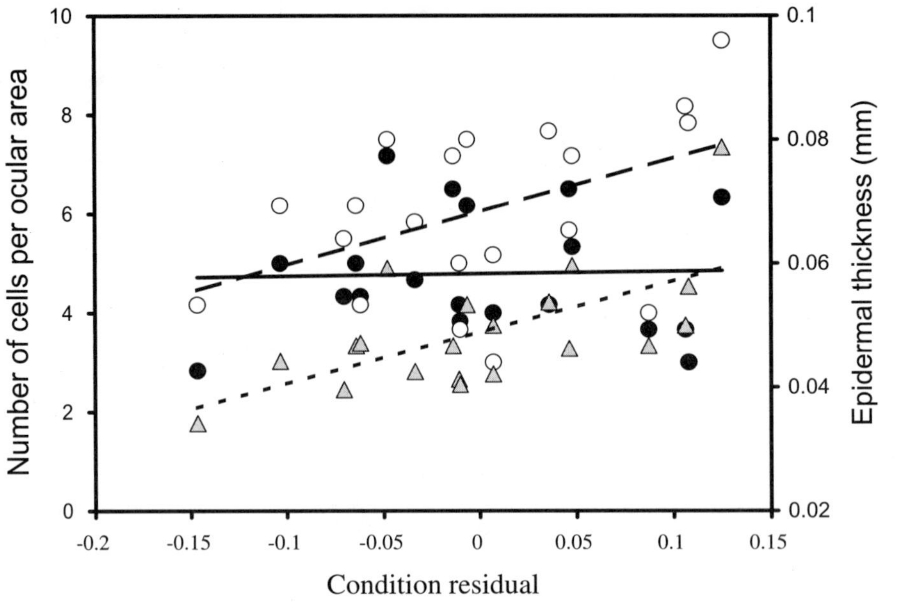

FIGURE 12.2. Number of alarm substance cells (open circles, long-dashed line), mucus cells (solid circles, solid line), and epidermal thickness (shaded triangles, small-dashed line) as a function of physical condition in fathead minnows. Condition was estimated by calculating the regression residuals about a fitted line for Ln (Weight) versus Ln (Length). Positive residuals indicate a fish that is relatively heavy for its length (good physical condition), negative residuals indicate a fish that is relatively light for its length (poor physical condition). As physical condition increased, epidermal thickness increased ($P < 0.05$) and fathead minnows increased investment into alarm substance cells (ASC) ($P < 0.05$), but did not change investment in mucus cells. Proliferation of ASCs in good-condition fishes indicates that the metabolic cost for ASCs is met only by individuals with abundant resources. (After Wisenden and Smith, 1997.)

It is difficult to dispute that injury-released cues from minnow skin, including club cells, play a role in mediating predator–prey interactions. Similarly, it is difficult to defend the assertion that alarm signaling must be the only or original function of these cells. Ostariophysan cells have been implicated in antipathogen defence (Chapman and Johnson, 1997; Poulin et al., 1999), wound healing (Al-Hassan et al., 1985), and absorbance of ultraviolet radiation (Lam et al., 1974; Blazer et al., 1997).

Epidermal club cells with similar staining characteristics to ostariophysan alarm substance cells occur in non-ostariophysans such as percids (darters: Smith, 1982; walleye: B.D. Wisenden, M. Rzaszutak, T. Overbo, and T. Thiel, unpublished), cottids (Hugie et al., 1991; Bryer et al., 2001), and poeciliids (Bryant, 1987) (Fig. 12.3).

Little is known about the chemical nature of alarm cues. Widespread use of peptides as a source of environmental information suggests proteins as a general starting point for alarm cue chemistry (Atema and Stenzler, 1977; Rittschof, 1990). Early work by Huttel and Sprengling (1943, as cited in Lebedeva et al., 1975) concluded that the alarm substance in cyprinids is a pterin similar to isoxanthopterin, with a molecular weight of 253. Subsequent study failed to demonstrate biological activity (Schutz, 1956, as cited in Lebedeva et al., 1975). Hypoxanthine 3(N) oxide was named as the putative active ingredient of ostariophysan alarm cells in an unpublished PhD dis-

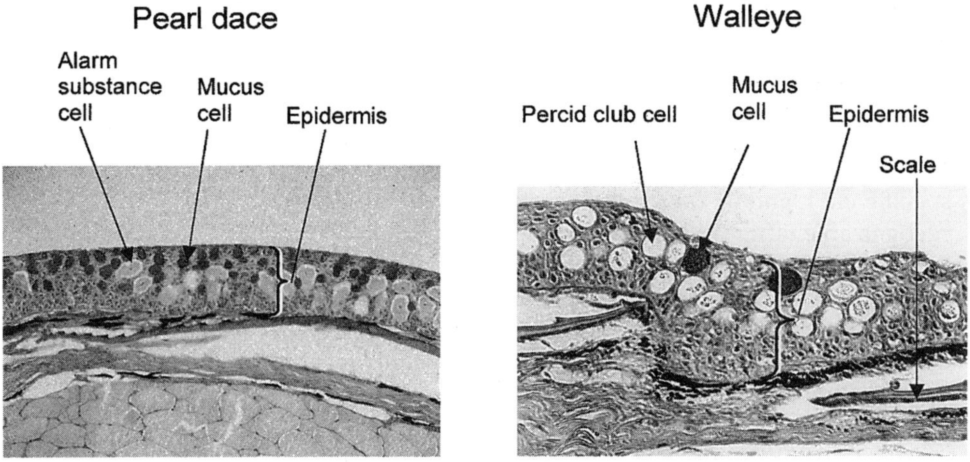

FIGURE 12.3. Histological sections of skin of a pearl dace (Ostariophysi, Cyprinidae: *Margariscus margarita* [photo by R.J.F. Smith]) and a 15-cm walleye (Acanthopterygii, Percidae: *Stizostedion vitreum* [photo by B.D. Wisenden]). The PAS stain adheres to carbohydrates. The presence of large club cells in the skin of Ostariophysans and Acanthopterygians suggests convergent evolution promoted by a common ecological selection gradient.

sertation by Argentini (1976, as cited in Smith, 1999). Subsequent experimental evidence using hypoxanthine 3(N) oxide confirm its biological activity in inducing alarm behavior (Pfeiffer et al., 1985; Brown et al., 2000; 2001). Other recent work showed that protein-reduced (heated) minnow skin extract loses its ability to invoke alarm behavior (N.L. Korpi, L.D. Louisiana, J.J. Provost, and B.D. Wisenden, unpublished). The involvement of a protein in minnow alarm cue is consistent with the observed decline in intensity of cross-species reactions among ostariophysans (Schutz, 1956, as cited in Smith, 1999) and among non-ostariophysan salmonids (Mirza and Chivers, 2001) with increasing phylogenetic distance.

2.4. Chemical Cues Released by Predators

Recognition of predator odors, or kairomones, is arguably the most prevalent mode by which chemical cues mediate predator–prey interactions (see Kats and Dill, 1998 for review). Kairomones are released continuously and independent of stage in the predation sequence.

The chemical nature of these cues is virtually unknown but notable exceptions exist. Stickleback and pike each release a freely dissociated molecule known to cause a diel vertical movement response in *Daphnia* (von Elert and Pohnert, 2000). Peptide-based kairomones are known for some species of amoebas (Kusch, 1999), ciliates (Wicklow, 1997), crustaceans (Rittschof, 1990), a mollusc (Rittschof, 1990), and amphibians (Lutterschimdt et al., 1994). Other kairomones do not seem to be peptide based. For example, the kairomone released by the *Daphnia* predator *Chaoborus* is resistant to a general peptidase (Parejko and Dodson, 1990). Heparin disaccharides evoke responses in marine gastropods similar to that of fish kairomones (Rahman et al., 2000). An alkaloid released by the seastar *Dermasterias imbricata* induces detachment and flight behavior in the anthozoid *Stomphia coccinea* (Elliott et al., 1989).

2.5. Cues from the Predator's Digestive Tract

Species-specific cues, and/or their metabolites, remain recognizable after passing through a predator's digestive tract (Chivers and Smith, 1998). Dietary cues are released at the postingestion stage of the predation sequence, but functionally, prey use dietary cues at the detection stage. Prey response to dietary alarm cues is known for amphibians (Wilson and Lefcort, 1993; Lefcort, 1996; Chivers et al., 1999; Madison et al., 1999; Wildy et al., 1999; Belden et al., 2000), fish (Gelowitz et al., 1993; Mathis and Smith, 1993b,c; Brown et al., 1995; Jachner, 1997; Godard et al., 1998; Hirvonen et al., 2000; Pettersson et al., 2000), aquatic insects (Chivers et al., 1996b; Huryn and Chivers, 1999), and crustaceans (Stirling, 1995; Mathis and Hoback, 1997; Jacobsen and Stabell, 1999). These studies show the importance of separating the effects of dietary alarm cues from those of kairomones in experiment design. Dietary cues are assumed to be chemically unchanged from the injury-released alarm cues, but this has not been verified, in large part because the chemistry of injury-released cues is so poorly understood.

3. Short-Term Behavioral Responses

Short-term behavioral responses ameliorate imminent threats of predation. Behavioral responses include increased vigilance, information gathering (predator inspection), and crypsis (reduced activity, movement out of the water column). Once the predator has been detected, and a predatory attack has begun or seems likely to begin, prey hide among densely structured objects or among conspecifics (shoaling) or flee the area. Flight may employ rapid and erratic swimming (skittering or dashing behavior) that makes it difficult for predators to visually isolate individual prey. The suite of behaviors evoked by chemical cues significantly reduces the probability of predation (Sih, 1986; Hews, 1988; Mathis and Smith, 1993d; Pijanowska, 1997; Stewart et al., 1999; Wisenden et al., 1999).

The duration of the response is usually matched to predator activity and may range from minutes to hours (e.g., Atema and Stenzler, 1977; Wisenden et al., 1995b), or even days (Dawidowicz et al., 1990; Ringelberg and van Gool, 1995; Christine et al., 1997; Spieler and Linsenmeir, 1999). The cost/benefit ratios influencing response duration have not been well explored.

Prey species in nature constantly trade-off foraging, risk avoidance, mate search, and other behaviors essential to life, and rarely engage in any one behavior without influence from others (Lima and Dill, 1990). Prey that are hungry (Smith, 1981; Vadas et al., 1994; Brown and Smith, 1996; Uiblein et al., 1996; Chivers et al., 2000; Gillette et al., 2000) or heavily parasitized (Jakobsen and Wedekind, 1998) do not respond overtly to chemical indicators of risk. When chemical cues indicate risk but visual verification of risk shows no signs of an approaching predator then an overt behavioral response may not occur (Magurran et al., 1996; Aabjörnsson et al., 1997; Hartman and Abrahams, 2000). Prey species vulnerable to visually oriented predators may respond to kairomones only during the day (Dawidowicz et al., 1990; Aabjörnsson et al., 1997; Cieri and Stearns, 1999; McIntosh and Peckarsky, 1999). Multifactor trade-offs involving chemical cues are probably more the norm than the exception in nature (Covich et al., 1994; Kerby and Kats, 1998; Huhta et al., 1999; Eklöv, 2000; Turner et al., 2000). This is clearly an area in which to direct future research.

4. Long-Term Responses

Chemical cues can induce life-historical and/or morphological changes in prey. Life history shifts occur in amphibians by way of altering time to hatching (Sih and Moore, 1993; Moore et al., 1996), increased time to, and size at, metamorphosis, representing a foraging cost of predator avoidance (Wildy et al., 1999;

Nicieza, 2000), or shorter time to maturation as a means to escape aquatic predators (Chivers et al., 1999). *Daphnia* life history responds to kairomones but different predator types invoke different responses. Larvae of the midge *Chaoborus* prey most effectively on small size classes of *Daphnia* and thus cause an increase in size at, and time to, maturation (Lüning, 1995; Tollrian, 1995; Repka and Pihlajamaa, 1996; Repka and Walls, 1998). Fishes selectively consume large *Daphnia*, thus, fish kairomones accelerate *Daphnia* maturation (Dawidowicz and Loose, 1992; Stibor, 1992; Reede, 1995), make smaller and more numerous neonates (Tollrian, 1994; Reede, 1995; Sakwinska, 1998), or stimulate formation of drought and freeze-tolerant (ephippial) eggs (Pijanoska and Stolpe, 1996).

Induced morphological responses are changes in shape or structure that render the prey species less vulnerable to predators (Tollrian and Harvell, 1999). Ciliates change structure within hours of exposure to kairomones (Kusch, 1993a; Wicklow, 1997) in order to reduce vulnerability to predation (Kusch, 1993b). *Daphnia* grow neck teeth in response to kairomones from *Chaoborus* (e.g., Dodson, 1989) and fishes (Tollrian, 1994) that deter predation (Dodson and Wagner, 1996). Induced morphology can be derived from maternal effects (Agrawal et al., 1999). *Daphnia* kairomones induce changes in the phytoplankton on which they graze (Lürling and Beekman, 1999). Grazing induces the synthesis of chemical deterrents in coral (e.g., Gochfeld, 1995) and *Fucus* (van Alstyne, 1988). Crucian carp (Brönmark and Miner, 1992; Stabell and Lwin, 1997) and goldfish (Mirza, 1998) exposed to alarm cues or kairomones change from the competitively superior elongate form (Holopainen et al., 1997; Pettersson and Brönmark, 1997) to a deep-bodied form. Greater body depth increases handling time for pike predators (Nilsson et al., 1995). Frog tadpoles change tail shape and body proportions to facilitate faster escape responses (van Buskirk and McCollum, 2000). Shell morphology in gastropods (Appleton and Palmer, 1988; DeWitt et al., 1999) and bivalves (Leonard et al., 1999) responds to chemical indicators of risk.

5. Learned Versus Genetic Antipredator Responses

Prey recognize predators by innate genetic programs and learned associations. Genetic programs confer a high speed of response but to a limited range of predator types. Learned associations confer great flexibility but require at least one naïve encounter with predator cues.

Genetically based responses to chemical cues occur in fishes (Miklósi et al., 1997; Hirvonen et al., 2000), amphibians (Griffiths et al., 1998; Laurila, 2000), insects (Huhta et al., 1999), and crustaceans (e.g., Schwartz, 1991; Appelberg et al., 1993; de Meester, 1993; Lüning, 1995). Prey that migrate from a location with predators to a location without predators may continue to exhibit maladaptive antipredator behavior (Storfer and Sih, 1998).

A number of species are known to associate novel stimuli with risk after a single simultaneous encounter of the novel stimulus and conspecific alarm cues (Chivers and Smith, 1998). Selection should favor learned recognition in habitats in which predator identity varies seasonally or spatially across populations. Acquired recognition of predator identity is evolutionarily ancient as demonstrated in a recent experiment with planaria (Wisenden and Millard, 2001; Fig. 12.4). Presentation of alarm cues and novel stimuli need not be simultaneous for learned associations to occur (Korpi and Wisenden, 2001). Naïve prey can acquire predator recognition by detecting dietary cues of conspecifics (Mathis and Smith, 1993c; Chivers et al., 1996b; Brown and Godin, 1999b). Similarly, prey come to recognize heterospecific alarm cues by detecting them among dietary cues of a predator that has fed on both prey species (R.S. Mirza, and D.P. Chivers, unpublished). So great is the fitness benefit for rapid acquisition of predator identity that fishes form aversive responses even to irrelevant and nonbiological stimuli (Suboski et al., 1990; Chivers and Smith, 1994b; Hall and Suboski, 1995; Yunker et al., 1999). One way that fishes limit stimuli to which they associate danger is to key in on motion (Wisenden and Harter, 2001), because motion is a reliable indicator of predator identity.

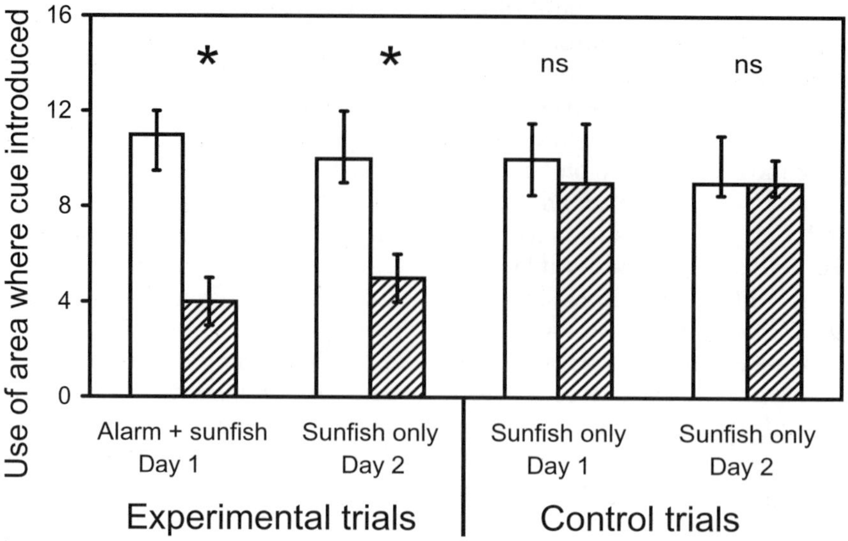

FIGURE 12.4. Median (±25 percentiles) number of times planaria (*Dugesia dorotocephala*) were observed in the test area before (open bars) and after (hatched bars) the addition of test stimuli. On Day 1 flatworms received sunfish odor plus alarm cues from injured conspecific planaria (experimental trials) or sunfish odor alone (control trials). All planaria were placed in a clean dish with fresh water and retested the next day (Day 2) with sunfish odor only. Planaria learned to associate predation risk with sunfish odor after a single simultaneous presentation of the two stimuli. Asterisk indicates $P < 0.05$, ns indicates $P > 0.05$. (After Wisenden and Millard, 2001.)

6. Future Directions

The proximate mechanisms of risk assessment are poorly understood, especially in freshwater habitats. I have emphasized chemistry in this review to provide some starting points for work in this direction. Interactions between ambient water chemistry and signaling molecules and their receptors are not well studied. Knowing something about the chemistry of the cues and the sensory physiology that processes them is key to understanding the evolution of these systems. Further exploration of the hierarchy and interactions among different sensory modalities should bring new insights. Interactions between sensory modality, context, and ontogeny seem particularly promising (Mathis and Vincent, 2000).

More field manipulation experiments are needed to verify laboratory findings and to test behavioral contingencies. Competing and conflicting ecological demands impact behavioral decision making (Lima and Dill, 1990). More complex and difficult to quantify is the role of individual antipredator responses in population parameters for predator and prey populations (Heckmann, 1995; Scheffer, 1997).

This review shows that all types of predator–prey interactions in aquatic environments are chemically mediated to some degree (Table 12.1). No other sensory modality provides the same subtlety, diversity, and richness of contextual cues to inform prey about predation risk. This is attributable to the length of evolutionary time over which these mechanisms have formed. Future investigations hold great promise in unraveling the remaining mysteries of chemical communication.

Acknowledgments. R. Jan F. Smith has had enormous impact on aquatic chemical ecology of predator–prey interactions through his own work and through those of us fortunate enough to have benefited from his tutelage. Much of the contents of this review, and the review itself, would not exist if not for Jan. I am grateful to R.S. Mirza for providing helpful comments and additional references that improved the quality of the manuscript.

References

Aabjörnsson, K., Wagner, B.M.A., Axelsson, A., Bjerselius, R., and Olsen. (1997). Responses of *Acilius sulcatus* (Coleoptera: Dytiscidae) to chemical cues from perch (*Perca fluviatilis*). *Oecologia* 111:166–171.

Adams, M.J., and Claeson, S. (1998). Field response of tadpoles to conspecific and heterospecific alarm. *Ethology* 104:955–961.

Agrawal, A.A., Laforsch, C., and Tollrian, R. (1999). Transgenerational induction of defences in animals and plants. *Nature* 401:60–63.

Al-Hassen, J.M., Thompson, M., Criddle, K.R., Summers, B., and Criddle, R.S. (1985). Catfish epidermal secretions in response to threat or injury: A novel defense response. *Mar. Biol.* 88: 117–123.

Appelberg, M., Soederbaeck, B., and Odelstroem, T. (1993). Predator detection and perception of predation risk in the crayfish *Astacus astacus* L. *Nord. J. Fresh. Res.* 68:55–62.

Appleton, R.D., and Palmer, A.R. (1988). Waterborne stimuli released by predatory crabs and damaged prey induce more predator-resistant shells in a marine gastropod. *Proc. Natl. Acad. Sci. USA* 85:4387–4391.

Atema, J., and Stenzler, D. (1977). Alarm substance of the marine mud snail, *Nassarius obsoletus*: Biological characterization and possible evolution. *J. Chem. Ecol.* 3:173–187.

Belden, L.K., Wildy, E.L., Hatch, A.C., and Blaustein, A.R. (2000). Juvenile western toads, *Bufo boreas*, avoid chemical cues of snakes fed juvenile, but not larval, conspecifics. *Anim. Behav.* 59:8 71–875.

Blazer, V.S., Fabacher, D.L., Little, E.E., Ewing, M.S., and Kocan, K.M. (1997). Effects of ultraviolet-B radiation on fish: Histological comparison of a UVB-sensitive and a UVB-tolerant species. *J. Aquat. Anim. Health.* 9:132–143.

Bolser, R.C., and Hay, M.E. (1996). Are tropical plants better defended? Palatability and defences of temperate vs. tropical seaweeds. *Ecology* 77: 2269–2286

Brönmark, C., and Miner, L.B. (1992). Predator-induced phenotypical change in body morphology in crucian carp. *Science* 258:1348–1350.

Brooks, W.R. (1989). Hermit crabs alter sea anemone placement patterns for shell balance and reduced predation. *J. Exp. Mar. Biol. Ecol.* 132:109–121.

Brown, G.E., and Godin, J.-G.J. (1997). Antipredator responses to conspecific and heterospecific skin extracts by threespine sticklebacks: Alarm pheromones revisited. *Behaviour* 134: 1123–1134.

Brown, G.E., and Godin, J.-G.J. (1999a). Chemical alarm signals in wild Trinidadian guppies (*Poecilia reticulata*). *Can. J. Zool.* 77:562–570.

Brown, G.E., and Godin, J.-G.J. (1999b). Who dares, learns: Chemical inspection behaviour and acquired predator recognition in a characin fish. *Anim. Behav.* 57:475–481.

Brown, G.E., and Smith, R.J.F. (1996). Foraging trade-offs in fathead minnows (*Pimephales promelas*, Osteichthyes, Cyprinidae): Acquired predator recognition in the absence of an alarm response. *Ethology* 102:776–785.

Brown, G.E., Adrian, J.C. Jr., and Shih, M.L. (2001). Behavioural responses of fathead minnows to hypoxanthine-3-N-oxide at varying concentrations. *J. Fish Biol.* 58:1465–1470.

Brown, G.E., Chivers, D.P., and Smith, R.J.F. (1995). Fathead minnows avoid conspecific and heterospecific alarm pheromones in the faeces of northern pike. *J. Fish Biol.* 47:387–393.

Brown, G.E., Adrian, J.C. Jr., Smyth, E., Leet, H., and Brennan, S. (2000). Ostariophysan alarm pheromones: Laboratory and field tests of the functional significance of nitrogen oxides. *J. Chem. Ecol.* 26:139–154.

Bryant, P.B. (1987). A study of the alarm system in selected fishes of Northern Mississippi. PhD dissertation, University of Mississippi.

Bryer, P.J., Mirza, R.S., and Chivers, D.P. (2001). Chemosensory assessment of predation risk by slimy sculpins (*Cottus cognatus*): Responses to alarm, disturbance, and predator cues. *J. Chem. Ecol.* 27:533–546.

Chapman, G.B., and Johnson, E.G. (1997). An electron microscope study of intrusions into alarm substance cells of the channel catfish. *J. Fish Biol.* 51:503–514.

Chivers, D.P., and Smith, R.J.F. (1994a). Intra- and interspecific avoidance of marked areas with skin extract from brook sticklebacks (*Culaea inconstans*). *J. Chem. Ecol.* 20:1517–1524.

Chivers, D.P., and Smith, R.J.F. (1994b). Fathead minnows, *Pimephales promelas*, acquire predator recognition when alarm substance is associated with the sight of unfamiliar fish. *Anim. Behav.* 48:597–605.

Chivers, D.P., and Smith, R.J.F. (1998). Chemical alarm signalling in aquatic predator–prey systems: A review and prospectus. *Ecoscience* 5:338–352.

Chivers, D.P., Brown, G.E., and Smith, R.J.F. (1996a). The evolution of chemical alarm signals: Attracting predators benefits alarm signal senders. *Am. Nat.* 148:649–659.

Chivers, D.P., Puttlitz, M.H., and Blaustein, A.R. (2000). Chemical alarm signaling by reticulate sculpins, *Cottus perplexus. Environ. Biol. Fish* 57:347–352.

Chivers, D.P., Wisenden, B.D., and Smith, R.J.F. (1995). The role of experience in the response of fathead minnows (*Pimephales promelas*) to skin extract of Iowa darters (*Etheostoma exile*). *Behaviour* 132:665–674.

Chivers, D.P., Wisenden, B.D., and Smith, R.J.F. (1996b). Damselfly larvae learn to recognize predators from chemical cues in the predator's diet. *Anim. Behav.* 52:315–320.

Chivers, D.P., Kiesecker, J.M., Wildy, E.L., Anderson, M.T., and Blaustein, A.R. (1997). Chemical alarm signalling in terrestrial salamanders: Intra- and interspecific responses. *Ethology* 103:599–613.

Chivers, D.P., Kiesecker, J.M., Marco, A., Wildy, E.L., and Blaustein, A.R. (1999). Shifts in life history as a response to predation in western toads (*Bufo boreas*). *J. Chem. Ecol.* 25:2455–2464.

Christine, A., Utne, W., and Bacchi, B. (1997). The influence of visual and chemical stimuli from cod *Gadus morhua* on the distribution of two-spotted goby *Gobiusculus flavescens* (Fabricius). *Sarsia* 82:129–135.

Cieri, M.D., and Stearns, D.E. (1999). Reduction of grazing activity of two estuarine copepods in response to the exudate of a visual predator. *Mar. Ecol. Prog. Ser.* 177:157–163.

Commens, A.M., and Mathis, A. (1999). Alarm pheromones of rainbow darters and responses to skin extracts of conspecifics and congeners. *J. Fish Biol.* 55:1359–1362.

Covich, A.P., Crowl, T.A., Alexander, J.E. Jr., and Vaughn, C.C. (1994). Predator-avoidance responses in freshwater decapod-gastropod interactions mediated by chemical stimuli. *J. North Am. Benthol. Soc.* 13:283–290.

Dawidowicz, P., and Loose, C.J. (1992). Metabolic costs during predator-induced diel vertical migration of *Daphnia. Limnol. Oceanogr.* 37: 1589–1595.

Dawidowicz, P., Pijanowska, J., and Ciechomski, K. (1990). Vertical migration of *Chaoborus* larvae is induced by the presence of fish. *Limnol. Oceanogr.* 35:1631–1637.

de Meester, L. (1993). Genotype, fish-mediated chemicals, and phototactic behavior in *Daphnia magna. Ecology* 74:1467–1474.

Dewitt, T.J., Sih, A., and Hucko, J.A. (1999). Trait compensation and cospecialization in a freshwater snail: Size, shape and antipredator behaviour. *Anim. Behav.* 58:397–407.

d'Heursel, A., and Haddad, C.F.B. (1999). Unpalatability of *Hyla semilineata* tadpoles (Anura) to captive and free-ranging vertebrate predators. *Ethol. Ecol. Evol.* 11:339–348.

Dodson, S.I. (1989). The ecological role of chemical stimuli for the zooplankton: Predator-induced morphology in *Daphnia. Oecologia* 78:361–367.

Dodson, S.I., and Wagner, A.E. (1996). Temperature affects selectivity of *Chaoborus* larvae eating *Daphnia. Hydrobiologia* 325:157–161.

Ebel, R., Marin, A., and Proksch, P. (1999). Organ-specific distribution of dietary alkaloids in the marine opisthobranch *Tylodina perversa. Biochem. Syst. Ecol.* 27:769–777.

Eklöv, P. (2000). Chemical cues from multiple predator-prey interactions induce changes in behavior and growth of anuran larvae. *Oecologia* 123:192–199.

Elliott, J.K., Ross, D.M., Pathirana, C., Miao, S., Andersen, R.J., Singer, P., Kokke, W.C.M.C., and Ayer, W.A. (1989). Induction of swimming in *Stomphia* (Anthozoa: Actiniaria) by imbricatine, a metabolite of the asteroid *Dermasterias imbricata. Biol. Bull. Mar. Biol. Lab. Woods Hole* 176:73–78.

Epifanio, R.A., Martins, D.L., Villaca, R., and Gabriel, R. (1999). Chemical defenses against fish predation in three Brazilian octocorals: 11 beta, 12 beta-epoxypukalide as a feeding deterrent in *Phyllogorgia dilatata. J. Chem. Ecol.* 25:2255–2266.

Gelowitz, C.M., Mathis, A., and Smith, R.J.F. (1993). Chemosensory recognition of northern pike (*Esox lucius*) by brook stickleback (*Culaea inconstans*): Population differences and the influence of predator diet. *Behaviour* 127:105–118.

Gillette, R., Huang, R., Hatcher, N., and Moroz, L.L. (2000). Cost-benefit analysis potential in feeding behavior of a predatory snail by integration of hunger, taste, and pain. *Proc. Natl. Acad. Sci. USA* 97:3585–3590.

Gochfeld, D. (1995). Evidence for predation-induced defenses in a hard coral. *Twenty-third Benthic Ecology Meeting*, Rutgers State Univ. Inst. Marine Coastal Sciences, New Brunswick, NJ.

Godard, R.D., Bowers, B.B., and Wannamaker, C. (1998). Responses of golden shiner minnows to chemical cues from snake predators. *Behaviour* 135:1213–1228.

Griffiths, R.A., Schley, L., Sharp, P.E., Dennis, J.L., and Roman, A. (1998). Behavioural responses of *Mallorcan* midwife toad tadpoles to natural and unnatural snake predators. *Anim. Behav.* 55:207–214.

Hall, D., and Suboski, M.D. (1995). Visual and olfactory stimuli in learned release of alarm reactions by zebra danio fish (*Brachydanio rerio*). *Neurobiol. Learn. Mem.* 63:229–240.

Hanifin, C.T., Yotsu-Yamashita, M., Yasumoto, T., Brodie, E.D. III, and Brodie, E.D. Jr. (1999). Toxicity of dangerous prey: Variation of tetrodotoxin levels within and among populations of the newt *Taricha granulose*. *J. Chem. Ecol.* 25: 2161–2176.

Hartman, E.J., and Abrahams, M.V. (2000). Sensory compensation and the detection of predators: The interaction between chemical and visual information. *Proc. R. Soc. Lond. B.* 267:571–575.

Harvell, C.D., West, J.M., and Griggs, C. (1996). Chemical defense of embryos and larvae of a West Indian gorgonian coral, *Briareum asbestinum*. *Invert. Reprod. Devel.* 30:239–246.

Hazlett, B.A. (1985). Disturbance pheromones in the crayfish *Orconectes virilis*. *J. Chem. Ecol.* 11: 1695–1711.

Hazlett, B.A. (1990a). Disturbance pheromone in the hermit crab *Calcinus laevimanus* (Randall, 1840). *Crustaceana* 58:314–316.

Hazlatt, B.A. (1990b). Source and nature of disturbance-chemical system in crayfish. *J. Chem. Ecol.* 16:2263–2275.

Heckmann, K. (1995). Predator-induced defences in Protozoa. *Naturwissenschaften* 82:107–116.

Hedberg, T.S. (1981). A possible stress-warning marker in ambystomatid salamanders. *J. Exp. Zool.* 216:349–355.

Hews, D.K. (1988). Alarm response in larval western toads, *Bufo boreas*, release of larval chemicals by a natural predator and its effect on predator captue efficiency. *Anim. Behav.* 36:125–133.

Hirvonen, H., Ranta, E., Piironen, J., Laurila, A., and Peuhkuri, N. (2000). Behavioural responses of naive Arctic charr young to chemical cues from salmonid and non-salmonid fish. *Oikos* 88: 191–199.

Holopainen, I.J., Aho, J., Vornanen, M., and Huuskonen, H. (1997). Phenotypic plasticity and predator effects on morphology and physiology of crucian carp in nature and in the laboratory. *J. Fish Biol.* 50:781–798.

Hugie, D.M., Thuringer, P.L., and Smith, R.J.F. (1991). The response of the tidepool sculpin *Oligocottus maculosus*, to chemical stimuli from injured conspecifics and alarm signalling in the Cottidae (Pisces). *Ethology* 89:322–334.

Huhta, A., Muotka, T., Juntunen, A., and Yrjoenen, M. (1999). Behavioural interactions in stream food webs: The case of drift-feeding fish, predatory invertebrates and grazing mayflies. *J. Anim. Ecol.* 68:917–927.

Huryn, A.D., and Chivers, D.P. (1999). Contrasting behavioral responses by detritivorous and predatory mayflies to chemicals released by injured conspecifics and their predators. *J. Chem. Ecol.* 25:2729–2740.

Jachner, A. (1997). The response of bleak to predator odour of unfed and recently fed pike. *J. Fish Biol.* 50:878–886.

Jacobsen, H.P., and Stabell, O.B. (1999). Predator-induced alarm responses in the common periwinkle, *Littorina littorea*: Dependence on season, light conditions, and chemical labelling of predators. *Mar. Biol.* 134:551–557.

Jakobsen, P.J., and Wedekind, C. (1998). Copepod reaction to odor stimuli influenced by cestode infection. *Behav. Ecol.* 9:414–418.

Jordão, L.C., and Volpato, G.L. (2000). Chemical transfer of warning information in non-injured fish. *Behaviour* 137:681–690.

Kats, L.B., and Dill, L.M. (1998). The scent of death: Chemosensory assessment of predation risk by prey animals. *Ecoscience* 5:361–394.

Kats, L.B., Petranka, J.W., and Sih, A. (1988). Antipredator defences and the persistence of amphibian larvae with fishes. *Ecology* 69:1865–1870.

Kerby, J.L., and Kats, L.B. (1998). Modified interactions between salamander life stages caused by wildfire-induced sedimentation. *Ecology* 79:740–745.

Kiesecker, J.M., Chivers, D.P., Marco, A., Quilchanos, C., Anderson, M.T., and Blaustein, A.R. (1999). Identification of a disturbance signal in larval red-legged frogs *Rana aurora*. *Anim. Behav.* 57:1295–1300.

Korpi, N.L., and Wisenden, B.D. (2001). Learned recognition of novel predator odour by zebra danios, *Danio rerio*, following time-shifted presentation of alarm cue and predator odour. *Environ. Biol. Fish* 61:205–211.

Kusch, J. (1993a). Induction of defensive morphological changes in ciliates. *Oecologia* 94:571–575.

Kusch, J. (1993b). Behavioural and morphological changes in ciliates induced by the predator *Amoeba proteus*. *Oecologia* 96:354–359.

Kusch, J. (1999). Self-recognition as the original function of an amoeban defence-inducing kairomone. *Ecology* 80:715–720.

Lam, F.L., Brown, G.B., and Parham, J.C. (1974). Purine N-oxides. LVI. Photoisomerization of 1-hydroxy- to 3-hydroxyxanthine: Photochemistry of related 1-hydroxypurines. *J. Org. Chem.* 39: 1391–1395.

Laurila, A. (2000). Behavioural responses to predator chemical cues and local variation in antipredator performance in *Rana temporaria* tadpoles. *Oikos* 88:159–168.

Lebedeva, N.Y., Malyukina, G.A., and Kasumyan, A.O. (1975). The natural repellent in the skin of cyprinids. *J. Ichthyol.* 15:472–480.

Lefcort, H. (1996). Adaptive, chemically mediated fright response in tadpoles of the southern leopard frog, *Rana utricularia*. *Copeia* 1996:455–459.

Leonard, G.H., Bertness, M.D., and Yund, P.O. (1999). Crab predation, waterborne cues, and inducible defenses in the blue mussel, *Mytilus edulis*. *Ecology* 80:1–14.

Lima, S.L., and Dill, L.M. (1990). Behavioral decisions made under the risk of predation: A review and prospectus. *Can. J. Zool.* 68:619–540.

Lüning, J. (1995). Life-history responses to *Chaoborus* of spined and unspined *Daphnia pulex*. *J. Plankton Res.* 17:71–84.

Lürling, M., and Beekman, W. (1999). Grazer-induced defenses in *Scenedesmus* (Chlorococcales; Chlorophyceae): Coenobium and spine formation. *Phycologia* 38:368–376.

Lutterschmidt, W.I., Marvin, G.A., and Hutchison, V.H. (1994). Alarm response by a plethodontid salamander (*Desmognathus ochrophaeus*): Conspecific and heterospecific "*Schreckstoff*." *J. Chem. Ecol.* 20:2751–2760.

McClintock, J.B., and Janssen, J. (1990). Pteropod abduction as a chemical defence in a pelagic Antarctic amphipod. *Nature* 346:462–464.

McIntosh, A.R., and Peckarsky, B.L. (1999). Criteria determining behavioural responses to multiple predators by a stream mayfly. *Oikos* 85:554–564.

Madison, D.M., Maerz, J.C., and McDarby, J.H. (1999). Optimization of predator avoidance by salamanders using chemical cues: Diet and diel effects. *Ethology* 105:1073–1086.

Magurran, A.E., Irving, P.W., and Henderson, P.A. (1996). Is there a fish alarm pheromone? A wild study and critique. *Proc. R. Soc. Lond. B.* 263: 1551–1556.

Marin, A., Lopez, M.D., Esteban, M.A., Meseguer, J., Munoz, J., and Fontana, A. (1998). Anatomical and ultrastructural studies of chemical defence in the sponge *Dysidea fragilis*. *Mar. Biol.* 131: 639–645.

Marvin, G.A., and Hutchison, V.H. (1995). Avoidance response by adult newts (*Cynops pyrrhogaster* and *Notophthalmus viridescens*) to chemical alarm cues. *Behaviour* 132:95–106.

Mathis, A., and Hoback, W.W. (1997). The influence of chemical stimuli from predators on precopulatory pairing by the amphipod, *Gammarus pseudolimnaeus*. *Ethology* 103:33–40.

Mathis, A., and Lancaster, D. (1998). Response of terrestrial salamanders to chemical stimuli from distressed conspecifics. *Amphibia-Reptilia* 19: 330–335.

Mathis, A., and Smith, R.J.F. (1992). Avoidance of areas marked with a chemical alarm substance by fathead minnows (*Pimephales promelas*) in a natural habitat. *Can. J. Zool.* 70:1473–1476.

Mathis, A., and Smith, R.J.F. (1993a). Intraspecific and cross-superorder responses to chemical alarm signals by brood stickleback. *Ecology* 74:2395–2404.

Mathis, A., and Smith, R.J.F. (1993b). Chemical labeling of northern pike (*Esox lucius*) by the alarm pheromone of fathead minnows (*Pimephales promelas*). *J. Chem. Ecol.* 19:1967–1979.

Mathis, A., and Smith, R.J.F. (1993c). Fathead minnows, *Pimephales promelas*, learn to recognize northern pike, *Esox lucius*, as predators on the basis of chemical stimuli from minnows in the pike's diet. *Anim. Behav.* 46:645–656.

Mathis, A., and Smith, R.J.F. (1993d). Chemical alarm signals increase the survival time of fathead minnows (*Pimephales promelas*) during encounters with northern pike (*Esox lucius*). *Behav. Ecol.* 4:260–265.

Mathis, A., and Vincent, F. (2000). Differential use of visual and chemical cues in predator recognition and threat-sensitive predator avoidance responses by larval newts (*Notophthalmus viridescens*). *Can. J. Zool.* 78:1646–1652.

Mathis, A., Chivers, D.P., and Smith, R.J.F. (1995). Chemical alarm signals: Predator deterrents or attractants? *Am. Nat.* 146:994–1005.

Miklósi, Á., Pongrácz, P., and Csányi, V. (1997). The ontogeny of antipredator behaviour in paradise fish larvae (*Macropodus opercularis*). II. The response to chemical stimuli of heterospecific fishes. *Behaviour* 134:391–413.

Mirza, R.S. (1998). Induced morphological changes in fishes mediated by chemical stimuli associated with predation. MS thesis, University of Saskatchewan, Canada.

Mirza, R.S., and Chivers, D.P. (2001). Are chemical alarm cues conserved within salmonid fishes? *J. Chem. Ecol.* 27:1641–1655.

Moore, R.D., Newton, B., and Sih, A. (1996). Delayed hatching as a response of streamside salamander eggs to chemical cues from predatory sunfish. *Oikos* 77:331–335.

Nicieza, A.G. (2000). Interacting effects of predation risk and food availability on larval anuran behaviour and development. *Oecologia* 123:497–505.

Nilsson, P.A., Brönmark, C., and Pettersson, L.B. (1995). Benefits of a predator-induced morphology in crucian carp. *Oecologia* 104:291–296.

Parejko, K., and Dodson, S. (1990). Progress towards characterization of a predator/prey kairomone: *Daphnia pulex* and *Chaoborus americanus*. *Hydrobiologia* 198:51–59.

Pettersson, L.B., and Brönmark, C. (1997). Density-dependent costs of an inducible morphological defense in crucian carp. *Ecology* 78:1805–1815.

Pettersson, L.B., Nilsson, P.A., and Brönmark, C. (2000). Predator recognition and defence strategies in crucian carp, *Carassius carassius*. *Oikos* 88:200–212.

Pfeiffer, W., (1977). The distribution of fright reaction and alarm substance cells in fishes. *Copeia* 1977:653–665.

Pfeiffer, W., Riegelbauer, G., Meir, G., and Scheibler, B. (1985). Effect of hypoxanthine-3(N)-oxide and hypoxanthine-1(N)-oxide on central nervous excitation of the black tetra *Gymnocorymbus ternetzi* (Characidae, Ostariophysi, Pisces) indicated by dorsal light response. *J. Chem. Ecol.* 11:507–524.

Pijanowska, J. (1997). Alarm signals in *Daphnia*? *Oecologia* 112:12–16.

Pijanowska, J., and Stolpe, G. (1996). Summer diapause in *Daphnia* as a reaction to the presence of fish. *J. Plankton Res.* 18:1407–1412.

Poulin, R., Marcogliese, D.J., and McLaughlin, J.D. (1999). Skin-penetrating parasites and the release of alarm substances in juvenile rainbow trout. *J. Fish Biol.* 55:47–53.

Powlik, J.J., Lewis, A.G., and Verma, N. (1997). The response of *Tigriopus californicus* to chlorophytic macroalgae, including *Cladophora trichotoma* Kuetzing. *Estuar. Coast Shelf Sci.* 44:327–337.

Rahman, Y.J., Forward, R.B. Jr., and Rittschof, D. (2000). Responses of mud snails and periwinkles to environmental odors and disaccharide mimics of fish odor. *J. Chem. Ecol.* 26:679–696.

Reede, T. (1995). Life history shifts in response to different levels of fish kairomones in *Daphnia*. *J. Plankton Res.* 17:1661–1667.

Repka, S., and Pihlajamaa, K. (1996). Predator-induced phenotypic plasticity in *Daphnia pulex*: Uncoupling morphological defenses and life history shifts. *Hydrobiologia* 339:67–71.

Repka, S., and Walls, M. (1998). Variation in the neonate size of *Daphnia pulex*: The effects of predator exposure and clonal origin. *Aquat. Ecol.* 32:203–209.

Ringelberg, J., and Van Gool, E. (1995). Migrating *Daphnia* have a memory for fish kairomones. *Mar. Freshwat. Behav. Physiol.* 26:249–257.

Rittschof, D. (1990). Peptide-mediated behaviors in marine organisms: Evidence for a common theme. *J. Chem. Ecol.* 16:261–272.

Sakwinska, O. (1998). Plasticity of *Daphnia* magna life history traits in response to temperature and information about a predator. *Freshwat. Biol.* 39:681–687.

Scheffer, M. (1997). On the implications of predator avoidance. *Aquat. Ecol.* 31:99–107.

Schwartz, S.S. (1991). Predator-induced alterations in *Daphnia* morphology. *J. Plankton Res.* 13: 1151–1161.

Scrimshaw, S., and Kerfoot, W.C. (1987). Chemical defenses of freshwater organisms: Beetles and bugs. In: *Predation: Direct and Indirect Impacts on Aquatic Communities* (Kerfoot, W.C., Sih, A., eds.), pp. 240–262. Hanover: UP of New England.

Sih, A. (1986). Antipredator responses and the perception of danger by mosquito larvae. *Ecology* 67:434–441.

Sih, A., and Moore, R.D. (1993). Delayed hatching of salamander eggs in response to enhanced larval predation risk. *Am. Nat.* 142:947–960.

Smith, R.J.F. (1981). Effect of food deprivation on the reaction of Iowa darters (*Etheostoma exile*) to skin extract. *Can. J. Zool.* 59:558–560.

Smith, R.J.F. (1982). Reaction of *Percina nigrofasciata*, *Ammocrypta beani* and *Etheostoma swaini* (Percidae, Pisces) to conspecific and intergeneric skin extracts. *Can. J. Zool.* 17:2253–2259.

Smith, R.J.F. (1992). Alarm signals in fishes. *Rev. Fish Biol.* 2:33–63.

Smith, R.J.F. (1999). What good is smelly stuff in the skin? Cross-function and cross taxa effects in fish "alarm substances." In: *Advances in Chemical Signals in Vertebrates* (Johnston, R.E., Müller-Schwarze, D., and Sorensen, P.W., eds.), pp. 475–487, New York: Kluwer Academic/Plenum Press.

Snyder, N.F.R. (1967). An alarm reaction of aquatic gastropods to intraspecific extract. Cornell University Agricultural Experiment Station,

New York State College of Agriculture, Ithaca, NY, Memoir 403:1–122.

Spieler, M., and Linsenmair, K.E. (1999). Aggregation behaviour of *Bufo maculatus* tadpoles as an antipredator mechanism. *Ethology* 105:665–686.

Stabell, O.B., and Lwin, M.S. (1997). Predator-induced phenotypic changes in crucian carp are caused by chemical signals from conspecifics. *Environ. Biol. Fishes* 49:145–149.

Stachowicz, J.J., and Hay, M.E. (1999). Mutualism and coral persistence: The role of herbivore resistance to algal chemical defense. *Ecology* 80:2085–2101.

Stewart, T.W., Gafford, J.C., Miner, J.G., and Lowe, R.L. (1999). *Dreissena*-shell habitat and antipredator behavior: Combined effects on survivorship of snails co-occurring with molluscivorous fish. *J. North Am. Benthol. Soc.* 18:274–283.

Stibor, H. (1992). Predator induced life-history shifts in a freshwater cladoceran. *Oecologia* 92:162–165.

Stirling, G. (1995). *Daphnia* behaviour as a bioassay of fish presence or predation. *Funct. Ecol.* 9:778–784.

Storfer, A., and Sih, A. (1998). Gene flow and ineffective antipredator behavior in a stream-breeding salamander. *Evolution* 52:558–565.

Suboski, M.D., Brian, S., Carty, A.E., McQuoid, L.M., Seelen, M.I., and Seifert, M. (1990). Alarm reaction in acquisition and social transmission of simulated predator recognition by zebra danio fish (*Brachydanio rerio*). *J. Comp. Psychol.* 104:101–112.

Tachibana, K., and Gruber, S.H. (1988). Shark repellent lipophilic constituents in the defence secretion of the Moses sole (*Pardachirus marmoratus*). *Toxicon* 26:839–853.

Tollrian, R. (1994). Fish-kairomone induced morphological changes in *Daphnia lumholtzi* (Sars) *Arch. Hydrobiol.* 130:69–75.

Tollrian, R. (1995). Predator-induced morphological defenses: Costs, life history shifts, and maternal effects in *Daphnia pulex*. *Ecology* 76:1691–1705.

Tollrian, R., and Harvell, C.D. (1999). The evolution of inducible defenses: Current ideas. In: *The Ecology and Evolution of Inducible Defenses* (Tollrian, R., and Harvell, C.D., eds.), pp. 306–321. Princeton, NJ: Princeton University Press.

Turner, A.M., Bernot, R.J., and Boes, C.M. (2000). Chemical cues modify species interactions: The ecological consequences of predator avoidance by freshwater snails. *Oikos* 88:148–158.

Uiblein, F., Roca, J.R., Baltanas, A., and Danielopol, D.L. (1996). Tradeoff between foraging and antipredator behaviour in a macrophyte dwelling ostracod. *Arch. Hydrobiol.* 137:119–133.

Vadas, R.L., Sr., Burrows, M.T., and Hughes, R.N. (1994). Foraging strategies of dogwhelks, *Nucella lapillus* (L.): Interacting effects of age, diet and chemical cues to the threat of predation. *Oecologia* 100:439–450.

van Alstyne, K.L. (1988). Herbivore grazing increases polyphenolic defences in the intertidal brown alga *Fucus distichus*. *Ecology* 69:655–663.

van Alstyne, K.L., Ehlig, J.M., and Whitman, S.L. (1999). Feeding preferences for juvenile and adult algae depend on algal stage and herbivore species. *Mar. Ecol. Prog. Ser.* 180:179–185.

van Buskirk, J., and McCollum, S.A. (2000). Functional mechanisms of an inducible defence in tadpoles: Morphology and behaviour influence mortality risk from predation. *J. Evol. Biol.* 13:336–347.

von Elert, E., and Pohnert, G. (2000). Predator specificity of kairomones in diel vertical migration of *Daphnia*: A chemical approach. *Oikos* 88:119–128.

Waddell, B., and Pawlik, J.R. (2000). Defences of Caribbean sponges against invertebrate predators. I. Assays with hermit crabs. *Mar. Ecol. Prog. Ser.* 195:125–132.

White, D.S. (1989). Defence mechanisms in riffle beetles (Coleoptera: Dryopoidea). *Ann. Entomol. Soc.* 82:237–241.

Wicklow, B.J. (1997). Signal-induced defensive phenotypic changes in ciliated protists: Morphological and ecological implications for predator and prey. *J. Eukaryot. Microbiol.* 44:176–188.

Wildy, E.L., Chivers, D.P., and Blaustein, A.R. (1999). Shifts in life-history traits as a response to cannibalism in larval long-toed salamanders (*Ambystoma macrodactylum*). *J. Chem. Ecol.* 25:2337–2346.

Wilson, D.J., and Lefcort, H. (1993). The effect of predator diet on the alarm response of red-legged frog, *Rana aurora*, tadpoles. *Anim. Behav.* 46:1017–1019.

Wilson, D.M., Puyana, M., Fenical, W., and Pawlik, J.R. (1999). Chemical defense of the Caribbean reef sponge *Axinella corrugata* against predatory fishes. *J. Chem. Ecol.* 25:2811–2824.

Wisenden, B.D. (2000). Scents of danger: The evolution of olfactory ornamentation in chemically mediated predator-prey interactions. In: *Animal Signals: Signalling and Signal Design in Animal Communication* (Espmark, Y., Amundsen, T., and Rosenqvist, G., eds.), pp. 365–386. Trondheim, Norway: Tapir Academic Press.

Wisenden, B.D., and Harter, K.R. (2001). Motion, not shape, facilitates association of predation risk with novel objects by fathead minnows (*Pimephales promelas*). *Ethology* 107:357–364.

Wisenden, B.D., and Millard, M.C. (2001). Aquatic flatworms use chemical cues from injured conspecifics to assess predation risk and to associate risk with novel cues. *Anim. Behav.* 62: 761–766.

Wisenden, B.D., and Smith, R.J.F. (1997). The effect of physical condition and shoalmate familiarity on proliferation of alarm substance cells in the epidermis of fathead minnows. *J. Fish Biol.* 50:799–808.

Wisenden, B.D., and Smith, R.J.F. (1998). A re-evaluation of the effect of shoalmate familiarity on the proliferation of alarm substance cells in ostariophysan fishes. *J. Fish Biol.* 53:841–846.

Wisenden, B.D., Chivers, D.P., and Smith, R.J.F. (1995a). Early warning in the predation sequence: A disturbance pheromone in Iowa darters (*Etheostoma exile*). *J. Chem. Ecol.* 21:1469–1480.

Wisenden, B.D., Chivers, D.P., and Smith, R.J.F. (1997). Learned recognition of predation risk by *Enallagma* damselfly larvae (Odonata, Zygoptera) on the basis of chemical cues. *J. Chem. Ecol.* 23:137–151.

Wisenden, B.D., Cline, A., and Sparkes, T.C. (1999). Survival benefit to antipredator behavior in the amphipod *Gammarus minus* (Crustacea: Amphipoda) in response to injury-released chemical cues from conspecifics and heterospecifics. *Ethology* 105:407–414.

Wisenden, B.D., Pohlman, S.G., and Watkin, E.E. (2001). Avoidance of conspecific injury-released cues by free-ranging *Gammarus lacustris* (Crustacea: Amphipoda). *J. Chem. Ecol.* 27:1249–1258.

Wisenden, B.D., Chivers, D.P., Brown, G.E., and Smith, R.J.F. (1995b). The role of experience in risk assessment: Avoidance of areas chemically labelled with fathead minnow alarm pheromone by conspecifics and heterospecifics. *Ecoscience* 2:115–122.

Yunker, W.K., Wein, D.E., and Wisenden, B.D. (1999). Conditioned alarm behavior in fathead minnows (*Pimephales promelas*) resulting from association of chemical alarm pheromone with a nonbiological visual stimulus. *J. Chem. Ecol.* 25:2677–1286.

Zulandt-Schneider, R.A., and Moore, P.A. (2000). Urine as a source of conspecific disturbance signals in the crayfish *Procambarus clarkii*. *J. Exp. Biol.* 203:765–771.

13 Mechanisms of Ultraviolet Polarization Vision in Fishes

Craig W. Hawryshyn

Abstract

Our understanding of polarization vision in vertebrates, especially in teleost fishes, has grown significantly over recent decades. Work in my laboratory has led to the development of a biophysical model indicating that the spatial arrangement of photoreceptors or geometry of the cone mosaic provides the basis for ultraviolet polarization sensitivity. This biophysical mechanism appears to be based on the selective reflection of axial polarized light by the partitioning membrane, formed along the contact zone between the members of the double cones, onto neighboring UV-sensitive cones. In this review, I discuss the historical development of this problem and present some interesting insights aimed at future work on the neuronal coding of e-vector.

1. Introduction

Animals often perceive the world in a manner that is different from that of humans. Each sense an animal possesses contributes to a mélange of sensations and overall perception that Niko Tinbergen described as the animal's *Merkvelt* or perceptual world (Tinbergen, 1951). Professor Tinbergen's observations were insightful for the time (1940–1950) since the idea that animals may see things that we may not presents a paradox with respect to how scientists go about investigating the sensory world of any animal. These constraints were very much in evidence in the initial efforts concerning the research on ultraviolet-polarization sensitivity in vertebrate animals (Hawryshyn and Beauchamp, 1985). This motivated a series of experiments examining different hypotheses regarding the characteristics of optical stimuli resulting in a clear demonstration of ultraviolet (UV) sensitivity (Hawryshyn and Beauchamp, 1982, 1985). Further experiments confirmed this suggestion that fishes and other vertebrate animals can see UV optical stimuli and that fish have cone photoreceptors that are primarily sensitive to UV light. Initially, ideas such as UV light not penetrating water and UV light being absorbed by the lens of fish eyes provided logical barriers that made UV vision, in any vertebrate, seem improbable. However, these logical barriers turned out to be nothing more than the human eye predicting what the fish could see. More recently, evidence has grown regarding the observation of ultraviolet visual sensitivity in vertebrate animals and the cone photoreceptors that mediate this aspect of vision, dissolving the issue of the scientist's bias (see *Vision Research*, 1994, vol. 34, Special Issue on the Biology of Ultraviolet Vision in Animals; *Journal of Experimental Biology*, 2001, vol. 204, Special Issue on the Biology of Ultraviolet and Polarization Vision.)

Questions surrounding UV vision have shifted from looking at the organisms that possess UV vision to defining the functional domain of UV vision in aquatic ecosystems. While numerous possibilities exist, the one most studied thus far is the role of UV vision in the detection of polarized light, (yet another visual attribute that humans do not appreciate). Many invertebrates use UV receptors to detect plane-polarized light (Waterman, 1981; Wehner, 1989), so we tested the hypothesis that cyprinids use ultraviolet vision to mediate polarization sensitivity (Hawryshyn and McFarland, 1987). Our hypothesis that there was a link between polarization sensitivity and UV vision in goldfish was confirmed. It is this link I would like to address in this chapter. What is the role of UV-polarized light vision in the behavior of a variety of species of fishes and how are UV-polarized light signals encoded in the visual system?

2. Underwater Polarized Light Field

To properly understand the role of UV polarization vision in orientation and navigation in fish behavior we must consider the characteristics of the light field in the aquatic environment. Celestial e-vector patterns have been extensively described (Brines and Gould, 1982; Wehner, 1989; Horvàth and Wehner, 1999). Unpolarized solar radiation is partially polarized by small particles in the atmosphere (Rayleigh scattering). This process produces a regular pattern of e-vectors throughout the celestial hemisphere. Two general observations describe the pattern of polarized light in the celestial hemisphere: (1) The dominant e-vector observed at a particular point in the sky is perpendicular to the plane of incidence (a plane comprised of the position of the sun, the position of the observer, and the point observed in the sky). (2) The maximum band of polarization is 90° away from the sun, and thus if the sun is on the horizon, the maximum band of polarization would appear as a band across the sky at the zenith and have an orientation perpendicular to the solar meridian. As the sun's elevation changes throughout the day, the maximum band of polarization and the pattern of e-vector distributions rotate (see Wehner, 1989 and Hawryshyn, 1992 for graphical illustrations).

Extensive measurements of underwater-polarized light fields have also been described (Waterman, 1981; Loew and McFarland, 1990). Unlike the hemispherical field in the terrestrial situation, the polarization field underwater is spherical with respect to the observer's point in space. Most of the polarization of underwater light on cloudy days results from scattering by water molecules with little contribution from the sky light polarization, but on clear days there maybe a significant celestial contribution, especially during crepuscular periods (Waterman, 1981). The precise description of the polarized light field is based on three parameters: I, the intensity of the e-vector, p, the percent polarization of a point in the polarized light field, and the e-vector orientation (Loew and McFarland 1990). These factors are highly dependent on the altitude of the sun, as shown in Figure 13.1A.

Underwater polarization occurs through three processes: (1) direct transfer of skyborne polarization, (2) reflection at the air–water interface (Horvàth and Varjú, 1997), and (3) scattering by water molecules and very small particles in the water column. The degree of underwater polarization generated by these mechanisms is subject to variance due to surface wave action and atmospheric conditions (cloud cover). Solar altitude greatly affects the band of maximum polarization underwater, as it does in the terrestrial environment. When the sun is at zenith, the band of

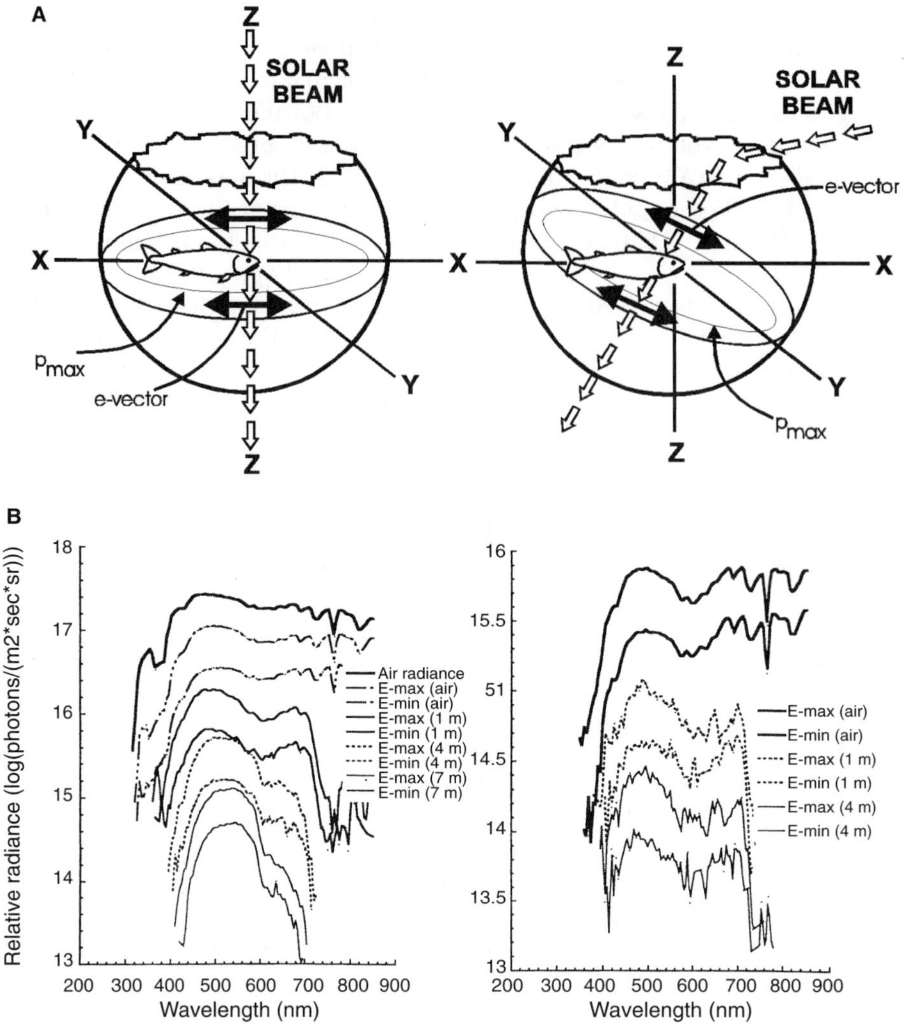

FIGURE 13.1. (A) Distribution of the degree of polarization of the e-vector in the underwater visual sphere of a fish at zenith and low solar altitude. The open arrows represent the solar beam in air and underwater, dark arrows the orientation of the e-vector relative to the horizontal plane, and stippling the distribution of maximal polarization. The upper circular areas indicate the extent of the aerial view through Snell's window. In a flat calm (left), the angle is slightly greater than 97°; for a disturbed surface (right), Snell's window is distorted by waves and is seen from below as an expanded flashing window of bright light. The maximal angle to the horizontal plane assumed by the e-vector approaches 49° as a result of refraction across the air–water interface. (After Loew and McFarland, 1990.) (B) Spectral characteristics of the radiance and polarized light fields during dusk in (*left*) Lake Cowichan (sun just above the horizon); and (*right*) Ogden Point Breakwater (sun just below the horizon, sunset at 21:02 hr Pacific Standard Time). Percent polarization: Lake Cowichan—74.2% (air), 65.4% (1 m), 63% (4 m), 52.3% (7 m); Ogden Point Breakwater—72.7% (air), 67.2% (1 m), 65.4% (4 m). Radiance is always the highest in air and diminishes with depth. E-max, E-min curves for a given depth have the same trace but E-max is always the highest. For clarity, 1 log unit was added to all air and 1 m scans in *left*, and 0.5 log unit was added to the 4 m scans. In *right*, 0.5 log unit was added to the air and 1 m scans. (After Novales-Flamarique and Hawryshyn, 1997a.)

maximum polarization is arranged perpendicular to the solar beam, but at dawn or dusk, when the altitude of the sun is much lower, the band of maximum polarization tilts and assumes an oblique orientation due to refraction of the solar beam by the surface water.

The underwater-polarized light fields of freshwater and coastal marine habitats have been recently examined by Novales-Flamarique and Hawryshyn (1997a). The spectral distribution of underwater-polarized light fields was measured in the upper photic zone of moderately productive marine and freshwater. Percent polarization levels during the day were lower than 40% but during crepuscular periods rose to close to 70%. The spectral characteristics of underwater light are shown for a coastal marine habitat during the day and dusk (Novales-Flamarique and Hawryshyn, 1997a). When the sun is higher in the daylight sky, the average percent polarization is much lower than when the sun is located lower in the sky, such as at dusk (Fig. 13.1B).

3. Polarization-Sensitive Retinal Mechanisms

Investigators working with birds have used electroretinogram (ERG) recording to demonstrate polarized light discrimination (Delius et al., 1976). However, these experiments did not clearly identify the receptor mechanism(s) that mediate polarization sensitivity. Experiments by Waterman and associates looked for receptor mechanisms by varying the test wavelength of the polarized light stimulus. Although different test wavelengths were used, because of a combination of factors, such as not using chromatic adaptation and not using UV linearly polarized stimuli, their results revealed just one class of polarization-sensitive receptor mechanism (UV photoreception in goldfish was reported in Hawryshyn and Beauchamp, 1982, 1985 well after the studies by Waterman and associates). More recent studies on goldfish (Hawryshyn and McFarland, 1987) and salmonids (Parkyn and Hawryshyn, 1993, 2000) have indicated that a UV-sensitive cone mechanism exhibits polarization sensitivity orthogonal to the green- and red-sensitive cones (Fig. 13.2). These observations indicate that differential polarization sensitivity between cone mechanisms can provide the potential for e-vector discrimination.

The blue-sensitive cones do not exhibit polarization sensitivity and, interestingly, this is the part of the spectrum in which the underwa-

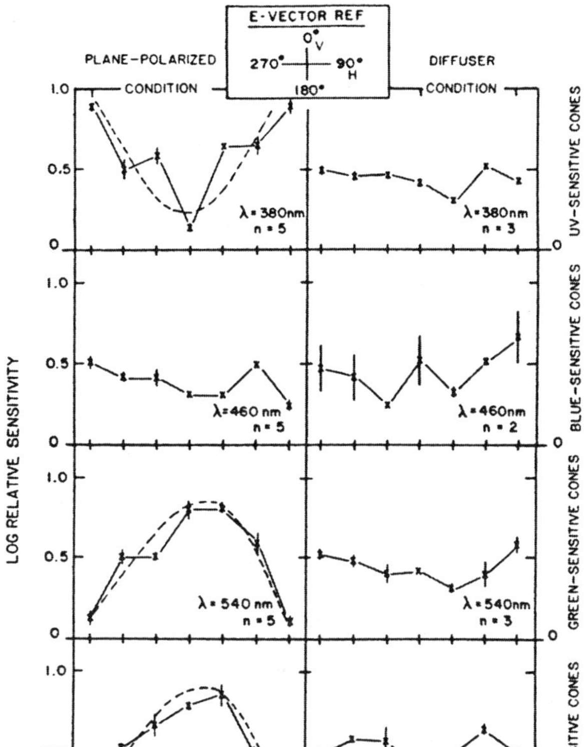

FIGURE 13.2. Polarization sensitivity curves for the cone photoreceptor mechanisms of goldfish. Test condition on the left side, control condition on the right. Each curve represents the mean relative polarization sensitivity (one standard error bar centered on the mean). The dashed line drawn for the UV-, green-, and red-sensitive cone receptor curves (left panel) represents the best fit by eye. (After Hawryshyn and McFarland, 1987.)

ter light field has the lowest degree of polarization underwater (Ivanoff and Waterman, 1958; Novales-Flamarique and Hawryshyn, 1997a). Dark-adapted trout do not exhibit polarization sensitivity (Parkyn and Hawryshyn, 1993). Furthermore, our examination of the dynamic range of sensitivity of different cone mechanisms (Hawryshyn, 1991) reveals that the UV-sensitive mechanism operates over a range of ambient intensity, which could be considered "mesopic," and thus functional polarization vision may be restricted to this window of ambient intensity. These are important considerations for visual ecologists who may be assuming that polarization vision may be functional throughout the whole diurnal period.

Like color vision, polarization vision depends on the possession of at least two differentially sensitive receptor (cone) mechanisms with respect to e-vector (Bernard and Wehner, 1997). In addition, to enable inter-receptor comparisons, the cone mechanisms must have overlapping regions of spectral sensitivity to minimize spectral confounds. For instance, in the case of the honeybee, those receptors that mediate polarization sensitivity are exclusively restricted to the UV spectrum (Wehner et al., 1975). In the case of cyprinid (Hawryshyn and Hárosi, 1991) and salmonid fishes (Hawryshyn and Hárosi, 1994; Hawryshyn et al., 2001), all cone absorption spectra overlap in the UV spectrum. Without this capability, fishes would not be able to make discriminations of e-vector independent of brightness or hue differences.

It has long been assumed the α-absorption band of a cone mechanism defines the range of the spectrum over which the cone mechanism is sensitive and the β-absorption band of the cone mechanism adds little utility to the responsivity of the cone mechanism. While this may be true for color vision, our research suggests this is not the case for polarization vision (Parkyn and Hawryshyn, 1993). The β-absorption band of the green- and red-sensitive cones overlaps with the α-band of the UV sensitive cones close to 360 nm. It is the β-absorption band of the green- and red-sensitive cones that renders these photoreceptors capable of exhibiting the polarization sensitivity (Parkyn and Hawryshyn, 1993). Therefore, under the appropriate photic conditions, UV stimuli can stimulate the UV (through α-band stimulation), green-, and red-sensitive mechanisms (through β-band stimulation), producing orthogonal polarization sensitivity. Two differentially sensitive polarization cone mechanisms operating in the same spectral range (UV) provide a convincing basis for neural coding and hence the potential for discrimination of e-vector.

4. Central Nervous System Processing of Polarization Information

Studies have demonstrated the presence of tectal cells sensitive to e-vector in fishes (Waterman and Aoki, 1974; Waterman and Hashimoto, 1974). Recent electrophysiological experiments have demonstrated neuronal processing underlying polarization vision in salmonid fishes (Coughlin and Hawryshyn, 1995). Using extracellular, single-unit recording techniques, we recorded from neurons in the *torus semicircularis* (ts), a subtectal central nervous system (CNS) structure associated with multimodal dependent orientation behavior. These neurons exhibited e-vector tuning. The experiments identified UV/red-sensitive cone opponent units, such as the unit shown in Figure 13.3A (UV-on and red-off). The spectral sensitivity characteristics of the color-opponent unit permit one to identify the origin of the cone mechanism response. We then confirmed orthogonal polarization sensitivity in the color-opponent pair, by demonstrating that the single-unit interneuron, which receives these inputs, exhibits responses of opposite polarity for orthogonal presentations of e-vector in the UV spectrum (Fig. 13.3B). This observation explains why fishes fail to show correct behavioral orientation when the polarized light field does not contain UV light. Without two differentially sensitive polarization receptors, the potential for discrimination on the basis of e-vector orientation is not possible (Coughlin and Hawryshyn, 1995). While the neurons in the *torus semicircularis* can make this basic discrimination, their acuity for finer-scale angular

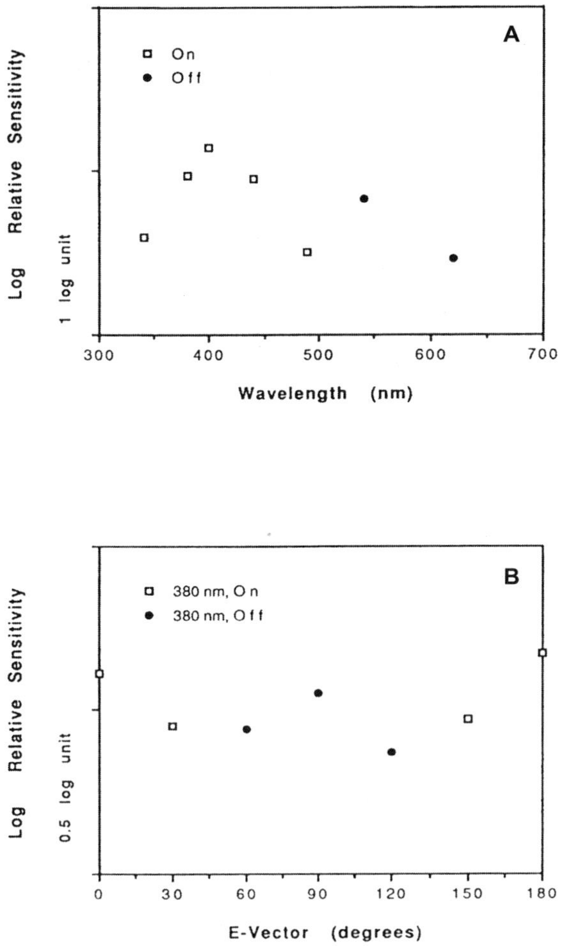

FIGURE 13.3. Spectral and polarization curves of a biphasic unit recorded from the *torus semicircularis* in rainbow trout. (A) Spectral sensitivity of a UV-red color-coded unit. A weak yellow background was used to achieve sensitivity balance in the UV- (on-response shown by the open squares) and red-sensitive (off-response shown by the filled circles) cone mechanisms. (B) Polarization sensitivity of color-coded unit to a 380 nm polarized stimulus. The unit shows coding for e-vector of a 380-nm polarized stimulus, with vertical e-vectors generating threshold on-responses and horizontal e-vectors generating threshold off-responses. When spectral sensitivity was later measured under a strong yellow background, it was a biphasic unit as described above. (After Coughlin and Hawryshyn, 1995.)

differences in e-vector seems questionable and this has been corroborated by behavioral experiments that evaluate the e-vector discrimination of rainbow trout (Degner and Hawryshyn, 2001).

5. Biophysical Models of Photoreceptor Polarization Sensitivity

Our knowledge of the specific mechanisms mediating the process of polarization vision is more thoroughly understood for invertebrates than for vertebrates. The organization and orientation of visual pigment–bearing membrane in photoreceptors is well known for invertebrates. Hence the biophysical basis for the preferential absorption of e-vector by invertebrate photoreceptors has been convincingly established (Wehner et al., 1975). The orientation of visual pigment molecules within a given rhabdomeric microvillus is aligned in one axis, permitting preferential absorption of plane-polarized light.

The biophysical mechanism for polarization sensitivity of different cone types in fishes (Hawryshyn and McFarland, 1987) remains unclear, although a number of hypotheses have been suggested: an intraphotoreceptor dichroic filter, chromophore alignment, light obliquely striking the outer segment, receptor wave-guiding, and receptor oil droplet refraction. Our observations that salmonids may use different cone mechanisms to accomplish the task of polarization vision has presented an interesting challenge for determining the biophysical characteristics of photoreceptors, which generate cone selective sensitivity to e-vector. While Hawryshyn and McFarland (1987) showed that UV-sensitive cones prefer the vertical e-vector orientation and the green- and red-sensitive cones prefer the horizontal e-vector, single-unit recordings have shown that e-vector coding occurs only in the UV spectrum and only when UV-sensitive cones are active (Coughlin and Hawryshyn, 1995). Therefore, neuronal coding of the e-vector appears to occur as a result of differential responses in UV-sensitive cones alone, within the cone mosaic, with respect to e-vector. How can differential e-vector coding occur only within the UV spectrum?

The spatial pattern of the cone mosaic and the ultrastructural properties of double-cones appears to play a major role in UV polarization vision. Novales-Flamarique et al. (1998) demonstrated that double cones possess a partitioning membrane that separates the two elements of the double cone and this membrane tilts (roughly 15°) at the distal end of the inner

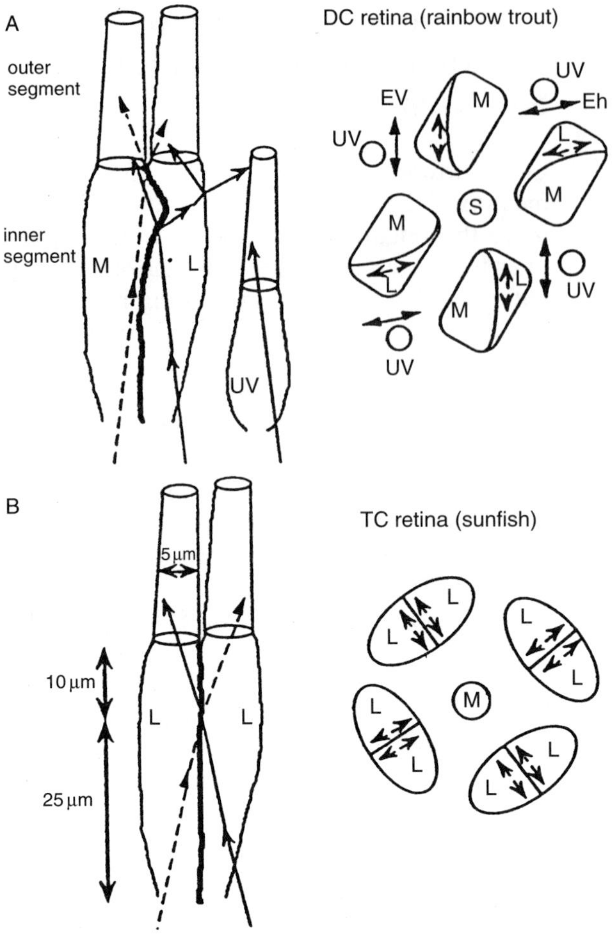

FIGURE 13.4. Schematic representation of anisotropic transfer of linearly polarized light, assuming that the partitions act as dielectric mirrors. (A) Rainbow trout double- and (B) green sunfish twin-cone mosaic units. Letter designations indicate long or red-sensitive (L), medium or green-sensitive (M), short or blue-sensitive (S), and ultraviolet-sensitive (UV) cone outer segments. Double arrows indicate the predominant polarization direction after reflection from the partition. The TC dimensions indicated in B are approximately valid for the corresponding DC ones in A. (After Novales-Flamarique et al., 1998.)

segments. The tilt is oriented toward neighboring UV cones and thus polarized light is reflected from this tilted surface onto adjacent UV cones at an angle of attack transverse to the axial orientation of cone photoreceptors, facilitating dichroic absorption of incident polarized light. The axial reflection of polarized light off the partitioning membrane of a double cone onto an adjacent UV cone is illustrated in Figure 13.4. The right panel of Figure 13.4A shows a tangential view of the cone mosaic, and in this perspective note that the axial reflection from double cone partitioning membrane surfaces is directed in two directions as double cones are arranged in the square cone mosaic. The two directions of reflection are arranged orthogonal to one another. Hence, reflection in orthogonal directions would result in a two-detector polarization system based mainly on the contributions of the UV cones. Within a square cone mosaic unit: (1) vertically oriented polarized light would strike the outer segments of two corner UV cones, situated at oblique positions in the corners of the cone mosaic; and (2) horizontally oriented polarized light would strike UV cones, obliquely positioned but in opposite corners of the cone mosaic. Thus double cones play an important role in the translation of the polarization orientation information into intensity modulation of the UV cones. While the UV cones receive the transverse reflection, the outer segments of the double cones receive the residual polarization distal to the partition membrane reflection. This is consistent with what is normally observed with respect to polarization sensitivity characteristics of chromatically adapted and isolated cones (Hawryshyn and McFarland,

1987). Blue-sensitive cones, which are the central cones in the square mosaic unit, do not receive transverse reflections from the double cone partitioning membranes and hence do not exhibit polarization sensitivity.

Figure 13.4B (left panel) shows the mosaic pattern and radial view of green sunfish double cones. Green sunfish have been reported to possess polarization sensitivity (Cameron and Pugh, 1991); however, attempts to replicate the data have not been successful (Novales-Flamarique and Hawryshyn, 1997b). Figure 13.4B shows that double cones of green sunfish lack the characteristic tilt in the partitioning membrane seen in salmonids, and thus cannot generate the transverse reflection onto neighboring cones. Furthermore, attempts to identify birefringent inclusions in the inner segment of these cones have been unsuccessful (Novales-Flamarique et al., 1995). Therefore, it seems sunfish lack the structural characteristics of photoreceptors for polarization sensitivity, which we identify in salmonids. Other species such as the common white sucker possess UV cones yet lack polarization sensitivity. Histological examination of common white sucker photoreceptor organization shows a random mosaic of cone types and hence there is no geometric plan for the cone-specific reflection patterns observed in cyprinids and salmonids (Novales-Flamarique and Hawryshyn, 1998). Thus, the critical factor determining polarization vision, at least in cyprinid and salmonids, is the geometric pattern of the cone mosaic and the tilted partitioning membrane of the double cones. Our current research efforts will extend this work by examining the interneuronal processing in the retina, especially the role of horizontal and bipolar cells in coding e-vector information.

6. Ontogenetic Changes in the Spatial Pattern of Cone Mosaics and Polarization Sensitivity

Rainbow trout (*Oncorhynchus mykiss*) lose UV photosensitivity during the course of development (Hawryshyn et al., 1989; Beaudet et al., 1993). Figure 13.5 illustrates the spectral

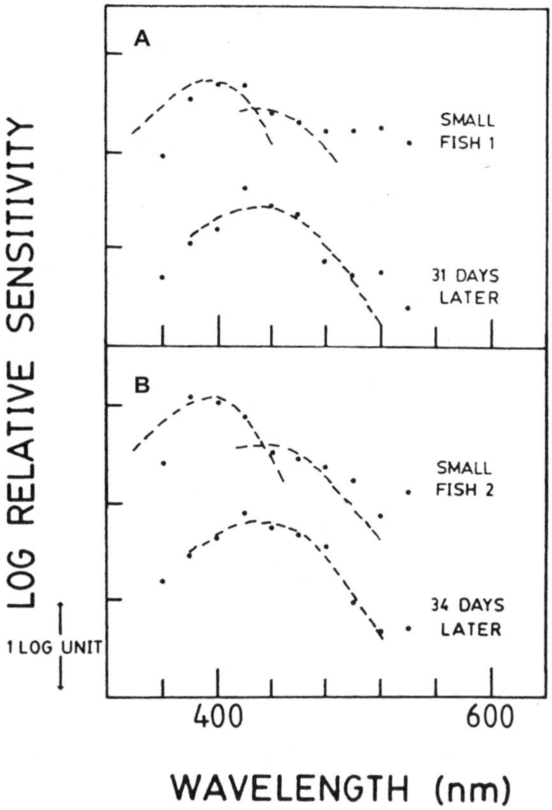

FIGURE 13.5. Changes in UV sensitivity within individuals. The spectral sensitivity of two rainbow trout was measured and then remeasured about one month later. A yellow-plus-blue background was used to isolate the UV peak for both panels A and B. (A) The top curve shows the spectral sensitivity of a rainbow trout weighing 44 g. The 390-nm and 430-nm λ_{max} absorption curves were compared to the spectral-sensitivity points. Fish no. 1 exhibited an increase in body weight of 17 g over a 31-d period and the retest spectral-sensitivity curve is shown in the lower curve. A 430-nm λ_{max} absorption curve was compared to the lower spectral-sensitivity curve. (B) The top curve shows the spectral sensitivity of rainbow trout weighing 60 g. The 390-nm and 430-nm λ_{max} absorption curves were compared to the spectral-sensitivity curve. Fish no. 2 exhibited an increase in body weight of 9 g over a 34-d period and the retest spectral-sensitivity curve is shown in the lower curve. A 430-nm λ_{max} absorption curve was compared to the lower spectral-sensitivity curve. *Note*: Curves in this figure are arbitrarily arranged on the ordinate for clarity. (After Hawryshyn et al., 1989.)

sensitivity of two juvenile trout exhibiting UV sensitivity with a peak sensitivity of 390 nm and the same trout about 25% larger showing a characteristic loss of UV sensitivity. This observation has been shown for both within and between fish measurements (a similar conclusion was made by Bowmaker and Kunz, 1987, for brown trout). Evidence from Kunz et al. (1994) indicated that the disappearance of UV photosensitivity in salmonids results from programmed cell death (apoptosis) of UV cones soon after transition to the adult stage. This inference was further supported by the premature loss of UV sensitivity in rainbow trout treated with thyroxine (Browman and Hawryshyn, 1992, 1994), a hormone known to play a key role in the development of all vertebrates. Histological examinations of tangential sections of salmonid retina were examined to determine the spatial distribution of photoreceptors and a full square mosaic was found in the majority of the retina. The salmonid retina consists of a square mosaic made up of double cones (green- and red-sensitive), blue-sensitive single cones, and UV-sensitive single corner cones in parr salmonids (freshwater juvenile stage), but subsequent to the parr-smolt metamorphosis salmonids lose their UV cones (see Figs. 13.6A and B for an example of the changes in the cone mosaic of rainbow trout).

In a recent study, Beaudet et al. (1997) found that within the cone mosaic of a sexually mature salmonid fish, the accessory corner cones were again present, suggesting the potential for UV cone regeneration at a point later in the fish's life history. Mechanisms for the regeneration of UV cones are not fully understood. UV-sensitive corner cones may regenerate from pluripotent cells within the retina or it is possible that rod precursor cells are differentiating into UV corner cones; these possibilities are now being investigated. It is important to note, however, that there are important changes in the pattern of the square cone mosaic during the life history of salmonids that have important implications for the detection of UV-polarized stimuli.

Our current understanding of the ontogenetic changes in the cone mosaic of salmonids provides an opportunity to sketch a model of life-history events and visual function in salmonid fishes: (1) Juvenile salmonids (freshwater parr) possess UV-sensitive cones and hence polarization sensitivity. (2) At the transition to the marine environment, commonly referred to as smoltification, UV-sensitive cones disappear from the retina and this appears to be mediated through elevated levels of serum thyroxine (Browman and Hawryshyn, 1992, 1994). With the loss of UV cones, polarization vision also disappears since polarization sensitivity is now limited to one polarization-sensitive channel (double cones) and hence is univariant. (3) At the point of sexual maturity, salmonids exhibit regeneration of UV cones, which may be triggered by an elevation of serum thyroxine (Browman and Hawryshyn 1994; Beaudet et al., 1997; Deutschlander et al., 2001). Thus, in the life-history stage leading up to and during sexual maturity, salmonids may possess the capability of UV-polarization vision (Beaudet and Hawryshyn, 1999; see Fig. 13.6C). The presence of UV-polarization vision could help salmon navigate during migrations.

7. Behavioral Evidence of the Use of UV Polarization Vision in Navigation

Studies have shown that fishes exhibit free-swimming spatial orientation to polarized light fields. Two behavioral techniques have been used to examine the response of fishes to linearly polarized light:

1. *Innate/unconditioned studies*, whereby an e-vector field is imposed on a circular orientation tank and the angular orientation of the fish is scored, relative to the e-vector orientation. While data from these studies show considerable variability in angular response, they nonetheless exhibit statistically significant orientation to the e-vector (Waterman and Forward, 1970; Forward and Waterman, 1973). Studies have established that this orientation results from the use of e-vector as a cue (i.e.,

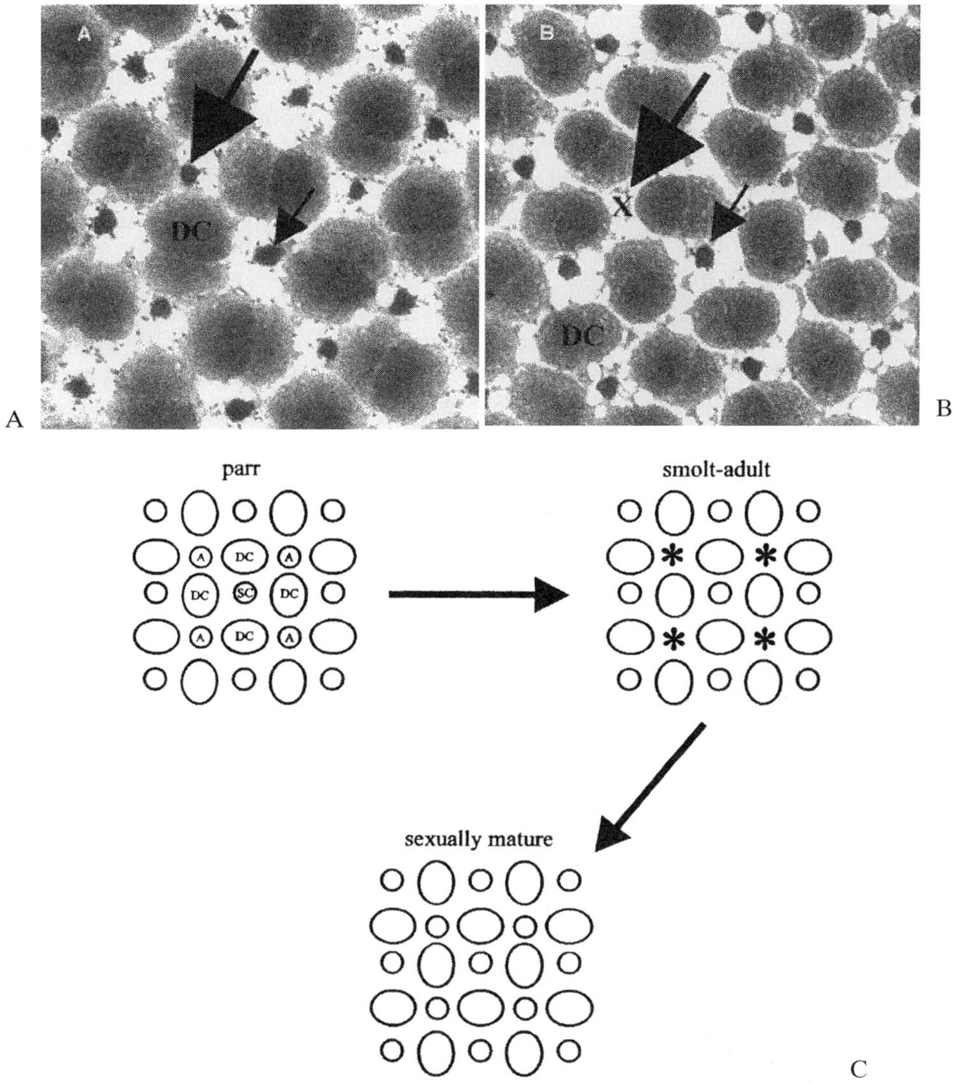

FIGURE 13.6. Cone mosaic pattern observed in salmonid fishes. (A) A complete square cone mosaic sampled from rainbow trout. The small arrow points at a central cone (blue-sensitive cone), which is quadrilaterally surrounded by four double cones labeled DC (green- and red-sensitive cones). Note that the partitioning membrane of each double cone is aligned to the central cone. The large arrow points to a corner cone (UV-sensitive cone); these are arranged in the corners of the square cone mosaic. (B) An incomplete square mosaic sampled from rainbow trout. In this illustration, the small arrow points at the center cone and the large arrow points to an X, marking the location where one would expect a corner cone. (Hawryshyn, C.W. et al., unpublished.) (C) Schematic representation of the ontogenetic sequence of the disappearance and reappearance of UV cones in the retina of Pacific salmonids. When sectioned tangentially at the level of the photoreceptors, the fish retina normally exhibits arrays of cone cells organized in a crystal-like fashion. At the time of seaward migration, the accessory corner cone (A) is lost from the retina of parr and only double (DC) and central single cones (SC) remain. This process is reversed later in life, at the time of sexual maturation and after the return to freshwater. (After Beaudet and Hawryshyn, 1999.)

polarotaxis) rather than from differential brightness patterns (i.e., phototaxis) generated from the polarized light reflecting off the internal surfaces of the test tank (Forward and Waterman, 1973). A completely different technique was used by Kawamura et al. (1981) in which innate responses to flashes of polarized light from above were monitored in terms of changes in unconditioned heart rate activity (bradycardia). A number of species of fishes including cichlids and trout were shown to exhibit innate cardiac responses to polarized light.

2. *Conditioned response studies*, whereby a fish is trained to swim to a particular location within a tank under an imposed polarized light field. The fish is given a food reward for swimming to a target location, and after a number of trials, the fish's movement to the target location becomes consistent. Once trained, a fish is released to the center of a test tank and tested for angular orientation to a randomly imposed e-vector orientation. The advantage of this technique is that it reduces the variance in angular orientation both within and between individuals and this in turn permits the investigator to reliably conduct longitudinal studies. A summary of the experiments that we have conducted on rainbow trout (*Onchorhyncus mykiss*) is given in Table 13.1. These experiments show a number of important features of the polarization vision-mediated orientation behavior in trout: (i) The orientation accuracy of test fishes is highly dependent on the spectral composition of the polarized light field. A polarized light field containing UV radiation is necessary for orientation, since the angular orientation performance of trout is notably compromised when the UV-sensitive cones are not stimulated. (ii) Partially polarized light can affect the orientation accuracy when the degree of polarization is less than 60–70% (Hawryshyn and Bolger, 1991; recently confirmed by Novales-Flamarique and Hawryshyn, 1997a, using electrophysiological recording).

Because the degree of polarization in natural marine and freshwater ecosystems is variable, fishes using the polarized light cues may require behavioral adjustments, such as position, in the water column to facilitate the acquisition of the polarized light cue. Our most recent behavioral orientation experiments, using a field situation for testing, indeed show that salmonids (sockeye, rainbow trout, steelhead salmon) are capable of orienting to the prevailing e-vector down to as low as 45% (Parkyn, 1998). The difference in performance of salmonids in labora-

TABLE 13.1. Summary of experiments on orientation of trout to polarized light.

Photic conditions	Proportion of trout statistically oriented to e-vector
(I) *Spectral content*	
(i) UV plus visible-spectrum polarized light field	12/14
(ii) No UV; only visible-spectrum polarized light field	3/11
(iii) UV polarized light field	12/15
(II) *Degree of polarization*	
UV plus visible-spectrum partially polarized light field	17/19
(i) 90%	11/14
(ii) 83%	12/13
(iii) 77%	10/17
(iv) 75%	3/13
(v) 68%	2/17
(vi) 65%	0/15
(vii) 63%	
(III) *Ontogeny*	
(i) Immature (possesses UV cones)	3/3
(ii) Mature (no UV cones present)	0/3

tory (tungsten-halogen light source) and field conditions maybe related to proportionately higher content of UV radiation in natural sunlight. (iii) Salmonids exhibit an ontogenetic loss of UV sensitivity (Bowmaker and Kunz, 1987; Hawryshyn et al., 1989; Browman and Hawryshyn, 1992, 1994) that can affect orientation accuracy (Hawryshyn et al., 1990).

8. Future Directions

1. *How life-history strategy influences visual function in salmonids.* Pacific salmonids exhibit an impressive array of life-history strategies, from landlocked (freshwater) to obligatory anadromous forms (distinct freshwater and marine phases). While many species of Pacific salmon are anadromous, there are divergent life-history strategies within species such as *Onchorhyncus mykiss* (rainbow trout—land locked; steelhead—anadromous) and *Onchorhyncus nerka* (kokanee—land locked; steelhead—anadromous). Yet other species, such as cutthroat and brook trout, exhibit facultative anadromy, moving frequently between the freshwater and marine habitat. Our knowledge of natural plasticity of the cone mosaic of salmonids would suggest that there may be interesting intraspecific and interspecific differences in the visual capabilities in Pacific salmonids. Changes in the structure of the cone mosaic translate directly into the functional utility of UV-polarization vision of salmonid fishes, in this spectrum of photic environments.

2. *Open-ocean utilization of polarization cues by salmonids assessed through ultrasonic telemetry.* Piloting, compass orientation, and bicoordinate navigation have been proposed as conceptual models for navigation in salmonids but the nature of the stimuli used in the open-ocean environment remains to be described. It has been well established that salmon are capable of using polarized light to guide their orientation behavior but the step toward examining this behavior depends on the development of ultrasonic telemetry systems that are capable of reliably tracking salmon and examining the directional behavior of salmon while experimentally manipulating polarized light cues. Ultrasonic tags have been successfully used to track salmonids in the open-ocean and coastal marine ecosystems; however, the technology currently available has restricted the number of fishes that could be tracked at any given time and the duration that any salmon could be tracked. A towed-array system now under development would permit the acquisition of tracking data from a number of salmon. Placing hydrophones at greater depth and using array gain to increase the range at which fishes can be tracked will allow us to use ultrasonic telemetry simultaneously in several salmon implanted with ultrasonic tags. The system would provide the horizontal position of each fish in addition to recording the time series of measurements from each tag (e.g., temperature, depth, heart rate). Optically active eye covers can modify or degrade the polarization field incident on the eye. The relative movements of salmon that differ in the imposed optical characteristics of the eye, in relation to the local meteorological and oceanographic conditions, will be used to assess the potential use of UV-polarized light cues as guidance mechanisms.

Acknowledgments. I would like to thank Theodore Haimberger and Jim Plant for their assistance in assembling the illustrative material. Garnet Martens kindly provided the photomicrographs of the cone mosaics used in Figures 6A and B. I am extremely grateful to the Natural Sciences and Engineering Research Council of Canada, which has funded my research program since 1984. Finally, I would like to thank all my graduate students and postdoctoral fellows who have made important contributions to this research.

References

Beaudet, L., and Hawryshyn, C.W. (1999). Ecological aspects of vertebrate visual ontogeny. In: *Adaptive Mechanisms in the Ecology of Vision* (Archer, S.N., Djamgoz, M.B.A., Loew, E.R., Partridge, J.C., and Vallerga, S., eds.), pp. 413–437. Great Britan, Kluwer Academic Publishers.

Beaudet, L., Browman, H.I., and Hawryshyn, C.W. (1993). Optic nerve response and the retinal structure in rainbow trout of different sizes. *Vision Res.* 33:1739–1746.

Beaudet, L., Novales-Flamarique, I., and Hawryshyn, C.W. (1997). Cone photoreceptor topography in the retina of sexually mature Pacific salmonid fishes. *J. Comp. Neurol.* 383:49–59.

Bernard, G.D., and Wehner, R. (1977). Functional similarities between polarisation and color vision. *Vision Res.* 17:1019–1028.

Bowmaker, J.M., and Kunz, Y.W. (1987). Ultraviolet receptors, tetrachromatic colour vision and retinal mosaics in the brown trout (*Salmo trutta*): Age-dependent changes. *Vision Res.* 27:2102–2108.

Brines, J.L., and Gould, J.L. (1982). Skylight polarisation patterns and animal orientation. *J. Exp. Biol.* 96:69–91.

Browman, H.I., and Hawryshyn, C.W. (1992). Thyroxine induces a precocial loss of ultraviolet photosensitivity in rainbow trout (*Oncorhynchus mykiss*). *Vision Res.* 32:2303.

Browman, H.I., and Hawryshyn, C.W. (1994). The developmental trajectory of ultraviolet photosensitivity in the rainbow trout is altered by thryoxin. *Vision Res.* 34:1397–1406.

Cameron, D., and Pugh, E. (1991). Double cones as a basis for a new type of polarisation vision in vertebrates. *Nature* 353:161–164.

Coughlin, D.J., and Hawryshyn, C.W. (1995). A cellular basis for polarised-light vision in rainbow trout. *J. Comp. Physiol. A.* 176:261–272.

Degner, S.L., and Hawryshyn, C.W. (2001). Orientation of rainbow trout (*Oncorhynchus mykiss*) to multiple patches of linearly polarized light. *Can. J. Zool.* 79:407–415.

Delius, J.D., Perchard, R.J., and Emmerton, J.A. (1976). Polarised light discrimination by pigeons and an electroretinographic correlate. *J. Comp. Physiol. Psychol.* 70:560–571.

Deutschlander, M.E., Greaves, D.K., Haimberger, T.J., and Hawryshyn, C.W. (2001). Functional mapping of UV photosensitivity during metamorphic transitions in a salmonid fish, *Oncorhynchus mykiss*. *J. Exp. Biol.* 204:2401–2413.

Forward, R.B., and Waterman, T.H. (1973). Evidence for e-vector and light intensity pattern discrimination by the teleost *Dermogenys. J. Comp. Physiol. A.* 87:189–202.

Hawryshyn, C.W. (1991). Light adaptation properties of the ultraviolet-sensitive cone mechanisms in comparison to the other receptor mechanisms of goldfish. *Vis. Neurosci.* 6:293–301.

Hawryshyn, C.W. (1992). Polarisation vision in fish. *Am. Sci.* 80:164–175.

Hawryshyn, C.W., and Beauchamp, R.D. (1982). Aberrant high blue sensitivity in goldfish. *Invest. Opthalmol. Vis. Sci. Suppl.* 22:282.

Hawryshyn, C.W., and Beauchamp, R.D. (1985). Ultraviolet photosensitivity in goldfish: An independent UV retinal mechanism. *Vision Res.* 25: 11–20.

Hawryshyn, C.W., and Bolger, A. (1991). Spatial orientation of rainbow trout: Effects of the degree of polarisation of the polarised light field. *J. Comp. Physiol. A.* 167:691–697.

Hawryshyn, C.W., and Hárosi, F.I. (1991). Ultraviolet photoreception in carp: Microspectrophotometry and behaviorally determined action spectra. *Vision Res.* 31:567–576.

Hawryshyn, C.W., and Hárosi, F.I. (1994). Spectral characteristics of the visual pigments in rainbow trout (*Oncorhyncus mykiss*). *Vision Res.* 34: 1385–1392.

Hawryshyn, C.W., and McFarland, W.N. (1987). Cone mechanisms and the detection of polarised light in fish. *J. Comp. Physiol. A.* 160:459–465.

Hawryshyn, C.W., Haimberger, T.J., and Deutschlander, M.E. (2001). Microspectrophotometric measurements of vertebrate photoreceptors using CCD-based detection technology. *J. Exp. Biol.* 204:2431–2438.

Hawryshyn, C.W., Arnold, M.G., Bowering, E., and Cole, R.L. (1990). Spatial orientation of rainbow trout to plan-polarised light: The ontogeny of e-vector discrimination and spectral sensitivity characteristics. *J. Comp. Physiol. A.* 166:565–574.

Hawryshyn, C.W., Arnold, M.G., Chaisson, D.J., and Martin, P.C. (1989). The ontogeny of ultraviolet photosensitivity in rainbow trout. *Vis. Neurosci.* 2:247–254.

Horváth, G., and Varjú, D. (1997). Polarisation pattern of freshwater habitats recorded by video polarimetry in red, green and blue spectral ranges and its relevance for water detection by aquatic insects. *J. Exp. Biol.* 200:1155–1163.

Horváth, G., and Wehner, R. (1999). Skylight polarisation as perceived by desert ants and measured by video polarimetry. *J. Comp. Physiol. A.* 184:1–7.

Ivanoff, A., and Waterman, T.H. (1958). Factors, mainly depth and wavelength, affecting the degree of underwater polarisation. *J. Mar. Res.* 16:283–307.

Kawamura, G., Shigata, A., and Yonemori, T. (1981). Response of teleost to the plane of polarised light as determined by the heartbeat rate. *Bull. Jpn. Soc. Sci. Fish.* 47:727–729.

Kunz, Y.W., Wildenburg, G., Goodrich, L., and Callaghan, E. (1994). The fate of ultraviolet receptors in the retina of the Atlantic salmon (*Salmo salar*). *Vision Res.* 34:375–383.

Loew, E.R., and McFarland, W.N. (1990). The underwater visual environment. In: *Vision in Fishes* (Douglas, R., and Djamgoz, M.B.A., eds.), pp. 1–43. London: Chapman & Hall.

Novales-Flamarique, I., and Hawryshyn, C.W. (1997a). Is vertebrate use of underwater polarised light restricted to crepuscular time periods? *Vision Res.* 37:975–989.

Novales-Flamarique, I., and Hawryshyn, C.W. (1997b). No evidence for polarisation sensitivity in freshwater sunfish from multi-unit optic nerve recordings. *Vision Res.* 37:967–973.

Novales-Flamarique, I., and Hawryshyn, C.W. (1998). The common white sucker: A fish with ultraviolet sensitivity that lacks polarisation sensitivity. *J. Comp. Physiol. A.* 182:331–341.

Novales-Flamarique, I., Hawryshyn, C.W., and Hárosi, F.I. (1998). Double cone internal reflection as a basis for polarised light detection. *J. Opt. Soc. Am. A.* 15:349–358.

Novales-Flamarique, I., Oldenbourg, R., and Hárosi, F.I. (1995). Transmission of polarised light through sunfish double cones reveals minute optical anisotropies. *Biol. Bull.* 189(2):220–222.

Parkyn, D.C. (1998). Visual biology of salmonids with special reference to polarised light sensitivity. PhD dissertation. University of Victoria. Canada.

Parkyn, D.C., and Hawryshyn, C.W. (1993). Polarised-light sensitivity in rainbow trout (*Oncorhychus mykiss*): Characterization from multi-unit responses in the optic nerve. *J. Comp. Physiol. A.* 172:493–500.

Parkyn, D.C., and Hawryshyn, C.W. (2000). Spectral and polarisation of sensitivity of salmonids: A comparative study. *J. Exp. Biol.* 203:173–1191.

Tinbergen, N. (1951). *The Study of Instinct.* New York: Oxford University Press.

Waterman, T.H. (1981). Polarisation sensitivity. In: Comparative physiology and evolution of vision in invertebrates. B. Invertebrate visual centers and behavior. In: *Handbook of Sensory Physiology*, VII/6B (Autrum, J., ed.), pp. 281–469. New York: Springer.

Waterman, T.H., and Aoki, K. (1974). E-vector sensitivity patterns in the goldfish optic tectum. *J. Comp. Physiol. A.* 95:13–27.

Waterman, T.H., and Forward, R.B. (1970). Field evidence for polarised light sensitivity of fish *Zenarchopterus. Nature* 228:85–87.

Waterman, T.H., and Hashimoto, H. (1974). E-vector discrimination by the goldfish optic tectum. *J. Comp. Physiol. A.* 95:1–12.

Wehner, R. (1989). Neurobiology of polarisation vision. *Trends Neurosci.* 12:353–359.

Wehner, R., Bernard, G.D., and Gieger, E. (1975). Twisted and non-twisted rhabdoms and their significance for polarisation detection in the bee. *J. Comp. Physiol. A.* 104:225–245.

14 Aspects of the Sensory Ecology of Cephalopods

Roger T. Hanlon and Nadav Shashar

Abstract

This chapter views cephalopod sensory capabilities in the context of behavioral ecology. Some extensive field studies and controlled laboratory experiments have recently shed new light on (1) camouflage and defence, (2) foraging and feeding, and (3) reproductive behavior. There have been exciting discoveries on mechanisms and functions of vision and olfaction. Cephalopods (excluding *Nautilus*) are highly visual animals, have keen visual acuity, and see well under highly varying light conditions. Nearshore cephalopods, such as the cuttlefish, have remarkable abilities to camouflage themselves on diverse substrates using visual cues alone. None of these cues include color because cuttlefishes are color-blind, and monochromatic vision may be the norm for many cephalopods. Cuttlefishes respond to the size, contrast, number, and area of light objects in the background to switch on disruptive coloration. Polarization sensitivity (PS) enables squid and cuttlefishes to detect transparent as well as silvery-reflecting prey more effectively, and they may use PS for intraspecific communication. Foraging octopuses use visual cues to camouflage themselves, but when movement negates camouflage, they become highly conspicuous, change appearance with remarkable frequency, and even mimic fishes. In the sea, hearing is a better long-distance sensor for predator detection: Cephalopods have low-frequency sensitivity and a lateral line (analogous to fishes) that may perform this function. Olfaction may play a role in mate choice of squid and cuttlefishes. In male squid, contact chemoreception of egg capsules stimulates highly aggressive behavior on spawning grounds. Octopuses seem capable of detecting food odor at a distance, and olfaction may improve predation on crabs by cuttlefishes. *Nautilus* olfaction is used for distant food odor detection and location, and perhaps for mate choice. Future emphasis on an integrative approach to the sensory ecology of cephalopods is likely to explain many other facets of their complex behavior.

1. Introduction

This chapter is about *what* cephalopods sense and *why*, not about *how* they sense the environment. Previous assessments of cephalopod sensory systems have depended mostly on anatomical, physiological, or biochemical data and made extrapolations concerning putative behavioral function; two reviews by Budelmann (1994, 1996) are especially noteworthy. Brief synopses have been provided by Wells (1978), Williamson (1995), and Hanlon and Messenger (1996). Vision, in particular, has been reviewed several times partly because more studies on vision have been conducted in cephalopods (e.g., Messenger, 1981, 1991; Hartline and Lange, 1984; Land, 1984; Saibil, 1990; Young, 1991; Muntz, 1999). There is particular need to learn more about chemical and mechanical stimuli that may be influencing cephalopod behavior (reviewed by Hanlon and Messenger, 1996).

We take a different approach here, one that is stimulated partly by several recent discoveries about cephalopod sense organs, but more so by greater advances in understanding the behavior of cephalopods. We take the liberty, from thousands of hours of direct observation and study of cephalopods in the natural environment and in large laboratory tanks, to speculate on ways that cephalopods use sensory information for a variety of behavioral decisions in their everyday lives. Thus, we attempt to blend the two to provide a "behavioral-ecological" view of cephalopod sensory function. Each of the following sections begins with a brief synopsis of its contents.

2. Camouflage and Other Forms of Defense

Cephalopods are perhaps the masters of camouflage but, surprisingly, achieve this in the absence of color vision. Cuttlefishes, in particular, produce highly disruptive body patterns on certain backgrounds, and use visual cues (size, contrast, number, and area) from the light-colored objects in the background to switch on the disruptive coloration. There is the possibility that cuttlefishes and squid can remain camouflaged to fishes while simultaneously signaling one another with polarization signals. Octopus camouflage fails when they move too quickly. In response to this problem, one species in Indonesia mimics flatfishes when it swims across open sand areas. Hearing is a better long-distance sensory modality for predator detection and cephalopods have low-frequency sensitivity and a lateral line (analogous to the lateral line in fishes) that may perform this function.

2.1. Matching Substrate Patterns Without Color Vision

Cephalopods are spectacular in their ability to match the color, brightness, texture, and pattern of a very wide variety of surrounding substrates (Fig. 14.1A,B; see color plate). A long-standing question is how they match color if, as a number of previous publications suggest, they are colorblind (cf. Messenger, 1991). Marshall and Messenger (1996) demonstrated that cuttlefishes do not respond to different wavelengths in the substrate (Fig. 14.1C). Instead, different body patterns were generated when they perceived different intensities in the substrate. The higher the contrast in the background the bolder and more high-contrast the body pattern. This experiment, which has since been repeated in *Octopus vulgaris* (Shashar and Lerner, unpublished observations), is a compelling demonstration of one aspect of "colorblind camouflage" since it provides insight into how cephalopods might make some visual discriminations of substrate values (intensity in this case). This type of experiment raises questions about how cephalopods might make other fine adjustments of color when they use mechanisms of crypsis as different as "general resemblance" and "disruptive coloration." In Figure 14.1A,B note the sophistication of the match. How do cephalopods determine other aspects of pattern? How do they match their skin with the texture of the surrounding algae, stones, or sand? Cuttlefishes, at least, match texture visually without tactile feedback (Hanlon and Messenger, 1988). It is also

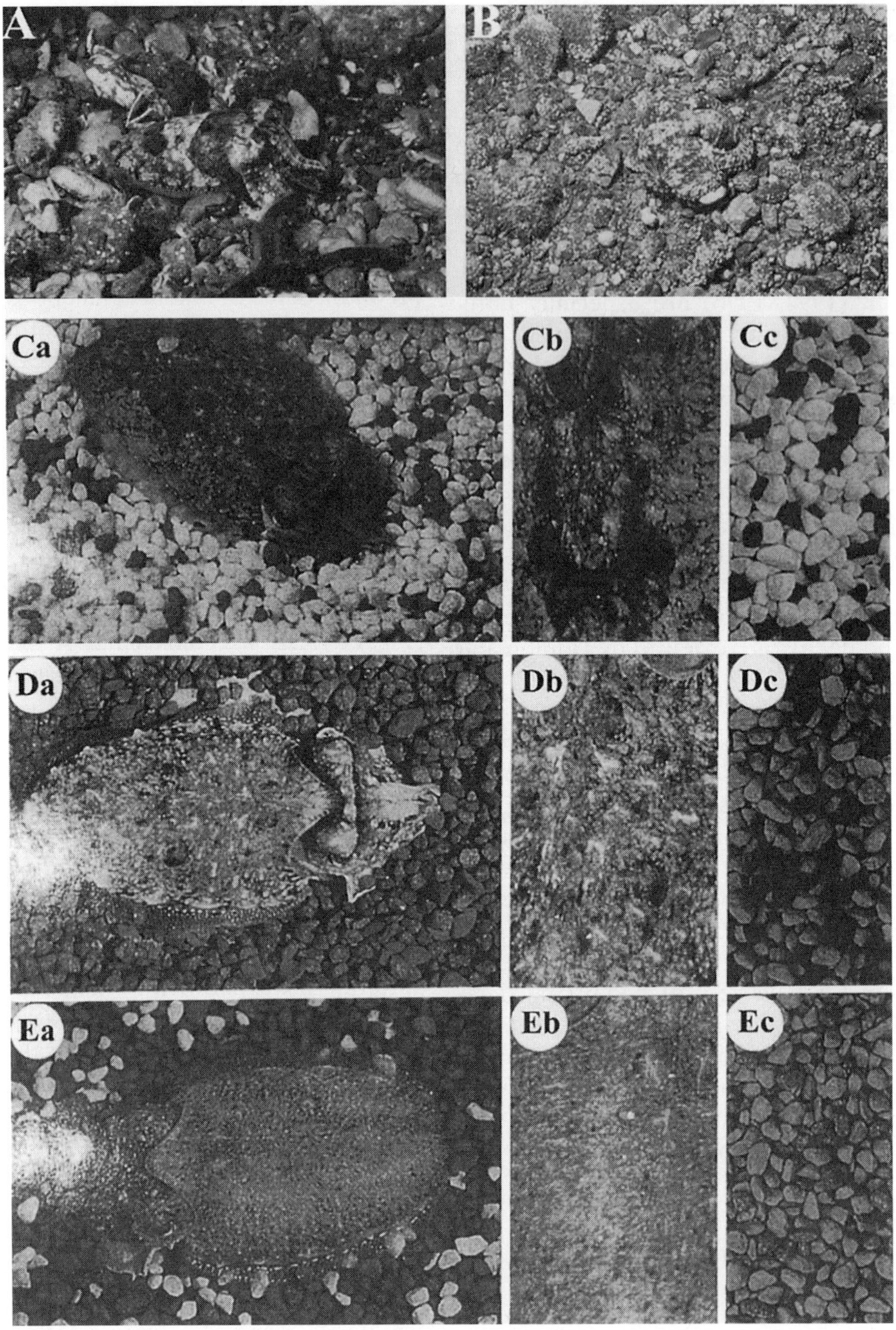

FIGURE 14.1. Camouflage by a color-blind cuttlefish. (A, B) Adult *Sepia officinalis* matching the color, brightness, texture, and pattern of the substrate in the Mediterranean Sea, and *Octopus bimaculatus* matching features of natural substrates off Catalina Island, California. (C–E) Subadult *S. officinalis* (100-mm mantle length) showing its color-blindness. Ca: photographed in daylight on a prepared background—red gravel on white; Cb: detail of mantle skin of the same animal; Cc: the same background photographed through a green interference filter. The D series is the same as C but with red gravel on blue; in E series, with yellow gravel on blue. The animals are not matching body color to that of the gravel background since they perceive it as uniform. (From Marshall and Messenger, 1996. Reprinted with permission from *Nature*. Copyright © 1996 Macmillan Magazines Limited.) (See color plate)

mysterious how an octopus or cuttlefish can match so many local substrate colors when their chromatophore pigments only cover longer wavelengths (browns, reds, yellows). Octopuses and cuttlefishes have many broadband reflectors in the skin (leucophores and iridophores) lying below the chromatophores. Messenger (summarized in 2001) has speculated that these are important in making color matches by reflecting local background colors. This remarkably adaptive "chromatophore system" requires considerably more research before it is understood from both the sensory and motor sides.

2.2. Cuttlefishes Respond to White Objects in the Substrate for Disruptive Camouflage

Unlike most animals that can use one or a few mechanisms of crypsis (Cott, 1940), cephalopods are known to switch swiftly (within fractions of a second) among general resemblance, disruptive coloration, countershading, deceptive resemblance, mimicry, rarity through neurally controlled polyphenism, and cryptic behavior (Hanlon and Messenger, 1996; Hanlon et al., 1999a). However, little is known about the visual features of the substrate that are used as sensory cues to switch from one mechanism of crypsis (e.g., general resemblance) to another (e.g., disruptive coloration).

How do cuttlefishes determine when to camouflage themselves by means of disruptive coloration (e.g., Figs. 14.1A and 14.2g) rather than, for example, general resemblance to the substrate? Chiao and Hanlon (2001a,b) addressed this question by presenting cuttlefishes with computer-generated substrates whose features could be controlled and quantified. They demonstrated that disruptive coloration was "switched on" visually by light objects in the background and that the size, contrast, and number of white squares in a checkerboard pattern determined whether cuttlefishes became disruptive in coloration. The size of the white checkerboard squares had to be close in size to the "white square" skin component on their dorsal mantles (Hanlon and Messenger, 1988) before they would show it. Note in Figure 14.2a,b that the squares are too large and thus the cuttlefish remain uniformly patterned. It was then determined (Figs. 14.2c–h) that cuttlefish responded to the *area* of white objects in the background, rather than their shape, to switch on its own "white square" (Chiao and Hanlon, 2001b). It was unexpected that a thin oval (Fig. 14.2h) would produce disruptive coloration (its shape is nothing like the "white square" of the animal), but this finding is consistent with their hypothesis that cuttlefishes are responding to the area of the object and not the shape. By testing like this with computer-generated patterns, it should now be possible to investigate other details of the visual cues that benthic cephalopods use to adjust their coloration in an adaptive manner.

2.3. Can Cephalopods Remain Camouflaged and Yet Still Communicate with One Another Using Polarization Signals?

Cuttlefishes possess a prominent polarization pattern on the arm and head region (Shashar et al., 1996). This pattern was not present during all behaviors, yet the responses of cuttlefishes (7 animals, both sexes; 10 presentations each) to their own images in a mirror changed when the polarization patterns of the reflected images were distorted. These initial behavioral data suggest that cuttlefishes might communicate with one another through a polarization visual channel. No definitive proof is yet available for this novel idea; however, *Octopus*, *Loligo*, and *Euprymna* can also produce and control polarization signals that emanate from iridophores in their arms (Cronin et al., 1995; Shashar and Hanlon, 1997; Hanlon et al., 1999b; Mathger and Denton, 2001; Shashar et al., 2001). Acetylcholine has been implicated in controlling the appearance of polarization patterns in the arms of squid (Shashar et al., 2001). The hypothesis of polarization communication has been examined further by Boal, Shashar, Hanlon, and collaborators (personal communication and unpublished observations). Significant findings were that a cuttlefish's polarization patterning changed when another

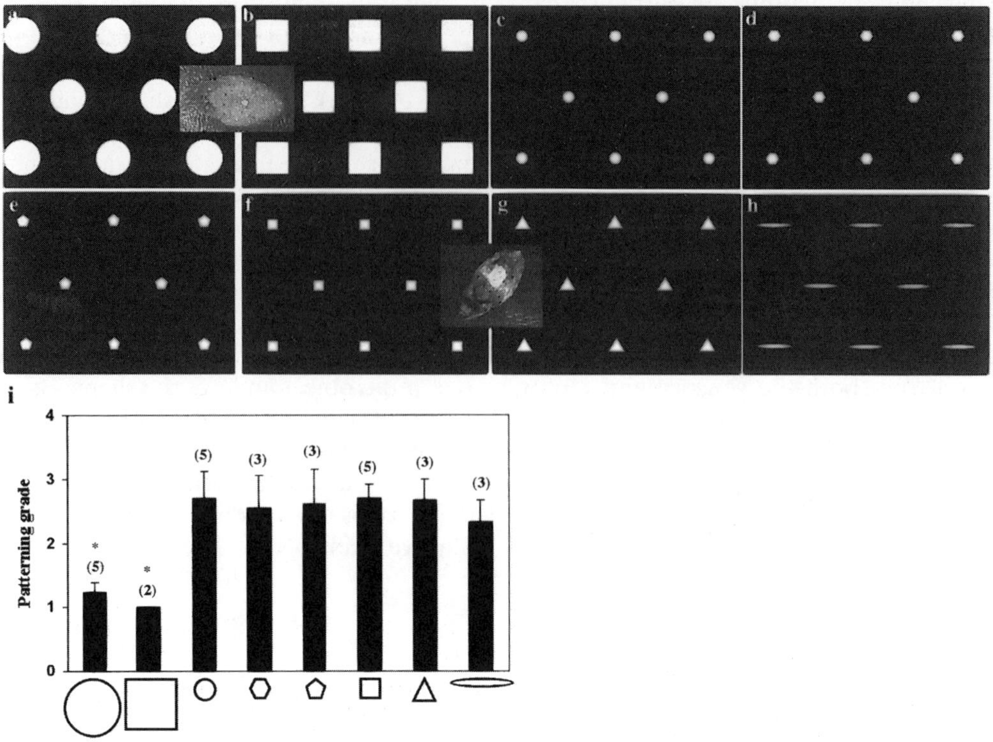

FIGURE 14.2. Cuttlefish (*Sepia pharaonis*) cueing on the *area* of light background objects to turn on their disruptive coloration. (a, b) Two large control images on which the cuttlefish (centred) did not show the white square on its mantle. For five images (c–g), cuttlefishes were expected to—and did—show white square disruptive coloration. On one image (h) they were not expected to elicit white square due to its highly different aspect ratio, yet they did because the area was the same. (i) A summary of results from all eight images. Patterning grade 3 is disruptive. The number in parentheses indicates the number of cuttlefishes tested. Results from a and b were significantly different from c–h ($p < 0.00001$). (From Chiao and Hanlon, 2001b.)

cuttlefish was watching it and females do not recognize other females when the latter's polarization patterns are distorted. Although somewhat circumstantial, these results suggest some sort of communication function. Polarization vision, and the potential for polarization-based signaling, is not limited to cephalopods (Marshall et al., 1999). If, in the future, a social function of polarization sensitivity can be proved (i.e., a communication channel "hidden" from predators without polarization sensitivity [PS]), it would reveal a previously unknown mode of communication in the animal kingdom.

Predator avoidance may be another function of PS. Since PS is used to detect and selectively attack transparent and opaque silvery reflecting prey (Section 3 below), it is reasonable to predict that cephalopods could identify polarization reflections of their predators (Shashar et al., 2000).

2.4. Octopuses Mimic Flatfishes Behaviorally

Foraging octopuses are often camouflaged except when they move quickly. *Octopus cyanea* on Pacific coral reefs uses the tactic of changing appearance hundreds of times per hour when moving, and by being highly conspicuous in coloration (Hanlon et al., 1999a). A different tactic has been discovered in an Indonesian octopus that lives on barren volcanic sand plains: It mimics locally abundant flatfishes (Norman et al., 2001). Hanlon, Conroy, and Forsythe (unpublished observations) showed that this "mimic octopus" stays camouflaged when still or slowly moving, and only mimics flatfishes when the octopus moves more rapidly or swims. Thus, when movement would betray camouflage, the octopus switches tactics and looks like a fish. Vision is the obvious modality for this behavioral decision. There is speculation that this octopus may mimic several other local species.

2.5. Can Squid "Hear" Predators Approaching?

A lateral line analogue was proposed by Hanlon and Budelmann (1987) and proven by Budelmann and Bleckmann (1988), who demonstrated that cuttlefishes and squid respond to water movements as low as 0.06 μm (Fig. 14.3B), which is equivalent to the threshold of the hair cells in fish lateral lines (Bleckmann et al., 1991). Biologically, this sensitivity means that cuttlefishes and squid are able to detect a moving fish of 1 m in body length from a distance of about 30 m (Budelmann, 1994). Hanlon and Budelmann (1987) mentioned field anecdotes in which squid and cuttlefishes reacted to predator fishes well before they appeared visually on clear coral reefs; this would correspond to distances of 25–30 m. Budelmann et al. (1991) suggested that juvenile cuttlefishes used the lateral line to capture shrimp in total darkness, and it is conceivable that schooling squid might use the lateral line to maintain school structure.

Low-frequency hearing was demonstrated by Packard et al. (1990), who used classical conditioning to show that cuttlefishes, squid, and octopus responded behaviorally (alteration in the breathing rhythms) to frequencies of about 1–100 Hz. They did not test higher frequencies, so additional research is needed on this subject. Packard et al. (1990) surmised that the statocyst organ may be the likeliest organ to sense these frequencies by detecting the particle motion rather than sound pressure.

3. Foraging and Feeding

Cephalopods are voracious carnivores that rely upon stealth, speed, and thus keen senses to obtain prey. High visual acuity combined with PS enables them to detect organisms camouflaged by transparency or silvery reflection. Given the novelty of those feats, might it be possible for midwater squid to use PS to detect counterilluminating prey? Distance chemore-

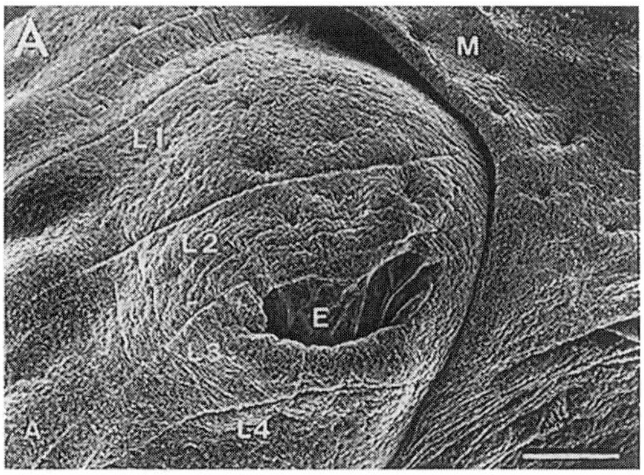

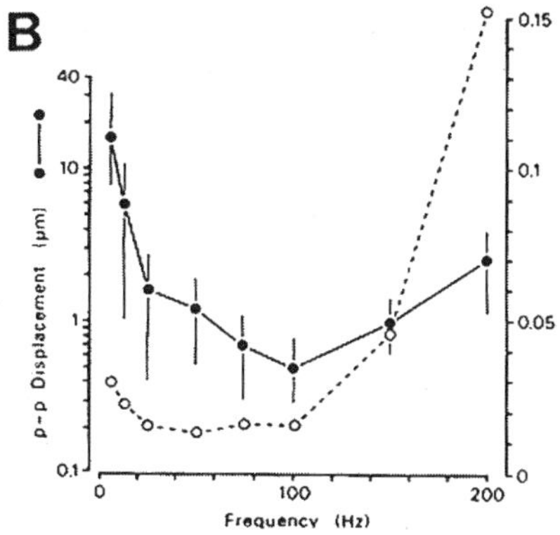

FIGURE 14.3. (A) Lateral line analogue in a 14-day-old cuttlefish, *Sepia officinalis*. Oblique lateral view to show the four epidermal lines (L1–L4) that run in anterior/posterior direction. Some lines continue onto an arm, A. E is the eye, and M is the anterior edge of the mantle. Scale bar is 0.5 mm. (From Budelmann, 1994. Copyright © 1994 from *Marine and Freshwater Behavior and Physiology*. Reproduced by permission of Taylor & Francis, Inc., http://www.routledge-ny.com) (B) Thresholds of the dorsal head lines to local water movements. The peak-to-peak water displacement (closed circles, solid line, left ordinate) and velocity (open circles, dashed line, right ordinate) are shown as a function of stimulus frequency. (From Budelmann and Bleckmann, 1988. Copyright © 1988 Springer-Verlag.)

ception may help octopuses to find hidden food organisms, and cuttlefishes to improve their first predation on dangerous prey such as crabs. *Nautilus* has long been known to be adept at distance chemoreception, yet recent experiments indicate that they can track odor plumes in three dimensions using the bilateral rhinophores.

3.1. Squid Use Polarization Sensitivity to Detect Transparent Prey

Transparency enables aquatic organisms to avoid detection by visual predators, and most zooplankton in the oceans are transparent (Fig. 14.4A). However, Shashar et al. (1998) inspected the tissues of 20 zooplankton species and found that various tissues such as muscles, cuticles, and photoreceptors are polarization active. Thus, a predator with PS would tend to see these zooplankton in high contrast (Fig. 14.4A, right). Shashar et al. (1998) demonstrated that squid hatchlings detected zooplankton prey at 70% greater distance when viewed under linearly polarized light, which could increase their foraging volume by nearly five times while in this planktonic phase of their paralarval life. Since zooplankton partially depolarized the light passing through them, PS may also provide a cue as to the location of zooplankton patches to the squid. These authors then demonstrated how adult squid might use PS. They examined the choices of adult squid when presented with small transparent glass beads that imitated prey items; some of the beads were polarization active and others were not (indicated by *PA* and *Reg* in Fig. 14.4B). Fifty squid were tested, and it was clear that they preferred polarization-active beads over regular beads (Fig. 14.4B, part G; PA = 50, Reg = 16; $x^2 = 17.52$, $p < 0.0001$). This latter experiment explains curious field observations that adult squid often appear to feed on tiny prey items only millimeters in length. Apparently, squid can see them better when contrast is enhanced by PS. It is strange and unexpected that squid would feed on prey so small, but mariculture experiments support the preferences of squid to feed on an extremely large size range of prey throughout their life cycle. Squid are mostly crepuscular and nocturnal feeders, and it may also be noteworthy that partially polarized light is at its maximum during crepuscular periods (Novales-Flamarique and Hawryshyn, 1997).

This begs another question: How can cephalopods see well in the full range of light conditions? In addition to high visual acuity—often similar to that of humans—cephalopods are sensitive to a wide range of light intensities. In their retinas, the lengths of the photoreceptors' outer segments are typically in the range of 200–400 μm in both diurnal and nocturnal cephalopods, including *Nautilus* (Muntz, 1991; Saibil and Hewat, 1987; Young, 1963; Zonana, 1961), suggesting that even animals that are most active in bright light conditions maintain high sensitivity for dim illumination. This outer segment length can shorten in response to light by approximately 25%, and movement of screening pigments is also part of the light adaptations (Young, 1963, 1971). In dim light conditions, when photoreceptors are longer and lack screening pigments, and when the pupil of the eye is wide open, blurring due to off-axis illumination can occur and somewhat reduce visual acuity (Muntz, 1999). However, the orthogonal arrangement and polarization sensitivity of neighboring photoreceptors may compensate for this blurring to maintain visual acuity, as well as PS, in a wide range of illumination conditions.

3.2. Cuttlefishes Use Polarization Sensitivity to Detect Opaque, Silvery Prey

Cephalopods often need to break the camouflage of (i.e., detect) silvery light-reflecting fishes. Does PS play a role in this detection? The concept behind silvery reflective scales is intensity and wavelength matching, which camouflages fishes by reducing the dark shaded areas on their bodies so that they match the background illumination against which they might be seen (Denton, 1970). Using an imaging polarimeter, Shashar et al. (2001) showed that a target fish produced areas that reflected partially linearly polarized light, which a PS predator would detect as high

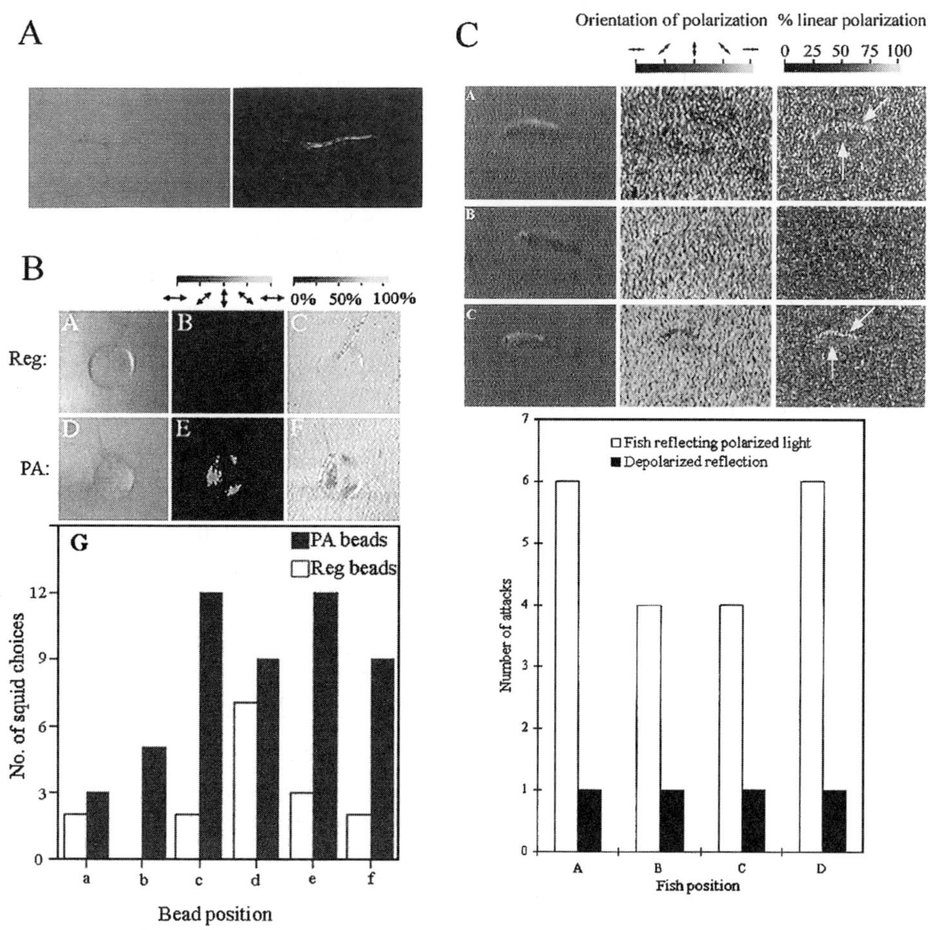

FIGURE 14.4. Overcoming camouflage of prey with polarization sensitivity. (A) A transparent *Lucifer* sp. shrimp camouflaged in normal light (left) and the same shrimp viewed between crossed polarizing filters, demonstrating contrast created by its polarization activity. (B) Adult squid attack choices of glass beads that imitated transparent prey. A–F: Glass beads (1-cm diameter, equal transparency in the 400–650 nm range) appear identical to a human observer. Using a submersible imaging polarimeter, the light intensity, A, D—left, e-vector orientation, B, E–center, and percentage of linear polarization, C, F—right, were measured. Some beads, D, E, F, were heat-stressed during manufacture and hence were polarization-active (PA), whereas others, A, B, C, were not stressed and did not affect the light's polarization (Reg). G: In 66 bead choices by squid, there was a significant preference for the PA beads independent of the bead's position (A–F, across the tank, 20 cm from each side, and 16 cm between them). (From Shashar et al., 1998. Reprinted with permission from *Nature*. Copyright © 1998 Macmillan Magazines Limited.) (C) Adult cuttlefish using PS to prey on silvery fishes. A–C: Target fish, *Peprilus triacanthus*, as observed underwater with an imaging polarimeter. A: Fish without any filter. B: Fish with a filter distorting linear polarization attached to its side. C: Fish with a polarization-inert filter. *Left*: A black/white image of the fish. *Center*: Orientation of polarization, where horizontal polarization is coded into white or black, and vertical polarization into 50% gray. *Right*: Partial polarization image where black represents unpolarized light (0%) and white codes for full linear polarization (100%). Two areas on fish that were main sources of reflected polarized light are indicated by arrows; the lateral side of the fish reflected light that was 30–50% linearly polarized and the dorsal area reflected light with 70–95% linear polarization. *Bottom panel*: cuttlefish significantly preferred fishes having regular reflection through a polarization-inert filter over fishes with a depolarized reflection ($x^2 = 17.3$, $p < 0.0001$). (From Shashar et al., 2000. Reprinted from *Vision Research*, Vol. 40, Shashar et al., Cuttlefish use polarization sensitivity in predation on silvery fish, pp. 71–75, Copyright © 2000, with permission from Elsevier Science.)

contrast against the seawater background (Fig. 14.4C). The question became: Do cuttlefish use PS in predation, and will they selectively attack such silvery reflecting fish? As shown in Figure 14.4C (bottom), adult *Sepia officinalis* preyed preferentially on fishes with normal polarization reflection compared with those that were presented with filters that greatly reduced polarization reflection (20 fishes with polarization, 4 without; $x^2 = 17.3$, $p < 0.0001$). As in the squid experiments (Section 3.1), cuttlefishes preferred prey targets that reflected polarized light, even though they (like squid) could probably see the other fishes presented simultaneously during these experiments.

Octopuses could use PS for predation too, as pointed out by Shashar and Cronin (1996), who demonstrated that octopuses could be trained to select objects (such as prey) based solely on the presence/absence of a pattern produced by polarization contrast within the target. Theirs was the first paper to demonstrate a visual discrimination based on polarization light patterns. Even more striking was the finding that octopuses could generalize the concept of contrast, no matter what the orientations or intensity of the center or surround were in the target.

3.3. Do Deepwater Squid Detect Prey Using PS?

Light at night as well as light at depth is plane polarized, although the extent and nature of the polarization are not known in full detail (cf. Waterman, 1955). Shashar, Milbury, and Hanlon (in press) recently demonstrated that at least some deep-sea squid have the orthogonal arrangement of microvilli in the retina required to detect PS.

What function could this fulfill? We propose that squid predators use PS when viewing upward to see prey silhouetted against downwelling light. This idea is based largely on the elegant work of Richard Young and colleagues (e.g., Young et al., 1979), who demonstrated that some midwater squid detect downwelling light and then produce counterillumination on their ventral surfaces so that their shadow is altered or eliminated. Presumably this reduces predation from below, since many animals look for shadows of prey against downwelling light. The counterillumination is often produced by ventral photophores, and some (if not many) of these photophores contain iridophores that reflect light before the light is "broadcast" (e.g., Arnold et al., 1974; Herring, 1994). Some of this light is likely to be linearly polarized due to the reflection process, and could potentially be distinguished from surrounding illumination by a predator with PS. The same mechanism could aid a cephalopod predator in shallow waters at night. Given the complex and diverse visual adaptations of aquatic animals (cf. Archer et al., 1999), it would not be very surprising that midwater squid might use this sensory specialty for predation. Testing this idea would be difficult but potentially rewarding.

3.4. Octopus Predation: Vision versus Contact and Distance Chemoreception

Octopuses hide in dens most of the time and emerge to forage. Clearly they can use their vision to locate and attack prey, and this behavior has been the basis for hundreds of visual discrimination and learning studies (Wells, 1978). Field studies, however, seem to tell a different story—one in which octopuses appear to be mostly tactile feeders, using the numerous suckers (eight arms, each with about 200 suckers) to sense and grasp a variety of molluscs and crustaceans (e.g., Mather, 1991; Forsythe and Hanlon, 1997). Contact chemoreception (mainly the suckers) is well developed and the octopus brain has separate brain centers for touch and taste learning (Wells, 1978; Hanlon and Messenger, 1996). But what about distance chemoreception? The experiments of Chase and Wells (1986), Boyle (1983), and Lee (1992) all show the ability of octopuses to detect and react to chemical cues (even single amino acids) from prey items. Collectively, these field and laboratory studies suggest that distance chemoreception may be more important in guiding octopuses to their food than has been assumed hitherto. The relative importance of suckers, lips, and "olfactory organs" (e.g., Lucero et al., 1992; Hanlon and

Messenger, 1996) in such responses remains to be established.

3.5. Cuttlefishes May Use Olfactory Cues to Improve First Predation on Crabs

Highly unexpected results were obtained by Boal et al. (2000), who were attempting to corroborate the finding of Fiorito and Scotto (1992) and demonstrate observational learning in cuttlefishes. These experiments were based on vision, specifically that cuttlefishes might more quickly learn how to attack a crab successfully (i.e., without being pinched) by first watching other cuttlefishes successfully attack a crab. However, one of the control experiments involved placing a cuttlefish in the experimental tank, although preventing visual contact. These "control" cuttlefishes (all of which were reared in the laboratory and had never seen or smelled a crab) were inadvertently exposed to crab odor due to the construction of the experimental tank, and they (as well as the cuttlefishes that observed predation) performed better in their first attacks on crabs. These results (which are unique in the observational learning literature; Robert, 1990) suggest that odor may serve as a primer for cuttlefish predatory attack behavior, perhaps by enhancing food arousal and improving attention.

3.6. *Nautilus* Detect and Find Food Sources by Distance Chemoreception

It has long been suspected that ancient *Nautilus* uses chemoreception as its primary sense. Only recently, however, has experimental evidence been provided. *Nautilus* is principally a scavenger. Basil et al. (2000), using a large flume under dark conditions, demonstrated that *Nautilus pompilius* could consistently detect and follow turbulent odor plumes to the source over distances up to 10 m. Intact or sham-operated *Nautilus*, upon detection of waterborne odor, swam toward the source; examples from six animals are shown in Figures 14.5A,B. These *Nautilus* usually swam more or less toward the source and adjusted their path when they swam out of the odor plume, even when they swam below or above the path, indicating an ability to sense in three dimensions (Fig. 14.5A, right). About a third of the animals performed wide casting movements across the plume, indicating that they have two types of orientation behavior. The paired rhinophores were necessary for orientation behavior: When they were blocked temporarily either uni- or bilaterally, *Nautilus* detected odor but could not track the plume and locate the source (Figs. 14.5C,D). The heading angles of normal *Nautilus* quickly became more accurate and stabilized in typical runs, but all intact animals showed poor heading angles as they approached within 60 cm of the source (Fig. 14.5B, right), perhaps indicating a gradual switch to a different behavior, possibly involving the tentacles. These laboratory findings corroborate telemetry data (e.g., O'Dor et al., 1993) from *Nautilus* in nature: Vertically foraging animals could potentially detect quantities of odor from horizontal cross-currents along the coral reef slopes and sample them in all dimensions even against a moderately strong flow. In the complex environment of a coral-reef/oceanic interface, detecting and following wisps of odor would be conducive to the "smelling and groping" mechanism of a vertical scavenger. The role of the 90 thin tentacles remains enigmatic, although they detect odor. Vision may still be highly important in foraging and daily vertical migrations by keeping sight of the bottom, because *Nautilus* does not migrate straight up and down, but rather along the slopes. Muntz (1999) also suggests that the pinhole camera eye may be able to detect bioluminescence at several meters' distance, and this may guide *Nautilus* to decaying food, which is often bioluminescent or attracts bioluminescent organisms.

4. Reproductive Behavior

Cephalopods are short-lived and fecund, and detailed study of some squid and cuttlefishes illustrate complex mating systems in which both males and females have multiple behavioral tactics. Male visual displays during agonistic bouts are elaborate, and females also have a

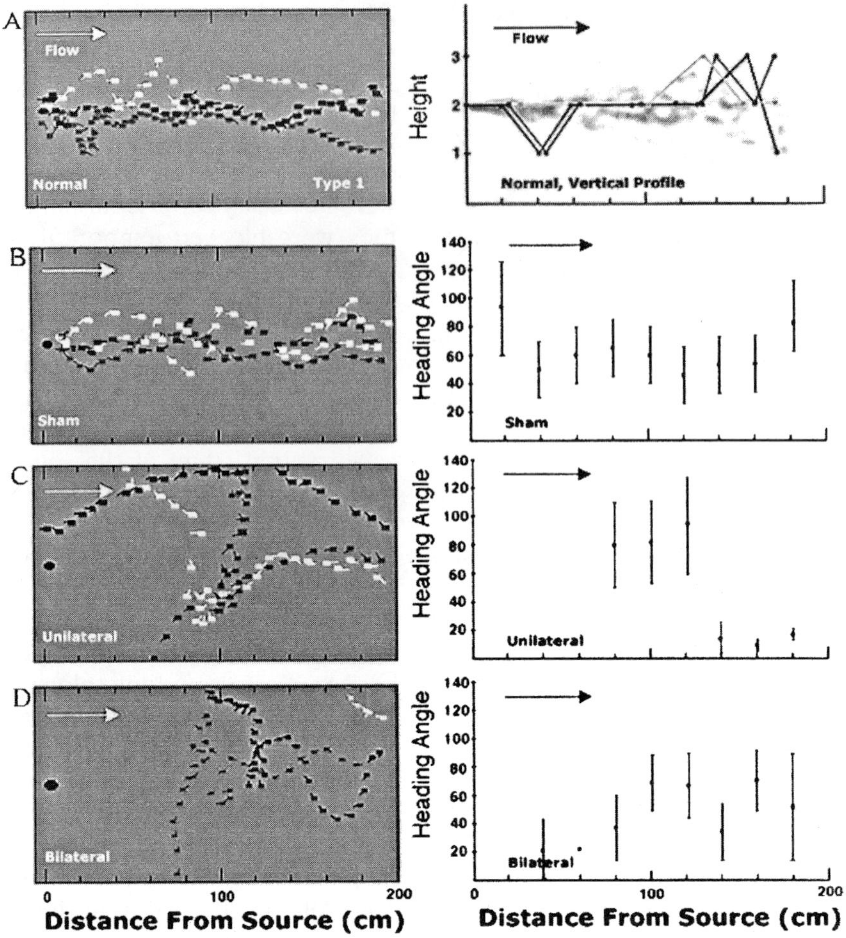

FIGURE 14.5. Chemotactic behavior in *Nautilus pompilius*. (A) *Left*: Source-directed tracks of three nautiluses (moving right to left) do a distant odor source (black circle on left). Each symbol represents the position of an individual animal at each second. *Right*: Vertical excursions of three nautiluses and their compensation as they lose the odor plume (shaded area). Height was scored as below the odor plume (1), in the plume (2), or above the plume (3). (B) *Left*: Sham-operated nautiluses found the odor source. *Right*: Heading angles for normal and sham-operated animals were poor at first, then improved and stabilized until they reached the source. (C, D) *Left*: Unilaterally or bilaterally blocked animals (i.e., Vaseline plug applied to rhinophores) were unable to locate the source. *Right*: Heading angles were erratic during search. (From Basil et al., 2000. Reprinted by permission of Company of Biologists Ltd.)

repertoire of visual signals used during mate choice. It is no surprise that vision is important, but it is peculiar that cuttlefish males have developed a distinctive visual signal that portends their intention to lose a fight. It was unexpected that chemical sensing plays a role in cuttlefish females' choice of mates, and that male squid use chemical cues in egg capsules to initiate highly aggressive behavior related to competition for mates. The finding that *Nautilus* may use chemical cues for locating mates seems appropriate, given the limitations of their vision.

4.1. Cuttlefish Males Adjust Fighting Behavior by Visual Signals

Agonistic contests in cuttlefishes and squid are mostly visual battles with some minor physical tests of strength (Moynihan and Rodaniche, 1982; DiMarco and Hanlon, 1997; summarized by Hanlon and Messenger, 1996). Adamo and Hanlon (1996) discovered an odd behavior during cuttlefish fights—males signaled their intentions with the highly visual signal called "dark face." Lighter-faced males eventually withdrew from the contest, essentially signaling their intent to lose; this is contrary to predictions from traditional game theory (Maynard-Smith, 1982). The explanation relates to the behavioral ecology of reproduction in cuttlefishes, which are not social and have unsophisticated methods of sexual recognition (Boal, 1996). Males signal "maleness" to other males (via the zebra striping that males use in fights) lest they be mated, but they apparently evolved a secondary signal of dark face that signifies the drive or intent to escalate the contest; not showing dark face decreases the possibility of escalation once the fight has been initiated with the zebra patterning.

4.2. Cuttlefish Females Choose Mates by Olfactory Cues

Partly because cuttlefish males use dramatic visual signals during sexual selection, it was thought that females would evaluate males based on their visual displays or their size and dominance (i.e., the winner of a bout). Boal (1996, 1997) tested these ideas and found surprising results. Males did not recognize their own mate from another mated female, but they did discriminate between mated and unmated females, suggesting the use of chemical cues. Female preference (Boal, 1997) was not related to known indices of male dominance, it was not related to the relative darkness of the males' faces (see Section 4.1), and it was not influenced by viewing male–male interactions. In addition, male zebra striping (the major visual feature of the agonistic display) appeared to repel rather than attract females. Quite by accident, Boal noticed from videotapes that some females were seemingly making their choices of males before the visual barrier was removed in each trial. Post-hoc analysis allowed her to correlate female choice with a small subset of males that were chosen repeatedly as mates in these trials. That is, females were able to detect which males has previously mated and how recently they had mated, and they preferred to approach those males regardless of whether those males were winners of fights, the largest or most aggressive, or most attractive visual displaying animals. Chemical cues (pheromones) provide a plausible explanation for female choice in these experiments because females were directly downstream from males and had olfactory information available to them (which explains the "choice" before the visual barrier was lifted). In subsequent experiments, Boal and Marsh (1998) used a Y maze to test for social recognition using chemical cues, but their data did not support this hypothesis statistically. They stated that their inconclusive results were suggestive of chemical communication, and that future experimentation with modified methodology was needed. Overall, these experiments are the first to indicate how chemical cues are used by a cephalopod for important reproductive behaviors. In this respect, they may not be so different from other molluscan classes that use chemical cues for a variety of behaviors.

4.3. Squid Males Become Aggressive and Initiate Sexual Selection Based on Visual and Chemical Cues Present in Deposited Egg Capsules

How is sexual behaviour initiated in squid mating aggregations? *Loligo pealeii*, a widespread species in the northwest Atlantic, migrates annually to nearshore spawning grounds where it sometimes—not always—constructs large communal egg beds. As in many taxa, there are sex-specific markings and behaviors that initiate courtship and competition for mates (Hanlon et al., 1999b). In *L. pealeii*, however, King et al. (1999) have demonstrated that males will engage in the *highest* level of fighting behavior within two minutes of seeing and touching eggs—even in the absence of females. In an extensive set of laboratory trials, they demonstrated first that males were attracted visually to eggs (or facsimiles of eggs), but this alone was not sufficient to evoke the full set of agonistic behaviors. If the males were allowed to approach the eggs, they would always touch them and blow water on them, which was followed within two minutes by exceptionally high levels of fighting. King et al. (op. cit.) measured five primary behaviors, the most aggressive of which was forward lunge grab (FLG), in which one male jetted forward and seized the other squid and attempted to bite it. This is an extreme behavior for cephalopods, which usually use only visual signaling to conduct and resolve fights (Hanlon and Messenger, 1996). Squid performed up to 100 chases and 2 to 6 FLGs per two minutes during these trials, and as long as eggs were left in the tank, they did not habituate these behaviors within the 50-min duration of the trials. If egg capsules were removed from the tank, the strongest agonistic behaviors declined within minutes. When eggs were hidden under an opaque but porous cover that allowed waterborne stimuli to circulate in the tank, squid were not attracted to egg capsules nor did they show increased agonistic behavior. The results imply a dual sensory input (visual together with tactile or chemosensory) that immediately stimulates intrasexual male competition. These male behaviors (egg interactions followed by fighting) have been seen on field video tapes of *L. pealeii*, *Loligo opalescens* (California), and *Loligo vulgaris reynaudii* (S. Africa). Once again, experimenters have been surprised to see chemical cues playing a part in behaviors that seem at first to be largely or exclusively visual.

4.4. *Nautilus* Females Discriminate Conspecifics by Odor

Following the feeding experiments reported in Section 3.6 above, Basil et al. (in press) tested the idea that *Nautilus* may use chemical cues for reproduction. Using a Y maze, they found that *Nautilus* left the start box of the maze only in the presence of conspecific odor. Females were attracted to the odor of males, but not females, both in terms of their first choice and the time they spent in the arm with the male odor. Conversely, males did not seem to prefer odor of one sex or the other. The level of arousal of males and females when making a choice of odor differed as well, with females extending their tentacles fully when tracking an odor. Female preference also emerged when viewing the results from the perspective that females might be avoiding same-sex odor, because they chose no odor over same-sex odor. Overall, their results suggest that mate detection by female *Nautilus pompilius* may be guided initially by olfaction. These findings may relate to the natural habitat of *Nautilus*, which are solitary, nocturnal organisms and that must somehow find mates. There is some evidence (Haven, 1977) that females migrate from cooler to warmer (shallower) waters during the late spring and summer, and they could potentially use olfaction to find potential mates and lay eggs, which are known to be laid at temperatures of 21–24°C, and in the laboratory only when temperatures are slowly raised. Much has yet to be learned about this mating system, but it seems obvious that *Nautilus* represents the ancient cephalopods in which olfaction played a major role rather than vision in the modern coleoids.

5. Other Senses and Behaviors

There are photosensitive vesicles in octopods (attached to the stellate ganglion in the mantle) and in squid and cuttlefishes (lying close to the olfactory lobe in the brain). These structures have the ultrastructure and biochemistry of photoreceptors and there is good physiological evidence that they respond to light. In squid they aid in counterillumination (Young et al., 1979), and there is speculation that they may aid diel vertical migration in many cephalopods. There is no evidence for an electric sense or a magnetic sense but no one has looked carefully. There are many mechanoreceptors and chemoreceptors in the suckers, and probably throughout the whole body surface, because all cephalopods seem sensitive to touch, but the body receptors have not been studied (Wells, 1978). It is likely that cephalopods can sense pain, although no specific "nociceptors" have been identified. Statocysts provide information about gravity and acceleration and their anatomy and physiology vary according to the behavioral habits of the species. The statocysts also control the countershading reflex (Ferguson and Messenger, 1991) and perhaps function in hearing (Packard et al., 1990). Neck hair cells and the associated nuchal cartilage structures of squid form a neck receptor organ that, together with statocysts and visual inputs, controls the position of the animal's head and body (Preuss and Budelmann, 1995a). Such a sense organ helps stabilize a free-swimming, high-speed jetting visual attacker such as a squid. It also contributes to the dorsal light reflex in squid (Preuss and Budelmann, 1995b).

6. Conclusion

Every line of evidence still indicates that vision is the primary sensory modality for the major behaviors involved in defence, feeding, and reproduction. But such a view has obscured the roles of other senses, and recent discoveries (both experimental and anecdotal) have shown their critical role.

Probably few, if any, researchers can appreciate how well—and what—cephalopods see, especially under varying light levels. Polarization sensitivity enhances contrast vision for feeding under a variety of light conditions, and it is possible that PS also imparts some of the capabilities that color vision provides other animals (Bernard and Wehner, 1977). The idea that some cephalopods may use PS to communicate is novel and exciting but requires more definitive behavioral experimentation. There is still a great deal to learn about visual capabilities of cephalopods.

More surprising is the role of chemoreception. While octopuses might be expected to use contact chemoreception due to their tactile behaviors, anatomy, and life styles, recent experiments show that cuttlefishes and squid use chemical information for elements of reproductive behavior. Predictably, *Nautilus* have a very well-developed sense of smell, but it is intriguing that they have bilateral sensors (the rhinophores) that seem to sense in three dimensions. From the ecological viewpoint, sensing in three dimensions would seem mandatory for an animal that lives such a vertical life style in a habitat with dynamic water patterns.

We in the cephalopod research community are just matriculating into the world of sensory ecology. There are now extensive field observations of selected cephalopods, and there has been substantial progress in moving from descriptions of behavior to controlled experiments in the laboratory as well as structured and quantitative field studies. It is time to combine sensory biology with behavioral ecology to begin to provide a more coherent and accurate picture of how cephalopods survive and thrive in natural environments, and this is a major challenge for the coming generation of cephalopod researchers.

Acknowledgments. We are indebted to many colleagues over the years for discussions that have helped formulate some of the ideas in this chapter. Among these we especially mention John Messenger, Tom Cronin, Jean Boal, Ulli Budelmann, C.-C. Chiao, Jennifer Basil, Jelle Atema, and John Forsythe. We thank John Messenger and Jelle Atema for reviewing the manuscript. We also thank the National Science Foundation, the National Oceanographic and Atmospheric Administration (particularly WHOI Sea Grant and the West Coast National Underseas Research Program), the US/Israel Bi-national Science Foundation, and the Sholley Foundation for funding portions of this research. We dedicate this chapter to the memory of Ellen Grass, whose passion for sensory biology and marine invertebrates was matched by few, and will be missed by many.

References

Adamo, S.A., and Hanlon, R.T. (1996). Do cuttlefish (Cephalopoda) signal their intentions to conspecifics during agonistic encounters? *Anim. Behav.* 52:73–81.

Archer, S.N., Djamgoz, M.B.A., Loew, E.R., Partridge, J.C., and Vallerga, S. (eds.). (1999). *Adaptive Mechanisms in the Ecology of Vision.* Dordrecht: Kluwer Academic Publishers.

Arnold, J.M., Young, R.E., and King, M.V. (1974). Ultrastructure of a cephalopod photophore. II. Iridophores as reflectors and transmitters. *Biol. Bull.* 147:522–534.

Basil, J.A., Hanlon, R.T., Sheikh, S.I., and Atema, J. (2000). Three-dimensional odor tracking by *Nautilus pompilius*. *J. Exp. Biol.* 203:1409–1414.

Basil, J.A., Lazenby, G.B., Nakanuku, L., and Hanlon, R.T. (in press) Female *Nautilus* are attracted to male conspecific odor. *Bull. Mar. Sci.*

Bernard, G.D., and Wehner, R. (1977). Functional similarities between polarization vision and color vision. *Vision Res.* 17:1019–1028.

Bleckmann, H., Budelmann, B.U., and Bullock, T.H. (1991). Peripheral and central nervous responses evoked by small water movements in a cephalopod. *J. Comp. Physiol. A.* 168:247–257.

Boal, J.G. (1996). Absence of social recognition in laboratory-reared cuttlefish, *Sepia officinalis* L. (Mollusca: Cephalopoda). *Anim. Behav.* 52:529–537.

Boal, J.G. (1997). Female choice of males in cuttlefish (Mollusca: Cephalopoda). *Behaviour* 134:975–988.

Boal, J.G., and Marsh, S.E. (1998). Social recognition using chemical cues in cuttlefish (*Sepia officinalis* Linnaeus, 1758). *J. Exp. Mar. Biol. Ecol.* 230: 183–192.

Boal, J.G., Wittenberg, K.M., and Hanlon, R.T. (2000). Observational learning does not explain improvement in predation tactics by cuttlefish (Mollusca: Cephalopoda). *Behav. Process.* 52:141–153.

Boyle, P. (1983). Ventilation rate and arousal in the octopus. *J. Exp. Mar. Biol. Ecol.* 69:129–136.

Budelmann, B.U. (1994). Cephalopod sense organs, nerves and the brain: Adaptations for high performance and life style. *Mar. Freshwat. Behav. Physiol.* 25:13–33.

Budelmann, B.U. (1996). Active marine predators: the sensory world of cephalopods. *Mar. Freshwat. Behav. Physiol.* 27:59–75.

Budelmann, B.U., and Bleckmann, H. (1988). A lateral line analogue in cephalopods: Water waves generate microphonic potentials in the epidermal head lines of *Sepia* and *Lolliguncula*. *J. Comp. Physiol. A.* 164:1–5.

Budelmann, B.U., Riese, U., and Bleckmann, H. (1991). Structure, function, biological significance of the cuttlefish "lateral lines." In: *The Cuttlefish: First International Symposium on the Cuttlefish* Sepia (Boucaud-Camou, E., ed.), pp. 201–209. Caen: Centre de Publications de l'Universite de Caen.

Chase, R., and Wells., M.J. (1986). Chemotactic behaviour in *Octopus*. *J. Comp. Physiol. A.* 158: 375–381.

Chiao, C.-C., and Hanlon, R.T. (2001a). Cuttlefish camouflage: Visual perception of size, contrast and number of white squares on artificial substrata initiates disruptive coloration. *J. Exp. Biol.* 204: 2119–2125.

Chiao, C.-C., and Hanlon, R.T. (2001b). Cuttlefish cue visually on area—not shape or aspect ratio—of light objects in the substrate to produce disruptive body patterns for camouflage. *Biol. Bull.* 201: 269–270.

Cott, H.B. (1940). *Adaptive Coloration in Animals.* London: Methuen.

Cronin, T.W., Shashar, N., and Wolff, L. (1995). Imaging technology reveals the polarized light fields that exist in nature. *Biophotonics Int.* 2:38–41.

Denton, E.J. (1970). On the organization of reflecting surfaces in some marine animals. *Philos. Trans. R. Soc. Lond. B. Biol. Sci.* 258:285–313.

DiMarco, F.P., and Hanlon, R.T. (1997). Agonistic behavior in the squid Loligo plei (Loliginidae, Teuthoidea): Fighting tactics and the effects of size and resource value. *Ethology* 103:89–108.

Ferguson, G.P., and Messenger, J.B. (1991). A countershading reflex in cephalopods. *Proc. R. Soc. Lond. B. Biol. Sci.* 243:63–67.

Fiorito, G., and Scotto, P. (1992). Observational learning in *Octopus vulgaris*. *Science* 256:545–547.

Forsythe, J.W., and Hanlon, R.T. (1997). Foraging and associated behavior by *Octopus cyanea* Gray, 1849 on a coral atoll, French Polynesia. *J. Exp. Mar. Biol. Ecol.* 209:15–31.

Hanlon, R.T., and Budelmann, B.-U. (1987). Why cephalopods are probably not "deaf." *Am. Nat.* 129:312–317.

Hanlon, R.T., and Messenger, J.B. (1988). Adaptive coloration in young cuttlefish (*Sepia officinalis* L.): The morphology and development of body patterns and their relation to behaviour. *Philos. Trans. R. Soc. Lond. B. Biol. Sci.* 320:437–487.

Hanlon, R.T., and Messenger, J.B. (1996). *Cephalopod Behaviour.* Cambridge: Cambridge University Press.

Hanlon, R.T., Forsythe, J.W., and Joneschild, D.E. (1999a). Crypsis, conspicuousness, mimicry and polyphenism as antipredator defences of foraging octopuses on Indo-Pacific coral reefs, with a method of quantifying crypsis from video tapes. *Biol. J. Linn. Soc.* 66:1–22.

Hanlon, R.T., Maxwell, M.R., Shashar, N., Loew, E.R., and Boyle, K.-L. (1999b). An ethogram of body patterning behavior in the biomedically and commercially valuable squid *Loligo pealei* off Cape Cod, Massachusetts. *Biol. Bull.* 197:49–62.

Hartline, P.H., and Lange, G.D. (1984). Visual systems of cephalopods. *Comp. Physiol. Sensory. Syst.* 335–356.

Haven, N. (1977). The reproductive biology of *Nautilus pompilius* in the Philippines. *Mar. Biol.* 42:177–184.

Herring, P.J. (1994). Reflective systems in aquatic animals. *Comp. Biochem. Physiol.* 109:513–546.

King, A.J., Adamo, S.A., and Hanlon, R.T. (1999). Contact with squid egg capsules increases agonistic behavior in male squid (*Loligo pealei*). *Biol. Bull.* 197:256.

King, A.J., Adamo, S.A., and Hanlon, R.T. (in press). Agonistic behaviour between male squid: Squid egg mops provide sensory cues for swift conflict escalation. *Anim. Behav.*

Land, M.F. (1984). Molluscs. In: *Photoreception and Vision in Invertebrates* (All, M.A., ed.), pp. 699–725. New York: Plenum Press.

Lee, P.G. (1992). Chemotaxis by *Octopus maya* Voss et Solis in a Y-maze. *J. Exp. Mar. Biol. Ecol.* 156: 53–67.

Lucero, M.T., Horrigan, F.T., and Gilly, W.F. (1992). Electrical responses to chemical stimulation of squid olfactory receptor cells. *J. Exp. Biol.* 162: 231–249.

Marshall, N.J., and Messenger, J.B. (1996). Colour-blind camouflage. *Nature* 382:408–409.

Marshall, N.J., Cronin, T.W., Shashar, N., and Land, M. (1999). Behavioural evidence for polarisation vision in stomatopods reveals a potential channel for communication. *Curr. Biol.* 9:755–758.

Mather, J.A. (1991). Foraging, feeding, and prey remains in middens of juvenile *Octopus vulgaris* (Mollusca: Cephalopoda). *J. Zool.* 224:27–39.

Mathger, L.M., and Denton, E.J. (2001). Reflective properties of iridophores and fluorescent "eyespots" in the loliginid squid *Alloteuthis subulata* and *Loligo vulgaris*. *J. Exp. Biol.* 204: 2103–118.

Maynard-Smith, J. (1982). Evolution and the theory of games. Oxford: Oxford University Press.

Messenger, J.B. (1981). Comparative physiology of vision in Molluscs. In: *Handbook of Sensory Physiology, Vol. VII/6C: Comparative Physiology and Evolution of Vision in Invertebrates* (Autrum, H., ed.), pp. 93–200. Berlin: Springer-Verlag.

Messenger, J.B. (1991). Photoreception and vision in molluscs. In: *Evolution of the Eye and Visual System* (Cronly-Dillon, J.R., and Gregory, R.L., eds.), pp. 364–367. London: Macmillan Press.

Messenger, J.B. (2001). Cephalopod chromatophores: neurobiology and natural history. *Biol. Rev.* 76:473–528.

Moynihan, M., and Rodaniche, A.F. (1982). The behavior and natural history of the Caribbean reef squid *Sepioteuthis sepioidea*, with a consideration of social, signal and defensive patterns for difficult and dangerous environments. *Adv. Ethol.* 25:1–150.

Muntz, W.R.A. (1991). Anatomical and behavioural studies on vision in *Nautilus* and *Octopus*. *Am. Malac. Bull.* 9:69–74.

Muntz, W.R.A. (1999). Visual systems, behaviour, and environment in cephalopods. In: *Adaptive Mechanisms in the Ecology of Vision* (Archer, S.N., Djamgoz, M.B.A., Loew, E.R., Partridge, J.C., and Vallerga, S., eds.), pp. 467–483. Dordrecht: Kluwer Academic Publishers.

Norman, M.D., Finn, J., and Tregenza, T. (2001). Dynamic mimicry in an Indo-Malayan octopus. *Proc. R. Soc. Lond. B. Biol. Sci.* 268:1755–1758.

Novales-Flamarique, I., and Hawryshyn, C.W. (1997). Is the use of underwater polarized light by fish restricted to crepuscular time periods? *Vision Res.* 37:975–989.

O'Dor, R.K., Forsythe, J., Webber, D.M., Wells, J., and Wells, M.J. (1993). Activity levels of *Nautilus* in the wild. *Nature* 362:626–628.

Packard, A., Karlsen, H.E., and Sand, O. (1990). Low frequency hearing in cephalopods. *J. Comp. Physiol. A.* 166:501–505.

Preuss, T., and Budelmann, B.U. (1995a). Proprioceptive hair cells on the neck of the squid *Lolliguncula brevis*: A sense organ in cephalopods for the control of head-to-body position. *Philos. Trans. R. Soc. Lond. B. Biol. Sci.* 349:153–178.

Preuss, T., and Budelmann, B.U. (1995b). A dorsal light reflex in a squid. *J. Exp. Biol.* 198:1157–1159.

Robert, M. (1990). Observational learning in fish, bird, and mammals: A classified bibliography spanning over 100 years of research. *Psychol. Rec.* 40:289–311.

Saibil, H.R. (1990). Photoreception in squid. In: *Light and Life in the Sea* (Herring, P.J., Campbell, A.K., Whitfield, M., and Maddock, L., eds.), pp. 199–207. Cambridge: Cambridge University Press.

Saibil, H., and Hewat, E. (1987). Ordered transmembrane and extracellular structure in squid photoreceptor microvilli. *J. Cell. Biol.* 105:19–28.

Shashar, N., and Cronin, T.W. (1996). Polarization contrast vision in *Octopus*. *J. Exp. Biol.* 199:999–1004.

Shashar, N., and Hanlon, R.T. (1997). Squid (*Loligo pealei* and *Euprymna scolopes*) can exhibit polarized light patterns produced by their skin. *Biol. Bull.* 193:207–208.

Shashar, N., Hanlon, R.T., and Petz, A. de M. (1998). Polarization vision helps detect transparent prey. *Nature* 393:222–223.

Shashar, N., Milbury, C.A., and Hanlon, R.T. (in press). Polarization vision in cephalopods: Neuroanatomical and behavioral features that illustrate aspects of form and function. *Mar. Freshwat. Behav. Physiol.*

Shashar, N., Rutledge, P.S., and Cronin, T.W. (1996). Polarization vision in cuttlefish: A concealed communication channel? *J. Exp. Biol.* 199:2077–2084.

Shashar, N., Hagan, R., Boal, J.G., and Hanlon, R.T. (2000). Cuttlefish use polarization sensitivity in predation on silvery fish. *Vision Res.* 40:71–75.

Shashar, N., Borst, D.T., Ament, S.A., Saidel, W.M., Smolowitz, R.M., and Hanlon, R.T. (2001). Polarization reflecting iridophores in the arms of the squid *Loligo pealeii*. *Biol. Bull.* 201:267–268.

Waterman, T.H. (1955). Polarization of scattered sunlight in deep water. *Deep Sea Res.* 3(Suppl.): 426–434.

Wells, M.J. (1978). *Octopus: Physiology and Behaviour of an Advanced Invertebrate.* London: Chapman & Hall.

Williamson, R. (1995). A sensory basis for orientation in cephalopods. *J. Mar. Biol. Assoc. U.K.* 75:83–92.

Young, J.Z. (1963). Light and dark adaptations in the eyes of some cephalopods. *Proc. Zool. Soc. Lond.* 140:255–270.

Young, J.Z. (1971). *The Anatomy of the Nervous System of* Octopus vulgaris. Oxford: Clarendon Press.

Young, J.Z. (1991). Light has many meanings for cephalopods. *Vis. Neurosci.* 7:1–12.

Young, R.E., Roper, C.F.E., and Walters, J.F. (1979). Eyes and extraocular photoreceptors in midwater cephalopods and fishes: Their roles in detecting downwelling light for counterillumination. *Mar. Biol.* 51:371–380.

Zonana, H.V. (1961). Fine structure of the squid retina. *Johns Hopkins Hosp. Bull.* 109:185–205.

15
Recent Progress in Aquatic Vertebrate Olfaction

H. Peter Zippel and Lars G.C. Lüthje

Abstract

This chapter presents physiological data including in vivo extracellular recordings from small single epithelial sensory neurons and single relay neurons in the goldfish olfactory bulb. In contrast to other vertebrates, biologically relevant stimuli such as preovulatory and ovulatory pheromones, the alarm pheromone, and food components such as amino acids have been identified structurally in fishes. These stimuli can be dissolved in water and applied in defined molarities and at low (10^{-7}–10^{-9}M) concentrations. A distinct advantage of this model is its experimental accessibility where the olfactory epithelium lies peripherally and is covered only by a fold of skin and the pedunculated olfactory bulb can easily be exposed. In the olfactory epithelium, cells were found to respond to either one of the applied stimuli ("specific") or to many different stimuli ("general"). In the olfactory bulb, responses from both mitral and ruffed cells (or bulbar relay neurons) can be recorded simultaneously and with a single electrode. They can be easily discriminated anatomically and physiologically. Similar recordings are not possible in other vertebrates because the respective relay neurons (mitral cells and tufted cells) are located in different bulbar layers. Mitral cells and ruffed cells in goldfish are spontaneously active during the interstimulus phases when a laminar flow of water passes over the epithelium. Stimulus application can result in no response, excitation, or inhibition. During excitation of mitral cells (which only have an epithelial input from sensory neurons), ruffed cells are simultaneously inhibited via granule cells. During inhibition of mitral cells, the activity of ruffed cells is increased because inhibitory granule cells are no longer activated.

The activity of mitral cells heavily depends on the activity of epithelial neurons. When the flow of water over the epithelium is interrupted, approximately 80% of epithelial neurons, in addition to the mitral cells, decrease their impulse rates. During anaesthesia of the epithelium, the mitral cell activity in the bulb decreases to zero and the ruffed cells become rhythmically active. The olfactory system of fishes is both chemo- and mechanosensitive. This has also been shown for olfactory bulb neurons of spontaneously breezing mammals. Cross-adaptation experiments demonstrate that, during permanent application of an odor, the sensitivity for a related odor decreases. We have not been able to show full or partial cross-adaptations during application of stimuli selected from the respective electro-olfactogram. It is thought that cross-adaptation phenomena depend on sensory neurons in the olfactory epithelium, relay neurons in the olfactory bulb, and, hypothetically, also on processing by central nuclei.

1. Introduction

The most distinct difference between aquatic and land animals with respect to chemical senses is the stimulus-delivery medium. Volatility is an essential characteristic for olfactory stimuli of air-breathers, whereas water solubility of stimuli or adsorption onto suspended particles are critical for both olfactory and gustatory stimuli in fishes. However, in both terrestrial and aquatic animals, stimulus solubility in the mucus overlying the chemical sensory neurons is necessary. The gustatory sense of fishes is, as a result of the aquatic medium, not limited to being a contact sense, but may also function as a distance sense. Furthermore, it is relatively highly sensitive to a great number of specific compounds after operative elimination of olfaction (Zippel, 1993).

This chapter will review mainly the recent progress in the physiology of olfaction in fishes. In contrast to land vertebrates, many types of biologically relevant chemical sensory stimuli have been both identified and structurally characterized in fishes (especially for the cyprinidae) along with their effective concentrations (see below). Another great advantage of fishes as experimental models for the study of olfaction is their slightly less differentiated anatomy, including a number of unique experimental prerequisites: The olfactory epithelium is easily accessible and contains both microvillous and ciliated receptor neurons (Hansen et al., 1999); in contrast to terrestrial vertebrates, different types of receptor neurons are more equally distributed over single lamellae and the midline raphe; odors can be applied in a constant flow to the epithelium while the experimental animal is artificially respirated through the mouth and the gills; the cyprinid olfactory bulb is pedunculated and, like the bulb of terrestrial vertebrates, contains two kinds of relay neurons; in contrast to land vertebrates, the bulbar relay neurons in fishes lie close to each other in the external plexiform layer (Kosaka and Hama, 1979, 1981, 1982, 1982–1983; Kosaka, 1980). In contrast to mitral cells, ruffed cells contain no synaptic inputs from the olfactory epithelium, and structural peculiarities of ruffed cells innervating great numbers of inhibitory granule cells make it easy to discriminate the potentials of both types of relay neurons during simultaneous recordings with a single electrode (Zippel et al., 1999, 2000).

Different aspects of chemoreception in fishes have been examined previously (Hara, 1982, 1992), including a number of chapters in the precursor of this book, *Sensory Biology of Aquatic Animals* (Atema et al., 1988). Functional properties of vertebrate olfactory receptor neurons were published by Getchell (1986) and transduction mechanisms in vertebrate olfactory receptor cells were recently reviewed by Schild and Restrepo (1998). This review of the physiology of chemoreception in the fish olfactory system will concentrate on two main points: (1) function and stimulus discrimination recorded in vivo from single olfactory receptor neurons, and (2) information processing in the main olfactory bulb.

2. In Vivo Recordings from Single Neurons

In vivo extracellular recordings from small single receptor neurons were made in the goldfish olfactory epithelium using platinum-black electrodes and from mitral and ruffed cells in the olfactory bulb (Zippel et al., 1999, 2000) with single tungsten microelectrodes (5–7 MΩ, A–M systems). Anatomical differences characterizing mitral cells and ruffed cells were published by Kosaka and Hama (1979, 1981, 1982, 1982–1983) in three teleost species. In goldfish, physiological responses from the two different types of relay neurons were recorded extracellularly and simultaneously. Ruffed cells can be clearly distinguished from mitral cells morphologically (Kosaka and Hama, 1979, 1981, 1982, 1982–1983) and physiologically (Zippel et al., 1999, 2000). During interstimulus intervals, mitral cells responded with higher, frequently burst-like impulse rates triggered by the activity of the epithelial receptor neurons. The mitral cell activity could be totally suppressed during local anesthesia of the olfactory epithelium (Zippel et al., 1999). Ruffed cell impulse rates were low, and each action potential triggered a long-lasting (3–5 ms), continuously variable, summed granule cell potential. In contrast to mitral cells, blockade of epithelial receptor cells significantly increased the activity of ruffed cells. The ruffed cells, which have no input from the olfactory epithelium, are spontaneously active, and are laterally inhibited by granule cells activated by mitral cells. An extremely slow ($0.95\,ml{\cdot}s^{-1}$) and laminar flow of water during 180-second interstimulus phases and during 15-s stimulus applications enabled recordings of responses to stimuli of more than one hour in the epithelium and for several hours in the olfactory bulb. Responses to biologically relevant stimuli such as preovulatory and ovulatory pheromones, a probable alarm pheromone, and amino acids from food sources were investigated to determine whether pheromones and nonpheromone stimuli are recognized by membrane receptors of different or the same individual sensory neurons. Slow laminar flow of water allowed investigation of the mechanosensitivity of the epithelial neurons, which also responded to the onset and termination of waterflow. Responses recorded from single epithelial olfactory sensory neurons were compared with recordings obtained from olfactory bulb relay neurons using similar experimental procedures and data evaluation.

In goldfish, many types of biologically relevant chemical sensory stimuli have been identified and structurally characterized along with their effective concentrations. Amino acids, relevant food stimuli in goldfish (Zippel et al., 1993; von Rekowski et al., 1994), were applied at 10^{-7} M, which elicited no observed electro-olfactogram (EOG) responses (Zippel et al., 1997a). Preovulatory (17α,20β-dihydroxy-4-pregnen-3-one, 10^{-9} M) and ovulatory (prostaglandin F_{2a} and a metabolite 15-ketoprostaglandin F_{2a}, 10^{-7} M; for citations see Zippel et al., 2000) pheromones were applied in concentrations, which elicited maximal EOG potentials. The probable alarm pheromone, hypoxanthine-3(N)-oxide (Pfeiffer, 1982), which does not evoke a recordable EOG response at any concentration (P.W. Sorensen, personal communication), was applied at 10^{-9} M.

In other vertebrates, significant effects of pheromones are behaviorally evident, although the respective chemical structures are unknown at present. The olfactory system of mammalian vertebrates is divided into two subsystems, the "general" olfactory system relevant for the majority of olfactory stimuli, and the accessory (vomeronasal) olfactory system, mainly responsible for pheromonal stimuli. In mammals, different structures (i.e., the olfactory and vomeronasal organs, respectively) are responsible for the majority of olfactory stimuli and pheromones (McLean and Shipley, 1992). The olfactory epithelium of fishes possesses both types of olfactory receptor neurons, microvillous and ciliated cells, which, in other vertebrates, are more discretely separated into the main olfactory system (ciliated receptor cells) and the accessory olfactory system (microvillous cells). In mammals, central neural pathways exist from the accessory olfactory system to the accessory olfactory bulb and to

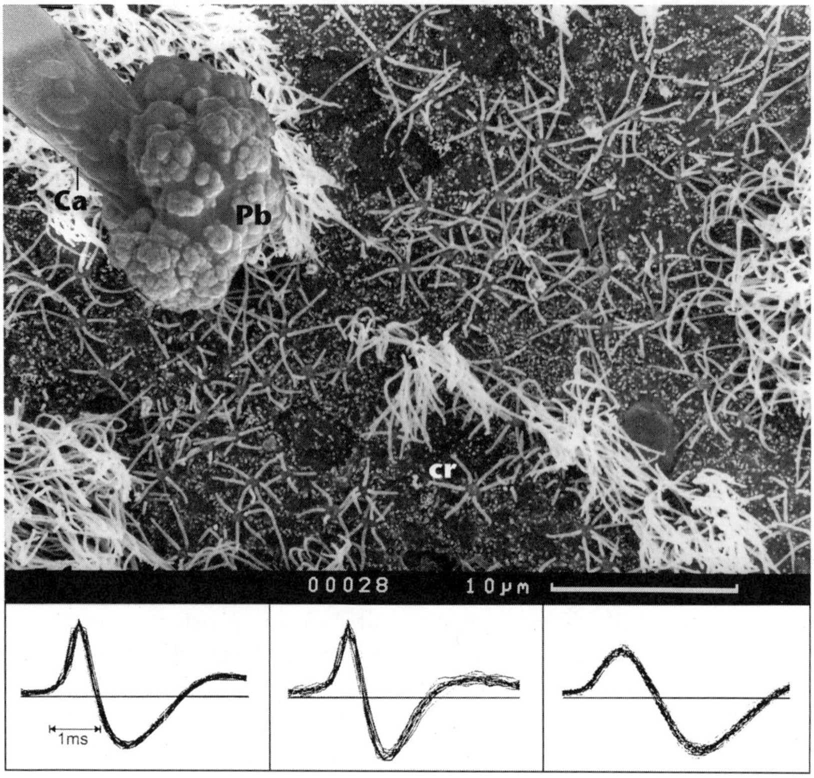

FIGURE 15.1. Photomontage presenting the recording procedure in the olfactory epithelium of goldfish. The SEM (scanning electron microscopic micrograph) from a less densely innervated olfactory lamella of a fish 33 days after olfactory nerve axotomy was made by A. Hansen (Zoologisches Institut und Museum, Universität Hamburg). The SEM of the tip of a platinum-black (Pb) electrode was made by P. Schwartz (Anatomisches Institut, Universität Göttingen). Below the photomontage, superimposed action potentials (15–30 each) are shown from 3 single in vivo extracellularly recorded receptor neurons. Ca = bumped tip of a glass capillary; cr = ciliated receptor neurons with 3–5-µm-long terminal cilia; cell body (not present in figure) approximately 1–2 µm in diameter. For details see Hansen et al. (1999) and text.

different central nuclei, while in fishes no accessory olfactory system is present.

Figure 15.1 demonstrates in a photomontage the experimental procedure in the epithelium. Platinum-black electrodes were made in the laboratory and methods used have been described previously (for details see: Gesteland, 1975; Caprio, 1995). Figure 15.1 also shows three superimposed potentials from single olfactory receptor neurons. The goldfish olfactory epithelium is exquisitely structured and the distribution of different cell types known (Hansen et al., 1999). In the present experiments, rather low concentrations of amino acids were applied that can be readily discriminated in training experiments (Zippel et al., 1993; von Rekowski et al., 1994).

2.1. Single Cell Recordings

The present experiments show that olfactory receptor neurons in goldfish can respond to application of a comparatively great number of pheromones and nonpheromonal stimuli (Fig. 15.2). Recent electrophysiological (EOG), anatomical, and behavioral (Zippel et al., 1997b) results suggest that ciliated receptor neurons respond preferentially to amino acids and to other nonpheromonal stimuli, whereas findings that microvillous receptors respond to pheromonal stimuli cannot be supported. In contrast to more "generalized" receptor neurons (Fig. 15.2), epithelial neurons were present, where the effects of applied stimuli were found to be much more limited (see Fig. 15.4). In recently published abstracts, contradictory findings on microvillar sensory neurons were reported for zebrafish (Lipschitz and Michel, 2000) and goldfish (Sorensen and Caprio, 2000). In zebrafish, the data suggested that amino acids preferentially stimulated microvillar olfactory sensory neurons, and in goldfish the majority of pheromone-sensitive locations were found in lamellar locations containing a higher density of microvillar sensory neurons. During pheromone and amino acid application in goldfish (this chapter) and during application of only amino acids in catfish (Kang and Caprio, 1995, 1997; Koce and Valentinčič, 2000; Valentinčič and Koce, 2000), quantitative differences were recorded in inhibitory and excitatory effects on the olfactory sensory neurons. Whether one type of sensory neuron (ciliated or microvillous) is responding to a greater number of different experimental odors is thus far unknown. Molecular studies suggest that only a very small subset (~1%) of olfactory sensory neurons express a specific or a small number of receptor proteins in the catfish olfactory epithelium (Ngai et al., 1993a,b), but this does not exclude the possibility that the "specificity" of these proteins may be greater or "narrower."

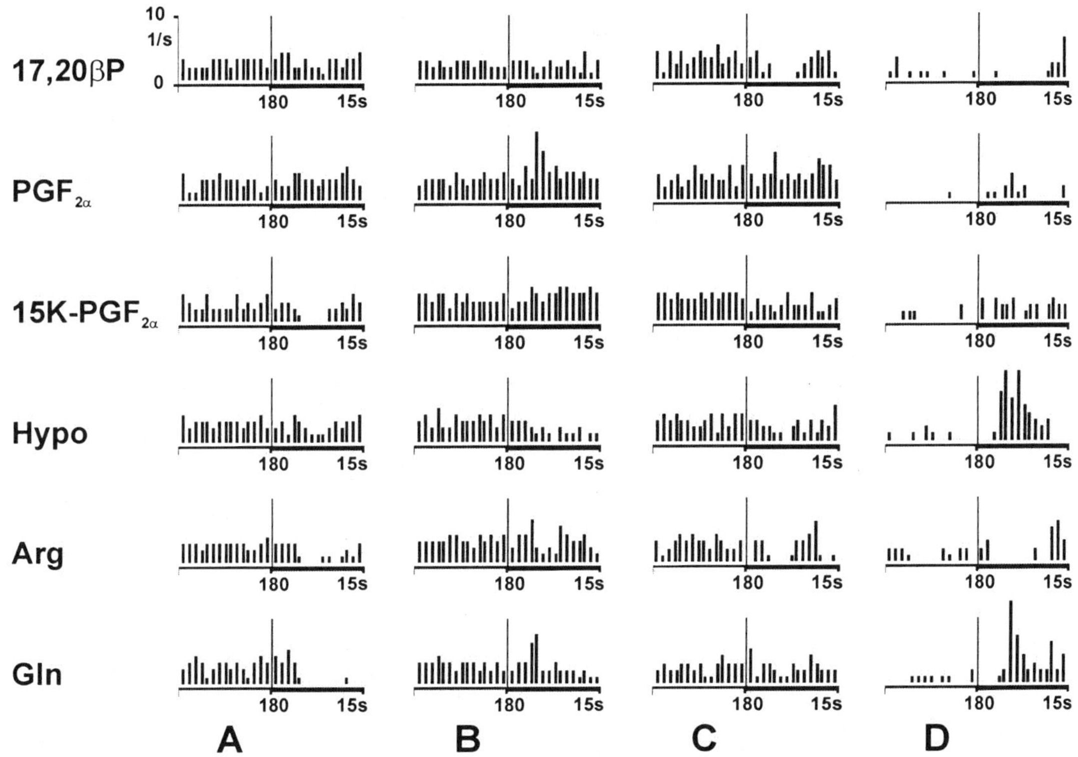

FIGURE 15.2. In vivo recordings from four epithelial receptor neurons during application of pheromones and amino acids. (A) Phasic inhibition by 15K-PGF$_{2\alpha}$ and Arg, and tonic inhibition by Gln. (B) Phasic-tonic excitation by PGF$_{2\alpha}$, tonic excitation by 15K-PGF$_{2\alpha}$, slight inhibition by Hypo, slight excitation by Arg, and phasic excitation by Gln. (C) Phasic inhibition by 17,20β-P and Arg; slight excitation by PGF$_{2\alpha}$. (D) Inhibitions by 17,20β-P, and Arg; slight excitation by 15K-PGF$_{2\alpha}$; strong excitation by Hypo and Gln. 17,20βP = 17α,20β-dihydroxyprogesterone; PGF$_{2\alpha}$ = prostaglandin F$_{2\alpha}$; 15K-PGF$_{2\alpha}$ = 15-keto-prostaglandin F$_{2\alpha}$; Hypo = hypoxanthine-3(N)-oxide; Arg = L-arg; Gln = L-gln. Recordings present the last 15 s of the 180 s interstimulus phase and the 15 s of stimulus application. For further details see text.

A great variety of effects has also been recorded from olfactory bulb mitral and ruffed cells during application of various experimental odors (Zippel et al., 1999, 2000). During olfactory stimulation, contrasting interactions were frequently present, which resulted in a drastic intensification of centrally transmitted information; that is, during simultaneous recordings from interacting pairs of mitral and ruffed cells the activity is modulated via inhibitory interneurons (granule cells). If the activity of the mitral cells is high (e.g., during application of an excitatory olfactory stimulus), granule cells activated by the mitral cells laterally inhibit ruffed cells. If the activity of the mitral cells is low, or becomes very low during application of an inhibitory olfactory stimulus, the activity of the ruffed cells is increased because lateral inhibition is no longer present. The increased activity of the ruffed cells results in a simultaneous activation of pools of granule cells, which obviously support the inhibition of the mitral cells' activity.

During stimulus application, no responses were noted when the activity of mitral cells and ruffed cells remained similar (±25% in comparison to interstimulus activity). Figure 15.3 shows responses of four pairs of relay neurons during the application of preovulatory and ovulatory pheromones, and amino acids. Neurons in A show different responses to amino acids and rather similar responses during preovulatory and ovulatory pheromone application. Neurons shown in B and C revealed a great variety of activity patterns during application of preovulatory and ovulatory pheromones, and amino acid application. Neurons in D show rather similar responses to the different classes of natural stimuli. The mitral cell activity is totally inhibited and the ruffed cell activity simultaneously increased. In general, application of amino acids and preovulatory and ovulatory pheromones results in a great variety of responses of olfactory bulb relay neurons to different classes of stimuli. These data are in good accordance with behavioral findings. In contrast to "nonfamiliar stimuli" like amyl acetate and α-ionone, ovulatory pheromones elicit courtship behavior (Zippel et al., 1997b) and amino acids and mixtures of amino acids are rapidly learned and discriminated (Zippel et al., 1993; von Rekowski et al., 1994). Recently, detailed behavioral and EOG studies have demonstrated the discriminative ability of binary mixtures of amino acids and their individual components by catfish (Valentinčič et al., 2000a,b).

In comparison to epithelial sensory neurons, olfactory bulb relay neurons respond to greater numbers of olfactory stimuli. During application of L-arginine and five structural analogues (Fig. 15.4), approximately 32% of the epithelial receptor neurons and only 9% of the bulbar relay neurons did not respond to any stimulus. Roughly 30% of sensory neurons responded during application of one or two of the six stimuli, while 20–25% of bulbar relay neurons responded during application of two, three, or four stimuli. In contrast to EOG recordings and labeling experiments in zebrafish (Lipschitz and Michel, 1999), we were unable to provide evidence that the olfactory system of goldfish has the potential to discriminate between amino acid odorants with modified carboxyl and amino groups (Conze and Zippel, 2001; Möller-de Beer and Zippel, 2001).

During amino acid application in goldfish, significantly different EOG amplitudes were recorded only when high (10^{-3} and 10^{-4} M) concentrations were applied to the olfactory epithelium (Zippel et al., 1997a). Application of 10^{-6} M concentrations resulted in only a small amplitude EOG. Application of lower concentrations of amino acids produced no EOGs (Zippel et al., 1997a), although concentrations of 10^{-7} M and lower are still learned and discriminated (Zippel et al., 1993; von Rekowski et al., 1994). During application of low (10^{-7} M) concentrations, responses were recorded from olfactory bulb relay neurons (Willms and Zippel, 1999) in which no clearly recognizable differences in the responses to the different amino acids could be found. During recordings from single epithelial sensory neurons, application of six amino acids at concentrations of 10^{-7} M (and 10^{-9} M, see below) showed no recognizable differences in activity in goldfish. In general, 50% of olfactory sensory neurons showed no responses (27% excitatory and 23% inhibitory). The high number of inhibitory

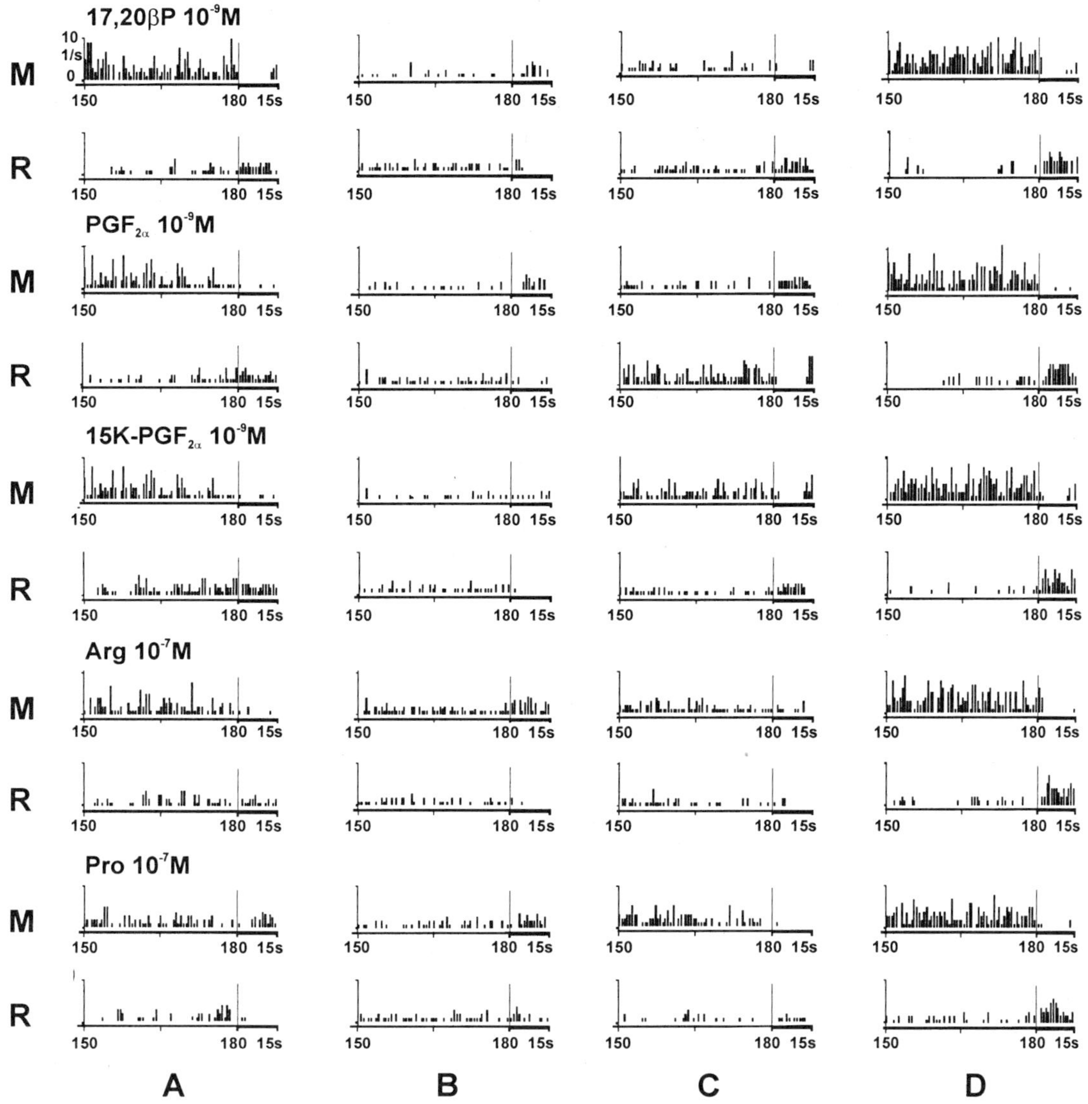

FIGURE 15.3. Activity patterns simultaneously recorded from four pairs of mitral cells (M) and ruffed cells (R) during stimulation with pheromones and amino acids. In all the pairs contrasting interactions between the relay neurons are quite evident. (A) M cell is inhibited by pheromones and Arg and the activity of R released; Pro slightly excites M, which laterally (viva granule cells) inhibits R. (B) Quantitatively different excitations recorded from M, which laterally inhibited R. (C) A great variety of effects released by the various stimuli is evident. (D) The high interstimulus activity of M is totally blocked by all the stimuli while a high excitatory activity is released in R. Recordings present the last 30 s of the interstimulus phase and the 15 s of stimulus application. Pro = L-proline; for the rest of stimuli see Figure 15.2.

responses indicated for catfish (Kang and Caprio, 1995; Valentinčič and Koce, 2000; Koce and Valentinčič, 2000) may be either a species difference or the result of application of rather high concentrations of stimuli.

2.2. Dose-Response Characteristics

Dose-response tests resulted in a greater variety of response characteristics by olfactory bulb mitral (and ruffed) cells than by single

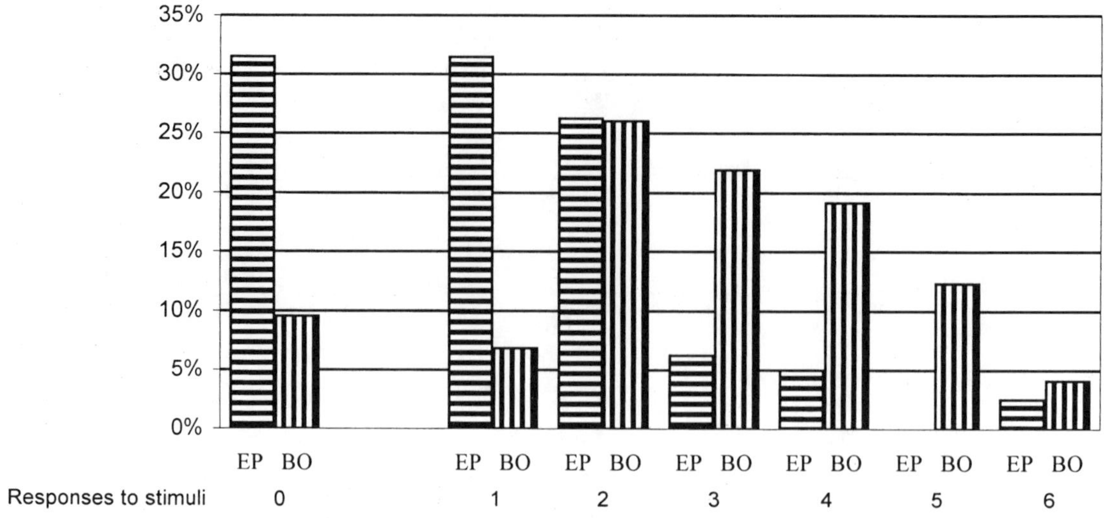

FIGURE 15.4. Responses of 53 epithelial receptor neurons (EP) and 65 olfactory bulb mitral cells (BO) during application of L-arginine and five structural analogs: 32% of receptor neurons respond to none or to one of the six stimuli; 27% of receptor neurons respond during application of two of the six stimuli; modulations of the interstimulus activity during application of three to six stimuli were found in 5% or lower percentages. Due to glomerular convergence of receptor axons only 9% of mitral cells did not respond to any of the six stimuli. In contrast to epithelial receptor neurons, in 20–25% of olfactory bulb mitral cells effects were recorded during application of two to four of the six stimuli, and above 10% of relay neurons still changed activity during application of five of the six stimuli.

epithelial receptor neurons (Figs. 15.5 and 15.6). Dose-response characteristics were only recorded from sensory olfactory neurons when at least the application of one of the six amino acids tested resulted in a clear response at 10^{-7} M. The examples selected from four different epithelial receptor neurons (Fig. 15.5) show two response characteristics, either similar responses during application of different molarities or weak or no response during application of the lowest (10^{-9}) molarity. Neurons A and B show similar tonic excitations, whereas

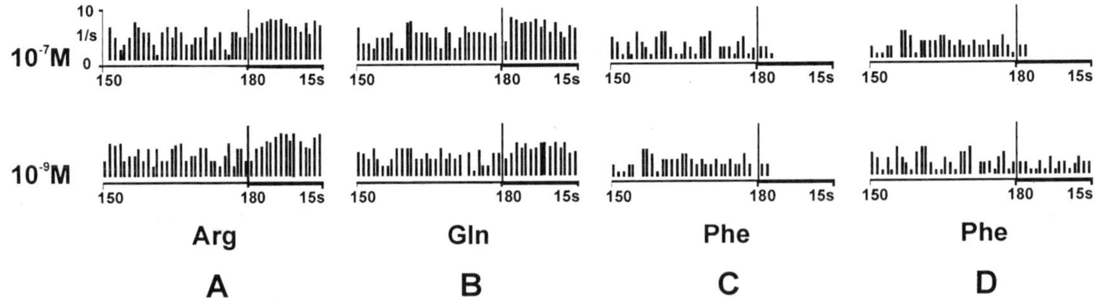

FIGURE 15.5. Dose-response patterns in vivo recorded from four single olfactory receptor neurons during amino acid application. Number of action potentials per second is shown for the last 30s of the 180s interstimulus water phases and during the 15s stimulus applications. Neurons A and B show similar-tonic excitation responses during 10^{-7} and 10^{-9} molar concentrations. The activity of neuron C is tonically suppressed. In neuron D, a tonic suppression was found during application of the high and no effect recorded during application of the low molarity. For abbreviations see Figure 15.2; Phe = L-phenylalanine.

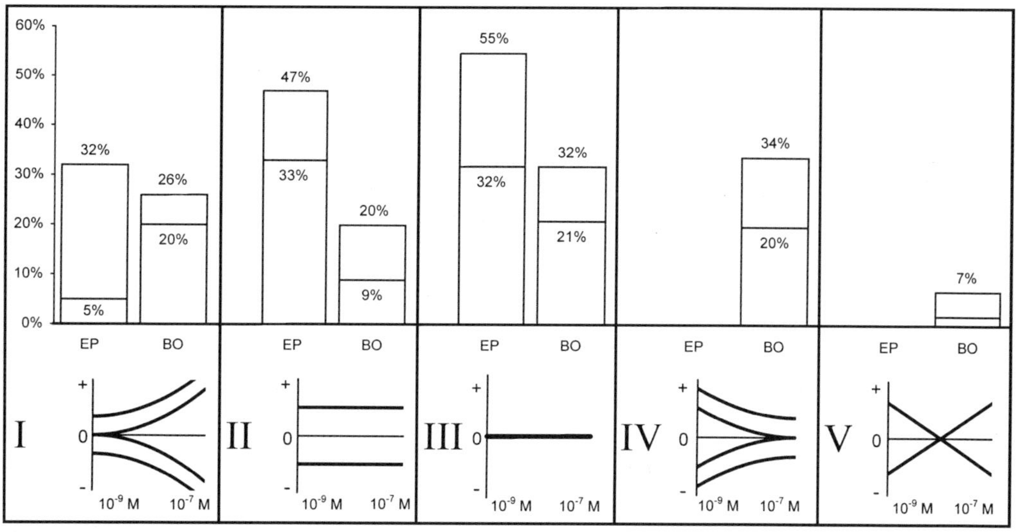

FIGURE 15.6. Schematic representation of different dose-response patterns in vivo recorded from 65 olfactory bulb mitral cells (BO) and 51 epithelial sensory neurons (EP). (I) Decreasing excitations and inhibitions; (II) Excitations and inhibitions remain similar. (III) No response to any concentration = indifferent. (IV) Increasing excitations and inhibitions. (V) Excitation (or inhibition) during application of low, and reverse patterns recorded during application of high concentration. Patterns IV and V so far were never recorded from single olfactory receptor neurons. Minimum and maximum values inserted. For details see text.

neuron C shows tonic inhibition during application of 10^{-7} and 10^{-9} M L-Phenylalanine. Neuron D, however, showed no response to 10^{-9} M L-Phenylalanine. Figure 15.6 summarizes the responses recorded from 51 epithelial sensory neurons and from 65 olfactory bulb mitral cells. The distribution of the various effects and the variety of response patterns remain relatively similar for olfactory bulb relay neurons, and are independent from the variety of stimuli applied (Zippel et al., 1999). No response was recorded from a large number of epithelial sensory neurons during application of different amino acid concentrations. From Figure 15.6, it is evident that similar responses during application of different stimulus concentrations were more frequently recorded than weaker (or no) response during application of stimuli in the lower concentration.

The variety of response patterns recorded from mitral cells is the result of either intrabulbar processes or of bulbopetal central pathways. Stronger or opposite (inhibition instead of excitation and vice versa) effects recorded from bulbar relay neurons during application of lower stimulus concentrations (Fig. 15.6) were not found during recordings from epithelial receptor neurons.

2.4. Bimodality

In contrast to olfactory stimulation, the majority of olfactory receptor neurons in the epithelium exhibited significant modulation in activity when the slow and laminar water flow over the epithelium was interrupted. The extremely high mechanosensitivity of olfactory receptor neurons can be clearly seen after interruption of the laminar flow (Fig. 15.7). The majority (77%) of the 72 receptor neurons in the epithelium significantly decreased their activity after interruption of the flow (Figs. 15.7A,B,G), whereas in a minority of receptor neurons (14%), activity increased after interruption of the laminar water flow (Figs. 15.3C,G). A few receptor neurons (9%) either did not change the activity (Figs. 15.7D,G) or gradually returned to the activity recorded during

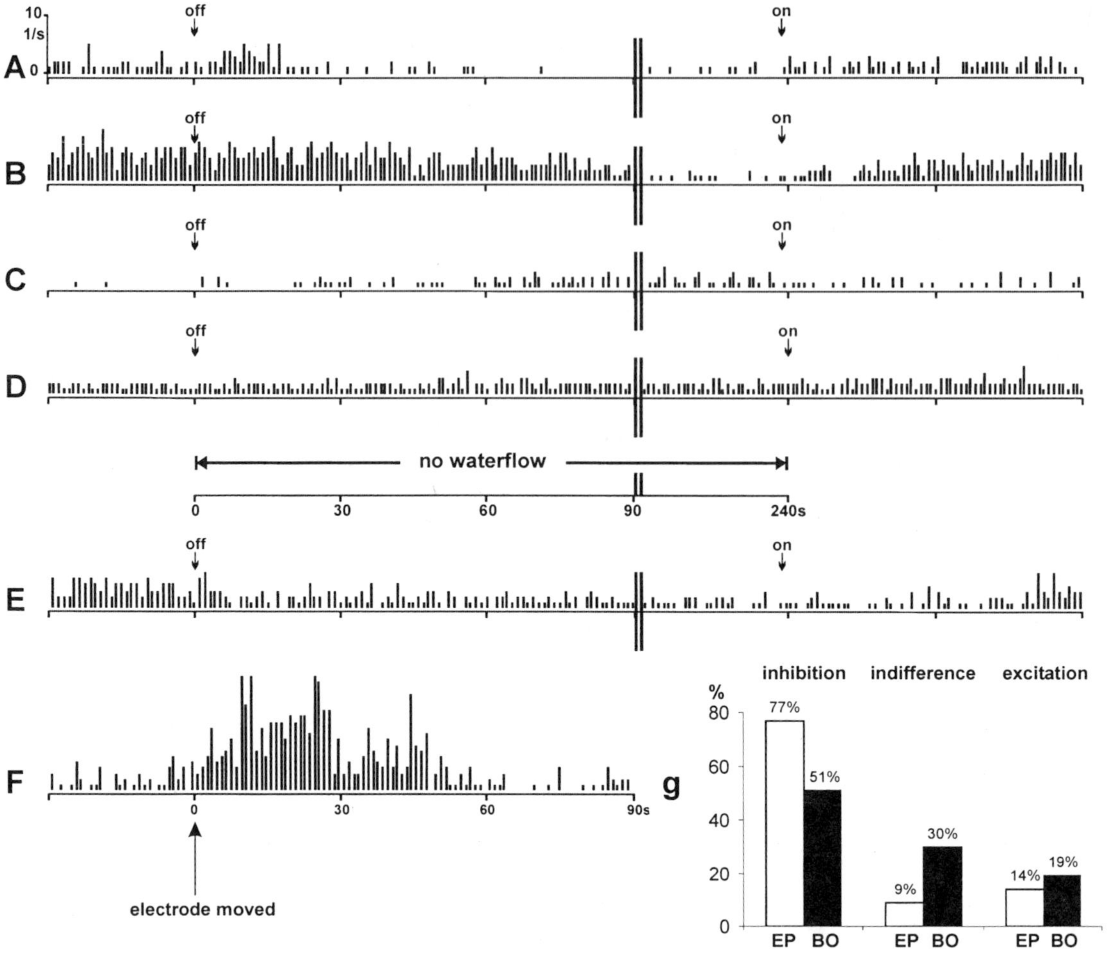

FIGURE 15.7. Various water-"off" and water-"on" responses recorded from epithelial receptor neurons and olfactory bulb relay neurons. A–E show effects recorded from single cells after termination of epithelial water flow for 4 min (90 s after termination and the last 30 s before turning on water flow again). Neural activity 30 s before and 60 s after termination of the laminar water flow are shown. A–D recorded from epithelial receptor neurons and E from an olfactory bulb mitral cell. (A, B) An initial increase of activity was frequently recorded during the 30 s after termination of the waterflow. Thereafter an immediate (A) or a delayed (B) decrease of activity was recorded. After onset of the water flow the activity rapidly returned to values recorded before termination of water flow. (C, D) Rarely an excitation was recorded after termination of water flow, or no significant (D) change of activity. (E) Inhibition of activity of a mitral cell recorded after termination of the epithelial water flow. (F) Strong excitation recorded from an epithelial receptor neuron after a slight movement of the recording electrode. (G) Summary of effects recorded from 105 olfactory bulb mitral cells and 72 receptor neurons in the olfactory epithelium. Inhibition was recorded in the majority of relay (51%) and sensory (77%) neurons 4 min after termination of the water flow. No significant change (indifference) was found in 30% of relay neurons in the olfactory bulb and 9% of epithelial sensory neurons. Excitation was recorded in 19% of bulbar relay neurons and in 14% of epithelial receptor neurons.

laminar application of tap water. Figure 15.7E shows surprisingly similar effects recorded from olfactory bulb mitral cells during interruption of the laminar water flow through the epithelium. Figure 15.7G summarizes effects recorded from 105 mitral cells and 72 epithelial receptor neurons during interruption of the laminar water flow.

Adequate mechanical stimuli, such as laminar water currents, probably cause turbulence and induce slight movements of the mucus and the cilia of the receptor cells. The slightest movements of the recording electrode frequently resulted in a significant increase in activity of the receptor neurons (Fig. 15.7F). The mechanosensitivity of the sensory neurons often made it possible to optimally position the recording electrode. Before stimulus application, and for determination of the spontaneous activity, the recording electrode was left in its final position for 3–5 minutes.

The obvious mechanosensitivity of olfactory bulb relay neurons described for spontaneously breezing mammals (Chaput, 1983; Chaput and Lankheet, 1986; Reinken and Schmidt, 1986; Ravel et al., 1987; Chaput et al., 1992) and for fishes (Zippel et al., 1995) is mediated by olfactory sensory neurons and not by mechanosensitive receptor neurons in the olfactory epithelium. In fishes, the flow rate through the epithelium might, in combination with the concentration of olfactory stimuli, enable the calculation of the distance of a relevant olfactory stimulus.

In "classical" in vivo recordings from frog olfactory receptor neurons (Duchamp et al., 1974; Getchell, 1986), the extremely low spontaneous activity of the neurons resulted in no significant inhibitory responses observed during stimulus application. In goldfish, the interstimulus activity during laminar water current is surprisingly high (mean value: $2.3 \cdot s^{-1}$), and only slightly lower than in olfactory bulb mitral cells (Zippel et al., 1999, 2000). Three to four minutes after the termination of the waterflow, a significantly lower (mean value: $.82 \cdot s^{-1}$) activity was recorded from single epithelial receptor neurons. In recordings from catfish, an even higher interstimulus activity probably is the result of the high flow rate through the epithelium (Kang and Caprio, 1995, 1997). In in vitro recordings, the spontaneous activity of receptor neurons either is not present, or is experimentally induced by depolarization of the receptor neurons (Reisert and Mathews, 1999; Vogler and Schild, 1999). A comparison of effects to naturally important stimuli was, in the aforementioned studies, not possible because naturally important stimuli were not known for the investigated species.

2.5. Cross-Adaptation Experiments

For cross-adaptation (CA) experiments, olfactory bulb relay neurons were selected where epithelial stimulus application resulted in clearly evident responses. Recordings from single relay neurons lasted 1.5 to 4.0 hours. After recording responses of olfactory bulbar neurons to stimuli dissolved in water (with 180-s interstimulus water phase intervals), the same stimuli were tested subsequent to one of the stimuli (i.e., the adapting stimulus) being presented to the olfactory mucosa instead of water during the 180-s interstimulus phases. The application of a stimulus instead of water during the interstimulus intervals resulted in either short-lasting (1–2 min) or an extensive (<30 min) alteration of the interstimulus activity recorded during the preceding water phases. During the 15-s stimulus applications, various effects were recorded in comparison to the effects recorded after application of tap water: (1) no response during stimulus application (i.e., a full cross-adaptation); (2) a significantly reduced response (i.e., a partial cross adaptation); (3) response similar to the response after the water-interstimulus-phase (i.e., no cross adaptation); (4) larger effects than those recorded before the cross-adaptation experiment; and (5) opposing response in comparison to response recorded after water (excitation instead of inhibition, and vice versa). Response characteristics mentioned under 4 and 5 have never been reported in EOG recordings.

During CA experiments, individual mitral cells showed a great variety of responses (response types 1 to 5 mentioned above) (Fig.

15.8). EOG findings could, in part, be confirmed since prostaglandins PGF and 15K-PGF mainly bind to different receptor molecules. The previous finding, that 21-C steroids bind to one and the same class of receptor molecules (Sorensen et al., 1990) could not be confirmed (Fig. 15.8). In general, each mitral cell seems to have a rather specific input from various types of

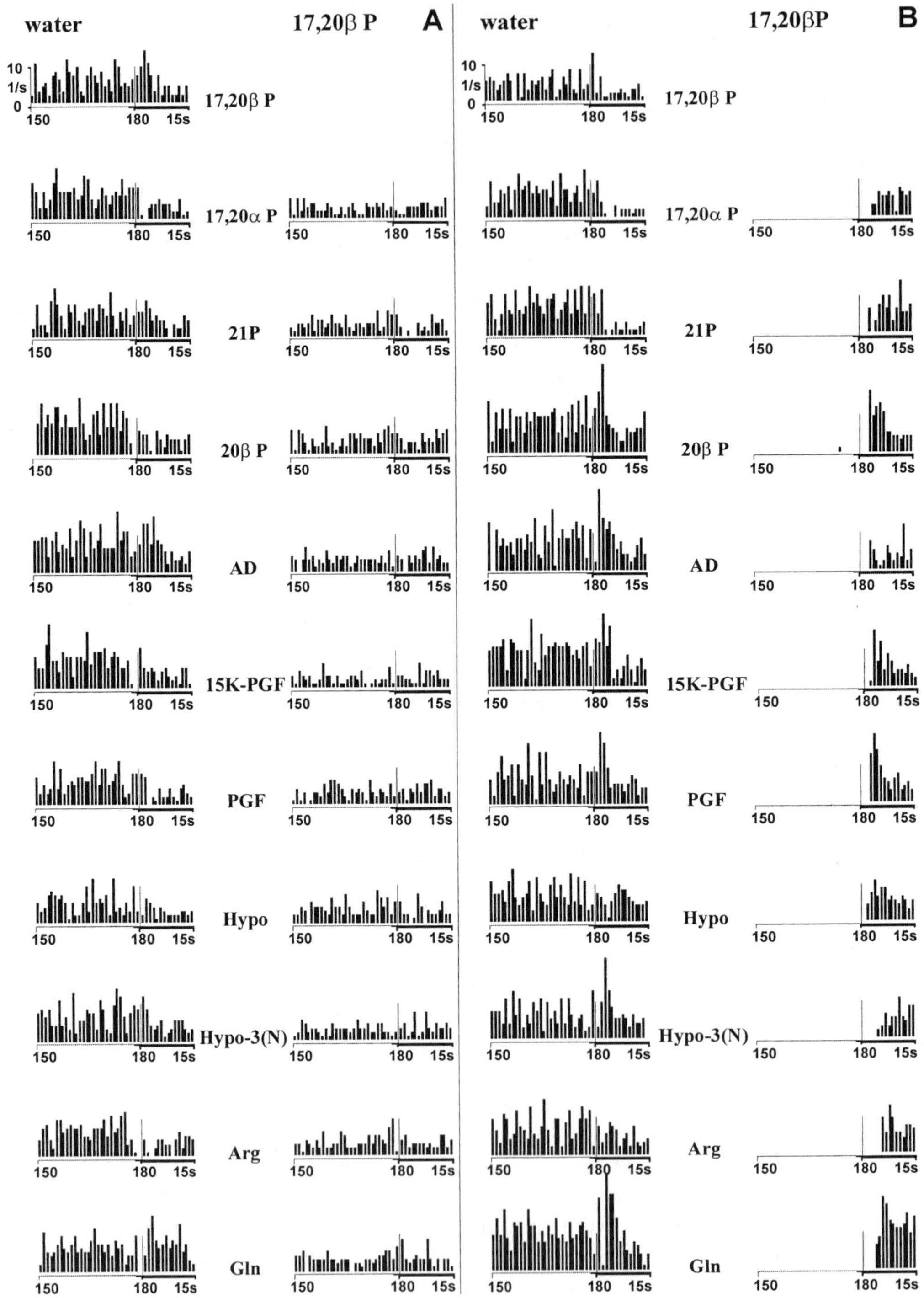

epithelial sensory neurons and, during application of a great variety of stimuli, can show a CA or a partial CA, show similar responses, react stronger, or even show opposing effects. Summarizing the effects recorded from a representative number of mitral cells, however, showed a more or less equal distribution of the abovementioned effects independent of the CA stimulus applied (Fig. 15.9).

Physiological (Caprio and Byrd, 1984; Ohno et al., 1984; Caprio et al., 1989; Sveinsson and Hara, 1990; Kang and Caprio, 1991; Michel and Derbridge, 1997) and biochemical (Cagan and Zeiger, 1978; Bruch and Rulli, 1988) investigations demonstrate a number of different types of receptor molecules for amino acids (AA). CA experiments during application of 11 amino acids showed a similarly large range of response characteristics and a similar distribution of the abovementioned effects, especially during application of pheromones and control stimuli. In goldfish, there are more than four relatively independent transduction mechanisms (acidic AA, basic AA, short-chain neutral AA, and long-chain neutral AA) described for catfish on the basis of EOG recordings (Caprio and Byrd, 1984).

In another CA series, structurally similar and structurally dissimilar, nonfamiliar stimuli were investigated. During water application throughout the 180-s interstimulus-phases, stimulation with structurally similar stimuli like α- and β-ionone or α-damascone and damascenone showed different effects from approximately 40% to 50% of bulbar relay neurons. During CA experiments, mixtures of the nonfamiliar stimuli were applied during the interstimulus phases because application of pure stimuli (see above, and Figs. 15.8 and 15.9) did not result in the expected numbers of cross adaptations. Effects and modulations of effects, however, were again similar to data recorded in the CA experiments with pheromones and amino acids. Mean values for the various modulations during CA experiments again were surprisingly similar in comparison to pheromone and amino acid experiments. During CA application of (for example) a mixture of α and β-ionone the numbers of cross adaptations and partial cross adaptations were not significantly greater during application of α-ionone and β-ionone than during application of the rest of stimuli (Kokemüller and Zippel, 2001).

From recent CA experiments in goldfish in which recordings were made on the basis of EOG experiments (Caprio and Byrd, 1984; Sorensen et al., 1990) and studies on single sensory neurons in the epithelium, it is evident that a large variety of responses (response types 1 to 5 mentioned above) also can be recorded from single sensory neurons (U. Jakob, and H.P. Zippel, unpublished; X. Cretsi, and H.P. Zippel, unpublished). The cross-adaptation effects recorded from relay olfactory bulb neurons and epithelial sensory neurons demonstrate significant differences in comparison to EOG recordings and an incredible variety of response characteristics even during application of related and structurally similar stimuli. The heterogeneity of responses recorded from olfactory bulb relay neurons and the unpredictability of cross adaptations or other effects shows that the glomerular convergence of sensory axons is rather variable. A

FIGURE 15.8. Cross-adaptation experiment with 17,20βP, the preovulatory pheromone. During application of water in the interstimulus phase mitral cell A is quantitatively differently inhibited by all the stimuli except Gln, which is excitatory. Cell B is also inhibited by all the stimuli. The inhibition in cell B, however, is preceded by a phasic excitation during application of 6 stimuli. In both cells, application of 17,20βP during the 180 s interstimulus phases results in a permanent, and in cell B, total inhibition of the activity. Stimulation of the epithelial sensors with the remaining 10 stimuli in cell A results in a full cross adaptation (except 21P, which again elicits an inhibition), and in massive opposing effects (excitation instead of inhibition) in cell B. For further details see text. 17,20βP = 17α,20β-dihydroxyprogesterone; 17,20αP = 17α,20α-dihydroxyprogesterone; 21P = 11-deoxycorticosterone; 20βP = 4-pregnen-20β-ol-3-one; AD = androstendione; 15K-PGF$_{2_\alpha}$ = 15-ketoprostaglandin; PGF = prostaglandin F$_{2_\alpha}$; Hypo = hypoxanthine; Hypo-3(N) = hypoxanthine-3(N)-oxide; Arg = L-arg; Gln = L-gln.

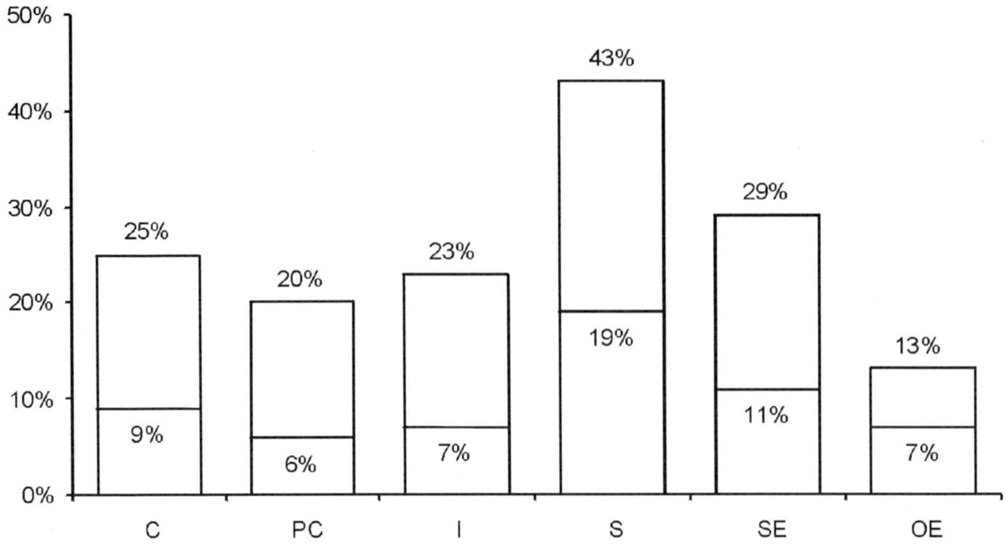

FIGURE 15.9. Summary of effects recorded from olfactory bulb mitral cells during cross-adaptation experiments with the preovulatory pheromone 17,20βP. In contrast to EOG recordings, we have not been able to show significantly greater amounts of cross adaptation during application of three different 4-pregnen 21-C steroids during permanent application of 17,20βP instead of water during the interstimulus phase. Nearly similar amounts and a great variety of effects recorded during application of 17,20βP instead of water during the interstimulus phase. Nearly similar amounts recorded during application of the other 4-pregnen 21-steroids and of the seven different control stimuli (one 19-C steroide (androstendione), two ovulatory pheromones, the probable alarm pheromone hypoxanthine-3(N)-oxide and hypoxanthine, and finally two amino acids; for details see text). In 54 cross-adaptation experiments with 17,20βP, a large variety of effects was recorded from bulbar mitral cells, which receive a rather variable input from different epithelial receptor neurons. C = cross-adaptation (i.e., no response during stimulus application); PC = partial cross-adaptation (i.e., a reduced response); I = indifferent response during water and during cross adaptation experiment; S = response remains similar; SE = stronger effect during cross-adaptation experiment; OE = opposing effect during cross-adaptation (excitation instead of inhibition and vice versa). For stimulus abbreviations see Figure 15.8, and for further details see text.

local projection from specific olfactory neurons to circumscribed narrow bulbar positions in zebrafish (e.g., Baier and Korsching, 1994; Baier et al., 1994; Friedrich and Korsching, 1997) and salmonid fishes (Hara and Zhang, 1998) does not necessarily mean that only a specific population of bulbar mitral cells located in the vicinity of respective glomeruli gets the input from only specific populations of epithelial sensory neurons. First, recordings from different areas of the olfactory bulb show no significant responses to a specific odor (e.g., amino acids or pheromones). In contrast, the discriminative ability of single relay neurons is incredibly sensitive and can drastically vary from neuron to neuron. Second, a number of different "specific" olfactory sensory neurons can converge onto one glomerulus. Third, the dendrites of the mitral cells divide and can terminate in several glomeruli (Oka, 1983) located every 100 μm throughout the diameter of the olfactory bulb (1,000 μm) of goldfish. For telencephalic analysis, it is sufficient that preferential projection of more specific information about food stimuli, the alarm pheromone, and other pheromones mainly projects to telencephalic nuclei via different olfactory subtracts (Hamadani et al., 2000).

3. Perspectives

The present review has concentrated on recent physiological data. We also have shown the great advantages of using fishes in olfactory research where biologically relevant stimuli are known and there is a unique anatomical arrangement of different types of ("neighboring") relay neurons. A disadvantage is the fact that in the olfactory epithelium both types of sensory neurons (ciliated and microvillous cells) are present. From recordings with platinum-black electrodes, we can say only that a great variety of responses exists during application of various stimuli, but unfortunately not which type of sensor is responding. After recordings from single sensory neurons following olfactory nerve axotomy, which results in a more rapid regeneration of ciliated sensory neurons and a slightly delayed regeneration of microvillous cells, we unfortunately found no significant differences in the response properties of the two types of neurons during application of pheromone and amino acid food stimuli. We urgently need investigations in species with a separated olfactory system where the main olfactory system possesses ciliated sensory neurons and the accessory (vomeronasal) olfactory system possesses microvillous sensory neurons. At present, it is rather difficult to interpret findings recorded from different species with a great variety of techniques. Odorants can be detected by animals in the micro- and even picomolar range. In patch-clamp or slice preparations, however, only high concentrations, and probably not in the "physiological range," have been applied and often elicit nonspecific effects. When experiments are conducted in vitro, the normal external milieu of the sensory cells is completely lost. The use of micromolar (or even higher molarity) concentrations may be essential to obtain a reasonable number of observable cell responses.

It seems essential to carefully and more systematically investigate the bimodality of the olfactory system. This might explain the differences in response characteristics (e.g., no or almost no inhibitory responses during stimulus application), and depend on methodological concerns such as the velocity and the duration of the stimulus application. Finally, physiological recordings should be made from central nuclei. So far, we have no idea about how and where the analysis of chemical stimuli really takes place. Fishes continue to be an excellent experimental model due to the existence of detailed anatomical studies.

Acknowledgments. The authors gratefully acknowledge the support of A. Hansen and D.L. Lipschitz, and the critical and helpful comments on the manuscript provided by J Caprio. Supported by DFG Zi-112/7-3.

References

Atema, J., Fay, R.R., Popper, A.N., and Tavola, W.N. (1988). *Sensory Biology of Aquatic Animals*. New York: Springer.

Baier, H., and Korsching, S. (1994). Olfactory glomeruli in the zebrafish form an invariant pattern and are identifiable across animals. *J. Neurosci.* 14:219–230.

Baier, H., Rotter, S., and Korsching, S. (1994). Connectional topography in the zebrafish olfactory system: Random positions but regular spacing of sensory neurons projecting to an individual glomerulus. *Proc. Natl. Acad. Sci. USA* 91: 11646–11650.

Bruch, R.C., and Rulli, R.D. (1988). Ligand binding specificity of a neutral L-amino acid olfactory receptor. *Comp. Biochem. Physiol. B.* 91:535–540.

Cagan, R.H., and Zeiger, W.N. (1978). Biochemical studies of olfaction: Binding specificity of radioactively labelled stimuli to an isolated olfactory preparation from rainbow trout (*Salmo gairdneri*). *Proc. Natl. Acad. Sci. USA* 75:4679–4683.

Caprio, J. (1995). Olfactory and taste recordings in fish. In: *Experimental Cell Biology of Taste and Olfaction Current Techniques and Protocols* (Spielman, A.I., and Brand, J.G., eds.), pp. 251–261. Boca Raton: CRC Press.

Caprio, J., and Byrd, R.P. Jr. (1984). Electrophysiological evidence for acidic, basic, and neutral amino acid olfactory receptor sites in the catfish. *J. Gen. Physiol.* 84:404–422.

Caprio, J., Dudek, J., Robinson, J.J. II (1989). Electro-olfactogram and multiunit olfactory receptor responses to binary and trinary mixtures of amino acids in the channel catfish, *Ictalurus punctatus*. *J. Gen. Physiol.* 93:245–262.

Chaput, M. (1983). Effects of olfactory peduncle sectioning on the single unit responses of olfactory bulb neurons to odor presentation in awake rabbits. *Chem. Senses.* 8:161–177.

Chaput, M.A., and Lankheet, M.J. (1986). Influence of stimulus intensity on the categories of single-unit responses recorded from olfactory bulb neurons in awake freely breathing rabbits. *Physiol. Behav.* 40:453–462.

Chaput, M.A., Buonviso, N., and Berthommier, F. (1992). Temporal patterns in spontaneous and odour-evoked mitral cell discharges recorded in anaesthetized freely breathing animals. *Eur. J. Neurosci.* 4:813–822.

Conze, C., and Zippel, H.P. (2001). Cross-adaptation experiments in goldfish (*Carassius auratus*) olfactory bulb relay neurons demonstrate a great variety of response patterns during application of L-arg and structural analogues. In: *Proc. 4th Meeting of the German Neurosci. Soc. and the 28th Neurobiol. Conf.* (Elsner, N., and Kreutzberg, G.W., eds.), p. 460. Thieme: Stuttgart.

Duchamp, A., Revial, M.F., Holley, A., and MacLeod, P. (1974). Odor discrimination by frog olfactory receptors. *Chem. Senses Flavor* 1:213–233.

Friedrich, R.W., and Korsching, S.I. (1997). Combinatorial and chemotopic odorant coding in the zebrafish olfactory bulb visualized by optical imaging. *Neuron* 18:737–752.

Gesteland, R.C. (1975). Techniques for investigating single unit activity in the vertebrate olfactory epithelium. In: *Methods in Olfactory Research* (Moulton, D.G., Tur, A., and Johnston, J.W. Jr., eds.), pp. 269–322. London: Academic Press.

Getchell, T.V. (1986). Functional properties of vertebrate olfactory receptor neurons. *Physiol. Rev.* 66: 772–817.

Hamdani, E.H., Stabell, O.B., Alexander, G., and Døving, K.B. (2000). Alarm reaction in the crucian carp is mediated by the medial bundle of the medial olfactory tract. *Chem. Senses* 25:103–109.

Hansen, A., Zippel, H.P., Sorensen, P.W., and Caprio, J. (1999). Ultrastructure of the olfactory epithelium in intact, axotomized, and bulbectomized goldfish, *Carassius auratus*. *Micros. Res. Tech.* 45: 325–338.

Hara, T.J. (1982). *Chemoreception in Fishes.* Amsterdam, Oxford, New York: Elsevier.

Hara, T.J. (1992). *Fish Chemoreception.* London, Glasgow, New York, Tokyo, Melbourne, Madras: Chapman & Hall.

Hara, T.J., and Zhang, C. (1998). Topographic bulbar projections and dual neural pathways of the primary olfactory neurons in salmonid fishes. *Neuroscience* 82:301–313.

Kang, J.S., and Caprio, J. (1991). Electro-olfactogram and multiunit olfactory receptor responses to complex mixtures of amino acids in the channel catfish, *Ictalurus punctatus*. *J. Gen. Physiol.* 98: 699–721.

Kang, J.S., and Caprio, J. (1995). In vivo responses of single olfactory receptor neurons in the channel catfish, *Ictalurus punctatus*. *J. Neurophysiol.* 73: 172–177.

Kang, J.S., and Caprio, J. (1997). In vivo responses of single olfactory receptor neurons of channel catfish to binary mixtures of amino acids. *J. Neurophysiol.* 77:1–8.

Kosaka, T. (1980). Ruffed cell: A new type of neuron with a distinctive initial unmyelinated portion of the axon in the olfactory bulb of the goldfish (*Carassius auratus*). II. Fine structure of the ruffed cell. *J. Comp. Neurol.* 193:119–145.

Kosaka, T., and Hama, K. (1979). Ruffed cell: A new type of neuron with a distinctive initial unmyelinated portion of the axon in the olfactory bulb of the goldfish (*Carassius auratus*). I. Golgi impregnation and serial thin sectioning studies. *J. Comp. Neurol.* 186:301–320.

Kosaka, T., and Hama, K. (1981). Ruffed cell: A new type of neuron with a distinctive initial unmyelinated portion of the axon in the olfactory bulb of the goldfish (*Carassius auratus*). III. Three-dimensional structure of the ruffed cell dendrite. *J. Comp. Neurol.* 201:571–587.

Kosaka, T., and Hama, K. (1982). Structure of the mitral cell in the olfactory bulb of the goldfish (*Carassius auratus*). *J. Comp. Neurol.* 212:365–384.

Kosaka, T, and Hama, K. (1982–1983). Synaptic organization in the teleost olfactory bulb. *J. Physiol.* (Paris) 78:707–719.

Koce, A., and Valentinčič, T. (2000). The amplitude of the electroolfactogram in catfish correlates with the proportion of responding ORNs. *Pflugers Arch.* 439:171–172.

Kokemüller, A., and Zippel, H.P. (2001). Stimulus discrimination and cross-adaptation experiments in goldfish (*Carassius auratus*) olfactory bulb relay neurons during application of very similar non-familiar stimuli. In: *Proc 4th Meeting of the German Neurosci. Soc. and the 28th Neurobiol. Conf.* (Elsner, N., and Kreutzberg, G.W. eds.), p. 466. Thieme: Stuttgart.

Lipschitz, D.L., and Michel, W.C. (1999). Physiological evidence for the discrimination of L-arginine from structural analogs by the zebrafish olfactory system. *J. Neurophysiol.* 82:3160–3167.

Lipschitz, D.L., and Michel, W.C. (2000). Do amino acids stimulate ciliated or microvillar olfactory sensory neurons? *Chem. Senses* 25:669–670.

McLean, J.H., and Shipley, M.T. (1992). Neuroanatomical substrates of olfaction. In: *Science of Olfaction* (Serby, M.J., and Chobor, K.L., eds.), pp. 126–171. New York: Springer.

Michel, W.C., and Derbridge, D.S. (1997). Evidence of distinct amino acid and bile salt receptors in the olfactory system of the zebrafish, *Danio rerio*. *Brain Res.* 764:179–187.

Möller-de Beer, A.M., and Zippel, H.P. (2001). *In-vivo*-recordings from single olfactory sensory neurons in goldfish (*Carassius auratus*) demonstrate their biomodality. In: *Proc. 4th Meeting of the German Neurosci. Soc. and the 28th Neurobiol. Conf.* (Elsner, N., and Kreutzberg, G.W., eds.). p. 464. Thieme: Stuttgart.

Ngai, J., Dowling, M.M., Buck, L., Axel, R., and Chress, A. (1993a). The family of genes encoding odorant receptors in the channel catfish. *Cell* 72:857–666.

Ngai, J., Chess, A., Dowling, M.M., Necles, N., Macagno, E.R., and Axel, R. (1993b). Coding of olfactory information: Topography of odorant receptor expression in the catfish olfactory epithelium. *Cell* 72:667–680.

Oka, Y. (1983). Golgi electron-microscopic and combined Golgi-electron-microscopic studies of the mitral cells in the goldfish olfactory bulb. *Neuroscience* 8:723–742.

Ohno, T., Yoshii, K., and Kurihara, K. (1984). Multiple receptor types for amino acids in the carp olfactory cells revealed by quantitative cross-adaptation method. *Brain Res.* 310:13–21.

Pfeiffer, W. (1982). Chemical signals in communication. In: *Chemoreception in Fishes* (Hara, T.J., ed.), pp. 307–326. New York: Elsevier.

Ravel, N., Caille, D., and Pager, J.A. (1987). A centrifugal respiratory modulation of olfactory bulb unit activity: A study on acute rat preparation. *Exp. Brain Res.* 65:623–628.

Reinken, U., and Schmidt, U. (1986). Reactions of olfactory bulb neurons to different stimulus intensities in laboratory mice. *Exp. Brain Res.* 63:151–157.

Reisert, J., and Matthews, H.R. (1999). Adaptation of the odour-induced response in frog olfactory receptor cells. *J. Physiol. (Lond.)* 519:801–813.

Schild, D., and Restrepo, D. (1998). Transduction mechanisms in vertebrate olfactory cells. *Physiol. Rev.* 78:429–466.

Sorensen, P.W., and Caprio, J. (2000). Restricted distribution and chemospecificity of pheromone-sensitive olfactory receptor neurons in the goldfish olfactory epithelium. *Chem. Senses* 25:670.

Sorensen, P.W., Hara, T.J., Stacey, N.E., and Dulka, J.G. (1990). Extreme olfactory specificity of male goldfish to the preovulatory steroidal pheromone 17α,20β-dihydroxy-4-pregnen-3-one. *J. Comp. Physiol. A.* 166:73–383.

Sveinsson, T., and Hara, T.J. (1990). Multiple olfactory receptors for amino acids in arctic char (*Salvelinus alpinus*) evidenced by cross-adaptation experiments. *Comp. Biochem. Physiol. A.* 97:289–293.

Valentinčič, T., and Koce, A. (2000). Coding principles in fish olfaction as revealed by single unit, EOG and behavioral studies. *Pflugers Arch.* 439:193–195.

Valentinčič, T., Kralj, J., Stenovec, M., Koce, A., and Caprio, J. (2000a). The behavioral detection of binary mixtures of amino acids and their individual components by catfish. *J. Exp. Biol.* 203: 307–3317.

Valentinčič, T., Metelko, J., Ota, D., Pirc, V., and Blejec, A. (2000b). Olfactory discrimination of amino acids in brown bullhead catfish. *Chem. Senses* 25:21–29.

Vogler, C., and Schild, D. (1999). Inhibitory and excitatory responses of olfactory receptor neurons of *Xenopus laevis* tadpoles to stimulation with amino acids. *J. Exp. Biol.* 202:997–1003.

von Rekowski, C., Schmilewski, J., and Zippel, H.P. (1994). Behavioral experiments in goldfish confirm amino acids as olfactory meaningful natural stimuli. *Adv. Biosci.* 93:491–496.

Willms, H.-G., and Zippel, H.P. (1999). Epithelial cross-adaptation experiments with amino acids result in a variety of different effects in goldfish olfactory bulb relay neurons. In: *From Molecular Neurobiology to Clinical Neuroscience* (Elsner, N., and Eysel, U. eds.), p. 375. Stuttgart: Thieme.

Zippel, H.P. (1993). Regeneration in the peripheral and the central olfactory system: A review of morphological, physiological and behavioral aspects. *J. Hirnforsch.* 34:207–229.

Zippel, H.P., Hansen, A., and Caprio, J. (1997a). Renewing olfactory receptor neurons in goldfish does not require contact with the olfactory bulb to develop normal chemical responsiveness. *J. Comp. Physiol. A.* 181:425–437.

Zippel, H., Sorensen, P.W., and Hansen, A. (1997b). High correlation between microvillous olfactory receptor cell abundance and sensitivity to pheromones in olfactory nerve-sectioned goldfish. *J. Comp. Physiol. A.* 180:39–52.

Zippel, H.P., Reschke, C., and Korff, V. (1999). Simultaneous recordings from two physiologically different types of relay neurons, mitral cells and ruffed cells, in the olfactory bulb of goldfish. *Cell. Mol. Biol.* 45:327–337.

Zippel, H.P., Voigt, R., Knaust, M., and Luan, Y. (1993). Spontaneous behavior, training and discrimination training in goldfish using chemosensory stimuli. *J. Comp. Physiol. A.* 172:81–90.

Zippel, H.P., Gloger, M., Lüthje, L., Nasser, S., and Wilcke, S. (2000). Pheromone discrimination ability of olfactory bulb mitral and ruffed cells in goldfish (*Carassius auratus*). *Chem. Senses* 25:339–349.

Zippel, H.P., Fiedler, W., Hager, D., Kühling-Thees, J., Maier, K., von Rekowski, C., Schumann, R., Tiedemann, W., Voigt, R., and Wächter, T. (1995). Responses of goldfish olfactory bulb relay neurons during mucosal application of various chemical, mechanical and temperature stimuli. *Biophysics* 40:171–192.

Part 4
Visual Adaptations to Limited Light Environments

Michael F. Land

Life in the sea provides visual opportunities and challenges that are very different from those on land. The very edges of the oceans are not so different from terrestrial environments: There is plenty of light, a range of available wavelengths, and the cover of rocks, reef, or plants where prey animals can hide and predators can lurk.

In the open ocean, however, there is no cover, and below the upper 100 m of the epipelagic zone the light available from daylight becomes weaker and weaker, and increasingly restricted in wavelength to a narrow band of blue light centred on 480 nm. Useable light from the sun has disappeared completely by 1,000 m (Denton, 1990), but some animals retain eyes, even down to the ocean floor several thousand meters deeper. This is because there is much bioluminescence at these depths. Animals emit light for communication, for defense, and for camouflage (Herring, 2000), and this background of flashes and glows continues to make vision useful for those animals that have retained eyes.

If we consider the life of a fish in the mesopelagic (twilight) zone between 100 and 1,000 meters, the contrast with terrestrial or shallow-water life becomes clear. There is no plant life, so the only sources of food are other animals or detritus that falls from above. The only ways to see food are to look upward, and sight it against the residual daylight (so many fishes and crustaceans have large, upward-pointing eyes), or to look for flashes of luminescence in the dark below (smaller, downward-pointing eyes are also common; Land, 2000). However, animals must take pains not to be seen by the predators below them, and this means that they must be suitably camouflaged. As Eric Denton showed, the sides of a fish can be well disguised with vertically oriented plane mirrors, because the light reflected from them is virtually identical to the light that would have come through the fish—the next best thing to being transparent. However, from directly below, the silhouette of an animal cannot be disguised in this way. The only way to achieve this is to make the underside of the animal match the light from above, and this can be done only by illuminating it with bioluminescent photophores. Thus downward-pointing light-organs are found in many fishes, crustaceans, and cephalopod molluscs. This brief account illustrates the visual arms race in the midocean: vision versus camouflage. With nowhere to shelter, all animals must play versions of the same game.

The three chapters in this section explore different aspects of vision in the various depth strata of the ocean. Warrant, Collin, and Locket (Chapter 16) examine eye structure in relation to the way the nature of the visual task changes with depth. In particular, they ask how eyes adapted to the task of detecting silhouettes against the downwelling daylight differ from those that are more concerned with the rather easier task of detecting luminescent flashes. By comparing bottom-dwelling with upper-mesopelagic fishes they show that differences in sensitivity, resolution, and the use of foveas all can be related to this shift in the type of light used to locate food. Douglas, Hunt, and Bowmaker (Chapter 17) consider the implications of the changes in the spectrum available to animals at different depths. As one would predict, most fishes from the lower-mesopelagic regions have only blue-sensitive visual pigments, but a substantial minority have two or even three visual pigments, apparently exploiting the small amount of shorter- or longer-wavelength light. Of greatest interest, however, are the stomiid dragon fishes, which, in spite of the fact that their world is blue, have visual pigments sensitive in the red. These fishes also have red-emitting photophores, so it seems that they have their own private wavelength, which could be used for communication, or simply to illuminate the surroundings in the search for prey.

With the chapter by Marshall Cronin, and Frank (Chapter 18) we return to the sunnier waters of the ocean's edge. After reviewing color vision in the crustaceans (most are violet/green dichromats), the authors concentrate on the unique and astonishing color-vision system of the reef-dwelling stomatopods (mantis shrimps). These have 12 visual pigments in their color-vision system (including 4 sensitive in the ultraviolet) and a further 4 in other parts of the eyes. They show that the spectral range of the pigments is adapted to the light environment, with deeper-living species having blue-shifted sensitivities. This sensitivity range can be shifted in the laboratory by subjecting young animals to different spectral regimes. The final part of the chapter deals with temporal resolution, and as one might expect, deeper-living animals have slower retinas (because it takes longer to generate an adequate photon count). The exceptions seem to be the euphausiids (krill), possibly because they are active hunters of copepods and so need faster vision, which may be made possible by their high-sensitivity superposition eyes.

References

Denton, E. (1990). Light and vision at depths greater than 200 metres. In: *Light and Life in the Sea* (Herring, P.J., Campbell, A.K., Whitfield, M., and Maddock, L., eds.), pp. 127–148. Cambridge: Cambridge University Press.

Herring, P.J. (2000). Species abundance, sexual encounter and bioluminescent signalling in the deep sea. *Phil. Trans. R. Soc. B.* 355:1273–1276.

Land, M.F. (2000). On the functions of double eyes in mid-water animals. *Phil. Trans. R. Soc. B.* 355: 1147–1150.

16
Eye Design and Vision in Deep-Sea Fishes

Eric J. Warrant, Shaun P. Collin, and N. Adam Locket

Abstract

The three great faunal zones of the deep-sea—the twilight mesopelagic zone, the numbingly dark bathypelagic zone, and the vast flat expanses of the benthic zone—are home to an enormous variety of fishes. In each of these zones, the light environment is unique and highly predictable. In the mesopelagic zone (150–1,000 m), the downwelling daylight creates an extended scene that becomes increasingly dimmer with depth. Bioluminescent point-source flashes, on the other hand, become increasingly visible. Upon entry to the bathypelagic zone at 1,000 m no daylight remains, and the scene becomes entirely dominated by point-like bioluminescence. This changing nature of visual scenes with depth—from extended source to point source—has had a profound effect on the eyes and visual capacities of deep-sea fishes. So too has the wide and dim horizon seen by benthic fishes as they cruise over the ocean floor. An exploration of these effects is the topic of this review. The eyes of fishes living at different depths in all three zones are compared, and their eye designs related to what is known of their natural histories. This analysis leads to the conclusion that the intensity, direction, and particularly the spatial organization of light into extended and point-like sources have been the dominant players in the evolution of deep-sea visual systems.

1. Introduction

The world's oceans comprise the largest habitat on Earth, 168 times larger than all terrestrial habitats combined (Cohen, 1994). The bright upper 100 m of the ocean—the *epipelagic zone*—is home to large agile predatory fishes and a rich plankton fauna. The deep-sea only starts below this zone, in the twilight world of the *mesopelagic zone*, which stretches down to the limit of daylight at 1,000 m. Beyond 1,000 m we find the vast *bathypelagic zone*, the largest and darkest habitat on Earth. The only light found at these depths is the bioluminescent light produced by animals themselves. Even though most ocean creatures live in the shallower illuminated depths, a remarkable variety also live in the vast darkness of the deep (Marshall, 1979). But at all depths the struggle to find food and mates is equally pressing and, in these contexts, the sensory systems of fishes are immensely important. The visual system is no exception.

With increasing depth in the mesopelagic zone, daylight becomes progressively dimmer and bluer. In contrast, bioluminescent flashes—drowned by daylight in the upper-mesopelagic zone—become more distinct with increasing depth. It is within this rather changeable light environment that visual adaptations show their greatest variation. Upon entrance into the bathypelagic zone daylight disappears all together, and bioluminescence is the only light that fishes can see, a stimulus that has largely constrained the designs of their eyes. Until recently, the relationship between the eye designs of fishes and depth in the ocean was perplexing. Sir John Murray and Johan Hjort—after several months of pioneering oceanographic work aboard the Norwegian steamer *Michael Sars*—exclaimed that "nothing seems more hopeless in biological oceanography than the attempt to explain the connection between the development of the eyes and the intensity of light at different depths in the ocean" (Murray and Hjort, 1912). During the last 15 years we have started to understand the reasons for their dilemma. It turns out that marine eyes are adapted not only to light intensity per se, but also to its spatial organization. In the ocean, light either is organized as an extended scene of visual features (as in the brighter shallower depths), or is point-like (as is typical of a bioluminescent flash). With increasing depth extended daylight illumination gradually gives way to bioluminescence, and at any given depth one or both types of light may be visible. Which visual adaptations evolve will depend on which type—point or extended or both—has the greatest ecological meaning, and this of course will vary widely from species to species. Thus, at any given depth the range of visual adaptations could be quite large.

The eyes and visual systems of deep-sea fishes, and indeed of all deep-sea animals, have thus evolved under two dominating influences: the changing nature of visual scenes with depth and the natural histories of individual species. These two influences—which form the thematic basis of this review—have led to some of the most remarkable eye designs found in the animal kingdom.

2. Light in the Sea

As light propagates through water it is absorbed and scattered, and this has a marked effect on its quality at different depths in the sea. The intensity of light, as well as its color, degree of polarization, and contrast, is significantly altered as it travels through the ocean.

2.1. The Optical Properties of Water

The intensity of light, even in the clearest oceans and lakes, reduces by approximately 1.5 orders of magnitude for every 100 m of depth. By 600–700 m during the day, light intensities reach starlight levels (Clarke and Denton, 1962). Almost no daylight remains upon entrance to the bathypelagic zone at 1,000 m, and it is very unlikely that marine animals can see daylight below this depth (Denton, 1990).

In a clear ocean, light also becomes progressively bluer with depth, with the orange-red part of the spectrum (beyond 550 nm) almost completely absorbed within the first 100 m (Tyler and Smith, 1970). Ultraviolet light is also absorbed, but not quite as effectively, with biologically relevant intensities remaining to at least 200 m (Frank and Widder, 1996; Losey et al., 1999). Below 200 m, the downwelling daylight is almost monochromatic (475 nm).

In the upper depths of the ocean, daylight is visible in all directions around the animal. An underwater "space light," incident from below and from the side, is due to the scatter of daylight from particles suspended in the water. In very clear water, light scatter—and thus the space light intensity—is lower than in cloudier water. However, with increasing depth the space light intensity declines, and the available daylight comes increasingly from above. This radiance distribution is dominated by the position of the sun in shallower water, but this dominance declines with depth, disappearing altogether below the so-called "asymptotic depth." Below this depth—which is about

400 m in clearest ocean water—the radiance distribution is vertically symmetric (Jerlov, 1976). Full stop only. Light originating laterally and from below is respectively about 40 times and 300 times dimmer than light originating directly from above.

No matter what its intensity, the space light has a detrimental effect on the contrast and visibility of objects in the sea. The scattering of the downwelling daylight by the interposing body of water will create a veiling "haze" of space light within which the object may disappear from sight. Such scattering greatly reduces the contrasts of objects, and limits the furthest distance at which they can be reliably detected (Lythgoe, 1979, 1988; Jagger and Muntz, 1993). In the brightest and clearest ocean water, the veiling space light zeros the contrast of a dark object at a range of about 40 m (Lythgoe, 1979).

2.2. Bioluminescence

Bioluminescence is a major light source in the sea, with many ecological meanings (Herring, 1978). Its contrast and visibility—just as for objects illuminated by daylight—are limited by the intensity of the surrounding space light. During the day in the brighter depths above 100 m, most bioluminescent signals are probably not visible (Denton, 1990). But below this depth, as the space light becomes progressively weaker, bioluminescent signals become increasingly visible. In the blackness of the bathypelagic zone they reach their greatest contrast.

Bioluminescent signals are usually point-source flashes, although in some cases they can be somewhat extended, as in the tunicate *Pyrosoma*, which may grow to the size of a bus (Herring, 2000). The length of a flash may vary from hundreds of milliseconds to several seconds, and their frequency in the sea can vary from 1 to 160 flashes from each steradian (solid angle) of water per minute. Below 1,000 m flash frequency drops considerably, and becomes very infrequent below 2,000 m (Clarke and Hubbard, 1959). The intensity and color of these flashes can vary considerably (see Herring, 1978), but a typical flash is blue and contains between 10^7 and 10^{13} photons, no doubt a highly visible stimulus in the darkness of the deep-sea.

2.3. The Changing Nature of Visual Scenes with Depth

The gradual decline in the downwelling daylight, and the increasing prominence of bioluminescent point sources, creates a visual scene that changes dramatically with depth. In the brighter upper depths, where even the sea floor may be visible, the visual scenes viewed by animals are *extended* in all directions. But at greater depths, the scene begins to be dominated by bioluminescent point sources, especially from below where the space light is dimmest. Upward, and even frontward, the scene may still be extended. At still deeper levels, bioluminescent point sources can be seen against the dim extended daylight in all directions. Below 1,000 m, beyond the penetration of daylight, the visual scene is entirely *point-like*.

It is this change in visual scene, from the bright sunny layers of the epipelagic to the fathomless darkness of the bathypelagic, which has arguably had the greatest influence on the evolution of aquatic vision. And this is no better seen than in the eyes of mesopelagic fishes, the topic to which we turn next.

3. The Eyes and Vision of Mesopelagic Fishes

Without doubt, the mesopelagic zone is the most varied from the point of view of vision. Within its 850 m, daylight intensities vary by over 13 log units and bioluminescent signals are more abundant here than in any other zone of the sea. Mesopelagic fishes therefore have access to the full range of visual scenes—from extended to point source. Not surprisingly, they also have the most varied visual adaptations found in the ocean.

3.1. Adaptations for Increased Sensitivity to a Dim Extended Scene

The features of eyes that maximize their sensitivity to dim *extended* scenes have been nicely summarized by Land (1981). In short, good sensitivity to an extended scene results from a pupil of large area ($\pi A^2/4$), and photoreceptors each viewing a large solid angle ($\pi d^2/4f^2$ steradians) of visual space and absorbing a substantial fraction of the incident light ($1 - e^{-kl}$ for monochromatic light). Here A is the diameter of the pupil, f the focal length of the eye, and d, l, and k the diameter, length, and absorption coefficient of the photoreceptors, respectively. If all lengths have units of μm, then k has units of μm^{-1}. The optical sensitivity S of the eye to an extended monochromatic scene (in units of μm^2 sr) is then simply given by the product of these three factors (Kirschfeld, 1974; Land, 1981):

$$S = \left(\frac{\pi}{4}\right)^2 A^2 \left(\frac{d}{f}\right)^2 (1 - e^{-kl}) \qquad (16.1)$$

Wider pupils, larger photoreceptors, or shorter focal lengths all increase S. These are common features of eyes that view objects in the dim extended space light of the mesopelagic zone. These and other adaptations for increased sensitivity in mesopelagic fishes have been the subjects of intense study for over 100 years. We will only summarize the highlights, but for a definitive treatment, the reviews of Marshall (1954, 1971), Locket (1977), Munk (1980), and Collin (1997) are recommended.

Larger eyes have room for larger pupils, and while no quantitative data exist, the eyes of fishes do tend to become larger relative to body size the deeper they live in the mesopelagic zone (Marshall, 1979). Upon entrance to the bathypelagic zone, eyes tend to become smaller again, although pupil size relative to eye size tends to become greater (Marshall, 1954; Munk and Frederiksen, 1974), which has important implications for detecting bioluminescent point sources, as we shall see later.

The eyes of mesopelagic fishes either have a "typical" form, with lateral placement on the head, large fields of view, and restricted binocular overlap, or they are "tubular" in form, with restricted fields of view and large binocular overlap (Munk, 1980; Fig. 16.1). Tubular eyes are mostly found in fishes from 11 families inhabiting the lower mesopelagic (Munk, 1966, 1980; Collin et al., 1997). They have a large distal spherical lens that focuses a sharp image onto the main retina that lines the almost-flat proximal base of the eye. The eye's tubular shape severely restricts the main retina's field of view (which is typically less than 50° wide), but their placement on the head usually ensures that both eyes effectively view the same region of space, thus collecting double the number of photons. Moreover, even though tubular eyes have a small retinal area, image brightness is far greater than in a spherical eye of similar axial length (see Fig. 12 in Locket, 1977). This is due to an increased effective aperture and a decreased retinal curvature (equivalent to an increased effective eye diameter). Thus, while tubular eyes sacrifice a wide field of view, they gain in sensitivity, but without the handicap of an excessively large eye. This sensitive retina is

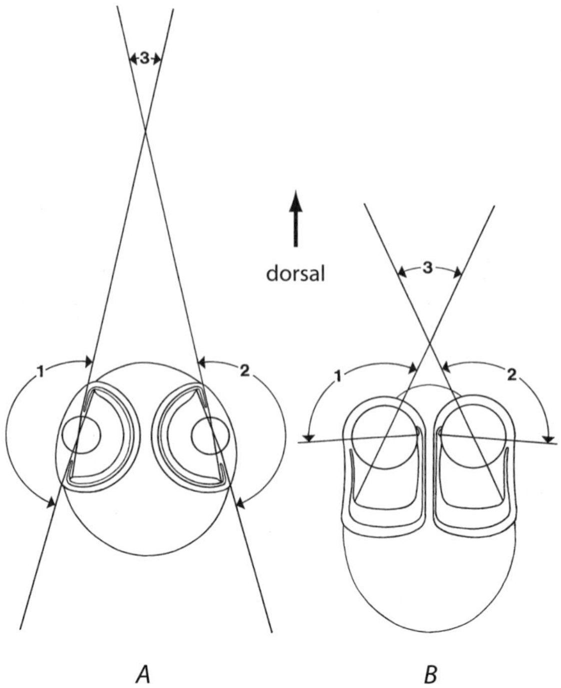

FIGURE 16.1. The two major eye designs of mesopelagic deep-sea fishes. (A) The typical form. (B) The dorsally directed tubular form. *1* = visual field of the left eye; *2* = visual field of the right eye; *3* = binocular visual field. (Adapted from Munk, 1980.)

then directed at the most important region of the mesopelagic light field.

Tubular eyes either point dorsally, to scan for prey silhouetted against the dim downwelling daylight (as in scopelarchids and hatchet fishes; Fig. 16.1B), or they point forward, possibly to chase prey swimming directly ahead (as in the telescope fish *Giganturus chuni*). There is some evidence, however, that fishes like *Giganturus* may even orient vertically in the ocean so that the tubular eyes are still directed dorsally (Marshall, 1979, p. 387; Ellis, 1996, p. 258; Justin Marshall and Bruce Robison personal communication). One rarely seen tubular-eyed fish, *Winteria*, even has the ability to rotate its tubular eyes to switch from a frontal view (at least along the nasocaudal body axis) to a position where they look directly upward (Justin Marshall and Peter Herring, personal communication).

The narrow visual fields of the main retinae of tubular eyes severely mask potential dangers, or even food sources, lurking in unseen parts of the fish's immediate surroundings. To overcome this problem, many tubular eyes possess special adaptations (Locket, 1977, 2000; Munk, 1980), such as a second "accessory retina" (*1* in Fig. 16.2A). Despite lying too close to the lens to receive focused images, the accessory retina can nevertheless detect light signals—such as a movement or a bioluminescent flash—from a greatly extended lateral visual field. Some fishes also have a small eye-like "retinal diverticulum" that bulges outward from the lateral side of the tubular eye, and can detect light originating up to 60° ventrally (Fig. 16.2A). Light is reflected from a sheet of guanine crystals, and strikes the photoreceptors of the retinal diverticulum through a clear epidermal window. "Lens pads"—common in scopelarchids—also extend the lateral and ventral visual field (*lp* in Fig. 16.2B). Constructed of corneal lamellae, the lens pad is a transparent light-guiding structure that directs light, originating from up to 70° ventral, to the dorsal-most part of the accessory retina.

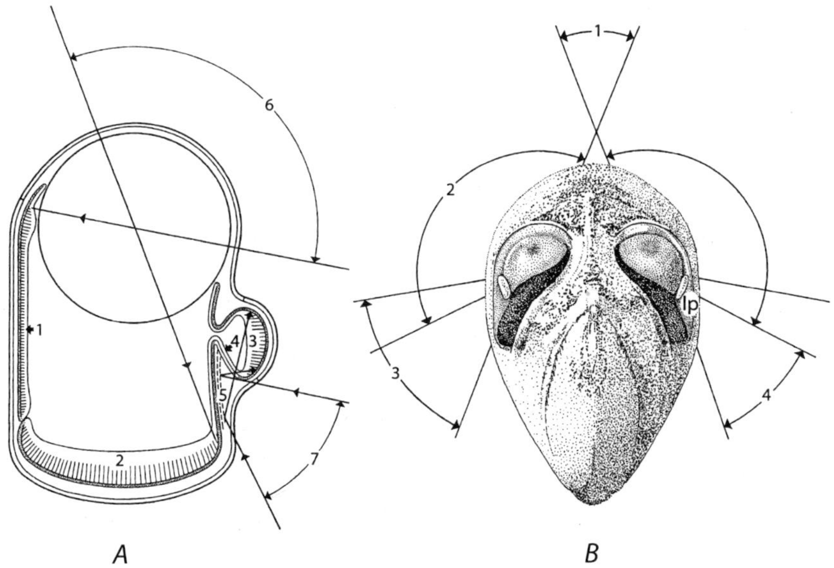

FIGURE 16.2. The visual fields of dorsal tubular eyes, retinal diverticuli, and lens pads. (A) A dorsal tubular eye with a laterally placed retinal diverticulum. *1* = accessory retina; *2* = main retina; *3* = retina of the retinal diverticulum; *4* = epidermal window; *5* = reflective sheet of guanine crystals; *6* = the eye's visual field; *7* = the retinal diverticulum's visual field. (B) Frontal view of the head of a scopelarchid showing the visual fields of the two eyes. *1* = dorsal binocular visual field; *2* = visual field of the right eye; *3* = total visual field for light reaching the accessory retina through the lens pad; *4* = ventral extension of the visual field provided by the lens pad; *lp* = lens pad. (Adapted from Munk, 1980.)

Mesopelagic fishes with eyes of "typical" form also have adaptations for increasing the visual field, and thus the sensitivity of their eyes. Their laterally placed eyes tend to restrict the frontal field of view, an important direction in space for the detection and pursuit of prey. To overcome this, many deep-sea fishes have a large "rostral aphakic gap," a frontal elongation of the pupil far beyond the margin of the lens (Fig. 16.3A; Munk and Frederiksen, 1974). Together with sighting grooves along the snout, the aphakic gap allows the full diameter of the lens to collect light frontally and to focus it onto the temporal (i.e., posterior) part of the retina (Fig. 16.3B), a retinal region where high-acuity visual processing commonly occurs. In deeper water, where maximum capture of light in *any* direction may be critical, fishes often have a "circumlental aphakic gap" (Fig. 16.3B,C; Munk and Frederiksen, 1974), a pupil that is larger than the lens in all directions (not just rostrally). Aphakic gaps greatly increase sensitivity (Fig. 16.3D), but only at the expense of spatial resolution: light that misses the lens and leaks through the gap will contaminate the retina with an image-degrading background haze.

The optical adaptations we have described so far all improve sensitivity by making the aperture through which light reaches the retina larger, a benefit for both point and extended source detection. This is equivalent to increasing A in Eq. 16.1. But this is not the only strategy available. For extended source detection (Eq. 16.1), a larger visual channel receptive field ($\pi d^2/4f^2$ steradians), or longer (l) and more absorbent (k) photoreceptors, will also improve sensitivity. Mesopleagic fishes are highly specialized in these neural respects as well.

The retinae of deep-sea fishes tend to be exclusively constructed of tightly packed rods, the vertebrate photoreceptor class adapted for dim light. Their photoreceptive outer segments are typically four times longer than in man, and are rich in photopigments tuned to the blue downwelling light (see chapter 17 by Partridge et al.). They also possess an absorption coefficient k at least twice as large as rods in animals from brighter habitats (Partridge et al., 1989; Warrant and Nilsson, 1998). The rods are sometimes also stacked several layers deep (Fig. 16.4A). Six layers are not uncommon in these so-called "multibank retinae," and such an arrangement would increase the path length (l in Eq. 16.1) for light absorption by a factor of six, thereby increasing sensitivity. However, even a single layer would have absorbed almost all of the incident light, so it is not immediately obvious what the benefit of a multibank retina actually is. Nevertheless, more layers are found in the most important parts of the retina (such as the temporal foveae), and in one extreme species, the alepocephalid *Bajacalifornia drakei*, there are no less than 28 (Locket, 1985)! In many fishes, the absorption path length is further increased—effectively doubled—by a reflective tapetum of guanine crystals lining the back of the eye. In certain circumstances, this solution may be more effective than increasing photoreceptor length (Warrant and Nilsson, 1998).

The final adaptation for increasing sensitivity is to widen the receptive fields of the visual channels (Eq. 16.1). The rods constitute the finest array of visual channels available, but these are very slender, typically only 2 or 3 μm wide. In many mesopelgaic fishes—such as the scopelarchids—groups of more than 20 rods are bundled into round cups of pigment epithelial cells filled with reflective guanine crystals (Fig. 16.4B; Locket, 1971, 1977, 1999; Collin et al., 1998). These reflective cups effectively turn the rod bundle into a kind of "macroreceptor," with a much wider receptive field than a single rod, and up to 50 times the sensitivity. Of course, this spatial summation of rods comes only at the cost of resolution.

Spatial summation of rods need not be as obvious as this. Via the bipolar cells, rods in all vertebrate retinae converge in large numbers onto single ganglion cells, and the size of the converging rod pool sets the size of the ganglion cell's receptive field. It also sets its potential sensitivity and the local spatial resolution. Denser packing of ganglion cells in the retina implies smaller and less sensitive receptive fields, but higher spatial resolution. A retina designed to see a dim extended source should thus have less densely packed ganglion cells. With large and sensitive receptive fields, these ganglion cells sacrifice spatial resolution in

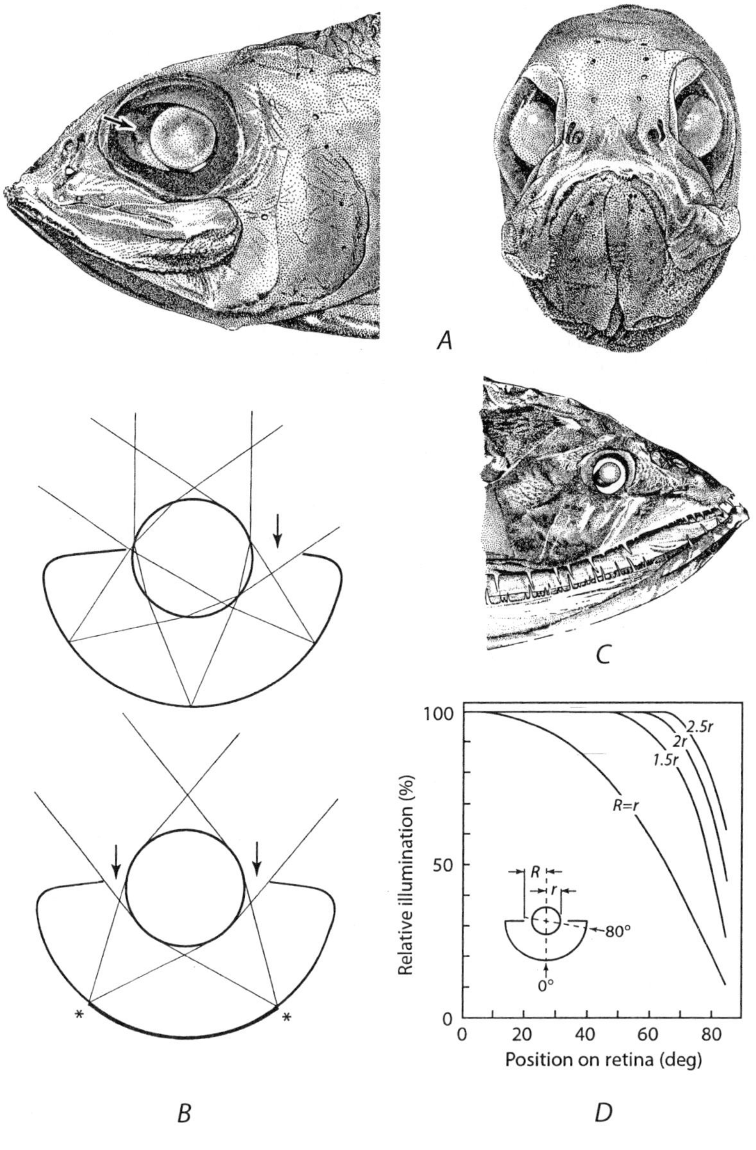

FIGURE 16.3. Aphakic gaps in the eyes of deep-sea fishes. (A) A rostral aphakic gap (*arrow*) in the eye of *Grimatroctes microlepis*. Together with sighting grooves in the snout, the rostral aphakic gaps entirely expose the lens of each eye in the frontal visual field. The temporally placed foveae that view the same frontal field are then assured of maximum light capture. (From Munk, 1980.) (B) The function of rostral (*upper*) and circumlental (*lower*) aphakic gaps (*arrows*). A small circumlental gap allows the retinal area delimited by the *stars* to be illuminated by the full aperture of the lens. A rostral gap allows full illumination of only the central and temporal areas of the retina. Unfortunately, the nasal areas of the retina also receive unfocused light. (Adapted from Locket, 1977.) (C) Circumlental aphakic gaps surround the entire lens, as shown here in *Gonostoma bathyphilum*. (From Munk and Frederiksen, 1974.) (D) The effect of circular circumlental aphakic gaps on illumination at different angular positions on the retina. As the radius of the gap (R) increases relative to the radius of the lens (r), the illumination of the retina increases, especially in the periphery. (Redrawn from Munk, 1980).

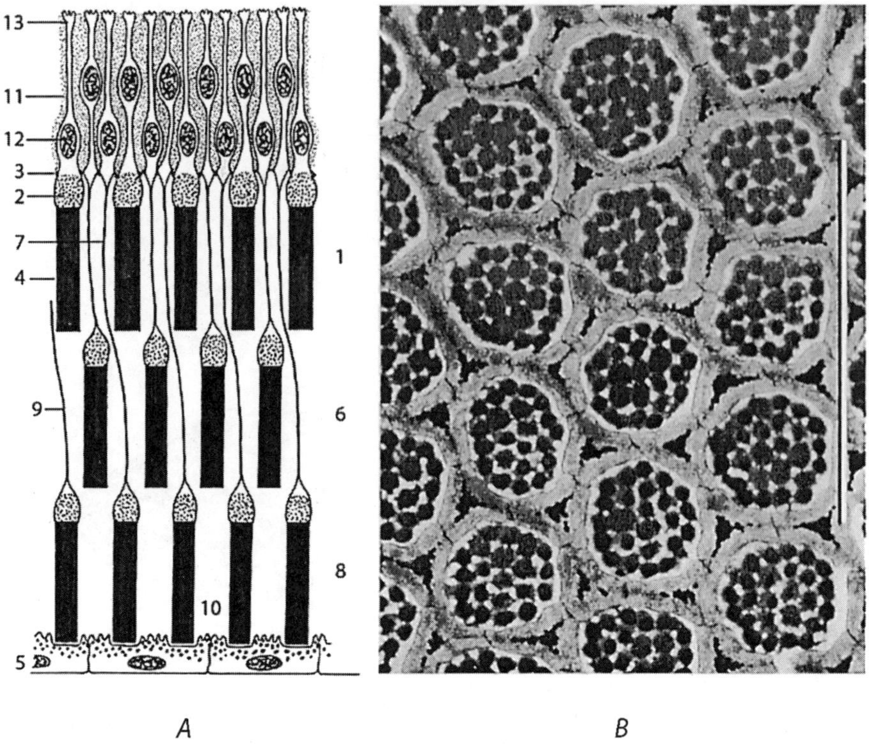

FIGURE 16.4. Retinal adaptations for increased sensitivity to extended sources in deep-sea fishes. (A) A multibank retina consisting of 3 banks (*1*, *6*, *8*). *2* = ellipsoid; *3* = outer limiting membrane; *4* = outer segment; *5* = pigment epithelium; *7* & *9* = filamentous myoids, *10* = ventricular space; *11* = radial fibre cytoplasm; *12* = rod nuclei; *13* = synaptic spherules. (B) The grouped retina of *Scopelarchus guntheri*, sectioned tangentially at the level of the inner-outer segment junctions. Scale bar = 50 µm. (From Locket, 1977.)

favor of photon catch. Such an "unspecialized" retina is found in lower mesopelagic lantern fishes of the genus *Lampanyctus* (Collin and Partridge, 1996; Wagner et al., 1998). These have a uniformly low density of ganglion cells, and presumably high rod convergence, throughout the retina. With only 1,000 to 3,000 cells mm^{-2}, visual acuity is no better than about 0.5°, a lot lower than in other deep-sea fishes with specialized retinae (see below). But using Eq. 16.1 it is easy to discover that the eyes of these fishes are extremely sensitive to the dim extended world of the lower mesopelagic zone. In *L. macdonaldi*, the pupil diameter A is 2,500 µm, the focal length f is $1.25A$ = 3,125 µm (Matthiessen's (1882) ratio), and the diameter of a ganglion cell receptive field is 25 µm (assuming 2,000 cells mm^{-2}). From Eq. 16.1, the sensitivity S of the eye to an extended source is then 247 µm^2 sr, a very high value for a vertebrate eye, in fact about 100 times more sensitive than for a nocturnal toad (Warrant and Nilsson, 1998). The eyes of deep-sea fishes are up to 120 times more sensitive to an extended scene than the human eye (Munk, 1980; Denton, 1990).

3.2. Point Source or Extended Source?

Fishes living in the brighter upper depths of the mesopelagic zone see an extended world where other animals are clearly illuminated by the downwelling daylight. During the transition to the lower mesopelagic zone, point-source bioluminescence becomes more prevalent and eventually dominates the visual scene. Wide pupils and long absorbent photoreceptors improve sensitivity no matter what the source, point or extended. In contrast, large ganglion cell receptive fields, like those we have just discussed in *Lampanyctus*, only improve sensitivity to an extended source. To collect as much light as possible from a point source, the ganglion cell's receptive field need not be any larger than the point-like image itself. If it were larger, this would only degrade accurate localization of the point source. Deep-sea fishes that need to see point-source bioluminescence thus tend to have wide pupils and sharp spatial resolution (Warrant, 2000).

This transition from an extended to a point-source world is clearly seen in the eyes of deep-sea fishes (Fig. 16.5). In the mesopelagic zone, where extended space light and bioluminescent flashes coexist, the retinae of fishes can be adapted to one or both of these light sources. In the brighter waters of the upper mesopelagic zone we find many fishes that depend on the downwelling daylight to illuminate their prey. A good example is the deepwater bass *Howella sherborni*, a fish living at depths ranging from 25 to 300 m (Fig. 16.5A). Their "typical" eyes each have a retina that contains a large temporal area of higher resolution—the "area centralis"—which samples nearly the entire frontal field of view with at least 14,000 ganglion cells mm^{-2} (Collin and Partridge, 1996). A deep fovea, containing up to 24,500 ganglion cells mm^{-2}, is also present (Locket, 1992; Collin and Partridge, 1996). *Howella*'s retina is clearly designed for viewing extended scenes, including their prey, with high resolution. The searsid *Searsia koefoedi*, a fish with a similar eye design to *Howella* but that lives much deeper in the mesopelagic zone (450–1,000 m), has a very different retina (Fig. 16.5B; Collin and Partridge, 1996). Its region of high resolution, also located temporally, is much smaller and confined to a steep-sided convexiclivate fovea (ca. 24,000 cells mm^{-2}). As we will see below when we discuss the eyes of bathypelagic fishes, this design is ideal for detecting and localizing point-source bioluminescence. The lantern fish *Lampanyctus macdonaldi*, whose "unspecialized" but sensitive retina we discussed above, lives at similar depths to *Searsia*. However, unlike *Searsia*, its retina is unsuited to accurate point-source detection, but is ideal for viewing the dim extended world of the lower mesopelagic zone. A mesopelagic fish with a retina adapted to both types of light is the bigfin pearleye *Scopelarchus michaelsarsi* (Fig. 16.5C), whose dorsally directed tubular eyes have different retinal areas devoted separately to point sources and extended sources (Locket, 1971, 1977; Collin et al., 1998).

These observations can be made more quantitative by plotting the smallest angular separation of ganglion cells as a function of depth for some 20 species of deep-sea fishes (Warrant, 2000, using data given in Wagner et al., 1998). Greater anatomical acuity results from a narrower separation of ganglion cells. The resulting histogram (Fig. 16.5D) reveals two important features. First, the eyes of fishes *on average* become sharper with depth, with the eyes of bathypelagic fishes being the sharpest. These eyes typically have the potential to resolve details subtending just 5 minutes of arc, perfect for detecting point-source bioluminescence, the only light seen at these depths. Second, the variation across species in ganglion cell separation (and thus acuity) is large in the brighter upper levels (Fig. 16.5D: error bars), but gradually declines with depth. Minimal variation occurs in the bathypelagic zone (separation = 4.8 ± 2.9 arc min), which is easy to understand. Here the only light sources are point sources and the *only* useful strategy involves little summation and high acuity. The large variation in the mesopelagic zone reflects its semiextended nature. Some mesopelagic species are adapted to point sources, some to extended sources, and others to both (Warrant, 2000).

3.3. Adaptations for the Directionality of Deep-Sea Light

The drive to increase the sensitivity of the eyes has not been the only force in the evolution of vision in deep-sea fishes. The increasingly dorsal origin of daylight has also been a major force. Dorsal tubular eyes, with their narrow, highly overlapping dorsally directed visual

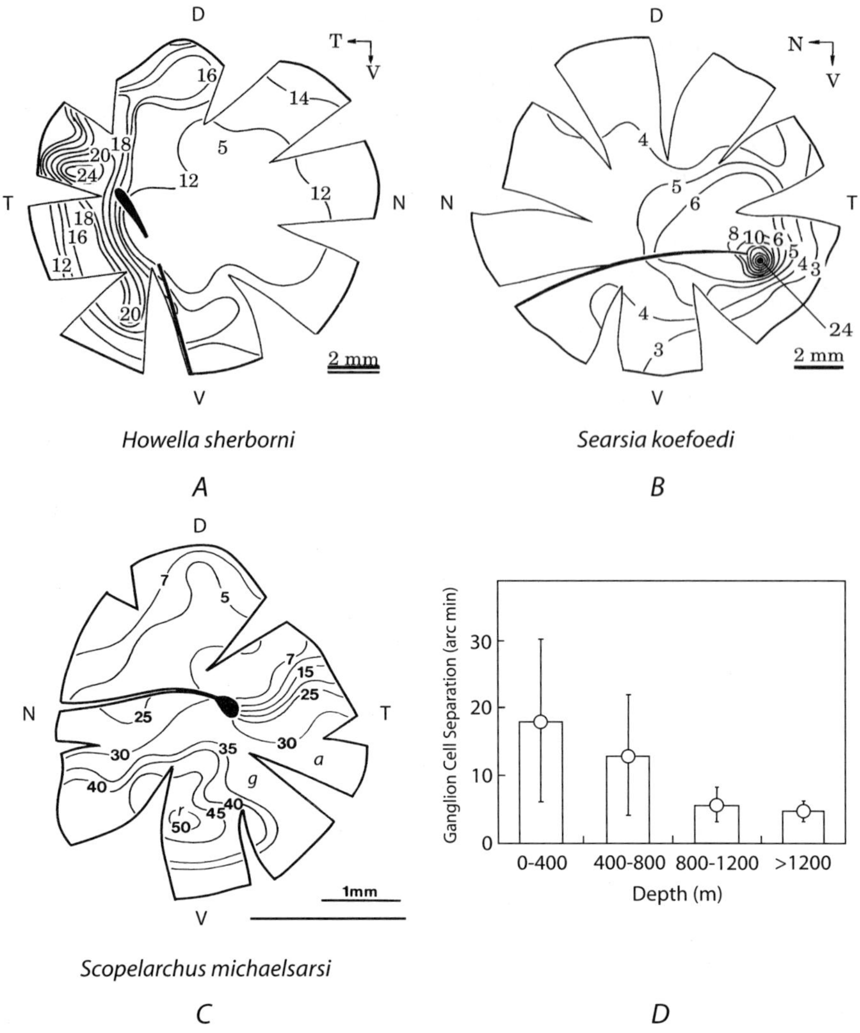

FIGURE 16.5. The transition from extended to point-source visual scenes as reflected in the retinal structure of deep-sea fishes. (A–C) Iso-density contour maps of the distribution of retinal ganglion cells. The retinae are shown as flat mounts, and densities are given in thousands of cells per mm^2. T = temporal (with frontal visual field); N = nasal (with posterior visual field); D = dorsal; V = ventral. Scale bars = 2 mm. (A) The deepwater bass *Howella sherborni*, living between 25 and 300 m, has a retina designed to see a bright extended scene with high resolution. (Adapted from Collin and Partridge, 1996.) (B) The searsid *Searsia koefoedi*, living between 450 and 1,000 m, has a retina optimally designed to see point-source bioluminescence. (Adapted from Collin and Partridge, 1996.) (C) *Scopelarchus michaelsarsi*, living between 250 and 1,000 m, has a retina designed to see both extended scenes (using a grouped retina subregion *g*, and an area of very large ganglion cells, the "area giganto cellularis" *a*), and point-sources (using a sharp area centralis *r*). (Adapted from Collin et al., 1997.) (D) The finest separation of ganglion cells found in a survey of 20 species of deep-sea fishes living at different depths, showing mean separation (in arc min) and standard deviation. Deeper-living fishes tend toward sharper retinae, a reflection of the increasing dominance of bioluminescent point-source illumination with depth. (From Warrant, 2000.)

fields, are a direct result of this influence. These eyes are designed to catch as much of the downwelling light as possible, and to enhance the discrimination of any small prey object that might be silhouetted against it. Their dorsal visual fields are typically analyzed by regions of enhanced spatial resolution in the retina, and this high acuity helps fishes to accurately localize small, silhouetted prey. The dorsal tubular eyes of many mesopelagic deep-sea fishes, like hatchet fishes and scopelarchids, provide classic examples (Fig. 16.2). The hatchetfish *Argyropelecus affinis*, which lives in the mid-mesopelagic zone, has well-developed dorsally oriented tubular eyes. The ventral part of its main retina has an area centralis of high ganglion cell density (up to 34,000 cells mm^{-1}) that samples a narrow dorsal field of view (Fig. 16.6A; Collin et al., 1997; Wagner et al., 1998).

For any edible mesopelagic animal, the possession of a distinct silhouette is a decided disadvantage. It is of little surprise, then, that many animals have evolved mechanisms for removing their own silhouettes, thus reducing the risk of becoming prey (see the review by Nilsson, 1997). All of these can reduce the contrast of an animal in the downwelling light considerably. Even modest reductions can make them very difficult to detect (Land, 2000). One such mechanism, practiced by many mesopelagic animals, notably fishes, shrimps, and cephalopods (see Chapter 14), is bioluminescent counterillumination. This requires a battery of photophores along the ventral surface of the body, each of which produces bioluminescence that mimics the color, intensity, and angular distribution of the surrounding downwelling daylight (Denton et al., 1972), thus effectively eliminating the animal's silhouette. However, the color of the

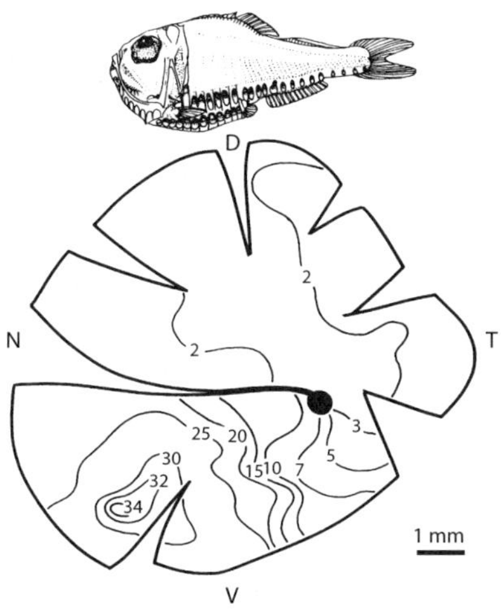

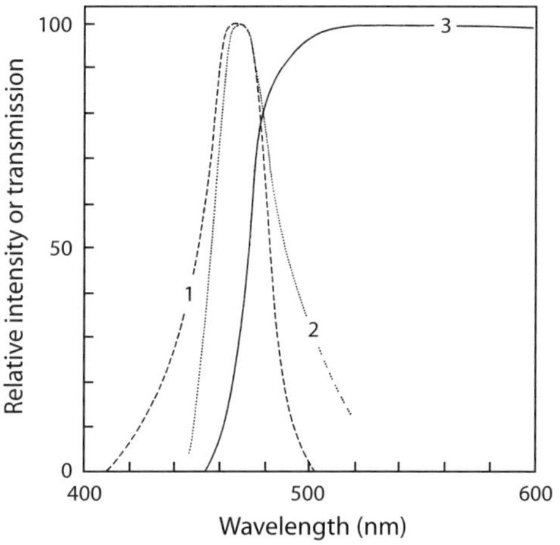

FIGURE 16.6. Visual adaptations for the directionality of light in the deep sea. (A) The dorsally directed tubular eyes of the mesopelagic hatchetfish *Argyropelecus affinis* have well-developed regions of higher resolution in the ventral retina. With its high density of ganglion cells, the ventral retina is ideally constructed to detect small silhouetted objects in the dorsal visual field. All conventions as in Figure 16.5. (Adapted from Wagner et al., 1998.) (B) Bioluminescent counterillumination and the yellow lenses of deep-sea fishes. Counterillumination produced by the euphausiid shrimp *Euphausia tenera* (*curve 2*) has a spectrum that is slightly greener than that of downwelling daylight at 500m (*curve 1*). The yellow lenses of many deep-sea fishes act as long-pass filters (*curve 3*) that cut out most of the downwelling blue light and accentuate the difference between the counterillumination and the remaining daylight. (Adapted from Munk, 1980.)

light produced by the photophores is not exactly the same as that of the surrounding daylight, having a spectrum slightly broader on the green side of the emission peak. The dorsal eyes of many deep-sea fishes can pick out this difference by having lenses colored by yellow pigment. These lenses selectively remove the blue part of the spectrum, and paradoxically cut out most of the light that could have reached the retina (Fig. 16.6B; Munk, 1980; Douglas and Thorpe, 1992). Their advantage, however, is that they enhance the slightly greener counterillumination (Muntz, 1976; Somiya, 1976). The unsuspecting prey, instead of being camouflaged, now shines like a torch. Chapter 17 contains further discussion of spectral sensitivities in the deep-sea.

4. Adaptations for Vision in the Bathypelagic Zone

"The bathypelagic zone, the largest environment on earth, is cold and dark (apart from fitful sparks of living light) and the most deserted life zone in the ocean, both in numbers of organisms and of species" (Marshall, 1979). These fathomless depths, beyond the penetration of daylight, are a dark world with little food, where most animals have been forced to adopt extremely low respiration rates to survive. Compared to their relatives from brighter, shallower waters, bathypelagic fishes have flimsy skeletons, weak and watery muscles, reduced internal organs, and very slow swimming speeds. Their eyes are frequently described as "small," "regressed," or "degenerate," no doubt casualties of the unsustainable energy cost of supporting large eyes (Laughlin et al., 1998). But as we shall see, the eyes of many bathypelagic animals are surprisingly sophisticated—despite their reputation.

4.1. The Visibility of Bioluminescent Point Sources in the Dark

Marshall's "fitful sparks" of bioluminescence—rare and often dim—are the only things bathypelagic eyes see. They reveal the presence of mates, predators, or sources of food, and in this impoverished realm such signals cannot afford to be missed. To accurately localize a flash requires reliable detection (i.e., good sensitivity to a point source), precise discrimination of the direction of the flash (i.e., good spatial resolution), and some mechanism for determining how far away the flash was (i.e., good depth discrimination).

The first of these—reliable detection—requires a large pupil. Imagine a point-source bioluminescent flash, containing a total of E photons at source, is located r meters from an eye of pupil diameter A. It is easy to show (Warrant, 2000) that the number of photons N that enter the eye, and are absorbed by a photoreceptor, is

$$N = \frac{EA^2}{16r^2} e^{-\alpha r} \qquad (16.2)$$

This equation assumes that the photoreceptors absorb all light incident upon them (which is a reasonable assumption in deep-sea fishes). The exponential term describes the attenuation of the bioluminescent flash due to the scattering and absorption of light by water, and α is the total attenuation coefficient (Lythgoe, 1979). For clear water and blue light, $\alpha = 0.05\,\mathrm{m}^{-1}$ (Denton, 1990). According to Eq. 16.2, smaller pupils, or bioluminescent flashes that are dimmer and further away, will reduce the photon catch and decrease the reliability of detection.

All dark-adapted photoreceptors suffer from intrinsic noise in the form of spontaneous isomerizations—or "dark noise"—that mimic the responses to real photons. These "false photon" responses set the ultimate limit to sensitivity (Aho et al., 1988). If it were not for dark noise, bathypelagic eyes could theoretically detect bioluminescent flashes infinitely far away. But even low rates of dark noise can easily swamp the real signals generated by photons arriving from a dim and/or distant flash.

How far away can a bathypelagic fish with a certain pupil size perceive a bioluminescent flash of given intensity? If the dark noise rate is X false photons per second, then following the logic of Land (1981), N photons entering the eye from a bioluminescent flash can be distinguished from the dark noise with 95% reliability when

$$N \geq 1.96\sqrt{2X} \quad (16.3)$$

For threshold detection, Eqs. 16.2 and 16.3 can now be equated:

$$\frac{EA^2}{16r^2} e^{-\alpha r} = 1.96\sqrt{2X} \quad (16.4)$$

In frogs, $X = 0.011\,\text{sec}^{-1}$ at 16°C. This rate falls by slightly more than a factor of 10 for every 10°C drop in temperature (Aho et al., 1993), so for a bathypelagic fish in the very cold waters of the deep-sea X could be around $0.0001\,\text{sec}^{-1}$. If we use this value and assume the fish has an average bathypelagic pupil diameter and sees a blue bioluminescent flash of average intensity (A = 7.3 mm, E = 10^{10} photons, and α = $0.05\,\text{m}^{-1}$: Warrant, 2000), Eq. 16.4 can be solved numerically to obtain r = 97 meters. In other words, a fish with a 7.3-mm-wide pupil will see the flash at distances up to 97 m away. If the rate of dark noise is 10 times larger, the maximum distance is reduced to about 80 meters. For the brightest flashes and the largest pupils found in the bathypelagic zone, the maximum range of visibility is about 230 m (assuming the same dark noise). Thus, where the intervening water quality allows, bathypelagic fishes have the potential to see bioluminescent flashes at considerable distances. As we shall see below, this may actually be to their disadvantage.

Once a flash is detected, its interception requires the fish to calculate its range, a task that is far from trivial. The cues that we use to judge distances to objects—parallax effects, perspective cues, and learned features of object size—are all absent in the pitch darkness of the bathypelagic zone. The only remaining cue is the slight disparity of the images in the two eyes. This requires a reasonable eye separation, a limitation in bathypelagic fishes, which are generally rather small. And because it is impossible to distinguish a dim flash nearby from a bright flash further away, flash brightness cannot be used as a cue either. How then do bathypelagic fishes judge the distance to bioluminescent flashes? This is still far from clear, but the specialized retinal adaptations discussed in the next section could provide the solution.

4.2. The Eyes of Bathypelagic Fishes

Relative to body size, bathypelagic fishes tend to have smaller eyes than their mesopelagic cousins. The tubular eye design, so common in lower mesopelagic species, has now basically disappeared. Some species, like whale fishes, have succumbed to the bathypelagic darkness, and possess clearly regressed eyes that lack lenses and have a degenerate retinae (Munk, 1966; Marshall, 1971). Others, like the alepocephalids, have large and highly developed eyes (Marshall, 1971). These fishes apparently rely much more on vision than on olfaction (Collin et al., 2000). In the bathypelagic zone the visual abilities of fishes are thus highly variable, and this certainly reflects wide differences in their natural histories, the details of which sadly remain rather sketchy.

As we saw above, even small bathypelagic eyes are capable of detecting a bioluminescent flash many tens of meters distant. Their wide pupils and long absorbent photoreceptors maximize sensitivity, and their presumably low rates of intrinsic dark noise (admittedly still unknown) probably maximize the range of detection (Eq. 16.4). Our calculation above suggested that an average eye could detect an average flash up to around 100 m away. But what does such a range actually mean for a bathypelagic fish?

For a start, a flash 100 meters away would be hopelessly out of range for most bathypelagic fishes. They would simply lack the energy to reach a flash this far distant. And with a swimming speed that is typically around one body length per second (Marshall, 1979), a bathypelagic fish of 10 cm length would require 17 min to make the journey. During this time, the producer of the flash could have moved impossibly far away. The fish, having expended massive amounts of precious energy, would have little chance of finding its potential mate or prey. Despite being able to see flashes so far away, bathypelagic fishes probably only react to those that occur at much closer range.

Paradoxically, a smaller and less sensitive eye might actually be a better adaptation for bathypelagic life. In addition to being energetically less expensive, smaller eyes reduce the range of bioluminescent visibility and act as a "distance filter." Compared to the average eye we discussed above (with a 7.3-mm pupil), a small eye with a small pupil (say, 1 mm wide) can see an average bioluminescent flash only half as far away (up to 47 m). Impossibly distant flashes, which might otherwise have been tempting to pursue, are simply beyond visual range.

If the size of the eye limits the maximum range where a flash can be detected, then other cues—such as the disparity of images formed in the two eyes—must be used to determine the actual distance to a flash. The very slight difference (or disparity) in the position of images formed on the left and right retinae (due to the finite separation of the eyes) can be used to calculate flash distance. Flashes that are closer produce greater disparity than flashes that are further away. An accurate determination of this disparity, which requires reasonable eye separation and good spatial resolution, can thus be used to infer the range of the flash. More accurate calculations of flash range result from greater eye separation and finer resolution. Are these characteristics of bathypelagic fishes?

Some bathypelagic fishes are large and have reasonably separated eyes, but this cannot be said of the very large number of species that are less than 20 cm in length. On the other hand, as we have already seen in Figure 16.5D, spatial resolution in bathypelagic fishes is surprisingly good, with foveal acuity in the vicinity of 5 minutes of arc across a broad spectrum of species. One such fish is *Rouleina attrita* (Fig. 16.7A), a fish living between 1,400 and 2,100 m (Wagner et al., 1998) that has rostral aphakic apertures, and deep convexiclivate temporal foveae possessing up to 27,000 ganglion cells mm^{-2}. The eyes of *Rouleina* (and many other similar species) are clearly designed for accurately localizing bioluminescent point sources in the frontal visual field.

Deep convexiclivate foveae (Fig. 16.7B) may also play a role in determining the range of bioluminescent flashes (Locket, 1985, 1992). The steep curvature and slightly higher refractive index of the radial fiber layer that lines the foveal pit has the potential to alter, and even distort, the focus of incoming light rays. Assuming that the axes of the two eyes are fixed and converging, the lines of sight of the foveal centers will cross at a particular point X in front of the fish. A point source located at X will be imaged symmetrically by the steep-sided foveal pit in each eye (Fig. 16.7C), resulting in perfectly round blur-circle images that are slightly magnified (the pit acts as a weak negative lens). If the point source is located closer to the fish than X (point B in Fig. 16.7C), the steep lateral sides of the pits distort the images, skewing them into comet-like shapes, with the "tail" of each comet pointing laterally across the retina. This image skewing is more pronounced for point sources closer to the fish. The opposite is true of a point source located further away than X (point A in Fig. 16.7C): the steep medial sides of the foveal pits skew the images such that their tails point medially, the extent of skewing increasing for point sources further away. If the fish can measure the extent and direction of image skew, it has a way of calculating the range of point sources.

Rostral aphakic apertures, deep convexiclivate foveae, sharp spatial acuity, and paradoxically small eyes are all adaptations that could help bathypelagic fishes to accurately localize points of bioluminescence in the immense darkness. Even with such adaptations the task seems overwhelming at ranges greater than a few tens of meters, and this could quite possibly set the ecologically meaningful range for bathypelagic visual interactions.

5. Adaptations for Vision in the Benthic Zone

The open expanses of the sea floor have had a major influence on the visual systems of the animals that live there. First, its vast plains have constrained their occupants to live in a flat, two-dimensional world, a living space quite unlike the entirely three-dimensional world of the pelagic depths above. Second, the sea floor is the repository of a continuous downward rain of organic matter that endows it with a rich

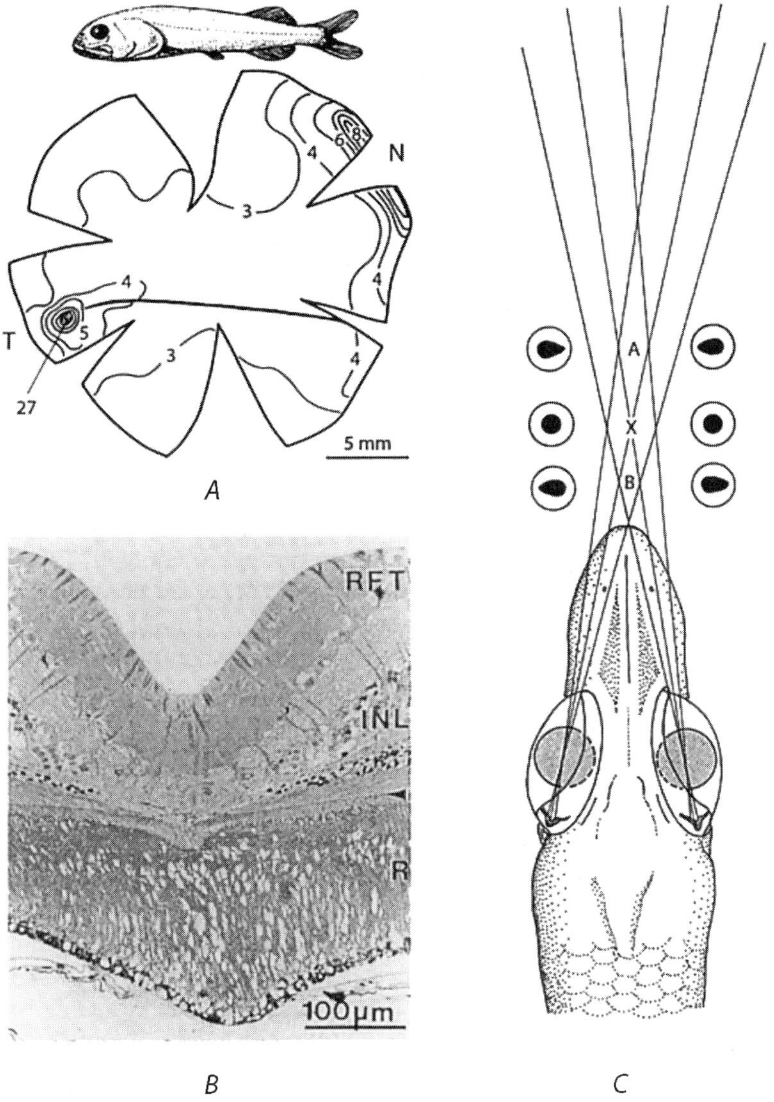

FIGURE 16.7. Visual adaptations in bathypelagic fishes. (A) *Rouleina attrita* has a retinal design that is common in bathypelagic fishes, with deep convexiclivate temporal foveae containing densely packed ganglion cells. This design is ideal for localizing bioluminescent point sources in the frontal visual field. All conventions as in Figure 16.5. (B) A transverse light microscope section through the deep convexiclivate fovea of the alepocephalid *Conocara macroptera*, a fish living in the bathypelagic zone down to 2.2 km. *RFT* = radial fiber layer; *INL* = inner nuclear layer; *R* = rod photoreceptors. (C) A possible role for deep convexiclivate foveae in determining the range of bioluminescent point sources. Range may be calculated using the extent and direction of image skewing within the steep sides of the foveal pit in each eye. See text for details. (All panels adapted from Wagner et al., 1998; C Adapted from Locket, 1985.)

nutrient level and fuels an active food web (Marshall, 1979). This better nutrition has resulted in fishes of hardier constitution that are much stronger swimmers than their cousins in the pelagic zones. Some have surprisingly large and well developed eyes, even at depths beyond the penetration of daylight.

5.1. The Visual World of the Benthic Zone

At depths where daylight penetrates, the sea floor is clearly visible to any animal with sufficiently sensitive eyes. A fish that cruises slowly over the bottom in search of food will be continuously aware of a single dominant visual feature: a horizon. In these fishes, as in many terrestrial hunters that scan the horizon for smaller animals upon which to prey, the eyes have special adaptations to see this horizon better. Even beyond the penetration of daylight, a sea-floor horizon may still be visible. Like the blinking lights of a sparsely populated landscape seen from an airplane at night, the wide-open sea bottom is home to an enormous variety of bioluminescent invertebrates. This expanse of blinking light, if it exists, would contrast strongly with the perfectly dark waters above, and thus create an obvious horizon. Oplophorid prawns, sepiolid squids, brittle stars, sea stars, sea lilies, sea cucumbers, sea spiders, polychaetes, sea pens, and sea pansies are but a few of the common benthic invertebrates that are bioluminescent (Marshall, 1979). Many of these—particularly the cnidarians—are thought to light up when an active swimmer brushes over them.

5.2. Benthic and Benthopelagic Vision in Fishes

As in pelagic fishes, those living in and over the sea floor show the same variations in eye development with depth (Marshall, 1954, 1971). Species inhabiting sea floors dimly lit by downwelling daylight have comparatively large eyes relative to head size, and retinae packed with rods, adaptations that are well suited to viewing a dim extended scene. In benthic species that live beyond the penetration of daylight, the trend reverses. With a few notable exceptions (see below), eyes tend to become smaller again, and at abyssal depths can become very small indeed, some little more than a small black point buried in the skin. Despite their tiny size, many of these eyes are fully functional. Others, however, are clearly degenerate, like those of the bottom-dweller *Ipnops* (Munk, 1959; Locket, 1977). This fish has two curious plate-like structures covering the top of its flattened head, with no discernible cornea, iris, or lens, and no possibility of obtaining a focused image. Each plate is made of a thin transparent membrane of bone that overlies a flat sheet of rods and a poorly differentiated layer of second- and third-order neurons devoid of ganglion cells. Such eyes could do little more than simply detect the presence or absence of bioluminescence. In these fishes, and many other visually impaired species, other senses presumably play an important role in the procurement of food. In the words of Walls (1942, p. 398): "Life on the bottom is largely life in one plane, and the finding of food by touch and chemoreception is vastly easier. Go far enough along the bottom (if you're a fish), and you're bound to bump into something good to eat."

But even at tremendous depths, the eyes of many benthic fishes are surprisingly large and well developed. Many strongly swimming rat-tails and alepocephalids cruise above sea floors as far down as 5 kilometers but have the large eyes and frontal receptive fields of an active visual predator (Marshall, 1954, 1971; Wagner et al., 1998). What are they looking at? Unfortunately one can only speculate on the basis of their eyes, which often possess deep convexiclivate temporal foveae of high spatial resolving power and a pupil that incorporates a frontal aphakic gap. Many alepocephalids, such as *Rouleina attrita* (Fig. 16.7A) and *Bathytroctes microlepis* (Fig. 16.8A), have eyes like this (Locket, 1992; Wagner et al., 1998). We noted earlier that this design is well suited to accurately localizing bioluminescent point sources in the frontal visual field. Curiously, their eyes are much larger than needed, being of a size much better suited to seeing dim extended scenes. But where can one find such scenes several kilometers below the penetration of daylight? One intriguing possibility—and admittedly one that is purely speculatory—is that the fishes themselves can either produce or induce bioluminescent illumination of the sea bottom and any potential prey upon it. Could rat-tails, with their colonies of bioluminescent bacteria—harbored within open glands along the belly (Marshall, 1971)—illuminate the sea floor if close enough? Could other fishes mechanically induce bioluminescence in bottom-dwelling invertebrates to achieve the same effect? We may never know, but the large eyes of many benthopelagic fishes could easily take advantage of it.

The visual horizon provided by the flat ocean floor—either by direct daylight illumination above 1,000 m, or by a bioluminescent host of seabed invertebrates below—has also had a striking effect on the design of deep-sea eyes. Just as in African veldt animals, like large predatory cats and their antelope prey, the eyes of benthic fishes often have a region of the retina with heightened visual performance directed along the horizon, the likeliest location for potential predators and prey. This height-

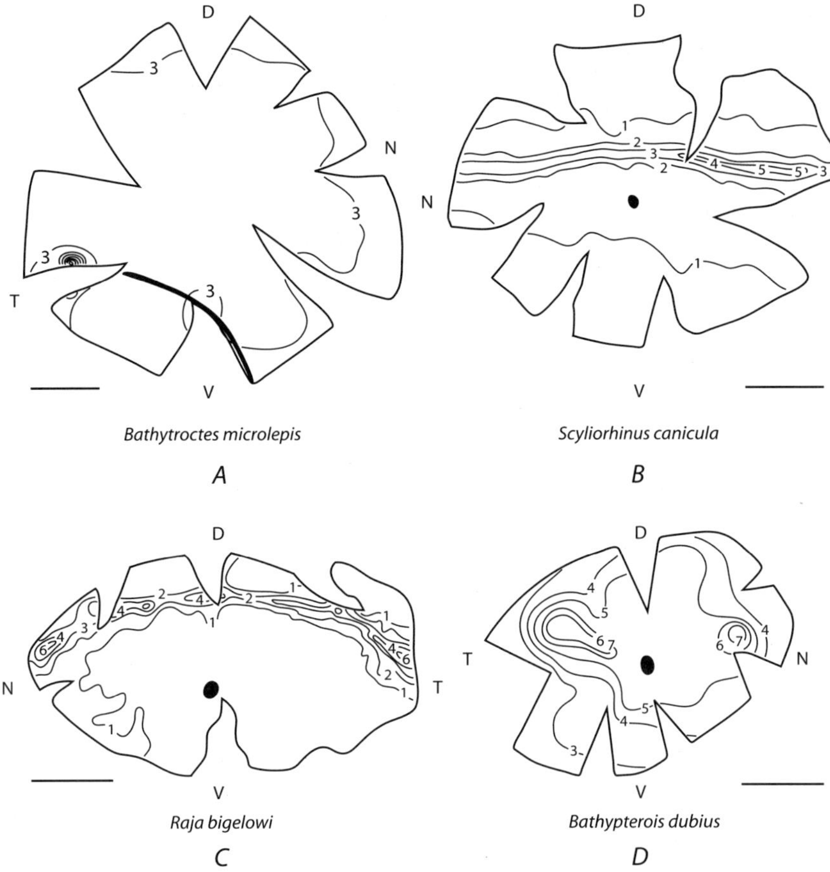

FIGURE 16.8. Visual adaptations in benthic and benthopelagic fishes. (A) The alepocephalid *Bathytroctes microlepis*, which cruises sea floors lying 1.1–2.7 km below the surface, clearly has eyes adapted for seeing bioluminescent point sources in the frontal visual field. (Redrawn from Collin and Partridge, 1996.) (B) The retina of *Scyliorhinus canicula*, a benthic shark that inhabits sea floors down to 400 m, has a horizontal visual streak aligned with the dimly illuminated visual horizon. (Redrawn from Bozzano and Collin, 2000.) (C) *Raja bigelowi*, a benthic shark that inhabits sea floors down to 2.2 km, also has a horizontal visual streak, possibly for viewing a sea bottom illuminated by bioluminescent animals. (Redrawn from Bozzano and Collin, 2000.) (D) The tripod fish *Bathypterois dubius*, which sits on sea floors down to 2 km, has a retina with two areae retinae, one pointing frontally, the other pointing backward. This may be a useful design for the tripod fish's sit-and-wait predatory lifestyle. (Redrawn from Collin and Partridge, 1996.) Scale bars = 5 mm (A–C), 1 mm (D). All conventions as in Figure 16.5.

ened performance is due to the presence of a horizontal "visual streak," an elongated strip of densely packed ganglion cells whose horizontal orientation is exactly aligned with the horizon. Visual steaks are common in shallow-water fishes living on bright sandy bottoms (Collin and Pettigrew, 1988). Until recently they were unknown in deep-sea fishes, but are now known in several species of bottom-cruising deep-sea sharks (Bozzano and Collin, 2000). These include species that inhabit sea floors both above (Fig. 16.8B) and below (Fig. 16.8C) the penetration of daylight. The peak ganglion cell density of these streaks is rather low, typically about 2,300 cells mm^{-2}. This is 12 times lower than that found in the point-source-detecting deep foveae of the bathypelagic fishes we discussed earlier (Fig. 16.7), suggesting a role in detecting dim extended horizons with high sensitivity. The slight dorsal placements of the visual streaks in Figure 16.8 suggest that deep-sea sharks view a somewhat ventrally located horizon, as might be seen from several meters above the bottom. In species like *Raja bigelowi* (Fig. 16.8C), which can cruise sea floors 2 km below the surface, the visual streak is presumably capable of sampling a bioluminescent horizon, and alerting the shark to other benthic fishes that unwittingly illuminate themselves by setting off the bioluminescent alarm systems of sea-floor invertebrates.

Not all fishes that inhabit the oozes of the sea floor possess a horizontal visual streak. Such is the tripod fish *Bathypterois dubius* (Fig. 16.8D), which orients against the oncoming current and props itself up on the bottom with the help of its stiffened pelvic and caudal fins. Its retina has an even and rather low ganglion cell density that increases slightly within two "areae retinae," one pointing frontally and slightly ventrally, the other pointing backward (Wagner et al., 1998). This retinal design is well suited to the tripod fish, a sit-and-wait predator that feeds on small bioluminescent copepods that wash by in the current. The temporal (frontally directed) areae retinae might be used to detect a copepod that suddenly appears in the frontal-ventral visual field, the part of the visual field where prey is expected to arrive.

6. Conclusion

Ninety years have passed since John Murray and Johan Hjort first puzzled over the apparent lack of correlation between the preferred depths of fishes in the ocean and the designs of their eyes. Since that time we have learned a lot about the nature of the deep-sea and the lives of the creatures that inhabit it, and many of these things have helped to ease their dilemma. One of these—the changing nature of visual scenes with depth, from extended to point source—has profoundly influenced the design of eyes in deep-sea fishes, and can explain many previously anomalous observations. However, many anomalies still remain. Now that we are beginning to venture into the depths to watch these elusive creatures at first hand, we may finally be able to understand what they are looking at, and what the limitations of their eyes might be. Until then, the deep-sea—our last natural frontier—remains the guardian of its secrets.

Acknowledgments. The authors would like to thank Dr. Kerstin Fritsches, Dr. Justin Marshall, and Anna Gislén for critically reading the manuscript. EJW is extremely grateful to the Swedish Research Council and STINT for their ongoing support. SPC acknowledges the support of the Natural Environment Research Council (NERC-UK), the Australian Research Council, the Jan Potter and George Alexander Foundation, the British Council, and the Alexander von Humboldt Foundation for their generous support.

References

Aho, A.-C., Donner, K., Helenius, S., Larsen, L.O., and Reuter, T. (1993). Visual performance of the toad (*Bufo bufo*) at low light levels: Retinal ganglion cell responses and prey-catching accuracy. *J. Comp. Physiol. A*. 172:671–682.

Aho, A.-C., Donner, K., Hydén, C., Larsen, L.O., and Reuter, T. (1988). Low retinal noise in animals with low body temperature allows high visual sensitivity. *Nature* 334:348–350.

Bozzano, A., and Collin, S.P. (2000). Retinal ganglion cell topography in elasmobranchs. *Brain. Behav. Evol.* 55:191–208.

Clarke, G.L., and Denton, E.J. (1962). Light and animal life. In: *The Sea* (Hill, M.N., ed.), pp. 456–468. New York, London: Wiley-Interscience.

Clarke, G.L., and Hubbard, C.J. (1959). Quantitative records of the luminescent flashing of oceanic animals at great depths. *Limn. Oceanog.* 4:163–180.

Cohen, J.E. (1994). Marine and continental food webs: Three paradoxes? *Phil. Trans. R. Soc. Lond. B.* 343:57–69.

Collin, S.P. (1997). Specialisations of the teleost visual system: Adaptive diversity from shallow-water to deep-sea. *Acta. Physiol. Scand.* 161, Suppl. 638:5–24.

Collin, S.P., and Partridge, J.C. (1996). Fish vision: Retinal specialisations in the eyes of deep-sea teleosts. *J. Fish Biol.* 49A:157–174.

Collin, S.P., and Pettigrew, J.D. (1988). Retinal topography in reef teleosts. II. Some species with prominent horizontal visual streaks and high density. *Brain Behav. Evol.* 31:283–295.

Collin, S.P., Hoskins, R.V., and Partridge, J.C. (1997). Tubular eyes of deep-sea fishes: A comparative study of retinal topography. *Brain Behav. Evol.* 50:335–357.

Collin, S.P., Hoskins, R.V., and Partridge, J.C. (1998). Seven retinal specialisations in the tubular eyes of the deep-sea pearleye *Scopelarchus michaelsarsi*: A case study in visual optimisation. *Brain Behav. Evol.* 51:291–314.

Collin, S.P., Lloyd, D.J., and Wagner, H.-J. (2000). Foveate vision in deep-sea teleosts: A comparison of primary visual and olfactory inputs. *Phil. Trans. R. Soc. Lond. B.* 355:1315–1320.

Denton, E.J. (1990). Light and vision at depths greater than 200 metres. In: *Light and Life in the Sea* (Herring, P.J., Campbell, A.K., Whitfield, M., and Maddock, L., eds.), pp. 127–148. Cambridge: Cambridge University Press.

Denton, E.J., Gilpin-Brown, J.B., and Wright, P.G. (1972). The angular distribution of the light produced by some mesopelagic fish in relation to their camouflage. *Proc. R. Soc. Lond. B.* 182:145–158.

Douglas, R.H., and Thorpe, A. (1992). Shortwave absorbing pigments in the ocular lenses of deep-sea teleosts. *J. Mar. Biol. Assoc. U.K.* 72:93–112.

Ellis, R. (1996). *Deep Atlantic*. New York: Lyons Press.

Frank, T.M., and Widder, E.A. (1996). UV light in the deep sea: *In situ* measurements of downwelling irradiance in relation to the visual threshold sensitivity of UV-sensitive crustaceans. *Mar. Fresh. Behav. Physiol.* 27:189–197.

Herring, P.J. (1978). Bioluminescence in invertebrates other than insects. In: *Bioluminescence in Action* (Herring, P.F., ed.), pp. 199–200. London: Academic Press.

Herring, P.J. (2000). Species abundance, sexual encounter and bioluminescent signalling in the deep sea. *Phil. Trans. R. Soc. Lond. B.* 355:1273–1276.

Jagger, W.S., and Muntz, W.R.A. (1993). Aquatic vision and the modulation transfer properties of unlighted and diffusely lighted natural waters. *Vision Res.* 33:1755–1763.

Jerlov, N.G. (1976). *Marine Optics*. Amsterdam: Elsevier Scientific.

Kirschfeld, K. (1974). The absolute sensitivity of lens and compound eyes. *Z. Naturforsch.* 29C:592–596.

Land, M.F. (1981). Optics and vision in invertebrates. In: *Handbook of Sensory Physiology, Vol. VII/6B* (Autrum, H., ed.), pp. 471–592. Berlin, Heidelberg, New York: Springer.

Land, M.F. (2000). On the functions of double eyes in midwater animals. *Phil. Trans. R. Soc. Lond. B.* 355:1147–1150.

Laughlin, S.B., de Ruyter van Steveninck, R.R., and Anderson, J.C. (1998). The metabolic cost of neural information. *Nature Neurosci.* 1:36–41.

Locket, N.A. (1971). Retinal anatomy in some scopelarchid deep-sea fishes. *Proc. R. Soc. Lond. B.* 178:161–184.

Locket, N.A. (1977). Adaptations to the deep-sea environment. In: *Handbook of Sensory Physiology, Vol. VII/5* (Crescitelli, F., ed.), pp. 67–192. Berlin, Heidelberg, New York: Springer.

Locket, N.A. (1985). The multiple bank rod foveae of *Bajacalifornia drakei*, an alepocephalid deep-sea teleost. *Proc. R. Soc. Lond. B.* 224:7–22.

Locket, N.A. (1992). Problems of deep foveas. *Aust. NZ. J. Ophthalm.* 20:281–295.

Locket, N.A. (1999). Vertebrate photoreceptors. In: *Adaptive Mechanisms in the Ecology of Vision* (Archer, S.N., Djamgoz, M.B.A., Loew, E.R., Partridge, J.C., and Vallerga, S., eds.), pp. 51–71. Dordrecht, Boston, London: Kluwer Academic.

Locket, N.A. (2000). On the lens pad of *Benthalbella infans*, a scopelarchid deep-sea teleost. *Phil. Trans. R. Soc. Lond. B.* 355:1167–1169.

Losey, G.S., Cronin, T.W., Goldsmith, T.H., Hyde, D., Marshall, N.J., and McFarland, W.N. (1999). The UV visual world of fishes: A review. *J. Fish. Biol.* 54:921–943.

Lythgoe, J.N. (1979). *The Ecology of Vision*. Oxford: Clarendon Press.

Lythgoe, J.N. (1988). Light and vision in the aquatic environment. In: *Sensory Biology of Aquatic Animals* (Atema, J., Fay, R.R., Popper, A.N., and Tavolga, W.N., eds.), pp. 57–82. Berlin, Heidelberg, New York: Springer.

Marshall, N.B. (1954). *Aspects of Deep-Sea Biology*. London: Hutchinson.

Marshall, N.B. (1971). *Explorations in the Life of Fishes*. Cambridge, MA: Harvard University Press.

Marshall, N.B. (1979). *Developments in Deep-Sea Biology*. Poole: Blandford.

Matthiessen, L. (1882). Über die Beziehungen, welche zwischen dem Brechungsindex des Kernzentrums der Krystalllinse und den Dimensionen des Auges bestehen. *Pflügers Arch. ges. Physiol.* 27:510–523.

Munk, O. (1959). The eyes of *Ipnops murrayi* Gunther 1887. *Galathea Rep.* 3:79–87.

Munk, O. (1966). Ocular anatomy of some deep-sea teleosts. *Dana Rep.* 70:1–62.

Munk, O. (1980). Hvirveldyrøjet: Bygning, funktion og tilpasning. Copenhagen: Berlingske Forlag.

Munk, O., and Frederiksen, R.D. (1974). On the function of aphakic apertures in teleosts. *Vidensk Meddr. Dansk. Naturh. Foren.* 137:65–94.

Muntz, W.R.A. (1976). On yellow lenses in mesopelagic animals. *J. Mar. Biol. Assoc. U.K.* 56:963–976.

Murray, J., and Hjort, J. (1912). *The Depths of the Ocean*. London: Macmillan.

Nilsson, D.-E. (1997). Eye design, vision and invisibility in planktonic invertebrates. In: *Zooplankton: Sensory Ecology and Physiology* (Lenz, P.H., Hartline, D.K., and Purcell, J.E, eds.), pp. 149–162. Newark: Gordon & Breach.

Partridge, J.C., Shand, J., Archer, S.N., Lythgoe, J.N., and van Groningen-Luyben, W.A.H.M. (1989). Interspecific variation in the visual pigments of deep-sea fishes. *J. Comp. Physiol. A.* 164:513–529.

Somiya, H. (1976). Functional significance of the yellow lens in the eyes of *Argyropelecus affinis*. *Mar. Biol.* 34:93–99.

Tyler, J.E., and Smith, R.C. (1970). *Measurement of Spectral Irradiance Underwater*. New York: Gordon & Breach.

Wagner, H.-J., Fröhlich, E., Negishi, K., and Collin, S.P. (1998). The eyes of deep sea fish. II. Functional morphology of the retina. *Prog. Retin. Eye. Res.* 17:637–685.

Walls, G.L. (1942). *The Vertebrate Eye and its Adaptive Radiation*. Bloomfield Hills, Michigan: Cranbrook Press.

Warrant, E.J. (2000). The eyes of deep-sea fishes and the changing nature of visual scenes with depth. *Phil. Trans. R. Soc. Lond. B.* 355:1155–1159.

Warrant, E.J., and Nilsson, D.-E. (1998). Absorption of white light in photoreceptors. *Vision Res.* 38: 195–207.

17
Spectral Sensitivity Tuning in the Deep-Sea

Ronald H. Douglas, David M. Hunt, and James K. Bowmaker

Abstract

Deep-sea fishes can be exposed to both dim residual sunlight and bioluminescence, both of which are usually most intense in the blue/green part of the spectrum. Not surprisingly, therefore, the vast majority of deep-sea fishes possess only one visual pigment with maximal absorbance (λ_{max}) between 475 and 490 nm. While a single visual pigment with maximal sensitivity approximately "matched" to both the downwelling sunlight and bioluminescence is clearly the norm within the deep-sea, 21 of the 200 species examined to date have two visual pigments within their retina, while 3 have at least three. The λ_{max} values of such multiple visual pigments cover a wider range of the spectrum than those found in single pigment species, allowing enhanced perception of more unusual signals. Although sunlight penetrating the water column is most intense around a relatively narrow, blue/green, part of the spectrum, it can contain significant amounts of shorter-wave radiation. Similarly, bioluminescent emissions often extend beyond 500 nm. In three genera of stomiid dragon fishes with more than one visual pigment the link between visual pigment absorption and photic stimulus is very clear. These animals, in addition to normal blue/green bioluminescence, also produce extreme-longwave light. Two genera, *Aristostomias* and *Pachystomias*, have evolved very red-sensitive visual pigments, while *Malacosteus* possesses a dietary-derived chlorophyll-based photosensitizing pigment, allowing perception of this "secret" waveband hitherto thought undetectable by other inhabitants of the deep-sea. Preliminary evidence, however, indicates that some myctophids, a group that are preyed upon by the stomiids, may also have evolved a photosensitizing mechanism to allow perception of the longwave bioluminescence produced by the dragon fish. The sequences of rod opsin genes from 28 species show that only nine sites are involved in the tuning of rod pigments in deep-sea fishes. The phylogenetic distribution of amino acid substitutions in the different species indicates that animals with shortwave-shifted rod visual pigments were present at or near the base of the Euteleost lineage, indicating that they were adapted to the same visual environment as present-day deep-sea fishes.

1. Introduction

The deep ocean, which accounts for around 97% of the habitable space on the planet, and is therefore the Earth's "typical" environment, is often thought of as a zone of perpetual darkness. However, even a cursory examination of the animals, especially the fishes, inhabiting this region shows this cannot be the case, as most have large, prominent eyes. Although the eyes are reduced in some species, this is the exception rather than the rule, even in deep demersal species (Douglas et al., 1995).

The deep-sea is unusual as there are two sources of illumination: downwelling sunlight, which is rapidly attenuated with depth so that no usable light penetrates below 1,000 m in even the clearest oceans (Denton, 1990); and bioluminescence, produced by over 80% of the animals inhabiting the deep ocean (Widder, 1999).

The visual systems of all vertebrates function in a similar way. Light is focused by the lens and cornea (although the cornea is usually optically insignificant underwater) onto a photosensitive retina composed of several distinct types of neurone and glial cells arranged into a well-defined series of layers (Fig. 17.1). Light is absorbed by visual pigments housed within the membranes of discs located in the photoreceptor outer segments. On absorption of a photon, these visual pigments "bleach," undergoing a number of conformational changes that activate an enzyme cascade. This leads to a hyperpolarization of the cell membrane and, in turn, causes a reduction in photoreceptor transmitter release. The ensuing neurobiological signal is then processed by the rest of the retina and central nervous system and results in visual perception.

While the eyes of all vertebrates conform to the same basic plan, those of deep-sea fishes have become specialized in a number of ways to make best use of the relatively few photons available. Thus they often have large and/or upwardly pointing tubular eyes (Fig. 17.1A) and retinae containing photoreceptors with unusually long outer segments or arranged into multiple layers (Fig. 17.1B) (Munk, 1966; Locket, 1977; Wagner et al., 1998). Furthermore, most vertebrates have two types of photoreceptors within their retinae: cones subserving high light level (photopic) vision, which allow the perception of detail and color, and rods, which become active in lower (scotopic) levels of illumination and are adapted for optimizing sensitivity. Although some deep-sea fishes do possess cones (e.g., Munk, 1965, 1981, 1984; Fröhlich et al., 1995), the majority have retinae containing just a single class of rod.

The wavelengths that an animal perceives are, in the first instance, determined by its visual pigments (although absorption by the ocular media and processing by the central nervous system (CNS) obviously have a role to play). All visual pigments have the same basic bell-shaped absorption profile with a well-defined point of greatest sensitivity (λ_{max}) where photon absorption is maximal (Fig. 17.2). Among vertebrates, the λ_{max} values of visual pigments range from 350 to 635 nm (Bowmaker, 1990, 1991, 1995; Douglas, 2001 for reviews), reflecting the diversity of the photic environment. Within deep-sea inhabitants, however, as we shall see below, the range of pigments is more restricted, reflecting the more conservative nature of the available stimuli (Douglas et al., 1998a for review).

2. Visual Pigment Structure

Photon capture within photoreceptors is thus the role of the photosensitive visual pigments. These pigments are members of the superfamily of G protein–coupled receptors that function through the activation of a guanine nucleotide–binding protein (G protein) and an

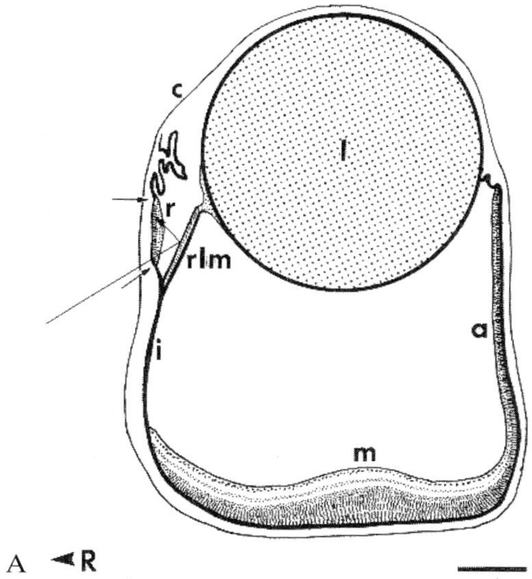

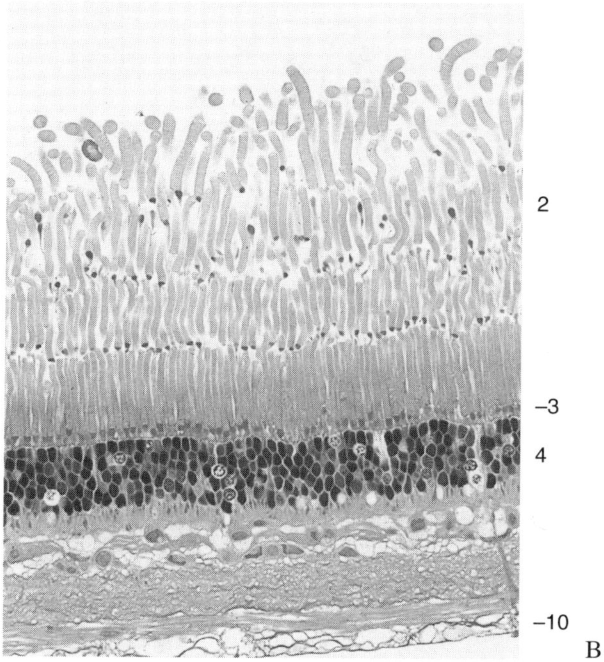

FIGURE 17.1. (A) Camera lucida drawing of a resin section of the tubular eye of *Stylephorus chordatus*. Apart from entering the eye through the cornea (c) from above and being focused by the lens (l) onto the main retina (m), light also projects from the ventral visual field onto a retinal diverticulum (r). Light entering the eye via this latter route (indicated by the arrows) is reflected by deposits in the retractor lentis muscle (rlm) and impinges on an accessory retina (a). i = iris; R = rostral; scale bar = 1.0 mm. (From Collin et al., 1998. Reprinted by permission of Karger, Basel.) (B) Transverse sections through the retinae of *Notacanthus chemnitzii*. The numbers indicate the traditional labels given to layers of the retina. The retinal pigment epithelium has been removed. 2—layer composed of rod outer and inner segments, note the 4 banks of rods presumably endowing this species with greater sensitivity than one possessing a conventional single tier retina. 3—external limiting membrane. 4—outer nuclear layer comprising the cell bodies of the photoreceptors. 10—internal limiting membrane. (Courtesy H.-J. Wagner.)

effector enzyme that changes the level of a second messenger in the cell cytoplasm. Each visual pigment is based on the same basic structure of a chromophore attached to an opsin protein. In vertebrates, the chromophore is either 11-*cis*-retinal or 11-*cis*-3,4-dehydroretinal, derivatives of vitamin A_1 and A_2 respectively, to give either rhodopsin (a term often incorrectly used to indicate rod pigments) or porphyropsin pigments. By themselves, retinal and 3,4-dehydroretinal absorb maximally at around 380 and 400 nm respectively. However, when combined with opsin protein, the visual pigment can have an absorbance maximum anywhere within a wide spectral range.

Opsins consist of a single polypeptide chain of 340–500 amino acids that form seven α helical transmembrane (TM) regions connected by cytoplasmic and extracellular loops (Dratz and Hargrave, 1983; Findlay and Pappin, 1986). In the tertiary structure (Fig. 17.3), the α helices form a bundle within the membrane, creating a hollow cavity on the extracellular side, the chromophore-binding pocket (Schertler and Hargrave, 1995; Palczewski et al., 2000). All visual pigments possess a lysine residue in the

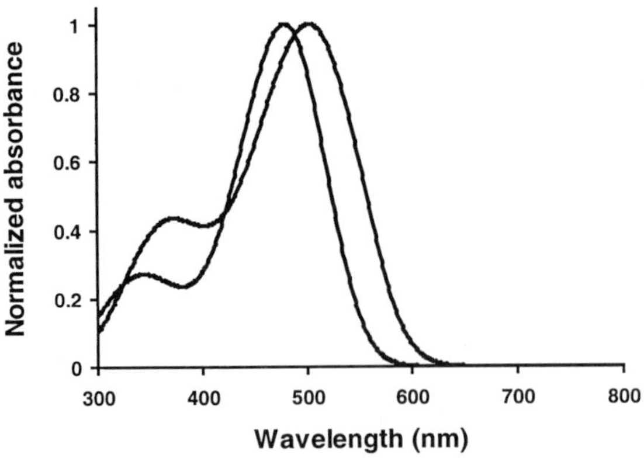

FIGURE 17.2. Visual pigment templates for a theoretical rhodopsin visual pigment with λ_{max} 480 nm and a porphyropsin pigment based on the same opsin (λ_{max} 530 nm) forming a pigment "pair." The relationship used to calculate the λ_{max} of the porphyropsin was derived by Whitmore and Bowmaker (1989). The templates fitted are those of Govardovskii et al. (2000).

seventh transmembrane region at site 296 (bovine rod opsin numbering) that is covalently linked to the chromophore via a Schiff base (SB). In vertebrates, this SB is generally thought to be protonated, with the negatively charged residue at site 113 (usually glutamate) acting as a counterion to electrostatically stabilize the proton of the SB (Nathans, 1990). Absorption of light causes the isomerization of the chromophore from 11-*cis*- to all-*trans*-retinal and this in turn causes major structural changes that include the displacement of the positively charged SB from its interaction with Glu113 (Zvyaga et al., 1994; Shieh et al., 1997). Low stability of the uncompensated, positively charged group in the hydrophobic environment of the retinal binding pocket leads to deprotonation and the production of the photointermediate metarhodopsin II (MII). These structural changes enable MII to activate the G protein transducin.

The particular λ_{max} of a visual pigment is thought to depend on a number of interactions (Kochendoerfer et al., 1999), although their relative importance may vary from pigment to pigment. The strength of the interaction between the glutamate counterion and the protonated Schiff base (PSB) is critical, since a strong interaction will prevent delocalization of the charge on the PSB along the chromophore, thereby stabilizing the ground state and resulting in a shorter wave-sensitive pigment. Photoexcitation of 11-*cis*-retinal induces a significant increase in electron delocalization, and a corresponding change in its dipole moment (Kropf and Hubbard, 1958; Mathies and Stryer, 1976). Interactions of charged, polar, or polarizable residues that alter delocalization will lead to a change in the energy difference between ground and excited states. Increases in delocalization will result in longwave shifts in the absorbance spectrum, whereas decreases will lead to shortwave shifts. Constriction of the chromophore binding pocket by bulky residues particularly at site 121 may also affect spectral tuning by planarization of the polyene chain of the chromophore due to stearic interactions with the opsin (Shieh et al., 1997).

The λ_{max} of a visual pigment will depend therefore on the combination of opsin and chromophore. Rhodopsin and porphyropsin pigments based on the same opsin protein will differ in λ_{max}, with the latter longwave shifted, particularly at longer wavelengths (Fig. 17.2). These two pigments are known as a rhodopsin/porphyropsin pigment "pair." Several empirically derived formulas relate the λ_{max} values of pigments based on the two chromophores within such pairs (e.g., Whitmore and Bowmaker, 1989; Parry and Bowmaker, 2000).

Most vertebrate visual pigments are rhodopsins. Porphyropsins are largely confined to fishes inhabiting longwave-rich freshwater, some amphibians and turtles. Although, even some terrestrial reptiles also possess A_2 pigments (Provincio et al., 1991; Bowmaker et al., 2000). Some species have pigments using both chromophores within their retina and some, such as those migrating between fresh- and saltwater, actually change their chromophore to suit the environment (Beatty, 1984). The balance of A_1/A_2 pigments within a retina is influenced by many factors, some of which remain unclear, but include temperature, day length, and reproductive status (Dartnall et al., 1961; Bridges, 1972; Loew and Dartnall, 1976; Muntz and Mouat, 1984).

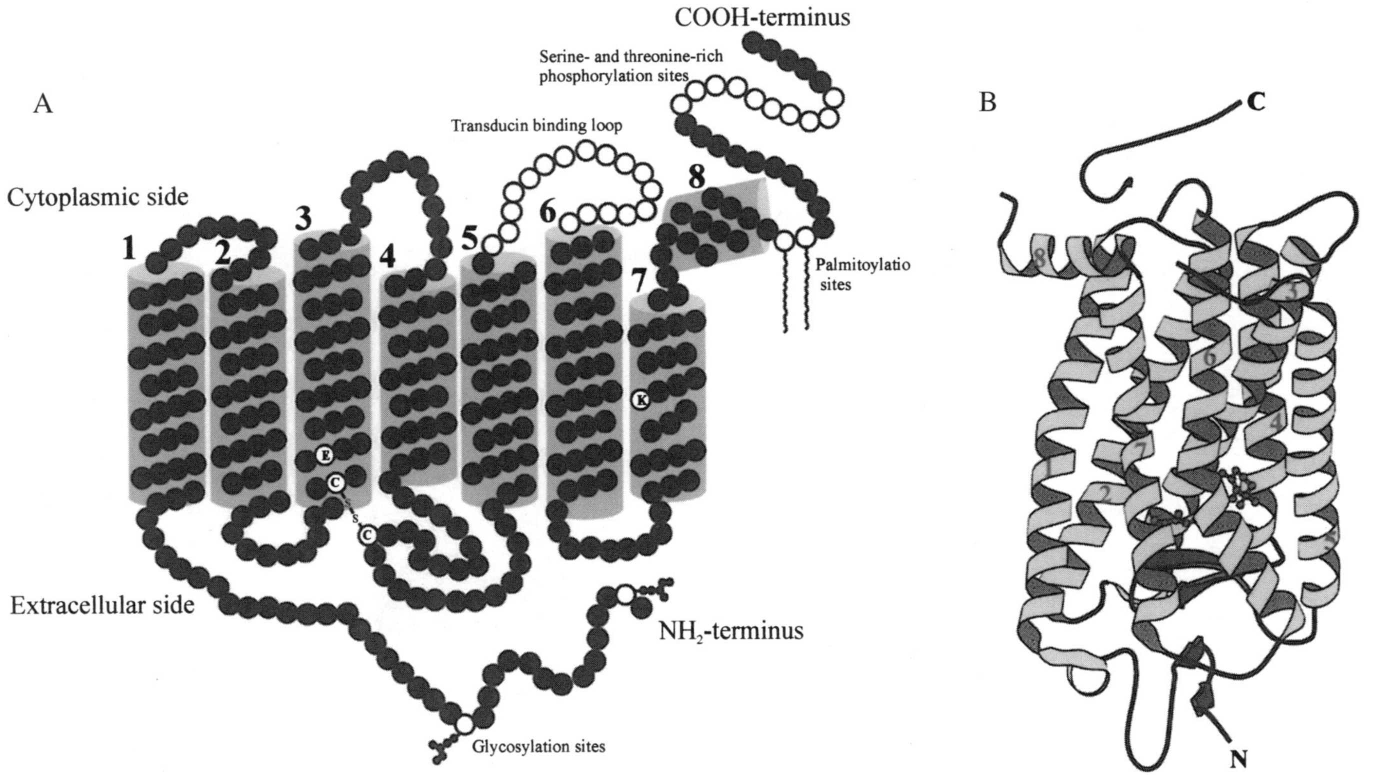

FIGURE 17.3. Structure of a visual pigment. (A) Diagram of an opsin molecule showing seven α-helical transmembrane regions connected by intra- and extracellular loops. The positions of key amino acid residues and protein domains are indicated. (Redrawn from Palczewski et al., 2000.) (B) Ribbon drawing showing the crystal structure of bovine rhodopsin, viewed parallel to the plane of the membrane. Helices 1 to 7 are transmembrane and helix 8 is cytoplasmic. The retinal chromophore is shown bonded by a Schiff base linkage to lysine at site 296. This figure is derived from crystallographic data (Palczewski et al., 2000).

3. Spectral Characteristics of Deep-Sea Visual Pigments

To date, visual pigment absorption spectra have been collected, using a variety of methods, for 200 of the estimated 2,500 species of deep-sea fishes; 176 of these have retinae containing a single visual pigment (Fig. 17.4A), with the remainder possessing two or more pigments (Fig. 17.4B).

3.1. Retinae Containing a Single Visual Pigment

Not only do the vast majority of deep-sea fishes have only a single visual pigment within their retinae, but in all cases so far described, they use retinal as their chromophore (e.g., Fig. 17.5). Ninety percent of these pigments have a λ_{max} between 475 and 490 nm and even those whose sensitivity maxima are at the extremes of the range (451 nm and 494 nm) are not far removed from the average (Fig. 17.4A).

Figure 17.4 confirms what has long been suggested (Bayliss et al., 1936; Clarke, 1936), that, in comparison to shallow-water animals whose rods absorb maximally round 500 nm (Lythgoe, 1972), the spectra of deep-sea pigments are hypsochromatically shifted toward shorter wavelengths. At a superficial level this makes perfect sense as an adaptation to the spectral filtering by the water. Sunlight is selectively attenuated with depth, so that after a few hundred meters (substantially less in some bodies of water), light primarily consists of a narrow band of radiation between 470 and 480 nm (Kampa, 1970; Jerlov, 1976; Kirk, 1983).

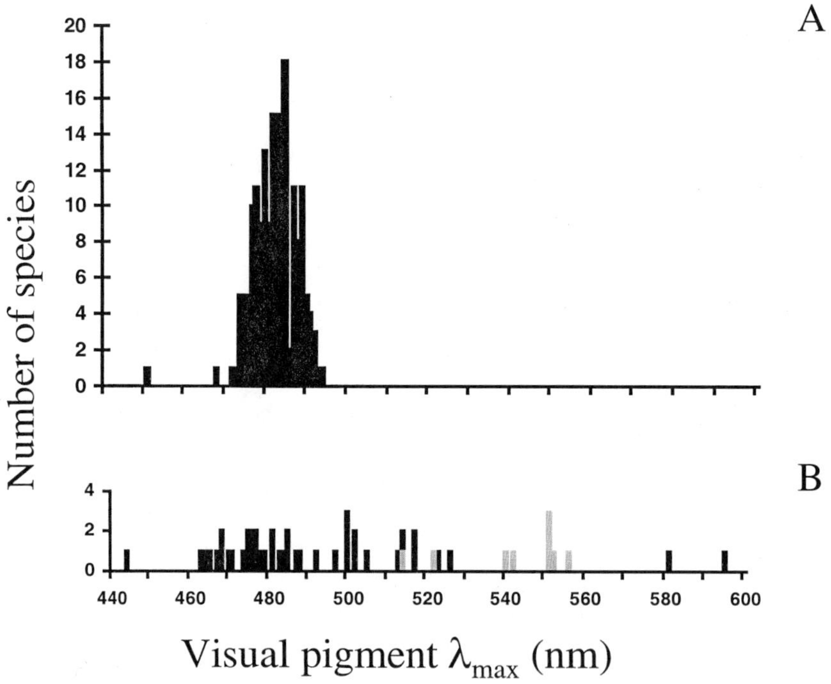

FIGURE 17.4. Distribution of visual pigment λ_{max} in 200 species of deep-sea fishes. (A) 176 species with a single visual pigment within their retinae. (B) 24 species containing multiple visual pigments. The values shown are mainly from published sources but also include 21 species for whom data are unpublished. The majority of the pigments were analyzed by extraction but some data derived from microspectrophotometry and retinal wholemounts are also included. Vitamin A_1-based pigments are shown in black and Vitamin A_2-based porphyropsins are represented in gray.

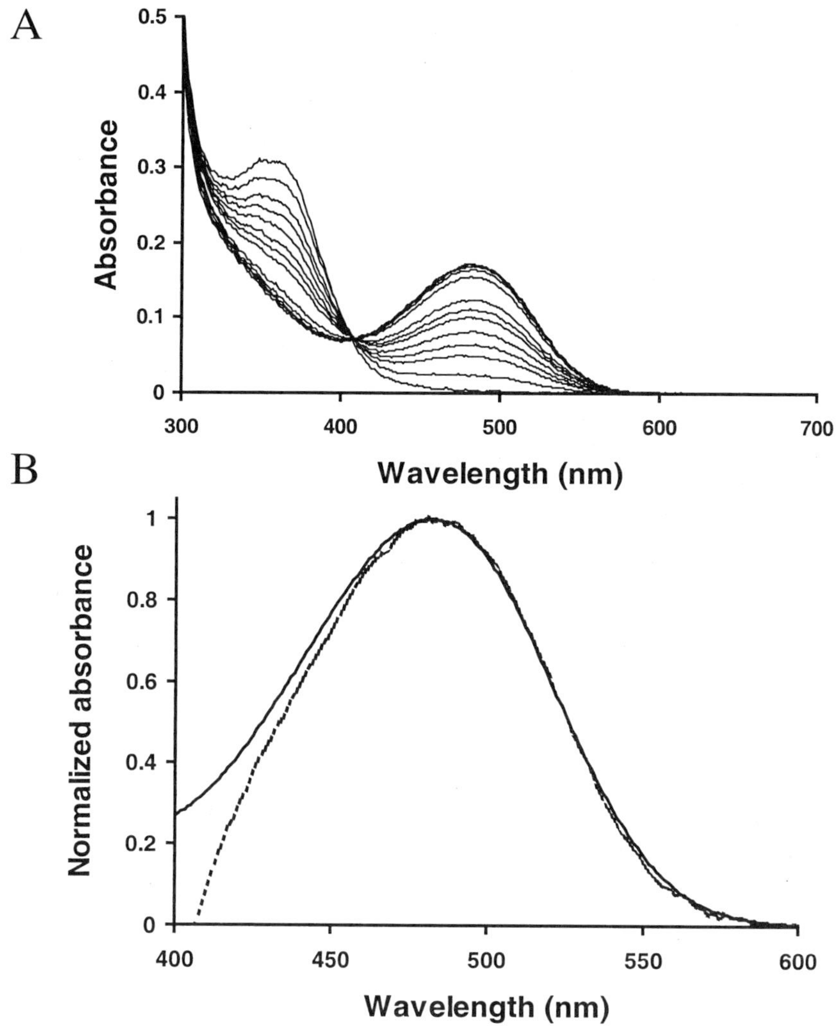

FIGURE 17.5. Detergent extract from the retina of an individual *Avocettina infans*, a typical deep-sea fish with only a single visual pigment. (A) Partial bleach with individual curves derived following monochromatic bleaches of various durations. In order of decreasing absorbance at 500 nm these were: (i) initial measurement of the unexposed extract 10 min after the addition of 100 μl 1 M hydroxylamine to 1 ml of extract; (ii) after 5 min exposure to monochromatic light of 666 nm; (iii) 5 min 650 nm; (iv) 5 min 634 nm; (v) 2 min 609 nm; (vi) 2 min 586 nm; (vii) 30 sec 576 nm; (viii) another 30 secs 576 nm; (ix) another 1 min 576 nm; (x) 30 secs 565 nm; (xi) another 30 secs 565 nm; (xii) 30 sec 552 nm; (xiii) 2 min 501 nm. (B) All difference spectra constructed from the individual bleaches in A had a λ_{max} in the same part of the spectrum, indicating the presence of only one visual pigment. The difference spectrum (dotted line) obtained by subtracting the fully bleached spectrum (curve xiii) from that of the unexposed extract (curve i), which was best fit by a rhodospsin template with λ_{max} 482 nm (solid line), was therefore used to describe the pigment (Douglas unpublished). The templates used here, and throughout this chapter for previously unpublished data, are those described by Govardovskii et al. (2000) fitted using methods outlined by Hart et al. (2000).

Similarly, the bioluminescence of most organisms also has peak emissions at around these wavelengths (Herring, 1983; Widder et al., 1983, 1989; Latz et al., 1988; Haddock and Case, 1999). Thus the absorption maxima of the visual pigments are close to the wavelengths of maximal downwelling irradiation and the emission maxima of most bioluminescence. However, theoretical modeling shows that the requirements for seeing downwelling illumination and biolouminescence are somewhat different and most deep-sea visual pigments are probably adapted for the detection of bioluminescence rather than background sunlight (Partridge, personal communication; Douglas et al., 1998a).

It has been argued that the precise location of the λ_{max} of deep-sea fish visual pigments is of little consequence due to the relatively broad absorption spectrum resulting from their unusually high concentration within the outer segments (Munz, 1965; Bowmaker, 1976; Crescitelli, 1991). However, in a photon-limited environment such as the deep-sea, it is likely that there will be a significant adaptive advantage in any gain in visual sensitivity, and there is thus likely to be selection for visual pigments that confer the highest possible sensitivity. It is, however, possible that in some instances the position of the λ_{max} is not only related to the available photic stimuli but is also determined by the need to maximize the signal-to-noise ratio and thermal stability, the effects of ambient pressure, or phylogenetic constraints (Douglas et al., 1998a for review). Such arguments receive some support from studies of the cottoid species flock in lake Baikal, in which deep-dwelling species have λ_{max} values similar to those of deep-sea fishes (Bowmaker et al., 1994; Hunt et al., 1996). This body of freshwater is, however, likely to contain significantly more longwave radiation than the open ocean and bioluminescence is not known to occur here. Given the quite different photic conditions in lake Baikal and the open ocean, it is difficult to understand why deep-dwelling species in both environments would have spectrally similar visual pigments if light conditions were the only determining factor.

3.2. Retinae Containing Multiple Visual Pigments

The possession of a single visual pigment with λ_{max} around 475–490 nm approximately matched to the predominant wavelengths of bioluminescence and residual sunlight is clearly the norm within the deep-sea. However, a significant number of animals have more than one visual pigment within their retinae; 21 species have two visual pigments, while 3 have at least three. The λ_{max} values of such multiple visual pigments are clearly more variable than those found in single-pigment species (Fig. 17.4B).

Maximal sensitivity outside of the narrow waveband displayed by single-pigment species could be beneficial for a number of reasons. First, the residual sunlight in the upper 1,000 m of the water column is not as restricted as sometimes imagined. For example, some deep-sea oplophorid shrimps have been shown, both electrophysiologically (Frank and Case, 1988) and behaviorally (Frank and Widder, 1994, 1996), to be sensitive to ultraviolet (UV)/violet light, and to possess a shortwave-sensitive visual pigment (λ_{max} 410 nm; Cronin and Frank, 1996). Furthermore, measurements have shown that enough UV reaches depths of up to 600 m during the daytime to be perceived by these animals (Frank and Widder, 1996). Although such shortwave sensitivity has not yet been demonstrated in deep-sea fishes, it does indicate that the range of wavelengths potentially available is quite large.

Similarly, although most bioluminescence peaks around 450–490, the emission spectra are quite variable between species and can include significant amounts of longer- and shortwave radiation (Herring, 1983; Widder et al., 1983, 1989; Latz et al., 1988; Haddock and Case, 1999). A special case are three genera of stomiid dragon fishes that produce far-red bioluminescence, providing one of the best examples of an adaptation of visual pigments to a specific signal (see below).

Perhaps the simplest way of making two visual pigments is to have a single opsin bound to the two different chromophores, forming a rhodopsin/porphyropsin pigment "pair" (see above). However, only three rather unusual deep-sea dragon fishes appear to have adopted this strategy (see Figs. 17.8 and 17.9). A further two species, *Bonapartia pedaliota* (Douglas and Partridge, 1997) and *Myctophum nitidulum* (Fig. 17.6), have also been shown, using partial bleaching of detergent extracts, to have two visual pigments within their retinae based on the two different chromphores. However, these have λ_{max} values too far separated to be considered a pair. Therefore they most likely represent the two different chromphores bound to two different opsins.

In general, vitamin A_2–based pigments are relatively rare in deep-sea fishes (accounting for only 9 of the 227 pigments so far described) and most animals with two pigments within their retina appear to use different opsins, both of which are linked to retinal (e.g., Fig. 17.7).

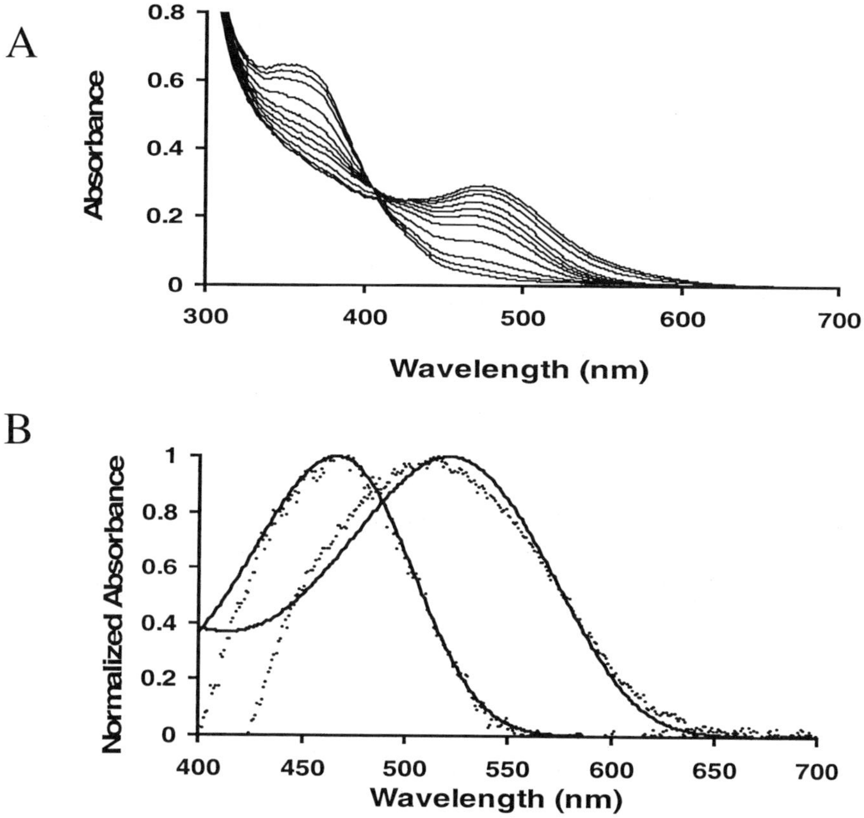

FIGURE 17.6. Detergent extract from the retina of an individual *Myctophum nitidulum*. (A) Partial bleach with individual curves derived following monochromatic bleaches of various durations. In order of decreasing absorbance at 500 nm these were: (i) initial measurement of the unexposed extract 10 min after the addition of 100 μl 1 M hydroxylamine to 1 ml of extract; (ii) after 5 min exposure to monochromatic light of 666 nm; (iii) 5 min 650 nm; (iv) 5 min 634 nm; (v) 3 min 609 nm; (vi) 2 min 586 nm; (vii) 2 min 576 nm; (viii) 2 min 565 nm; (ix) 1 min 552 nm; (x) 1 min 544 nm; (xi) 2 mins 501 nm. (B) Difference spectra constructed from successive bleaches resulted in λ_{max} values shifted progressively toward shorter wavelengths, indicating the presence of more than one bleachable pigment. The normalized difference spectra (dotted lines) of the extremes of long and shortwave bleaching (obtained by subtracting curves i–iii and x–xi respectively) were best fit by a porphyropsin template with λ_{max} 521 nm and a rhodospsin template with λ_{max} 466 nm (solid lines). The pigments of two individuals were analyzed and, on average, best fit by templates with λ_{max} 522 and 468 nm (Douglas, unpublished.)

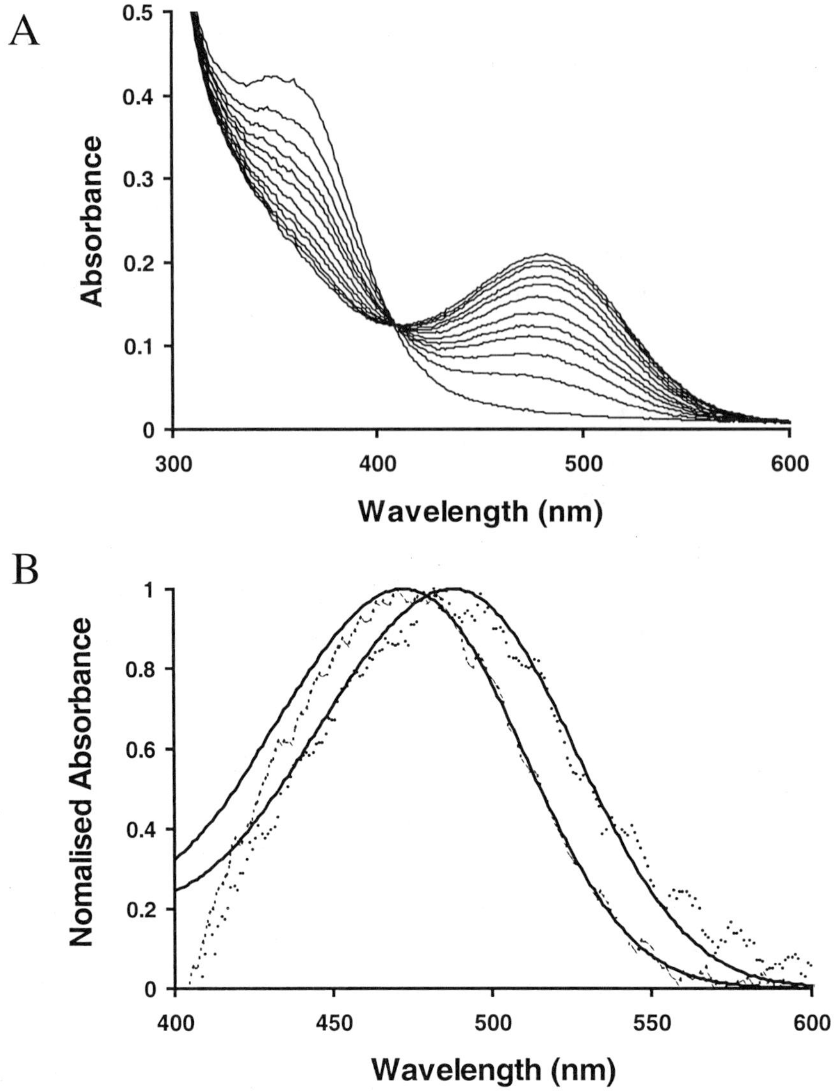

FIGURE 17.7. Detergent extract from the retina of an individual *Ceratoscopelus warmingii.* (A) Partial bleach with individual curves derived following monochromatic beaches of various durations. In order of decreasing absorbance at 500 nm these were: (i) initial measurement of the unexposed extract 10 min after the addition of 100 µl 1 M hydroxylamine to 1 ml of extract; (ii) after 5 min exposure to monochromatic light of 666 nm; (iii) 4.5 min 650 nm; (iv) 5 min 634 nm; (v) 1 min 609 nm; (vi) another 2 min 609 nm; (vii) 1 min 586 nm; (viii) 30 sec 576 nm; (ix) another 30 sec 576 nm; (x) 30 sec 565 nm; (xi) 30 sec 552; (xii) 2 mins 501 nm. (B) The normalized difference spectra (obtained by subtracting curves i–iv and xi–xii respectively, dotted lines) were best fit by rhodopsin templates with λ_{max} 472 and 488 nm (solid lines). The pigments of a total of six individuals were analyzed and on average were best fit by templates with λ_{max} 468 and 488 nm (Douglas, unpublished.)

Three species of deep-sea fish have at least three visual pigments within their retinae. Two of these (*Aristostomias titmanni*, *Pachystomias microdon*) are discussed below as they couple the emission of longwave bioluminescence with the possession of extremely red-shifted visual pigments. Apart from these, only *Scopelarchus analis*, which produces conventional blue/green bioluminescence, has been shown to have three visual pigments (rhodopsins with λ_{max} values 444, 479, and 505 nm; Partridge et al., 1992). Not surprisingly, members of the genus *Scopelarchus* have highly complex tubular eyes (Munk, 1966; Locket, 1971, 1977; Collin et al., 1998; Wagner et al., 1998) with seven distinct retinal regions spread over a main retina on the ventral surface of the eye, as well as a medial accessory retina, and a diverticular retina. Furthermore, the retina contains cones and both grouped and ungrouped rods as well as morphologically distinct subtypes of ganglion cell. Most significantly, the rod outer segment containing the 507 nm visual pigment was always found to also possess the 445 nm pigment in the part of the outer segment nearest the inner segment, indicating the photoreceptor was in the process of changing from one opsin to another. A similar transition occurs in a few other fishes such as eels (Wood et al., 1992; Wood and Partridge, 1993; Archer et al., 1995). In all other deep-sea fishes with multiple visual pigments in which MSP has been performed, the different pigments have always been shown to be located in separate photoreceptors (Partridge et al., 1988, 1989, 1992).

Perhaps the most obvious advantage to possessing more than one visual pigment is that it simply extends the spectral range of the animal's sensitivity compared to an animal with just one pigment. Conceivably, the different pigments might be tuned to different stimuli, such as bioluminescence and background spacelight. If the output of the receptors is simply summed, the absolute sensitivity will inevitably be increased by the possession of more than one pigment (Douglas and Partridge, 1997).

Intriguingly, however, the output from the various photoreceptors could also be used as the basis for a form of color vision, albeit mediated by rods rather than cones (Douglas and Partridge, 1997; Douglas et al., 1998a). In fact, Denton and Locket (1989) have suggested that in species with several banks of rods, a situation not uncommon in deep-sea fishes (Fig. 17.1; Fröhlich and Wagner, 1995), the spectral sensitivities of the various banks will be different due to screening of the more scleral layers of photoreceptors by the more vitreal cells. Color vision would thus be theoretically possible even if the animal contained just a single visual pigment. Color vision, whatever its underlying receptors, could, for instance, be used to identify species on the basis of differences in bioluminescent hue. Unfortunately, since most deep-sea species are at best moribund when captured and as the working environment in which they are caught is unsuitable for most forms of physiological experimentation, nothing is known about either their behavioral or electrophysiological responses to different wavelengths. Thus, for the time being, evidence for color vision in deep-sea fish is absent.

3.3. Extreme Longwave Sensitivity

In three genera of stomiid dragon fishes (*Aristostomias*, *Pachystomias* and *Malacosteus*) the link between visual pigment absorption and photic stimulus is an obvious one. These animals possess two distinct types of light-producing photophores, a postorbital one producing conventional blue/green bioluminescence, and a larger suborbital photophore producing far-red illumination peaking sharply at over 700 nm (Figs. 17.8 and 17.9; Denton et al., 1970, 1985; Widder et al., 1984). To enable them to see their own far-red bioluminescence, they were originally shown by extract spectrophotometry and MSP (Bowmaker et al., 1988; Crescitelli, 1989, 1991; O'Day and Fernandez, 1974; Knowles and Dartnall, 1977; Partridge et al., 1989) to each possess a rhodopsin/porphyropsin pigment pair separated into two spectrally distinct populations of rods, whose λ_{max} values are longwave shifted compared to the pigments of other deep-sea fishes (λ_{max} ca. 515–525, 540–550; Douglas et al., 1998a for review). Subsequent work using fresh retinal wholemounts, which conserves pigments destroyed by extraction or preservation, has shown that *Aristostomias* (Partridge and Douglas, 1995) and *Pachystomias* (Douglas et al., 1998a) have an additional visual pigment absorbing at longer wavelengths composed of a longer-wave opsin bound to retinal (*Aristostomias tittmanni*, λ_{max} ca. 585 nm; *Pachystomias microdon*, λ_{max} ca. 580–595 nm). Although it has

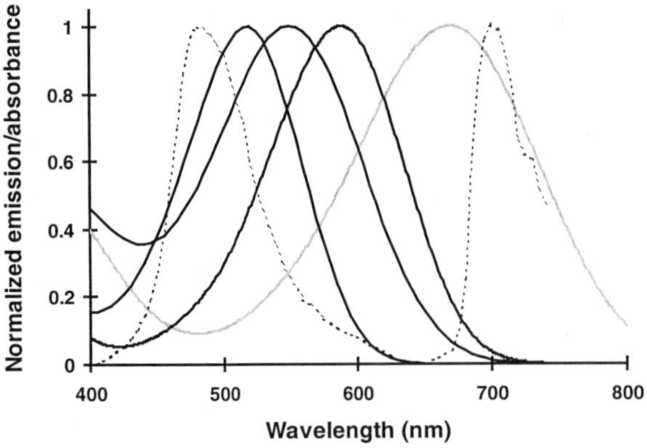

FIGURE 17.8. Bioluminescence of *Aristostomias tittmanni* (dotted line; Widder et al., 1984), and the best fit templates (solid lines) of the three visual pigments so far identified in its retinae (a rhodopsin/porphyropsin pigment pair with λ_{max} values 520nm and 551nm and a rhodopsin with λ_{max} 588nm). The gray line represents a theoretical porphyropsin with λ_{max} 669nm, which is the "partner" of the longwave-sensitive rhodopsin (calculated using the formula of Whitmore and Bowmaker, 1989). (Douglas et al., 1998a.)

never been convincingly isolated, due most probably to the use of dim red light during tissue handling, one might theoretically expect these animals to possess a fourth visual pigment consisting of the longwave opsin bound to 3,4-dehydroretinal (λ_{max} ca. 670nm; Douglas et al., 1998a; Fig. 17.8).

Although *Malacosteus niger* possesses a pigment pair similar to that based on the shortwave opsins of *Aristostomias* and *Pachystomias* (λ_{max} ca. 520 and 540nm; Fig. 17.9), it does not have the extreme red-sensitive pigments within its retinae displayed by the other two red-light-emitting stomiids, even when intact outer segments are examined (Douglas et al., 1998b, 1999). Instead it uses a photostable pigment, consisting of a mixture of demetallated and defarnesylated derivatives of bacteriochlorophylls *c* and *d*, located within the outer segments, with an absorption maximum around 670nm, as a photosensitizing pigment. Thus longwave bioluminescence is absorbed by the photosensiting pigment and indirectly bleaches the shorter-wave-absorbing visual pigments (Bowmaker et al., 1988; Douglas et al., 1998b, 1999; Fig. 17.9).

Since these stomiids are sensitive to their own far-red bioluminescence, which other animals in the deep-sea with conventional visual pigments cannot see, they have what could be regarded as a "private" waveband that could be used for intraspecific signaling immune from detection by potential predators, or for the covert illumination of prey (Partridge and Douglas, 1995). This would seem to give them a significant selective advantage. Perhaps not surprisingly, we have recently found evidence that at least one member of the myctophidae, a family that is heavily preyed on by the stomiid dragon fish (Sutton and Hopkins, 1996), may also have evolved a method for longwave sensitivity. Partial bleaching of extracts from *Bolinichthys longipes* revealed a single "conventional" deep-sea visual pigment absorbing maximally at 488 nm (Fig. 17.10). It is clear from examination of this figure, however, that retinal extracts of this species are far from conventional, as they not only show the bleachable visual pigment but also a photostable pigment absorbing maximally at long wavelengths around 670–671nm. The absorption spectrum of this pigment is similar to that of the presumed photosensitizer in the *Malacosteus niger* retina (Fig. 17.9). Preliminary data from fluorescence emission and excitation spectra indicate that this pigment, like that of *Malacosteus*, is a chlorophyll/bacteriochlorophyll derivative. Limited microspectrophotometer (MSP) evidence also indicates this pigment may in fact be linked to a second, longer-wave-absorbing visual pigment with a λ_{max} around 553nm.

Although we have no direct evidence, these data suggest that *B. longipes* may be using a chlorophyll-based pigment in a way similar to *M. niger*, as a photosensitizer to enhance the visibility of longwave illumination. To date the only species known to either produce or sense longwave bioluminescence in the deep-sea are some stomiids. The above data suggest that *B. longipes* may also be able to see this light. Although *Bolinichthys* possesses many photophores producing "normal" blue-green bioluminescence, it is not known to produce far-red

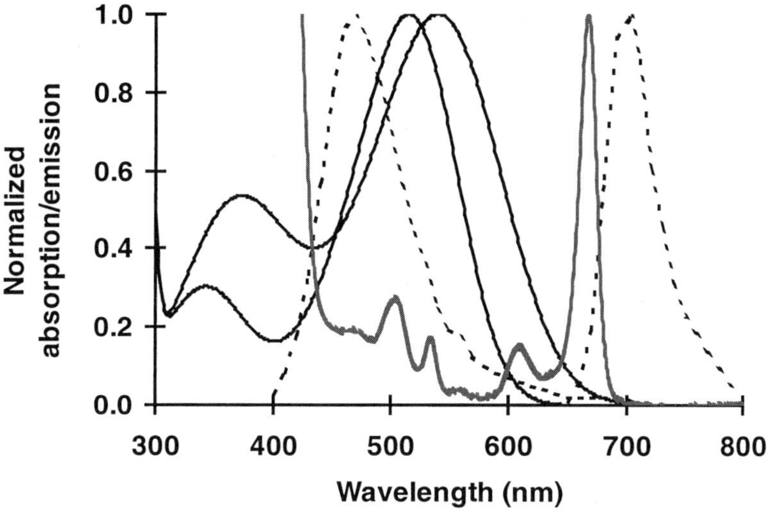

FIGURE 17.9. (B) Summary of absorbance spectra of *Malacosteus niger* retinal pigments and emission spectra of its bioluminescence. The solid black curves represent the two visual pigment templates that most closely fit the extract derived difference spectra: a rhodopsin with λ_{max} 515 nm and a porphyropsin with λ_{max} 540 nm, respectively. The emission spectra for the shortwave postorbital photophore and the long-wave-emitting suborbital photophore (dotted lines) are taken from Widder et al. (1984). The absorption spectrum of the presumed photosensitizer (gray line) is represented by a purified diethyl ether extract of a retinal suspension. (Modified from Douglas et al., 1999.)

light (Herring, personal communication). It is, however, possible that *Bolinichthys* may be involved in a sensory arms race with the stomiids, having evolved a system for seeing the far-red light of the dragon fish and thereby affording them a degree of protection from predation.

We have speculated that *M. niger* obtains its photosensitizer from a diet of copepods (Douglas et al., 2000). Interestingly, 70% of the diet of *B. longipes* is also composed of copepods (Kinzer et al., 1993).

It would be intriguing to know if other myctophid retinae, apart from *Bolinichthys longipes*, also contain a potential photosensitizing pigment. It is also worth noting that outside of the stomiids the most longwave pigment isolated by extraction is also found in a myctophid (Fig. 17.6; *Myctophum nitidulum*; λ_{max} 522 nm). One might speculate that if these retinae were examined using retinal wholemount techniques, they too, like *Aristostomias* and *Pachystomias*, may be shown to have additional red-sensitive visual pigments.

4. Molecular Basis of Deep-Sea Visual Pigment Tuning

The question of tuning of rod pigments in deep-sea fishes has been addressed by Hope et al. (1997) and Hunt et al. (2001) by comparing the sequences of the rod opsin gene from 28 species of deep-sea fishes drawn from seven different orders of the Euteleostei. The species together with the λ_{max} values of their visual pigments are listed in Table 17.1. With the exception of *Aristostomias tittmanni*, *Malacosteus niger*, and *Ceratoscopelus warmingii*, all the species studied possess only a single rhodopsin pigment in the retina (Ali et al., 1976; Partridge et al.,

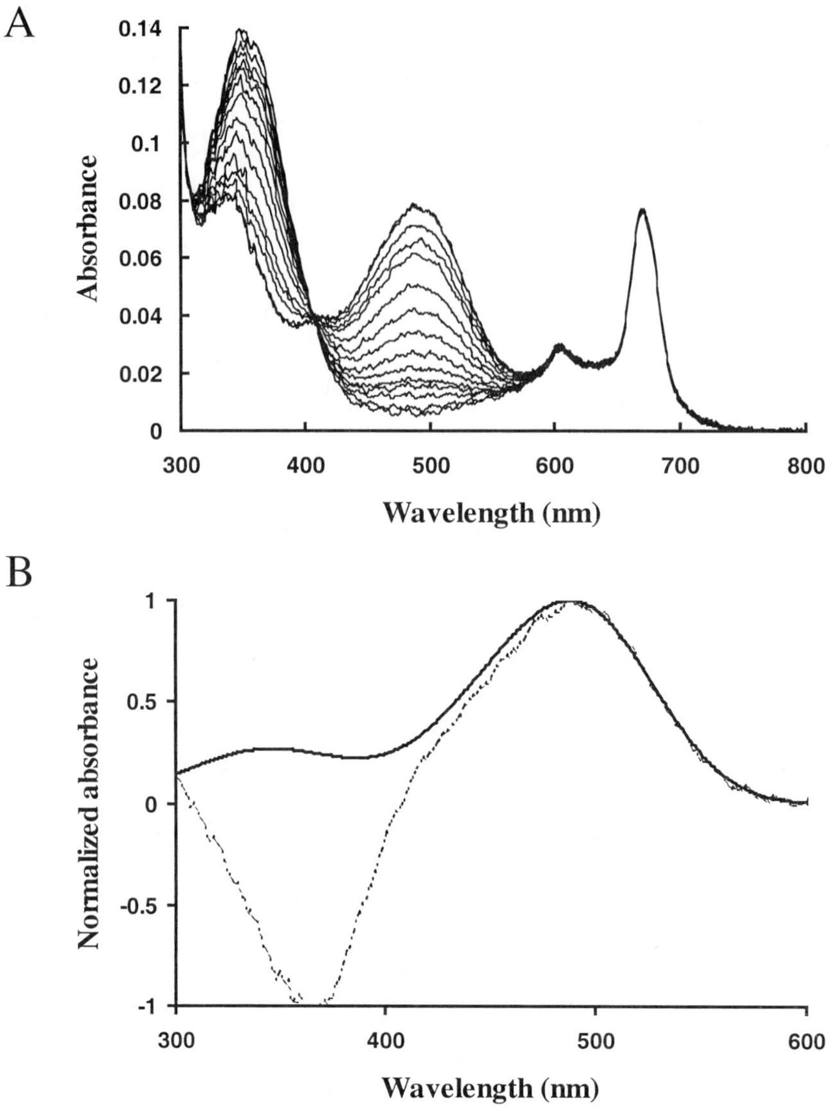

FIGURE 17.10. *Bolinichthys longipes* visual pigment in detergent extracts. (A) Absorbance spectra of a retinal extract of *B. longipes* following various durations of exposure to monochromatic light. In order of decreasing absorbance at 500 nm the absorbance spectra are: (i) initial measurement of the unexposed extract 10 min after the addition of 30 µl 1 M hydroxylamine to ca. 600 µl of extract; (ii) after 2 min exposure to monochromatic light of 709 nm; (iii) 10 min 634 nm; (iv) another 10 min 634 nm; (v) another 10 min 634 nm; (vi) 5 min 609 nm; (vii) another 5 min 609 nm; (viii) another 5 min 609 nm; (ix) another 5 min 609 nm; (x) another 5 min 609 nm; (xi) another 5 min 609 nm; (xii) another 5 min 609 nm; (xiii) 1.5 mins 576 nm; (xiv) another 5 mins 576 nm; (xv) 2 mins white light. (B) Normalized difference spectrum (dashed line) constructed by subtracting the absorbance spectrum of the totally bleached extract (curve xv in A) from the unbleached spectrum (curve i in A). This was best fit by a template (solid line) of a rhodopsin with λ_{max} 488 nm. Similar data were obtained from a further two extracts and by MSP (Douglas and Bowmaker, unpublished).

TABLE 17.1. Classification and λ_{max} values of visual pigments whose rod opsin genes have been sequenced.

Superorder: Stenopterygii Order: Stomiformes	λ_{max} **(nm)**[a]	**Superorder: Scopelomorpha Order: Myctophiformes**	λ_{max} **(nm)**[a]	**Superorder: Protacanthopterygii Order: Osmeriformes**	λ_{max} **(nm)**[a]
Aristostomias tittmani	523/551/581	*Diaphus rafinesquei*	489	*Conocara salmonea*	480
Malacosteus niger	517/542	*Bolinichthys indicus*	489	*Alepocephalus bairdii*	476
Photostomias guernei	483	*Ceratoscopelus warmingii*	468/488		
Chauliodus danae	484	*Lampanyctus alatus*	485	**Superorder: Acanthopterygii**	
Chauliodus sloani	485	*Benthosema suborbitale*	487	**Order: Beryciformes**	
Idiacanthus fasciola	485			*Anoplogaster cornuta*	485
Stomias boa	489	**Superorder:**		*Hoplostethus mediterraneus*	479
Ichthyocossus ovatus	489	**Paracanthopterygii**			
Vinciguerria nimbaria	477	**Order: Gadiformes**			
Argyropelecus aculeatus	477	*Phycis blennoides*	494	**Superorder: Cyclosquamata**	
Argyropelecus gigas	477	*Coryphaenoides guntheri*	479	**Order: Aulopiformes**	
Gonostoma bathyphilium	481	**Order: Ophidiiformes**		*Bathysaurus ferox*	481
Gonostoma elongatum	483	*Bassozetus compresis*	476	*Bathysaurus mollis*	479
		Cataetyx laticeps	468		

[a] λ_{max} values from Douglas (unpublished), Douglas and Partridge (1997), Douglas et al. (1999), Partridge (1989), and Partridge et al. (1988, 1989).

1988, 1989; Douglas et al., 1995; Fröhlich et al., 1995), indicating that only a single rod opsin is present in the photoreceptors.

Except for species from the same genus, the amino acid divergence between the rod opsin sequences of these deep-sea fishes species is generally >20%. In order therefore to identify candidate sites for spectral tuning, two overlapping approaches were used. The sequences were mapped onto a model based on conserved residues across >500 G protein–linked receptor proteins (Baldwin, 1993; Baldwin et al., 1997). This model not only identifies the seven helices but, together with the three-dimensional map of frog rhodopsin determined by electron cryomicroscopy (Schertler and Hargrave, 1995), also provides a framework for orientating each helix with respect to the exterior lipid membrane and the central hydrophilic retinal-binding pocket. Only substitutions that result in a change in either charge or polarity (Nathans, 1990; Nakayama and Khorana, 1991) in residues that are located in the transmembrane helical regions, and are either adjacent to this pocket or face another helix, appear to be important for the spectral tuning of the resulting pigment (Merbs and Nathans, 1993; Asenjo et al., 1994; Hunt et al., 1996; Hope et al., 1997). The number of candidate tuning sites was then extended to include additional sites identified from the crystal structure of bovine rhodopsin (Palczewski et al., 2000) that are again adjacent to either the retinal binding pocket or Schiff base.

A summary of the results is shown in Table 17.2. *Aristostomias tittmani* and *Malacosteus niger* from the order Stomiiformes, as detailed earlier, are exceptional among deep-sea fishes in producing longwave bioluminescence and both possess red-shifted rod pigments with λ_{max} values around 520 nm. Both species are also unique among the species studied in possessing phenylalanine rather than tyrosine at site 261 and isoleucine rather than serine at site 292. Substitutions at both sites have been implicated in longwave shifts in other cone and rod pigments (Asenjo et al., 1994; Yokoyama et al., 1995; Hunt et al., 1996; Sun et al., 1997; Fasick et al., 1998). These two substitutions would therefore appear to act together in producing longwave-shifted rod pigments in these two related species of deep-sea fishes, with substitution at site 208 responsible for the 5 nm difference in λ_{max} between the two species.

A further seven substitutions (Table 17.2) are all concerned with shortwave shifts, giving an overall total of only nine sites that are involved in the tuning of rod pigments in deep-sea fishes. Of these nine sites, however, only four, at residues 83, 122, 124, and 292, are commonly substituted in the different species. Substitutions at site 83 were first noted by Hope et al.

TABLE 17.2. Candidate tuning sites for spectral shifts in rod pigments.

λ_{max}	Candidate tuning sites								
Stomiiformes	83	122	124	132	208	261	292	299	300
>500 nm	Asn	Glu	Ala	Ala	Phe/Tyr	Tyr	Ile	Ala	Leu
489–481 nm	Asn	Glu	Ala/Gly	Ala	Phe	Phe	Ser	Ala	Leu/Ile
<480 nm	Asn	Gln	Ala	Ala	Phe	Phe	Ser	Ala	Leu
Myctophiformes									
489–487 mm	Asn/Asp	Gln	Ala/Gly	Ser	Phe	Phe	Ala	Ala	Ile
Gadiformes									
492 nm	Asp	Glu	Ala	Ala	Phe	Phe	Ala	Ala	Ile
479 nm	Asp	Val	Ala	Ala	Phe	Phe	Ser	Ala	Ile
Ophidiiformes									
476 nm	Asn	Glu	Ser	Ala	Phe	Phe	Ser	Ala	Thr
468 nm	Asn	Glu	Ser	Ala	Phe	Phe	Ser	Ala	Ile
Osmeriformes									
480 nm	Asp	Gln	Ala	Ser	Phe	Phe	Ser	Ala	Ile
476 nm	Asp	Gln	Val	Val	Phe	Phe	Ser	Ala	Ile
Beryciformes									
485 nm	Asn	Glu	Ala	Ala	Phe	Phe	Ser	Ser	Ile
479 nm	Asn	Glu	Ala	Ala	Phe	Phe	Ser	Ala	Ile
Aulopiformes									
481–479 nm	Asn	Glu	Ser	Ala	Phe	Phe	Ser	Thr	Ile

(1997) to be one of the main mechanisms for the tuning of rod pigments of deep-sea fishes to λ_{max} values of <490 nm, and site 292, which is commonly occupied by polar serine, is responsible in most species for a substantial part of the shortwave shift. The major exception is in the myctophids, where alanine is present at 292; in this group, the shortwave shift is achieved instead by polar glutamine at site 122. The shortwave-shifted pigments of stomiiforms and osmeriforms with λ_{max} values of 480 nm and below possess glutamine at site 122 and serine at site 292. In contrast, similar shortwave shifts seen in the aulopiformes are achieved by the pairing of serine at 124 and 292.

Dartnall and Lythgoe (1965) and Bridges (1965) were the first to note that the λ_{max} values of vertebrate visual pigments cluster around certain points in the spectrum. This is particularly apparent in primate M and L cone pigments, with each spectral location attributable to a particular combination of amino acid residues at three main sites (Neitz et al., 1991; Williams et al., 1992; Hunt et al., 1998; Dulai et

TABLE 17.3. Amino acid residues at four sites in species from different orders but with pigments that spectrally locate at three cluster points.

Cluster point		$H\lambda_{max}$ (nm)	Amino acid sites			
			83	122	124	292
1	**Stomiiformes**	489	Asn	Glu	Ala	Ser
	Myctophiformes	489	Asp	Gln	Ala	Ala
2	**Stomiiformes**	482	Asn	Glu	Ala	Ser
	Gadiformes	482	Asp	Val	Ala	Ser
	Aulopiformes	481	Asn	Glu	Ser	Ser
3	**Stomiiformes**	477	Asn	Gln	Ala	Ser
	Osmeriformes	477	Asn	Glu	Ala	Ser
	Ophidiiformes	476	Asn	Glu	Ser	Ser

al., 1999). The same phenomenon is seen in the rod pigments of deep-sea fishes (Partridge et al., 1989, 1992) and a similar explanation has been advanced, namely that the spectral location of pigments belonging to the same cluster group arises from a common set of amino acid substitutions (Douglas et al., 1998a). The pattern of substitutions across the 28 species included in this study indicates, however, that for the four most commonly used sites this is true only for species from the same order. In general, therefore, the presence of cluster points in deep-sea fish rod opsins cannot be attributed to the selection of a particular set of amino acid substitutions. Rather, it is achieved by different combinations of amino acids at these sites that produce the same net spectral shift. This is summarized in Table 17.3 where representative species from different orders with the same or very similar λ_{max} values are listed. For each cluster group, there are at least two different combinations of residues at the four sites.

Finally, the phylogenetic distribution of amino acid substitutions in the different species of deep-sea fishes (Hunt et al., 2001) indicates that the rod opsin of the ancestral species would have possessed serine at 292 and that the resulting pigment would have been shortwave shifted to around 485 nm or 480 nm, depending on whether glutamine or glutamate was present at site 122. This implies that species with shortwave-shifted rod visual pigments were present at or near the base of the Euteleost lineage, indicating that they were adapted to the same visual environment as present-day deep-sea fishes.

References

Ali, M.A., and Anctil, M. (1976). *Retinas of Fishes: An Atlas*. Berlin: Springer.

Archer, S.N., Hope, A., and Partridge, J.C. (1995). The molecular basis for the blue-green sensitivity shift in the rod visual pigments of the European eel. *Proc. Roy. Soc. Lond. B.* 262:289–295.

Asenjo, A.B., Rim, J., and Oprian, D.D. (1994). Molecular determinants of human red/green color discrimination. *Neuron* 12:1131–1138.

Baldwin, J.M. (1993). The probable arrangement of the helices in G protein-coupled receptors. *EMBO J.* 12:1693–1703.

Baldwin, J.M., Schertler, G.F.X., and Unger, V.M. (1997). An alpha-carbon template for the transmembrane helices in the rhodopsin family of G-protein-coupled receptors. *J. Mol. Biol.* 272: 44–164.

Bayliss, L.E., Lythgoe, J.N., and Tansley, K. (1936). Some forms of visual purple in sea fishes with a note on the visual cells of origin. *Proc. Roy. Soc. Lond. B.* 120:95–114.

Bowmaker, J.K. (1976). Vision in pelagic animals. In: *Adaptation to Environment: Essays on the Physiology of Marine Animals* (Newell, R.C., ed.), pp. 430–479. London: Butterworths.

Bowmaker, J.K. (1990). Visual pigments of fishes. In: *The Visual System of Fishes* (Douglas, R.H., and Djamgoz, M.B.A., eds.), pp. 81–107. London: Chapman & Hall.

Bowmaker, J.K. (1991). The evolution of visual pigments and photoreceptors. In: *Evolution of the Eye and Visual System* (Gregory, R.L., and Cronly-Dillon, J.R., eds.), pp. 63–81. London: Macmillan.

Bowmaker, J.K. (1995). The visual pigments of fish. *Prog. Ret. Eye. Res.* 15:1–31.

Bowmaker, J.K., Dartnall, H.J.A., and Herring, P.J. (1988). Longwave-sensitive visual pigments in some deep-sea fishes: Segregation of "paired" rhodopsins and porphyropsins. *J. Comp. Phys. A.* 163:685–698.

Bowmaker, J.K., Loew, E.R., and Ott, M. (2000). Porphyropsins and rhodopsins in chameleons, *Chamaeleo dilepis* and *Furcifer lateralis*. *Invest. Ophthalmol. Vis. Sci.* 41:3177.

Bowmaker, J.K., Thorpe, A., and Douglas, R.H. (1991). Ultraviolet sensitive cones of the goldfish. *Vision Res.* 31:349–352.

Bowmaker, J.K., Govardovskii, V.I., Shukolyukov, S.A., Zueva, L.V., Hunt, D.M., Sideleva, V.G., and Smirnova, O.G. (1994). Visual pigments and the photic environment: The cottoid fish of lake Baikal. *Vision Res.* 34:591–605.

Bridges, C.D.B. (1965). The grouping of visual pigments about preferred positions in the spectrum. *Vision Res.* 5:223–238.

Bridges, C.D.B. (1972). The rhodopsin-porphyropsin visual system. In: *Photochemistry of Vision* (Dartnall, H.J.A., ed.), pp. 417–480. Berlin: Springer.

Clarke, R.L. (1936). On the depth at which fishes can see. *Ecology* 17:452–456.

Collin, S., and Partridge, J.P. (1997). Retinal specializations in the eyes of deep-sea teleosts. *J. Fish. Biol.* 49 (Suppl. A):157–174.

Collin, S.P., Hoskins, R.V., and Partridge, J.C. (1998). Seven retinal specializations in the tubular eyes of the deep-sea pearleye, *Scopelarchus michaelsarsi*: A case study in visual optimisation. *Brain Behav. Evol.* 50:335–357.

Crescitelli, F. (1989). The visual pigments of a deep-water Malacosteid fish. *J. Mar. Biol. Assoc. U.K.* 69:43–51.

Crescitelli, F. (1991). Adaptations of visual pigments to the photic environment of the deep-sea. *J. Exp. Zool. (Suppl.)* 5:66–75.

Cronin, T.W., and Frank, T. (1996). A short-wavelength photoreceptor class in a deep-sea shrimp. *Proc. Roy. Soc. Lond. B.* 263:861–865.

Dartnall, H.J.A., and Lythgoe, J.N. (1965). The spectral clustering of visual pigments. *Vision Res.* 5:81–100.

Dartnall, H.J.A., Lander, M.R., and Munz, F.W. (1961). Periodic changes in the visual pigments of fish. In: *Progress in Photobiology* (Christensen, B., and Buchmann, B., eds.), pp. 203–213. Amsterdam: Elsevier.

Denton, E.J. (1990). Light and vision at depths greater than 200 metres. In: *Light and Life in the Sea* (Herring, P.J., Campbell, A.K., Whitfield, M., and Maddock, L., eds.), pp. 127–148. Cambridge: Cambridge University Press.

Denton, E.J., and Locket, N.A. (1989). Possible wavelength discrimination by multibank retinae in deep-sea fishes. *J. Mar. Biol. Assoc. U.K.* 69:409–435.

Denton, E.J., Gilpin-Brown, J.B., and Wright, P.G. (1970). On the "filters" in the photophores of mesopelagic fish and on a fish emitting red light and especially sensitive to red light. *J. Physiol.* 208:72–73P.

Denton, E.J., Herring, P.J., Widder, E.A., Latz, M.F., and Case, J.F. (1985). The roles of filters in the photophores of oceanic animals and their relation to vision in the oceanic environment. *Proc. Roy. Soc. Lond. B.* 225:63–97.

Douglas, R.H. (2001). The ecology of teleost fish visual pigments: A good example of sensory adaptation to the environment? In: *Ecology of Sensing* (Barth, F.G., Schmid, A., eds.), pp. 215–235 Berlin: Springer Verlag.

Douglas, R.H., and Partridge, J.C. (1997). On the visual pigments of deep-sea fish. *J. Fish. Biol.* 50: 68–85.

Douglas, R.H., Mullineaux, C.W., and Partridge, J.C. (2000). Long-wave sensitivity in deep-sea stomiid dragonfish with far-red bioluminescence: Evidence for a dietary origin of the chlorophyll-derived retinal photosensitizer of *Malacosteus niger*. *Phil. Trans. Roy. Soc. Lond. B.* 355:1269–1272.

Douglas, R.H., Partridge, J.C., and Hope, A.J. (1995). Visual and lenticular pigments in the eyes of demersal deep-sea fishes. *J. Comp. Phys. A.* 177:111–122.

Douglas, R.H., Partridge, J.C., and Marshall, N.J. (1998a). The visual systems of deep-sea fish. I. Optics, tapeta, visual and lenticular pigmentation. *Prog. Ret. Eye Res.* 17(4):597–636.

Douglas, R.H., Partridge, J.C., Dulai, K.S., Hunt, D.M., Mullineaux, C.W., and Hynninen, P.H. (1999). Enhanced retinal longwave sensitivity using a chlorophyll-derived photosensitiser in *Malacosteus niger*, a deep-sea dragon fish with far red bioluminescence. *Vision Res.* 39:2817–2832.

Douglas, R.H., Partridge, J.C., Dulai, K., Hunt, D., Mullineaux, C.W., Tauber, A., and Hynninen, P.H. (1998b). Dragon fish see using chlorophyll. *Nature* 393:423–424.

Dratz, E.A., and Hargrave, P.A. (1983). The structure of rhodopsin and the outer segment disc membrane. *Trends Biochem. Sci.* 8:128–131.

Dulai, K.S., von Dornum, M., Mollon, J.D., and Hunt, D.M. (1999). The evolution of trichromatic colour vision by opsin gene duplication in New World and Old World primates. *Genome Res.* 9:629–638.

Fasick, J.I., and Robsinson, P.R. (1998). Mechanism of spectral tuning in the dolphin visual pigments. *Biochemistry* 37:433–438.

Findlay, J.B., and Pappin, D.J. (1986). The opsin family of proteins. *Biochem J.* 238:625–642.

Frank, T.M., and Case, J.F. (1988). Visual spectral sensitivities of bioluminescent deep-sea crustacea. *Biol. Bull.* 175:261–273.

Frank, T.M., and Widder, E.A. (1994). Evidence for behavioural sensitivity to near-UV light in the deep-sea crustacean *Systellaspis sebilis*. *Mar. Biol.* 118:279–284.

Frank, T.M., and Widder, E.A. (1996). UV light in the deep-sea: *In situ* measurements of downwelling irradiance in relation to the visual threshold sensitivity of UV-sensitive crustaceans. *Mar. Fresh. Behav. Physiol.* 27:189–197.

Fröhlich, E., and Wagner, H.-J. (1995). Patterns of rod proliferation in deep-sea fish retinae. *Vision Res.* 35:1799–1811.

Fröhlich, E., Negishi, K., and Wagner, H.-J. (1995). The occurrence of dopaminergic interplexiform cells correlates with the presence of cones in the retinae of fish. *Vis. Neurosci.* 12:359–369.

Govardovskii, V.I., Fyhrquist, N., Reuter, T., Kuzmin, D.G., and Donner, K. (2000). In search of the visual pigment template. *Vis. Neurosci.* 17:509–528.

Haddock, S.H.D., and Case, J.F. (1999). Bioluminescence spectra of shallow and deep-sea gelatinous zooplankton: Ctenophores, medusae and siphonophores. *Mar. Biol.* 133:571–582.

Hart, N.S., Partridge, J.C., Bennett, A.T.D., and Cuthill, I.C. (2000). Visual pigments, cone oil droplets and ocular media in four species of estridid finch. *J. Comp. Physiol. A.* 186:681–694.

Herring, P.J. (1983). The spectral characteristics of luminous marine organisms. *Proc. Roy. Soc. Lond. B.* 220:183–217.

Hope, A.J., Partridge, J.C., Dulai, K.S., and Hunt, D.M. (1997). Mechanisms of wavelength tuning in the rod opsins of deep-sea fishes. *Proc. Roy. Soc. Lond. B.* 264:155–163.

Hunt, D.M., Fitzgibbon, J., Slobodyanyuk, S., and Bowmaker, J.K. (1996). Spectral tuning and molecular evolution of rod visual pigments in the species flock of cottoid fish in Lake Baikal. *Vision Res.* 36:1217–1224.

Hunt, D.M., Dulai, K.S., Cottrill, P.B., Partridge, J.C., and Bowmaker, J.K. (2001). The molecular basis for spectral tuning of rod visual pigments in deep sea fish. *J. Exp. Biol.* 204:3333–3344.

Hunt, D.M., Dulai, K.S., Cowing, J.A., Julliot, C., Mollon, J.D., Bowmaker, J.K., Li, W.-H., and Hewett-Emmett, D. (1998). Molecular evolution of trichromacy in primates. *Vision Res.* 38:3299–3306.

Jerlov, N.G. (1976). *Marine Optics*. Amsterdam: Elsevier Scientific.

Kampa, E.M. (1970). Underwater daylight and moonlight measurements in the Eastern North Atlantic. *J. Mar. Biol. Assoc. U.K.* 50:391–420.

Kinzer, J., Bottgerschnack, R., and Schulz, K. (1993). Aspects of horizontal distribution and diet of myctophid fish in the Arabian Sea with reference to the deep-water oxygen deficiency. Deep-sea research part II: *Topical Studies in Oceanography* 40:783–800.

Kirk, J.T.O. (1983). *Light and Photosynthesis in Aquatic Ecosystems*. Cambridge: Cambridge University Press.

Knowles, A., and Dartnall, H.J.A. (1977). *The Photobiology of Vision*. New York: Academic Press.

Kochendoerfer, G.G., Lin, S.W., Sakmar, T.P., and Mathies, R.A. (1999). How color visual pigments are tuned. *TIBS* 24:300–305.

Kropf, A., and Hubbard, R. (1958). The mechanism of bleaching rhodopsin. *Ann. N.Y. Acad. Sci.* 74: 266–280.

Latz, M.I., Frank, T.M., and Case, J.F. (1988). Spectral composition of bioluminescence of epipelagic organisms from the Sargasso Sea. *Mar. Biol.* 98: 441–446.

Locket, N.A. (1971). Retinal anatomy in some scopelarchid deep-sea fishes. *Proc. Roy. Soc. Lond. B.* 178:161–184.

Locket, N.A. (1977). Adaptations to the deep-sea environment. In: *Handbook of Sensory Physiology, VII/5: The Visual System in Vertebrates* (Crescitelli, F., ed.), pp. 67–192. Berlin: Springer.

Loew, E.R., and Dartnall, H.J.A. (1976). Vitamin A1/A2-based visual pigment mixtures in cones of the rudd. *Vision Res.* 16:891–896.

Lythgoe, J.N. (1972). List of vertebrate visual pigments. In: *Handbook of Sensory Physiology, VII/1: The Photochemistry of Vision* (Dartnall, H.J.A., ed.), pp. 604–624. Berlin: Springer.

Mathies, R., and Stryer, L. (1976). Retinal has a highly dipolar vertically excited singlet state: Implications for vision. *Proc. Natl. Acad. Sci. USA* 73:2169–2173.

Merbs, S.L., and Nathans, J. (1993). Role of hydroxyl-bearing amino acids in differentially tuning the absorption spectra of the human red and green cone pigments. *Photochem. Photobiol.* 58:706–710.

Munk, O. (1965). *Omosudis lowei* (Günther, 1887): A bathypelagic deep-sea fish with an almost pure-cone retina. *Vidensk Medd. Dansk Naturh. Foren.* 128:341–355.

Munk, O. (1966). Ocular anatomy of some deep-sea teleosts. *Dana. Rep.* 70:1–62.

Munk, O. (1981). On the cones of the mesopelagic teleost *Trachipterus trachypterus* (Gmelin, 1789). *Vidensk Medd. Dansk Naturh. Foren.* 143:101–111.

Munk, O. (1984). Duplex retina in the mesopelagic teleost *Radiicephalus elongates* (Osorio, 1917). *Vidensk Medd. Dansk. Naturh. Foren.* 145:183–199.

Muntz, W.R.A., and Mouat, G.S.V. (1984). Annual variations in the visual pigments of brown trout inhabiting lochs providing different light environments. *Vision Res.* 24:1575–1580.

Munz, F.W. (1965). Adaptation of visual pigments to the photic environment. In: *Ciba Foundation Symposium on Physiology and Experimental Psychology of Colour Vision* (Wolstenholme, G.E.W., and Knight, J., eds.), pp. 27–45. London: Churchill.

Nakayama, T.A., and Khorana, H.G. (1991). Mapping of the amino acids in membrane-embedded helices that interact with the retinal chromophore in the bovine rhodopsin. *J. Biol. Chem.* 266:4269–4275.

Nathans, J. (1990). Determinants of visual pigment absorbance: Role of charged amino acids in the putative transmembrane segments. *Biochemistry* 29:937–942.

Neitz, M., Neitz, J., and Jacobs, G.H. (1991). Spectral tuning of pigments underlying red-green color vision. *Science* 252:971–974.

O'Day, W.T., and Fernandez, H.R. (1974). *Aristostomias scintillans* (Malacosteidae): A deep-sea fish with visual pigments apparently adapted to its own bioluminescence. *Vision Res.* 14:545–550.

Palczewski, K., Kumasaka, T., Hori, T., Behnke, C.A., Motoshima, H., Fox, B.A., Le Trong, I., Teller, D.C., Okada, T., Stenkamp, R.E., Yamamoto, M., and Miyano, M. (2000). Crystal structure of rhodopsin: A G protein-coupled receptor. *Science* 289:739–745.

Parry, J.W.L., and Bowmaker, J.K. (2000). Visual pigment reconstitution in intact goldfish retina using synthetic retinaldehyde isomers. *Vision Res.* 40:2241–2247.

Partridge, J.C., and Douglas, R.H. (1995). Far-red sensitivity of dragon fish. *Nature* 375:21–22.

Partridge, J.C., Archer, S.N., and Lythgoe, J.N. (1988). Visual pigments in the individual rods of deep-sea fishes. *J. Comp. Phys. A.* 162:543–550.

Partridge, J.C., Archer, S.N., and van Oostrum, J. (1992). Single and multiple visual pigments in deep-sea fishes. *J. Mar. Biol. Assoc. U.K.* 72:113–130.

Partridge, J.C., Shand, J., Archer, S.N., Lythgoe, J.N., and van Groningen-Luyben, W.A.H.M. (1989). Interspecific variation in the visual pigments of deep-sea fishes. *J. Comp. Physiol. A.* 164:513–529.

Provencio, I., Loew, E.R., and Foster, R.G. (1992). Vitamin A_2-based visual pigments in fully terrestrial vertebrates. *Vision Res.* 32:2201–2208.

Schertler, G.F., and Hargrave, P.A. (1995). Projection structure of frog rhodopsin in two crystal forms. *Proc. Natl. Acad. Sci. USA* 92:11578–11582.

Shieh, T., Han, M., Sakmar, T.P., and Smith, S.O. (1997). The steric trigger in rhodopsin activation. *J. Mol. Biol.* 269:373–384.

Sun, H., Macke, J.P., and Nathans, J. (1997). Mechanisms of spectral tuning in the mouse green cone pigment. *Proc. Natl. Acad. Sci. USA* 94:8860–8865.

Sutton, T.T., and Hopkins, T.L. (1996). Trophic ecology of the stomiid (Pisces: Stomiidae) fish assemblage of the eastern Gulf of Mexico: Strategies, selectivity and impact of a top mesopelagic predator group. *Mar. Biol.* 127:179–192.

Wagner, H.-J., Fröhlich, E., Negishi, K., and Collin, S.P. (1998). The eyes of deep-sea fish, II. Functional morphology of the retina. *Prog. Ret. Eye Res.* 17:637–685.

Whitmore, A.V., and Bowmaker, J.K. (1989). Seasonal variation in cone sensitivity and shortwave absorbing visual pigments in the rudd, *Scardinius erythrophthalmus*. *J. Comp. Physiol. A.* 166:103–115.

Widder, E.A. (1999). Bioluminsecence. In: *Adaptive Mechanisms in the Ecology of Vision* (Archer, S.N., Djamgoz, M.B.A., Loew, E.R., Partridge, J.C., and Vallerga, S., eds.), pp. 555–581. Dordrecht: Kluwer.

Widder, E.A., Latz, M.I., and Case, J.F. (1983). Marine bioluminescence spectra measured with an optical multichannel detection system. *Biol. Bull.* 165:791–810.

Widder, E.A., Latz, M.I., Herring, P.J., and Case, J.F. (1984). Far red bioluminescence from two deep-sea fishes. *Science* 225:512–514.

Widder, E.A., Bernstein, S.A., Bracher, D.F., Case, J.F., Reisenbichler, K.R., Torres, J.J., and Robison, B.H. (1989). Bioluminescence in the Monterey Submarine Canyon: Image analysis of video recordings from a midwater submersible. *Mar. Biol.* 100:541–551.

Williams, A.J., Hunt, D.M., Bowmaker, J.K., and Mollon, J.D. (1992). The polymorphic photopigments of the marmoset: Spectral tuning and genetic basis. *EMBO J.* 11:2039–2045.

Wood, P., and Partridge, J.C. (1993). New opsin induced in the retinal rods of the eel (*Anguilla anguilla* L.). *Proc. Roy. Soc. Lond. B.* 254:227–232.

Wood, P., Partridge, J.C., and DeGrip, W.J. (1992). Rod visual pigment changes in the elver of the eel *Anguilla anguilla* (L) measured by microspectrophotometry. *J. Fish Biol.* 41:601–611.

Yokoyama, R., Knox, B.E., and Yokoyama, S. (1995). Rhodopsin from the fish, *Astyanax*: Role of tyrosine 261 in the red shift. *Invest. Ophthalmol. Vis. Sci.* 36:939–945.

Zvyaga, T.A., Fahmy, K., and Sakmar, T.P. (1994). Characterization of rhodopsin-transducin interaction: A mutant rhodopsin photoproduct with a protonated Schiff base activates transducin. *Biochemistry* 33:9753–9761.

18 Visual Adaptations in Crustaceans: Chromatic, Developmental, and Temporal Aspects

N. Justin Marshall, Thomas W. Cronin, and Tamara M. Frank

Abstract

Crustaceans possess a huge variety of body plans and inhabit most regions of Earth, specializing in the aquatic realm. Their diversity of form and living space has resulted in equally diverse eye designs. This chapter reviews the latest state of knowledge in crustacean vision concentrating on three areas: spectral sensitivities, ontogenetic development of spectral sensitivity, and the temporal properties of photoreceptors from different environments. Visual ecology is a binding element of the chapter and within this framework the astonishing variety of stomatopod (mantis shrimp) spectral sensitivities and the environmental pressures molding them are examined in some detail. The quantity and spectral content of light changes dramatically with depth and water type and, as might be expected, many adaptations in crustacean photoreceptor design are related to this governing environmental factor. Spectral and temporal tuning may be more influenced by bioluminescence in the deep ocean, and the spectral quality of light at dawn and dusk is probably a critical feature in the visual worlds of many shallow-water crustaceans. Plasticity in photoreceptor tuning is a recently emerging theme both in crustaceans and other animals. The seasonal variation in crayfish spectral sensitivity and spectral sensitivity change in single stomatopod species from different depths provide two examples of this. Other oddities such as the need to see the heat from hydrothermal vents, color dances in water-fleas, and the possible influences of temperature on the spectral tuning of visual pigments are also discussed.

1. Introduction

Crustaceans possess a greater variety of eye types than any other invertebrate group (Land, 1984; Fig. 18.1 [see color plate]). They inhabit all the zones of the aquatic realm and some, such as the woodlice (*Armadillidium* sp.), have ventured permanently onto land. The optical design of their eyes includes "simple" or "camera" eyes like our own; compound eyes (apposition, refracting superposition) similar to the insects; eyes with reflecting superposition or

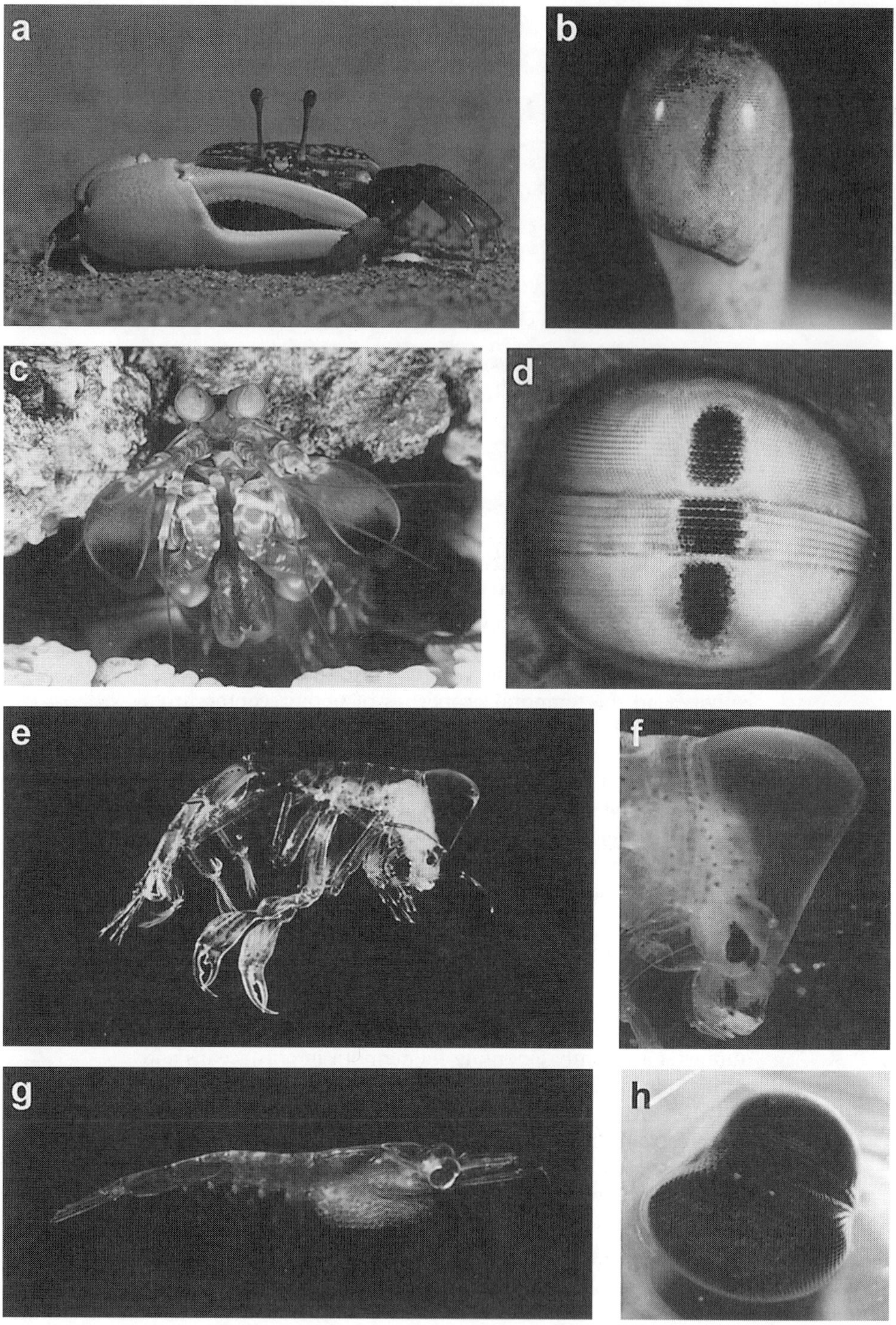

FIGURE 18.1. Crustaceans and their eyes. The fiddler crab *Uca polita* and closeup view of front of eye (a, b), (a—photograph by Jochen Zeil.) The stomatopod *Odontodactylus scyllarus* and closeup view of front of eye (c, d). The hyperiid amphipod *Phronima sedentaria* and closeup view of lateral aspect of head and eyes (e, f), (e—photograph by Mike Land.) The euphausiid *Nematobrachion megalops* and closeup view of lateral aspect of eye (g, h). Note the eye subdivisions in the last three examples. (See color plate)

mirrored optics that may just gather light or form perfect images; and even eyes that possess no optics at all (Land, 1981a, 1984; Land and Nilsson, 2002). This diversity of light-gathering mechanisms partially reflects the range of light intensities that crustaceans experience (10^{13} to 10^{1} photons $cm^{-1} sr^{-1} nm^{-1}$) but is to some extent phylogenetic (Land, 1981a; Cronin, 1986; Marshall et al., 1999a,b). A cursory glance at the spread of peak spectral sensitivities (315–710 nm, Table 18.1; Fig. 18.2) suggests a similar diversity in visual pigments and color vision potential. However, much of the variety seen comes from our knowledge of an isopod (*Ligia*) (Hariyama et al., 1993), a water-flea (*Daphnia*) (Smith and Macagno, 1990), and the mantis shrimps (stomatopods) (Cronin and Marshall, 1989a,b), which possess tri-, tetra-, and polychromatic color vision, respectively. Other crustaceans examined possess one or two spectral mechanisms, usually with one placed close to 400 nm and the other between 450 nm and 550 nm depending on habitat.

It is clear that our knowledge of crustacean spectral mechanisms still has many holes in it (Table 18.2) and there are likely to be many interesting surprises for anyone exploring these fascinating animals. One purpose of this chapter is to rekindle some of the interest in crustacean visual mechanisms, a declining discipline in these days of more applied biology. Interested readers are directed to two recent reviews in this area, Marshall et al. (1999b) and Cronin and Hariyama (2002).

The relationship between eye design constraints, such as the habitat light environment of different species, and their resulting spectral sensitivities, forms a major theme of this chapter. We also examine changes in spectral sensitivity during the development of single species, as many crustaceans shift from shallow-living planktonic larval stages to deeper-living adults and this involves substantial changes in habitat and therefore light regime and way of life. By way of comparison, and within the same framework of visual ecology, our knowledge of the temporal properties of crustacean photoreceptors is also reviewed. All photoreceptors face the dual task of catching enough photons and processing the information fast enough to eat or avoid being eaten. In light-limited habitats, such as deep-oceans or turbid waters, this results in compromises in temporal and optical aspects of photoreceptor design that mirror changes in spectral sensitivity (Land, 1981a,b; Warrant, 1999).

2. A Note on Methods

Five technical methods have been used to gather the data reviewed here. Spectrophotometry of retinal extracts and microspectrophotometry (MSP) of intact photoreceptors and optical filters are the two most common methods of determining spectral sensitivity. Extraction of visual pigment from the retina and its cellular environment can result in changes in the position of the sensitivity maximum (λ_{max}), so these results need to be treated with the caution of hindsight. Intracellular or extracellular electrophysiological recording from intact eyes provide two even more precise measures of eye spectral sensitivity, and these methods are also the ones used most extensively for determining temporal properties of photoreceptors. While recording from individual cells is desirable, for practical reasons (such as the need to conduct experiments onboard vibrating ships in order to work on live animals), the extracellular electroretinogram (ERG) is often used. Finally, in this scale of visual system "dissection," behavioral determination of both spectral sensitivity and flicker fusion frequency (a measure of photoreceptor temporal properties—see Section 5) gives the closest approximation of the parameters of an animal's visual system. Again, for a variety of practical reasons, this may not be possible. One result of this variety of techniques is that some caution is needed in cross comparisons between data gathered with different methods.

Molecular genetics is now being used to examine fish (Carleton et al., 2000) and mammalian (Jacobs and Deegan, 2000) visual pigments, for example, with fascinating results. The target molecule of this method is the opsin or protein component of the visual pigment. All visual pigments are composed of a

TABLE 18.1. Spectral sensitivities in crustaceans.

Species	λ_{max} (MSP) S-max /nm	Method of study	Author(s)	Year
Subphylum Crustacea				
Class Branchiopoda				
Subclass Diplostraca				
Order Cladocera				
Daphnia magna	348,434,525, & 608	EON	Smith & Macagno	1990
Subclass Sarsostraca				
OrderAnostraca				
Artemia salina	410	ERG	Hertel	1972
Class Maxillopoda				
Subclass Copepoda				
Acartia tonsa	450–520	BP	Stearns & Forward	1984
Subclass Cirripedia				
Order Thoracica				
Suborder Balanomorpha				
Balanus amphitrite	532	ERG	Hillman, Dodge, Hochstein, Knight & Minke	1973
	532	MSP	Minke & Kirschfield	1978
Balanus balanoides (nauplius)	510–530	BP	Barnes & Klepal	1972
Balanus eburneus	532	ERG	Hillman, Dodge, Hochstein, Knight & Minke	1973
	532	MSP	Minke & Kirschfield	1978
Class Malacostraca				
Subclass Eumalacostraca				
Superorder Holpocarida				
Order Stomatopoda				
Superfamily Squilloidea				
Cloridopsis dubia	510	MSP	Cronin, Marshall & Caldwell	1994b
Squilla empusa	517	MSP	Cronin, Marshall & Caldwell	1994b
	360 & 500	OP	Cronin	1985
Squilla mantis	535–555	IC	Schiff	1963
Superfamily Gonodactyloidea				
Gonodactylus oerstedii	400–551 (11)	MSP	Cronin & Marshall	1989a&b
	325	MSP	Cronin, Marshall, Quinn, & King	1994d
	400–695 (11)	CS	Cronin, Marshall, Caldwell & Shashar	1994c
	312–710	IC	Marshall & Oberwinkler	1993
	312–410	IC	Marshall & Oberwinkler	1999
Gonodactylaceus aloha	400–551 (11)	MSP	Cronin, Marshall & Caldwell	1996
Neogonodactylus curacaoensis	400–520 (8)	MSP	Cronin, Marshall & Caldwell	1996
Psuedosquilla ciliata	400–539 (11)	MSP	Cronin & Marshall	1989b
	400–680 (11)	CS	Cronin, Marshall, Caldwell & Shashar	1994c
Hemisquilla ensigera	414–535 (11)	MSP	Cronin, Marshall, Caldwell & Shashar	1994c
	440–625 (11)	CS	Cronin, Marshall, Caldwell & Shashar	1994c
	330	MSP	Cronin, Marshall, Quinn, & King	1994d
Odontodactylus brevirostris	402–535 (11)	MSP	Cronin, Marshall, Caldwell & Shashar	1994c
	400–623 (11)	CS	Cronin, Marshall, Caldwell & Shashar	1994c
Odontodactylus scyllarus	400–546 (11)	MSP	Cronin, Marshall & Caldwell	1994b
	425–655 (11)	CS	Cronin, Marshall & Caldwell	1994b
Odontodactylus havanensis	407–520 (8)	MSP	Cronin, Marshall & Caldwell	1996
Superfamily Lysiosquilloidea				
Coronis scolopendra	407–533 (11)	MSP	Cronin, Marshall, Caldwell & Shashar	1994c
	425–620 (11)	CS	Cronin, Marshall, Caldwell & Shashar	1994c
Lysiosquilla maculata	330	MSP	Cronin, Marshall, Quinn, & King	1994d
Lysiosquilla sulcata	397–538 (11)	MSP	Cronin, Marshall, Caldwell & Shashar	1994c
	410–620 (11)	CS	Cronin, Marshall, Caldwell & Shashar	1994c

TABLE 18.1. *Continued*

Species	λ_{max} (MSP) S-max /nm	Method of study	Author(s)	Year
Superorder Eucarida				
Order Euphausiacea				
Euphausia pacifica	462	EX	Kampa	1955
	483	MSP	Widder, Hiller-Adams & Case	1987
Euphausia superba	462	EX	Denys	1982
	485	EX	Denys & Brown	1982
	487	ERG	Frank and Widder	1999
Meganyctiphanes norvegica	460–465	EX	Fisher & Goldie	1959
	460–515, 490	ERG	Boden, Kampa & Abbott	1961
	488	MSP	Denys & Brown	1982
	490	ERG	Frank and Widder	1999
Nematobrachion sexpinosus	478	ERG	Frank and Widder	1999
Nematobrachion boopis	487	ERG	Frank and Widder	1999
Nematoscelis megalops	465	EX	Fisher & Goldie	1961
	490	ERG	Frank	unpubl.
Stylocheiron maximum	470	EX	Fisher & Goldie	1961
	479	ERG	Frank and Widder	1999
Thysanoessa raschii	460–465	EX	Fisher & Goldie	1961
Thysanopoda acutifrons	480	EX	Fisher & Goldie	1961
Thysanopoda orientalis	478	ERG	Frank and Widder	1999
Suborder Pleocyemata				
Order Decapoda				
Suborder Dendrobranchiata				
Penaeus duorarum	516	EX	Fernandez	1965
Family Oplophoridae				
Acanthephyra curtirostris	510	ERG	Frank & Case	1988a
	485	MSP	Hiller-Adams, Widder & Case	1988
Acanthephyra smithi	510	ERG	Frank & Case	1988a
	491	MSP	Hiller-Adams, Widder & Case	1988
Janicella spinacauda	400 & 500	ERG	Frank & Case	1988a
Notostomus gibbosus	490	ERG	Frank & Case	1988a
Notostomus elegans	490	ERG	Frank & Case	1988a
Oplophorus gracilirostris	400 & 500	ERG	Frank & Case	1988a
Oplophorus spinosus	400 & 500	ERG	Frank & Case	1988a
Systellaspis debilis	400 & 500	ERG	Frank & Case	1988a
	493	MSP	Hiller-Adams, Widder & Case	1988
	410 & 498	MSP	Cronin & Frank	1996
Family Pasiphaedae				
Paciphaea multidentata	497	ERG	Frank and Widder	1999
Family Penaeidae				
Funchalia villosa	489	ERG	Frank and Widder	1999
Faimly Sergestidae				
Sergestes tenuiremis	495	MSP	Hiller-Adams, Widder & Case	1988
Sergestes similis	495	MSP	Lindsay et al	1999
Sergestes arcticus	495	ERG	Frank and Widder	1999
Sergestes corniculum	500	ERG	Frank and Widder	1999
Sergia grandis	500	ERG	Frank and Widder	1999
Suborder Pleocyemata				
Infraorder Caridea				
Family Palaemonidae				
Palaemonetes palludosus	539	EX	Fernandez	1965
Palaemonetes vulgaris	390 & 540	ERG	Wald & Seldin	1968
Family Bresiliidae				
Rimicaris exoculata	500	EX	vanDover, Szuts, Chamberlain, Cann	1989

TABLE 18.1. *Continued*

Species	λ_{max} (MSP) S-max /nm	Method of study	Author(s)	Year
Infraorder Astacidea				
Astacus fluviatilis	530	MSP	Hamacher & Kohl	1981
Astacus leptodactylus	530	MSP	Hamacher & Stieve	1984
Homarus americanus	515	EX	Wald & Hubbard	1957
	515	MSP	Bruno, Barnes & Goldsmith	1977
Orconectes rusticus	530	MSP	Goldsmith	1978
	535	MSP	Cronin & Goldsmith	1982
Procambarus clarkii	640	IC	Nosaki	1969
	530	MSP	Cronin & Goldsmith	1982
	567 (A2)	MSP	Zeiger & Goldsmith	1989
	560,600,640	MSP	Cronin & Hariyama	2002
Procambarus milleri	440 & 530	IC	Cummins & Goldsmith	1981
Camberellus schufeldtii	522	MSP	Crandall and Cronin	1995
Cambarus ludovicianus	526	MSP	Crandall and Cronin	1995
Engaeus cunicularius	529	MSP	Crandall and Cronin	1995
Nephrops norvegicus	522	MSP	Crandall and Cronin	1995
Homarus gammarus	496	MSP	Marshall, Kent & Cronin	1999b
Infraorder Palinura				
Panulirus argus	504	EX	Fernandez	1965
Infraorder Anomura				
Pleuroncodes planipes	503	EX	Fernandez	1973
	523	IC	Fernandez	1973
Superfamily Paguridea				
Family Diogenidae				
Clibanarius vittatus	510	MSP	Cronin & Forward	1988
Dardanus fucosus	511	MSP	Cronin & Forward	1988
Petrochirus diogenes	508	MSP	Cronin & Forward	1988
Family Coenobitidae				
Coenobita clypeatus	508	MSP	Cronin & Forward	1988
Coenobita rugosa	491	MSP	Cronin & Forward	1988
Family Paguridae				
Pagurus annulipes	495	MSP	Cronin & Forward	1988
Pagurus longicarpus	515	MSP	Cronin & Forward	1988
Pagurus pollicaris	515	MSP	Cronin & Forward	1988
Superfamily Galatheoidea				
Family Porcellanidae				
Polyonyx gibbesi	—	MSP	Cronin & Forward	1988
Superfamily Hippoidea				
Family Hippoidae				
Emerita talpoida	—	MSP	Cronin & Forward	1988
Infraorder Brachyura				
Hemigrapsus edwardsii	513	EX	Briggs	1961
Leptograpsus variegtus	513	EX	Briggs	1961
	484	IC	Stowe	1980
Section Oxystomata				
Family Calappidae				
Calappa flammea	486	MSP	Cronin & Forward	1988
Hepatus epheliticus	487	MSP	Cronin & Forward	1988
Section Oxyrhyncha				
Family Majidae				
Libinia dubia	489	MSP	Cronin & Forward	1988
Libinia emarginata	493	MSP	Briggs	1961
	493	MSP	Hays & Goldsmith	1969

TABLE 18.1. *Continued*

Species	λ_{max} (MSP) S-max /nm	Method of study	Author(s)	Year
Section Cancridea				
Cancer irroratus	496	MSP	Cronin & Forward	1988
Pandalus borealis (nauplius)	475–500	ERG	Eaton and Brown	1970
Section Brachyrhyncha				
Family Portunidae				
Arenaeus cribrarius	498	MSP	Cronin & Forward	1988
Callinectes ornatus	501	MSP	Cronin & Forward	1988
Callinectes sapidus	477	EX	Bruno & Goldsmith	1973
	440 & 508	IC	Martin & Mote	1982
Carcinus maenas	503	MSP	Cronin & Forward	1988
	508	B	Horridge	1967
	508	MSP	Bruno, Mote & Goldsmith	1973
Ovipales stephensoni	440 & 508	IC	Martin & Mote	1982
Portunus spinimanus	505	MSP	Cronin & Forward	1988
	483	MSP	Cronin & Forward	1988
Family Xanthidae				
Eurypanopeous depressus	490	MSP	Cronin & Forward	1988
Menippe mercenaria	494	MSP	Cronin & Forward	1988
Panopeus herbstii	493	MSP	Cronin & Forward	1988
Panopeous obesus	493	MSP	Cronin & Forward	1988
Pilumnus sayi	489	MSP	Cronin & Forward	1988
Rhithropanopeus harrisii	495	MSP	Cronin & Forward	1988
Family Geryonidae				
Geryon quinquendens	473	MSP	Cronin & Forward	1988
Family Grapsidae				
Sesarma cinereum	492	MSP	Cronin & Forward	1988
Sesarma reticulatum	508	IC	Scott & Mote	1973
	493	MSP	Cronin & Forward	1988
Family Ocypodidae				
Uca pugilator	508	IC	Scott & Mote	1973
Uca pugnax	508	IC	Scott & Mote	1973
Family Gecarcinidae				
Gecarcinus lateralis	510	ERG	Lall & Cronin	1987
	487	MSP	Cronin & Forward	1988
Superorder Peracaridia				
Order Mysida				
Gnathophausia ingens	490 & 520	ERG	Frank & Case	1988b
Order Isopoda				
Ligia	330,460 & 520	ERG	Hariyama, Tsukahara & Meyer-Rochow	1993
Order Amphipoda				
Suborder Hyperiidae				
Phronima sedentaria	470	ERG	Frank and Widder	1999
	480	MSP	Cronin and Marshall	Unpubl
Recent Reviews			Marshall, Kent Cronin	1999a
			Cronin & Hariyama	2002
			Cronin & Marshall	In Press

λ_{max} from MSP and S-max from a variety of methods. For the stomatopoda, where known, both l-max and S-max calculated from λ_{max} and intrarhabdomal filtering are given. Numbers in brackets are total numbers of visual pigments. Abbreviations as follows: MSP—microspectrophotometry, EX—visual pigment extract, OP—optical physiology, ERG—electroretinogram, EON—extracellular/optic nerve, IC—intracellular electrophysiology, B—behavioral, BP—behavioral/phototaxis, CS—calculated sensitivity.

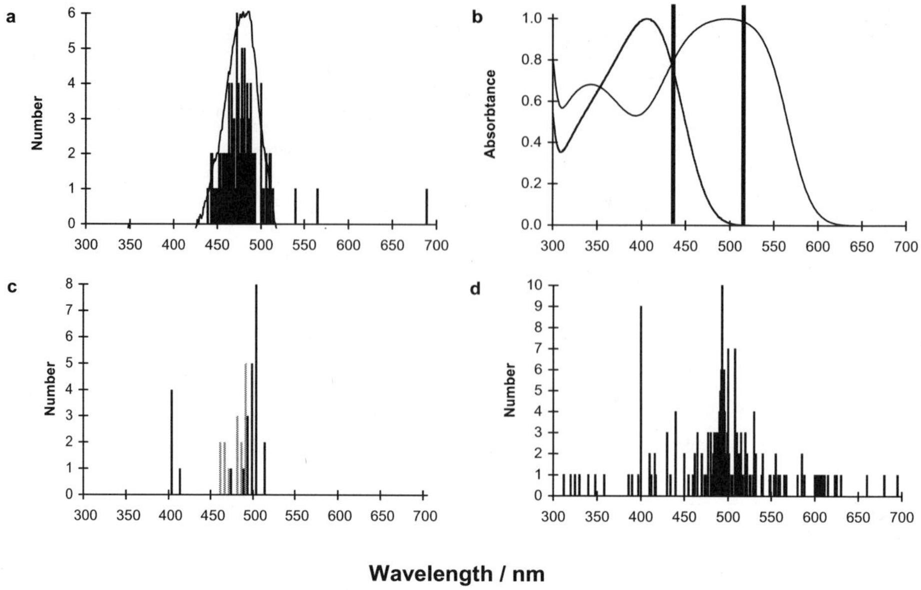

FIGURE 18.2. Spectral sensitivities in crustaceans and the light they may try to capture. (a) The peak wavelengths of bioluminescent emissions plotted as a histogram. (Data from Latz et al., 1988.) Also shown is irradiance at 335m in clear ocean water near Cuba (solid line—normalized to 6 to fit existing axis). (Widder and Frank, unpublished.) (b) Spectral sensitivities of the midwater oplophorid shrimp *Systellaspis debilis*. Curves are calculated absorptance and therefore take into account photo pigment density and photoreceptor length. (Marshall et al., 1999b.) The vertical bars encompass the spectral region in which most bioluminescent emissions are found (see a). (c) Peak sensitivities of photoreceptors in euphausiids (gray) compared to oplophorids and sergestids (black). (d) Spectral sensitivity maxima of all crustaceans studied thus far. Most of the apparent diversity is from stomatopods. (Data summarized in Marshall et al., 1999b.)

chromophore embedded in the center of a 7-helix transmembrane protein. Changes in either of these two subunits can affect spectral tuning (λ_{max}) and a brief examination of this with respect to chromophore changes is included in Section 3.4. Changes in opsin amino acid sequence have been linked with λ_{max} for some time (e.g., Nathans, 1989); however, little molecular work has been attempted with crustaceans (Sacamoto et al., 1996). Interestingly, in common with some vertebrates, the crab *Hemigrapsus sanguinensis* shows multiple expression of opsin types in one cell type (each of the R1–7 cells; Sacamoto et al., 1996; Shand et al., 2001; Chapter 8). Until more complete genomic libraries and expression systems are developed, the significance of this, and the other revelations this powerful technique is likely to bring, remain in the background for crustaceans. Especially in light of the "natural molecular laboratory" offered by the 16 spectral sensitivities of stomatopods, each possibly based on different opsins (Section 3.5), this is an area in desperate need of further research.

3. Crustaceans' Spectral Sensitivity

Before we examine the array of spectral sensitivities crustaceans exhibit and the evolutionary pressures, environmental and otherwise, which mold them, some discussion of the photoreceptor types in crustacean eyes is worthwhile. Many crustacean compound eyes, including those in most decapod, mysid, and stomatopod species, contain two anatomically distinct photoreceptor classes. Seven retinular cells (therefore called R1–7), construct the main portion of the rhabdom, and a single cell, R8, forms a smaller, usually distally placed part (e.g., Eguchi and Waterman, 1967; Marshall et al., 1991a,b). Both R8 and R1–7 cells contribute microvilli, or membrane tubes, to the rhabdom, and it is within these that the light-absorbing visual pigment molecule is located. All R8 cells known contain relatively short-wavelength, UV- or violet-sensitive visual pigment and R1–7 cells generally contain visual pigments maximally sensitive in the blue/green region of the spectrum. This pattern is found in many decapods including crabs (Martin and Mote, 1982; Cronin and Forward, 1988; Forward et al., 1988), crayfish (Cummins and Goldsmith, 1981), lobsters (Cummins et al., 1984), and decapod shrimps (Frank and Case 1988a; Frank and Widder, 1994a,b; Cronin and Frank, 1996).

R8 cells in some of the deep-sea decapods and other crustaceans are lost or greatly reduced, as the increasingly blue light from the surface or bioluminescent sources becomes the only light available, resulting in these species having monochromatic visual systems (Table 18.1; Fig. 18.2). In addition, R8 cells are small, and many studies including MSP and electrophysiology have probably missed their contribution to the retina, and erroneously concluded that these species have monochromatic visual systems. The crabs and crayfish are good examples, as almost all examined anatomically so far possess R8 cells (Eguchi and Waterman, 1967; Eguchi, 1973; Stowe et al., 1986; Marshall, unpublished; and see brief review in Marshall et al., 1991). The intracellular study of Martin and Mote (1982) found functional R8 cells in two species of crab, *Ovipales stephensoni* and *Callinectes sapidus*, and another intracellular investigation (Cummins and Goldsmith, 1981) revealed its presence in the crayfish *Procambarus milleri* (Table 18.1). Functional ultraviolet (UV)/violet-sensitive R8 cells are likely to be present in many crabs and crayfish and this is not obvious from Table 18.1.

Where examined, the nondecapod crustaceans with the typical R8/R1–7 cell arrangement, such as some squilloid stomatopods, sometimes possess this basic UV-violet + blue-green spectral sensitivity pattern (Cronin, 1985; J. Marshall and T.W. Cronin, unpublished). Seemingly dependent on their habitat, other squilloids reduce or lose their R8 cells (Schiff, 1963; Cronin, 1985). In the gonodactyloid and lysiosquilloid stomatopods, the basic theme is still present but has been greatly expanded upon (Section 3.4). ERG has demonstrated the presence of two spectral sensitivities, peaking at 490 nm and 520 nm, in the deep-sea mysid *Gnathophausia ingens* (Frank and Case, 1988b). Although at least some mysids do possess R8 cells (Hallberg, 1977), the precise anatomical positions of the two spectral mechanisms in *G. ingens*, along with reasons for their rather long-wavelength sensitivity for a deep-sea animal, are yet to be determined (Frank and Case, 1988b).

One other large crustacean order, which has previously received some attention, is the Euphausiacea. Where examined, these oceanic and often deep-living (mesopelagic) shrimp have only seven retinular cells, all with the one spectral sensitivity. It is unknown whether R8 cells in this group were secondarily lost. The flurry of other sensitivities found in *Meganyctiphanes norvegica* in the late 1950s and early 1960s (Table 18.1) is probably due to methodological problems (Frank and Widder, 1999). Isopods and amphipod eyes examined contain five or seven retinular cells, generally in a monomorphic population (Edwards, 1969; Ball, 1977; Hallberg et al., 1980; Hallberg and Nilsson, 1983).

3.1. Spectral Sensitivities in Different Habitats: Comparisons of Freshwater, Shoreline, and Open Ocean

Where spectral sensitivities lie in clusters (Fig. 18.2), functional reasons are often sought as an explanation. Ecological explanations behind λ_{max} positions in several animals, including crustaceans, center around optimization of the photoreceptors for maximum sensitivity according to habitat light regime (Bayliss et al., 1936;

TABLE 18.2. A phylogenetic analysis of current knowledge of crustacean spectral sensitivities.

Class	Subclass	Superorder	Order	Suborder	Infraorder	Spectral sensitivities	Number of spectral sensitivities	Example reference (See Table 18.1 for full listing)
Remipedia			Nectiopoda					
			Enantiopoda					
Cephalocardia								
Branchiopoda	Sarsostraca		Anostraca			*	1	Hertel (1972)
	Calmanostraca		Notostraca					
	Diplostraca		Cladocera			*	4	Smith and Macagno (1990)
			Conchostraca					
Maxillopoda	Ostracoda	Myodocopa						
		Podacopa						
		Palaeocopa						
	Mystacocarida							
	Copepoda		Calanoida			*	1	Stearns and Forward (1984)
			Harpacticoida					
			Cyclopoida					
			Poecilostomatoida					
			Siphonostomatoida					
			Monstrilloida					
			Misophirioda					
			Mormonilloida					
	Branchiura							
	Cirripedia		Thoracica			*	1	Hillman et al. (1973)
			Ascothoracia					
			Acrothoracica					
			Rhizocephala					
	Tantulocarida							
Malacostraca	Phyllocarida		Leptostraca					
	Eumalacostraca	Hoploarida	Stomatopoda			**	1,2,16	Cronin and Marshall (1989a,b)
		Syncarida	Bathynellacea					
			Anaspidacea					

Eucarida	Euphausiacea			**	1,(3)	Boden et al. (1961), Frank and Widder (1999)
	Amphionidacea					
	Decapoda	Dendrobranchiata		*	1	Fernandez (1965)
		Pleocyemata	Caridea	**	1,2	Frank and Case (1988a), Marshall et al. (1999b)
			Stenopodidea			
			Thalassinidea			
			Astacidea	*	1,2,4	Cummins and Goldsmith (1981)
			Palinura	*	2	Cummins et al. (1984)
			Anomura	**	1	Cronin and Forward (1988)
			Brachyura	**	1,2	Cronin and Forward (1988)
Peracarida	Mysida			*	2	Frank and Case (1988a), Marshall et al. (1999b)
	Lophogastrida					
	Cumacea					
	Tanaidacea					
	Mictacea					
	Spelaeogriphacea					
	Thermosbaenacea					
	Isopoda	Gnathiidea				
		Anthuridea				
		Flabellifera				
		Oniscidea		*	3	Hariyama et al. (1993)
		Valvifera				
		Phreatoicidea				
		Asellota				
		Epicaridea				
		Calabozoidea				
	Amphipoda	Gammaridea				
		Hyperiidea		*	1	Frank and Widder (1999)
		Caprellidea				
		Ingofiellidea				

*Species examined.
**More than 10 species examined.
Numbers in brackets indicate doubtful results (Frank and Widder, 1999).
Crustacean groups down to infraorder are tabulated and, where known, spectral sensitivities scored. Knowledge exists of spectral sensitivities in only 9 of the existing 36 orders and often these data are from one species only. Best known are the Decapoda, Stomatopoda, and Euphausiacea.

Clarke, 1936; Lythgoe, 1979; Cronin and Forward, 1988; Forward et al., 1988; Marshall et al., 1999b). Light penetrating clear, deep ocean waters is maximally bright, close to 475 nm (Figs. 18.2 and 18.3; Jerlov, 1976; Frank and Widder, 1996). This blue-water environment generally shifts to green closer to the coast, and in fresh waters, yellow light may penetrate further than other wavelengths (Fig. 18.3; Jerlov, 1976; Lythgoe, 1979). As with fishes (Lythgoe, 1979), freshwater crustaceans generally possess longer-wavelength sensitivity than their marine counterparts (Fig. 18.3; Table 18.1). Crustacean R1–7 cell sensitivities in the marine environment lie mostly between 460 and 525 nm, while the range in freshwater crustaceans is 510–567 nm but may go up to 640 nm in winter crayfish (Table 18.1; Fig. 18.3; and see Section 3.3).

Optimization, or at least correlation of spectral sensitivity to habitat, is found within the marine habitat. Crabs are often intertidal coast dwellers or reef dwellers, experiencing the greener water of these habitats (Jerlov, 1976). Their R1–7 spectral sensitivities, having maxima between 473 and 523 nm (average 499 nm, Table 18.1; Figs. 18.2 and 18.3), lie toward the long-wavelength side of those known in crustaceans. The euphausiid shrimps (krill), on the other hand, are open-water, often deep-living crustaceans and their sensitivities peak between 460 and 490 nm (average 475 nm; Table

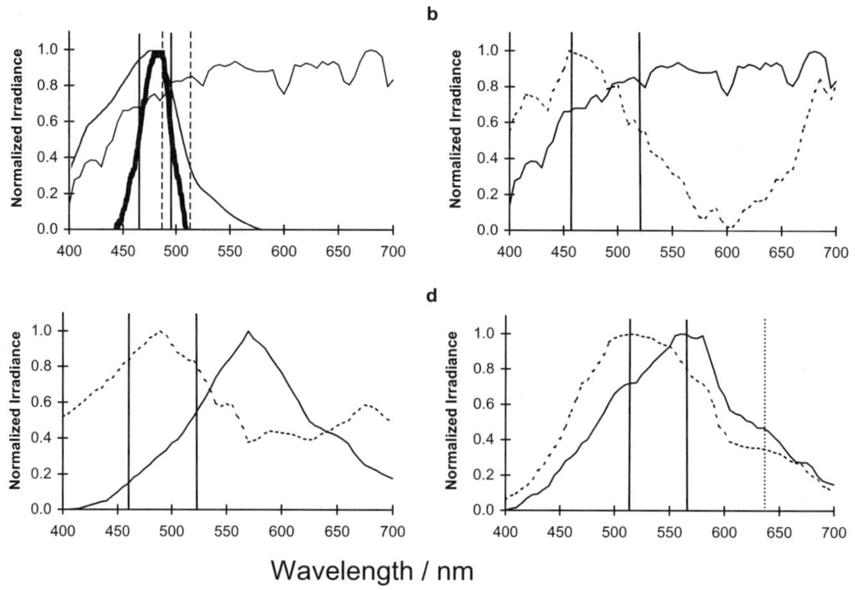

FIGURE 18.3. Light environments and spectral sensitivity ranges of crustaceans in different habitats. (a) Normalized irradiance at 3 m, 60 m, and 335 m (progressively thicker lines) in clear oceanic water. (Deep-water data from the ocean close to Cuba, Widder and Frank, unpublished; 3 m data from Munz and McFarland, 1973.) Solid vertical lines encompass spectral sensitivity maxima of euphausiids; dotted vertical lines encompass spectral sensitivity maxima of oplophorid and sergestid shrimps. (b) Irradiance at 3 m in clear oceanic reef water at noon (solid line) and after sunset (dotted line). (Data from Eniwetok atoll in the Pacific ocean—Munz and McFarland, 1973.) Solid vertical lines encompass all R1–7 cell sensitivities in most monochromatic and dichromatic marine crustaceans. (c) Light at noon (solid line) and just after dark (dotted line) in an estuary. Crab R1–7 spectral sensitivities lie between the vertical lines. (Light data from Munz and McFarland, 1977) (d) Freshwater crustacean light environment (Lake Michigan, from Forward et al., 1988) at noon (solid) and dusk (dotted) and spectral sensitivity range of most R1–7 cells (between vertical solid lines). The vertical dotted line marks the peak sensitivity in winter Japanese crayfish *Procambarus clarkii*.

18.1; Figs. 18.2 and 18.3), closely matching the blue peak transmission of such waters. The non-euphausiid ocean dwellers seem to be exceptions to the "sensitivity rule" for reasons discussed in Section 3.2.

The R1–7 photoreceptors of shallow-living marine crustaceans are perhaps best adapted to the crepuscular light environment (Fig. 18.3; Marshall et al., 1999b) rather than the daytime spectral world of their respective habitats. This is certainly true for coastal crustaceans, such as crabs (Fig. 18.2; Cronin and Forward, 1988), whose spectral sensitivities are better matched to the spectrum of light reaching them at dawn and dusk (Fig. 18.3). This trend, an extension of the sensitivity hypothesis, is also apparent in marine fishes (McFarland and Munz, 1975; McFarland, 1991), and it may be that for many shallow-living marine animals, it is most critical to have good vision during this time of day. On coral reefs it is a period of intense predation (McFarland, 1991), and a variety of crustaceans become more active during these periods (Forward et al., 1988). The day and night light environments in freshwater are more similar to each other than in shallow marine habitats (Fig. 18.3). In freshwater the overriding factor in light quality is the organic and suspended particle content of the water (Jerlov, 1976).

Shallow-water crustaceans, in any habitat other than the most turbid waters, have broad-spectrum light available for vision (Fig. 18.3). In theory, they could place spectral sensitivities almost anywhere and indeed, one group, the stomatopods, have literally done this (Section 3.4). Aside from this exception, however, R1–7 cell sensitivities peak between 460 and 525 nm, a spectral window containing almost 45% of all light at dawn and dusk. It seems that being closely tuned to the light experienced during crepuscular periods is of utmost importance to many crustaceans (Fig. 18.3).

The reason why R8 cell sensitivities are clustered around 400 nm is obscure. For crustaceans in shallow environments, there are certainly plenty of photons available in this spectral region and, in fact, even in clear oceanic waters there may be enough UV photons for effective vision several hundred meters down (Frank and Widder, 1994a,b, 1996; Section 3.2). One advantage of two spectrally distinct sensitivities is that they can form the basis of a color vision system. However, good evidence for this has yet to be conclusively shown in any dichromatic crustacean. Initial suggestions of the existence of color vision in fiddler crabs have not been satisfactorily replicated (Hyatt, 1975). As we shall see in Section 3.4, color vision has been mastered in one crustacean group, the stomatopods.

Finally in this section, the littoral isopod *Ligia exotica* deserves a mention, for this woodlouse-like crustacean is trichromatic, with sensitivities at 330, 460, and 520 nm (Hariyama et al., 1993). *L. exotica* certainly has broad-spectrum light in its habitat, but reasons for its spectral diversity in behavioral or ecological terms remain a tantalizing mystery.

3.2. Spectral Sensitivities in the Open Ocean and Deep-Sea

As we have just seen, the euphausiids seem to have adapted to their blue-water world by maximizing the sensitivity of their eyes to this environment. The spectral sensitivity range of their visual systems is also a good match to bioluminescent sources of light (Fig. 18.2). As shown a number of years ago, the most effective spectral sensitivity position for detecting small spots of light may not be an exact match to the light emitted, but this is critically dependent on background contrast (Lythgoe, 1966, 1968; Loew and Lythgoe, 1978). A spectral sensitivity matched to downwelling light is good for spotting dark targets against the lighter background, such as, for example, looking up from 400 m during the day. A spectral sensitivity offset from the maximum irradiance is more useful for the detection of light objects against a darker background (Lythgoe, 1966). The target could be a bioluminescent flash, which exceeds the intensity of the downwelling light. That many euphausiids seem to favor the former strategy may suggest that, for them at least, detecting small silhouettes overhead, whether they be predator, prey, or conspecifics, is important. Alternatively, at night, when the background has no photons in it, bioluminescence will always be best detected by a spectral sensitivity matched to it (Land, 1981a; Warrant, 1999; and see Chapter 16). It also seems possible that euphausiid vision may be good for this task; however, other than knowing that they undergo diurnal vertical migration in huge swarms, more needs to be known about euphausiid biology before answering these questions (Frank and Widder, 1999).

It is interesting that one other group of mesopelagic crustaceans, the hyperiid amphipods, also match their spectral sensitivity to the downwelling light or bioluminescence (Table 18.1; Frank and Widder, 1999; T.W. Cronin and J. Marshall, unpublished). Intriguingly, both euphausiids and hyperiids often have split or double eyes with one-half looking up and the other half down (Land et al., 1979; Land, 1981). It would make a nice story if upper and lower lobes in the eyes of either group had different spectral sensitivities, one for examining bright bioluminescent sources against the dim depths and one for spotting dark spots against downwelling light. All indications so far, however, are that, while the optics are certainly different (Land et al., 1979; Land, 1981), the spectral sensitivities are essentially identical (Frank and Widder, 1999; T.W. Cronin, unpublished). In fact, the spectral distribution of light in both upwelling and downwelling light in the lower part of the mesopelagic realm, where many of these animals live, is very similar (Jerlov, 1976; Denton, 1990), so, in this sense, similar spectral sensitivity over the whole eye, despite its optical subdivisions, is not a surprise.

Several of the other deep-sea crustaceans known are likely dichromats. This includes the oplophorids and mysids, while the sergestids seem anatomically more like the euphausiids in that no functional R8 system has been found. *G. ingens* is the only mysid characterized and, being the only representative of its order, and with fascinatingly orange sensitivity, it and other mysids must remain a wonderful project for the future. With λ_{max} ranging between 490 and 520 nm (average 497 nm), R1–7 cell sensitivity in both the oplophorids and sergestids is longer than would be expected if they were trying to catch all photons from either downwelling light or bioluminescence.

Four possible explanations for this are as follows.

1. R1–7 sensitivity in oplophorids and sergestids is offset from the downwelling light maximum and bioluminescent sources (Fig. 18.3) in order to best *detect* these bright sources against the relatively dimmer background (see discussion above in relation to euphausiids and Lythgoe, 1966).
2. Many mesopelagic crustaceans undergo diurnal vertical migrations of several hundred meters. During these movements, the shrimps face the problem of knowing their depth accurately, and these migrations are critical both for finding food and avoiding predators (Frank and Widder, 1994a,b). One suggested way of doing this visually is by ratiometric comparison of light at different wavelengths, thus overcoming the problem of varying light due to both depth and daytime differences (Frank and Case, 1988a,b). While this may work in the top few hundred meters, below the euphotic zone, there is little change in spectral shape with time or depth (Frank and Widder, 1996). In this oceanic region, therefore, this hypothesis is replaced with the idea that the dichromatic system provides a method for discriminating bioluminescent spectra, as explained in item 3 below (Cronin and Frank, 1996).
3. The R1–7 spectral sensitivity (average 497 nm) and R8 cell sensitivity (usually close to 400 nm) of dichromatic mesopelagic crustaceans is ideal for *discrimination* of bioluminescent sources (Fig. 18.2). Most bioluminescent sources known have peak emissions between 440 nm and 515 nm (Latz et al., 1988). For *Systellaspis debilis*, a mesopelagic dichromat, this represents the region of maximum signal difference between R8 and R1–7 cells (Fig. 18.2). As a result, different bioluminescent emissions or indeed spectral differences between bioluminescence and background will be encoded by large R8/R1–7 cell signal differences. It is not known if such signal comparisons are made by

any deep-sea crustacean or indeed by any malacostracan. However, as bioluminescence is perhaps "the major visual stimulus in the deep-sea" (Frank and Widder, 1999), it is an attractive speculation that the spectral sensitivities of R8 and R1–7 cells have evolved in positions providing maximum interpretation of their meaning (Cronin and Frank, 1996).

4. The exact position of λ_{max} does not matter in deep-sea crustaceans, within limits, as the sensitivity of their R1–7 cells, set by the large dimensions and high visual pigment density, is such that they have effectively equal sensitivity over a broad range (Crescitelli et al., 1985; Hiller-Adams et al., 1988). The R1–7 cell sensitivity for *S. debilis* plotted in Figure 18.2, for example, is calculated according to these parameters (Marshall et al., 1999b) and is *almost* equally sensitive from 470 to 520 nm. Whether *almost* is good enough in the photon-starved depths is by no means certain.

An intriguing idea for the relatively long-wavelength sensitivity in the vent shrimp *Rimicaris exoculata*, and its close relatives, is worth mentioning. It has been suggested that these animals are straining to see light produced from the heat of the deep-water hot vents they inhabit rather than bioluminescence (most vents are below the photic zone—van Dover et al., 1989; Pelli and Chamberlain, 1989). While there may be just enough sensitivity in their 500 nm visual pigment to see the light emitted from the superheated water, this idea has yet to be satisfactorily proven (Land, 1989).

3.3. Freshwater Reexamined: Season, Temperature, and Color Dances

Two unrelated oddities are described here, the tetrachromatic water-flea and seasonal changes in spectral sensitivity in crayfish.

As described above, all visual pigments are constructed from two parts, a chromophore and surrounding opsin or protein part. In crayfish, two different chromophores have been found, retinal (producing A_1 visual pigment) and 3-dehydroretinal (producing A_2 visual pigment). A_2 based visual pigments absorb at longer wavelengths than A_1 equivalents, were first described in the invertebrates in crayfish (Suzuki et al., 1984), and are known in many freshwater vertebrates (Bridges, 1972; Knowles and Dartnall, 1977). A_2 based visual pigments are common in freshwater, evidently to shift spectral sensitivities toward the longer wavelengths of light that characterize such habitats (Lythgoe, 1979). Aquatic animals that migrate from freshwater to seawater and back (such as salmonids or anguilliforms), often switch between the A_1/A_2 systems for this reason (Lythgoe, 1979). Combinations of A_1 and A_2 are known within the same cells of vertebrates and indeed crustaceans (Sakamoto et al., 1996).

The crayfish, *Procambarus clarkii*, changes the 3-dehydroretinal content of its retina seasonally, apparently in response to both light environment and temperature and this has resulted in the variety of spectral sensitivities previously reported (Table 18.1). Recently, Cronin and Hariyama (2002) reported that summer crayfish eyes only contain A_1-based, visual pigment and a single R1–7 spectral sensitivity maximum at 600 nm. Winter crayfish eyes contain both A_1- and A_2-based visual pigments and four different spectral sensitivities at 560 nm (narrow), 600 nm (narrow), 600 nm (broad), and 640 nm (broad). The broad sensitivities are probably A_2-based and the narrow sensitivities A_1-based, and in order to explain the wintertime proliferation a second opsin must be expressed. Two opsins and two chromophore types will result in the four sensitivities seen.

Seasonal changes in mating behavior and water quality may underlie this change in the crayfish's visual world. Mating increases during the Japanese winter and the different spectral sensitivities may help discriminate between the differently colored males and females in muddy winter water (Cronin and Hariyama,

2002). Similar arguments are raised to explain seasonal variation in spectral sensitivity and color in stickleback (Cronly-Dillon and Sharma, 1968) and guppies (Houde and Endler, 1990).

The tiny freshwater water-flea *Daphnia magna* possesses small (22-ommatidia) compound eyes but has a surprising four spectral sensitivities located in its eight cell rhabdoms (Smith and Macagno, 1990). These peak at 348, 434, 525, and 608 nm, but instead of constructing a four-dimensional, tetrachromatic color space, it seems likely that each spectral sensitivity drives a certain behavior and these have been described as color dances by Smith and Baylor (1953). Blue light makes *D. magna* dance in an agitated manner and red light induces calm. How such behavior relates to cladoceran survival is unknown but such wavelength-specific behaviors are also described in lepidopteran insects (Scherer and Kolb, 1987; Kelber, 1996).

3.4. Polychromatic Vision: The Stomatopod Crustaceans

It is in the stomatopods (or mantis shrimps, a very ancient crustacean lineage) that spectral mechanisms reach their greatest complexity in crustaceans and perhaps in all animals. Some stomatopods in the superfamily Squilloidea are like most other crustaceans and have only one or two photoreceptor classes based on no more than a pair of visual pigment types (Cronin, 1985; Cronin et al., 1993). But species in the superfamilies Lysiosquilloidea and Gonodactyloidea express up to 16 different spectral types of photoreceptors in a single retina (Cronin and Marshall, 1989a,b; Cronin et al., 1993, 1994a,c, 1996; Marshall and Oberwinkler, 1999). Stomatopod vision is so different from that of other species that it will receive a rather extended discussion in this chapter.

3.4.1. Stomatopod Retinas, Visual Pigments, and Filter Pigments

To understand the nature of stomatopod vision, a brief discussion of their ocular anatomy is necessary, focusing on the eyes of lysiosquilloids and gonodactyloids (see Horridge, 1978; Manning et al., 1984; Schiff et al., 1986; Marshall et al., 1991a,b; Marshall and Land, 1993a,b; Cronin and Marshall, 2001). Stomatopod compound eyes are typically composed of three ommatidial regions. Two of these form the nearly symmetrical dorsal and ventral halves of the eye, while the third region separates them and forms an equatorial midband, usually with six parallel rows of ommatidia. The dorsal and ventral hemispheric ommatidial arrays resemble those of typical decapod crustaceans. Here, a main rhabdom, formed from a ring of seven retinular cells and containing a middle-wavelength visual pigment, is topped by an eighth retinular cell having an ultraviolet-sensitive rhodopsin (Cronin, 1994; Marshall and Oberwinkler, 1999). In any given species, visual pigments in all these ommatidia are identical, and thus provide, at best, dichromatic vision specialized for polarizational (Marshall et al., 1991a) and spatial (Horridge, 1978; Marshall and Land, 1993a) analysis.

The complex spectral system is confined to the four most dorsal ommatidial rows of the six-row midband. Here, rhabdoms are divided into a series of three to five tiers, depending on the particular retinal location and species being considered (Marshall, 1988; Cronin and Marshall, 1989a,b, 2001, 2002; Marshall et al., 1991a,b). In all lysiosquilloids and gonodactyloids, rhabdoms in Row 1 (the most dorsal) and Row 4 have three tiers: the R8 rhabdomere at the top followed by a main rhabdom that is subdivided into an upper tier formed from three retinular cells overlying a deeper one with rhabdomeres of the four remaining cells. Each tier of the main rhabdom contains a different visual pigment based on a retinal$_1$ chromophore (Goldsmith and Cronin, 1993). The pigment of the distal tier absorbs at shorter wavelengths than that of the proximal tier by about 25 to 50 nm (Cronin and Marshall, 1989a,b, 2001, 2002; Cronin et al., 1993, 1994a, 1996), thereby partially blocking its normal short-wavelength absorption and sharpening its spectral tuning (see Cronin, 1994). This effect can be seen in Figure 18.4 by comparing the panels labeled "Visual Pigments" with the corresponding panels for "Spectral Sensitivities"—the sensitivities are more sharply tuned and are slightly more separated than the absorption spectra of the visual pigments alone in Rows 1 and 4.

The other two "chromatic" midband rows take tuning and filtering further (see Cronin and Marshall, 2001). In Rows 2 and 3, the junctions between the rhabdomere of R8 and the topmost tier of the main rhabdom are separated by a section of photostable filter material, generally containing a carotenoid pigment that transmits medium to long wavelengths (Figs. 18.4 and 18.5). By removing light that would normally stimulate the main absorption bands of underlying visual pigments, these unique filters limit the photoreceptor's sensitivity to medium to long wavelengths (Fig. 18.1). In many species (and almost all gonodactyloids) a second filter, transmitting at yet-longer wavelengths, is located between the distal and proximal tiers of the main rhabdom, further tuning spectral sensitivity functions of the most proximal receptor tier. Due to all this filtering, photoreceptors in Rows 2 and 3 of the midband are spectrally specialized to detect light of longer wavelengths (Fig. 18.4), sometimes out to nearly 700 nm.

Together, main rhabdoms of the dorsal four midband rows smother the spectrum from near 400 nm to about 600 or 700 nm with eight narrow, closely spaced receptor classes. Each row specializes in the analysis of a particular region within this range. The pattern illustrated in Figure 18.4 is found in all 16 stomatopod species with six ommatidial rows in the midband region that have been described. In order of increasing wavelength, spectral coverage of each row is as follows: Row 1, Row 4, Row 2, and Row 3. The remaining two rows, the most ventral, are specialized not for spectral analysis of light but for its polarizational analysis (Marshall, 1988; Marshall et al., 1991a, 1999a,b).

3.4.2. *Color Vision in Stomatopods*

With their complex retinas, visual pigment arrays, and diversity of spectral sensitivity mechanisms, it is no surprise that the stomatopods have true color vision; they can visually discriminate among objects solely on the basis of spectral reflectance. This ability was first suspected by Caldwell and Dingle (1975), who noted that body colors associated with visual displays tend to be more intense in the most aggressive species. Subsequently, Hazlett (1979) found that mantis shrimps evaluate colors of the meral spot, a prominent signal mark displayed during aggressive encounters. Most recently, Marshall et al. (1996) trained stomatopods to discriminate stimuli based on their spectral reflectances, a conclusive demonstration that the animals possess color vision. Mantis shrimps themselves are beautifully colored, and it is possible that some of their signaling colors are spectrally shaped for easy discrimination by other members of their species (Chiao et al., 2000).

3.4.3. *Tuning of Color Vision*

Stomatopods are unique among animals (not to mention crustaceans) in producing narrowly tuned photoreceptor classes by means of serially arranged, photostable filters in a single photoreceptor unit (Cronin et al., 1994a,d; Douglas and Marshall, 1999; Cronin and Marshall, 2001). The numbers and spectral positions of filter classes found in a single retina vary among species and habitats (Fig. 18.5). Lysiosquilloids (e.g., *Lysiosquilla maculata*, Fig. 18.5, top left) typically have fewer classes (two or three) than do the gonodacytyloids, most of which possess four filter types per retina, two in Row 2 and two in Row 3 (remaining panels of Fig. 18.5).

The benefit of narrow tuning is easily realized in shallow marine waters, where incident light is bright and spectrally broad. Here, the very high photon densities overcome the greatly diminished sensitivity that results from all the filtering (Cronin et al., 1994a). This is not the case with species that live deeper in the water column, where absorption by the overlying water diminishes intensities at all wavelengths. Long wavelengths in particular, beyond 550 nm, are severely attenuated even in the clearest marine waters. Therefore, receptors

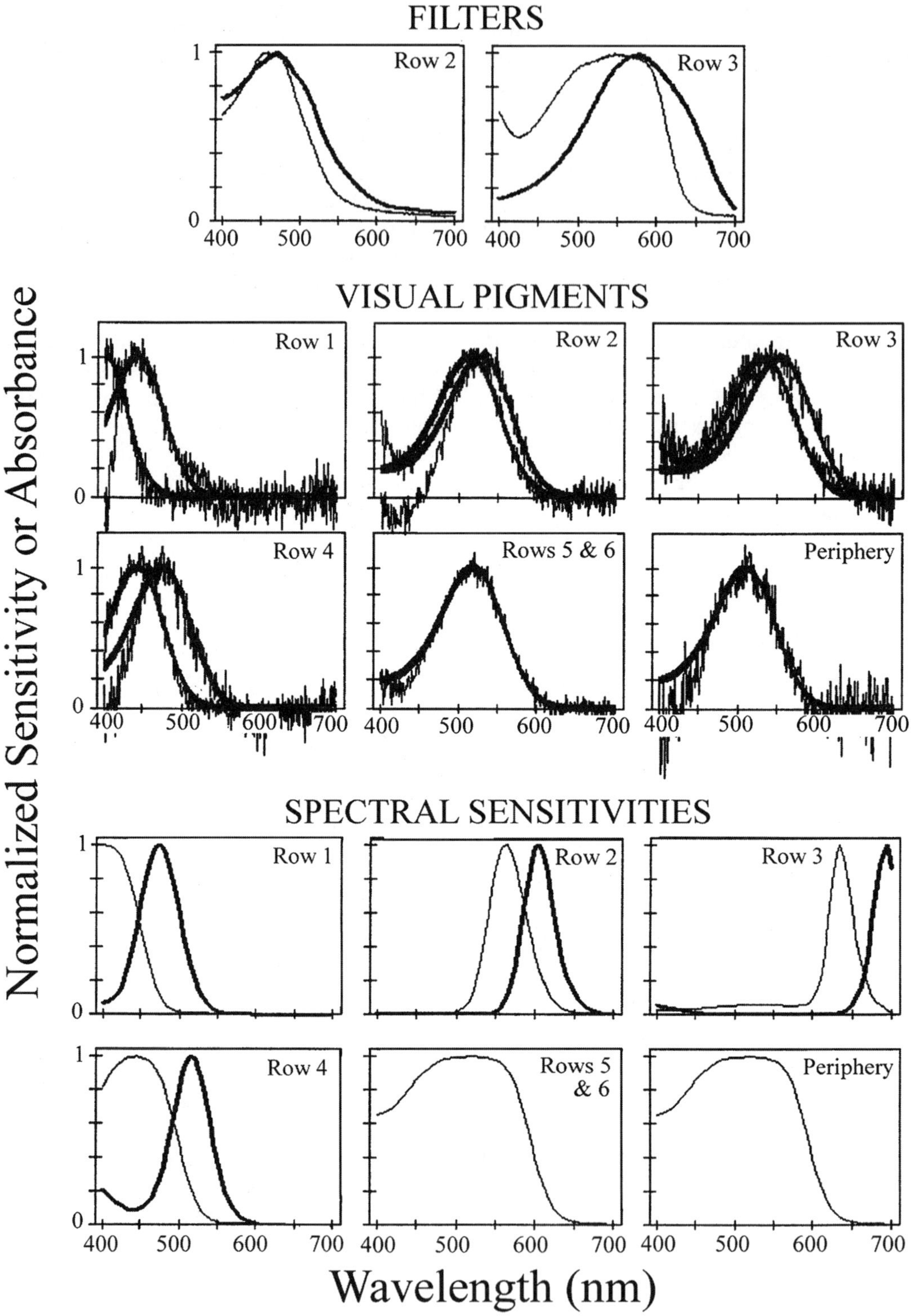
FILTERS
Row 2
Row 3
VISUAL PIGMENTS
Row 1
Row 2
Row 3
Row 4
Rows 5 & 6
Periphery
SPECTRAL SENSITIVITIES
Row 1
Row 2
Row 3
Row 4
Rows 5 & 6
Periphery
1
0
400
500
600
700
Normalized Sensitivity or Absorbance
Wavelength (nm)

like those of Row 3 in Figure 18.4 would probably be nonfunctional at depths exceeding a few meters.

Nevertheless, many stomatopod species live in deeper water, or are active at night. Their retinas include filter sets that are blue-shifted relative to those of shallower species (Cronin et al., 1994a). This shift in filtering greatly enhances sensitivities of the deeper receptors, because the longest-wavelength receptors (Figs. 18.4 and 18.6) are based on visual pigments that absorb maximally near 550 nm. The peak sensitivities at much longer wavelengths (>650 nm) therefore rely on the merest absorption by the long-wavelength tails of these pigments. Even a small shift to shorter wavelengths in the filter's transmission can produce a huge increase in photon capture, and thus absolute sensitivity, in the underlying photoreceptors.

Filters of *Gonodactyaceus mutatus* (Fig. 18.4) or *Gonodactylus smithii* (Fig. 18.5, middle left panel) typify those of shallow-water gonodactyloids, while those of *Pseudosquilla ciliata* or *Gonodactylellus affinis* (Fig. 18.5, top and middle right) characterize related species from deep water. Note that the filters of Row 3 in the species that live deeper are blue-shifted by 50 nm or more, compared to the corresponding classes in shallow-water retinas. This level of change in the transmission spectra of the filters profoundly affects spectral sensitivities of long-wavelength receptors.

Somewhat surprisingly, the visual pigments of deeper-living species do not show indications of tuning to the photic environment (Fig. 18.6, top). Within each retinal region, the spectral placement of visual pigment λ_{max} certainly varies, but the pattern of this change is not related in any simple way to depth. In the top part of Figure 18.6, λ_{max} positions of visual pigments in main rhabdoms of corresponding retinal regions of all lysiosquilloid and gonodactyloid species characterized to date (16 species) are plotted. The data have been placed into four classes according to the depths at which each species normally lives, from intertidal to deep. In the midband rows devoted to color vision (Rows 1–4), there is a general pattern of increasing λ_{max} in the order of Row 1, 4, 2, and 3, and the proximal tier's pigment visual pigment is always sensitive to longer wavelengths than the distal tier's. While λ_{max} may vary by 15 to 50 nm in any given region across species, there is no particular pattern of change with habitat depth; the ranges tend to overlap nearly completely in species from different habitats. For contrast, note that the visual pigments of the peripheral retina (i.e., the retinal hemispheres) do show the typical

FIGURE 18.4. Filter pigments, visual pigments, and spectral sensitivities of photoreceptors in the retina of the stomatopod crustacean, *Gonodactylaceus mutatus. Top:* Normalized absorbance spectra of intrarhabdomal filters in Rows 2 and 3 of midbands of compound eyes. Each panel illustrates filters from one row (light trace, distal filter; dark trace, proximal filter). *Middle:* Normalized absorbance spectra (light, jagged traces) and best-fit templates (dark, smooth traces) of visual pigments in main rhabdoms of each retinal region of compound eyes. Rows 1 through 6 refer to the midband region (Rows 5 and 6 contain the same visual pigment); "periphery" refers to the peripheral retina. Main rhabdoms in Rows 1 through 4 have two tiers, each with a different visual pigment; the visual pigment of the proximal tier always absorbs at longer wavelengths than that of the distal tier. Best-fit template spectra have the following maxima: Row 1 distal tier, 400 nm; Row 1 proximal tier, 443 nm; Row 2 distal tier, 513 nm; Row 2 proximal tier, 527 nm; Row 3 distal tier, 532 nm; Row 3 proximal tier, 553 nm; Row 4 distal tier, 443 nm; Row 4 proximal tier, 475 nm; Rows 5 and 6, 518 nm; peripheral retina, 510 nm. *Bottom:* Computed spectral sensitivities of all main rhabdoms or tiers. Each panel shows the sensitivity of one retinal region, as for visual pigments (above). In panels for Rows 1 through 4, light traces illustrate sensitivities of distal tiers; dark traces illustrate proximal tiers. Sensitivity maxima (rounded to the nearest 5 nm) are as follows: Row 1 distal tier, 400 nm; Row 1 proximal tier, 465 nm; Row 2 distal tier, 560 nm; Row 2 proximal tier, 605 nm; Row 3 distal tier, 635 nm; Row 3 proximal tier, 695 nm; Row 4 distal tier, 445 nm; Row 4 proximal tier, 515 nm; Rows 5 and 6, 520 nm; peripheral retina, 510 nm.

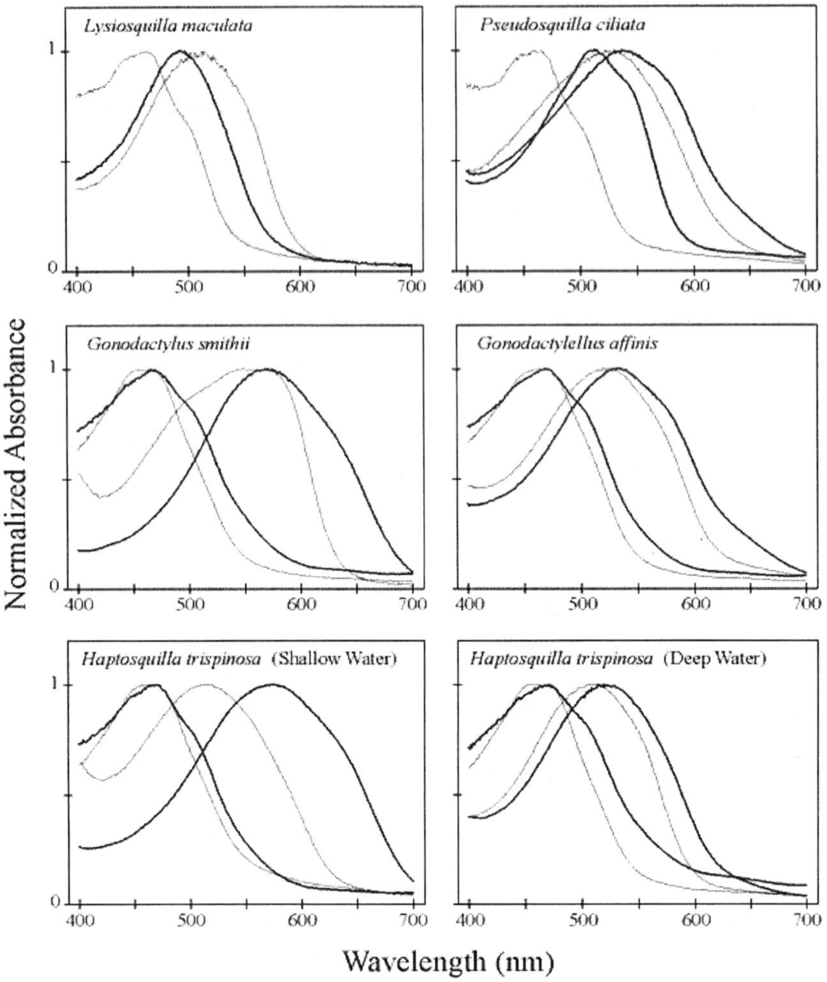

FIGURE 18.5. Normalized absorbance spectra of all filter classes in retinas of selected lysiosquilloid and gonodactyloid stomatopods. The species name is given in each panel. The lysiosquilloid *L. maculata* has only three filter types, while the gonodactyloids (*P. ciliata*, *G. smithii*, *G. affinis*, and *H. trispinosa*) have four types. Distal filters are plotted as thin traces, proximal filters as thick traces. In each panel, the wavelength of maximum absorption proceeds from left to right in the following sequence: Distal filter, Row 2; proximal filter, Row 2; distal filter, Row 3; proximal filter, Row 3. The "shallow-water" filter set was used to compute the sensitivity maxima plotted for *H. trispinosa* in Figure 18.3. See text for further details.

pattern of decreasing λ_{max} with increasing depth (see also Cronin and Marshall, 2002).

The lower half of Figure 18.6 shows how filtering, by visual pigments alone (proximal tiers of Rows 1 and 4) or by the combined effects of visual pigments and filters (Rows 2 and 3), affects peak spectral sensitivity. The effect is by far the greatest in Row 3, where peak spectral sensitivity can be shifted by over 100 nm to longer wavelengths than the visual pigment's λ_{max}. Also, the very-long-wavelength receptors of Row 3 (particularly in the proximal tier), which arise from the use of very long-wavelength-transmissive filters, exist only in gonodactyloid stomatopods that occupy very shallow water. Filtering is much less extreme in gonodactyloids living in deeper water and all lysiosquilloid species (Figs. 18.5 and 18.6).

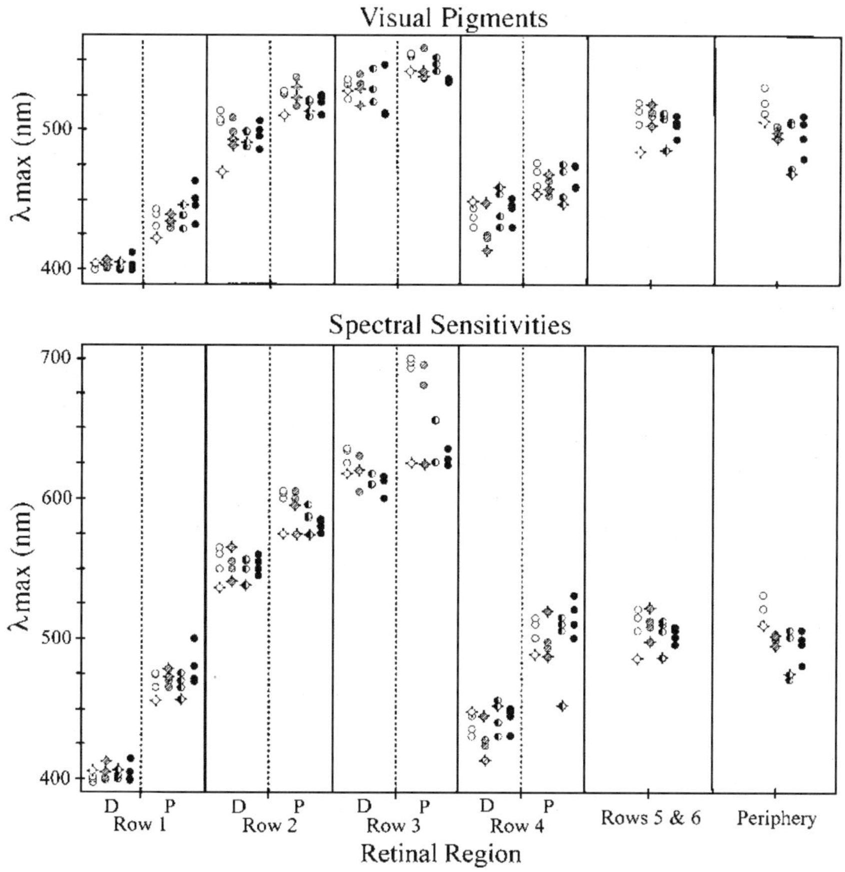

FIGURE 18.6. Spectral maxima of visual pigment absorbances (*top*) and spectral sensitivities (*bottom*) in 4 lysiosquilloid and 12 gonodactyloid species of stomatopod crustaceans from four classes of photic environments. Each panel of the figure includes data from main rhabdoms of all retinal regions, with each plotted point representing data from one receptor class of one species. Each retinal region is plotted in an enclosed rectangle; in the tiered Rows 1 to 4, this area is subdivided by a dashed vertical line so each tier is plotted separately (D, distal tier; P, proximal tier). Species have been assigned to four photic environmental classes, based on their most common depths of occurrence: open circles, intertidal; hatched circles, shallow subtidal (<5 m); half-filled circles, inhabiting a depth range from shallow to deep; filled circles, deep (5–50 m). Gonodactyloid species and their environmental class assignments are as follows: *Gonodactylellus affinis* (range of depths), *Gonodactylaceous mutatus* (intertidal), *Gonodactylus smithii* (intertidal), *Gonodactylopsis spongicola* (range of depths), *Haptosquilla trispinosa* (shallow subtidal), *Hemisquilla ensigera* (deep), *Neogonodactylus curacaoensis* (range of depths), *N. oerstedii* (intertidal), *Odontodactylus brevirostris* (deep), *O. havanensis* (deep), *O. scyllarus* (range of depths), and *Pseudosquilla ciliata* (shallow subtidal). Symbols for lysiosquilloid species are marked with a cross, and are as follows: *Coronis scolopendra* (shallow subtidal), *Lysiosquilla sulcata* (shallow subtidal), *Pullosquilla litoralis* (intertidal), and *P. thomassini* (range of depths).

Among gonodactyloids, there is a clear progression to shorter sensitivity maxima in Row 3 receptors of successively deeper-living species, despite the overall similarities in their visual pigments in this retinal region (Fig. 18.6).

A recent, unexpected finding is that individuals *within* a single stomatopod species have retinas that are tuned to specific habitats in their overall range of photic environments (Cronin et al., 2001, 2002; Cronin and Caldwell, 2002). In *Haptosquilla trispinosa*, for instance, filters of Row 3 vary among individuals, depending on where each lives (Cronin et al., 2001). This is illustrated in the bottom row of panels in Figure 18.5; note that the filters of deep-water *H. trispinosa* (bottom right) are severely blue-shifted compared to those of the shallow-water individuals (bottom left). The specific tuning to habitat occurs during development, and can be induced in the laboratory by raising young *H. trispinosa* in lighting that mimics illumination in shallow or deep water (Cronin et al., 2001). Recent results suggest that the filters can be altered at any time throughout the life of an individual, illustrating just how flexible the tuning of spectral sensitivity can be.

3.4.4. The Ultraviolet System

Ultraviolet (UV) sensitivity is common among arthropods in general, and crustaceans in particular, but stomatopods sample this spectral range most thoroughly (Cronin et al., 1994b; Marshall and Oberwinkler, 1999). At least four UV receptor types, with λ_{max} from 315 to 380 nm, can be present. Here again, the theme of spectral tuning by filtering recurs. The UV classes all have sensitivity functions that are too narrow for visual pigments alone, and they instead seem to be formed by filtering various UV visual pigments, with λ_{max} ranging from 300 to 370 nm, with nonvisual pigments probably located in the cornea or crystalline cone (Marshall and Oberwinkler, 1999).

While there is as yet no behavioral evidence that stomatopods can discriminate among stimuli based solely on their ultraviolet reflectances, it is likely that their color vision operates throughout a 400-nm spectral range, from ~300 nm to beyond 700 nm, the broadest coverage of any group of animals. Throughout the entire span, receptors remain narrowly tuned by the judicious combination of visual pigments with appropriate filters, matching the series of λ_{max} functions steadily from shortest to longest wavelengths while preserving discrimination at every point. The design promotes outstanding color constancy (Osorio et al., 1997), and may lead to a color vision sense that is defined by a series of wavebands rather than broad opponencies, much as the auditory system samples frequency space.

4. Development of Crustacean Spectral Sensitivity Mechanisms

Many marine crustacean species pass through larval stages, metamorphosing to the adult form. Not only is the sensory equipment of larvae and adults likely to be different, but they often inhabit different sensory worlds with divergent sensory requirements. For instance, almost all marine crustaceans are planktonic as larvae, but the adults are often nektonic or benthic. Therefore, changes in eye size, and frequently optical design as well (Nilsson, 1983; Nilsson et al., 1986), are often accompanied by metamorphosis of spectral mechanisms. Crustacean eye development has been recently reviewed (Cronin and Hariyama, 2002; Cronin and Jinks, 2002), and only a very brief account is provided here.

Larval crustaceans probably have the typical dichromatic spectral sensitivity design, with a short-wavelength visual pigment housed in R8 overlying a main rhabdom of 7 retinular cells packed with a middle-wavelength pigment. This is conjectural, however, since all larval visual pigments described to date are from the main rhabdoms (Cronin et al., 1995; Jutte et al., 1998; Cronin and Jinks, 2002), with spectral maxima ranging from 447 to 509 nm.

While data concerning changes in visual pigments are sparse, it appears that many species retain the larval pigment as adults; this is the case with the crab *Callinectes sapidus* (Cronin et al., 1995) or the squilloid stomatopod *Squilla empusa* (Cronin and Jinks, 2002). However, in polychromatic stomatopod species the retina completely turns over at metamorphosis; the larval retina is rejected, and an entirely new adult-type retina appears (Williams et al., 1985; Cronin et al., 1995). In these cases, the larval visual pigments apparently do not survive metamorphosis, and the adult retina not only introduces the photostable retinal filters but also expresses a new set of visual pigments. As in other aspects of their visual physiology, compound eye development in gonodactyloid and lysiosquilloid stomatopods is unusual.

5. Temporal Aspects of Crustacean Vision

Animals are challenged with the task of detecting movement and predator/prey locations under a wide range of light intensities. This requires balancing high sensitivity against the competing demand of high resolution in order to obtain the most comprehensive picture under each lighting condition.

One mechanism for dramatically improving photon capture is to increase the integration time of the photoreceptor (Pirenne and Denton, 1952; Pirenne, 1967; Lythgoe, 1979; Snyder, 1979), that is, increasing the temporal summation. This would have the same effect as leaving the shutter open longer on a camera in dim light (Lythgoe, 1979), and comes at the cost of temporal resolution. However, an analysis by Warrant (1999) demonstrated that, in the absence of summation (both spatial and temporal), the ability to resolve detail is lost at much higher intensities than when summation is occurring. Therefore, he concluded that the improvement in photon capture in dim light due to summation far outweighs losses in spatial and temporal resolution. In addition, his model predicts that animals trying to see small, slowly moving objects are better off if temporal summation is employed as a strategy for improving photon capture rather than spatial summation, and that for the detection of point sources (which are common sources of light in the marine environment where bioluminescence is so prevalent), spatial summation is not very effective (Chapter 16). Srinivasan and Bernard (1975) determined that the spatial resolution of objects moving at angular velocities above a critical value is also dependent on photoreceptor dynamics in the compound eye. These last two points are of particular importance for crustaceans: The vast majority of crustacean species are aquatic, so either they or their prey must be constantly moving, and for deep-living marine species, bioluminescent point sources are the prevailing light stimuli. Therefore, any comprehensive analysis of the visual systems of crustaceans should include the temporal characteristics of the photoreceptor.

Classic studies by Autrum (1950, 1958) on intact eyes of terrestrial arthropods gave rise to the idea that the temporal response characteristics of photoreceptors match the habitat and lifestyle of the organism. More recent studies on the intracellular responses of single cells (Howard et al., 1984; de Souza and Ventura, 1989; Laughlin and Weckstrom, 1993) support this idea in that the response characteristics of photoreceptor cells from slow-moving nocturnal species are considerably slower than those of fast-moving, diurnal species. One would predict the same type of match for crustaceans, with the photoreceptors of those species inhabiting dim-light environments showing slower response dynamics than those inhabiting bright-light environments.

Determination of the maximum flicker fusion frequency (FFF), which is the maximum flicker rate the eye is capable of following at any light intensity, has been used most often in comparative studies of temporal resolution in crustaceans. Comparing results from these studies (Fig. 18.7 and Table 18.3) shows that, in general, the trend observed for insects is also present for crustaceans. For terrestrial crustaceans, those species that inhabit a bright environment during the day (the day-active terrestrial isopods *Ligia occidentalis* and *L. italica* and the hermit crab *Pagurus*) have considerably higher flicker fusion frequencies (50–120 Hz) than the wood lice (*Porcellio* and *Armadillo*), which are nocturnal and hide beneath stones during the day (24–32 Hz). For aquatic species, the shallow-living species (water louse *Asellus*, rock lobster *Jasus edwarsii*, and crayfish *Cambarus*) have considerably higher flicker fusion frequencies (50–60 Hz) than the deeper-living marine species (17–32 Hz), as one would anticipate due to differences in their ambient daytime light fields.

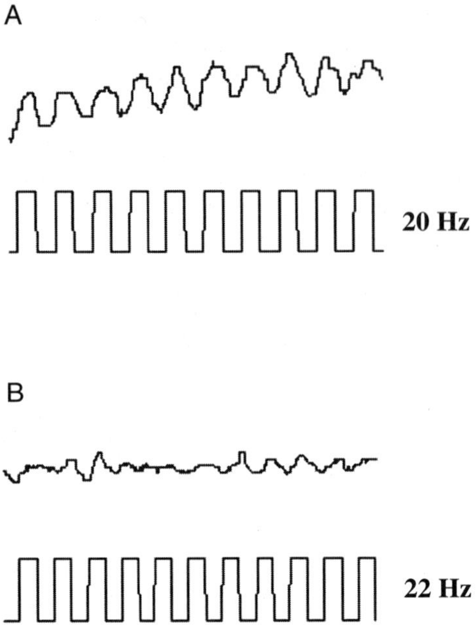

Figure 18.7. Representative example of flicker fusion frequency. Top trace is the electroretinogram (ERG) response recorded from the eye of the penaeid shrimp, *Funchalia villosa*. Lower trace is the flickering light stimulus. Data shown are 0.5 s of a 2-s stimulus pulse. (A) The eye was able to follow the stimulus light at 20 Hz. (B) The eye was not able to follow the stimulus light at 22 Hz. The critical flicker fusion frequency for the specimen is 20 Hz, as this was the highest frequency, at any irradiance, at which the ERG of this specimen was able accurately to match the phase of the stimulus light.

The ambient light field in the marine environment at a depth of 400 m in Jerlov's Type 1 or 1A water (the clearest ocean water; Jerlov, 1976) is equivalent to that provided by moonlight for nocturnal terrestrial species (Munz and McFarland, 1973; Land, 1981a,b), and therefore one might anticipate that the maximum FFF of these deeper-living marine species would be equivalent to that of the two terrestrial nocturnal species that have been studied. This is indeed the case, with some rather notable exceptions. The euphausiid crustaceans (*Stylocheiron* and *Nematobrachion*), which all have daytime depths below 250 m, have flicker fusion frequencies (40–50 Hz) approaching those of the shallow-living species. This brings up the issue that for coastal and open-ocean crustacean species, matching temporal resolution to the ambient light field takes on added complexity due to the prevalence of bioluminescence in the marine environment. The euphausiids, which appear to possess photoreceptors with flicker fusion frequencies too fast for their dim ambient light environment, feed primarily on rapidly moving, bioluminescent copepods. Since their prey might be visible as a rapidly moving bright light against a dark background, the advantages of higher temporal resolution might outweigh the disadvantages of lower sensitivity.

Two other studies have been conducted on the temporal characteristics of crustacean photoreceptors, utilizing intermediate light levels to look at flicker fusion frequencies (Moeller and Case, 1995) and frequency responses to sinusoidal light stimuli (Johnson et al., 2000). Because temporal resolution is so strongly dependent on light intensity (Bröcker, 1935; Crozier and Wolf, 1939; Crozier et al., 1939; Frank, 2000), results from these studies cannot be directly compared with results from the studies utilizing the invariant maximum flicker fusion frequency. However, the same general trends are apparent in these two studies, in that the deeper-living and/or slower-moving species from dimmer light environments have slower-responding eyes than the shallower-living and/or more active species from brighter habitats.

6. Conclusion

Crustaceans have long served as model organisms for the study of visual function. Often this has been with the primary concern focused on basic aspects of visual pigment photochemistry or visual processing, or on the applicability of their modular eyes to artificial imaging designs or autonomous systems. While retaining much of this traditional significance, today they are studied mainly either for their own sakes or for their ability to contribute to our understanding of how the visual systems of organisms are specialized for function in challenging environments or for unusual and exotic visual tasks. With their huge diversity of optical designs,

TABLE 18.3. Maximum flicker fusion frequencies and habitat/depth distributions.

Species	Temporal	Habitat/daytime depth
Order Isopoda		
Ligia occidentalis[1]	120	terrestrial, rocky shores
Ligia italica[2]	78	terrestrial, rocky shores
Asellus communis[3]	52	shallow freshwater
Armadillo officinalis[2]	24	terrestrial, nocturnal
Porcellio loewis[2]	32	terrestrial, nocturnal
Order Decapoda		
Pagurus bernhardus[4]	56	terrestrial, rocky shores
Cambarus bartoni[5]	50	shallow freshwater
Jasus edwarsii[6]	60	coastal, sublittoral to 50 m
Funchalia villosa[7,8]	21	open ocean, 300–550 m
Janicella spinacauda[7]	31	open ocean, 500–600 m
Oplophorus gracilirostris[7]	32	open ocean, 500–650 m
Systellaspis debilis[7]	21	open ocean, 600–900 m
Pasiphaea multidentata[8]	17	open ocean, 600–960 m
Sergestes arcticus[8]	24	open ocean, 600–960 m
Sergia filictum[7]	24	open ocean, 600–900 m
Sergia grandis[8]	22	open ocean, 600–900 m
Order Amphipoda		
Phronima sedenteria[8]	27	open ocean, 120–600 m
Order Euphausiacea		
Stylocheiron maximum[7,8]	40	open ocean, 250–500 m
Nematobrachion sexspinosus[7]	50	open ocean, 400–600 m
Nematobrachion flexipes[7]	44	open ocean, 450–600 m

[1] Ruck and Jahn (1954).
[2] Benguerrah and Carricaburu (1976).
[3] Crozier and Zerrahan-Wolf (1939).
[4] Bröcker (1935).
[5] Crozier and Wolf (1939).
[6] Meyer-Rochow and Tiang (1984).
[7] Frank (1999).
[8] Frank (2000).

body plans, habitats, and ecological niches, the crustaceans continue to fascinate students of visual or sensory physiology. They have much to teach us about the evolution, function, and design of sensory systems.

References

Autrum, H. (1950). Die Belichtungspotentiale and das Sehen der Insekten (Untersuchungen an *Calliphora* und *Dixippus*). *Z. Vergleich. Physiol.* 32:176–227.

Autrum, H. (1958). Electrophysiological analysis of the visual systems in insects. *Exp. Cell. Res.* 5:426–439.

Ball, E.E. (1977). Fine structure of the compound eyes of the midwater amphipod *Phronima* in relation to behavior and habitat. *Tissue Cell* 9:521–536.

Barnes, H., and Klepal, W. (1972). Phototaxis in stage I nauplius larvae of two cirripedes. *J. Exp. Mar. Ecol.* 10:267–273.

Bayliss, L.E., Lythgoe, J.N., and Tansley, K. (1936). Some forms of visual purple in sea fishes with a note on the visual cells of origin. *Proc. R. Soc. Lond. B.* 120:95–114.

Benguerrah, A., and Carricaburu, P. (1976). L'electroretinogramme chez les crustacea isopodes. *Vision Res.* 16:1173–1177.

Boden, B.P., Kampa, E.M., and Abott, B.C. (1961). Photoreception of a planktonic crustacean in relation to light penetration in the sea. In: *Progress in Photobiology* (Christensen, B., and Buchamann, B., eds.), pp. 189–196. New York: Elsevier North-Holland.

Bridges, C.D.B. (1972). The rhodopsin-porphyropsin visual system, In: *Handbook of Sensory Physiology, Vol. VII/1: Photochemistry of Vision* (Dartnall, H.J.A., ed.), Berlin, Heidelberg, New York: Springer-Verlag.

Briggs, M.H. (1961). Visual pigment of grapsoid crabs. *Nature* 190:784–786.

Bröcker, H. (1935). Untersuchungen über das Sehvermögen der Einsiedlerkrebse. *Zool. Jahrb. Abt. Allgem. Zool. Physiol. Tiere* 55:399–430.

Bruno, M.S., Mote, M.I., and Goldsmith, T.M. (1974). Spectral absorption and sensitivity measurements in single ommatidia of the green crab, *Carcinus. J. Comp. Physiol.* 82:151–163.

Bruno, M.S., Barnes, S.N., and Goldsmith, T.H. (1977). The visual pigment and visual cycle of the lobster, *Homarus. J. Comp. Physiol.* 120:123–142.

Caldwell, R.L., and Dingle, H. (1975). Ecology and evolution of agonistic behavior in stomatopods. *Naturwissenschaften* 62:214–222.

Carleton, K.L., Harosi, F.I., and Kocher, T.D. (2000). Visual pigments of African cichlid fishes: Evidence for ultraviolet vision from microspectrophotometry and DNA sequences. *Vision Res.* 40:879–890.

Chiao, C.C., Cronin, T.W., and Marshall, J. (2000). Eye design and color signaling in a stomatopod crustacean, *Gonodactylus smithii. Brain Behav. Evol.* 56:107–122.

Clarke, G.L. (1936). On the depth at which fishes can see. *Ecology* 17:452–456.

Crandall, K.A., and Cronin, T.W. (1995). Opsin evolution in crayfishes: Effects of functional constraints. In: *Current Topics on Molecular Evolution* (Nei, M., and Takahata, N., eds.), pp. 19–27. Institute of Molecular Evolutionary Genetics, Pennsylvania State University.

Crescitelli, F., McFall-Ngai, M., and Horwitz, J. (1985). The visual pigment sensitivity hypothesis: Further evidence from fishes of varying habitats. *J. Comp. Physiol.* 157:323–333.

Cronin, T.W. (1985). The visual pigment of a stomatopod crustacean, *Squilla empusa. J. Comp. Physiol.* 156:679–687.

Cronin, T.W. (1986). Optical design and evolutionary adaptation in crustacean compound eyes. *J. Crust. Biol.* 6:1–23.

Cronin, T.W. (1994). Polychromatic vision in mantis shrimps. *Sensorie Systemy* 8:95–106.

Cronin, T.W., and Caldwell, R.L. (2002). Tuning of photoreceptor function in mantis shrimp species occupying a range of depths. II. Filter pigments. *J. Comp. Physiol. A.* 188:187–197.

Cronin, T.W., and Forward, R.B. Jr. (1988). The visual pigments of crabs. I. Spectral characteristics. *J. Comp. Physiol. A.* 162:463–478.

Cronin, T.W., and Frank, T.F. (1996). A short-wavelength photoreceptor class in a deep-sea shrimp. *Proc. Roy. Soc. Lond. B.* 263:861–865.

Cronin, T.W., and Goldsmith, T.H. (1982). Quantum efficiency and photosensitivity of the rhodopsin-metarhodopsin conversion in crayfish photoreceptors. *Photochem. Photobiol.* 36:447–454.

Cronin, T.W., and Hariyama, Y. (2002). Spectral sensitivity in crustaceans. In: *The Crustacean Nervous System* (Wiese, K., ed.), pp. 499–511. Berlin, Heidelberg, New York: Springer.

Cronin, T.W., and Jinks, R.N. (2002). Ontogeny of vision in marine crustaceans. *Am. Zoologist* 41: 1098–1107.

Cronin, T.W., and Marshall, N.J. (1989a). Multiple spectral classes of photoreceptors in the retinas of gonodactyloid stomatopod crustaceans. *J. Comp. Physiol. A.* 166:267–275.

Cronin, T.W., and Marshall, N.J. (1989b). A retina with at least ten spectral types of photoreceptors in a stomatopod crustacean. *Nature* 339:137–140.

Cronin, T.W., and Marshall, J. (2001). Parallel processing and image analysis in the eyes of mantis shrimps. *Biol. Bull.* 200:177–183.

Cronin, T.W., and Marshall, J. (in press). The visual world of mantis shrimps. In: *Complex Worlds from Simpler Nervous Systems* (Prete, F.R., ed.), Boston: MIT Press.

Cronin, T.W., Caldwell, R.L., and Erdmann, M.V. (2002). Tuning of photoreceptor function in mantis shrimp species occupying a range of depths. I. Visual pigments. *J. Comp. Physiol. A.* 188:179–186.

Cronin, T.W., Caldwell, R.L., and Marshall, J. (2001). Tunable colour vision in a mantis shrimp. *Nature* 411:547–548.

Cronin, T.W., Marshall, N.J., and Caldwell, R.L. (1993). Photoreceptor spectral diversity in the retinas of squilloid and lysiosquilloid stom-atopod crustaceans. *J. Comp. Physiol. A.* 172:339–350.

Cronin, T.W., Marshall, N.J., and Caldwell, R.L. (1994a). The intrarhabdomal filters in the retinas of mantis shrimps. *Vision Res.* 34:279–291.

Cronin, T.W., Marshall, N.J., and Caldwell, R.L. (1994b). The retinas of mantis shrimps from low-light environments (Crustacea; Stomatopoda; Gonodactylidae). *J. Comp. Physiol. A.* 174:607–619.

Cronin, T.W., Marshall, N.J., and Caldwell, R.L. (1996). Visual pigment diversity in two genera of mantis shrimps implies rapid evolution. *J. Comp. Physiol. A.* 179:371–384.

Cronin, T.W., Marshall, N.J., Caldwell, R.L., and Pales, D. (1995). Compound eyes and ocular pigments of crustacean larvae (Stomatopoda and Decapoda, Brachyura). *Mar. Freshwater Behav. Physiol.* 26:219–231.

Cronin, T.W., Marshall, N.J., Caldwell, R.L., and Shashar N. (1994c). Specialization of retinal function in the compound eyes of mantis shrimps. *Vision Res.* 34:2639–2656.

Cronin, T.W., Marshall, N.J., Quinn, C.A., and King, C.A. (1994d). Ultraviolet photoreception in mantis shrimp. *Vision Res.* 34:1443–1452.

Cronly-Dillon, J., and Sharma, S.C. (1968). Effect of season and sex on the photopic spectral sensitivity of the threespined stickleback. *J. Exp. Biol.* 49:679–687.

Crozier, W.J., and Wolf, E. (1939). The flicker response contour for the crayfish. *J. Gen. Physiol.* 23:1–10.

Crozier, W.J., Wolf, E., and Zerrahn-Wolf, G. (1939). The flicker response contour for the isopod *Asellus*. *J. Gen. Physiol.* 22:451–462.

Cummins, D., and Goldsmith, T.H. (1981). Cellular identification of the violet receptor in the crayfish eye. *J. Comp. Physiol.* 142:199–202.

Cummins, D., Chen, D.-M.-Y., and Goldsmith, T.H. (1984). Spectral sensitivity of the spiny lobster, *Panulirus argus*. *Biol. Bull.* 166:269–276.

de Souza, J.M., and Ventura, D.F. (1989). Comparative study of temporal summation and response form in hymenopteran photoreceptors. *J. Comp. Physiol. A.* 165:237–245.

Denton, E.J. (1990). Light and vision at depths greater than 200 metres. In: *Light and Life in the Sea* (Herring, P.J., Campell, A.K., Whitfield, M., and Maddock, L., eds.), pp. 127–148. Cambridge: Cambridge University Press.

Denys, C.J. (1982). Ommachrome pigments in the eyes of *Euphausia superba* (Crustacea, Euphausiacea). *Polar Biol.* 1:1–10.

Denys, C.J., and Brown, P.K. (1982). The rhodopsins of *Euphausia superba* and *Meganyctiphanes norvegica* (Crustacea, Euphausiacea). *J. Gen. Physiol.* 80:451–472.

Douglas, R.H., and Marshall, N.J. (1999). A review of vertebrate and invertebrate ocular filters. In: *Adaptive Mechanisms in the Ecology of Vision* (Archer, S., Djamgoz, M.B.A., Loew, E., Partridge, J.C., and Vallerga, S., eds.), pp. 95–162. Dordrecht: Kluwer Academic.

Eaton, B.P., and Brown, C.M. (1970). Photoreception in the nauplius eye of *Pandalus borealis* Kröger: Decapoda, Crustacea. *Canad. J. Zool.* 48:119–121.

Edwards, A.S. (1969). The structure of the eye of *Ligea oceanica* L. *Tissue Cell* 1:217–228.

Eguchi, E. (1973). Orthogonal microvillus pattern in the eighth rhabdomere of the rock crab *Grapsus*. *Z. Zellforsch.* 137:145–157.

Eguchi, E., and Waterman, T.H. (1967). Changes in the retinal fine structure induced in the crab *Libinia* by light and dark adaptation. *Z. Zellforsch.* 79:209–229.

Fernandez, H.R. (1965). A survey of the visual pigments of decapod crustacea of South Florida. PhD dissertation. Coral Gables, FL: University of Miami.

Fernandez, H.R. (1973). Spectral sensitivity and visual pigment of the compound eye of the galatheid crab *Pleuroncodes planipes*. *Mar. Biol.* 20:148–153.

Fisher, L.R., and Goldie, E.H. (1959). The eye pigments of a euphausiid crustacean, *Meganyctiphanes norvegica* (M. Sars). *Proc. XV Intern. Congr. Zool. Lond.* pp. 533–535.

Fisher, L.R., and Goldie, E.H. (1961). Pigments of compound eyes. In: *Progress in Photobiology* (Christensen, B.B., and Buchamann, B., eds.), pp. 153–154. New York: Elsevier North-Holland.

Forward, R.B.J., Cronin, T.W., and Douglass, J.K. (1988). The visual pigments of crabs. II. Environmental adaptations. *J. Comp. Physiol. A.* 162: 479–490.

Frank, T.M. (1999). Comparative study of temporal resolution in the visual systems of mesopelagic crustaceans. *Biol. Bull.* 196:137–144.

Frank, T.M. (2000). Temporal resolution in mesopelagic crustaceans. *Phil. Trans. Roy. Soc. Lond. B.* 355:1195–1198.

Frank, T.M., and Case, J.F. (1988a). Visual spectral sensitivities of bioluminescent deep-sea crustaceans. *Biol. Bull.* 175:261–273.

Frank, T.M., and Case, J.F. (1988b). Visual spectral sensitivity of the bioluminescent deep-sea mysid, *Gnathophausia ingens*. *Biol. Bull.* 175:1–10.

Frank, T.M., and Widder, E.A. (1994a). Comparative study of behavioural-sensitivity thresholds to near-UV and blue-green light in deep-sea crustaceans. *Mar. Biol.* 121:229–235.

Frank, T.M., and Widder, E.A. (1994b). Evidence for behavioural sensitivity to near-UV light in the deep-sea crustacean *Systellaspis debilis*. *Mar. Biol.* 118:279–284.

Frank, T.M., and Widder, E.A. (1996). UV light in the deep-sea: *In situ* measurements of downwelling irradiance in relation to the visual threshold sensitivity of UV-sensitive crustaceans. *Mar. Fresh. Behav. Physiol.* 27:189–197.

Frank, T.M., and Widder, E.A. (1999). Comparative study of the spectral sensitivities of mesopelagic crustaceans. *J. Comp. Physiol. A.* 185:255–265.

Goldsmith, T.H. (1978). The spectral absorption of crayfish rhabdoms: Pigment photoproduct and pH sensitivity. *Vision Res.* 18:463–473.

Goldsmith, T.H., and Cronin, T.W. (1993). The retinoids of seven species of mantis shrimp. *Visual Neurosci.* 10:915–920.

Hallberg, E. (1977). The fine structure of the compound eyes of mysids (Crustacea: Mysidacea). *Cell Tiss. Res.* 184:45–65.

Hallberg, E., and Nilsson, D.E. (1983). The euphausid compound eye: A morphological reinvestigation (Crustacea: Euphausidacea). *Zoomorphologie* 103:59–66.

Hallberg, E., Nilsson, H.L., and Elofsson, R. (1980). Classification of amphipod compound eyes: The fine structure of the ommatidial units (Crustacea, Amphipoda). *Zoomorphologie* 94:279–306.

Hamacher, K.J., and Kohl, K.D. (1981). Spectroscopical studies of the *Astacus* visual pigment. *Biophys. Struct. Mech.* 7:338–343.

Hamacher, K.J., and Stieve, H. (1984). Spectral properties of the rhodopsin-system of the crayfish *Astacus leptodactylus*. *Photochem. Photobiol.* 39: 379–390.

Hariyama, T., Tsukahara, Y., and Meyer-Rochow, V.B. (1993). Spectral responses, including a UV-sensitive cell type, in the eye of the isopod *Ligia exotica*. *Naturwissenschaffen* 80:233–235.

Hays, D., and Goldsmith, T.H. (1969). Microspectrophotometry of the visual pigment of the spider crab *Libinia emarginata*. *Z. Vergl. Physio.* 65:218–232.

Hazlett, B.A. (1979). The meral spot of *Gonodactylus oerstedii* Hansen as a visual stimulus (Stomatopoda, Gonodactylidae). *Crustaceana* 36: 196–198.

Hertel, H. (1972). Aspekte zur photorezeption von *Artemia salina*. *Verh. Dtsch. Ges. Zool. Erlangen*. 14–78.

Hiller-Adams, P., Widder, E.A., and Case, J.F. (1988). The visual pigments of four deep-sea crustacean species. *J. Comp. Physiol. A.* 163:63–72.

Hillman, P., Dodge, F.A., Hochstein, S., Knight, B.W., and Minke, B. (1973). Rapid dark recovery of the invertebrate early receptor potential. *J. Gen. Physiol.* 62:77–86.

Horridge, G.A. (1967). Perception of polarization plane, colour, and movement in two dimensions by the crab, *Carcinus*. *Z. Vergl. Physiol.* 55:207–224.

Horridge, G.A. (1978). The separation of visual axes in apposition compound eyes. *Phil. Trans. Roy. Soc. Lond. B.* 285:1–59.

Houde, A.E., and Endler, J.A. (1990). Correlated evolution of female mating preferences and male color patterns in the guppy *Poecilia reticulata*. *Science* 14:1405–1408.

Howard, J., Dubs, A., and Payne, R. (1984). The dynamics of phototransduction in insects. *J. Comp. Physiol.* 154:707–718.

Hyatt, G.W. (1975). Physiological and behavioural evidence for colour discrimination by fiddler crabs (Brachyura, Ocypodidae, genus *Uca*). In: *Physiological Ecology of Estuarine Organisms* (Vernberg, F.J., ed.), pp. 333–365. Columbia, SC: University of South Carolina Press.

Jacobs, G.H., and Deegan, J.F.I. (2000). Photopigments and colour vision in New World monkeys from the family Atelidae. *Proc. Roy. Soc. Lond. B.* 268:695–702.

Jerlov, N.G. (1976). *Marine Optics*. Amsterdam: Elsevier.

Johnson, M.L., Shelton P.M.J., and Gaten, E. (2000). Temporal resolution in the eyes of marine decapods from coastal and deep-sea habitats. *Mar. Biol.* 136:243–248.

Jutte, P.A., Cronin, T.W., and Caldwell, R.L. (1998). Retinal function in the planktonic larvae of two species of *Pullosquilla*, a lysiosquilloid stomatopod crustacean. *J. Exp. Biol.* 201:2481–2487.

Kampa, E.M. (1955). Euphausiopsin: A new photosensitive pigment from the eyes of euphausiid crustaceans. *Nature* 175:996–998.

Kelber, A. (1996). Colour learning in the hawkmoth *Macroglossum stellatarum*. *J. Exp. Biol.* 199: 1127–1131.

Knowles, A., and Dartnall, H.J.A. (1977). Habitat and visual pigments. In: *The Eye, Vol. 2B: The Photobiology of Vision* (Davson, H., ed.), pp. 581–641. New York: Academic Press.

Lall, A.B., and Cronin, T.W. (1987). Spectral sensitivity of the compound eyes in the purple land crab *Gecarcinus lateralis* (Freminville). *Biol. Bull.* 173:398–406.

Land, M.F. (1981a). Optics and vision in invertebrates. In: *Handbook of Sensory Physiology, Vol. VII/6B* (Autrum, H., ed.), pp. 471–592. Berlin: Springer-Verlag.

Land, M.F. (1981b). Optics of the eyes of *Phronima* and other deep-sea amphipods. *J. Comp. Physiol.* 145:209–226.

Land, M.F. (1984). Crustacea. In: *Photoreception and Vision in Invertebrates* (Ali, M.A., ed.), pp. 401–438. New York: Plenum.

Land, M.F. (1989). The sight of deep wet heat. *Nature* 337:404.

Land, M.F., and Nilsson, D.-E. (2002). *Animal Eyes*. Oxford, New York: Oxford University Press.

Land, M.F., Burton, F.A., and Meyer-Rochow, V.B. (1979). The optical geometry of euphausiid eyes. *J. Comp. Physiol.* 130:49–62.

Laughlin, S.B., and Weckstrom, M. (1993). Fast and slow photoreceptors: A comparative study of the functional diversity of coding and conductances in the *Diptera*. *J. Comp. Physiol.* 172:593–609.

Latz, M.I., Frank, T.M., and Case, J.F. (1988). Spectral composition of bioluminescence of epipelagic organisms from the Sargasso Sea. *Mar. Biol.* 98:441–446.

Lindsay, S.M., Frank, T.M., Kent, J., Partridge, J.C., and Latz, M.I. (1999). Spectral sensitivity of vision and bioluminescence in the mid-water shrimp, *Sergestes similis*. *Biol. Bull.* 1:1–11.

Loew, E.R., and Lythgoe, J.N. (1978). The ecology of cone pigments in teleost fishes. *Vision Res.* 18: 715–722.

Lythgoe, J.N. (1966). Visual pigments and underwater vision. In: *Light as an Ecological Factor* (Bainbridge, R., Evans, G.C., and Rackham, O., eds.), Oxford: Blackwell.

Lythgoe, J.N. (1968). Visual pigments and visual range underwater. *Vision Res.* 8:997–1012.

Lythgoe, J.N. (1979). *The Ecology of Vision*. Oxford: Clarendon Press.

Manning, R.B., Schiff, H., and Abbott, B.C. (1984). Eye structure and classification of stomatopod crustaceans. *Zool. Scripta* 13:41–44.

Marshall, N.J. (1988). A unique colour and polarization vision system in mantis shrimps. *Nature* 333:557–560.

Marshall, N.J., and Land, M.F. (1993a). Some optical features of the eyes of stomatopods. I. Eye shape, optical axes and resolution. *J. Comp. Physiol. A.* 173:565–582.

Marshall, N.J., and Land, M.F. (1993b). Some optical features of the eyes of stomatopods. II. Ommatidial design, sensitivity and habitat. *J. Comp. Physiol. A.* 173:583–594.

Marshall, N.J., and Oberwinkler, J. (1993). Electrophysiology of mantis shrimp retina. *Abstracts of the British Photobiology Society*.

Marshall, J., and Oberwinkler, J. (1999). The colourful world of the mantis shrimp. *Nature* 401: 873–874.

Marshall, N.J., Cronin, T.W., and Shashar, N. (1999a). Behavioural evidence for polarisation vision in stomatopods reveals a potential channel for communication. *Current Biology* 9:755–758.

Marshall, N.J., Jones, J.P., and Cronin, T.W. (1996). Behavioural evidence for color vision in stomatopod crustaceans. *J. Comp. Physiol. A.* 179:473–481.

Marshall, N.J., Kent, K., and Cronin, T.W. (1999b). Pigments in eyes of marine crustaceans. In: *Adaptive Mechanisms in the Ecology of Vision* (Archer, S.N., Djamgoz, M.B.A., Loew, E.R., Partridge, J.C., and Vallerga, S., eds.), pp. 285–327. Dordrecht: Kluwer Academic.

Marshall, N.J., Land, M.F., King, C.A., and Cronin, T.W. (1991a). The compound eyes of mantis shrimps (Crustacea, Hoplocarida, Stomatopoda). I. Compound eye structure: The detection of polarised light. *Phil. Trans. R. Soc. Ser. B.* 334:33–56.

Marshall, N.J., Land, M.F., King, C.A., and Cronin, T.W. (1991b). The compound eyes of mantis shrimps (Crustacea, Hoplocarida, Stomatopoda). II. Colour pigments in the eyes of Stomatopod crustaceans: Polychromatic vision by serial and lateral filtering. *Phil. Trans. R. Soc. B.* 334:57–84.

Martin, F.G., and Mote, M.I. (1982). Colour receptors in marine crustaceans: A second class of retinular cell in the compound eyes of *Callinectes* and *Carcinus*. *J. Comp. Physiol.* 145:549–554.

McFarland, W.N. (1991). The visual world of coral reef fishes. In: *The Ecology of Fishes on Coral Reefs* (Sale, P.F., ed.), pp. 16–38. San Diego: Academic Press.

McFarland, W.N., and Munz, F.W. (1975). Part II: The photopic environment of clear tropical seas during the day. *Vision Res.* 15:1063–1070.

Meyer-Rochow, V.B., and Tiang, K.M. (1984). The eye of *Jasus edwardsii* (Crustacea, Decapoda, Palinuridae): Electrophysiology, histology and behaviour. *Zoologica* 134:1–61.

Minke, B., and Kirschfield, K. (1978). Microspectrophotometric evidence for two photointerconvertible states of visual pigment in the barnacle lateral eye. *J. Gen. Physiol.* 71:37–45.

Moeller, J.F., and Case, J.F. (1995). Temporal adaptations in visual systems of deep-sea crustaceans. *Mar. Biol.* 123:47–54.

Munz, F.W., and McFarland, W.N. (1973). The significance of spectral position in the rhodopsins of tropical marine fishes. *Vision Res.* 13:1829–1874.

Munz, F.W., and McFarland, W.N. (1977). Evolutionary adaptations of fishes to the photic environment. In: *The Visual System in Vertebrates* (Crescitelli, F., ed.), pp. 193–274. Berlin, Heidelberg, New York, Springer.

Nathans, J. (1989). The genes for colour vision. *Sci. Amer.* 1989:28–35.

Nilsson, D.E. (1983). Evolutionary links between apposition and superposition optics in crustacean eyes. *Nature* 302:818–821.

Nilsson, D.E., Hallberg, E., and Elofsson, E. (1986). The ontogenetic development of refracting superposition eyes in crustaceans: transformation of optical design. *Tissue Cell* 18:509–519.

Nosaki, A. (1969). Electrophysiological study of color encoding in the compound eye of crayfish, *Procambarus clarkii. Z. Vergl. Physiologie.* 64: 318–323.

Osorio, D., Marshall, N.J., and Cronin, T.W. (1997). Stomatopod photoreceptor spectral tuning as an adaptation for colour constancy in water. *Vision Res.* 37:3299–3309.

Pelli, D.G., and Chamberlain, S.C. (1989). The visibility of 350°C black-body radiation by the shrimp *Rimicaris exoculata* and man. *Nature* 337:460–461.

Pirenne, M.H. (1967). *Vision and the Eye*. London: Chapman & Hall.

Pirenne, M.H., and Denton, E.J. (1952). Accuracy and sensitivity of the human eye. *Nature* 170: 1039–1042.

Ruck, P., and Jahn, T.L. (1954). Electrical studies on the compound eye of *Ligia occidentalis* Dana (Crustacea: Isopoda). *J. Gen. Physiol.* 37:825–849.

Sakamoto, K., Hisatomi, O., Tokunaga, F., and Eguchi, E. (1996). Two opsins from the compound eye of the crab *Hemigrapsus sanguinensis*. *J. Exp. Biol.* 199:441–450.

Scherer, C., and Kolb, G. (1987). The influence of colour stimuli on visually controlled behaviour in *Aglais urticae* L. and *Parage aegeria* L. (Lepidoptera). *J. Comp. Physiol. A.* 161:891–898.

Schiff, H. (1963). Dim light vision of *Squilla mantis* L. *Am. J. Physiol.* 205:927–940.

Schiff, H., Manning, R.B., and Abbott, B.C. (1986). Structure and optics of ommatidia from eyes of stomatopod crustaceans from different luminous habitats. *Biol. Bull.* 170:461–480.

Scott, S., and Mote, M.I. (1973). Spectral sensitivity in some marine crustacea. *Vision Res.* 14:659–663.

Shand, J., Hart, N.S., Thomas, N., and Partridge, J.C. (2001b). Developmental changes in the visual pigments of black bream. *Invest. Ophthalmol. Visual Sci.* 42:S362.

Smith, F.E., and Baylor, E.R. (1953). Colour responses in the cladocera and their ecological significance. *Am. Nat.* 87:49–55.

Smith, K.C., and Macagno, E.R. (1990). UV photoreceptors in the compound eye of *Daphnia magna* (Crustacea, Branchiopoda): A fourth spectral class in a single ommatidia. *J. Comp. Physiol. A.* 166:597–606.

Snyder, A.W. (1979). The physics of vision in compound eyes. In: *Handbook of Sensory Physiology: Comparative Physiology and Evolution of Vision in Invertebrates* (Autrum, H., ed.), pp. 225–313. Berlin, Heidelberg, New York: Springer-Verlag.

Srinivasan, M.V., and Bernard, G.D. (1975). The effect of motion on visual acuity of the compound eye: A theoretical analysis. *Vision Res.* 15:515–525.

Stearns, D.E., and Forward, R.B., Jr. (1984). Photosensitivity of the calanoid copepod, *Acartia tonsa*. *Mar. Biol.* 82:85–90.

Stowe, S. (1980). Rapid synthesis of photoreceptor membrane and assembly of new microvilli in a crab at dusk. *Cell Tiss. Res.* 211:419–440.

Stowe, S., Fukudome, H., and Tanaka, K. (1986). Membrane turnover in crab photoreceptors studied by high-resolution scanning electron microscopy and by a new technique of thick-section transmission electron microscopy. *Cell Tiss. Res.* 245:51–60.

Suzuki, T., Makino-Tasaka, M., and Eguchi, E. (1984). 3-dehydroretinal (vitamin A_2 aldehyde) in crayfish eye. *Vision Res.* 24:783–787.

van Dover, C.L., Szuts, E.Z., Chamberlain, S.C., and Cann, J.R. (1989). A novel eye in "eyeless" shrimp from hydrothermal vents of the mid-Atlantic ridge. *Nature* 337:458–460.

Wald, G., and Hubbard, R. (1957). Visual pigment of a decapod crustacean: The lobster. *Nature* 180:278–280.

Wald, G., and Seldin, E.B. (1968). Spectral sensitivity of the common prawn, *Palaemonetes vulgaris*. *J. Gen. Physiol.* 51:694–700.

Warrant, E.J. (1999). Seeing better at night: Life style, eye design and the optimum strategy of spatial and temporal summation. *Vision Res.* 39:1611–1630.

Widder, E.A., Hiller-Adams, P., and Case, J.F. (1987). A multichannel microspectrophotometer for visual pigment investigations. *Vision Res.* 27: 1047–1055.

Williams, B.G., Greenwood, J.G., and Jillet, J.B. (1985). Seasonality and duration of the developmental stages of *Heterosquilla tricarinata* (Claus, 1871) (Crustacea: Stomatopoda) and the replacement of the larval eye at metamorphosis. *Bull. Mar. Sci.* 36:104–114.

Zeiger, J., and Goldsmith, T.H. (1989). Spectral properties of porphyropsin from an invertebrate. *Vision Res.* 29:519–527.

Part 5
Central Coordination and Evolution of Sensory Inputs

John Montgomery and John D. Pettigrew

Sensory processing of the aquatic environment is carried out by a huge diversity of animals, each with their own array of sensors and the signals that they detect. Parts 1 to 4 cover the wide range of peripheral sensory mechanisms and the behaviors they mediate. Part 5 ventures into the "hard" territory of the evolution of sensory inputs and their central coordination. The aquatic environment was the site of the evolution of the vertebrate brain, in particular its machinery for processing sensory input and distinguishing self-induced changes from real-world-induced changes. Accordingly, nature's evolutionary experiments in sensory processing are often elucidated by an examination of neural adaptations to the aquatic niche. This principle is well-illustrated by each of the chapters in this section.

Chapter 19 tackles the tough task of understanding the evolution of a rostral brain in bilaterial animals from a comparative perspective, drawing parallels between the way in which enlarged brains and elaborate neural circuitry have arisen in cephalopod molluscs, arthropods, and craniates. Dr. Butler proposes a model of enlargement of parts of the brain, along with paired eyes, as serving a preadaptation for later elaboration of the peripheral nervous system. Similarities can, in part, be explained by the novel concept of generative homology, where fundamental similarities in rostral brain development are due to shared genetic basis inherited from a common ancestor.

Chapter 20 provides an overview of a fundamental problem of central coordination of sensory inputs—that is, the stimulation of sensory systems by the animal's own activity. Electrosensory and mechanosensory systems are particularly prone to self-generated noise. Bioelectric potentials produced by the animal itself drive low-frequency electroreceptors over their entire dynamic range. Mechanosensory systems such as the lateral line detect differential movement of the body of the animal and the surrounding water, whether those movements are due to movements of the animal itself, or to external sources. Self-generated stimulation can also affect other sensory systems, for example, the visual system. As Bell et al. (1997) have stated, "Many aspects of sensory processing may be understood as the generation of expectations or predictions about sensory input and the removal of such expectations from the current sensory flow." Chapter 20 concentrates on how behaviorally important signals are extracted from noise in the elasmobranch electrosensory system and details the elegant cerebellar-like structure that performs this task. However, the issues raised in the chapter cut right across the sensory spectrum.

Sorting out signal/noise issues close to the periphery is one thing; understanding of sensory coordination at the level of the telencephalon is quite another. Chapter 21 makes an incursion into this challenging area, exploring connectivity and function of the teleost dorsal telencephalon, or pallium. It argues the case for the involvement of the pallium in primary visual processing and vision-related multisensory integration and sensorimotor control. The main anatomical substrates for these functions are proposed to be a reciprocal pallial-tectal loop, and multisensory thalamic pathways. It is speculated that the pallial-tectal loop is involved in habituation, spatial orientation, and complex visual integrative functions, whereas thalamic pathways are more involved in social and other species-typical behaviors.

On an evolutionary theme, Chapter 22 explores the remarkable similarities of the electroreceptive rostral bill organ in platypus and paddlefish. Convergent evolution provides us with some of our strongest insights into the evolutionary process and the structure/function relationship. At the level of fine structure, the electroreceptors of these two animals are quite different, just reinforcing our understanding of the independent evolution of these systems. Behavioral sensitivity for both the platypus and the paddlefish is much higher than the sensitivity of individual electroreceptors in each animal, with the implication that both perform complex signal processing on the inputs from large arrays of electroreceptors. The most remarkable convergence between the systems is the similar shape of the rostrum and the bill, with the concentration of electroreceptors along the edges. This produces an axis of greatest sensitivity to electrical stimuli that originates laterally, and a remarkably similar prey capture style. In both species, prey are captured following a short latency, lateral head saccade. The head saccades of both these taxa involve a very rapid computation of the details of the electric field produced by the prey and can be contrasted with the elasmobranch approach algorithm suggested by Kalmijn (1997). It is surprising that the flattened bill shape along with its computation of target position has evolved twice, and suggests that it would be very fruitful to pursue the underlying neural basis of target location in each of these cases.

These papers on the evolution and central coordination of sensory inputs may raise more questions than they answer. However, central coordination of sensory input at all levels of the neuraxis, and an appreciation of fundamental processes of evolution and development have an exciting part to play in generating a mechanistic understanding of sensory processing in the aquatic environment.

References

Bell, C., Bodznick, D., Montgomery, J., and Bastian, J. (1997). The generation and subtraction of sensory expectations within cerebellar-like structures. *Brain Behav. Evol. Suppl* 1:17–31.

Kalmijn, A.J. (1997). Electric and near-field acoustic detection: A comparative study. *Acta Phyiol. Scand.* 161; Suppl 638:25–38.

19
Sensory Systems and Brain Evolution Across the Bilateria: Commonalities and Constraints

Ann B. Butler

Abstract

Enlarged brains with relatively elaborate neural circuitry have arisen multiple times across bilaterally symmetrical animals, particularly in cephalopod molluscs, arthropods, and craniates (hagfishes and vertebrates). The transition from noncraniate chordate to craniate was marked by a substantial enlargement of both the central and peripheral nervous systems. A plausible model for this transition posits that enlargement of parts of the brain, along with paired eyes, occurred first and served as a preadaptation for later elaboration of the peripheral nervous system. This model is consistent with the sequence of evolutionary events that are postulated to have occurred in the elaboration of the brain within various arthropod lineages. The nervous systems of the earliest bilaterally symmetrical animals are deemed to have had at least the genetic basis for a rostral brain that included a hypothalamic region with paired eyes or eye spots, a hindbrain region, and descending nerve cord(s). Specializations present in the brains of both arthropods and craniates can be accounted for as elaborations on this basic plan.

1. Introduction

Sensory systems require analysis of stimuli by more or less elaborate pools of neurons if they are to be exploited for adaptive advantages. Enlarged, elaborated brains with paired eyes have arisen multiple times across the radiation of bilaterally symmetrical animals, which comprises protostomes and deuterostomes. Within deuterostomes, the origin of craniates (hagfishes and vertebrates) was marked by developmental changes that resulted in elaboration of the central nervous system and gain of many new structures. Similar neural elaborations occurred in protostome lineages, particularly arthropods and molluscs.

Recent studies of gene expression patterns provide exciting new insights into nervous system development across chordates (Holland and Holland, 1999) and also invite intriguing comparisons between deuterostome chordates and protostomes. Many shared characters can now be recognized as products of the same shared genetic bases and, as such, are examples of generative homology, or syngeny (Butler and Saidel, 2000). This new concept permits acknowledgment of "sameness" for a given character due to inheritance of the generative process(es) from a common ancestor, whether or not the character also meets criteria for phylogenetic (historical) (Wiley, 1981) or biological (Wagner, 1989) homology.

Butler (2000a,b) recently proposed a model for a transitional morphotype between the ancestral chordate condition and the definitive craniate condition, called a "cephalate." When this "cephalate" is compared with independently derived morphotypes of nervous systems of ancestral protostomes (Arendt and Nübler-Jung, 1996; Strausfeld, 1998; Wullimann, 2001) fundamental similarities are revealed. This chapter considers the emergence of the earliest craniates, recent work on the nervous system in extant cephalochordates (lancelets), and a scenario for the transition to definitive craniates. Brain evolution in some protostomes and comparisons of the cephalate morphotype with ancestral protostome morphotypes are then discussed. The initial basis for nervous system elaboration across all bilaterally symmetrical animals can be recognized: a forebrain including hypothalamus with paired eye structures and infundibulum, a midbrain/isthmus region, a hindbrain, and a nerve cord.

2. Chordate Nervous Systems and the Craniate Transition

Extant deuterostomes comprise echinoderms, hemichordates, and chordates. The latter encompass urochordates (tunicates, or ascidians), cephalochordates (lancelets, commonly called amphioxus), and craniates. In fact, lancelets may represent the outgroup to two sister radiations: the small, early Cambrian chordate *Haikouella* and the craniates (Holland and Chen, 2001). Among deuterostomes, a definitively elaborated brain is unique to craniates. While protostomes and nonchordate deuterostomes exhibit a ventral nerve cord that runs from the rostral brain region down the length of the body, chordates exhibit a dorsal nerve cord. Recent gene expression evidence (Holley et al., 1995; De Robertis and Sasai, 1996) supports the hypothesis broached by Geoffroy-Saint-Hilaire in 1822 that chordates are essentially inverted protostomes. Nübler-Jung and Arendt (1996) and Nielsen (1999) have argued that the inversion of the dorsoventral body axis took place in the transition to chordates rather than at the origin of deuterostomes. The chordate central nervous system of brain and dorsal nerve (spinal) cord is thus directly comparable to nonchordate nervous systems with the dorsoventral body axis reversed.

2.1. Craniate Nervous Systems

The ancestral craniate morphotype (Northcutt, 1985, 1996a,b; Wicht and Northcutt, 1992; Fritzsch and Northcutt, 1993; Braun, 1996; Wicht, 1996; Wullimann, 2001) had a forebrain with two divisions, telencephalon and diencephalons, plus midbrain and hindbrain (Fig. 19.1). Paired olfactory bulbs projected to most or all of the telencephalic pallium. The diencephalon comprised the epithalamus (at least habenular nuclei), dorsal thalamus, ventral thalamus, and a hypothalamus with paired lateral eyes. The midbrain consisted of a dorsal tectum and a ventral tegmentum. A major neural coordinating and motor center, the reticular formation, was present within the hindbrain. A cerebellum was lacking. Extant vertebrates additionally have a median, dorsal part of the epithalamus, the epiphysis (the pineal/parietal structure), that hagfishes lack. Assignment of an epiphysis to the common craniate ancestral morphotype is indicated by outgroup comparison with lancelets, as discussed below.

The peripheral nervous system of the earliest craniates comprised up to 16 cranial nerves (see Butler and Hodos, 1996). Of these, the optic and epiphyseal nerves differ in that they are not in fact "nerves" or parts of the peripheral nervous system but are rather tracts of the central nervous system, since their components all derive embryologically from the neural tube (see Eagleson and Harris, 1990; Eagleson et al., 1995). Most of the sensory system receptors and all of the bipolar neurons of the peripheral nervous system are derived embryologically from two tissues of crucial importance for the evolution of craniates, migratory neural crest

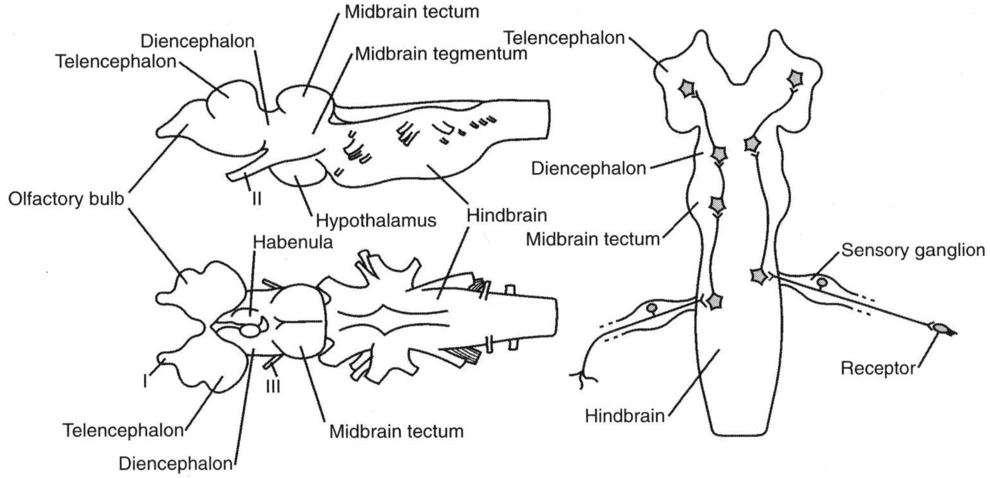

FIGURE 19.1. Lateral and dorsal views of ancestral craniate morphotype (left) constructed after Northcutt (1985), Wicht and Northcutt (1992), Braun (1996), and Wicht (1996). Cranial nerves IV–X are shown on brainstem unlabeled. Rostral is toward left. Schematic of central sensory pathways in craniates (right) with rostral toward the top.

and neurogenic placodes (see Northcutt, 1996b). The former arises from the neurectoderm at the lateral edges of the infolding neural tube, while the latter are thickened areas of ectoderm on the surface of the head. Migratory neural crest also gives rise to a host of other structures, including most of the cranium.

The diverse array of peripheral receptor systems for the true cranial nerves contrasts with the striking similarity of sensory pathway organization within the central nervous system (Fig. 19.1) (Butler and Hodos, 1996; Hodos and Butler, 1997, 2001). In all sensory systems, the bipolar neurons project to a set of first-order multipolar neurons (FOMs) in the central nervous system that, in turn, projects to other sets of multipolar neurons in the diencephalon or midbrain (Butler, 2000a,b). In the visual system, the FOMs are the retinal ganglion cells that receive input from the retinal bipolar cells. These FOMs give rise to two distinct pathways to sets of multipolar neurons within the diencephalon, one that is direct and one that involves an intermediate synapse in the midbrain tectum.

This basic pattern of direct diencephalic projections from the FOMs and/or a second pathway via the midbrain roof applies to most sensory systems (Butler, 1995; Butler and Hodos, 1996), the only exceptions being the two telencephalic cranial nerves, the olfactory and terminal nerves. The diencephalic target of each pathway then in turn projects to the telencephalon. No set of FOMs in the brainstem fails to give rise to an ascending pathway or projects directly to the telencephalon.

2.2. The Nervous System in Lancelets

Larval lancelets have many of the major brain parts that craniates do (Lacalli, 1994, 1996a,b,c; Lacalli et al., 1994; Lacalli and Kelly, 2000). The cerebral vesicle (Fig. 19.2A) appears to be homologous to the craniate diencephalon. One can compare its median, unpaired frontal organ of pigment cells and associated neurons with the paired eyes of craniates; its median, unpaired lamellar body with the craniate epiphysis; ventrally lying infundibular cells with the craniate infundibulum; and possibly its so-called balance organ with the craniate hypothalamus. In this regard, it is noteworthy that the craniate eyes arise embryologically by an initial midline outgrowth that then splits to form the bilateral pair of optic evaginations (Lacalli, 1994). This outgrowth is from the preoptic area at the rostral pole of the

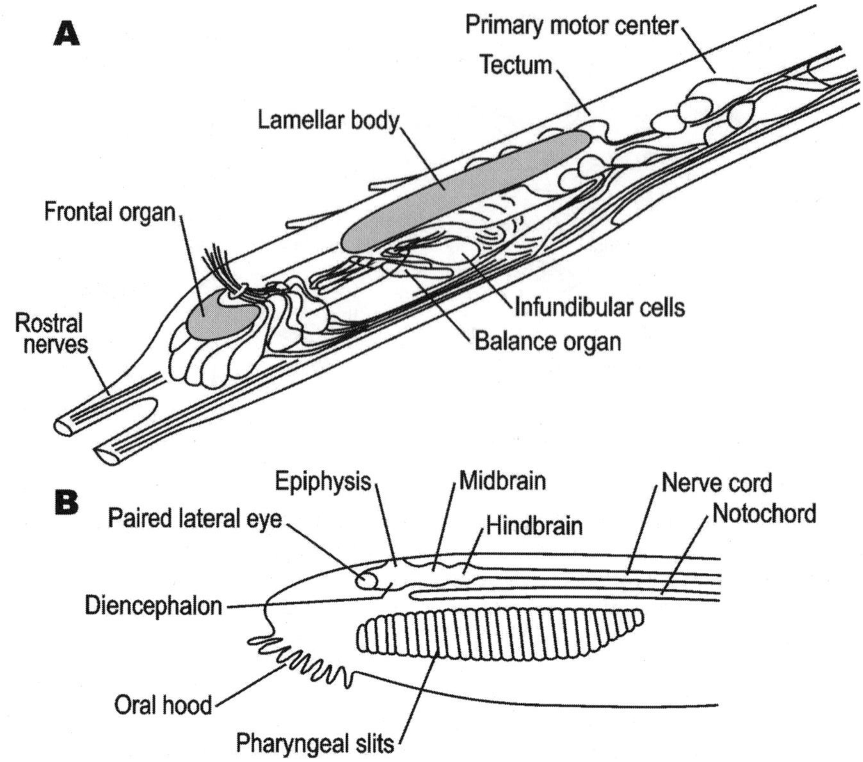

FIGURE 19.2. (A) Dorsolateral view of the lancelet brain, after Lacalli (1996b), with rostral toward lower left. (B) Lateral view of the cephalate morphotype, after Butler (2000a,b), with rostral toward left.

hypothalamus (Bulfone et al., 1993; Puelles and Rubenstein, 1993; Shimamura et al., 1997). A putative homologue of either the dorsal thalamus or posterior tuberculum, the major parts of the diencephalon used for ascending sensory relay in craniates, is not apparent in lancelets or in larval tunicate brains, so this crucial part of the craniate brain was probably absent in the common craniate-cephalochordate ancestor.

A small tectal region caudal to the cerebral vesicle is a putative homologue of the craniate midbrain (Lacalli, 1996a,b,c). It contains part of a column of motor neurons that extends caudally into the hindbrain region, the primary motor center (PMC). The PMC includes some reticular formation-like neurons (Fritzsch, 1996) that are presumably involved in the production of startle responses and undulatory swimming movements (Lacalli, 1996a,b,c). Lacalli (1996a,b,c) has identified three descending pathways from the frontal organ to tectal cells and from there to the PMC for such responses, one of which is a direct receptor cell–tectal connection. As discussed below, enhancement of this descending pathway may have been a key part of the transition from the ancestral chordate condition to definitive craniates.

Recent gene expression studies strongly support comparison of the lancelet cerebral vesicle to the craniate diencephalon (Holland and Chen, 2001). *AmphiOtx* is expressed in this region in lancelets (Holland and Garcia-Fernàndez, 1996; Holland et al., 1996; Williams and Holland, 1996), and its homologue *orthodenticle*, or *Otx*, in the forebrain and midbrain of craniates (Simeone et al., 1993). Similarly, the related gene *Hroth* is expressed in the anterior two-thirds of the sensory vesicle of tunicate larvae (Katsuyama et al., 1996; Wada et al., 1998), a structure that lies at the rostral end of the neuraxis and appears to correspond to the lancelet cerebral vesicle.

Lancelets have a pair of rostral nerves in the head region (Lacalli et al., 1994; Northcutt, 1996b), various sensory cells (Demski et al., 1996; Fritzsch, 1996), and several small dorsal roots (Lacalli et al., 1994; Fritzsch, 1996). Fritzsch (1996) noted a possible comparison of the second sensory nerve of lancelets with either the trigeminal or the facial nerve of craniates. In addition, dorsal sensory cells are present in the lancelet spinal cord. Thus, lancelets may have at least the precursors of two tissues crucial to craniate evolution, neurogenic placodes and migratory neural crest (Fritzsch and Northcutt, 1993; Fritzsch, 1996). Recent evidence supports the view that such tissues were present in the earliest chordates (Holland and Chen, 2001). The cupular organs of tunicates, a possible homologue of the vertebrate acousticolateralis system (Bone and Ryan, 1978), develop from epidermal thickenings and express *Pax* gene homologues in common with the developing vertebrate ear (Wada et al., 1998). Similarly, the so-called neurohypophyseal duct of tunicates, which develops caudal to the sensory vesicle in association with the pharynx and a large visceral ganglion, appears to be homologous to the olfactory/adenohypophyseal/hypothalamic placode of vertebrates (Nozake and Gorbman, 1992; Manni et al., 1999; and see Graham and Begbie, 2000): this structure gives rise to migratory cells that form a cerebral ganglion and to a neural complex of cells that are immunoreactive to pituitary hormones (Manni et al., 1999). Whittaker (1997) has pointed out that melanocytes, which are derived from neural crest in craniates, are present in these chordates in a series of photoreceptor units called the cup organs of Hesse that lie in the lateroventral part of the neural tube.

2.3. A Common Cephalochordate-Craniate Morphotype

We can thus deduce that a common cephalochordate-craniate ancestor had a rostral neural specialization that comprised a homologue of the craniate diencephalon with at least epiphyseal and hypothalamic components, possibly a homologue of the craniate midbrain, a column of motor cells that included a homologue of the craniate reticular formation, and a spinal cord with some sensory and motor elements. The common ancestor lacked the craniate-type of elaborate peripheral sensory system with bipolar neurons and receptor cells derived from migratory neural crest and neurogenic placodes, although precursors of these tissues were present.

The common ancestor may or may not have had paired eyes. As Lacalli (1994) has discussed, both single and paired apical organs (corresponding to the lancelet frontal organ) occur across various species of echinoderm larvae. These paired apical organs develop from an initial midline outgrowth that then splits, as is also the case for the formation of the paired eyes of craniates. While the phylogenetic distribution of paired eyes does not permit assignment of this character state to the common cephalochordate-craniate ancestor, one can deduce that the genetic bases for paired eyes were present, particularly a homolog of the gene *sonic hedgehog* (*Shh*). This gene contributes to dorsoventral differentiation of the neural tube and the development of paired eyes (see Rubenstein and Beachy, 1998; Rubenstein et al., 1998). In fact, interruption of the Shh signal results in a single, median, cyclopic eye (Chiang et al., 1996).

2.4. The Transition to Craniates

The transition from the common ancestral condition to definitive craniates was both rapid and extremely complex. It involved the achievement of a definitive brain with paired eyes and a bloom of peripheral nervous system components. Holland and Graham (1995) proposed that these events were approximately concurrent but did not discuss how, while Northcutt (1996b) proposed that the origin of craniates was rapid and "without transitional forms." The alternative possibility of serial transformation also has been proposed (Thomson, 1988; Lacalli, 1996b; Williams and Holland, 1996). Butler (2000a,b) recently offered a more detailed scenario for serial transformation with elaboration of the brain and paired eyes preceding elaboration of the placodal and migratory neural crest sensory systems and with the initial forebrain elaboration being restricted to the diencephalon. Recent evidence indicates that two genomic duplication events occurred during the transition (see Meyer and Schart, 1999), so one possibility is that a transitional condition arose with the first of these duplications and full-blown craniates with the second. Elaboration of a brain preceding that of the neural crest is consistent with fossil evidence of the chordate *Haikouella*, which may be the most proximal sister group to the craniates. *Haikouella* appears to have a brain with two or three lobes and possibly paired eyes (Holland and Chen, 2001).

A major adaptive benefit from elaboration of the peripheral nervous system in the absence of brain elaboration is difficult to envision and is not exhibited by any known animal, but the reciprocal scenario of brain elaboration occurring first is attractive. All instances of brain elaboration in bilaterally symmetrical animals include elaboration of paired eyes, and often the eyes are significantly enlarged. Preferential up-regulation of the genes that affect development of the craniate neural plate and paired eyes, such as *Shh* (see Rubenstein and Beachy, 1998), could have resulted in this grade of organization (Fig. 19.2B). Subsequent elaboration of migratory neural crest and placodes would have been immediately successful, since the basic template for sensory pathway organization would have been in place to be colonized and exploited by these new sensory systems, as discussed below.

As Northcutt (1996a) has discussed, the derivatives of the neural plate in craniates are a close match for the neural components present in cephalochordates, as revealed by fate mapping of the neural plate in the frog *Xenopus* (Eagleson and Harris, 1990; Eagleson et al., 1995). The neural plate is bordered by neural folds, lateral ridges along the lateral edges of the neural plate, and an anterior ridge rostrally. The most medial and proximal part of the anterior neural ridge and the adjacent anterior part of the neural plate give rise to the paired retinae, the ventral part of the hypothalamus, and to the infundibulum and posterior pituitary. The more caudal parts of the neural plate give rise to the major portions of the midbrain and the hindbrain tegmentum. Most of the components of the craniate telencephalon, as well as the anterior part of the pituitary, are derived from the midlateral part of the anterior neural ridge. The most anterior part of the lateral ridge gives rise to the epiphysis and the habenular nuclei, and the more caudal parts of the lateral ridge give rise to other structures, including the cerebellum (which is lacking in hagfishes and very small in lampreys). Thus, the postulated cephalate transitional brain maps onto the neural plate and the proximal and most medial part of the anterior neural ridge, along with a small portion of the rostral part of the lateral ridge from which the epiphysis develops. Northcutt (1996a,b) discussed these relationships in detail and noted that the craniate brain is "in part, an elaboration of a lancelet-like level of organization. . . ."

Neural plate elaboration in the cephalate morphotype (Butler, 2000a,b) implies that existing descending visuomotor pathways would have been fortified (Fig. 19.3A,B) and then could have served as a template for central pathways of other sensory systems. The newly elaborated dorsal thalamus would have received direct retinal projections and become an additional contributor to the descending pathways. Such pathways, while minor, are present in at least some craniates (Northcutt and Wullimann, 1988; Neary and Wilczynski, 1979, 1980). Subsequent addition of other sensory systems derived from migratory neural crest and placodes could have then followed the same pattern, and the telencephalon would have been "tacked on" as an additional, ultimate, rostral relay for these sensory pathways (Fig. 19.3C).

Initial visual dedication of the diencephalon and midbrain is consistent with the fact that no population of first-order multipolar neurons for any craniate neural crest or placodal sensory systems occurs there (see Butler and Hodos, 1996). The craniate peripheral sensory systems have FOMs in the telencephalon, hindbrain, and spinal cord, but only the visual systems

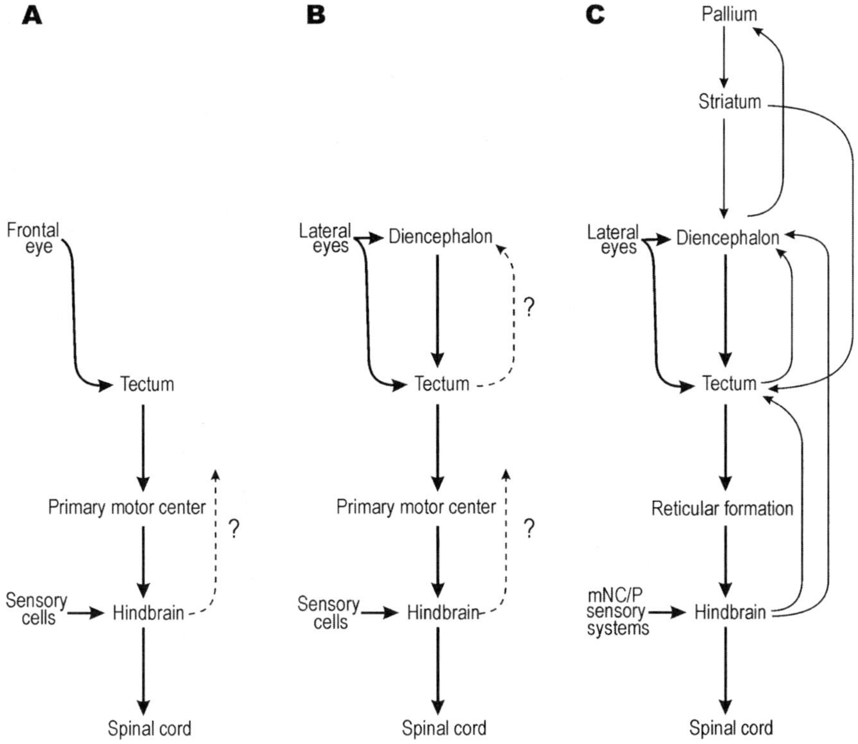

FIGURE 19.3. Proposed evolutionary sequence for central sensory pathways in the ancestral chordate-cephalate-craniate transition, after Butler (2000a,b). (A) Descending visual pathway in lancelet. (B) Postulated ancestral "cephalate" transitional condition with paired lateral eyes and augmentation of descending visual pathway. (C) Craniate condition with telencephalon and elaborated migratory neural crest/placodal (mNC/P) sensory systems.

have FOMs in the diencephalon, while no FOMs exist in the midbrain. Prior "colonization" of the diencephalon by visual FOMs could thus account for the lack of neural crest–placodal system FOMs in this region.

A mediolateral shift of expression gradients of some regulatory genes, with later up-regulation of genes that specify more lateral tissues, could have produced the serial transformation. Neural tissue development is antagonized by bone morphogenetic proteins (BMPs), which exert a repressive influence and promote epidermal fate (see Lumsden and Krumlauf, 1996; Kerszberg and Changeux, 1998; Rubenstein et al., 1998). Expression of BMPs in the medial region that gives rise to the neural plate is inhibited by expression of *chordin*, which is produced by cells in the underlying notochord. *Chordin* expression thus allows formation of the neural plate. Increased *chordin* expression could thus contribute to elaboration of the nerve cord, such as is present in protostomes with elaborated brains, and also to the hypothesized cephalate transitional form in the chordate line to craniates.

For later elaboration of placodal systems, up-regulation of BMPs lateral to the nerve cord could result in placode formation. For example, the signaling molecule BMP7, which is produced by pharyngeal endoderm, mediates induction of epibranchial placodes for the taste cranial nerves of vertebrates (Graham and Begbie, 2000). Different sets of placodes are induced by different signals, which suggests that various placodes may have independent evolutionary histories and may not have all been gained and/or elaborated at the same time (Graham and Begbie, 2000). Further, induction of a number of placodes is mediated by initial signals from the central nervous system itself, such as the olfactory and otic placodes, which depend on early signals from mesoderm and endoderm followed by signals derived from forebrain and isthmal (midbrain-hindbrain junction) tissue, respectively; the same brain region that induces a particular placode contains the FOMs that receive input from it (see Graham and Begbie, 2000). This observation strongly supports the idea that some elaboration of the central nervous system of craniates preceded that of the placodal sensory systems, since the latter depend on the former to develop.

3. Protostomes: Shared Generative Bases for the Same Brains

Across the Bilateria, many shared structures of like topology share at least some genetic specification. Such "sameness" has been called homology in some instances despite lack of phylogenetic congruity (Patterson, 1982). Sandeman and Scholtz (1995) noted that "some astounding examples of complex similarities . . . in vertebrates and arthropods . . . cannot possibly have been derived from a common ancestral character." On the other hand, it is now indisputable that, in many cases, the fundamental similarities of such characters are due to shared genetic bases inherited from a common ancestor. The concept of generative homology, or syngeny, has thus been proposed by Butler and Saidel (2000) for all characters with shared generative pathways regardless of phylogenetic distribution of the character itself.

Flatworms may be members of a sister group to the rest of the Bilateria, the Eubilateria (Reuter and Gustafsson, 1995). The rostral part of the nervous system consists of a rostral concentration of nerve cells, or brain, with associated cerebral lobes (Brusca and Brusca, 1990; Reuter and Gustafsson, 1995). Paired eye spots, or ocelli, are located on the dorsal surface of the head and are connected to the cerebral lobes (Brusca and Brusca, 1990; Kiyokazu et al., 1998). As with other Bilateria, *Otx*-related homeobox genes are expressed in the cerebral lobes, including in the region where the visual axons terminate (Yoshihiko et al., 1999). Also of interest, growth hormone releasing factor is predominantly localized in the brain region (Reuter and Gustafsson, 1995). The nervous system of the body consists of longitudinal nerve cords that are interconnected by commissures, creating a ladder-like configuration, the "orthogon."

In contrast to the flatworms, highly elaborated nervous systems occur in several taxa of protostomes, including the arthropod lineage, which is considered here. The arthropod brain (see Brusca and Brusca, 1990) consists of two to three ganglia, a protocerebrum, in some cases a deutocerebrum, and a tritocerebrum (Fig. 19.4A). The protocerebrum lies dorsal to the deutocerebrum and tritocerebrum. The tritocerebrum has connectives that encircle the esophagus and interconnect with a set of other head ganglia collectively referred to as the subesophageal ganglion. In insects, a hypocerebral ganglion is present in the region between the brain and the gut (Fig. 19.4A). It is associated with two pairs of glandular structures that are connected with neurosecretory cells in the protocerebrum and that are involved in neuroendocrine regulatory functions (Brusca and Brusca, 1990). Arendt and Nübler-Jung (1996, 1999) compared the medial zone of the brain in insects to the hypothalamic infundibular region of vertebrates. They aligned the non-hypothalamic parts of the prosencephalon of vertebrates with the more lateral parts of the protocerebrum and deutocerebrum on the basis of gene expression and axonal scaffolding data. Single or double ventral nerve cords extend caudally from the rostral part of the nervous system and have segmental ganglia along their length.

In crustaceans and insects (Sandeman and Scholtz, 1995; Strausfeld et al., 1995; Strausfeld, 1998) (Fig. 19.5A,B), the optic lobe structures lie within eye stalks and receive retinal input. They develop from *anlagen* of the protocerebrum, which also contains neuropils that span the midline: the fan-shaped body and (in some

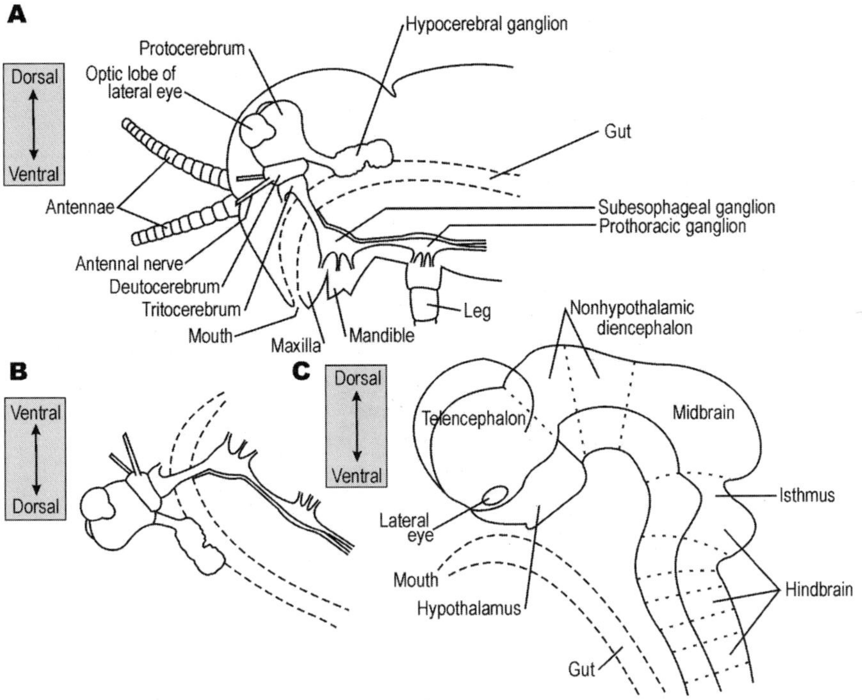

FIGURE 19.4. (A) Lateral view of the brain within the head of a generalized insect, after Brusca and Brusca (1990), showing the protostome position of the mouth. (B) Dorsoventrally inverted lateral view of generalized insect brain and gut from A for comparison with C. (C) Lateral view of generalized vertebrate neural tube, after Rubenstein and Beachy (1998), with the deuterostome mouth position. Rostral is toward left and dorsal toward top.

genera) a protocerebral bridge. In crustaceans, the protocerebrum also contains olfactory-recipient antennule lobes and related structures (including an accessory lobe and hemiellipsoid body). In insects, a central body complex is present in the protocerebrum, of which the fan-shaped body is one of several components. Antennal lobes lie within the deutocerebrum and tritocerebrum. In crustaceans, these lobes are only mechanoreceptive, but in insects they are both mechanosensory and olfactory and project to a large structure in the protocerebrum, the mushroom body, which has elaborate cytoarchitecture, receives visual input, and has been implicated in learning and memory.

A common ancestral crustacean-insect morphotype (Strausfeld, 1998) had the three preoral neuromeres, protocerebrum, deutocerebrum, and tritocerebrum, with lateralized neuropils (Fig. 19.5C). It lacked a protocerebral bridge but had at least one median neuropil, the fan-shaped body in the protocerebrum, and optic lobes with two consecutive neuropils (lamina and medulla) for input from the paired compound eyes. Mechanosensory antennal lobes in the deutocerebrum received input from a single pair of antennae. A mushroom body that received visual and mechanosensory projections from the respective sensory lobes may have been present in the protocerebrum. The common ancestral crustacean-insect morphotype also exhibited rostrocaudal scaffolding created by the longitudinal connectives during development.

Since many crustaceans exhibit an unpaired, median eye in addition to the paired compound eyes (Williamson, 1992), this character may also have been present in the common crustacean-insect ancestor. Some insects (honeybees and locusts, for example) exhibit one or more ocelli

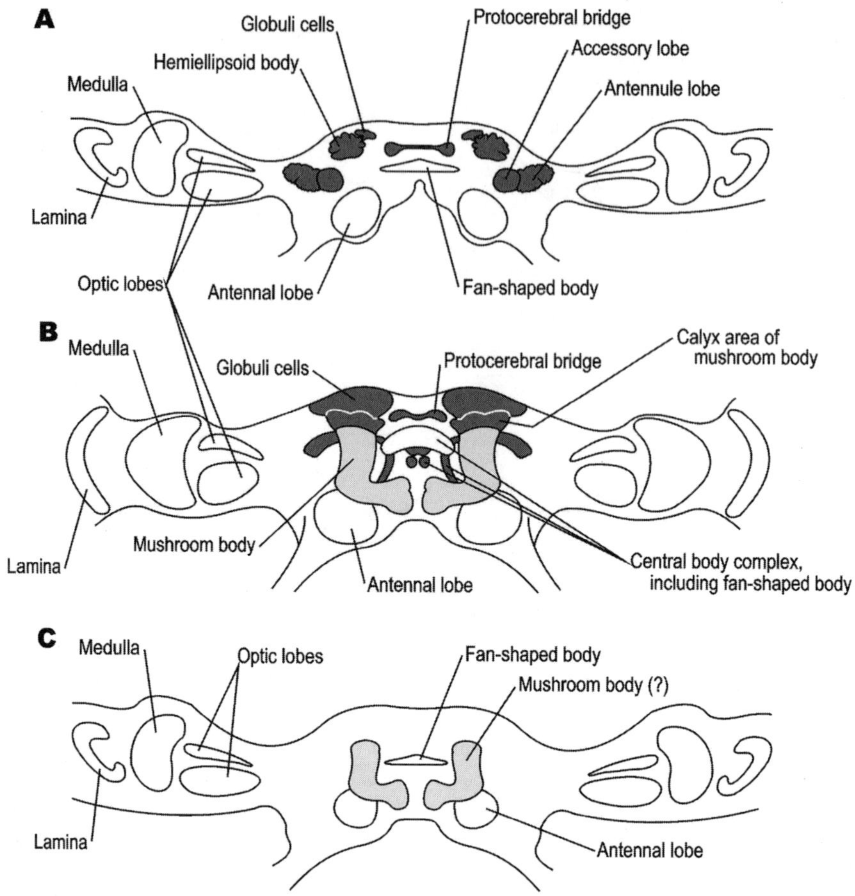

FIGURE 19.5. Dorsal views of the brain of a crustacean (A), an insect (B), and an ancestral crustacean-insect morphotype (C). A and B are after Strausfeld (1998) and C is constructed from A and B and from text discussion (Strausfeld, 1988). Apomorphic characters are indicated by dark shading and plesiomorphic characters are in white. The mushroom body in B and C is lightly shaded, since it may or may not have been in the common ancestor.

in the median region of the head (see Brusca and Brusca, 1990).

As Strausfeld's (1998) analysis indicates, each descendant taxon acquired different specializations. Insects added olfactory chemoreception to mechanoreception for their single pair of antennae, as well as calyces (a capped sensory-reception region) for the mushroom bodies and an elaborated central body complex. In crustaceans, the mushroom body was lost (if it was present ancestrally), and a second pair of head appendages for olfaction, with their associated antennule lobes (accessory lobes), and hemiellipsoid bodies were gained. Thus, in both cases, paired eyes were inherited from the common ancestor, whereas an olfactory system was secondarily acquired.

4. Cephalates and Protostome Ancestral Morphotypes: Commonalities of Design for Brain Elaboration

The finding that nervous systems share patterning genes across the Bilateria allows us to reconstruct a morphotype for the deep ancestry of elaborated nervous systems. Whether the common ancestor actually expressed this phenotype cannot be determined, but the generative bases for the shared characters were indeed present in that ancestor. This morphotype incorporates characters that are syngenous (Butler and Saidel, 2000), whether or not they are historically homologous.

The morphotype for the common ancestor of crustaceans and insects (Strausfeld, 1998) and the proposed cephalate morphotype (Butler, 2000a,b) for the transition to craniates both have a brain region with a rostral portion that includes paired lateral sets of neuropils in receipt of input from paired eyes (or eye spots). Neither have a sensory system for olfaction. Both have lateralization of brain neuropils (nuclei) and commissural connections.

In its plesiomorphic condition, the rostralmost part of the brain in protostomes, the protocerebrum, can be directly compared with the hypothalamus of chordates. A brain region that lies at the rostral pole of the neuraxis, gives rise to paired eyes (or eye spots), and is involved in neuroendocrine functions appears to be a fundamental feature of bilaterally symmetrical nervous systems. The unelaborated protostome protocerebrum and the chordate hypothalamus are thus postulated to be syngenogues. Due to flexure of the neuraxis, the protocerebrum in some protostomes lies dorsal to the rest of the brain neuromeres (Fig. 19.4A,B), just as the hypothalamus lies ventral to the rest of the prosencephalon in vertebrates (Fig. 19.4C).

The ancestral crustacean-insect and cephalate morphotypes share the lack of a syngenogue of the vertebrate telencephalon with olfactory innervation. The presence of a telencephalon thus appears to represent a later specialization in the process of brain elaboration (see Holland and Holland, 1999). The mushroom body calyces and some of the central body complex neuropils in insects and the hemiellipsoid bodies and antennule lobes in crustaceans (Strausfeld et al., 1995; Strausfeld and Hildebrand, 1999) are among the elaborated structures that may represent craniate telencephalic syngenogues.

The deutocerebrum and tritocerebrum align with the more caudal part of the vertebrate prosencephalon and the mesencephalon (Fig. 19.4) on the basis of gene expression data (Arendt and Nübler-Jung, 1996), although the specific boundaries of the mesencephalic region are not yet fully clarified (see Arendt and Nübler-Jung, 1999). Part of the deutocerebrum, and possibly the tritocerebrum, thus appear to be syngenous to the nonhypothalamic portions of the diencephalon and the small midbrain region of both cephalochordates and vertebrates, specifically at least an epiphysis and a tectal region. Continuing caudally, gene expression data indicate that the subesophageal ganglia are comparable to the rhombomeres of the vertebrate hindbrain (Fig. 19.4) (Arendt and Nübler-Jung, 1996; Hirth et al., 1998). The same topological relationships thus apply to both protostome and deuterostome brains, with the hypothalamus constituting the rostralmost end of the neuraxis.

The primitive brain morphotype, or "plesiocephalon," for bilaterally symmetrical metazoa thus has a diencephalon with paired eyes (or eye spots) that arise from the hypothalamic region at the rostral neural pole, an infundibulum, and a median eye structure. It also has a small midbrain region and descending visuomotor pathways to the hindbrain and nerve cord(s). Other sensory systems, such as mechanoreception, with receptor neurons that arise from simple ectoderm may project into the more caudal part of the neuraxis. At a later evolutionary stage in various lineages, augmented and/or additional sensory systems were then added to this basic template.

Acknowledgments. I thank Shaun Collin and Justin Marshall for inviting me to the first International Conference on Sensory Processing of the Aquatic Environment, and the conference participants, particularly Robert Barlow and William Saidel, for their valuable feedback. This work was supported by NSF Grant IBN-9728155.

References

Arendt, D., and Nübler-Jung, K. (1996). Common ground plans in early brain development in mice and flies. *BioEssays* 18:255–259.

Arendt, D., and Nübler-Jung, K. (1999). Comparison of early nerve cord development in insects and vertebrates. *Develop.* 126:2309–2325.

Bone, Q., and Ryan, K.P. (1978). Cupular sense organs in *Ciona* (Tunicata: Ascidiacea). *J. Zool. Lond.* 186:417–429.

Braun, C.B. (1996). The sensory biology of the living jawless fishes: A phylogenetic assessment. *Brain Behav. Evol.* 48:262–276.

Brusca, R.C., and Brusca, G.J. (1990). *Invertebrates*. Sunderland, MA: Sinauer Associates Inc.

Bulfone, A., Puelles, L., Porteus, M.H., Frohman, M.A., Martin, G.R., and Rubenstein, J.L.R. (1993). Spatially restricted expression of *Dlx-1*, *Dlx-2* (*Tes-1*), *Gbx-2*, and *Wnt-3* in the embryonic day 12.5 mouse forebrain defines potential transverse and longitudinal segmental boundaries. *J. Neurosci.* 13:3155–3172.

Butler, A.B. (1995). The dorsal thalamus of jawed vertebrates: A comparative viewpoint. *Brain Behav. Evol.* 46:209–223.

Butler, A.B. (2000a). Chordate evolution and the origin of craniates: An old brain in a new head. *Anat. Rec.* [*New Anat.*] 261:111–125.

Butler, A.B. (2000b). Sensory system evolution at the origin of craniates. *Phil. Trans. Roy. Soc. Lond.* 355:1309–1313.

Butler, A.B., and Hodos, W. (1996). *Comparative Vertebrate Neuroanatomy: Evolution and Adaptation*. New York: Wiley-Liss.

Butler, A.B., and Saidel, W.M. (2000). Defining sameness: Historical, biological, and generative homology. *BioEssays* 22:846–853.

Chiang, C., Litingtung, Y., Lee, E., Young, K.E., Corden, J.L., Westphal, H., and Beachy, P.A. (1996). Cyclopia and defective axial patterning in mice lacking *Sonic hedgehog* gene function. *Nature* 383:407–413.

Demski, L.S., Beaver, J.A., and Morrill, J.B. (1996). The cutaneous innervation of amphioxus: A review incorporating new observations with DiI tracing and scanning electron microscopy. *Israel J. Zool.* 42:S-117–S-129.

De Robertis, E.M., and Sasai, Y. (1996). A common plan for dorsoventral patterning in Bilateria. *Nature* 380:37–40.

Eagleson, G.W., and Harris, W.A. (1990). Mapping of the presumptive brain regions in the neural plate of *Xenopus laevis*. *J. Neurobiol.* 28:146–158.

Eagleson, G., Ferreiro, B., and Harris, W.A. (1995). Fate of the anterior neural ridge and the morphogenesis of the *Xenopus* forebrain. *J. Neurobiol.* 28:146–158.

Fritzsch, B. (1996). Similarities and differences in lancelet and craniate nervous systems. *Israel J. Zool.* 42:S-147–S-160.

Fritzsch, B., and Northcutt, R.G. (1993). Cranial and spinal nerve organization in amphioxus and lampreys: Evidence for an ancestral craniate pattern. *Acta Anat.* 148:96–109.

Graham, A., and Begbie, J. (2000). Neurogenic placodes: A common front. *Trends Neurosci.* 23: 313–316.

Hirth, F., Hartmann, B., and Reichert, H. (1998). Homeotic gene action in embryonic brain development of *Drosophila*. *Develop.* 125:1579–1589.

Hodos, W., and Butler, A.B. (1997). Evolution of sensory pathways in vertebrates. *Brain Behav. Evol.* 50:189–197.

Hodos, W., and Butler, A.B. (2001). Sensory system evolution in vertebrates. In: *Brain Evolution and Cognition* (Roth, G., and Wullimann, M.F., eds.), pp. 114–133. Heidelberg: Spektrum Akademischer Verlag.

Holland, L.Z., and Holland, N.D. (1999). Chordate origins of the vertebrate central nervous system. *Curr. Opin. Neurobiol.* 9:596–602.

Holland, N.D., and Chen, J. (2001). Origin and early evolution of the vertebrates: New insights from advances in molecular biology, anatomy, and palaeontology. *BioEssays* 23:142–151.

Holland, N.D., Panganiban, G., Henyey, E.L., and Holland, L.Z. (1996). Sequence and developmental expression of *AmphiDll*, an amphioxus *Distal-less* gene transcribed in the ectoderm, epidermis and nervous system: Insights into evolution of craniate forebrain and neural crest. *Develop.* 122:2911–2920.

Holland, P.W.H., and Garcia-Fernàndez, J. (1996). *Hox* genes and chordate evolution. *Dev. Biol.* 173:382–395.

Holland, P.W.H., and Graham, A. (1995). Evolution of regional identity in the vertebrate nervous system. *Perspect. Devel. Neurobiol.* 3:17–27.

Holley, S.A., Jackson, P.D., Sasai, Y., Lu, B., De Robertis, E.M., Hoffmann, F.M., and Ferguson, E.L. (1995). A conserved system for dorsal-ventral patterning in insects and vertebrates involving *sog* and *chordin*. *Nature* 376:249–253.

Katsuyama, Y., Wada, S., and Saiga, H. (1996). Homeobox genes exhibit evolutionary conserved regionalization in the central nervous system of an ascidian larva. *Zool. Sci.* 13:479–482.

Kerszberg, M., and Changeux, J.-P. (1998). A simple molecular model of neurulation. *BioEssays* 20:758–770.

Kiyokazu, A., Yukihiro, S., Kentaro, K., Chiyoko, K., Yoshihiko, U., and Kenji, W. (1998). Structure of the planarian central nervous system (CNS) revealed by neuronal cell markers. *Zool. Sci.* 15:433–440.

Lacalli, T.C. (1994). Apical organs, epithelial domains, and the origin of the chordate central nervous system. *Amer. Zool.* 34:533–541.

Lacalli, T.C. (1996a). Dorsoventral axis inversion: a phylogenetic perspective. *BioEssays* 18:251–254.

Lacalli, T.C. (1996b). Frontal eye circuitry, rostral sensory pathways and brain organization in amphioxus larvae: Evidence from 3D reconstructions. *Phil. Trans. Roy. Soc. Lond. B.* 351:243–263.

Lacalli, T.C. (1996c). Landmarks and subdomains in the larval brain of *Branchiostoma*: Vertebrate homologs and invertebrate antecedents. *Israel J. Zool.* 42:S-131–S-146.

Lacalli, T.C., and Kelly, S.J. (2000). The infundibular balance organ in amphioxus larvae and related aspects of cerebral vesicle organization. *Acta. Zool.* 81:37–48.

Lacalli, T.C., Holland, N.D., and West, J.E. (1994). Landmarks in the anterior central nervous system of amphioxus larvae. *Phil. Trans. Roy. Soc. Lond. B.* 344:165–185.

Lumsden, A., and Krumlauf, R. (1996). Patterning the vertebrate neuraxis. *Science* 274:1109–1115.

Manni, L., Lane, N.J., Sorrentino, M., Zaniolo, G., and Burighel, P. (1999). Mechanism of neurogenesis during embryonic development of a tunicate. *J. Comp. Neurol.* 412:527–541.

Meyer, A., and Schart, M. (1999). Gene and genome duplications in vertebrates: The one-to-four (-to-eight in fish) rule and the evolution of novel gene functions. *Curr. Opin. Neurobiol.* 11:699–704.

Neary, T.J., and Wilczynski, W. (1979). Anterior and posterior thalamic afferents in the bullfrog, *Rana catesbeiana*. *Neurosci. Abstr.* 5:144.

Neary, T.J., and Wilczynski, W. (1980). Descending inputs to the optic tectum in ranid frogs. *Neurosci. Abstr.* 6:629.

Nielsen, C. (1999). Origin of the chordate central nervous system—and the origin of chordates. *Dev. Genes. Evol.* 209:198–205.

Northcutt, R.G. (1985). The brain and sense organs of the earliest craniates: Reconstruction of a morphotype. In: *Evolutionary Biology of Primitive Fishes* (Forman, R.E., Gorbman, A., Dodd, J.M., and Olsson, R., eds.), pp. 81–112. New York: Plenum Press.

Northcutt, R.G. (1996a). The agnathan ark: The origin of craniate brains. *Brain Behav. Evol.* 48:237–247.

Northcutt, R.G. (1996b). The origin of craniates: neural crest, neurogenic placodes and homeobox genes. *Israel J. Zool.* 42:S-273–S-313.

Northcutt, R.G., and Wullimann, M.F. (1988). The visual system in teleost fishes: Morphological patterns and trends. In: *Sensory Biology of Aquatic Animals* (Atema, J., Fay, R.R., Popper, A.N., and Tavolga, W.N., eds.), pp. 515–552. Berlin: Springer-Verlag.

Nozake, M., and Gorbman, A. (1992). The question of functional homology of Hatschek's pit of amphioxus (*Branchiostoma belcheri*) and the vertebrate adenohypophysis. *Zool. Sci.* 9:387–395.

Nübler-Jung, K., and Arendt, D. (1996). Enteropneusts and chordate evolution. *Curr. Biol.* 6:352–353.

Patterson, C. (1982). Morphological characters and homology. In: *Problems of Phylogenetic Construction* (Joysey, K.A., and Friday, A.E., eds.), pp. 21–74. London: Academic Press.

Puelles, L., and Rubenstein, J.L.R. (1993). Expression patterns of homeobox and other putative regulatory genes in the embryonic mouse forebrain suggest a neuromeric organization. *Trends Neurosci.* 16:472–479.

Reuter, M., and Gustafsson, M.K.S. (1995). The flatworm nervous system: Pattern and phylogeny. In: *The Nervous System of Invertebrates: An Evolutionary and Comparative Approach* (Briedbach, O., and Kutsch, W., eds.), pp. 25–59. Basel: Brikhäuser Verlag.

Rubenstein, J.L.R., and Beachy, P.A. (1998). Patterning of the embryonic forebrain. *Curr. Opin. Neurobiol.* 8:18–26.

Rubenstein, J.L.R., Shimamura, K., Martinez, S., and Puelles, L. (1998). Regionalization of the prosencephalic neural plate. *Ann. Rev. Neurosci.* 21:445–477.

Sandeman, D., and Scholtz, G. (1995). Ground plans, evolutionary changes and homologies in decapod crustacean brains. In: *The Nervous System of Invertebrates: An Evolutionary and Comparative Approach* (Breidbach, O., and Kutsch, W., eds.), pp. 329–347. Basel: Birkhaüser Verlag.

Shimamura, K., Martinez, S., Puelles, L., and Rubenstein, J.L.R. (1997). Patterns of gene expression in the neural plate and neural tube subdivide the embryonic forebrain into transverse and longitudinal domains. *Dev. Neurosci.* 19:88–96.

Simeone, A., Acampora, D., Mallamaci, A., Stornaiuolo, A., D'Apice, M.R., Nigro, V., and Boncinelli, E. (1993). A vertebrate gene related to *orthodenticle* contains a homeodomain of the *bicoid* class and demarcates anterior neuroectoderm in the gastrulating mouse embryo. *EMBO. J.* 12:2735–2747.

Strausfeld, N.J. (1998). Crustacean-insect relationships: The use of brain characters to derive phylogeny amongst segmented invertebrates. *Brain Behav. Evol.* 52:186–206.

Strausfeld, N.J., and Hildebrand, J.G. (1999). Olfactory systems: Common design, uncommon origins? *Curr. Opin. Neurobiol.* 9:634–639.

Strausfeld, N.J., Buschbeck, E.K., and Gomez, R.S. (1995). The arthropod mushroom body: Its functional roles, evolutionary enigmas and mistaken identities. In: *The Nervous System of Invertebrates: An Evolutionary and Comparative Approach* (Breidbach, O., and Kutsch, W., eds.), pp. 349–381. Basel: Birkhaüser Verlag.

Thomson, K.S. (1988). *Morphogenesis and Evolution*. New York: Oxford University Press.

Wada, H., Saiga, H., Satoh, N., and Holland, P.W.H. (1998). Tripartite organization of the ancestral chordate brain and the antiquity of placodes: Insights from ascidian *Pax-2/5/8*, *Hox* and *Otx* genes. *Develop* 125:1113–1122.

Wagner, G.P. (1989). The biological homology concept. *Ann. Rev. Ecol. Syst.* 20:51–69.

Whittaker, J.R. (1997). Cephalochordates, the lancelets. In: *Embryology: Constructing the Organism* (Gilbert, S., and Raunio, A.M., eds.), pp. 365–381. Sunderland, MA: Sinauer Associates.

Wicht, H. (1996). The brains of lampreys and hagfishes: Characteristics, characters, and comparisons. *Brain Behav. Evol.* 48:248–261.

Wicht, H., and Northcutt, R.G. (1992). The forebrain of the Pacific hagfish: A cladistic reconstruction of the ancestral craniate forebrain. *Brain Behav. Evol.* 40:25–64.

Wiley, E.O. (1981). *Phylogenetics: The Theory and Practice of Phylogenetic Systematics*. New York: Wiley-Liss.

Williams, N.A., and Holland, P.W.H. (1996). Old head on young shoulders. *Nature* 383:490.

Williamson, D.I. (1992). *Larvae and Evolution: Toward a New Zoology*. New York: Chapman & Hall.

Wullimann, M.F. (2001). Brain phenotypes and early regulatory genes: The Bauplan of the metazoan central nervous system. In: *Brain Evolution and Cognition* (Roth, G., and Wullimann, M.F., eds.), pp, 11–40. Heidelberg: Spektrum Akademischer Verlag.

Yoshihiko, U., Kenji, W., and Kiyokazu, A. (1999). Distinct structural domains in the planarian brain defined by the expression of evolutionarily conserved homeobox genes. *Devel. Genes. Evol.* 209:31–39.

20 Electroreception: Extracting Behaviorally Important Signals from Noise

David Bodznick, John Montgomery, and Timothy C. Tricas

Abstract

Sharks, skates, and rays exploit a marine environment rich in bioelectric and motional electric fields, which are very weak but nevertheless very useful for prey capture, predator avoidance, social interactions, and orientation in the sea. The elasmobranchs' appreciation of these fields is made possible by two distinct specializations. The first of these is an array of extraordinarily sensitive receptors, the ampullae of Lorenzini, which derive much of their sensitivity from positive feedback mechanisms held delicately balanced at threshold. The second specialization is a sophisticated set of filter mechanisms in the brain for extracting the weak electrosensory signals from much stronger background noise. A large portion of this background noise is created by the fish's own movements. Recent experiments show that a remarkable adaptive filter mechanism implemented by the cerebellar-like circuitry of the medullary electrosensory nucleus accounts for much of the noise suppression. The specializations of receptors and CNS so well developed in these fishes allow us to recognize important general principles operating in other sensory systems and in other vertebrates.

1. Introduction

Despite its relatively recent discovery, electroreception is widespread in aquatic vertebrates. Furthermore, the distribution of electroreception indicates that an original system based on receptors like the ampullae of Lorenzini evolved very early in vertebrate phylogeny, and that the electrosense has been lost and reevolved several times in vertebrates (Bullock et al., 1982). This chapter reviews our knowledge of Lorenzinian electroreception as exemplified by elasmobranchs, with a particular emphasis on recent advances. Like all sensory systems, the electrosense has its own peculiar set of physical and biological potentialities and limitations, but unlike many other sensory systems, these lie outside our own sensory experience. So the particular challenge of electroreception is to understand and depict an electrosensory world unfamiliar to us. To do this requires an understanding of the behaviorally important signals in the environment and the ways in which these are detected, encoded, and processed by the sensory periphery and central nervous system. Perhaps more than in other sensory systems, the electrosense has the potential to be corrupted by environmental and self-generated noise, so recent studies have focused on the strategies and mechanisms that exist for the fish to extract the behaviorally important signals from this noise.

2. Behaviorally Important Signals

2.1. Predation and Mate Detection

The most significant source of behaviorally relevant signals for elasmobranch electroreception is other animals in the environment. All aquatic animals pump ions to maintain the integrity of their body fluids and in doing so separate ionic charges. This creates electrical fields that are short-circuited by the surrounding seawater, forming quasi-dipolar sources. Other currents are generated by muscle action potentials but these do not appear to be significant sources for the elasmobranch electrosense since their frequency composition is largely higher than the sensitivity range of the electroreceptors. The fact that natural local dipole sources in the marine environment are exclusively of animate origin may well be one of the most important features of the electrosense. In effect, the electrosense "sees" only other animals in the environment, and it is other animals that form the most salient behaviorally relevant stimuli as potential prey or predators, or as conspecifics in social interactions. In visual terms, this is analogous to the infrared detectors of snakes or military snipers, which pick out the body heat of their mammalian targets against a uniform or featureless background. That local dipole sources can be equated with the presence of animals simplifies the central processing tasks for the electrosense. It is not necessary to distinguish relevant electrical image from complex background (or figure from ground in the terminology of visual psychophysics), and discrimination can be relatively straightforward.

In line with this expectation, behavioral studies show that sharks, rays, and skates make just the same orienting responses to either artificially produced or naturally occurring dipole electric fields (Fig. 20.1A). Sharks attracted to a bait source exhibit well-directed attacks at an adjacent small dipole source, ignoring the visual stimulus of the bait (Kalmijn, 1982). The calculated fields at the point the shark turns toward the source provide the often-quoted threshold sensitivity of the system of 5 nV/cm. Swell sharks use their electrosensory system to detect small fish being washed toward them in the subtidal surge zone on the reef; the shark's response to an approaching dipole is simply to open its mouth (Tricas, 1982). During mating season male Atlantic stingrays in the Sea of Cortez approach and investigate a buried plastic model of a female when the model is equipped with a synthesized bioelectric field (Fig. 20.1B) (Tricas et al., 1995). As is clear from these examples, the electrosense plays a major role in predation and in mate location and probably also in other social behaviors. Social and reproductive interactions, in some skates, incorporate pulsed discharges of an electric organ in their tails (Mikhailenko, 1971; Mortenson and Whitaker, 1973; Bratton and Ayers, 1987), though the behavioral importance of these discharges is yet unknown. Finally, it seems likely that the electrosense is also involved in predator detection and avoidance. Even skate embryos still within the egg capsule pause their normal respiratory tail undulations in response to weakly electric fields, one source of which could be a predator nearby (Sisneros et al., 1998).

2.2. Orientation and Navigation

In addition to the relatively small-scale bioelectric fields, there are also large-scale electrical fields in the environment, and electrical fields produced by movements of the animal itself (Kalmijn, 1974). Both of these result from Faraday's Law of electromagnetic induction that specifies the electromotive force induced in a conductor moving within a magnetic field. In the case of large-scale environmental fields, these are induced by

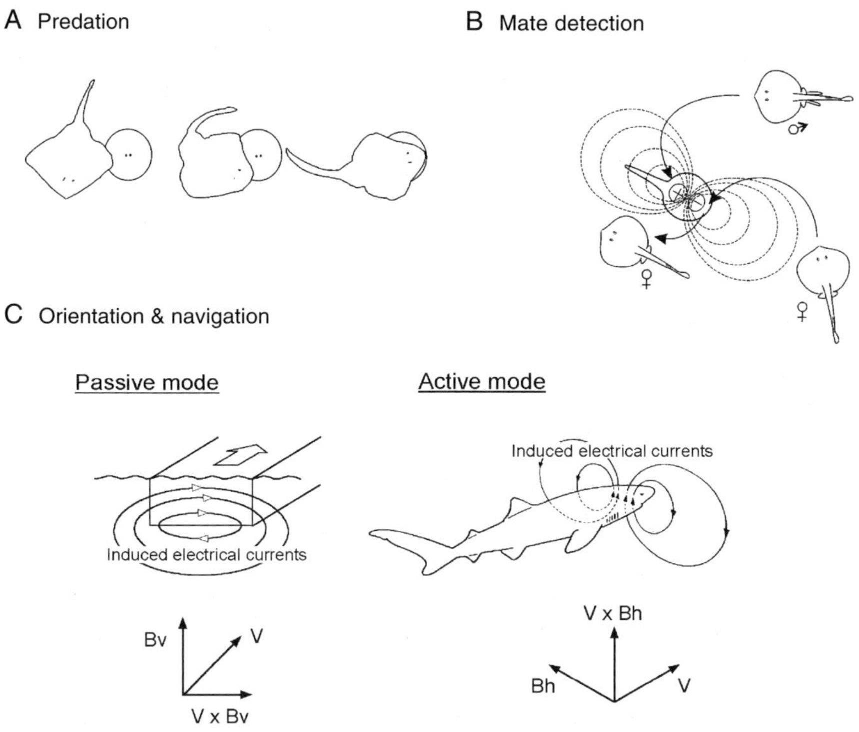

FIGURE 20.1. (A) Little skate, *Raja erinacea*, aroused by food odor, orienting to a DC dipole electrical field of 0.2 μV/cm at the perimeter of the test area (20 cm from electrodes buried in the sand). Images from video frames are in left-to-right order. (B) Mating stingrays, *Urolophus halleri*, in shallow water of the Sea of Cortez orient to a buried plastic model during playback of the low-frequency bioelectric field recorded from a female. Males approach, explore, and sometimes dig up buried models (as they do actual females) in an attempt to mate. Females also locate and approach the model and often bury next to it. (After Tricas et al., 1995.) (C) Diagrams illustrating motional electric fields available to elasmobranchs for navigation in the ocean. Large-scale fields associated with ocean streams may be used by the fish to set a heading with respect to the stream (passive mode), or the fish's own swimming movements within the Earth's magnetic field create fields whose polarity and intensity depends on the fish's compass heading and velocity (active mode). Abbreviations: Bh, Bv—Earth's magnetic field horizontal, vertical vector; V—velocity vector. (After Kalmijn, 1974.)

movements of oceanic and tidal streams in the Earth's magnetic field, and are well within the sensitivity of the elasmobranch electrosense. Because of boundary conditions at the water surface, the most relevant are horizontal fields produced by the horizontal movement of the stream through the vertical component of the earth's magnetic field (Fig. 20.1C). The relatively constant direction of these fields may provide a directional reference, allowing elasmobranchs to maintain a constant heading. Kalmijn (1997) refers to this as the passive mode of electronavigation. In the active mode, the animal's own movements induce an electric field across its body. Close to the surface, these self-induced fields could be distinguished from environmental fields by concentrating on the vertical fields induced by the horizontal movement of the animal through the horizontal component of the Earth's magnetic field. It is clear that as the horizontal heading changes, so too would the strength and polarity of the vertically induced field. As Murray (1962) suggested, this available cue could provide the basis of an electrosensory-mediated compass sense (Kalmijn, 1982). One complication is that the electroreceptors themselves do not encode purely DC potentials, but as Paulin (1995) has shown, the compass-heading cue could still be available from the modulated field produced by the shark's head undulations during swimming.

3. Electrosensory Periphery

3.1. Ampullae of Lorenzini and Distribution of Receptors

The ampullae of Lorenzini are the electrosensory organs in elasmobranchs. Thin-walled, conductive jelly-filled canals open onto the body surface at one end, and terminate in swellings, called ampullae, at the other (Fig. 20.2A,B). The electroreceptor cells line the walls of the ampullae, and each of these has a synaptic contact with a primary afferent electrosensory nerve fiber. There are 4–5 ampullary swellings at the base of each canal, and 5–10 afferents each receiving their input from several thousand electroreceptor cells. The canals are arranged in 3 or 4 clusters on each side, with the ampullary swellings close together, and the canals radiating outward various distances to their pore terminations on the body surface. As discussed below, the long canals of some of the receptors make a contribution to sensitivity in large-scale electric fields, but they also permit clustering of the ampullae of the different receptors so that they share a common internal reference potential (Kalmijn, 1974). This is important for electrosensory noise suppression in the brain. In skates and rays, there are generally many canal pores on the leading edge of the pectoral fin, and ventrally on the snout and around the mouth. In sharks, the receptor distribution is restricted to the head (for details see Chu and Wen, 1963; Bodznick and Boord, 1986). Although the general patterns of pore distributions of batoids and sharks are described, only recently have these been interpreted in relation to the fish's natural ecology and behavior. Raschi (1986) examined the distributions of ampullary pores in 40 species of skates and found that different patterns were associated with different diets and habitats. This "neuroecological" approach was used to analyze the projection vectors of canals from ampullary clusters in the white shark and barndoor skate by Tricas (2001). Distinct projection patterns were found both among and within ampullary groups (Fig. 20.2C), and were postulated to serve different biological functions, including aspects of prey capture and orientation to uniform electric fields.

3.2. Receptor Sensitivity and Frequency Response

The electroreceptors of elasmobranchs are the most sensitive known biological detectors of electrical potential, responding to small fractions of a microvolt. What factors might account for this extreme sensitivity? The long conductive canal of some receptors, in conjunction with the fish's relatively low skin resistance, allows sampling over widely spaced points in an external voltage gradient. The high resistance of the sensory epithelium compared to the negligible resistance along the canal length assures that most of the available voltage drop takes place directly across the receptor cells (Waltman, 1966). In an external uniform field of 5 nV/cm, the potential drop across the sensory epithelium of a receptor organ with a 20-cm canal may be nearly 0.1 μV. But this is still an exceedingly small potential and detecting it would seem to require much greater sensitivity than could be provided by the familiar voltage-gated ion channels of nerve or muscle cell membranes.

Positive feedback and spontaneous activity may be the keys to this puzzle. The sensory epithelium at rest shows regenerative electrical activity, and even the dissociated receptor cells exhibit spontaneous impulses in the absence of stimulation (Araneda and Bennett, 1993). The apical, lumen-facing membranes of the receptor cells bear voltage-gated calcium (Ca) channels. According to the model proposed first by Obara and Bennett (1972), the apical faces of the individual receptor cells in the unstimulated receptor organ are held by a bias current at a membrane potential on the brink of threshold for a regenerative Ca spike, occasionally exceeding threshold. Excitatory cathodal electrosensory stimuli then increase the amount of this regenerative activity, which in turn

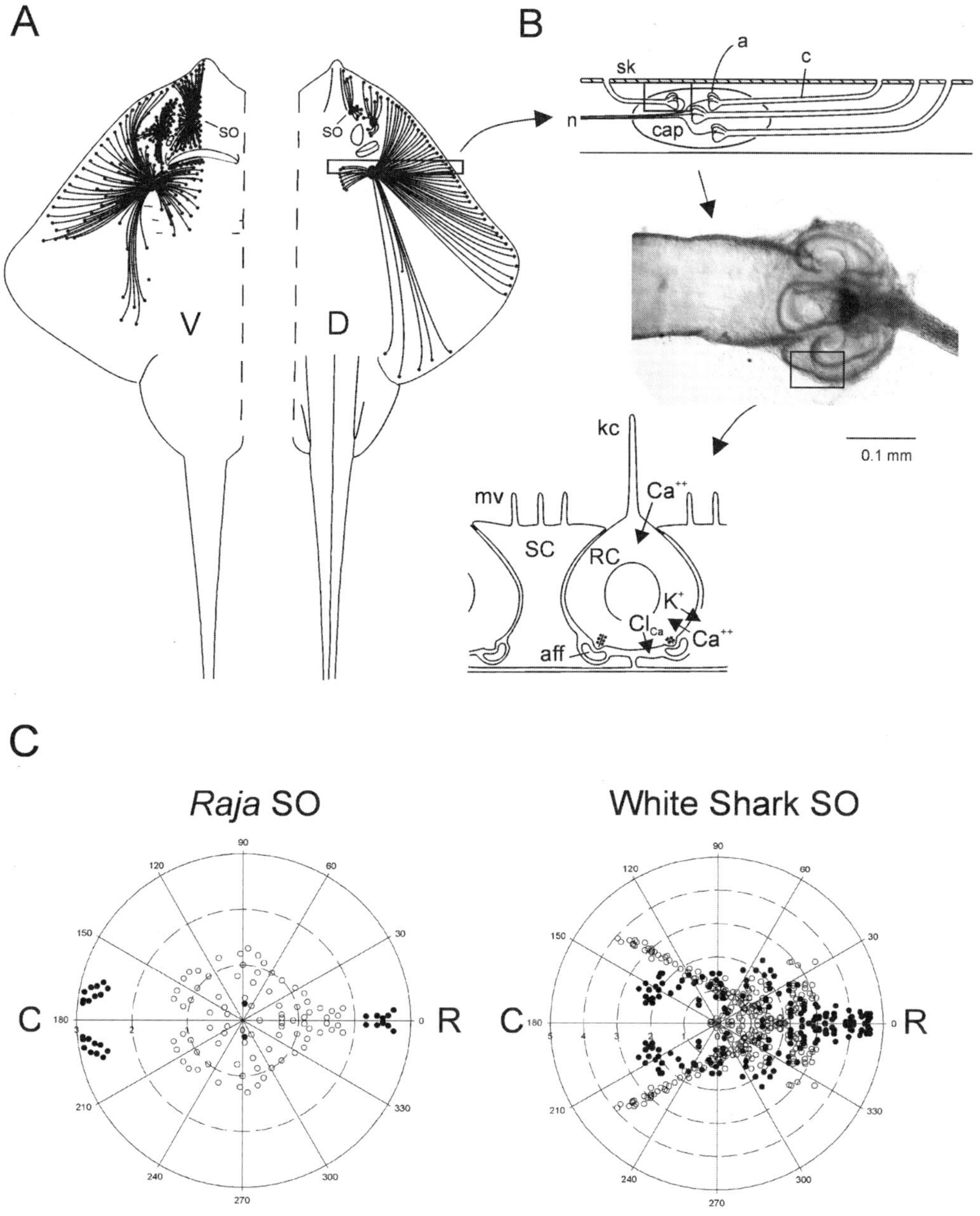

FIGURE 20.2. Ampullae of Lorenzini. (A) Distribution of receptors on the dorsal (D) and ventral (V) surface of *Raja erinacea*. (B) Schematic of receptor organs in cross section (*top*); high-contrast photo of a receptor ampulla and canal in wholemount (*lower left*); and electrosensory epithelium with known receptor cell ionic conductances (arrows) (*lower right*). (C) Polar plot of projection vectors of superficial ophthalmic (SO) receptor canals in the horizontal plane for the skate. Projections of dorsal group (filled circles) are distinct from those of canals that project ventrally (open circles). In skate, dorsal canals project in the rostral (R) and caudal (C) directions. Projections of canals on the ventral surface are omnidirectional. Dorsal superficial ophthalmic canals in the white shark project almost parallel with the rostrocaudal axis, whereas the ventral canals project obliquely toward the edges of the head. These different projections may subserve different behaviors such as orientation to uniform fields (long canals with projections parallel to the body axis) vs. detection of prey (short omnidirectional canals near the mouth). Radius units are centimeters. Data are from Raschi (1986) and Tricas (2001). Abbreviations: a—ampulla; aff—afferent nerve ending; c—canal; cap—capsule; kc—kinocilium; mv—microvilli; n—nerve; RC—receptor cell; SC—support cell; sk—skin; SO—superficial ophthalmic ampullae. (Fig. 20.2B is extensively modified after Murray, 1974).

increases the depolarization of the basal face, and consequently the level of ongoing synaptic transmitter release and afferent nerve discharge rate (Clusin and Bennett, 1979). Inhibitory anodal stimuli have the opposite effects. The basal face of the receptors exhibits a separate ongoing potential oscillation necessary for neurotransmitter release; the basal oscillation is generated by the interplay between inward current through L-type Ca channels and outward current through K channels and Ca-activated chloride (Cl) channels (see Fig. 20.2B) (Lu and Fishman, 1995a,b). Voltage clamp studies (Clusin and Bennett, 1979) indicate that the channels of the electroreceptor cells are just the same as those of other excitable cell membranes and show no extraordinary sensitivity. Instead, the great sensitivity of the individual electroreceptor cells is just that which is inherent in a positive feedback system poised at threshold.

Although positive feedback systems at threshold are extremely sensitive, they generally gain that sensitivity at the cost of nonlinearity. Nevertheless, the firing rate of the electrosensory afferent fibers is a linear function of stimulus intensity over the natural stimulus range and this is probably because the individual receptor cells in most circumstances operate independently and fire asynchronously (but see Lu and Fishman, 1994). The number of receptor cells that are concurrently active determines the afferent firing rate at any given time, and electrosensory stimuli increase or decrease this proportion of active receptor cells.

One final factor contributing to receptor sensitivity is the great convergence of hundreds to thousands of receptor cells onto each afferent fiber. This convergence can lead to a $\sqrt{N}$ improvement (ca. 30× for a 1,000:1 convergence ratio) in the signal-to-noise ratio in the afferents for stimuli generating coincident activity among receptor cells. In summary, the extreme sensitivity of the receptors is achieved by the length and passive electrical properties of the canals, the spontaneous activity of sensory epithelium and afferent fibers, and the high convergence ratio of receptor cells to afferents.

Elasmobranch electroreceptors accommodate to mV changes in DC level without losing their absolute incremental sensitivity (Bodznick et al., 1993). A dipole stimulus of 1 μV causes a similar change in afferent firing rate whether riding on a DC level of 2 mV or 0 mV. The value of this accommodative mechanism to the skate is likely to be in adjusting to its own transcutaneous DC potentials such as those generated by osmoregulatory ion pumping. These can be 0.2 mV or more in amplitude (i.e., about a thousand times the receptor threshold).

The physiology of elasmobranch electrosensory afferents has been extensively studied in a number of species, especially skates and stingrays. The afferents effectively respond to the voltage gradient developed between the pore opening on the skin and the interior of the animal adjacent to the base of the canal. As stated, the afferents are spontaneously active with an electrical gradient sensitivity of 1–25 I/s/μV/cm depending on species (Table 20.1). In the clearnose skate, afferent sensitivity increases as the animals grow (Sisneros et al., 1998). Since the afferents are spontaneously active, there is no discernible threshold in their response; however, the behaviorally determined threshold must correspond to a change in firing rate of only about a tenth of an impulse per second. Sinusoidal electric stimuli can be encoded by primary afferent neurons in the round stingray at levels as low as 10–20 nanovolts/cm (Tricas and New, 1998), which is near the behavioral response threshold of 1–5 nanovolts/cm. The enhanced sensitivity observed in the secondary processing cells known as ascending efferent neurons (AENs) of the dorsal nucleus may be at least partially due to convergence of primary afferents (Tricas and New, 1998).

The frequency sensitivity of the electroreceptors has also been determined for a number of different elasmobranch species (Fig. 20.3) and essentially exhibits a bandpass characteristic with a bandwidth (50% of maximal response) between about 0.1 and 10 Hz. The filter properties of the afferents appear to be determined largely by the properties of the receptor cells themselves, but canals longer than about 10 cm may contribute a low-pass characteristic to receptors (Waltman, 1966).

TABLE 20.1. Spontaneous activity (SA) and gain of elasmobranch electroreceptors and second-order neurons.

Animal	Cell type	SA (I/s)	Gain	Reference
P. triseriata	Afferent	18	4 I/s/μV/cm	Montgomery, 1984b
	Secondary cells	10	32 I/s/μV/cm	Montgomery, 1984b
R. erinacea	Afferent	14	0.9 I/s/μV/cm	Montgomery and Bodznick, 1993
	AENs	0	2.2 I/s/μV	Conley and Bodznick, 1994
U. halleri	Afferent	33	24 I/s/μV/cm	Tricas and New, 1998
R. eglanteria				Sisneros et al., 1998
Embryo	Afferent	12	2.5 I/s/μV/cm	
Juvenile		45	13 I/s/μV/cm	
Adult		45	19 I/s/μV/cm	

The frequency response curves of the clearnose skate change during ontogeny; the upper frequency range extends in juveniles, and then contracts in adults (Sisneros et al., 1998). The broad peak of frequency sensitivity in reproductively active adults is near the normal discharge rate of this species' electric organ. In Atlantic stingrays, increased sensitivity to low frequencies (0.5–2 Hz) takes place in wild males during the reproductive season and in the laboratory after exogenous testosterone treatment (Sisneros et al., 1998; Sisneros and Tricas, 2000), perhaps to enhance detectability of the female's standing or ventilatory potentials.

3.3. Reafference

In paralyzed animals (the default condition for most electrophysiological experimentation), the spontaneous activity of electrosensory afferents is relatively steady. This situation changes dramatically in freely breathing animals with the afferent nerves often driven over most of their dynamic range in time with the breathing movements (Fig. 20.4) (Montgomery, 1984a). This ventilatory self-stimulation (reafference) is driven by the standing potential that exists between the inside and the outside of the animal and that is modulated by changes in resistance pathways caused by opening and closing of the mouth and spiracle. The "inside/outside" origin of the potential determines that ventilatory reafference is very similar in all of the electroreceptors irrespective of where their pore openings are on the body surface. Thus ventilatory reafference is "common mode" across the afferent population. As we shall see, this has implications for the removal of ventilatory reafference by the electrosensory circuits of the hindbrain.

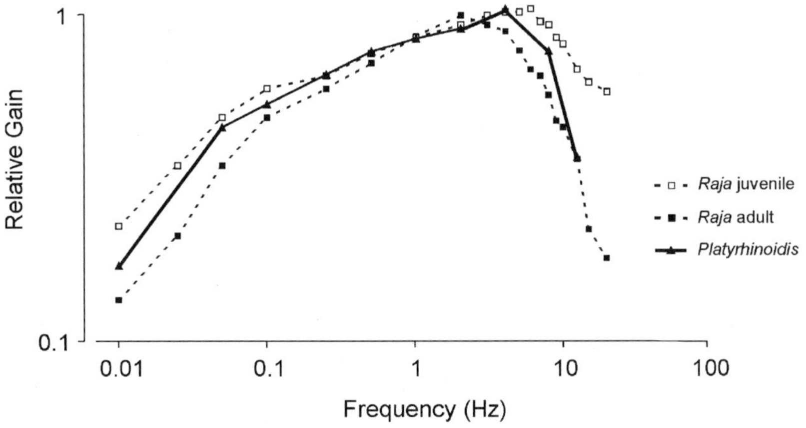

FIGURE 20.3. Frequency response functions of electrosensory primary afferents. Note the broader response in *Raja* juveniles than adults (Sisneros et al., 1998). *Platyrhinoidis* data from Montgomery (1984b).

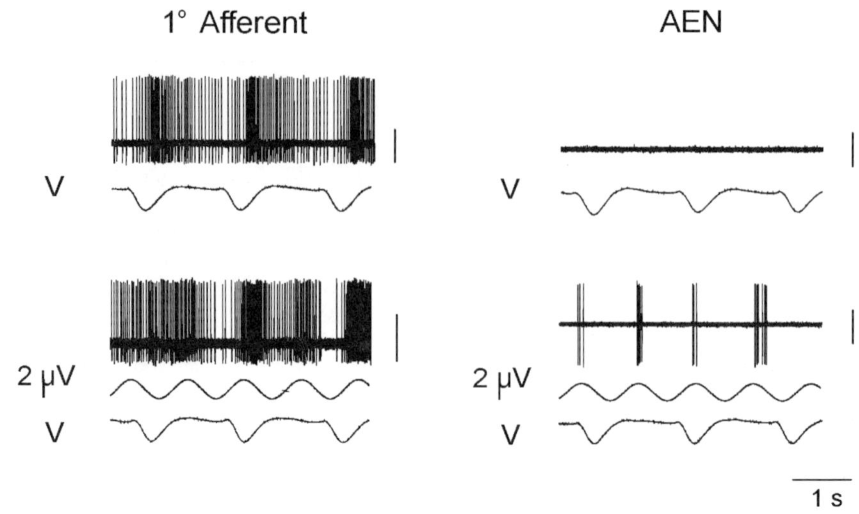

FIGURE 20.4. Reafference comparison for primary afferent and AEN. The ventilatory reafference in a primary afferent recorded from a freely breathing skate (top) masks the response of the same unit to a 2-μV, 1-Hz sinusoidal dipole field (bottom). AEN recorded about the same time was unaffected by ventilatory activity (*top*) but still very sensitive to the 2 μV dipole field (*bottom*). The traces just below the spike records in each case are the output of a force transducer monitoring ventilatory movements. The sinusoidal traces indicate the dipole field stimulus. (From Bodznick et al., 1999. Reprinted by permission of Company of Biologists Ltd.)

4. Central Processing

4.1. Hindbrain Circuits

The electrosensory nerves enter the dorsal octavolateral nucleus (DON) on the dorsolateral wall of the hindbrain. The DON has a distinct molecular layer cap composed of axons of granule cells from the auricular lobes of the cerebellum. Below the molecular layer, in what is termed the peripheral zone of DON, is a band of neuronal cell bodies, many of which are the so-called AENs. The rest of the nucleus is termed the central zone. The principal circuits of the DON are illustrated in Figure 20.5. Afferents enter the DON and course rostrally and caudally through the central zone of the nucleus. They make contact with the ventral dendrites of the AENs, and also with interneurons of the central zone. Many of these interneurons are GABAergic and in turn make inhibitory synaptic contacts with the AENs (Montgomery and Bodznick, 1993; Duman and Bodznick, 1996, 1997). AENs also have extensive dorsal dendrites that receive multiple inputs from parallel fibers and stellate interneu-

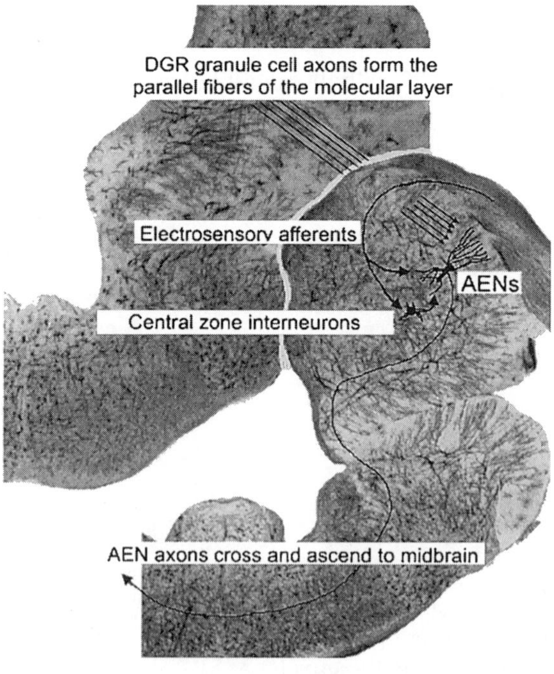

FIGURE 20.5. Principal neuron (AEN—ascending efferent neuron) and circuits of the dorsal octavolateralis nucleus (DON) are shown overlaying Golgi-stained transverse sections.

rons of the molecular layer. The stellate cells are inhibitory and GABAergic. The AEN axons cross the midline and ascend to the contralateral midbrain. Crossed inhibitory pathways (not illustrated in Fig. 20.5) also join the bilateral dorsal nuclei.

4.2. Receptive Fields and Common Mode Rejection

The responses of AENs differ from primary afferents in several respects. The spontaneous activity of AENs is much lower and less regular, they are more sensitive to extrinsic electrical fields (Table 20.1), they have complex receptive fields, and, most strikingly, they show very little response to the animal's own ventilatory movements. The low spontaneous rate and the lack of response to ventilation are illustrated in Figure 20.4. Whereas in the typical primary afferent it is difficult to see the spike-rate modulation of an external 2μV stimulus over the top of the ventilatory modulation, the typical AEN, with its zero spontaneous activity and no ventilatory modulation, provides a very clear modulation to the external stimulus. If one considers the ventilatory modulation to be "noise," then the clear modulation to the external stimulus coupled with the zero response to ventilation represents an effectively infinite improvement in the signal-to-noise ratio between the primary afferent and the AEN illustrated in Figure 20.4. One of the main factors in this dramatic improvement in the signal-to-noise ratio is that the AENs have balanced excitatory and inhibitory components to their receptive fields. Whereas primary afferents are activated by a stimulus applied to a single electrosensory canal pore, AENs have an excitatory receptive field consisting of one or several canal pores, and an inhibitory receptive field that can be either focal (one or several pores) or diffuse including components from the contralateral side. Because the ventilatory modulation in afferents is similar across the whole afferent population, balanced excitatory and inhibitory inputs onto the AEN can cancel most of the ventilation by simple subtraction (Montgomery, 1984b; New and Bodznick, 1990; Bodznick and Montgomery, 1992; Bodznick et al., 1992). This is the principle of common-mode rejection (CMR) that is used in differential electronic amplifiers to remove common-mode noise present at the positive and negative input terminals. Inhibitory receptive field input to AENs is mediated by GABAergic interneurons and commissural neurons of the DON (Duman and Bodznick, 1996, 1997).

Modeling studies show that CMR can remove most, but not all of the ventilatory reafference due to differences in the dynamics of the excitatory and inhibitory pathways (Nelson and Paulin, 1995). In addition, significant reafference from the fish's swimming and other movements is generally not common mode but rather affects different parts of the receptor array differently. Common mode rejection is ineffective against such noise. For these reasons, an additional more sophisticated filter is required to fully remove reafferent modulation in the AENs. This so-called adaptive filter is a property of the molecular layer of the DON and is described below.

Balanced excitatory and inhibitory receptive fields are required for CMR of ventilatory input. However, the precise nature of the receptive field structure also has a major impact on how external electric field stimuli are encoded at the level of the hindbrain (Salypongse et al., 1992). AENs with a focal excitatory receptive field and a diffuse inhibitory field do not respond well to uniform external fields. Widespread focal excitatory and inhibitory fields would be best suited to encode uniform fields. Kalmijn's (1997) approach algorithm and electronavigational hypothesis both require the animal to determine its angle with respect to the external field; however, in the little skate and in the carpet shark, widespread focal receptive fields appear to be the exception rather than the rule. From this we can conclude that, for these species, the spatial localization of small external dipoles, such as prey, is probably the most important function of the electrosense. As an aside, it is interesting that *R. erinacea* orienting to small dipoles ignore the vertical dimension. Dipoles held above the pectoral fin produce a positioning of the mouth to the position on the substrate below where the stimulus was presented (M. Jarnot and D.

Bodznick, unpublished). The skate has reduced source localization to a two-dimensional problem.

4.3. Adaptive Filter

The molecular layer of the DON provides the substrate for an adaptive filter that can cancel any self-generated noise. The presence of this additional filter is demonstrated in Figure 20.6. The experiment is to record from an AEN in a freely ventilating animal. A stimulus is presented to the excitatory receptive field triggered by the animal's own breathing movements (Fig. 20.6A). The AEN initially responds strongly to this stimulus, but over a time course of 10 minutes or so, the response declines (Fig. 20.6B). The AEN has learned to suppress the response to a stimulus that is correlated with its own movements. A control stimulus presented at about the same rate, but not triggered by ventilation, shows no decline over the same period. The mechanism underlying the cancellation of the ventilation-triggered stimulus is revealed when the stimulus is turned off; a negative image of the initial response remains (Fig. 20.6B, poststimulation). Where the AEN was initially excited it is now inhibited; where the AEN was inhibited, it is now excited. During the course of the coupling, a cancellation signal input to the AEN has developed that suppresses the AEN's response in an apparently additive fashion. After the stimulus has ended, the cancellation signal is removed with a similar time course to its development.

How could the AEN learn to cancel inputs associated with its own movements? What is the source of the cancellation signal, and how is it constructed and modified? The answers to these questions lie in the molecular layer circuitry. The parallel fibers of the molecular layer are the axons of granule cells from the auricular lobe of the cerebellum—specifically from an area termed the dorsal granular ridge (DGR). Recordings from this area show that DGR cells receive a variety of inputs, including efference copy signals from the motor control centers (Hjelmstad et al., 1996). In other words, the neural circuitry that drives ventilation, and presumably other movements, sends a copy of all the motor signals to the DGR. In addition, DGR cells receive information from proprioceptors activated by movement, and other inputs including descending electrosensory feedback. In effect, the molecular layer provides a rich matrix of information related to, among other things, self-generated movement. The parallel fibers of the molecular layer contact the spiny dendrites of the AENs, with many thousands of excitatory parallel fiber synapses on each AEN. In addition, parallel fibers contact stellate interneurons in the molecular layer that in turn make inhibitory connections with the AENs. So in theory, excitatory and inhibitory selections of the molecular layer information could be combined onto one AEN to generate the required cancellation signal in the molecular layer dendrites. The cancellation signal then would null the sensory reafference impinging on the ventral dendrites of the AEN. This description still begs the question as to how the selection of molecular layer information is made. How does each particular AEN, through experience, come to be selectively affected by just those molecular layer inputs that comprise an effective cancellation signal? It turns out that two simple learning rules are sufficient to make the selection of molecular layer inputs (Montgomery and Bodznick, 1994):

1. Each time the AEN is activated by the electrosensory inputs, turn down the gain of coincidentally active excitatory parallel fiber synapses (and/or conversely turn up the gain of coincidentally active inhibitory stellate cell synapses).
2. If parallel fiber synapses are active in the absence of AEN activity, increase their synaptic gain (and/or decrease the gain in the case of inhibitory synapses).

In effect, the changes in synaptic strength directed by these learning rules result in a reduction of molecular layer excitatory drive to the AEN to compensate for the increase in excitatory electrosensory input on the AEN's ventral dendrites. A net increase in the molecular layer excitatory drive compensates for an increased inhibitory electrosensory input ventrally.

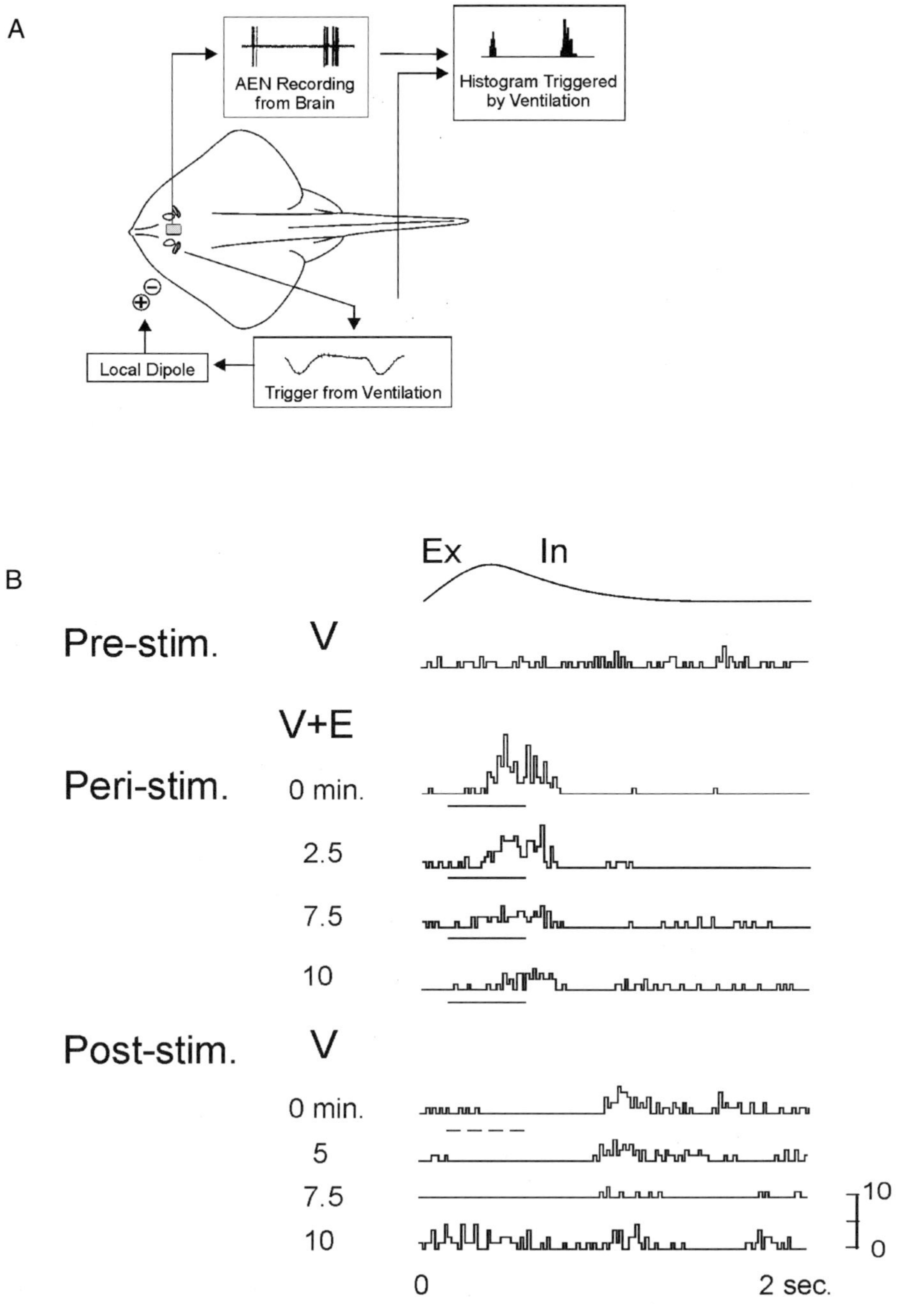

FIGURE 20.6. Adaptive filter experimental setup and results. (A) A monitor of branchial ventilatory movements of a partially immobilized and decerebrated skate triggers the computer-generated histogram of AEN spike activity recorded from the brain. The same trigger synchronizes presentation of a local dipole E field stimulus with the onset of ventilation. (B) Histograms of the AEN spiking activity during the ventilatory cycle recorded before (pre-stim.), during (peri-stim.), and after (post-stim.) the presentation of a 2-μV dipole time-linked with the ventilatory movements. Top trace schematically illustrates the ventilatory cycle, and the line beneath each trace indicates dipole stimulus timing. Note the decline in the E field response during the period of coupling (compare peri-stim. 10 vs. 0 min.), and the presence of a negative image of the initial response at stimulus offset (compare post-stim. 0 min. with pre-stim.) No such decline in response is observed if the stimulus is given at a similar rate but not time-linked to ventilation. Abbreviations: Ex—exhalation; In—inhalation; V—ventilatory movements alone; V+E—dipole field stimulus coupled with ventilation.

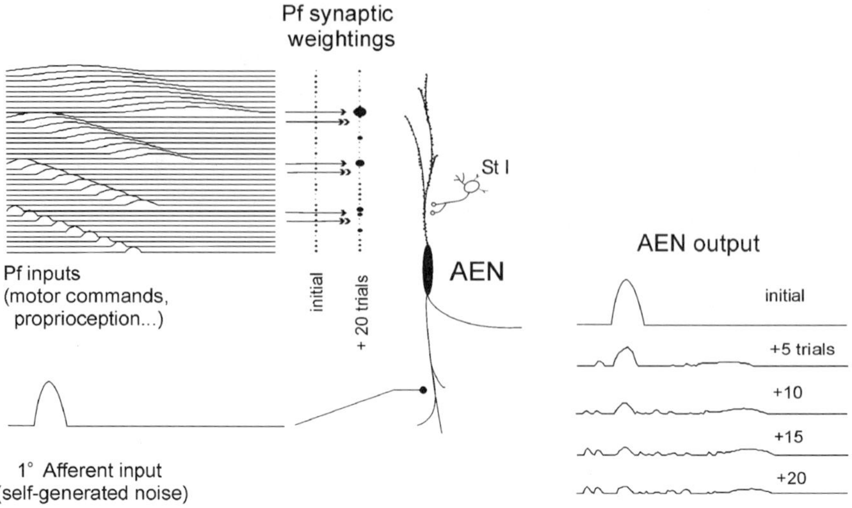

FIGURE 20.7. Adaptive filter model of Montgomery and Bodznick (1994). Following learning rules described in the text, the relative strengths (or weightings) of the excitatory parallel fiber (Pf) synapses to the apical dendrites of the AEN are altered through experience in order to construct a cancellation signal input that is the inverse of the expected self-generated noise brought in by the 1° afferents. The cancellation signal then adds with the 1° afferent input in order to eliminate the self-generated noise from the AEN output. In this model, molecular layer stellate interneurons (St I) provide a steady background of inhibition against which Pf excitation can be added or removed. Filled circles at the center of the figure denote by their size the strength of the synapse of each parallel fiber onto the AEN. Note that after 20 trials the learning rules result in a strengthening of the Pf synapses from those parallel fibers whose activity is consistently out of phase with the noise (single-headed arrows), and a reduction in the weighting of synapses of parallel fibers whose activity is consistently coincident with the noise (double-headed arrows). Please see text for additional details.

A simple computer model demonstrates the utility of these synaptic learning rules to generate reafference cancellation (Fig. 20.7). In this model, the parallel fiber signals are a series of different half-period sine waves staggered through the ventilatory cycle. Although these are clearly artificial, they can be thought of as the sequential motor signals that would drive the sequential movements of ventilation, along with the resulting propriosensory feedback. Below the parallel fiber signals is a representative half-sinusoid reafference, or self-generated noise, that has been constructed so as not to match any one of the parallel fiber signals. Initially the parallel fibers are connected with random, low synaptic weightings to the AEN, and the AEN output has a strong component of the self-generated noise. The computer program simply steps through the files applying the learning rules. If the AEN is silent, each of the parallel fiber files is examined to see which are active and which are not. Those that are active have their connection strengths incremented by a small amount (rule 2). If the AEN is active, coincidentally active parallel fibers have their connection strengths decreased (rule 1). With successive passes through the files (or with each ventilatory cycle), the "self-generated noise" in the AEN output is successfully canceled. In this simple model, stellate cell inhibitory synapses have been represented as fixed and they provide a constant background level of inhibition against which the parallel fiber excitatory inputs can be added or removed to achieve net shifts in the molecular layer excitatory and inhibitory input to the AEN. The anatomical arrangement with stellate interneurons receiving convergent input from thousands of different parallel fibers seems to suggest such a role for them in providing nonspecific background inhibition. However, models in which both inhibitory and excitatory molecular layer synapses are adjustable also work well (Nelson and Paulin, 1995).

As Figure 20.7 illustrates, this simple model is robust, and cancels arbitrary waveforms provided they are not too dissimilar to those in the parallel fibers. In essence, the model does a generalized inverse Fourier transform, synthesizing an arbitrary wave out of a series of quasi-sinusoids. The only reason that the filter does not also cancel extrinsic electrosensory signals is that they are not consistently linked with any of the molecular layer inputs; they are not predictable.

Since the learning rules are based on the activity of each individual AEN, the adaptive filter mechanism results in the development of a neuron-specific cancellation signal. Each AEN receives its own unique cancellation signal input that is fitted to the particular form of the reafference that the neuron experiences during a particular behavior (Fig. 20.8).

Experimental tests of the model have been conducted showing that the correlation of intracellular depolarization with breathing produces an appropriate cancellation signal, and that correlation of DGR stimulation with excitatory input to an individual AEN also effects a cancellation (Bodznick et al., 1999). Anti-Hebbian synaptic learning and a similar adaptive filter have also been clearly demonstrated in the electrosensory lateral lobe of weakly electric fishes (Bell et al., 1993, 1997; Bastian, 1995).

4.4. Ascending Projections and Midbrain Maps

Noise cancellation forms an important first stage to sensory processing in the nervous system. The AEN receptive field structures and adaptive filter precondition the signals before they are sent further up the neuraxis. Much less is known about these higher electrosensory areas in elasmobranchs. Ascending information is sent to a number of nuclei en route to the midbrain, but the functional correlates of these pathways is as yet unknown (Bodznick and Boord, 1986). At the midbrain level, electrosensory information is mapped into the regions of the tectum and the lateral mesencephalic nucleus (LMN) (Bodznick, 1991) and also into the anterior mesencephalic nucleus (AMN) in *Platyrhinoidis* (Schweitzer, 1986).

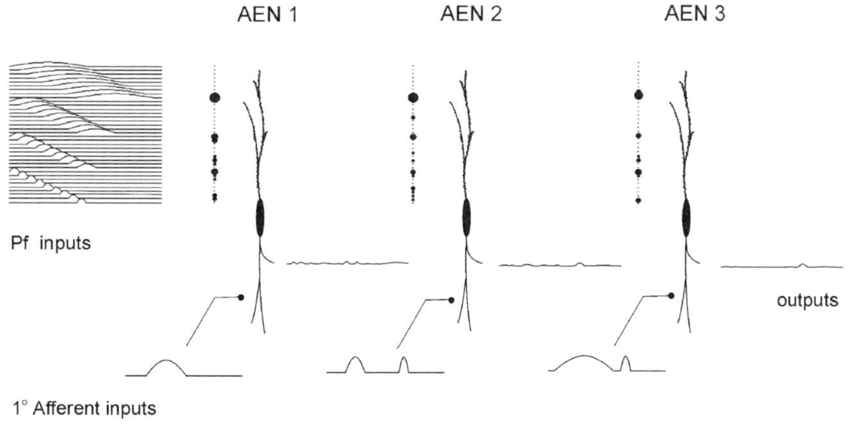

FIGURE 20.8. The adaptive filter is neuron-specific. The ouput of the model illustrates that since changes in the synaptic weightings are directed by the activity of each individual AEN, different patterns of reafference associated with a particular behavior result in a unique set of synaptic weightings for each AEN. That is, after the learning each AEN receives its own unique cancellation signal matched to eliminate the particular form of reafference it receives.

The electrosensory maps in the tectum in skates are in register with the visual maps, and both form an extensive overrepresentation of the lateral fin margin and the animal's horizon. Thus, like the skate's orienting responses to dipoles, the tectal sensory representation is largely a two-dimensional map of the horizontal plane. The electrosensory map of the LMN covers the ventral surface of the animal for which there is no visual equivalent from the dorsally placed eyes. Forebrain targets of the ascending electrosensory pathway include two separate nuclei in the diencephalon and at least one area in the medial pallium of the telencephalon (Schweitzer, 1983; Bodznick and Northcutt, 1984), but an examination of the physiology of these areas must await future studies.

References

Araneda, R.C., and Bennett, M.V.L. (1993). Electrical properties of electroreceptor cells isolated from skate ampullae of Lorenzini. *Biol. Bull.* 185:310–311.

Bastian, J. (1995). Pyramidal-cell plasticity in weakly electric fish: A mechanism for attenuating responses to reafferent electrosensory inputs. *J. Comp. Physiol. A.* 176:63–78.

Bell, C., Bodznick, D., Montgomery, J., and Bastian, J. (1997). The generation and subtraction of sensory expectations within cerebellum-like structures. *Brain Behav. Evol.* 50:17–31.

Bell, C.C., Caputi, A., Grant, K., and Serrier, J. (1993). Storage of a sensory pattern by anti-Hebbian synaptic plasticity in an electric fish. *Proc. Natl. Acad. Sci. USA* 90:4650–4654.

Bodznick, D. (1991). Elasmobranch vision: multimodal integration in the brain. *J. Exp. Zool. Suppl.* 5:108–116.

Bodznick, D., and Boord, R.L. (1986). Electroreception. New York: Springer-Verlag.

Bodznick, D., and Montgomery, J.C. (1992). Suppression of ventilatory reafference in the elasmobranch electrosensory system: Medullary neuron receptive fields support a common mode rejection mechanism. *J. Exp. Biol.* 171:127–137.

Bodznick, D., and Northcutt, R.G. (1984). An electrosensory area in the telencephalon of the little skate, *Raja erinacea*. *Brain Res.* 298:117–124.

Bodznick, D., Hjelmstad, G., and Bennett, M.V. (1993). Accommodation to maintained stimuli in the ampullae of Lorenzini: How an electroreceptive fish achieves sensitivity in a noisy world. *Jpn. J. Physiol.* 43:S231–S237.

Bodznick, D., Montgomery, J.C., and Bradley, D.J. (1992). Suppression of common mode signals within the electrosensory system of the Little Skate *Raja erinacea*. *J. Exp. Biol.* 17:107–125.

Bodznick, D., Montgomery, J.C., and Carey, M. (1999). Adaptive mechanisms in the elasmobranch hindbrain. *J. Exp. Biol.* 202:1357–1364.

Bratton, B.O., and Ayers, J.L. (1987). Observations on the electric discharge of two skate species (Chondrichthyes: Rajidae) and its relationship to behavior. *Environ. Biol. Fishes* 20:241–254.

Bullock, T.H., Northcutt, R.G., and Bodznick, D. (1982). Evolution of electroreception. *Trends Neurosci.* 5:50–53.

Chu, Y.T., and Wen, M.C. (1963). Monograph of fishes of China (no. 2): A study of the lateral-line canals system and that of Lorenzini ampullae and tubules of elasmobranchiate fishes of China. (In Chinese with English abstract.) Shanghai: Science and Technology Press.

Clusin, W.T., and Bennett, M.V. (1979). The ionic basis of oscillatory responses of skate electroreceptors. *J. Gen. Physiol.* 73:703–723.

Conley, R.A., and Bodznick, D. (1994). The cerebellar dorsal granular ridge in an elasmobranch has proprioceptive and electroreceptive representations and projects homotopically to the medullary electrosensory nucleus. *J. Comp. Physiol. A.* 174: 707–721.

Duman, C.H., and Bodznick, D. (1996). A role for GABAergic inhibition in electrosensory processing and common mode rejection in the dorsal nucleus of the little skate, *Raja erinacea*. *J. Comp. Physiol. A.* 179:797–807.

Duman, C.H., and Bodznick, D. (1997). Distinct but overlapping populations of commissural and GABAergic neurons in the dorsal nucleus of the little skate, *Raja erinacea*. *Brain Behav. Evol.* 49:99–109.

Hjelmstad, G., Parks, G., and Bodznick, D. (1996). Motor corollary discharge activity and sensory responses related to ventilation in the skate vestibulolateral cerebellum: Implications for electrosensory processing. *J. Exp. Biol.* 199:673–681.

Kalmijn, A.J. (1974). The detection of electric fields from inanimate and animate sources other than electric organs. In: *Handbook of Sensory Physiology, Vol. III/3* (Fessard, A., ed.), pp. 148–200. Berlin: Springer-Verlag.

Kalmijn, A.J. (1982). Electric and magnetic field detection in elasmobranch fishes. *Science* 218:916–918.

Kalmijn, A.J. (1997). Electric and near-field acoustic detection, a comparative study. *Acta Physiol. Scand. Suppl.* 638:25–38.

Lu, J., and Fishman, H.M. (1994). Interaction of apical and basal membrane ion channels underlies electroreception in ampullary epithelia of skates. *Biophys. J.* 67:1525–1533.

Lu, J., and Fishman, H.M. (1995a). Ion channels and transporters in the electroreceptive ampullary epithelium from skates. *Biophys. J.* 69:2467–2475.

Lu, J., and Fishman, H.M. (1995b). Localization and function of the electrical oscillation in electroreceptive ampullary epithelium from skates. *Biophys. J.* 69:2458–2466.

Mikhailenko, N.A. (1971). Biological significance and dynamics of electrical discharges in weak electric fishes of the Black Sea (in Russian). *Zool. Zh.* 50:1347–1352.

Montgomery, J.C. (1984a). Noise cancellation in the electrosensory system of the thornback ray: Common mode rejection of input produced by the animal's own ventilatory movement. *J. Comp. Physiol. A.* 155:103–111.

Montgomery, J.C. (1984b). Frequency response characteristics of primary and secondary neurons in the electrosensory system of the thornback ray. *Comp. Biochem. Physiol. A.* 79:189–195.

Montgomery, J.C., and Bodznick, D. (1993). Hindbrain circuitry mediating common mode suppression of ventilatory reafference in the electrosensory system of the little skate *Raja erinacea*. *J. Exp. Biol.* 183:203–215.

Montgomery, J.C., and Bodznick, D. (1994). An adaptive filter that cancels self-induced noise in the electrosensory and lateral line mechanosensory systems of fish. *Neurosci. Lett.* 174:145–148.

Mortenson, J., and Whitaker, R.H. (1973). Electric discharge in free swimming female winter skates, *Raja ocellata*. *Am. Zool.* 13:1266.

Murray, R.W. (1962). The response of the ampullae of Lorenzini of elasmobranchs to electrical stimulation. *J. Exp. Biol.* 39:119–128.

Murray, R.W. (1974). The ampullae of Lorenzini. In: *Handbook of Sensory Physiology, Vol. III/3* (Fessard, A., ed.), pp. 125–146. Berlin: Springer-Verlag.

Nelson, M.E., and Paulin, M.G. (1995). Neural simulations of adaptive reafference suppression in the elasmobranch electrosensory system. *J. Comp. Physiol. A.* 177:723–736.

New, J.G., and Bodznick, D. (1990). Medullary electrosensory processing in the little skate. II. Suppression of self-generated electrosensory interference during respiration. *J. Comp. Physiol. A.* 167:295–307.

Obara, S., and Bennett, M.V. (1972). Mode of operation of ampullae of Lorenzini of the skate, *Raja*. *J. Gen. Physiol.* 60:534–557.

Paulin, M.G. (1995). Electroreception and the compass sense of sharks. *J. Theor. Biol.* 174:325–339.

Raschi, W. (1986). A morphological analysis of the ampullae of Lorenzini in selected skates (Pisces, Rajoidei). *J. Morph.* 189:225–247.

Salypongse, A., Hjelmstad, G., and Bodznick, D. (1992). Second-order electroreceptive cells in skates have response properties dependent on the configuration of their inhibitory receptive fields. *Biol. Bull.* 183:349.

Schweitzer, J. (1983). The physiological and anatomical localization of two electroreceptive diencephalic nuclei in the thornback, ray, *Platyrhinoidis triseriata*. *J. Comp. Physiol. A.* 153:331–341.

Schweitzer, J. (1986). Functional organization of the electroreceptive midbrain in an elasmobranch (*Platyrhinoidis triseriata*). *J. Comp. Physiol. A.* 158:43–58.

Sisneros, J.A., and Tricas, T.C. (2000). Androgen-induced changes in the response dynamics of ampullary electrosensory primary afferent neurons. *J. Neurosci.* 20:8586–8595.

Sisneros, J.A., Tricas, T.C., and Luer, C.A. (1998). Response properties and biological function of the skate electrosensory system during ontogeny. *J. Comp. Physiol. A.* 183:87–99.

Tricas, T.C. (1982). Bioelectric-mediated predation by swell sharks *Cephaloscyllium ventriosum*. *Copeia* 1982:948–952.

Tricas, T.C. (2001). The neuroecology of the elasmobranch electrosensory world: Why peripheral morphology shapes behavior. *Environ. Biol. Fishes* 60:77–92.

Tricas, T.C., and New, J.G. (1998). Sensitivity and response dynamics of elasmobranch electrosensory primary afferent neurons to near threshold fields. *J. Comp. Physiol. A.* 182:89–101.

Tricas, T.C., Michael, S.W., and Sisneros, J.A. (1995). Electrosensory optimization to conspecific phasic signals for mating. *Neurosci. Lett.* 202:129–132.

Waltman, B. (1966). Electrical properties and fine structure of the ampullary canals of Lorenzini. *Acta. Physiol. Scand. Suppl.* 264:1–60.

21
In a Fish's Mind's Eye: The Visual Pallium of Teleosts

Leo S. Demski

Abstract

This chapter reviews visual functions of the pallium of teleost fishes. Information derives from studies utilizing three major groups of fishes: the Osteoglossomorpha or bony tongues, which include mormyrid weakly electric fishes (most notably the elephant-nose, *Gnathonemus petersii*); the Otophysi, which include catfish, minnows, and gymnotid weakly electric fishes; and the Percomorpha, the most derived of the bony fishes. Comparisons between the groups are complicated by differences in the preglomerular and/or thalamic nuclei that function as links between the tectum and pallium. Nevertheless similarities are apparent. Ipsilateral tectal information is projected to the dorsal part of the lateral pallium (dDl) via preglomerular nuclei in representatives of all three groups of fishes. The most differentiated system is a pathway from the tectum to a highly developed preglomerular cell group, the nucleus prethalamicus (PTH), to an enlarged dDl in the squirrelfish. The PTH is probably homologous to part of the anterior preglomerular nucleus of other percomorphs and possibly other tectal recipient preglomerular nuclei in the two other groups of fishes. In some coral reef percomorphs (e.g., squirrelfishes and surgeonfishes) the dDl is highly differentiated, obtaining a "cortical-like" appearance with vertical (radial) columns as well as horizontal lamination. Recent electrophysiological studies in percomorphs confirm the visual input to the area. The major output pathway from the typical cortical-like cells of dDl (at least in certain percomorphs and otophysians) is to an adjacent large-celled dorsal portion of the central pallial nuclear complex (dDc). This area projects to several layers in the ipsilateral tectum. Stimulation of the dDc has a profound effect on tectal activity. The pathway in squirrelfish appears to be topographically organized. These animals also have an additional connection between the regions in the form of unique specialized "bridge cells" that have processes in both dDl and dDc. Thus, at least for advanced percomorphs, the current evidence supports a reciprocal pallial-tectal interconnection or "vision loop." Visual information is also projected to the dorsomedial pallium (dDm) and probably dDl by thalamic nuclei, most notably the ventromedial nucleus (VMN). The VMN pathways are probably multisensory in nature. Connections between the dDm and the dDl in a number of fishes may also provide substrates for integration of visual and other types of sensory information (e.g., acoustic and electrosensory).

1. Introduction

Over the past 30 years increased study of the teleost dorsal telencephalon (pallium) has produced new insights concerning pallial anatomy and behavioral physiology. No longer can the structure be considered merely an olfactory organ or an engine for general nonspecific arousal but rather an integrator of complex sensory and sensorimotor components as is the pallium of higher tetrapods (see Aronson, 1981, for background). Indeed, these functions must now be considered "featured attractions" rather than "limited possibilities." This chapter reviews and synthesizes information on the teleost pallium as it is involved in primary visual processing and vision-related multisensory integration and sensorimotor control in representatives of three major teleost groups (Osteoglossomorpha, Otophysi, and Percomorpha). The chapter derives from: (1) a comprehensive analysis of available anatomical, physiological, and behavioral data concerning pallial involvement in processing sensory information (including "downward" modulation by pallial structures) in several of the best-studied modalities (i.e., acousticolateralis systems, chemosenses, and vision); (2) unpublished electrophysiological, anatomical, and behavioral studies of a tectal-pallial "vision loop" in coral reef and other percomorphs (L.S. Demski, unpublished); and (3) a review of sensory and neural mechanisms of cognitive and related behavioral functions in teleosts (Demski and Beaver, 2001). Trends in the evolution of the systems are suggested and the implications of the findings for a better understanding of teleost cognition stressed.

1.1. Comments on the Biology and Evolutionary History of the Fish Groups

The following is a capsule summary of the phylogeny and natural history of the fish taxa included in this chapter (see also Lauder and Liem, 1983; Nelson, 1984; Paxton and Eschmeyer, 1998). Although the Osteoglossomorpha or bony tongues are somewhat basal in terms of evolutionary history, some of them (e.g., the mormyrid elephant-nose fishes) present remarkable sensory and neural specializations, including a highly developed electrosense and hypertrophy of the cerebellum and pallium (Braford, 1995; Meek and Nieuwenhuys, 1998). The Otophysi, which are the largest group of freshwater fishes, include the cyprinids (minnows), the characins (piranhas, tetras, etc.), the silurids (catfishes), and the closely related gymnotids (weakly electric knifefishes). These fishes offer a variety of peripheral and central nervous system (CNS) specializations for olfaction, gustation, hearing, and mechanosensory and electrosensory lateral line senses (Schellart, 1992; Meek and Nieuwenhuys, 1998). The highly derived Percomorpha is the largest group of the teleosts with many advanced morphological traits and behavioral capabilities. For the most part, the percomorphs are vision specialists (Schellart, 1992; Demski and Beaver, 2001). Of particular interest to this review are studies in the Beryciformes (squirrelfishes and soldierfishes), the Scorpaeniformes (scorpion-fishes), and the Perciformes (perch-like fishes). Many of the percomorphs inhabit reefs or reef-like environments that are spatially complex and biologically diverse. The three main taxa cover much of the diversity among higher ray-finned fishes and any generalizations made across the groups will likely pertain to most teleosts.

1.2. Comments on the Anatomy and Development of the Teleost Pallium and its Diencephalic Connections

While a detailed discussion of peculiar features of the anatomy of the telencephalon and its diencephalic relay nuclei is not the purpose of this chapter, a few words of explanation are necessary for orientation of readers not familiar with bony fish neurology. These ray-finned (actinopterygian) fishes present considerable divergence from the thalamic–telencephalic models most commonly associated with tetrapods. The dissimilarities are especially noticeable in two areas: (1) the general anatomy/development of the dorsal telencephalon and (2) the developmental origin of the major diencephalic cell groups associated with the dorsal telencephalon.

The telencephalon of ray-finned fishes develops through a process of eversion or thickening of the neural tube walls rather than an invagination to form a ventricle as in tetrapods (details in Northcutt and Braford, 1980; Northcutt and Davis, 1983; Nieuwenhuys and Meek, 1990; Braford, 1995; Northcutt, 1995; Meek and Nieuwenhuys, 1998). The difference in development makes comparison among types difficult and suggestions of homologies tenuous. For this reason, a system of nomenclature for telencephalic structure in ray-finned fishes, based on topographic position rather than implied homology to tetrapods, was adapted by Meek and Nieuwenhuys (1998). The system is followed in this review. The telencephalon is divided into dorsal and ventral tiers, which generally translate into pallial and subpallial structures, respectively. Five pallial subdivisions are recognized: the area dorsalis telencephali pars medialis (Dm), pars dorsalis (Dd), pars lateralis (Dl), and pars posterioris (Dp) form a thickened wall extending from an enlarged pial surface area. A division composed of larger cells, the area centralis (Dc), sits deep to these nuclei in the central part of the hemisphere. The structures are named according to their relative positions, which are usually sufficiently constant to permit identification (Fig. 21.1). Further subdivisions are often made. Dc has components that relate anatomically and probably functionally to the overlying areas (Northcutt and Braford, 1980; Northcutt and Davis, 1983; Demski and Beaver, 2001).

The diencephalic region with strongest connections to the pallium in ray-finned fishes is not the dorsal thalamus (although this area is represented) as in tetrapods but rather a group of nuclei derived from the posterior tuberculum (PT). The PT may be partly comparable to the subthalamus. Most of the PT cell groups are part of the nucleus preglomerulosus complex, which is variously subdivided into components based on topographical location and connections. Comparison of preglomerular nuclei in relatively unrelated taxa has been difficult and homologies among nuclei largely unresolved (see Braford and Northcutt, 1983; Braford, 1995; Meek and Nieuwenhuys, 1998). Therefore, while useful in constructing evolutionary scenarios, such comparisons must be viewed with caution.

2. Osteoglossomorphs

Prechtl and associates (1998) have carried out the most comprehensive study of telencephalic sensory processing in any teleost. Over 7,000 sites in the dorsal medial (Dm), dorsal (Dd), and a portion of the underlying dorsal central areas (Dc) were sampled for multiunit and compound field potentials in response to light flashes, tones, water movements, and electrosensory stimuli in the weakly electric mormyrid elephant-nose fish *Gnathonemus petersii*. Most visual responses, which were usually recorded at sites deeper than those for the other stimuli, consisted of a relatively slow positive potential with superimposed high-frequency (>100 Hz) multiunit activity. The

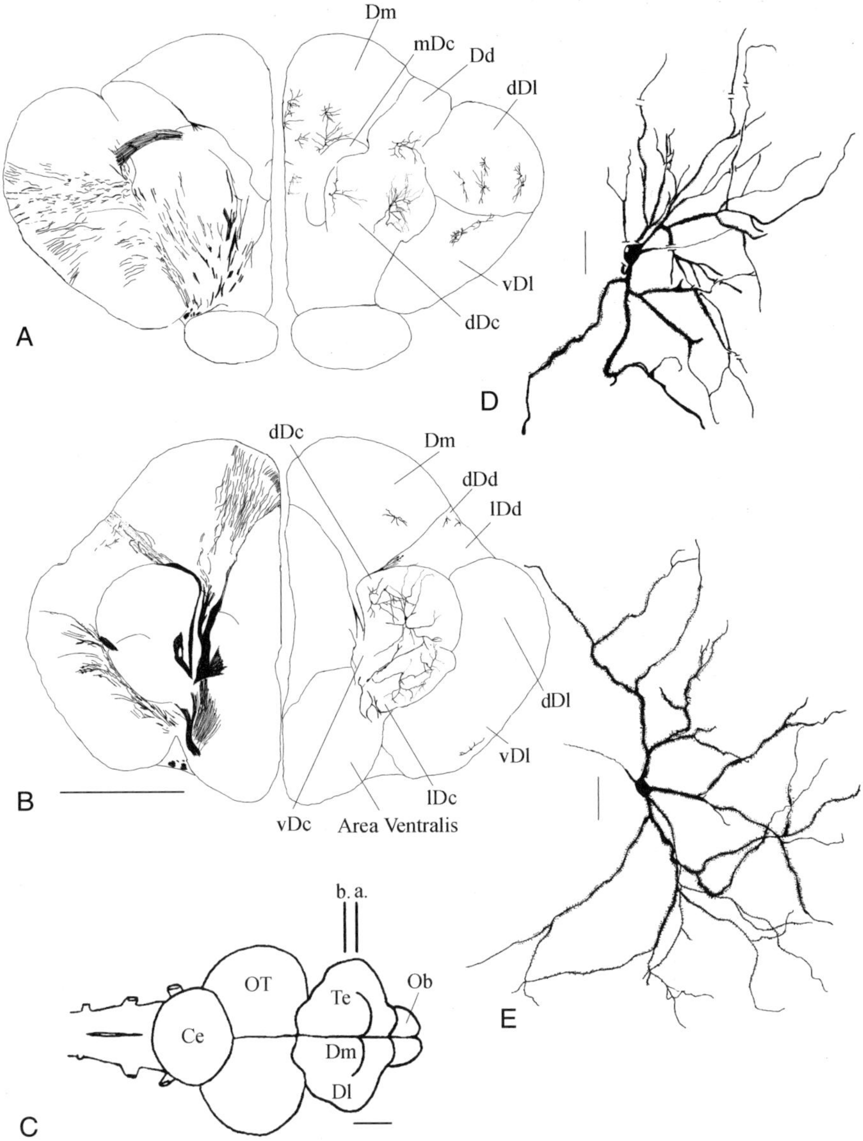

FIGURE 21.1. Cytoarchitecture of the central pallial telencephalon of the Texas cichlid based on rapid Golgi-stained material. Transverse sections A and B are through the center of the Dc and Dl, areas considered part of the "visual pallium." The rostrocaudal level of sections is indicated in C. Composites of myelinated fiber tracts (left side) and filled cells (right side) are illustrated. Note that the myelinated fibers, mostly associated with the lateral forebrain bundle, form a capsule around Dc. Enlargements of two of the spiny Dc neurons are depicted in D and E. See details in text. Abbreviations: area dorsalis: pars centralis, medial part (mDc), lateral part (lDc), ventral part (vDc): dorsal part (dDc = nucleus tectalis, see text); pars lateralis (Dl); dorsal part (dDl), ventral part (vDl); pars medialis (Dm); Ce, cerebellum; Ob, olfactory bulb; OT, optic tectum; Te, telencephalon. Scale = 1 mm in A–C, 50 μm in D and E. (From Mahoney, 2001.)

majority of responses were unimodal. Twenty-four percent of all the responses were multimodal, usually responding to two or three of the test stimuli; only 2% of the sites responded to all four types. Multimodal responses most often occurred near the boundaries between unimodal zones. The waveforms of multimodal responses were more varied than those for unimodal sites. In some cases, response to one stimulus would be the typical wave while the response to the second stimulus would be a later component of high-frequency spikey activity.

Sites responsive to the flash stimuli (unimodal) were the most scattered and it was likely that the primary area for vision was not adequately tested (see below). In contrast, data for the sites activated by the other stimuli were clearly localized in Dm and the associated part of Dc.

Visually receptive areas were found in parts of Dc, a result consistent with Dc connections to visual centers in *Gnathonemus* and other teleosts (see below). However, the study most likely did not adequately identify the primary visual centers of the pallium which, based on anatomical evidence in other Osteoglossomorphs (Braford, 1995), are probably centered in the lateral pallium (Dl). In the minimal sampling of the Dl carried out, only unimodal visual responses were recorded.

In *Gnathonemus* (Wullimann and Northcutt, 1990), the tectum projects to the dorsal preglomerular nucleus (PGd) and receives input from Dc. The PGd (sometimes referred to as "nucleus rotundus") is reciprocally connected with the Dl (Rooney and Szabo, 1991). In the African knifefish *Xenomystus nigri,* part of the Dl (Dla) receives a projection from the posterior preglomerular nucleus (PGp), which is reciprocally connected with the tectal torus longitudinalis and probably receives tectal input as well (Braford, 1995). Braford (1995) considers the PGp a likely homologue of the PGd of *Gnathonemus*. Thus, the areas most responsive to the flash stimuli are bidirectionally associated with the tectum via a preglomerular relay. This theme will be apparent in the other groups.

3. Otophysians

3.1. Visual Afferents to Pallium

Retinal and tectal recipient nuclei of the diencephalon have been identified in several otophysians. However, convincing evidence that any of these areas project to the telencephalon is unavailable. Various preglomerular nuclei project to the telencephalon in cyprinids, catfishes, and gymnotiforms, but contrary to the situation in the other groups, a strong tectal input to these areas has not been demonstrated (Sligar and Voneida, 1976; Ito and Kishida, 1977; Grover and Sharma, 1979, 1981; Echteler and Saidel, 1981; Luiten, 1981; Murakami et al., 1986b; Striedter, 1990, 1991, 1992; Zupanc, 1997; Corrêa et al., 1998; Wullimann, 1998; Corrêa and Hoffmann, 1999).

Nevertheless, striking cytoarchitectural similarities of Dl in otophysians and percomorphs (Northcutt and Davis, 1983; Meek and Nieuwenhuys, 1998; L.S. Demski, unpublished) are consistent with the possibility that the Dl of the otophysians receives a strong tectal-related input. Cellular changes in Dl in goldfishes in response to a visually based spatial learning task (Vargas et al., 2000) are also supportive of a visual projection to the area. Also consistent with the assumption is a report that bilateral ablation of telencephalon disrupts the "ventral substrate response" in the upside-down catfish, *Synodontis nigriventris*. The behavior is a visually mediated body orientation to surfaces observed in certain teleosts (see Section 4). These African catfishes typically rest ventral-side-up while under ledges (Meyer et al., 1976b).

It should also be noted that the Dl receives a strong projection from the PGl in cyprinids, gymnotids, characins, and catfishes (Murakami et al., 1986a; Striedter, 1990, 1991, 1992; Zupanc and Horschke, 1997) and that the PGl receives major afferents from acousticolateralis-related structures. Thus, it would not be surprising to find that Dl in these fishes may be dominated by eighth nerve and lateral line inputs with visual afferents (if they exist) of secondary importance.

3.2. Dc Projection to the Tectum

3.2.1. Anatomical Studies

Considerably more is known of the "downward" telencephalic visual pathways. Cells in a rostral part of Dc innervate several layers of the ipsilateral tectum in the catfish (Bass, 1981; Schlussman et al., 1990), carp (Luiten, 1981), goldfish (Grover and Sharma, 1981), piranha (Fiebig et al., 1983), and the weakly electric fish, *Gymnotus carapo* (Corrêa et al., 1998). Electron and light microscopic degeneration studies in carp (Marotte and Mark, 1975) and goldfish (Airhart and Kriebel, 1985) also report a projection to the contralateral tectum. The ipsilateral telencephalic tectal projection is similar to that of osteoglossomorphs (see above) and percomorphs (see below).

3.2.2. Physiological Studies

Lee and Bullock (1990) demonstrated that electrical stimulation of the rostral Dc (and not other pallial areas) in bullhead catfish (*Ictalurus nebulosus*) profoundly influences the ipsilateral tectum. A single shock to the area elicits a tectal field potential of two negative peaks N8 (8ms) and N25 (25ms) followed by a slow positive–negative wave P50–N95 (50–95ms). The N8 is probably due to the presynaptic activity of the telencephalic afferents while the N25 most likely relates to activation of deep tectal cells. The P50–N95 is probably a result of continued tectal activity (see below in discussion of system in percomorphs). It is especially noteworthy that the two "postsynaptic" waves were maximal with Dc stimulation at ~2Hz. This decrement with frequency may be a substrate for habituation to complex sensory stimuli (see below).

The Dc stimulation also triggered unit responses from cells in the deeper layers (i.e., at a depth of 300–400 μm). While the latency to activation was consistent across cells, their induced firing activity did not show any particular pattern. These cells either were unresponsive or at best demonstrated inconsistent activity to light flashes or optic tract stimulation. The unit response latency (20–25ms) corresponded with the N25. The units also responded best to repetitive stimulation of 2–3 Hz and failed to follow frequencies above 8Hz. Thus, there is good agreement in the time course of evoked field potentials as well as units. It is also significant that cells in the superficial tectal layers that have short latency to activation following optic nerve (ON) shock (reflected in the N6 peak, see below) do not respond to Dc stimulation.

The effects of Dc stimulation on tectal responses to ON stimulation were also studied. A conditioning shock to Dc decreases the typical N6 and P12 components of the ON shock. An interval between conditioning and test stimuli of about 90ms is maximally effective. The Dc shock also reduces and/or delays responses of certain tectal units to ON shock. These units are unresponsive to the Dc stimulation alone; they have a relatively long latency to the ON stimulation (20ms) and respond to moving stimuli moving in the horizontal direction.

3.2.3. Summary and Behavioral Implications

Dorsal central (DC) projections to the tectum can profoundly influence processing of retinal input as well as affect later components of tectal integration. Dc stimulation is complex; that is, it activates cells rather unresponsive to direct retinal input while modulating the responses of another class of tectal neuron that show directional sensitivity to visual stimuli. The electrophysiological findings are consistent with the anatomical observations of the distribution of telencephalic afferents in several layers with highest densities in lower parts of the tectum (see above). The Dc/tectal system in catfishes appears physiologically and anatomically similar to that of percomorphs (see below).

Electrical stimulation of the tectum of teleosts evokes eye and body movements (Akert, 1949; Meyer et al., 1970; Demski, 1983; Al-Akel et al., 1986; Herrero et al., 1998). Fiebig and coworkers (1985) demonstrated that stimulation in Dc and the related area of Dm elicits coordinated eye movements in spinally transected unanaesthetized piranhas. Movements of the jaw, gill covers, and pectoral fins often accompanied the eye movements. The results suggest that Dc may activate tectal sensorimotor circuits. Studies of Dc stimulation combined with tectal ablation are needed to test the hypothesis.

4. Percomorphs

The modern perch-like fishes are generally considered to be visual specialists based on behavior, color changes, and highly derived pathways associated with vision in the retina, tectum, diencephalon, and telencephalon (Collin and Pettigrew, 1989 and review by Demski and Beaver, 2001). Wullimann (1997) provides compelling evidence that the visual enhancements and adaptive radiation of these fishes were associated with the colonization of early bivalve and coelenterate reefs by ancestors of the group (ctenothrissiforms) and continued life in reef-like environments. There is little doubt that the modern coral reefs are the most visually rich of all aquatic habits. Thus, it is not surprising that certain reef fishes are useful subjects for understanding telencephalic visual functions. Historically, studies center on the squirrelfish (Holocentridae).

4.1. Studies in Squirrelfishes

These fishes have very large eyes and although they are active both day and night, they generally behave differently at these times. At dusk, the fishes emerge from their hiding places in the crevices of the coral rock and proceed to feed over large areas of the reef; at dawn they move back into their refuges (Hobson, 1974, 1975; Paxton and Eschmeyer, 1998; L.S. Demski, unpublished). Capture and release of tagged fishes have confirmed that they can return to their home base after being removed from the water and transported to adjacent areas of reef (L.S. Demski unpublished). Thus, the squirrelfishes probably have a visually based cognitive map of a considerable area surrounding their home base and the ability to use it under differing luminosities. Such functions may be mediated by interconnected enlarged areas in the tectum, diencephalon, and pallium (see below).

4.1.1. The Anatomically Defined Tectum–Nucleus Prethalamicus (PTH)/Dl/Dc/Tectum "Vision Loop"

In 1934, Meader studied retinal projections in the squirrelfish. He also identified a large non-retinal target nucleus situated between the optic tract and the lateral forebrain bundle, which, based on its location, was designated the nucleus prethalamicus (PTH). Subsequent investigations have indicated that the PTH receives a robust tectal input and projects to the pallial Dl (Ebbesson, 1972; Ebbesson and Vanegas, 1976; Ebbesson, 1980; Ito et al., 1980; Ito and Vanegas, 1983; Demski and Beaver, 1997, 2001; Demski et al., 1999). The tectal afferents are derived from retinal-recipient cells (Ito and Vanegas, 1984). Projections from Dl, dDm, and Dc to PTH have also been reported (Ito and Vanegas, 1983). These "downward" connections could provide pathways for regulating the flow of tectal information to Dl.

The tectal/PTH/Dl projections are topographically organized (Ebbesson, 1980; Demski and Beaver, 1997). Retinal connections to the tectum have not been mapped in the squirrelfish, hence it cannot be assumed that the tectal input to PTH is retinotopic. However, precise retinal mapping to the tectum is the condition in many other teleosts including percomorphs (Schwassmann, 1975; Vanegas, 1983; Vanegas and Ito, 1983; Meek and Nieuwenhuys, 1998) and can be expected in squirrelfishes as well. Thus, it is likely that retinotopic information is conveyed to the lateral pallium via PTH. A "simple" diencephalic relay function seems unlikely since the PTH has widespread inputs from the pallium and dorsal thalamus. Regarding comparison to other percomorphs, the PTH is most likely a rostral (anterior) component of the preglomerular complex (aPG; Northcutt and Wullimann, 1988; Meek and Nieuwenhuys, 1998).

The dorsal part of Dl (dDl) is hypertrophied and forms a cortex-like mantle over a greatly enlarged dorsal component of Dc (dDc) (Aronson, 1963; Demski and Beaver, 1996, 2001). The dDl has both a columnar as well as laminar organization. dDl projects topographically to dDc. The dDc is expanded dorsomedially to the extent that it fuses across the midline and attains a near-superficial (i.e., periventricular or dorsal) position. Of the four Dc areas identified only dDc projects to the tectum. The dDc-to-tectum pathway is also topographical (Ebbesson, 1980; Demski and Beaver, 1997, 2001). "Bridge cells," which may be unique to holocentrids, also interconnect dDl and dDc. The large sparsely distributed neurons extend "thin" dendrites and probably a branching axon into dDl cells while sending a long, "thick" dendritic-like process into dDc in areas where axons of the dDl cells ramify. The bridge cells could provide a short feedback loop between the structures.

A "vision loop" between the tectum and the hypertrophied dorsal pallium (dDl and dDc) is indicated. Shorter loops or potential feedback circuits are also identified within the larger system. A retinotopic representation is probably maintained throughout the system.

4.1.2. Other Anatomical Inputs to the "Vision Loop"

An especially strong input to dDl derives from part of the dorsal Dm that forms a superficial or ribbon-like structure (sdDm) on the medial surface of dDc (Demski et al., 1998; Demski and Beaver, 2001). Preliminary results from anterograde transport of tracer injected into sdDm suggest a projection to the hypertrophied cerebellum-like torus longitudinalis (TL) of the tectum (Demski and Beaver, 2001). The TL is especially well-developed in the squirrelfish (Kishida, 1979; Vanegas et al., 1979; Vanegas, 1983). In squirrelfish and goldfish TL units respond to either eye movements (corollary discharges) or dimming of ambient light (e.g., negative contrast stimuli) (Northmore et al., 1983). Experimental evidence in goldfish suggests that the TL functions as a broadband luminance processing system (Gibbs and Northmore, 1998) necessary for orientation as measured by the dorsal light reflex (Gibbs and Northmore, 1996). Thus, the proposed connections of sdDm to both TL and dDl could link the "visual telencephalon" with cerebellar-brainstem mechanisms controlling eye movements and perhaps visual tracking mechanisms, particularly to dark objects (e.g., shadows, etc.).

Based on dDm connections in other fishes (see above and Demski and Beaver, 2001), the sdDm pathway may also relay acousticolateralis, gustatory, and/or multisensory information to the dDl. The situation would allow for integration of sensory representations in a variety of modalities with topographical maps of visual space.

4.1.3. Physiological and Behavioral Studies on the Squirrelfish "Vision Loop"

Electrophysiological studies confirm that PTH cells respond to visual stimulation (Williams and Vanegas, 1982). Visual activity has also been recorded in dDl and dDc (L.S. Demski, unpublished). Light flashes triggered both multiphasic field potentials (Fig. 21.2A) as well as short unit bursts (~0.5 s), which also occurred spontaneously. Cells in dDl were also activated by moving-spot stimuli.

Large lesions in the dDl/Dc area reduce or block a normal transfer of habituation to a visual stimulus presented first in one eye and then in the other (Laming, 1987). In addition, squirrelfishes with extensive damage to the dorsal part of these areas have considerable difficulty in returning to their hiding places in their home aquaria. The observations are consistent with a deficit in spatial mapping and/or recognition of their surroundings (Demski and Beaver, 2001).

An immunocytochemical localization of four neurochemicals has been recently carried out in the squirrelfish (Herbsman et al., 2000). Findings of particular importance to the "vision loop" are summarized below. Gonadotropin-releasing hormone (GnRH), serotonin (5HT), tyrosine hydroxylase (TH), and substance P (SP) immunoreactive (ir) fibers were present in synaptic layers of the tectum that also receive projections from dDc. Scattered GnRH, 5HT, and TH-ir axons were present in PTH. A few GnRH-ir fibers, a sparse plexus of TH-ir axons, and a dense plexus of fine 5HT-ir fibers were localized in dDl. Only occasional GnRH and TH-ir fibers were found in dDc. The results, which are consistent with findings in other

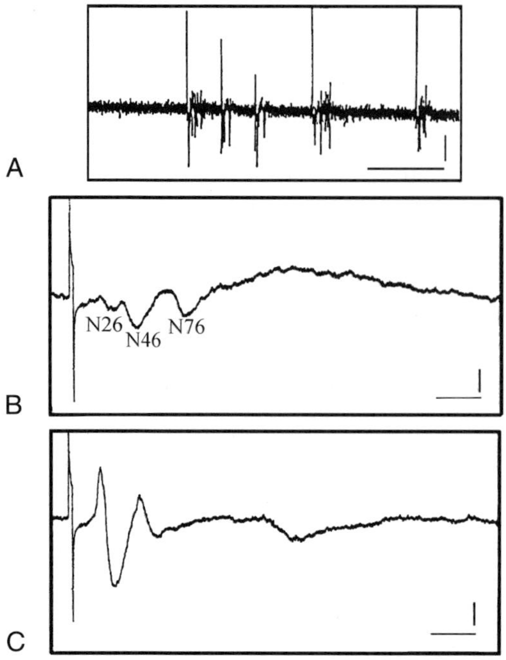

FIGURE 21.2. Recordings of responses to light flashes from a squirrelfish (A) and a pinfish (B, C). A is a digitized single sweep trace from the dorsolateral pallium (Dl). The five large spikes are artifacts caused by the firing of the strobe flash. A burst of unit activity consistently followed each visual stimulation. The decrement in tests 2 and 3 was caused by a decrease in intensity of the not fully recharged flash (as apparent by the decrease in artifact amplitude). The units were also responsive to moving stimuli. B represents an averaged evoked potential to a light flash recorded from the dorsolateral pallium (Dl) while C is a similar averaged recording of responses (over 30) from the optic tectum. Both records are aligned on the same time scale for direct comparison. The N26 wave is an artifact spread from the optic nerves. The other waves were generated locally. Notice that major tectal potential shifts slightly precede those of Dl. This most likely represents a projection from the tectum to the diencephalon (anterior preglomerular nucleus) to pallium (see text for further details). Time scale = 1 s in A and 30 ms in B and C. Vertical scale = ~20 μV in A, ~3 μV in B, and ~4 μV in C. (From Demski, unpublished; Morgan, 2001.)

teleosts, indicate that while GnRH, SP, and one or more of the TH-related neurotransmitters could influence processing in the "loop" at one or perhaps two loci, serotonin most likely exerts a widespread modulation of its functions. The locus coeruleus and raphé both project to the pallium in the squirrelfish (Ito et al., 1982; Demski and Beaver, 2001) and thus are likely sources of the TH-ir and 5HT-ir afferents, respectively.

4.2. PTH/Dl/Dc/Tectum Connection in Other Fishes

4.2.1. DC/Tectal Connections and Physiology

Tracer experiments in several reef and non-reef percomorphs (bluegill sunfishes, *Lepomis macrochirus*; white grunts, *Haemulon plumieri*; pinfishes, *Lagodon rhomboides*; Texas cichlids, *Cichlasoma cyanoguttatum*; ocean surgeonfishes, *Acanthurus bahianus*; and pop-eye catalufa, *Pristigenys serula*; Demski et al., 1999; Demski and Beaver, 2001; L.S. Demiski, unpublished; *Sebastiscus*, Murakami et al., 1983) confirm that a dorsal area of Dc projects to the tectum much as it does in squirrelfishes and the otophysians. The region in *Lepomis* corresponds to Dc-4 using the nomenclature of Northcutt and Davis (1983). It has been referred to as the nucleus tectalis (nT) to differentiate it from other components of the Dc complex (Demski and Beaver, 2001). The nT is small in the sunfish, intermediate in size in the pinfish, grunts, and Texas cichlids (Fig. 21.1), and highly developed in the reef-dwelling surgeonfish (almost to the extent found in the squirrelfish; Demski and Beaver, 2001). On a daily basis, schools of surgeonfishes forage across large areas of coral reefs grazing on the substrate. At dusk, the schools break up only to reform at dawn (Mazeroll and Montgomery, 1995; L.S. Demski, unpublished). The behavior of these fishes and the hypertrophy of the dDc are consistent with the structure being involved in visually based spatial mapping as suggested for the squirrelfish. Lesion studies in a sea bass (*Serranus scriba*) (Lazarević et al., 1989) and *Sebastiscus* (Murakami et al., 1983) also demonstrate the Dc/tectal projection. A report that electrical activity is evoked in a central telencephalic locus (presumed to be Dc) in response to tectal or optic nerve stimulation in *S. scriba* (Rakich et al., 1979) is also consistent with the suggested visual functions of part of Dc.

Electrical stimulation of dDc in largemouth bass (*Micropterus salmoides*) evokes complex activity) in tectal units (i.e., usually excitation followed by more prolonged inhibition but sometimes oscillations of bursts occur, Friedlander, 1983). The minimum 25 ms latency of early activation and varied response patterns are both consistent with the results of similar studies in catfish (Lee and Bullock, 1990; see above). dDc stimulation in pinfish L.S. Demski, unpublished) evokes a compound field potential also similar to that found for catfish (Lee and Bullock, 1990; see above), that is, an N20, P40, and N55 complex with maximum response at 2 Hz stimulation. The frequency sensitivity of the Dc modulation observed may help to explain observations that only low-frequency stimulation (3–10 Hz) of the Dc was effective in evoking stereotyped nestbuilding or "sweeping" responses in the free-swimming bluegill sunfish (Demski and Knigge, 1971; Demski, 1983). Presumably the Dc output triggered activity in frequency-sensitive lower areas, probably the preoptic area and possibly the tectum, which resulted in the organized behaviors.

4.2.2. PTH/aPG Connections

A PTH has been identified in *Sebastiscus* (Murakami et al., 1983; Yemane et al., 1996). Although it is less well-developed than its counterpart in the squirrelfish, the area is in the same location and has similar connections to DL (here separated into dDl and vDl) and other pallial areas (afferents from Dd, Dc, and dDm). Additionally, the nucleus receives afferents from the ventromedial nucleus of the thalamus (VMT) (Ito et al., 1986). Cytoarchitectural and tracer studies in our laboratory indicate that the lateral part of the anterior (rostral) preglomerular nucleus (aPG) in several species has anatomical relationships supporting a homology to PTH. The region is only moderately developed in bluegill sunfishes, pinfishes, and Texas cichlids but is large and laminar in the surgeonfish, much as it appears in the squirrelfish. Tracer studies performed on bluegill sunfishes, pinfishes, Texas cichlids, and pop-eye catalufas confirm that aPG receives afferents from the tectum and projects to dDl (Demski et al., 1999; Demski and Beaver, 2001). The comparative data support the aPG (PTH)/dDl/dDc/tectum as the primary pallial visual pathway in percomorphs.

4.2.3. dDL in Other Fishes

The dDl in *Sebastiscus* has the PTH and Dc connections reported for the squirrelfish (see above) as well as projections to the tectum, vDl, the inferior lobe, and nucleus paracommissuralis (a precerebellar relay) and input from most subpallial regions, locus coeruleus and raphé (Yemane et al., 1996). dDl projections to the Dc have also been observed in pinfishes, Texas cichlids, and bluegill sunfishes (Demski et al., 1999; L.S. Demski and J.A. Beaver, unpublished). In these cases, even small tracer implants in dDl result in anterogradely filled axons that cover the entire nT, whereas similar implants in the dDl of squirrelfishes reveal projections to only a small part of nT (less than 25%). The comparative result suggests a higher resolution in mapping topography for the squirrelfish system. Studies are needed to determine if this is also the case for other reef fishes with enlarged nT areas (e.g., surgeonfishes).

Recent electrophysiological studies on the pinfish confirm the visual projections to dDl (Morgan, 2001). Evoked potentials to a light flash were recorded in dDl. An early negative peak (N26) most likely resulted from spread from the optic nerve; however, two later negative potentials (N46 and N76) followed by a slow positive wave (about 120 ms in duration) were generated from the dDl (Fig. 21.2B,C). Similarly, a dorsal pallial locus (most likely Dl) for activity in response to tectal or optic nerve stimulation has also been reported for *S. scriba* (Rakich et al., 1979). In the "downward" direction, stimulation of Dl in *Micropterus* has a profound effect on the responses of light-flash sensitive units in several thalamic (see below) and pretectal areas. The evidence that dDl is an important, if not the primary, "vision center" beyond the tectum in percomorphs is compelling.

Neurobehavioral studies also support this idea. Certain percomorphs, especially those that live under ledges or in caves, demonstrate a strong visually guided orientation response to nearby surfaces (Meyer et al., 1976b, 1977). This "ventral substrate response" is abolished by bilateral removal of the telencephalon in the black croaker, *Cheilotrema saturnum* (Meyer et al., 1976b). The response is guided by complex visual cues and requires measurement of the plane of the substrate, typically under conditions of much irregularity. Based on the above discussion, among pallial regions, dDl and perhaps its associated nT are the most likely mediators of this high-level integrative activity.

4.3. Thalamic–Telencephalic Visual Pathways

Anatomical evidence supports thalamopallial visual pathway(s). The elaborate connections of the VMT have been extensively studied in *Sebastiscus* (Ito et al., 1986) and to a lesser extent in the cichlid *Oreochromis (Tilapia) niloticus* (Yoshimoto et al., 1998; Sawai et al., 2000). Regarding possible visual functions, the nucleus in *Sebastiscus* receives contralateral retinal, ipsilateral tectal, and Dc inputs and projects ipsilaterally to Dc, Dd, and Dm and bilaterally to the dorsomedial thalamic nucleus (DMT), the PTH, and the tectum. Connections to several structures related to other primary sensory systems and or multimodal-sensorimotor integration centers (i.e., cerebellum, torus semicircularis, spinal cord, and reticular formation) are also reported. The VMT most likely mediates multimodal integration, which includes a strong visual component. It may provide the main route for visual information to reach the Dm and thus provide for interactions with projections to the area of acousticolateralis and gustatory-related activity (see Murakami et al., 1986a; Yoshimoto et al., 1998; Demski and Beaver, 2001).

Physiological studies are consistent with the above suggestions and, in addition, indicate "downward" modulation of the "visual" thalamus. VMT cells respond to light flash and or tactile stimuli (other modalities were not tested) in *Micropterus* (Friedlander, 1983). The response to the flash is inhibited by conditioning stimulation in Dl. Stimulation in other areas of the pallium was not effective. DMT cells were also activated by light flashes. Pallial stimulation in Dm resulted in antidromic activation of the cells, thus indicating that they project to the Dm region. As mentioned above, in *Sebastiscus* the DMT receives afferents from VMT.

Neurobehavioral experiments implicate thalamic centers in complex visually related behaviors. Electrical stimulation in the area of both VMT and DMT in simultaneous hermaphroditic sea bass (*Serranus subligarious*) results in color changes associated with different sexual roles (Demski and Dulka, 1986; Demski, 1987). The color patterns clearly signal male- or female-related motivation levels as the partners trade eggs for fertilizations. The natural behaviors involve visual as well as tactile cues. The thalamic area is considered a sex hormone–sensitive focus for sensorimotor integration of the sexually relevant visual and tactile information.

4.4. Hypothetical Circuitry for Visual Processing in Percomorphs

The following discussion synthesizes information from a variety of species (as discussed above) into a putative "pallial visual system" for percomorphs. Its speculative nature is hopefully balanced by its heuristic value. The dorsal lateral pallium (dDl) receives retinotopic information from the tectum via PTH and multimodal input from VMT and/or Dm and in turn modulates visual (and/or perhaps multimodal) processing in the tectum, PTH, and VMT, either through direct connections or indirectly via its topographically organized connections to nT (dDc). The size, degree of differentiation, and most likely the resolution of the topographical mapping in the dDl/nT system reaches its epitome in reef-related species. The anatomical observations and results from limited neurobehavioral studies suggest involvement of the pathway in habituation to visual stimuli, spatial orientation, and mapping and complex visual integrative functions. Modulation of activity in the "system" via long ascending projections from the locus coeruleus (TH-related) and the raphé (5HT) is likely. Details of the integrative functions of the "visual pallium" await study.

Ventromedial nucleus of the thalamus (VMT) projections to Dm either directly or via relay in DMT provide multimodal, including visual, input to the medial pallium. The major multisensory integration of acousticolateralis, gustatory, visual, and probably somatosensory input most likely occurs in Dm, as it does in Osteoglossomorphs (see above).

5. Conclusion

Regarding projection of visual information to the pallium, most of the information concerns percomorphs (as summarized above). The tectal/preglomerular/prethalamic–Dl pathway has been established in this group by both anatomical and physiological results. The Osteoglossomorph rostral preglomerular projection to the pallium may be a homologous pathway but such suggestions are tenuous without more information. Clearly, there appear to be physiological and/or anatomical similarities in the systems in both groups and, even if they are homoplastic, comparisons will be interesting from an adaptationist point of view. It seems likely that a preglomerular/Dl pathway for tectal information is also present in the Otophysi but electrophysiological and detailed tracer studies are needed to settle the issue. Projection of visual information via the thalamus to Dm and other pallial regions probably occurs in all three groups. The nature of the information (e.g., is it multisensory?; is it qualitatively different from that of the preglomerular system?) remains to be determined. Behavioral studies suggest that the thalamic system may be more involved in key feature analysis in social and other species-typical behaviors rather than spatial orientation and possibly movement control. Regardless of the details, it seems reasonable to postulate that at least two visual pathways to pallium exist in the three groups.

Concerning downward modulation of the tectum, similar Dc-tectum systems are present in all three of the groups. Indeed, the pathway is probably plesiomorphic for ray-finned fishes as it appears to be present in gars (Northcutt, 1982). In percomorphs, at least, the Dl may also modulate activity in the tectum, PTH, and thalamic and pretectal visual centers.

Acknowledgments. The author wishes to thank Joel Beaver, Peter Mahoney, and Josh Morgan for their contributions to work cited in this paper. Financial support from National Science Foundation Grant IBN9724144RUI and the Leonard S. Florsheim Sr. endowment to the New College Foundation, and the use of facilities and help of the staff at the Institute of Marine Science, Roatan, Honduras, are greatly appreciated.

References

Airhart, M.J., and Kriebel, R.M. (1985). Telencephalic terminals in the major retinal synaptic lamina of the goldfish optic tectum. *Brain Res.* 336:363–367.

Akert, K. (1949). Der visuelle Greifreflex. *Helv. Physiol. Acta* 7:112–134.

Al-Akel, A.S., Guthrie, D.M., and Banks, J.R. (1986). Motor responses to localized stimulation of the tectum in the freshwater perch (*Perca fluviatilis*). *Neuroscience* 19:1381–1391.

Aronson, L.R. (1963). The central nervous system of sharks and bony fishes with special reference to sensory and integrative mechanisms. In: *Sharks and Survival* (Gilbert, P.W., ed.), pp. 165–241. Boston: D.C. Heath.

Aronson, L.R. (1981). Evolution of telencephalic function in lower vertebrates. In: *Brain Mechanisms of Behavior in Lower Vertebrates* (Laming, P., ed.), pp. 33–58. Cambridge: Cambridge University Press.

Bass, A.H. (1981). Telencephalic efferents in the channel catfish, *Ictalurus punctatus*: Projections to the olfactory bulb and optic tectum. *Brain Behav. Evol.* 19:1–16.

Braford, M.R. Jr. (1995). Comparative aspects of forebrain organization in the ray-finned fishes: Touchstones or not? *Brain Behav. Evol.* 46:259–274.

Braford, M.R. Jr., and Northcutt, R.G. (1983). Organization of the diencephalon and pretectum of the ray-finned fishes. In: *Fish Neurobiology, Vol. 2: Higher Brain Areas and Functions* (Davies, R.E., and Northcutt, R.G., eds.), pp. 117–163. Ann Arbor, MI: University of Michigan.

Collin, S.P., and Pettigrew, J.D. (1989). Quantitative comparison of the limits on visual spatial resolution set by the ganglion cell layer in twelve species of reef teleosts. *Brain Behav. Evol.* 34:184–192.

Corrêa, S.A.L., and Hoffmann, A. (1999). Reciprocal connections between the preglomerular complex and the dorsolateral telencephalon in the weakly electric fish, *Gymnotus carapo. Neurosci. Lett.* 261:131–134.

Corrêa, S.A.L., Grant, K., and Hoffmann, A. (1998). Afferent and efferent connections of the dorsolateral telencephalon in an electrosensory teleost, *Gymnotus carapo. Brain Behav. Evol.* 52:81–98.

Demski, L.S. (1983). Behavioral effects of electrical stimulation of the brain. In: *Fish Neurobiology, Vol. 2: Higher Brain Areas and Functions* (Davies, R.E., and Northcutt, R.G., eds.), pp. 317–359. Ann Arbor, MI: University of Michigan.

Demski, L.S. (1987). Diversity in reproductive patterns and behavior in teleost fishes. In: *Psychobiology of Reproductive Behavior: An Evolutionary Perspective* (Crews, D., ed.), pp. 1–27. Englewood Cliffs, NJ: Prentice-Hall.

Demski, L.S., and Beaver, J.A. (1996). Evolution of cognition in teleosts: Cytoarchitecture of the hypertrophied dorsal telencephalon of squirrelfish. *Soc. Neurosci. Abstr.* 22:135 (60.6).

Demski, L.S., and Beaver, J.A. (1997). Visually related pathways of the hypertrophied dorsal telencephalon of squirrelfish. *Soc. Neurosci. Abstr.* 23:2380 (927.7).

Demski, L.S., and Beaver, J.A. (2001). Brain and cognitive function in fishes. In: *Brain Evolution and Cognition* (Roth, G., and Wullimann, M.F., eds.), pp. 297–332. New York: Wiley; and Heidelberg: Spectrum Akademischer Verlag.

Demski, L.S., and Dulka, J.G. (1986). Thalamic stimulation evokes sex-color change and gamete release in a vertebrate hermaphrodite. *Experientia* 42:1285–1287.

Demski, L.S., and Knigge, K.M. (1971). The telencephalon and hypothalamus of the bluegill (*Lepomis macrochirus*). Evoked feeding, aggressive and reproductive behavior with representative frontal sections. *J. Comp. Neurol.* 143:1–16.

Demski, L.S., Beaver, J.A., Deignan, J.T., and Lucas, C.L. (1998). Intrinsic connections of the visually related dorsal telencephalon of squirrelfish. *Soc. Neurosci. Abstr.* 24:157 (64.3).

Demski, L.S., Beaver, J.A., Barlow, B., Funderburk, M.J., and Teelucksingh, K.H. (1999). The visual telencephalon of percomorph fishes: Tectal and cerebellar associations. *Soc. Neurosci. Abstr.* 25:103 (47.4).

Ebbesson, S.O.E. (1972). A proposal for a common nomenclature for some optic nuclei in vertebrates and the evidence for a common origin of two such cell groups. *Brain Behav. Evol.* 6:75–91.

Ebbesson, S.O.E. (1980). A visual thalamo-telencephalic pathway in a teleost fish (*Holocentrus rufus*). *Cell Tissue Res.* 213:505–508.

Ebbesson, S.O.E., and Vanegas, H. (1976). Projections of the optic tectum in two teleost species. *J. Comp. Neurol.* 165:161–180.

Echteler, S.M., and Saidel, W.M. (1981). Forebrain connections in the goldfish support telencephalic homologies with land vertebrates. *Science* 212: 683–685.

Fiebig, E., Ebbesson, S.O.E., and Meyer, D.L. (1983). Afferent connections of the optic tectum in the piranha (*Serrasalmus nattereri*). *Cell Tissue Res.* 231:55–72.

Fiebig, E., Meyer, D.L., and Ebbesson, S.O.E. (1985). Eye movements evoked from telencephalic stimulations in the piranha (*Serrasalmus nattereri*). *Comp. Biochem. Physiol.* 81A:67–70.

Friedlander, M.J. (1983). The visual prosencephalon of teleosts. In: *Fish Neurobiology, Vol. 2: Higher Brain Areas and Functions* (Davies, R.E., and Northcutt, R.G., eds.), pp. 91–115. Ann Arbor, MI: University of Michigan.

Gibbs, M.A., and Northmore, D.P.M. (1996). The role of the torus longitudinalis in equilibrium orientation measured with the dorsal light reflex. *Brain Behav. Evol.* 48:115–120.

Gibbs, M.A., and Northmore, D.P.M. (1998). Spectral sensitivity of the goldfish torus longitudinalis. *Visual Neurosci.* 15:859–865.

Grover, B.G., and Sharma, S.C. (1979). Tectal projections in the goldfish (*Carassius auratus*): A degeneration study. *J. Comp. Neurol.* 184:435–454.

Grover, B.G., and Sharma, S.C. (1981). Organization of external tectal connections in goldfish (*Carassius auratus*). *J. Comp. Neurol.* 196:471–488.

Herbsman, D.M., Beaver, J.A., and Demski, L.S. (2000). Distribution of GnRH, serotonin, tyrosine hydroxylase and substance P-like immunoreactivity in the brain of squirrelfish, with particular regard to hypertrophied telencephalic-tectal pathways. *Soc. Neurosci. Abstr.* 26:1240 (462.8).

Herrero, L., Rodríguez, F., Salas, C., and Torres, B. (1998). Tail and eye movements evoked by electrical microstimulation of the optic tectum in goldfish. *Exp. Brain Res.* 120:291–305.

Hobson, E.S. (1974). Feeding relationships of teleostean fishes on coral reefs in Kona Hawaii. *Fish Bull.* 72:915–1031.

Hobson, E.S. (1975). Feeding patterns among tropical reef fishes. *Amer. Sci.* 63:382–392.

Ito, H., and Kishida, R. (1977). Tectal afferent neurons identified by the retrograde HRP method in carp telencephalon. *Brain Res.* 130:142–145.

Ito, H., and Vanegas, H. (1983). Cytoarchitecture and ultrastructure of the nucleus prethalamicus, with special reference to degenerating afferents from optic tectum and telencephalon, in a teleost (*Holocentrus ascensionis*). *J. Comp. Neurol.* 221:401–415.

Ito, H., and Vanegas, H. (1984). Visual receptive thalamopetal neurons in the optic tectum of teleosts (Holocentridae). *Brain Res.* 290:201–210.

Ito, H., Murakami, T., and Morita, Y. (1982). An indirect telencephalo-cerebellar pathway and its relay nucleus in teleosts. *Brain Res.* 249:1–13.

Ito, H., Morita, Y., Sakamoto, N., and Ueda, S. (1980). Possibility of telencephalic visual projection in teleosts, Holocentridae. *Brain Res.* 197:219–222.

Ito, H., Murakami, T., Fukuoka, T., and Kishida, R. (1986). Thalamic fiber connections in a teleost (*Sebastiscus marmoratus*): Visual, somatosensory, octavolateral, and cerebellar relay region to the telencephalon. *J. Comp. Neurol.* 250:215–227.

Kishida, R. (1979). Comparative study of the teleostean optic tectum: Lamination and cytoarchitecture. *J. Hirnforsch.* 20:57–67.

Laming, P.R. (1987). Behavioural arousal and its habituation in the squirrel fish, *Holocentrus rufus*: The role of the telencephalon. *Behav. Neural. Biol.* 47:80–104.

Lauder, G.V., and Liem, K.F. (1983). The evolution and interrelationships of the actinopterygian fishes. *Bull. Mus. Comp. Zool.* 150:95–197.

Lazarević, L., Rakić, L., Gojković, M., and Drasković, Z. (1989). Striatal projections in teleost *Serranus scriba*. *Yugoslav Physiol. Pharmacol. Acta* 25: 53–63.

Lee, L.T., and Bullock, T.H. (1990). Responses of the optic tectum to telencephalic stimulation in catfish. *Brain Behav. Evol.* 35:313–334.

Luiten, P.G.M. (1981). Afferent and efferent connections of the optic tectum in the carp (*Cyprinus carpio* L.). *Brain Res.* 220:51–65.

Mahoney, P. (2001). The cytoarchitecture of the telencephalon of a cichlid fish: *Cichlasoma cyanoguttatum*. Honor's thesis. Sarasota: New College of Florida.

Marotte, L.R., and Mark, R.F. (1975). Ultrastructural localization of synaptic input to the optic lobe of carp (*Carassius carassius*). *Exp. Neurol.* 49:772–789.

Mazeroll, A.I., and Montgomery, W.L. (1995). Structure and organization of local migrations in brown surgeonfish (*Acanthurus nigrofuscus*). *Ethology* 99:89–106.

Meader, R.G. (1934). The optic system of the teleost, *Holocentrus*. 1. The primary optic pathways and the corpus geniculatum complex. *J. Comp. Neurol.* 60:361–407.

Meek, J., and Nieuwenhuys, R. (1998). Holosteans and teleosts. In: *The Central Nervous System of Vertebrates, Vol. 2* (Nieuwenhuys, R., ten Donkelaar, H.J., and Nicholson, C., eds.), pp. 759–937. Berlin and Heidelberg: Springer Verlag.

Meyer, D.L., Becker, R., and Graf, W. (1977). The ventral substrate response of fishes: Comparative investigation of the VSR about the roll and the pitch axis. *J. Comp. Physiol.* 117:209–217.

Meyer, D.L., Heiligenberg, W., and Bullock, T.H. (1976a). The ventral substrate response: A new postural control mechanism in fishes. *J. Comp. Physiol.* 109:59–68.

Meyer, D.L., Platt, C., and Distel, H.-J. (1976b). Postural control mechanisms in the upside-down catfish (*Synodontis nigriventris*). *J. Comp. Physiol.* 100:323–331.

Meyer, D.L., Schott, D., and Schaefer, K.-P. (1970). Reizversuche im Tectum opticum freischschwimmender Kabeljaue bzw. Dorsche (*Gadus morrhua* L.). *Pflügers Arch. Eur. J. Physiol.* 314:240–252.

Morgan, J.L. (2001). Preliminary electrophysiology of the tecto-telencephalo-tectal pathway in *Lagodon rhomboidies*. Honor's thesis. Sarasota: New College of Florida.

Murakami, T., Fukuoka, T., and Ito, H. (1986a). Telencephalic ascending acousticolateral system in a teleost (*Sebastiscus marmoratus*), with special reference to the fiber connections of the nucleus preglomerulosus. *J. Comp. Neurol.* 247:383–397.

Murakami, T., Ito, H., and Morita, Y. (1986b). Telencephalic afferent nuclei in the carp diencephalon, with special reference to fiber connections of the nucleus preglomerulosus pars lateralis. *Brain Res.* 382:97–103.

Murakami, T., Morita, Y., and Ito, H. (1983). Extrinsic and intrinsic fiber connections of the telencephalon in a teleost, *Sebastiscus marmoratus*. *J. Comp. Neurol.* 216:115–131.

Nelson, J.S. (1984). *Fishes of the World*, 2nd ed. New York: Wiley.

Nieuwenhuys, R., and Meek, J. (1990). The telencephalon of actinopterygian fishes. In: *Cerebral Cortex, Vol. 8A: Comparative Structure and Evolution of Cerebral Cortex, Part 1* (Jones, E.J., and Peters, A., eds.), pp. 31–73. New York: Plenum Press.

Northcutt, R.G. (1982). Localization of neurons afferent to the optic tectum in longnose gars. *J. Comp. Neurol.* 204:325–335.

Northcutt, R.G. (1995). The forebrain of gnathostomes: in search of a morphotype. *Brain Behav. Evol.* 46:275–318.

Northcutt, R.G., and Braford, M.R. Jr. (1980). New observations on the organization and evolution of the telencephalon of actinopterygian fishes. In: *Comparative Neurology of the Telencephalon* (Ebbesson, S.O.E., ed.), pp. 41–98.

Northcutt, R.G., and Davis, R.E. (1983). Telencephalic organization in ray-finned fishes. In: *Fish Neurobiology, Vol. 2: Higher Brain Areas and Functions* (Davies, R.E., and Northcutt, R.G., eds.), pp. 203–236. Ann Arbor, MI: University of Michigan.

Northcutt, R.G., and Wullimann, M.F. (1988). The visual system in teleost fishes: Morphological patterns and trends. In: *Sensory Biology of Aquatic Animals* (Atema, J., Fay, R.R., Popper, A.N., and Tavolga, W.N., eds.), pp. 515–552.

Northmore, D.P.M., Williams, B., and Vanegas, H. (1983). The teleostean torus longitudinalis: Responses related to eye movements, visuotopic mapping, and functional relations with the optic tectum. *J. Comp. Physiol.* 150:39–50.

Paxton, J.R., and Eschmeyer, W.N. (1998). *Encyclopedia of Fishes*, 2nd ed. San Diego: Academic Press.

Prechtl, J.C., von der Emde, G., Wolfart, J., Karamürsel, S., Akoev, G.N., Andrianov, Y.N., and Bullock, T.H. (1998). Sensory processing in the pallium of a mormyrid fish. *J. Neurosci.* 15:7381–7393.

Rakich, L., Belekova, M.G., and Konevich, D. (1979). Visual projections into the telencephalon and diencephalon of the fish *Serranus scriba* (electrophysiologic study). *Zh. Evol. Biokim. Fiziol.* 15:357–367. [The original article is in Russian. The information obtained for this paper is from a translated abstract (PubMed search).]

Rooney, D.J., and Szabo, T. (1991). Reciprocal connections between the "nucleus rotundus" and the lateral telencephalon in the weakly electric fish *Gnathonemus petersii*. *Brain Res.* 543:153–156.

Sawai, N., Yamamoto, N., Yoshimoto, M., and Ito, H. (2000). Fiber connections of the corpus mamillare in a percomorph teleost, Tilapia *Oreochromis niloticus*. *Brain Behav. Evol.* 55:1–13.

Schellart, N.A.M. (1992). Interrelations between the auditory, the visual and the lateral line systems of teleosts: A minireview of modeling sensory capabilities. *Neth. J. Zool.* 42:459–477.

Schlussman, S.D., Kobylack, M.A., Dunn-Meynell, A.A., and Sharma, S.C. (1990). Afferent connections of the optic tectum in channel catfish *Ictalurus punctatus*. *Cell Tissue Res.* 262:531–541.

Schwassmann, H.O. (1975). Central projections of the retina and vision. In: *Vision in Fishes: New Approaches in Research* (Ali, M.A., ed.), pp. 113–126. New York: Plenum Press.

Sligar, C.M., and Voneida, T.J. (1976). Tectal afferents in the blind cave fish *Astyanax hubbsi*. *J. Comp. Neurol.* 165:107–124.

Striedter, G.F. (1990). The diencephalon of the channel catfish, *Ictalurus punctatus*. II. Retinal, tectal, cerebellar and telencephalic connections. *Brain Behav. Evol.* 36:355–377.

Striedter, G.F. (1991). Auditory, electrosensory, and lateral line pathways through the forebrain in channel catfishes. *J. Comp. Neurol.* 312:311–331.

Striedter, G.F. (1992). Phylogenetic changes in the connections of the lateral preglomerular nucleus in ostariophysan teleosts: A pluralistic view of brain evolution. *Brain Behav. Evol.* 39:329–357.

Vanegas, H. (1983). Organization and physiology of the teleostean optic tectum. In: *Fish Neurobiology, Vol. 2: Higher Brain Areas and Functions* (Davies, R.E., and Northcutt, R.G., eds.), pp. 43–90. Ann Arbor, MI: University of Michigan.

Vanegas, H., and Ito, H. (1983). Morphological aspects of the teleostean visual system: A review. *Brain Res. Rev.* 6:117–137.

Vanegas, H., Williams, B., and Freeman, J.A. (1979). Responses to stimulation of marginal fibers in the teleostean optic tectum. *Exp. Brain Res.* 34:335–349.

Vargas, J.P., Rodríguez, F., López, J.C., Arias, J.L., and Salas, C. (2000). Spatial learning-induced increase in argyrophilic nucleolar organizer region of dorsolateral telencephalic neurons in goldfish. *Brain Res.* 865:77–84.

Williams, B., and Vanegas, H. (1982). Tectal projections in teleosts: Responses of some target nuclei to direct tectal stimulation. *Brain Res.* 242:3–9.

Wullimann, M.F. (1997). Major patterns of visual brain organization in teleosts and their relation to prehistoric events and the paleontological record. *Paleobiology* 23:101–114.

Wullimann, M.F. (1998). The nervous system. In: *The Physiology of Fishes*, 2nd ed. (Evans, D.H., ed.), pp. 245–282. Boca Raton, FL: CRC Press.

Wullimann, M.F., and Northcutt, R.G. (1990). Visual and electrosensory circuits of the diencephalon in mormyrids: An evolutionary perspective. *J. Comp. Neurol.* 297:537–552.

Yamane, Y., Yoshimoto, M., and Ito, H. (1996). Area dorsalis pars lateralis of the telencephalon in a teleost (*Sebastiscus marmoratus*) can be divided into dorsal and ventral regions. *Brain Behav. Evol.* 48:338–349.

Yoshimoto, M., Albert, J.S., Sawai, N., Shimizu, M., Yamamoto, N., and Ito, H. (1998). Telencephalic ascending gustatory system in a cichlid fish, *Oreochromis* (*Tilapia*) *niloticus. J. Comp. Neurol.* 392:209–226.

Zupanc, G.K.H. (1997). The preglomerular nucleus of gymnotiform fish: Relay station for conveying information between the telencephalon and diencephalon. *Brain Res.* 761:179–191.

Zupanc, G.K.H., and Horschke, I. (1997). Reciprocal connections between the preglomerular nucleus and the central posterior/prepacemaker nucleus in the diencephalon of weakly electric fish, *Apteronotus leptorhynchus*. *Neuroscience* 80:653–667.

22
Paddlefish and Platypus: Parallel Evolution of Passive Electroreception in a Rostral Bill Organ

John D. Pettigrew and Lon Wilkens

Abstract

A comparison is made between a mammalian, monotreme species and an actinopterygian fish that have each, indepen dently, evolved a similar, spoonbill-shaped rostral bill organ whose array of electroreceptors provides sufficient spatial information for prey capture in a freshwater environment without the need for visual cues. The platypus, *Ornithorhyncus anatinus* (Monotremata, Mammalia), has approximately 40,000 electroreceptors arranged in parasagittal rows on the bill organ. By means of behavioral and electrophysiological recording experiments in platypus, it has been shown that this array of electroreceptors can trigger an accurately directed head saccade to intersect aquatic prey that emit electrical signals. The threshold field strength for prey detection by platypus is 50 microvolts/cm, two orders of magnitude more sensitive than individual electroreceptors. The paddlefish, *Polyodon spathula* (Osteichthyes, Actinopterygii), can similarly execute a lateral head saccade to intersect prey, with a threshold field strength around 10 microvolts/cm, considerably more sensitive than the presumed sensitivity of individual electroreceptors. The remarkable anatomical and behavioral similarities between these two independent electroreceptive systems are described and discussed. Major differences between the two bill-organ systems include the mechanism of transduction at the electroreceptors and a prominent cooperative role played by 60,000 mechanoreceptors that are interdigitated among the electroreceptors in the platypus bill but not in the paddlefish.

1. Introduction

This paper grew out of a mutual recognition by the authors of the striking similarities between the rostral bill apparatus of the paddlefish and the platypus. This fish and this mammal each use the unusual sensory abilities of their bills to locate and catch small freshwater prey. Although neither of us has worked on the other's species, our knowledge has grown as we interacted. As a result we decided to try to present a comparison that illustrates the convergent similarities and important differences between these extraordinary animals that have independently evolved the use of passive electroreception by an elaborate rostral bill organ to catch aquatic prey.

We begin with a brief account of the biology of the two species and place them each in evolutionary context. We then give a detailed description of the electroreceptors and their arrangement in the bill organ in each species, with an account of the similarities and differences. We describe the lateral head saccade that is directed toward an electrically active prey in both species and show how this behavior has been used by investigators of both species to define the directionality and the threshold sensitivity of the electrosensory system. Since both species show far greater sensitivity to electrical stimuli at the whole animal level than at the level of individual receptors, we discuss this commonly observed discrepancy assuming that there is signal processing of the combined activity of large numbers of receptors. We show how both species avoid DC fields and finish with a discussion of the role of underwater mechanoreception as an adjunct to electroreception. This emerging bimodal viewpoint that considers waterborne mechanical stimuli arises from the extensive system of interdigitated electroreceptive and mechanoreceptive cortical neurons known in platypus, along with the mechanosensitivity of electrosensory ampullae in paddlefishes. Finally, we discuss the similarities and differences between the underwater prey detection systems of paddlefishes and platypus in the context of conservation issues.

2. Biographical Sketches of the Animals

2.1. Paddlefishes

The paddlefish, *Polyodon spathula*, is among the most primitive of bony-finned fishes (Osteichthyes, Actinopterygii), and together with sturgeon comprises an order of secondarily cartilaginous fishes, the Acipenseriformes. The North American paddlefish is one of only two extant species in the family Polyodontidae whose origins extend back into the Cretaceous period, over 100 mya; the sister species *Psephurus* is native to China. Paddlefishes once ranged throughout the turbid midwestern rivers of the United States.

By virtue of their elongated rostrum (i.e., bill or paddle), paddlefishes are unique among the approximately 25,000 species of living fishes. Unlike other fishes with long noses and jaws, the paddlefish rostrum is an elongation of its cranium. In contrast, the platypus upper bill is maxillary and is fused to the cranium. The prominent rostrum comprises one-third the length of the fish, even more in small fishes, and is responsible for the common functional misrepresentation indicated by the synonymic names "spoonbill cat" and "shovelnose cat." Paddlefishes also are notable for their large size, once known to reach 75 kg and 2 m, which they achieve by feeding on a rich zooplankton resource at the base of the food chain. As filter feeders, and one of the largest riverine species of fishes, they are the ecological equivalent in freshwater of baleen whales and large filter-feeding marine fishes such as the whale and basking shark.

Only recently (Wilkens et al., 1997) has the evolutionary significance of the paddle been correlated with the fish's planktonic diet and its muddy, nonvisual environment. In feeding experiments, small paddlefishes readily capture tiny plankton (*Daphnia*) without benefit of their visual, chemical, or hydrodynamic senses (Wilkens et al., 2001). This suggests the existence of an alternate sensory mechanism. Like the platypus bill, the paddlefish rostrum is covered by an extensive array of electroreceptors, here the ampullae of Lorenzini (Jørgensen et al., 1972). Thus, the rostrum and its ampullary electroreceptors comprise the functional equivalent of an antenna, a sensory device with sufficient sensitivity to detect the electric fields of their planktonic prey (Wilkens et al., 1997; Russell et al., 1999).

While we have established the role of electroreception in planktivorous feeding in small juvenile paddlefishes, the sensory mechanisms used by large adult fishes have not been determined. Unlike juveniles that feed selectively by capturing individual plankton, large fishes feed nonselectively by sieving plankton from the water using comb-like gill rakers not fully developed until fishes reach 22–23 cm in length (Rosen and Hales, 1981). However, large healthy fishes occasionally are found missing their paddle, presumably the result of mutilation by boat propellers. Since juvenile fishes have not been observed in the wild except with their paddles intact, it is presumed that the electrosensory system is an adaptation primarily benefiting young paddlefishes when feeding selectively (i.e., capturing plankton one at a time).

2.2. Platypus

The platypus, *Ornithorhyncus anatinus*, is one of three living species of monotremes (egg-laying mammals with a single opening, the cloaca, for both excretory and reproductive functions) (Griffith, 1978). The only other two extant monotremes (*Tachyglossus aculeatus* and *Zaglossus brujnii*) are land-dwelling, spine-covered echidnas that arose relatively recently (20–30 myr) from an ancient platypus lineage that can be traced back more than 100 myr (Pettigrew, 1999). The first specimen of a platypus to reach the West was believed at first to be a fake, as it was thought that the duck-like bill had been cunningly stitched to the luxuriant fur (Griffith, 1998). Platypus live in a long burrow with an entrance close to, or below, water level. The tunnel can be a tight fit, particularly in the breeding female's burrow, which is much longer than usual. The length and fit ensure that the platypus is dry by the time it reaches the nest area at the end, even when there is a high frequency of hunting sorties to the water that would tend to wet the tunnel, as for the breeding female. Eyes, nostrils, and ears are all closed underwater and prey are located by a combination of electroreception and mechanoreception (see below). Prey are benthic and include crustaceans, insect larvae, worms, and fishes. On an average foraging trip, which is usually at night, a platypus may consume half its weight in prey, in total darkness. Juveniles may hunt in daylight as well, perhaps because they are smaller, with a greater metabolic rate, and/or because they are less skillful at finding prey and therefore need to spend more time in this activity (J.D. Pettigrew, unpublished). An extensive series of behavioral observations has clarified how platypus use electroreception, perhaps in conjunction with mechanoreception, to locate aquatic prey in three dimensions (Manger and Pettigrew, 1995; Manger et al., 1996; Pettigrew et al., 1998). The 40,000 electroreceptors on the bill form a directional antenna that allows instantaneous determination of the azimuth and elevation of a prey's electrical discharge so that a head saccade can be directed accurately toward it. By combining mechanosensory and electrosensory inputs together in an extraordinary cortical structure resembling the celebrated ocular dominance columns of primates, the platypus may be able to fix the third coordinate of its prey, distance, from the time-of-arrival difference between the electrical signal and its later mechanical wave produced by prey.

Early development after hatching takes place in the burrow, with the female providing milk that is secreted onto her fur rather than from a nipple (Griffiths, 1978). Electroreceptors develop on the bill at this very early stage, before mechanoreceptors, suggesting that electroreceptors may have a functional role, perhaps in suckling, before the juvenile enters the water for the first time (Manger et al., 1999).

3. Arrangement of Electroreceptors on Rostrum and Bill

3.1. Paddlefish Ampullae

The paddlefish skin covering the upper and lower surfaces of the paddle, the head, opercular flaps, and lower jaw is penetrated by numerous pores, 65–140 mm in diameter and just visible to the naked eye. An 80-cm fish was estimated to have 57,365 pores (Kistler, 1906), with numbers as high as 75,000 per fish (Nachtrieb, 1910). Pores occur in clusters of 5–25 that are

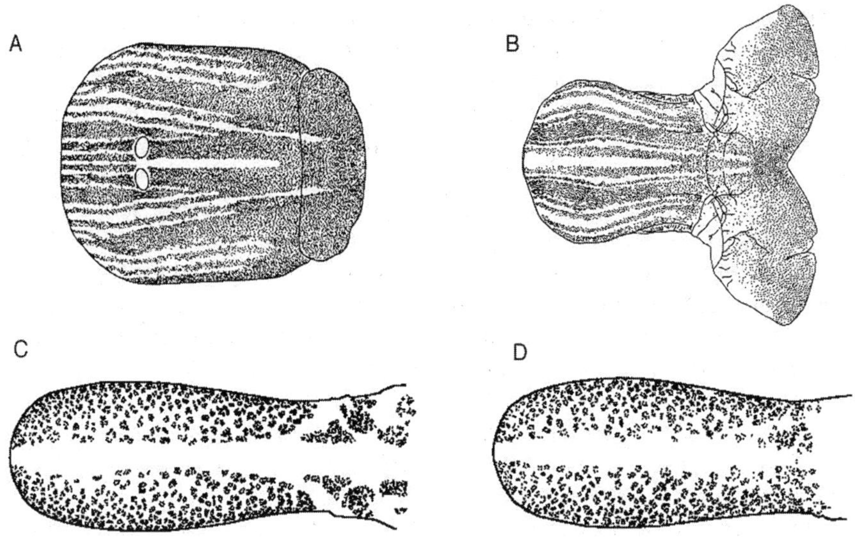

FIGURE 22.1. Electroreceptor arrays in platypus and paddlefishes. Views of upper (A, C) and lower (B, D) surfaces of the rostral bill organ of each species (there are also many electroreceptors on the inner surfaces of the upper and lower mandibles, not shown). Approximately 40,000 electroreceptors in platypus (mucous gland type) are arrayed in parasagittal rows, with a prominent raphe lacking in receptors along the midline. About 50,000–60,000 electroreceptors in paddlefishes (ampullae of Lorenzini type) are arranged in clusters with a prominent midline raphe where afferent nerve fibers travel.

more numerous near the lateral margins of the paddle, but absent from an 8–10-mm wide strip along the midline of both upper and lower surfaces (Fig. 22.1).

Each pore opens into the pear-shaped lumen of an ampullary organ 100–250 μm deep. The lower half of the ampulla is lined by a sensory epithelium of ciliated receptors and supporting cells (Jørgensen et al., 1972). Tight junctions near the luminar surface join the receptor and supporting cells. The receptor cells form synapses with the nerve endings of primary afferent neurons that converge into anterior lateral line nerves that travel to the brain. Each receptor synapses with several nerve terminals and each primary afferent innervates numerous receptor cells. Although characterized early on as sensory organs (Kistler, 1906), Nachtrieb (1910) argued vigorously that the ampullae were secretory organs owing to the gelatinous secretions inside the pits. Their electrosensory function as ampullae of Lorenzini was established unequivocally by the anatomical study of Jørgensen et al. (1972).

3.2. Platypus Electroreceptors

Electroreceptors are readily recognized in low-power micrographs (even in obliquely lit naked-eye views) because of the pores that penetrate the thick epidermis to enable the necessary flow of current to the nerve endings. There are two kinds of electroreceptor that can be easily recognized in low-power micrographs and SEM views of the bill skin because of differences in the size and shape of the flower-petal-like arrangement of keratin that surrounds the pore and that opens when the bill is immersed (Manger et al., 1996). These two kinds of electroreceptors are associated separately with serous and mucous glands. The role of the serous gland electroreceptors is still obscure and they will not be considered further here, except to note that they have a lower density and completely different distribution on the bill from the mucous gland electroreceptors and have been speculated to be involved in near-field or DC detection (Manger and Pettigrew, 1997). Mucous gland electroreceptors are much more common, with a total of 40,000 distributed over all the epidermal surfaces of the bill, with a marked concentration at the edges and a striking parasagittal stripe arrangement on the dorsum of the bill (Fig. 22.1). As their name suggests, mucous gland electroreceptors are modified mucous glands. The base of the gland duct is penetrated radially by about one dozen free nerve endings, originating from a daisy-chain-like arrangement of lateral connections between nerve terminals where it is suspected that there are gap junctions that would act to reduce noise (Manger and Pettigrew, 1998; Pettigrew et al., 1999). The free nerve endings pass through a barrier that is inferred to have very high impedance to lie in electrical contact with the medium at the center of the gland duct without any associated epithelial cell specialization (Fig. 22.2). This arrangement contrasts strongly with electroreceptors in the paddlefish where epithelial receptor cells provide the transducing interface between the nerve ending and electric current in the aqueous environment (Fig. 22.2).

The parasagittal stripe arrangement suggested to investigators that there might be integration between the output of different electroreceptors that lay with a parasagittal arrangement to each other. This was tested electrophysiologically and behaviorally. Receptive fields of electroreceptors recorded at the level of the neocortex show elongations that are roughly parallel to the sagittal axis, in accord with this suggestion (Krubitzer et al., 1991). Moreover, when the sensitivity of the whole bill is measured using the head saccade paradigm, the axis of maximal sensitivity is found lateral, slightly forward and down from an axis at right angle to the lateral edge of the bill (Manger and Pettigrew, 1997). Given the complex, slightly curved shape of the bill, this is what would be expected if there were a general

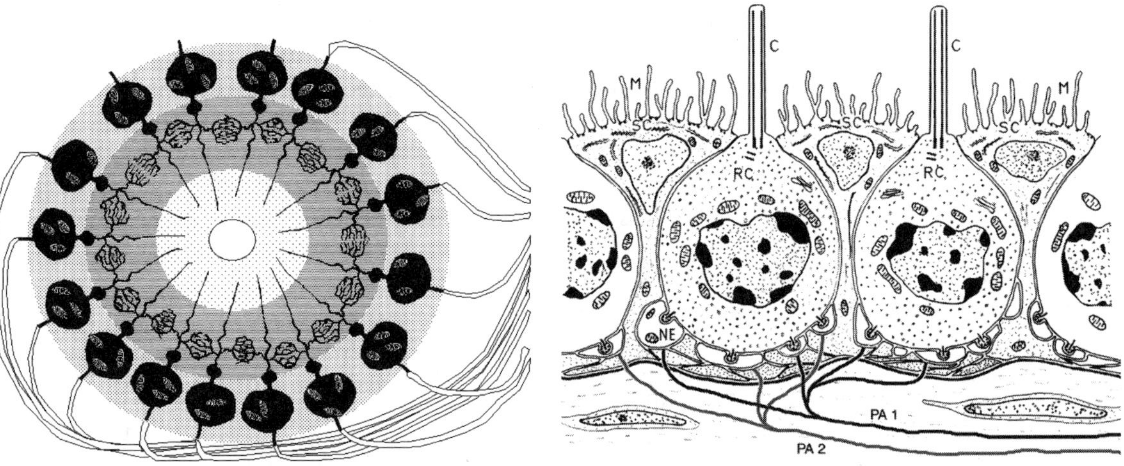

FIGURE 22.2. (A) Arrangement of free nerve endings in platypus mucous gland electroreceptor. Up to 12 myelinated fibers lose their myelin sheaths on entering the base of the gland duct. Lateral extensions of the free nerve endings originate from bulb-shaped expansions filled with mitochondria. The lateral extension joins all participating fibers in a daisy chain around the base of the duct, presumably to facilitate some form of common-mode rejection of electrical signals that are not shared in all terminals. A terminal filament pierces the electron-dense zone around the duct to lie free in the mucous-filled ambient. Note there there is no associated epithelial cell involved in electrical transduction at the terminal in the platypus, as there is in the paddlefish. (B) Generalized ampullary electroreceptors, as found in the paddlefish rostrum. Note that primary afferent nerve endings PA1 and PA2 innervate ciliated sensory epithelial cells that are responsible for transduction, in contrast to the free nerve endings seen in the mammalian electroreceptor. Modified from schematic of Jørgensen et al., 1972, by addition of primary afferents.

principle of integration that favored electroreceptors along the sagittal axis.

3.3. Comparison

The paddlefish lacks the tell-tale parasagittal stripe arrangement of electroreceptors that may reveal the underlying strategy by which the platypus integrates information from such a large number of receptors in its central nervous system. Yet, a parasagittal axis of integration is suggested in the paddlefish too, by the fact that paddlefish electroreceptors are distributed preferentially along the rostrum edges. A parasagittal organization is also suggested by the preference for lateral strikes that indicates that the paddlefish, like the platypus, is most sensitive to signals that emanate laterally, rather than frontally or from any other direction.

The fine structure of the electroreceptors in the two taxa, as might be expected from their separate phylogenetic derivations, is completely different. The trigeminal nerve endings of the platypus lack any associated epithelial cell specializations, instead showing an elaborate structure where a dozen or more nerve terminals are linked circumferentially around the pore, into which penetrate a dozen or more free nerve endings. This structure is unstudied physiologically and even microscopical information is scanty, but it seems very likely that the terminals are all linked by gap junctions that would form a pool of receptors that could act to reduce electrical noise from one pore, just as coupled pools of photoreceptors use simultaneity to reduce noise in low light levels where the signal-to-noise ratio is low. Under normal circumstances for the electroreceptive systems there is a high noise level, so it is easy to imagine a useful role played by a coupled pool of detectors that were all signaling the field strength in the same pore and performing some kind of common mode rejection operation. The absence of an epithelial cell to provide a receptor potential and to transduce the electrical signal could explain why platypus electrosensitivity is relatively low compared with the paddlefish, whose electroreceptors involve an epithelial receptor cell whose physiology is well-known from extensive studies of other passive electric fishes and which have lower thresholds than the platypus (Wilkens et al., 1997).

4. Lateral Head Saccade

4.1. Paddlefishes

The swimming undulations of the paddlefish trunk impart a lateral oscillating motion to the rostrum, with the cranium acting as the fulcrum. Paddlefishes are ram ventilators (Burggren and Bemis, 1992), so swimming is continuous. This saccade-like motion may serve to enhance prey detection by increasing the width of the electrical scan field. Rapid lateral saccades, a yaw-like motion, also characterize the motion sequences involved in prey capture, as dictated by the broad, flat shape of the rostrum. For example, as plankton drift by lateral to the rostrum the paddlefish makes sudden flexions that jerk the head and rostrum toward the prey as the mouth begins to open (Fig. 22.3). From the moment of detection the strike and gulp sequence takes 150–200 ms. For plankton encountered above or below the rostrum, a more complex motion sequence ensues. The fish initially rolls about its long axis, up to 90° or more, followed by the same lateral saccade, now directed in a vertical or oblique plane. Thus, each rapid strike, regardless of prey location, results in a slicing motion of the rostrum through the water that affords the least resistance. Plankton are captured by these accurate feeding strikes, up to 8–9 cm from the rostrum, some of which involve acrobatic combinations of rolls, lateral saccades, and complete turns if the plankton drifts past the mouth.

4.2 Platypus

The platypus lateral head saccade is elicited reflexly by a brief electrical stimulus. If the platypus is swimming, the head saccade causes resetting of the swimming rhythm of the body behind the head. In this way, the head

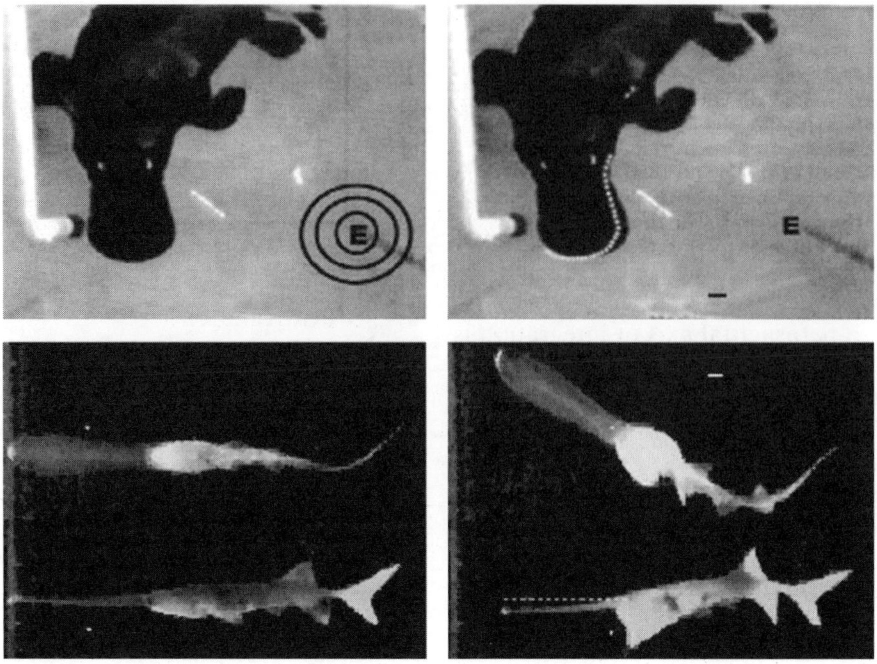

FIGURE 22.3. Head saccades to electrical stimuli by platypus and paddlefishes. *Stationary platypus*: The two video frames are taken 60 msec apart, with the platypus at rest in a tank (see Manger and Pettigrew, 1995 for more details). The electrical stimulus was a square-wave pulse from a unipolar electrode (E) that produced a field at the near side of the bill of around 200 mV/cm. The resulting saccade toward the stimulus had a small amplitude in this habituated platypus (saccades during patrolling and prey capture are larger), but can readily be seen by referring the right bill edge to the angled exhaust pipe on the left side of the figure and by reference to the dotted outline, which shows the position of the bill's left edge at 0 msec. In a swimming platypus, a saccade like this will act to reset the swimming direction behind the head so that the platypus changes direction and swims toward the prey. Saccades to electrical stimuli can follow to 15 Hz. (A fast sequence of saccades can be viewed at www.uq.eud.au/nuq/jack/platypus.html.) *Swimming paddlefish*: The prey item, *Daphnia*, is shown as a white highlight. The electrical discharge (no other sensory cues are available to the fish in this darkened tank illuminated with infrared) triggers a lateral and slightly downward-directed saccade, along with jaw opening, that will permit the paddlefish to swallow the prey. In both species, the spatial information that guides the accurate directionality of the saccade has been gathered almost instantaneously from the electrical field generated by the stimulus, rather than being the result of an algorithm to track the field strength as the animal moves through it (e.g., the Kalmijn algorithm). Instantaneous determination thereby implies sophisticated signal processing by the spatial array of electroreceptors on the bill. Scale = 1 cm in both cases.

saccade causes the platypus to swim toward the source of the stimulus (videotape in Manger, 1998). The head saccade is directed accurately toward the direction of the electrical source, whether this is above, below, or lateral to the bill, although it is much more sensitive to sources that emanate laterally, forward, and down, along the axis of the electroreceptive system. In captivity, platypus are intensely curious about gentle electrical stimuli, which are explored and mouthed. This exploratory behavior soon habituates, but the head saccade remains, with the platypus unable to suppress it after thousands of trials. The head saccade behavior can then form a useful basis for the exploration of the parameters of electroreception, such as threshold and directionality.

4.3 Comparison

Given the similar orientation of the rostrum and bill, it is perhaps not surprising that both paddlefish and platypus have evolved a short-latency head saccade with the same lateral direction to move through the water toward prey with least resistance (Fig. 22.4). Since both rostrum and bill have similar shapes and have the greatest concentration of electroreceptors along the edges, both would be expected to have axes of greatest sensitivity to electrical stimuli that originate laterally, rather than frontally or along some other axis, as has been shown for the platypus and as seems very likely for the paddlefish, too. The similarities in prey capture style can thus be attributed to several factors that have combined to ensure that an

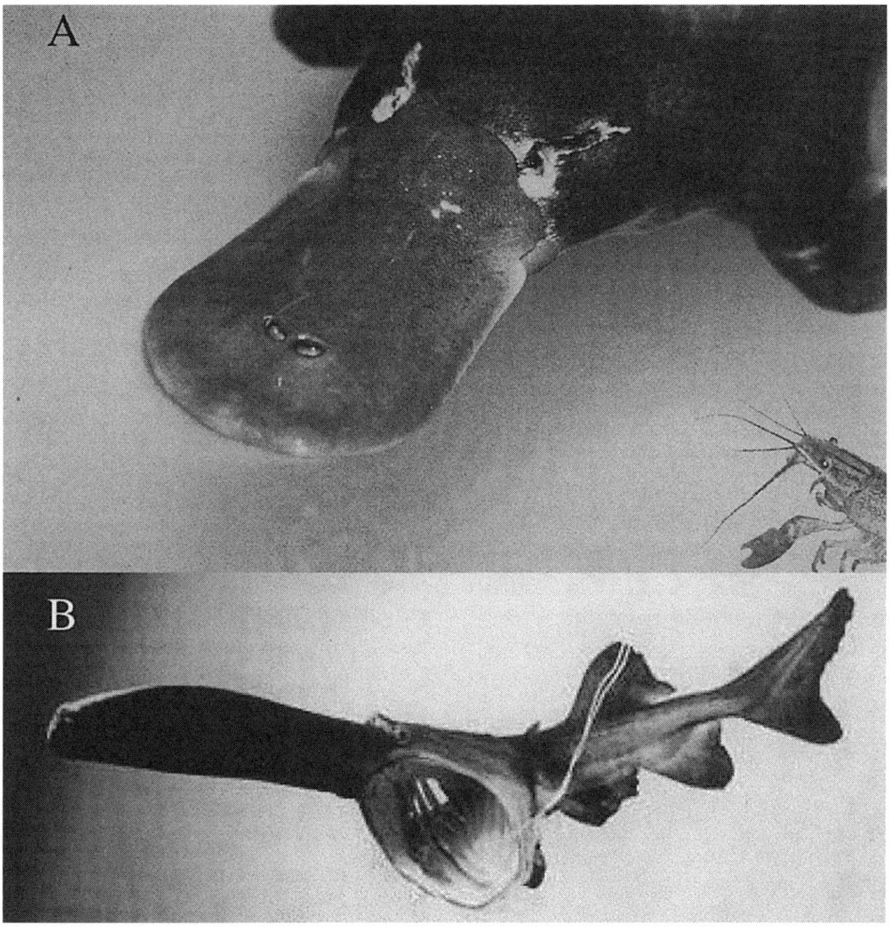

FIGURE 22.4. Comparison of platypus with paddlefish to show the similarity of the paddle-shaped bill organ and rostrum respectively. (A) The bill organ of the platypus (and one of its prey, the yabbie, *Cherax spp.*). About 40,000 electroreceptors and 60,000 mechanoreceptors form a sensory array that enables the capture of small benthic invertebrates, as well as some vertebrates such as fishes, without any assistance from vision, hearing, or olfaction. Note the numerous pits formed by the opening of the mucous gland electroreceptors on the bill; these pits are barely visible where they are in sharper focus on the more proximal region of the bill, near the eye. (B) Young paddlefish attacking the dipole wires as if trying to capture plankton: Its rostrum is innervated by about 60,000 ampullary electroreceptors that enable the capture, using the prey's electrical discharge, of free-swimming invertebrates such as *Daphnia*.

explosive strike by the bill has shortest latency when the bill orientation, neck muscle arrangement, and greatest sensitivity of the electroreceptor system are all optimal for a laterally directed saccade.

It is worth considering the underlying neural basis for a well-localized strike of this kind. Many passive electric fishes appear to use Kalmijn's "approach algorithm" to localize a dipole, but note that this cannot be true for the lateral head saccades of platypus and paddlefish, which are instantaneously directional and do not require successive approximation toward the target. Both platypus and paddlefish must "know" at very short latency, the shape of the electric field generated by the prey in order to localize it so accurately and quickly. Such a calculation would involve the simultaneous reporting of field strength from a large number of the electroreceptors distributed over the rostrum or bill. This calculation would be aided by the bill shape, whose flattened lateral dimension and high density of electroreceptors would help ensure that changes in field strength would be most detectable along just the axis most relevant to localizing the prey for a strike. The elaborate neural mechanisms that would provide a platypus with a "map" of field strength over the surface of the bill have been well-described (Krubitzer et al., 1996; Calford et al., 1998). We can only assume that there are similar mechanisms that subserve the high density of electroreceptors on the similarly shaped rostrum of the paddlefish.

5. Sensitivity

5.1. Paddlefishes

A lateral head strike, like the platypus's head saccade, can be elicited from a paddlefish by a *Daphnia* at 10 cm. Since *Daphnia* produce field strengths of around 1 μV/cm, the paddlefish's sensitivity is around 10 μV/cm, assuming a cubic fall-off in field strength from this circumscribed, tiny prey. Since a directed response would require the coordinated activity of many electroreceptors, we can assume that not all receptors will be stimulated with strengths as great as 10 μV/cm and that signal detection algorithms will be applied by the paddlefish nervous system to look at activation across the array. Little is known about the details of this process, which is responsible for the much greater sensitivity of the whole animal when compared with sensitivity of individual electroreceptors. The electrical threshold for detecting and avoiding a DC field, such as created by a metallic object underwater, is as much as 10 times lower in the paddle-fish. The difference between these two thresholds is not explained, but could derive from the punctate nature of stimulation with *Daphnia*, where one would expect different receptors in the array to be subject to markedly different stimulus strength according to their position in relation to the *Daphnia*, compared with the more uniform pattern of stimulation that one would expect from a large, submerged metal object.

5.2. Platypus

The most extensive studies of the electrosensory threshold of platypus were conducted using the lateral head saccade to a brief electrical stimulus (Manger and Pettigrew, 1998). If one adopts a one-to-one criterion for a head saccade following an electrical pulse, the threshold field strength is 50 μV/cm. This is 20–50× less than the 1–2 μV/cm threshold established for single nerve fibers emanating from the electroreceptors themselves (Gregory et al., 1987a,b; 1988), the difference being accountable in terms of signal detection in the massively enlarged bill-representation of the platypus brain. Little is known about the signal detection algorithms that are used by the platypus brain, but some idea of their sophistication can be gained by a study of time averaging that showed that a platypus can act as a computer of average transients and lower its threshold if the same electrical stimulus is presented repeatedly! (Farrebrandt et al., 1999). This phenomenon will be familiar to readers who have wondered whether their name may have been called in a noisy environment, only to have their suspicion confirmed when the call is repeated. If the same electrical stimulus is repeated many times, the platypus threshold falls to a low of 10 μV/cm.

The electrical threshold for detecting a DC source in platypus is not known, but from experiments with avoidance of metal objects we can assume that it is equal to, or less than, the threshold for a lateral head saccade.

5.3. Comparison

Both taxa have remarkable similarity in the sensitivity of the strike to electrical stimuli generated by prey, with the paddlefish system having two orders of magnitude greater sensitivity than the platypus, at 1 μV/cm. To strike at a *Daphnia*, the platypus would have to be 3 cm away, rather than the 10 cm at which a paddlefish can successfully strike its favorite prey. One would imagine that caddis-fly larvae, a favored platypus prey item with size and field strength in the range of *Daphnia*, would permit the closer approach needed by the platypus to elicit a strike. On the other hand, more mobile platypus prey that would require active pursuit and more distant detection, such as *Macrobrachium* and *Ateya*, have long bodies and strong fields compared with *Daphnia* and would be detectable by the platypus at 10 cm (Pettigrew et al., 1999). Whether these larger crustaceans would also be attractive to the paddlefish, as this comparison suggests, is an unanswered question.

In contrast to the systematic study of paddlefish avoidance of DC fields (Gurgens et al., 2000), there is no detailed study of the threshold for avoidance of DC fields by platypus. The greatly enhanced sensitivity of the paddlefish to DC fields raises the possibility that platypus may also be more sensitive to such fields than they are to AC fields such as those generated by prey, but the available data are agnostic on this point. It has been suggested that the less common serous gland electroreceptors mediate DC detection in platypus. This separation of function certainly echoes the dramatic difference in threshold between the detection of AC and DC in paddlefishes. For these reasons it might be profitable to look for a segregation of receptor sensitivity in paddlefishes and to undertake a more extensive study of DC thresholds in platypus.

6. Role of Mechanoreception

6.1. Paddlefishes

Any mechanical disturbance created by a *Daphnia* at the rostrum of the paddlefish when it is located 10 cm away is presumably very small and unlikely to be detectable compared with the threshold electrical stimulus of 1 μV/cm at this distance. For this reason, the mechanosensory properties of the ampullary receptors are not likely to play as important a role in detection of *Daphnia* as their electrosensory properties. Nevertheless, this aspect may be worthy of further investigation, both because the ampullary receptors do have acute mechanical sensitivity and because the near-field mechanical stimulation produced by *Daphnia* may cause surprises in the same way that the mechanical stimulation provided by platypus prey had unexpected effects on the sensory system for prey detection in the bill.

6.2. Platypus

There are 60,000 mechanoreceptors in the platypus bill, interspersed with the 40,000 electroreceptors. It was generally assumed that these mechanoreceptors were not involved in the electroreceptive detection of aquatic prey, but rather in processing the direct contact with prey that had been captured. That view has been overturned by a number of new observations: (1) When the bill is out of the water, the mechanoreceptors, "pushrods," are tightly enclosed within a pore that opens upon immersion of the bill to free the rod and thereby enable it to respond to small movements in the water. (2) At the neocortical level, there is an elaborate anatomical interdigitation of the mechanoreceptive and electroreceptive representations that enables them to cooperate in a sophisticated way so that individual neurons can signal the time of arrival difference between electrical and mechanical stimuli. (3) Finally, larger prey items, such as shrimp, generate substantial mechanical disturbance at the platypus bill, well above thresholds for the pushrods, when they escape with tail flicks at distances comparable to the threshold distance for electroreception (Pettigrew et al., 1999).

Taking these new observations all together, it seems that electroreception and mechanoreception cooperate intimately in the platypus. A huge cortical area is devoted specifically to this cooperation in the brain (Manger et al., 1998). Just as the shape of the electrical field produced by a prey item could be deciphered by the bill's electroreceptors, so the distribution of mechanical disturbance over the bill that is produced by prey movements could be deciphered by the distributed array of abundant pushrods and thereby provide information about the location of the prey. Since the arrival of the mechanical information will be delayed in relation to the distance of the prey, in contrast to the electrical information, which travels too close to the speed of light to provide any practical latency information, close coordination between the two sensory systems could lead to precise information about the distance of the prey (Pettigrew et al., 1999).

6.3. Comparison

Since the paddlefish also judges the distance of the prey quite accurately for a strike, one wonders if mechanoreception might play a role in paddlefish prey detection, as it does in platypus. The much greater electrosensitivity of the paddlefish system makes this unlikely. The much closer distances that a platypus must approach in order to detect the electrical activity of prey also mean that there is a greater likelihood that the mechanical stimuli from the same prey will be capable of exceeding the threshold of the pushrods. The fact that the platypus also has two completely separate sets of receptors, each one of which seems to be highly specialized for underwater mechanoreception and electroreception, respectively, argues that both may be significant in a way that is not true for the paddlefish, where any role of mechanoreception would have to be carried out by the same set of receptors that is also involved in electroreception.

7. Conservation Issues and Electroreception

Electrofishing may be having a deleterious impact on platypus in Australia. Forty thousand electroreceptive pores in the platypus bill lower its electrical impedance and make it vulnerable to the intense current generated by the electrofishing devices used to sample fish stocks in the waterways. Direct observations on rescued platypus indicate that they are immobilized in the vicinity of the probe for a number of minutes. Stress, to which platypus are particularly vulnerable, often causing death within a few hours of capture, and drowning would be possible consequences if the affected platypus were not rescued from the water.

There is less information about the impact of electrofishing on paddlefishes, but we predict that this will not be inconsequential because the ampullary electroreceptors provide a large number of parallel paths for the high-intensity current from the high-voltage probe, just as the pores in the platypus bill lower its impedance and increase its vulnerability. Because of the substantial resistance to the use of the platypus as an experimental animal, it is possible that the paddlefishes might be able to provide some baseline information about the effects of high voltages on bill electroreceptors.

Paddlefishes have been impacted by electrofishing, although no systematic study has been made to determine whether they are more or less susceptible than other fishes to high-voltage electrofishing, or to determine rates of recovery and/or mortality, although fishes have been observed to leap out of the water and break their rostrum in violent swimming collisions in response to current pulses. Other signs of stress, with limited recovery of normal swimming and latent mortality, have been reported (Scarnecchia et al., 1999), with physical damage including ruptured blood vessels and notochord. However, electrofishing has been touted recently as a successful sampling technique for paddlefishes, but with immediate mortality rates of 10% (Lein et al., 1997).

Paddlefish electrosensitivity presents an additional conservation issue with respect to migratory behavior. Fishes migrate upstream to spawn in smaller streams and tributaries (Purkett, 1961), and from river channels to flooded backwaters that provide seasonal planktonic food (Stockard 1907). Tracking studies document that fish migrations of several hundred kilometers are common (Russell, 1986; Pitman and Parks, 1994), in one case with recapture 2,000 km downstream (Rosen et al., 1982). During migration, paddlefishes face unnatural electrical fields produced by metal gates in locks and dams, and are reluctant to pass through partially open gates (Southall and Hubert, 1984). Sensitivity to metallic sources, first noted by Jørgensen et al. (1972), was recently quantified in an avoidance study by Gurgens et al. (2000). Fishes detected a 2.5-cm diameter bar at distances up to 38 cm, swimming away in an excited avoidance response. Field studies may eventually demonstrate the need to insulate these barriers in sensitive habitat areas.

8. Conclusion

Fundamental research on the platypus is strictly limited at present because of its status as a national icon. There are some ironies in this, since it was the neuroethological approach, using a combination of electrophysiological and behavioral investigations, that was responsible for the present knowledge of platypus electroreception and its beneficial application to both display and field contexts.

Display arenas for captive platypus are extremely popular because viewing this extraordinary mammal in the wild is made difficult by its nocturnal, secretive underwater habits. Captive displays had been associated in the past with high mortality from stress. These arenas now incorporate the new information about thresholds for electroreception into their design, with marked improvement of the welfare and longevity of the platypus. Design of screening and grounding for the aquarium to protect it from stray electrical fields caused by switches, Silicon-Controlled Rectifiers (SCRs), and electrical pumps, for example, can now be guided by the quantitative information about threshold sensitivity of the electrosensitive system. Such information became available only because of the carefully controlled neuroethological approach in the laboratory that involved capture and investigation of a small number of healthy animals. While purely behavioral observations in the field might now in hindsight be able to provide some observations on electroreceptive thresholds, the historical facts show that the key elements in these advances were laboratory-based and multidisciplinary, including electron microscopy of bill receptors and electrophysiology from individual nerve fibers and nerve cells. In other words, the very small numbers of platypus taken for these investigations is more than counterbalanced by the beneficial information about this species that was gained. This conclusion is supported also by a consideration of field conditions such as impact of water impedance on electroreception. Another potential area where precise information about platypus electroreception will be vital is the problem of the impact of electrofishing, which raises questions about platypus welfare that are potentially much more serious than the laboratory investigations that gave rise to the only relevant current source of information on the problem.

The quantitative laboratory-based information about electroreception is also relevant to field studies, in relation to the wide variation of salinity and water impedance in eastern Australian waterways. While it is clear that platypus can tolerate very low impedance conditions, such as in small waterways loaded with electrolytes that run off from cow pasture, it is unlikely that they could catch highly mobile prey, such as shrimp, under these conditions. This is because relatively pure, high-impedance water is necessary if the field generated by a mobile prey is sufficiently large to be detected by the platypus at a distance that would permit successful pursuit. In a small polluted stream, low water impedance may still be compatible with capture of prey if they are relatively immobile, like the caddis-fly larvae that are common in such situations. The platypus can then approach closely, inside the very small threshold radius that will obtain in such low-impedance situations. In contrast, rapidly moving prey, or rapidly moving water, or both, will severely impact on the platypus's ability to track prey in conditions of low water impedance. This may explain how platypus are so successful at catching fast-moving shrimp in pure streams, and perhaps illuminate the presently unexplained absence of platypus from increasingly saline streams on the western drainage of the Great Dividing Range.

One of the most extraordinary findings in these studies, the close coordination of electroreception and mechanoreception in a complex central array that is reminiscent of primate ocular dominance columns, was made just as laboratory-based platypus studies were being phased out. In the future, fruitful investigations would involve study of the combined action of mechanoreceptive and electroreceptive waterborne stimuli that originate from the same target. We predict that combined stimuli of this kind will be much more effective in eliciting platypus behavior than electrical or mechanical stimuli alone.

Acknowledgments. The work on platypus was supported by the Australian Electricity Supply Industry Research Board and an ARC Special Research Centre grant to JDP. Despite initial misgivings, Queensland Minister for the Environment, Pat Comben, gave important support that we believe has been vindicated by the wealth of new information gained on platypus electroreception that is relevant to both husbandry of captive animals and conservation in the wild. The work on paddlefishes was supported by the U.S. Office of Naval Research, National Science Foundation, and University of Missouri Research Board. This work could not have been done without the cooperation of the Missouri Department of Conservation, which has generously provided fishes from their Blind Pony Hatchery. Collecting juvenile fishes from the wild would have been prohibitively difficult.

References

Burggren, W.W., and Bemis, W.E. (1992). Metabolism and ram gill ventilation in juvenile paddlefish, *Polyodon spathula* (Chondrostei: Polyodontidae). *Physiol. Zool.* 65:515–539.

Fjällbant, T.T., Manger, P.R., and Pettigrew, J.D. (1998). Some related aspects of platypus electroreception: temporal integration behaviour, electroreceptive thresholds and directionality of the bill acting as an antenna. *Phil. Trans. Roy. Soc. B.* 353:1211–1219.

Gregory, J.E., Iggo, A., McIntyre, A.K., and Proske, U. (1987a). Electroreceptors in the platypus. *Nature* 326:386–387.

Gregory, J.E., Iggo, A., McIntyre, A.K., and Proske, U. (1987b). Responses of electroreceptors in the platypus bill to steady and alternating potentials. *J. Physiol. Lond.* 408:391–404.

Gregory, J.E., Iggo, A., McIntyre, A.K., and Proske, U. (1988). Receptors in the bill of the platypus. *J. Physiol.* 400:349–358.

Griffiths, M. (1978). *The Biology of the Monotremes.* Academic Press: New York.

Griffiths, M. (1998). Preface to platypus biology: Recent advances and reviews. *Phil. Trans. Roy. Soc. B.* 353:1057–1237.

Gurgens, C., Russell, D.F., and Wilkens, L.A. (2000). Electrosensory avoidance of metal obstacles by the paddlefish. *J. Fish. Biol.* 57:277–290.

Jørgensen, J.M., Flock, Å, and Wersäll, J.Z. (1972). The Lorenzinian ampullae of *Polyodon spathula*. *Z. Zellforsch.* 130:362–377.

Kistler, H.D. (1906). The primitive pores of *Polyodon spathula*. *J. Comp. Neurol.* 16:294–298.

Krubitzer, L., Manger, P., Pettigrew, J.D., and Calford, M. (1991). The organization of and connections of somatosensory cortex in monotremes: In search of a prototypical plan. *J. Comp. Neurol.* 351:261–306.

Manger, P.R., and Pettigrew, J.D. (1995). Electroreception and feeding behaviour of the platypus. *Phil. Trans. Roy. Soc. Lond. B.* 347:359–381.

Manger, P.R., and Pettigrew, J.D. (1996). Ultrastructure, number, distribution and innervation of electroreceptors and mechanoreceptor organs in the bill skin of the platypus, *Ornithorhynchus anatinus*. *Brain Behav. Evol.* 48:27–54.

Manger, P.R., Calford, M.B., and Pettigrew, J.D. (1996). Properties of electrosensory neurons in the cortex of the platypus (*Ornithorhyncus anatinus*): Implications for the processing of electrosensory stimuli. *Proc. Roy. Soc. Lond. B.* 263:611–617.

Musser, A.M., and Archer, M. (1998). New information about the skull and dentary of the Miocene platypus *Obduron dicksoni*, and a discussion of ornithorhynchid relationships. *Phil. Trans. R. Soc. Lond. B.* 353:1059–1061.

Nachtrieb, H.F. (1910). The primitive pores of *Polyodon spathula* (Walbaum). *J. Exp. Biol.* 9:455–468.

Pettigrew, J.D. (1999). Electroreception in monotremes. *J. Exp. Biol.* 202:1447–1454.

Pettigrew, J.D., Manger, P.R., and Fine, S.L.B. (1998). The sensory world of the platypus. *Phil. Trans. R. Soc. Lond. B.* 353:1199–1210.

Rosen, R.A., and Hales, D.C. (1981). Feeding of paddlefish, *Polyodon spathula*. *Copeia* 1981:441–455.

Russell, D.F., Wilkens, L.A., and Moss, F. (1999). Use of behavioural stochastic resonance by paddle fish for feeding. *Nature* 402:291–294.

Wilkens, L.A., Russell, D.F., Pei, X., and Gurgens, C. (1997). The paddlefish rostrum functions as an electrosensory antenna in plankton feeding. *Proc. Roy. Soc. Lond. B.* 264:1723–1729.

Wilkens, L.A., Wettring, B.A., Wagner, E., Wojtenek, W., and Russell, D.F. (2001). Prey detection in selective plankton feeding by the paddlefish: Is the electric sense sufficient? *J. Exp. Biol.* 204: 1381–1389.

Index